Handbook of Water and Wastewater Treatment Technology

Paul N. Cheremisinoff

New Jersey Institute of Technology
Newark, New Jersey

Marcel Dekker, Inc.   New York • Basel • Hong Kong

Library of Congress Cataloging-in-Publication Data

Cheremisinoff, Paul N.
Handbook of water and wastewater treatment technology / Paul N. Cheremisinoff.
p. cm.
Includes index.
ISBN 0-8247-9277-7 (acid-free paper)
1. Water—Purification. 2. Sewage—Purification. I. Title.
TD430.C4593 1995
628.1'62—dc20 94-33927
CIP

The publisher offers discounts on this book when ordered in bulk quantities. For more information, write to Special Sales/Professional Marketing at the address below.

This book is printed on acid-free paper.

MARCEL DEKKER, INC.
270 Madison Avenue, New York, New York 10016

Current printing (last digit):
10 9 8 7 6 5 4 3 2 1

PRINTED IN THE UNITED STATES OF AMERICA

Preface

This handbook is a collection of exact and useful information relating to the treatment of water and wastewater for municipal, sanitary, and industrial uses. The operations and processes implemented by users of water are numerous and an attempt has been made to reduce those described to the practical limits of a single volume. Preference has been given to those unit operations and processes which have the most general application and serve a broad range of users.

Treatment of material is primarily descriptive, although the text is liberally supplemented with diagrams and drawings. Wherever possible, the theory governing the processes and equipment described is briefly given to clarify the discussion. The intention was to give sufficient information to provide the thoughtful reader with a satisfactory understanding of the subject. This book should be useful to civil, environmental, manufacturing, petrochemical, and chemical engineers, as well as plant operators and technicians.

Paul N. Cheremisinoff

Contents

1

Water Characteristics

Both individuals and industry produce liquid and solid wastes. The liquid portion, wastewater, is essentially water supply after it has been fouled by use. From the standpoint of sources of generation, wastewater may be defined as a combination of the liquid of water-carried wastes removed from residences, institutions, and commercial and industrial establishments together with such groundwater, surface water, and storm water as may be present. The ultimate goal in wastewater management is the protection of the environment commensurate with economic, social, political, and health concerns.

With increasing density of population and industrial expansion, the need for treatment and disposal of waste has grown. The specific reasons for sewage and waste treatment are as follows.

HEALTH CONCERNS

Disease-producing organisms, especially those causing diarrhea, may be present in sewage. Little is known about the presence of toxic substances produced by bacterial decomposition of certain organic substances, although a variety of degradation products are formed. From a health standpoint, the greatest problem of pollution is its effect on public water supplies by overloading the treatment devices so that they pass intestinal organisms, and by producing intestinal irritants in water which are not removed by filtration. Sewage pollution of bathing waters and shellfish areas may result in epidemics such as typhoid or other intestinal diseases as well as contamination of the food chain. Sewage treatment which reduces the pollution and kills intestinal organisms assists in the production of a safe drinking water at a lower cost, prevents contamination of shellfish, and permits bathing, water sports, and recreation.

ESTHETIC CONCERNS

The discharge of sewage into streams and water courses produces odors and discoloration, results in nuisances from sludge, and interferes with bathing facilities and recreation.

PROPERTY DAMAGE

The discharge of sewage affects industrial water supplies by changing the character of the water. Odors and gases in sewage affect real estate by causing paints to discolor as well as damage to boats. Some treatment of wastewaters is usually necessary before disposal. The methods of treatment adopted must be sufficient to ensure the necessary degree of purification required to suit the means of disposal.

Most unit operations and processes used for wastewater treatment are constantly undergoing continual and intensive investigation from the standpoint of implementation and application. As a result, many modifications and new operations and processes have been developed and implemented; more need to be made to meet increasingly stringent requirements for environmental enhancement of water. In addition to the developments taking place with conventional treatment methods, alternative treatment systems and technologies are also being developed and introduced.

AVAILABLE TREATMENT SYSTEMS

The treatment process chosen is a function of several factors:

- Flow rate
- Waste strength and toxicity
- Availability of land
- Esthetics
- Discharge standards
- Climatic conditions
- Degree of permanence desired
- Costs

For example, in remote areas where land is inexpensive and climate is favorable, a percolation/evaporation pond may provide simple zero discharge solution, whereas in a suburban community in which the ultimate discharge enters surface waters, an esthetic and high-performance plant which may include some type of tertiary facility would be more appropriate. Table 1 shows the various wastewater treatment options in use. Materials removed during water/wastewater treatment is called sludge, and Table 2 lists the options for its management.

CHARACTERISTICS OF WASTEWATER

An understanding of the nature of wastewater is essential in the design and operation of collection, treatment, and disposal facilities and in the engineering management for environmental quality.

The physical properties and the chemical and biological constituents of wastewater and their sources are listed in Table 3. The important contaminants of interest in wastewater treatment are listed in Table 4. Wastewater characterization studies are conducted to determine the physical, biological, and chemical characteristics and the concentrations of constituents in the wastewater as the best means of reducing the pollutant concentrations.

EFFECTS OF POLLUTION

Effects of pollution can be manifested by many characteristics and variations in degree when pollution enters the aquatic environment. Specific environmental and ecological responses to a pollutant will depend largely on the volume and strength of the waste and the volume of water receiving it. Within each response there can be many changes in magnitude and degree. A classic response that has often been described is the effects of organic wastes that may be discharged from sewage-treatment plants and certain industries. As these wastes enter the receiving water, they create turbidity, decrease light penetration, and may settle to the bottom in substantial quantity to form sludge beds. Wastes are attacked by bacteria and this process of decomposition consumes oxygen from the water and liberates essential nutrients that in turn stimulate the production of some forms of aquatic life.

Upstream from the introduction of organic wastes is a clean water zone or one that is not affected by pollutants. At the point of waste discharge and for a short distance downstream there is formed a zone of degradation where wastes become mixed with the receiving waters and where the initial attack is made on the waste by bacteria and other organisms in the process of decomposition.

Following the zone of degradation there is a zone of active decomposition that may extend for miles or days of stream flow, which depends in large measure on the volume of the waste by the stream and the temperature of the water.

Table 1 Wastewater Treatment Options

		Primary treatment		Secondary treatment		
	Pretreatment	Chemical	Physical	Dissolved organics and colloidal material is removed	Suspended solids removal	Advance or tertiary treatment
Wastewater	Screening and Grit Removal	Neutralization	Flotation	Activated Sludge	Sedimentation	Coagulation Sedimentation
	Equalization and Storage	Coagulation	Sedimentation	Contact Stabilization		Filtration
	Oil Separation			Trickling Filter		Carbon Adsorption
		Hydrolysis		Aerated Lagoon		Ion Exchange
				Ozonation		Distillation
						Reverse Osmosis
						Electrodialysis

Table 2 Options for Sludge Management

Disposal	Sludge treatment	Sludge disposal
	Aerobic Digestion	
Chlorination Ozonation	Anaerobic Digestion	Incineration
Receiving Waters or Reuse	Wet Combustion	Land Fill
Controlled or Transported Discharge	Centrifugation	Soil Conditioning
Ocean Disposal	Thickening	Ocean Disposal
Surface Application or Ground Water Seepage	Vacuum Filtration Lagooning, or Drying Beds	
Evaporation + Incineration		

Biological processes that occur within this zone are similar in many respects to those that occur in a typical sewage treatment plant. Within this zone, waste products are decomposed and those products that are not settled as sludge are assimilated by organisms in life processes.

A recovery zone follows the zone of active decomposition. The recovery zone is essentially a stream reached in which water quality is gradually returned to that which existed prior to the entrance of pollutants. Water quality recovery is accomplished through physical, chemical, and biological interactions within the aquatic environment. The zone of recovery may also extend for many miles, and its extent will depend principally on morphometric features of the waterways. The zone of recovery will terminate in another zone of clean water or area unaffected by pollution that is similar in physical, chemical, and biological features to that which existed upstream from the pollution source.

Organic Wastes

The effects of organic wastes on the receiving stream often become confused with a specific stream because additional sources of pollution may enter the environment before the receiving water has been able to assimilate the entire effects of an initial source. When this occurs, the effects of subsequent introductions become superimposed on the initial source and the total effect may confine large reaches of stream to a particular zonal classification.

Effects of organic wastes in the static water environment, as opposed to the flowing water environment, are modified by the features of the receiving water. Zonal changes for flowing water do exist but may be compressed in great measure either laterally or vertically when the discharge is to a lake or estuary. Such compression may tend to decrease the severity of pollution that is often observed in the flowing water environment and, on the other hand, may increase

Table 3 Physical, Chemical, and Biological Characteristics of Wastewater and Their Sources

Characteristic	Sources
Physical Properties	
Color	Domestic and industrial wastes, natural decay of organic materials
Odor	Decomposing wastewater, industrial wastes
Solids	Domestic water supply, domestic and industrial wastes, soil erosion, inflow-infiltration
Temperature	Domestic and industrial wastes
Chemical Constituents	
Organics	
Carbohydrates	Domestic, commercial, industrial wastes
Fats, oils and grease	Domestic, commercial, industrial wastes
Pesticides	Agricultural wastes
Phenols	Industrial wastes
Proteins	Domestic and commercial wastes
Surfactants	Domestic and industrial wastes
Others	Natural decay of organic materials
Inorganics	
Alkalinity	Domestic wastes, domestic water supply, ground water infiltration
Chlorides	Domestic water supply, domestic wastes, ground water infiltration, water softeners
Heavy metals	Industrial wastes
Nitrogen	Domestic and agricultural wastes
pH	Industrial wastes
Phosphorus	Domestic and industrial wastes, natural runoff
Sulfur	Domestic water supply, domestic and industrial wastes
Toxic compounds	Industrial wastes
Gases	
Hydrogen sulfide	Decomposition of domestic wastes
Methane	Decomposition of domestic wastes
Oxygen	Domestic water supply, surface water infiltration
Biological Constituents	
Animals	Open watercourses and treatment plants
Plants	Open watercourses and treatment plants
Protista	Domestic wastes, treatment plants
Viruses	Domestic wastes

Table 4 Important Contaminants of Concern in Wastewater Treatment

Contaminants	Reason for importance
Suspended solids	Suspended solids can lead to the development of sludge deposits and anaerobic conditions when untreated wastewater is discharged in the aquatic environment.
Biodegradable organics	Composed principally of proteins, carbohydrates, and fats, biodegradable organics are measured most commonly in terms of BOD and COD. If discharged untreated to the environment, their biological stabilization can lead to the depletion of natural oxygen resources and to the development of septic conditions.
Pathogens	Communicable diseases can be transmitted by the pathogenic organisms in wastewater.
Nutrients	Both nitrogen and phosphorus, along with carbon, are essential nutrients for growth. When discharged to the aquatic environment, these nutrients can lead to the growth of undesirable aquatic life. When discharged in excessive amounts on land, they can also lead to the pollution of ground water.
Refractory organics	These organics tend to resist conventional methods of wastewater treatment. Typical examples include surfactants, phenols, and agricultural pesticides.
Heavy metals	Heavy metals are usually added to wastewater from commercial and industrial activities and may have to be removed if the wastewater is to be reused.
Dissolved inorganic solids	Inorganic constituents such as calcium, sodium, and sulfate are added to the original domestic water supply as a result of water use and may have to be removed if the wastewater is to be reused.

substantially the development of biotic nuisances such as algae or rooted aquatic plants that may develop from the nutrients released with and decomposed from the introduced organic materials.

Organism communities that may be related to pollution principally are those that are usually associated with the bed or bottom of the waterway; those that attach themselves to objects such as rocks, aquatic plants, brush, or debris submerged in the water; those that are essentially free floating and are transported by currents and wind, such as plankton and other microscopic forms; and those motile free-swimming organisms such as fish. Considering each of these common organism groups, a number of observations can be made on their reaction to the introduction of organic wastes to a flowing stream.

Upstream from waste sources such limiting factors as food and intense competition among organisms and among organism groups, predation, and available habitat for a particular species will limit organism populations to those that can be sustained by the particular environment. Most often the limiting factor will be available food. Within this population, however, there will exist a great number of organism species. Thus, the old biological axiom for an environment unaffected by pollution is one that supports a great number of species with the total population delimited largely by food supply.

Introduction of organic wastes causes conditions of existence for many organisms that become substantially degraded. Increased turbidity in the water reduces light penetration, which in turn will reduce the volume of water capable of supporting photosynthesizing plants. Particulate matter in settling will flocculate small floating animals and plants from the water. As the material settles, sludge beds are formed on the stream bed and many of the areas that formerly could have been inhabited by bottom-associated organisms become covered and uninhabitable.

The zone of degradation is the transition area between the clean water unaffected reach and a zone of decomposition of organic wastes. The dissolved oxygen may be diminished but not completely removed. Sludge deposits may be initiated but are not formed in maximum magnitude or extent. Conditions of existence become impaired, and typically there is a reduction in both the organism population and the number of species that can tolerate this environment.

Within the zone of active decomposition conditions of existence for aquatic life are at their worst. Breakdown of organic products by bacteria may have consumed available dissolved oxygen. Sludge deposits may have covered the stream bed and thus eliminate dwelling areas for the majority of bottom-associated organisms that could be found in an unaffected area. Fish spawning areas have been eliminated, but perhaps fish are no longer present because of diminished dissolved oxygen and substantially reduced available food. Here aquatic plants will not be found in large numbers, because they cannot survive on the soft shifting blanket of sludge. Turbidity may be high and floating plants and animals destroyed. Water color may be substantially affected. When organic materials are decomposed as the food supply is liberated for those particular organisms that are adapted to use this food source. Bacterial and certain protozoan populations may increase to extremely high levels. Bottom-associated organisms such as sludgeworms, bloodworms, and other wormlike animals may also increase to tremendous numbers, because they are adapted to burrowing within the sludge, deriving their food therefrom, and existing on sources and amounts of oxygen that may be essentially nondetectable by conventional field investigative methods. Within the zone of active decomposition, the organism species that can tolerate the environment are reduced to extremely low levels. Under some conditions, those bottom-associated animals that are visible to the unaided eye may be completely eliminated. Because of the tremendous quantity of food that is available to those organisms that are adapted to use it, the numbers of individuals of the surviving species may become great.

The zone of recovery is essentially the downstream transition zone between the zone of active decomposition and an environment that is unaffected by pollution. This zone features a gradual cleaning up of the environment, a reduction in those features that form adverse conditions for aquatic life, an increase in organism species, and a gradual decrease in organism population because of decreased food supply and the presence of some of the predators that are less sensitive individually to pollutional affects.

Because of variation in response among species to conditions of existence within the environment, and because of inherent difficulties in aquatic invertebrate taxonomy, the ecological evaluation of the total organism community is the acceptable approach in water pollution control studies. Investigators tend to place organisms in broad groups according to the general group response to pollutants in the environment. For example, the general group known as "sludgeworms" is found in both the unpolluted as well as the organically polluted environment. Value as a group lies in the fact that the numbers of individuals within the group are exceedingly low in unpolluted water, whereas in the organically polluted environment, its numbers may be very high. Examples of organisms that may inhabit both the unpolluted and polluted environments are listed in Table 5.

The converse of the effects of pollution on organisms is the effects of organisms on pollutants. Organic wastes, especially, the supply food which in turn produces an abundance of a few types of organisms produced in an unpolluted environment. When consuming organic wastes, the organisms stabilize the waste in a given number of feet or miles of horizontal stream in a manner similar to that in a vertical trickling filter that is designed especially for maximum stabilizing efficiency by the organisms.

As organic wastes become stabilized, other organism types predominate within the aquatic animal community. Midge larvae have been found to taint stream beds a brilliant red with their undulating bodies. Caddisfly larval populations greater than 1000 per square foot of stream bed or mayfly nymphs numbering more than 300 per square foot have been found. Figures 1–3 show representative stream bed–associated animals.

Inorganic Silts

Inorganic silts in the environment reduce severely both the types of organisms present and their populations. Particulate matter settling to the bottom can blanket the substrate and form undesirable physical environments for organisms that would normally occupy a habitat. Erosion silts change environments chiefly by screening out light, by changing heat radiation, by blanketing the stream bottom and destroying living spaces, and by retaining organic materials and other substances that can create unfavorable conditions. Developing eggs of fish and

Table 5 Examples of Organisms Sensitive and Tolerant to Polluted Water, Respectively

Clean water organisms		Clean or polluted water organisms	
Algae	*Cladophora* (green)	Iron Bacteria	*Sphaerotilus*
	Ulothrix (green)	Fungi	*Leptomitus*
	Navicula (diatom)	Algae	*Chlorella* (green)
Protozoa	*Trachelomonas*		*Chlamydomonas* (green)
Insects	*Plecoptera* (stoneflies)		Oscillatoria (blue-green)
	Negaloptera (hellgrammites, alderflies, and fishflies)		Phormidium (blue-green)
		Protozoa	*Carchesium* (stalked) colonial ciliate)
	Trichopetera (caddisflies)		*Colpidium* (noncolonial ciliate)
	Ephemeroptera (mayflies)	Segmented Worms	*Tubifex* (slugeworms)
			Limnodrilus (slugeworms)
Clams	*Unionidae* (pearl button)	Leeches	*Helobdeall stagnalis*
Fish	*Etheostoma* (darter)	Insects	*Culex pipiens* (mosquito)
	Notropis (shiner)		*Chironomus (Tendpipes) plumosus* (bloodworms)
	Chrosomus (dace)		*Tubifera (Eristalis tenax)* (rat-tailed maggot)
		Snail	*Physa integra*
		Clam	*Sphaerium* (fingernail clam)
		Fish	*Cyprinus carpio* (carp)

other organisms may be smothered by deposits of silt. Fish feeding may be hampered by silt deposits. Direct injury to fully developed fish, however, by nontoxic suspended matter occurs only when concentrations are higher than those commonly found in natural water or associated with pollution.

Toxic Metals

Wastes containing heavy metals, either individually or in combination, may be destructive to aquatic organisms and have a severe impact on the aquatic community. A severely toxic substance will eliminate aquatic biota until dilution, dissipation, or volatilization reduces the concentration below the toxic threshold. Generally toxic materials will reduce the aquatic biota except those species that are able to tolerate the observed concentration of the toxicant. Because toxic materials do not offer an increased food supply, such as organic wastes, there is no sharp increase in the population of those organisms that may tolerate a

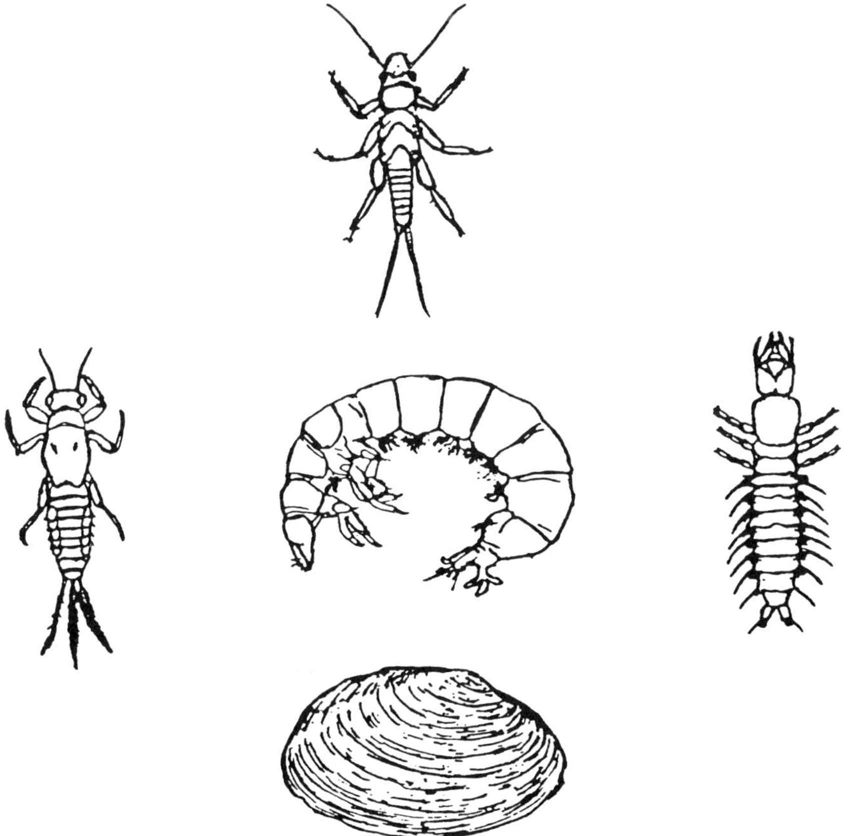

Figure 1 Clean water [sensitive] animals associated with stream beds include the stonefly nymph, mayfly naiad, caddisfly larva, and hellgrammite unionid clam.

specific concentration. The bioassay is an important tool in the investigation of these wastes, because the results from such a study indicate the degree of hazard to aquatic life of particular discharges; interpretations and recommendations can be made from these studies concerning the level of discharge that can be tolerated by the receiving aquatic community.

Temperature

Temperature is a regulator of natural processes within the water environment. It governs physiological functions in organisms, and acting directly or indirectly

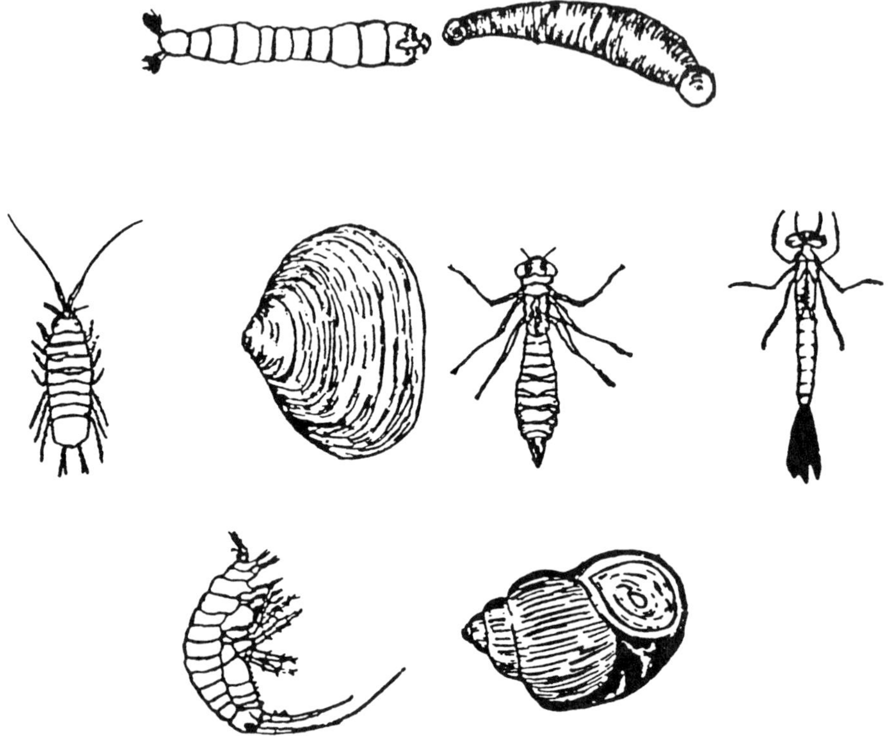

Figure 2 Intermediately tolerant animals associated with stream beds include the scud, sowbug, blackfly larvae, fingernail clam, damselfly nymph, dragonfly nymph, leech, and snail.

in combination with other water quality constituents, temperature affects aquatic life with each change. The effects of temperature changes include, for example, chemical reaction rates, enzymatic functions, molecular movements, and molecular exchanges between membranes within and between the physiological systems and organs of an animal. Because of the complex interactions involved, and often because of the lack of specific knowledge or facts, temperature effects as they pertain to an animal or plant are most efficiently assessed on the basis of net influence on the organism. Depending on the extent of environmental temperature change, organisms can be activated, depressed, restricted, or killed.

Temperature determines those aquatic species that may be present. It controls spawning and the hatching of young, regulates their activity, and stimulates or suppresses their growth and development. Temperature can attract and kill when the water becomes heated or chilled too suddenly. Colder water

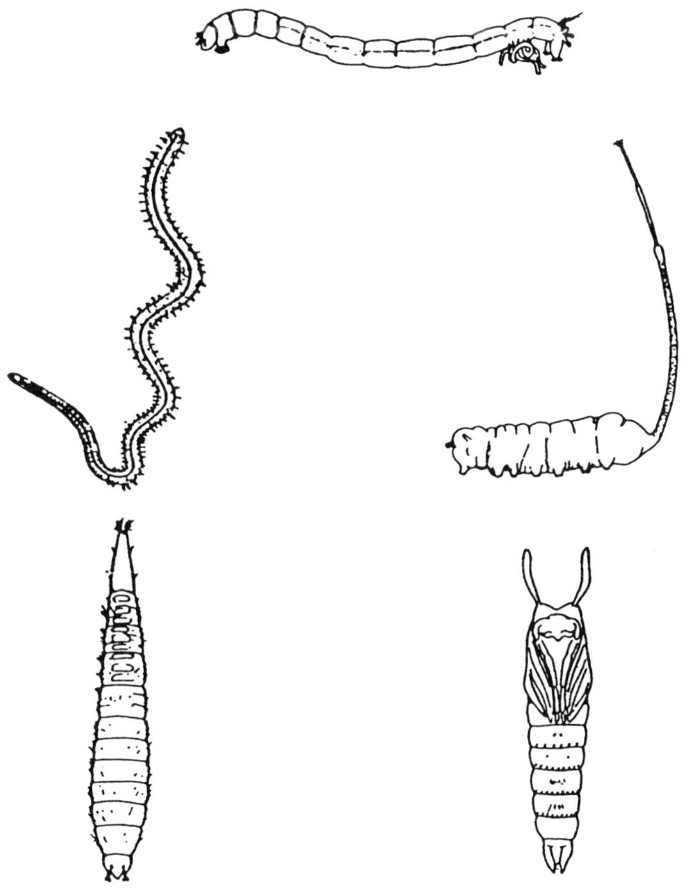

Figure 3 Very tolerant animals associated with stream beds include the bloodworm, midge larvae sludgeworm, rat-tailed maggot, sewage fly larva, and sewage fly pupa.

generally suppresses development and warmer water generally accelerates activity.

Temperature regulates molecular movement and thus determines the rate of metabolism and activity of all organisms, both those with a relatively constant body temperature and those whose body temperature is identical to, or follows closely, the environmental temperature. Because of its capacity to determine metabolic rate, temperature may be the most important single environmental entity to life and life processes.

Variations in temperature of streams, lakes, estuaries, and oceans are normal results of climatic and geological phenomena. Waters that support some form of aquatic life other than bacteria or viruses range in temperature from 26.6°F in polar sea waters to 185°F in thermal springs. Most aquatic organisms tolerate only those temperature changes that occur within a narrow range to which they are adapted whether it be high, intermediate, or low on this temperature scale.

Within the same species, the effects of a given temperature may differ in separate populations, in various life cycle stages, or between the sexes, and such effects may depend on the temperature history of the individual tested as well as on present or past effects of other environmental factors. Fresh water has the greatest density at 38°F; higher and lower temperatures result in waters with lower density. Seasonally induced temperature changes are greatest in the midlatitudes.

In lakes, insolation warms the surface waters in spring reducing densities compared with the deeper waters until eventually the density differences are sufficient to prevent the wind from mixing the body of water; thermal stratification then occurs. The warm upper layer is well mixed to a depth determined by waves and other wind-induced currents. The cool bottom waters (hypolimnion) become stagnant except for minor currents confined to this stratum. A stratum of sudden temperature changes (thermocline) separates these regions. In autumn, the lake radiates heat, surface temperatures decrease, surface water density increases, and water viscosity increases. Wind aided by reduced density differences between water layers mixes the surface with the bottom waters resulting in a homogeneous water mass. Depending on altitude and local climatic conditions, the lake continues to mix until the following spring in latitudes of less than about 40°F. In latitudes north of about 40°F, winter surface water temperatures are less than 38°F and these are superimposed over the water mass until they are cooled to freezing. An ice cover eliminates wind-induced mixing and stagnation occurs.

Thermal stratification may assume various patterns depending on geographical location, climatological conditions, depth, surface area, type of dam structure, penstock locations, and hydropower use. In general, large, deep impoundments will cool downstream waters in the summer and warm them in winter when withdrawal ports are deep; shallow, unstratified impoundments with large surface areas will warm downstream waters in the summer; water drawn from the surface of a reservoir will warm downstream waters; a reduction in normal flow downstream from an impoundment will cause marked warming in summer; and "runoff-river" impoundments, where the surface area has not been increased markedly over the normal river area, will produce only small changes in downstream water temperatures.

In the deep, stagnant, summer bottom waters as well as in ice-covered waters atmospheric reaeration is absent and oxygen from photosynthesis by plants is limited. Decomposing organisms (especially those settling to the bottom waters in summer) remove oxygen from the water and the gaseous by-products of decomposition are trapped. Undesirable soluble phosphorus, carbon dioxide, iron, and manganese concentrations increase in these stagnant waters. Designed thermal discharges can reduce some of these problems. Ice cover can be limited and thus allow wind and thermally induced currents to reduce winter stagnation. A deep-water summer discharge could warm hypolimnetic waters to decrease density and permit total water mass mixing where a cold water fishery would not be damaged by such action.

Stratification may occur in streams receiving heated effluents. There are three recognized forms of stream stratification: overflow, interflow, and underflow; the forms are determined by the relationship between the density of the influent and the density of the stream water.

Surface fresh waters in the United States vary from 32 to over 100°F according to the latitude, altitude, season, time of day, duration of flow, depth, and many other variables. Agents affecting natural water temperature are so numerous that no two water bodies, even in the same latitude, are likely to have the same thermal characteristics. Fish and other aquatic life occurring naturally in each body of water are those that have become adapted to the temperature conditions existing there. The interrelationships of species, length of daylight, and water temperature are so intimate that even a small change in temperature may have far-reaching effects. Inhabitants of a water body that seldom becomes warmer than 70°F are placed under stress, if not killed outright, by 90°F water. Even at 75–80°F, they may be unable to compete successfully with organisms for which 75–80°F is favorable. Similarly, the inhabitants of warmer waters are at a competitive disadvantage in cool water.

An animal's occurrence in a given habitat does not mean that it can tolerate the seasonal temperature extremes of that habitat at one time. The habitat must be cooled gradually in the fall if the animal is to become acclimatized to the cold water of winter and warmed gradually in the spring if it is to withstand summer heat. Some organisms might endure a temperature of 92–95°F for a few hours but not for days. Gradual change of water temperature with the season is important for other reasons. An increasing or decreasing temperature often triggers spawning, metamorphosis, and migration. The eggs of some freshwater organisms must be chilled before they will hatch properly.

The temperature range tolerated by many species is narrow during very early development; it increases somewhat during maturity and decreases again in old adults. Similarly, the tolerable temperature range is often more restrictive during the reproductive period than at other times during maturity. Upper lethal temperatures may be lower for animals from cold water than for closely related species from warm water. Many motile organisms such as fish, some zooplankton, certain algae, and some associated animals can avoid critical temperatures by vertical and horizontal migration into more suitable areas. However, some organisms may be attracted to areas with critical temperatures and succumb on arrival in these areas.

Changes in fish populations can result from many types of artificial cooling and heating of natural waters. These changes result from the discharge of condenser cooling water from thermal electric-generating plants, industrial waste-cooling waters, and other heated effluents and irrigation waters. Streams also are warmed by the sun when the shade from stream bank trees and other vegetation is eliminated. The discharge of cold water from stratified impoundments may provide an ideal habitat for trout and other cold-water fish when sufficient dissolved oxygen is present but not for the warm-water fish that inhabited the stream before impoundment.

For every 18°F increase in temperature, the chemical reaction rate is approximately doubled in an organism or in an environment. Life processes in the water are accelerated with temperature increases and slowed as the water cools. Solubility of gases, including oxygen, in water varies inversely with temperature. In fresh water, the solubility of atmospheric oxygen is decreased by about 55% as the temperature rises from 32 to 104°F under 1 atmosphere of pressure (760 mm Hg). Because all desirable living things are dependent on oxygen in one form or another to maintain the life processes that produce energy for growth and reproduction, dissolved oxygen is of imposing significance in the aquatic environment.

When organism metabolism increases because of higher temperatures, organism development is speeded and more dissolved oxygen is required to maintain existence. But bacterial action in the natural purification process to break down organic materials is also accelerated with increased temperatures, thus reducing the oxygen that could be available in the warmer water. When organisms use larger amounts of oxygen, and when oxygen has been reduced by temperature action and interaction, organisms may perish. Life stages that are especially vulnerable are the eggs and larvae. At higher temperatures, phytoplankton have been found to need greater amounts of certain growth factors such as vitamin B_{12}. For example, at between 96.8 and 98.2°F the vitamin requirement has been found to increase over 300 times for some species.

Fish and other mobile organisms seek a preferred temperature at which they can best survive, which is several degrees below a temperature that is lethal. Larger individuals tend to move out of areas that are too hot, but larvae and juveniles cannot often move fast enough to avoid a sudden temperature increase. Large fish and fish in schools avoid heated areas in summer but may be attracted to such areas in winter. This phenomenon may result in good fishing during the cooler months but an absence of this sport at other times.

Reproduction cycles may be changed significantly by increased temperature, because this function takes place under restricted temperature ranges.

Spawning may not occur at all if temperatures are too high. Thus, a fish population may exist in a heated area only by continued immigration. Disregarding the decreased reproductive potential, water temperatures need not reach lethal levels to wipe out a species. Temperatures that favor competitors, predators, parasites, and disease can destroy a species at levels far below those that are lethal.

Fish food organisms are altered severely when temperatures approach or exceed 90°F. Predominant algal species change, primary production is decreased, and bottom-associated organisms may be depleted or altered drastically in numbers and distribution. Increased water temperatures may cause aquatic plant nuisances when other environmental factors are favorable.

Synergistic actions of pollutants are more severe at higher water temperatures. Domestic sewage, refinery wastes, oils, tars, insecticides, detergents, and fertilizers more rapidly deplete oxygen in water at higher temperatures, and the respective toxicities are likewise increased.

Marine water temperatures do not change as rapidly or range as widely as those of fresh waters. Marine and estuarine fish, therefore, are less tolerant of temperature variation. Although this limited tolerance is greater in the estuarine than in the open water marine species, temperature changes are more important to those fish in estuaries and bays than to those in open marine areas. Marine surf-zone discharge from large-scale coastal power plants may be expected significantly to alter the shore environment for species of invertebrates and fish that are commonly found there.

DISSOLVED OXYGEN

Dissolved oxygen (DO) is a measure of water quality that in the right concentrations is essential not only to keep organisms living but also to sustain species reproduction, vigor, and the development of populations. Organisms undergo stress at reduced DO concentrations that makes them less competitive to sustain their species within the aquatic environment.

Oxygen enters the water by absorption directly from the atmosphere or by plant photosynthesis and is removed by respiration of organisms and by decomposition. Oxygen from the atmosphere may be by direct diffusion or by surface water agitation by wind and waves, which may also release dissolved oxygen under conditions of supersaturation.

In photosynthesis, aquatic plants utilize carbon dioxide and liberate dissolved and free gaseous oxygen at times of supersaturation. Since energy is required in the form of light, photosynthesis is limited to the photic zone where light is sufficient to facilitate this process.

During respiration and decomposition, animals and plants consume dissolved oxygen and liberate carbon dioxide at all depths where they occur. Because excreted and secreted products and dead animals and plants sink, most of the decomposition takes place in the hypolimnion; thus during lake stratification there is a gradual decrease of dissolved oxygen in this zone. After the dissolved oxygen is depleted, anaerobic decomposition continues with evolution of methane and hydrogen sulfide.

During thermal stratification, dissolved oxygen is usually abundant and is supplied by atmospheric aeration and photosynthesis. Phytoplankton are plentiful in fertile lakes and are responsible for most of the photosynthetic oxygen. The thermocline is a transition zone from the standpoint of dissolved oxygen as well as temperature. The water rapidly cools in this region, incident light is much reduced, and photosynthesis is usually decreased; if sufficient dissolved oxygen is present, some cold-water fish abound. As dead organisms that sink into the hypolimnion decompose, oxygen is utilized; consequently, the hypolimnion in fertile lakes may become devoid of dissolved oxygen following a spring overturn, and this zone may be unavailable to fish and most benthic invertebrates at this time. During the two brief periods in spring and fall when lake water circulates, temperature and dissolved oxygen are the same from top to bottom and fish use the entire water depth.

OTHER IMPURITIES IN WATER

In addition to dissolved mineral matter and dissolved gases, water may contain other impurities such as turbidity and sediment, color and organic matter, tastes and odors, and microorganisms.

Turbidity and Sediment

Finely divided, insoluble impurities that may be suspended in and compromise the clarity of a water are known as turbidity. Suspended impurities may be inorganic in nature, such as clay, rock flour, silt, calcium carbonate, silica, ferric hydroxide, and sulfur, or they may be organic in nature, such as finely divided vegetable, animal matter, oils, fats, greases, and microorganisms. Turbidity may be due to a single substance or more usually to a mixture of substances.

Suspended impurities may range in size from colloidal particles to coarse, sandy materials that can be kept in suspension only by mixing and turbulent flows. Material that is so coarse that it rapidly drops out of suspension is classified sediment. The dividing line between turbidity and sediment is not sharp, and there may be different ideas as to what constitutes turbidity or sediment. Additionally, agglomeration of smaller particles into larger ones is also possible. If the sediment content is sufficiently high, it may warrant preliminary sedimentation before coagulation.

Turbidity is a measure of the opacity of the water as compared with certain arbitrary standards and differs with varying materials and degrees of fineness.

Where the suspended matter is always of the same nature, it is possible to measure the divergence between the observed turbidity and the actual weight of the suspended matter and apply a correction factor to subsequent readings.

The standard method for the determination of turbidity is the Jackson candle method, or suspensions standardized by this method may be used, with or without dilution in other instruments. In the standard Jackson candle turbidimeter, the turbidities are determined from the depths of water through which a light disappears when viewed lengthwise through a tube. In practice, it is customary to use standard suspensions (which have been standardized and diluted by the addition of the calculated amounts of distilled water) for all measurements except extremely low ones.

The term *quiescent* as applied to the waters of lakes, ponds, and reservoirs is a relative one. That is, such bodies of water are relatively quiescent when compared with flowing streams, but wind and wave actions and temperature changes do effect movements in the water even if the inflow and outflow are small in comparison with the size of the body of water so that the effects may be practically negligible. Wind usually tends to push the upper layers of water in the direction to which it is blowing. The movement is greatest at the surface and diminishes downwardly until a plane is reached at which it is practically zero. Below this, the water may return in layers flowing in the opposite direction, slowly under this layer of nearly zero velocity, then increasing in speed in succeeding lower layers until a maximum is reached, after which it tapers off until it again reaches a layer of practically zero movement. Factors such as the shape of the shore line and the sheltering effect of islands may obviously create currents, so that much of the returning water is diverted to another course. Wind also creates waves which have a churning effect and tend to roll waters badly, especially where relatively shallow bodies of water are concerned. It is therefore common experience that many lake, pond, and reservoir waters which are usually clear may become very turbid during and for a period following high winds or storms.

Heavy rainfalls may also increase the turbidity of lakes, ponds, and reservoirs, especially that of the smaller ones. This may be due to dirt or clay being washed in from the banks or from tributaries swollen by heavy rains and carrying unduly heavy loads of turbidity. Obviously, the location of intakes in the proximity of such tributaries may greatly increase the turbidity load in the plant influent. Also, frequently plant intakes are located too close to sewer outlets, so that contamination from this source may occur under certain conditions of wind and currents. Temperature changes may also affect the turbidity of the water, with the greatest effect being noticed when the water is at uniform density, as this permits vertical circulation. Under such conditions, wind and wave action may greatly aggravate the pickup of turbidity from bottom sediments.

In bodies of water that are less than 25 feet in depth, vertical circulation is possible at all seasons except when the surface is frozen. In deeper bodies, vertical circulation usually occurs when the water is at or near its greatest density, which is at 39.2°F, and the effect on turbidity in certain lakes, ponds, and reservoirs is especially troublesome during the periods of spring and fall turnovers (or overturnings). In areas where freezing temperatures are encountered in the winter, the water on the bottom is stagnant at about the temperature of greatest density. Then in the spring, after the ice has melted, the surface layers increase in density on being warmed, and consequently sink until, when the temperature is about 39.2°F, all of the water is of uniform density and vertical circulation occurs. Then as the upper layers are warmed beyond this point, they become lighter; the body of water tends to stratify and the bottom and lower layers stagnate at about the point of maximum density. In the fall, the upper layers become heavier as they are chilled, until all of the water has reached about 39.27°F, when vertical circulation again is possible. On further cooling, the upper layers become lighter and the period of winter stagnation begins. With many bodies of water, the increase in turbidity at these periods is startling and may persist for several weeks, especially during the fall turnover.

Temperature has an effect on the growth of microorganisms and aquatic plants which, together with their decomposition products, may impart not only turbidity but also color, tastes, and odors to the water supply. Growth of some of these forms takes place in winter and even under the ice, although at greatly reduced rates compared with growth at summer temperatures. The worst contamination with living matter is experienced in small, shallow bodies of water with muddy bottoms.

It is therefore evident that lakes, ponds, and reservoirs, like rivers, must be individually studied as to their possible variations in turbidity and that it is not safe to assume that simply because a few random tests show a negligible turbidity that this will never be exceeded. Likewise, as has been pointed out, the location of the inlet in respect to its depth, distance from the shore, condition of the bottom, proximity to tributaries, and sewers, for example, must also be taken into account.

Turbidity in Ground Waters

Owing to the filtering action of the strata through which ground waters pass, most well and spring waters are free from turbidity. There are exceptions though. Some spring waters, for instance, may be clear the greater part of the time but may show appreciable turbidities after periods of heavy rains. This is especially the case in limestone country where there may be fairly large fissures and crevices through which water containing suspended material may travel. Obviously, any wells tapping such aquifers will show occasional turbidities. Shallow wells also may show turbidities at times; deep wells, with but few exceptions, yield very clear waters. In sandy strata, however, well waters may contain fine sand which must be removed by sand traps or other settling equipment.

Well waters may also be clear when drawn may develop turbidity on standing in contact with air. Iron-containing waters, for instance, may be sparklingly clear at first but may cloud on exposure to the air and finally deposit yellowish to reddish brown ferric hydroxide. Manganese-bearing waters may also, but usually not so readily, become turbid in contact with the air. Sulfur waters may deposit sulfur in contact with the atmosphere, but in many cases the sulfur is colloidal and is not always apparent in broad daylight. In a darkened room it will, however, show the Tyndall effect when struck by a small beam of light. Sulfur waters may also develop growths of sulfur bacteria, and iron- or manganese-bearing waters may develop growths of iron or manganese bacteria.

High turbidity in a water supply is undesirable for practically all uses except perhaps for certain types of surface condensers. The amount of turbidity which makes a water definitely objectionable varies with the nature of the turbidity and the purpose for which the water is intended. Therefore, tolerances for turbidity may vary.

Removal of turbidity and sediment is if a water supply has a high content of easily settled sediment, it may be advisable to remove a great portion of this by sedimentation tanks, basins, or reservoirs. On the other hand, if the content of heavy, easily settled sediment is small, the sediment and turbidity are usually taken out at the same time by:

- Coagulation and filtration
- Coagulation and settling
- Coagulation, settling, and filtration

Color and Organic Matter

Color is found mostly in surface waters, although some shallow well waters, a few springs, and an occasional deep well water may contain noticeable amounts of color. In general, though, deep well waters are practically colorless. The colors noted in water usually range from a very light straw color, through a yellowish brown, and up to a dark brown. In determining the color of a water, it is the true color that is of interest and not the apparent color. This is expressed as the color standard unit.

The true color of water is that only due to substances in solution; that is, it is the color of the water after the suspended matter has been removed. The accurate determination of color in water containing matter in suspension is impossible. The removal of suspended matter by centrifuging before the color observation is made gives the best results. No filter should be used, since filters may exert marked decolorizing action. The platinum-cobalt method of measuring color shall be considered as the standard, and the unit of color shall be that produced by 1 mg of platinum per liter.

The color of natural water supplies is usually organic in nature, and augmented in many cases by organic or colloidal iron or manganese. Arbitrary inorganic color standards measure only the relative depth of color of the water and not the mass of the actual coloring agent present. Color in water in large part exists in the form of colloidal suspensions of ultramicroscopic particles. Some of the color may be due to colloidal emulsions. A small part of the color is probably due to noncolloidal material, organic acids, and neutral salts in true solution. The colloidal coloring matter, whether suspensoids or emulsoids, carries an electrostatic charge. This charge may be positive or negative depending on the character and source of the water and varying in different waters. Since these particles carry an electrical charge and are in colloidal suspension, they obey the laws of cataphoresis when an electric current is sent through a colored water. The particles are attracted to the electrode of opposite sign from the charge which they carry. They are discharged, flocculate, and precipitate, with consequent reduction in the color of the water.

Removal of Color and Organic Matter

The soluble organic matter in water is a complex mixture of substances, some of which presumably have a high color and others of which have either a low color or are practically colorless. The intensity of color may be determined against an arbitrary scale, but this does not show the quantity of organic matter producing such a color. Oxygen consumption tests are of value in indicating whether or not excessive amounts of organic matter are present, but they are not translatable into definite amounts of organic matter. Such tests are merely a measure of the quantity of potassium permanganate, which is usually expressed as its available oxygen equivalent that is used up by a given volume of water under certain fixed conditions of time and temperature.

In general, removal or reduction of color and organic matter can be accomplished by coagulation, settling, and filtration. The most widely used coagulant is aluminum sulfate. Iron coagulants, ferric sulfate, and chlorinated copperas are also used. The coagulation is best carried out at the most favorable pH value, which should be determined by experiment: With aluminum sulfate, the optimum point varies over a range from about 5.5 to 6.8 for most waters, but this coagulant has been used at a pH below 5.0 in a few cases, and with waters having rather high mineral contents at a pH as high as 7.5. With ferric coagulants, the range from 3.5 to about 5.5 has been effective for highly colored waters. Ferric coagulants also precipitate well at pH values above 9.0, but a high pH usually tends to keep the color in solution. Addition of clay is valuable in coagulating colored waters which are low in turbidity, and this broadens the pH range at which good coagulation takes place. Activated silica is frequently useful as a coagulation aid in treating colored waters. Controlled agitation is of importance in securing good floc formation. Settling is of advantage and obviously lessens the load on filters, but settling of the coagulated water when not followed by filtration is often insufficient for good color removal.

Superchlorination is of value in removing color in some waters but has little effect (often below 20%) on the colors of other waters. The value of this method can only be made by experiment, and if it does not have the required effect when dosed to the breakpoint, there is no advantage to be gained by increasing the dosage beyond this. Activated carbon adsorbs some colors, but its adsorptive capacity for others is so uncertain that carbon use is confined almost entirely to taste and odor removal.

Odors are graded according to their nature and intensity. The terms used in the classification are descriptive. The intensity is often described as very faint, faint, distinct, decided, very strong, or by numbers, 0, 1, 2, 3, 4, 5, and so forth, referring to the number of successive dilutions perceptible, which is known as the odor threshold. To the trained observer, such odor determinations are of value in tracing the contamination, but noses are not equally sensitive, so that different observers may variously interpret the same odor. Therefore, both the classification and the intensity of the odor of a given sample of water may be different in the reports of various observers. This does not, by any means, imply that odor tests are worthless, for they are really very valuable, but it does mean that a certain amount of latitude has to be allowed in interpreting the reports of different observers.

Practically all of the odors in natural waters, with the exception of hydrogen sulfide, are organic in nature. Even the odors and tastes noticeable in many chlorinated waters are seldom due to the chlorine but rather to compounds formed by the action of the chlorine on organic matter present in the water. Some of these are so intense that as little as one part per billion is noticeable. These organic tastes and odors are usually confined to surface waters and are either absent or very low in amount in deep well waters.

Disagreeable odors and tastes render waters objectionable for many uses and industrial processes. They are intolerable in water used for beverages or food products, but it is not in these industries alone that odors are objectionable, for many wet-processed materials such as pulp, paper, and textile materials absorb odors.

The removal of inorganic tastes or odors due to hydrogen sulfide or iron has been discussed. Organic tastes and odors may be removed by means of activated carbon, aeration, or aeration followed by activated carbon. Activated carbon may be employed in powdered form in coagulation or settling basins or in granular form in filters. Partial chlorination may intensify certain odors, whereas breakpoint chlorination may destroy them. With surface waters requiring chlorination or superchlorination, it is often advisable to chlorinate first followed by activated carbon. In some instances, postchlorination in very small doses is then required.

Microorganisms

Microorganisms are common to surface waters but are usually either absent or present in small amounts in deep well waters. Contamination of deep well waters can occur by surface waters seeping down around the casing and, in limestone areas especially, crevices may furnish unobstructed passages for contamination. Shallow wells, some springs, and water from infiltration may also contain appreciable amounts of microorganisms and with iron- or manganese-bearing water iron or manganese bacteria. Deep well waters also may have microorganisms develop in them after they are drawn. Thus, deep well water which has been pumped into an open basin, tank, or reservoir often develops luxuriant growths of algae; or a deep well sulfur water may have clogging growths of sulfur bacteria develop on the trays of an aerator; or bacteria, which reduce sulfates to sulfides, may develop in water mains. In many instances, the point of entry of these organisms is not immediately obvious.

It is in surface waters that organic growths are found in the greater variety and profusion. There are literally many thousands of varieties. Some are visible to the naked eye, others become so at a fairly low magnification, and still others require the highest powers of the microscope. Some are plants and others are animals; in some cases, the exact status is yet unknown. Many of these plants look like animals; on the other hand, many of the animals look like plants.

Microorganisms differ greatly in form, color, and habits as well as in size. The variety of forms, shapes, and patterns is enormous. In one classification alone, the *Diatomacea*, it is estimated that there are over 10,000 species, each of which has its own distinctive shape, pattern, or design of silica shells. The colors of microorganisms also vary; for example, they may be green, yellow, red, pink, brown, blue-green, and in all shades, and some forms are quite transparent and colorless. Some microorganisms live only in sunlight, whereas others thrive in the dark. Some (aerobic) require dissolved air for their existence, whereas others (anaerobic) grow in its absence. They may be nonmotile or motile with various types of appendages that produce numerous methods of locomotion such as rolling, creeping, leaping and rowing, paddling, or swimming. Many of these microorganisms grow, although at reduced rates, in cold water and even under the ice. It is interesting to note that at the other extreme there are some blue-green algae which can live in hot water.

Discoloration and staining of wet-processed materials are frequently caused by microorganisms. Disagreeable tastes and odors, either from the living microorganisms or their decomposition products, may at times become troublesome. Decomposition of cellulose or other organic substances may be caused by microorganisms that are popularly lumped together as molds and slimes.

In many cooling waters, organic growths are a problem and treatment is necessary. Some of these organic growths are matted and fibrous in appearance, but others are so dense that at times they resemble an inorganic scale. Surface waters usually give the most trouble, especially during the warm summer months, but well waters which contain iron or manganese, usually from shallow wells or from infiltration galleries, frequently cause extremely troublesome and clogging growths of iron or manganese bacteria (*crenothrix* is the term commonly used collectively for these) in piping.

Sulfur waters frequently have threadlike growths that appear on aerator trays and in clear wells. Such waters contain elemental sulfur often in a finely divided, colloidal form. This sulfur may be due to the oxidation of the hydrogen sulfide to water and sulfur by dissolved oxygen, but since this reaction is rather slow, it is believed that the bacteria speed up the process. The sulfate-reducing bacteria, which reduce sulfates to sulfides, are often responsible for the production of shots of black water. These are obtained most frequently from dead ends in mains, and the black color is due to ferrous sulfide.

Tolerances for microorganism will differ according to the types and numbers present and the end use for the water in the various industries. Water that is to be used for drinking purposes should, of course, conform to the public health standards on drinking waters. Where other supplies are used in plants, cross-connections must be avoided and all dangers of back siphonage must be eliminated. Obviously, water that is to be used in the manufacture of beverages or food products must conform to strict standards. Even where products are sterilized in sealed containers, it is shortsighted to allow any water which is not above suspicion to come in contact with them.

The removal, destruction, or prevention of growths of microorganisms may be accomplished by various means. Algae and other chlorophyll-containing plants need sunlight in order to grow. Therefore, if the water can be stored in covered reservoirs, the growth of these organisms can be prevented. Floating rafts have sometimes been employed, but they are not recommended as they usually are not very satisfactory. In open settling basins, activated carbon has sometimes been employed for its light-screening action, but obviously this is somewhat limited in its applications. In large open reservoirs, treatments with measured dosages of copper sulfate are frequently employed. The dosages have to be carefully regulated in order to avoid killing fish.

In most industrial plants, killing and removal of organic growths is best accomplished in the settling basin in the water-treatment plant. This is done with chlorine and coagulation, settling, and filtration to remove the remains. Where chlorine is so used, the process is known as prechlorination, and it is frequently of value in reducing the dosages of coagulant required. In many cases, however, the bulk of the organic matter is removed by coagulation, settling, and filtration followed by postchlorination. In other cases, both prechlorination and postchlorination may be practiced.

Iron and manganese crenothrix growths are best prevented by removal of these metals followed by chlorination. In the case of sulfur waters, reduction of the hydrogen sulfide should first be carried out after which the water is chlorinated to remove the last traces of hydrogen sulfide and to kill any sulfate-reducing bacteria that may be present. Slimes and molds in such industries as pulp and paper mills are controlled by chlorination or by various other germicides. Where the amounts of water to be handled are relatively small and the content of microorganisms and turbidity is small, settling basins are frequently dispensed with and chlorine or a hypochlorite and a coagulant are added in the line leading to the filters. In these cases, especially if the water demand is not very large, a pressure reaction tank, usually known as a pressure settling tank, is of advantage before the filters. In plants where the cold lime or lime soda process of water softening is employed, the coagulation and settling may be effected in the equipment and used. If the treatment is carried to the caustic stage, a considerable reduction in the bacterial content may be effected, but postchlorination also is usually advisable.

WATER SUPPLIES

Water is used in such vast quantities that in sheer quantity and bulk it far overshadows all other materials. It requires over 250 tons of water to make one ton of steel, over 700 tons to make 1 ton of paper, over 100 tons to make 1 ton of aluminum, and the list could go on and on, for in practically every industry the greatest amount of any material used is water.

Sources may be a surface supply—river, creek, canal, pond, lake, or reservoir—or it may be a ground water supply—deep well, shallow well, spring, mine, or infiltration gallery. Whatever the source, the water will contain impurities, for no natural water supply consists of chemically pure water.

In different water supplies impurities will be varied both in character and amounts present. In fact, in many water supplies, they may at different periods vary over a rather wide range in the same supply. This is especially the case with flowing streams, which may show not only seasonal changes in composition but daily or hourly variations because of rainfalls. On the other hand, water from a large lake or a deep well may be remarkably constant in composition even over a period of many years.

Impurities that may be present in water supplies may be grouped generally as follows:

- Dissolved mineral matter
- Dissolved gases
- Turbidity and sediment
- Color and organic matter
- Tastes and odors
- Microorganisms

Whether or not these impurities are harmful depends on

- The nature and amounts of the impurities present
- The uses to which the water is intended
- The tolerances for various impurities for each use

Treated Water Supplies

The municipal water supply may be chlorinated, aerated, filtered, or softened. Most municipal water supplies, especially of surface origin, are chlorinated whether or not they receive any other treatment. In general, the dosages of chlorine used are very small, so that although the water is rendered safe for drinking purposes, there is little impact on the mineral matter in the water.

With some water supplies, usually surface supplies, chlorination results in objectionable tastes and odors. These chlorine tastes are due to the action of chlorine on certain organic impurities in the water, thereby forming compounds of such marked taste and odor that as little as 1 part in a billion parts of water may be noticeable.

Most filtered municipal water supplies are clear and low in color. If also treated so as to render the filter effluent comparatively nonaggressive, the suspended matter may be so slight as to be of little or no consequence for many industrial uses. In addition, a number of municipal water-treatment plants employ activated carbon for taste and odor removal, but for industrial uses where taste and odor control is important, it is advisable to install dedicated taste and odor removal equipment as an additional safeguard.

Aeration is practiced in many municipal water-treatment plants. The purpose is to reduce hydrogen sulfide content, to reduce tastes and odors, or, in combination with filtration, to remove iron or manganese. In such cases, lime may also be added, especially for manganese removal, to build up the pH value, and such additions may increase the hardness of the water unless partial softening is practiced.

Softened Municipal Water Supplies

Municipal water softening may best be considered separately from other forms of municipal water treatments, because it has so much greater an effect on the dissolved mineral content of the water. Over 1,000 municipalities in this country now partially soften their water supplies.

In general, municipalities do not attempt to soften the water completely but instead merely reduce the hardness to about 85 ppm. Some do not reduce it this far but instead reduce it to only 135 ppm or some intermediate point. Relatively few municipalities reduce it below 85 ppm, and there are even less that soften the water completely. Such partially softened water may be quite suitable for most household uses, but for most industrial uses further softening or treatment is typically required.

Fluoridated Municipal Water Supplies

A large number of municipalities fluoridate their water supplies as a dental health measure. The amount of fluoride added is very small, as the final concentration in the treated water is only 0.7–1.1 ppm, expressed as F (fluorine).

Where a municipal water supply is used in an industry, samples for analysis may be drawn from any convenient tap, and it is simply necessary to allow the water to run for several minutes before sampling and to observe the usual sampling precautions. In such cases, too, records are usually available that will show what variations in composition may be expected. All these records should be carefully studied, for the analysis of a single sample does not necessarily mean that it is representative of the quality of water that may be supplied at all times. Even if the records do not contain complete mineral analyses, they usually will give sufficient information to supplement the knowledge gained from the complete mineral analyses. In the case of most city waters, the manufacturer of water-treating equipment will usually have in its files a number of complete mineral analyses made at different periods.

If the municipality draws its water supply from a large lake or from deep wells tapping the same aquifer, the composition of the water may be remarkably constant. If the water instead is drawn from different well fields, from several sources, from various reservoirs, or from a river or stream variations in composition may be expected and the range of these variations should be known. If records of such variations are not available, a series of periodic analyses should be made or, if time does not permit such a course, allowances should be made for whatever variations in composition may reasonably be expected following a survey of the source or sources of supply.

Units Used in Water Analysis

There are many methods of expressing water analyses. Expressions can be quite complicated. In Table 6, the same quantity of calcium is shown in 28 different forms. It might also be expressed as equivalents per million (epm) or milliequivalents per liter (mEq/L), which are different ways of expressing the same numerical quantity, in this case 2 epm (or 2 mEq/L).

Table 6 The Same Amount of Calcium Expressed in Different Ways

Name	Formula	Parts per million	Parts per 100,000	Grains per US gallon	Grains per imperial gallon
Calcium	Ca	40	4	2.3	2.8
Calcium oxide	CaO	56	5.6	3.3	3.9
Calcium carbonate	$CaCO_3$	100	10	5.8	7.0
Calcium bicarbonate	$Ca(HCO_3)_2$	162	16.2	9.4	11.3
Calcium chloride	$CaCl_2$	111	11.1	6.5	7.8
Calcium sulfate	$CaSO_4$	136	13.6	8.0	9.5
Calcium nitrate	$Ca(NO_3)_2$	164	16.4	9.5	11.5

Water Analysis Units

Numerically there are only four basic units for water analyses, although there are six names for them. These are

- Parts per million (ppm): the number of parts of substance per million parts of water. This same unit is also known by two other names—milligrams per liter (mg/L) or grams per cubic meter (g/m^3). All three are numerically equal.
- Grains per US gallon (gpg): the number of grains of substance per 1 US gallon of water (1 grain = 1/7000 pound and 1 US gallon of water weighs 8.33 pounds)
- Parts per hundred thousand (pts/100,000): the number of parts of substance per 100,000 parts of water
- Grains per imperial gallon (gpg imp): the number of grains of substance per 1 British imperial gallon of water (1 grain = 1/7000 pound and 1 imperial gallon of water weighs 10 pounds)

Table 7 gives the interrelations between these basic units and other equivalents such as pounds per thousand gallons and per million gallons both for the US and imperial gallon. The most widely used basic unit is parts per million. Another unit that is commonly used in this country, especially in expressing hardness, is grains per US gallon, and one widely used for the same purpose in other English-speaking countries is grains per imperial gallon. The parts per hundred thousand unit is now much less widely used than formerly, having been largely displaced by parts per million or its equivalents.

In making analyses of fresh waters, the samples are measured by volume at the prevailing temperature and not by weight. Corrections for temperature or specific gravity are practically never made (e.g., boiler salines and brines are

Table 7 Water Analysis Units: Interrelations and Equivalents

1 Part per million (ppm)		
	= 1.0	milligrams per liter
	= 1.0	grams per cubic meter
	= 0.1	parts per 100,000
	= 0.0583	grains per US gallon
	= 0.00833	pounds per 1,000 US gallons
	= 8.33	pounds per 1,000,000 US gallons
	= 0.07	grains per imperial gallon
	= 0.01	pounds per 1,000 imperial gallons
	= 10.0	pounds per 1,000,000 imperial gallons
1 Grain per US gallon (gpg)		
	= 17.1	parts per million
	= 17.1	milligrams per liter
	= 17.1	grams per cubic meter
	= 1.71	parts per 100,000
	= 0.143	pounds per 1,000 US gallons
	=143.0	pounds per 1,000,000 US gallons
	= 1.2	grains per imperial gallon
	= 0.171	pounds per 1,000 imperial gallons
	=171.0	pounds per 1,000,000 imperial gallons
1 Part per hundred thousand (pts/100,000)		
	= 10.0	parts per million
	= 10.0	milligrams per liter
	= 10.0	grams per cubic meter
	= 0.583	grains per US gallon
	= 0.0833	pounds per 1,000 US gallons
	= 83.3	pounds per 1,000,000 US gallons
	= 0.7	grains per imperial gallon
	= 0.1	pounds per 1,000 imperial gallons
	=100.0	pounds per 1,000,000 imperial gallons
1 Grain per British imperial gallon		
	= 14.3	parts per million
	= 14.3	milligrams per liter
	= 14.3	grams per cubic meter
	= 1.43	parts per 100,000
	= 0.833	grains per US gallon
	= 0.119	pounds per 1,000 US gallons
	=119.0	pounds per 1,000,000 US gallons
	= 0.143	pounds per 1,000 imperial gallons
	=143.0	pouonds per 1,000,000 imperial gallons

exceptions), and it is assumed that 1 liter weighs one kilogram; that 1 US gallon weighs 8.33 pounds; and that 1 British imperial gallon weighs 10 pounds.

Water Analyses Expressed as Calcium Carbonate ($CaCO_3$) Equivalents

As noted in Table 6, even when one basic unit such as parts per million is used, it is possible to express part or all of the calcium in seven different ways, each one of which uses a different numerical figure to express the equivalent amount of calcium. This, of course, is an extreme example; for instance, one could hardly imagine anyone using all these values to show parts or all of the calcium present in the same water. However, in a series of analyses made by different analysts, it would be possible to find all forms, although each analysis might use only one, two, or possibly three of them.

If we listed the magnesium equivalents, there would be 7 of them, making a total of 14 different methods of expression for the two constituents that together account for the hardness of water. Since calculations for water treatment were so complicated, it was decided to express the total hardness as the calcium carbonate ($CaCO_3$) equivalent of all of the hardness constituents. This makes a convenient yardstick; moreover, calcium carbonate has the figure of 100 as its molecular weight, which is a convenient, round, easy number to remember and which is large enough so that practically all of the results appear as whole numbers. Thus, it is customary to express total hardness, calcium hardness, and magnesium hardness as their $CaCO_3$ equivalents. This equivalent extends to the three forms of alkalinity—bicarbonate, carbonate, and caustic (or hydroxide)—and is convenient to extend this to mineral acidity, which might also be shown as "negative alkalinity." It is common practice to express the cations of calcium, magnesium, and sodium; the total hardness; the bicarbonate, carbonate, and hydroxide alkalinity; the anions of sulfate, chloride, and nitrate; and mineral acidity in terms of their calcium carbonate ($CaCO_3$) equivalents in order to simplify the calculations used in water treatment. On the other hand, free carbon dioxide is usually expressed as CO_2 and iron, manganese, silica, and fluoride as Fe, Mn, SiO_2, and F, respectively.

Water Analyses Expressed as Equivalents per Million or Milliequivalents per Liter

The two units, equivalents per million (epm) and milliequivalents per liter (mEq/L), are numerically equal. To express the amount of a substance present in terms of either one, divide the number of parts per million of the substance present by its equivalent weight. The equivalent weights of the common cations and anions in water are Ca 20, Mg 12.2, Na 23, HCO_3 61, CO_3 30, OH 17, SO_4 48, Cl 35.5, and NO_3 62.

Although equivalents per million epm and milliequivalents per liter are used, they are not as widely used as the calcium carbonate equivalents. The objection is inconvenience: with these equivalents, many of the figures appear on the right hand of the decimal point, whereas with the calcium carbonate equivalents, especially when expressed in parts per million, most of the figures appear as whole numbers. The equivalents of most of the substances found in water work will be found in the Appendix.

Water Hardness: Methods of Expression

The method of expressing the hardness of a water is as its $CaCO_3$ equivalent, in parts per million, grains per US gallon, parts per hundred thousand, or grains per imperial gallon. Hardness is also expressed as $CaCO_3$ equivalents, in parts per million, grains per US gallon or to a limited extent in parts per hundred thousand.

In expressing results of a water analysis, no more than three significant figures should ever be given, for the methods employed do not warrant the use of a greater number. If the application of a factor results in such a figure as 214.64 ppm, it should be rounded off to an even 215 ppm. Even then, it should be understood that the accuracy of the determinations does not extend to +1 in the third significant figure. For certain determinations, two significant figures may be all that are warranted. The chemical factors and equivalents used in water-treatment calculations also need not consist of more than three significant figures. Thus the molecular weight of calcium chlorinate may be taken as 100 instead of 100.09, and the factor for converting Ca to $CaCO_3$ may be rounded off to 2.50 instead of the more exact 2.497.

WASTES AND WASTE EFFLUENT CHARACTERISTICS

Knowledge of the composition of sewage and industrial wastewaters allows a better assessment of methods of treatment that should be applied before discharge to a receiving body of water. Such knowledge helps to determine whether or not an industrial waste will attack the sewer, whether it should be treated alone or in admixture with sewage—and in what proportions, and whether a single-stage or multistage process should be used. A thorough study of the processes from which various components of an industrial waste arise, together with their composition, may lead to recovery of materials and water saving. Analysis of treated wastewaters is necessary to assess potential toxicity and disease hazards to humans and toxicity to fish and other biota. This is desirable, especially in light of the growing need for reuse and conservation of water, to help decide how to treat the effluents further for use as low-grade water for industrial purposes, and as potable water as well as to meet regulatory requirements of discharge.

The principal source of pollution in sewage is human excreta, with smaller contributions from food, washing, laundry, surface drainage, and so forth.

Industrial wastes, in general, consist of one or more strong spent liquors from specific processes together with comparatively weak waters from rinsing, washing, condensing, floor washing, and so forth. Typically analyses of the sewage itself have to be made for specific substances for particular purposes, such as indole for odor, detergents for foaming, or pesticides for health hazards.

Sampling

The value of analytical work can be reduced if the sampling techniques and programs adopted do not adequately take account of the nature of the wastewater and the wide, often rapid, fluctuations in flow, strength, and composition which occur in wastewaters and effluents. In purely domestic systems, the variations follow a fairly regular pattern, but where industrial wastes are discharged to the sewers other fluctuations will be observed. Continuous industrial processes often give rise to less variation in the wastewater produced than occurs in municipal sewage. Factories operating batch processes, or in which much washing water is used, often give rise to fluctuations in volume and composition of waste. Treatment of wastes by the biological processes reduces the fluctuations, and there is usually little systematic variation in the composition of the effluent.

Equipment used for sampling range from vessels hand operated such as beakers and buckets to sophisticated automatic devices which take constant volume samples at prescribed time intervals or which take volumes proportional to the flow of wastewater. The type of sampler must be suitable for the waste being sampled; for example, any tubes through which the liquid has to pass should be wide enough to prevent clogging by suspended solids. Sometimes it is necessary to analyze the subsamples; in other cases, it is important to know the total pollution load on a treatment plant or to be discharged. For this purpose a composite sample is made by mixing the subsamples in amounts proportional to the flow of liquid at the time of sampling. The time interval between subsamples is commonly 1 hr for sewage, but for industrial wastes more frequent samples may be required. To help decide the frequency of sampling, a knowledge of the manufacturing processes involved is invaluable. In view of postsampling changes, it may be necessary to make provision for preservation of subsamples by refrigeration or addition of a bacterial inhibitors.

The collection of samples for bacterial examination requires sterile conditions, especially where biological effluents are involved. The qualitative estimation of viruses is usually made by using the Moore "swab" technique, in which pads of sterile cotton gauze are immersed in the flowing liquid for a number of hours before being withdrawn for examination.

Sample Preservation

Because of changes which take place on standing, it is desirable that analyses be made immediately after collection. This may not be possible, but it is essential that for such determinations as dissolved gases, volatile substances, and bacterial numbers the analyses be made within a short time of collection. Of the changes which can occur when bacteria are present, the more important are the absorption of dissolved oxygen, growth and death of bacteria, hydrolysis of urea to ammonia, oxidation of ammonia to nitrite and nitrate, reduction of nitrite to gaseous nitrogen, disappearance of sugars, and formation of volatile fatty acids. Changes in the bacterial count on storage vary widely from sample to sample, probably due to unknown factors. The effect of storage on the 5-day biochemical oxygen demand (BOD) is consistently to lower the value obtained, but there is a wide scatter in the proportional reduction.

Other methods used to preserve samples do not allow the BOD or bacterial count to be estimated but are useful for determination of chemical and physical parameters. The addition of sulfuric acid keeps the suspended solids (SS) constant for 8 days and the chemical oxygen demand (COD) constant for at least 17 days, whereas the addition of the antibiotic polymyxin B, after heating the sample to 80°C for 1 hr, gave constant COD values for at least 6 months on storage at 22–27°C. Mercuric chloride, added at 3–50 mg/L, depending on the concentration of organic matter in the sample, preserved samples for at least 2 weeks for the determination of pH value, COD, total solids, volatile solids (VS), ammonia, nitrite, nitrate, and organic nitrogen but interfere in the determination of phosphate and phenol. It is undesirable to use mercuric chloride in gas-liquid chromatography methods and in the methods for organic carbon involving catalytic gaseous oxidation. Agents such as chloroform, formalin, thymol, and potassium cyanide are ineffective. The method chosen will depend on the situation, and two or more methods of preservation may have to be used for a single series of samples. Whatever method is chosen, the samples should be stored in the dark in full bottles.

Separation and Concentration

Some determinations can be carried out directly on the untreated sample, but because of the very low concentration of many constituents, and also sometimes for convenience, many samples are processed to separate or concentrate the constituent or to remove interfering compounds before analyses are made.

The constituents in suspended solids are best estimated on the dried solids, which are conveniently prepared by lyophilization (freeze drying) of the various solids fractions obtained by successive settlement, differential centrifugation, and ultrafiltration by membrane or Pasteur candle. For most constituents in solution, lyophilization is, again, a useful general concentrating process which prevents loss of heat-labile substances. The method has been used successfully with domestic sewage to yield fine, buff- or tan-colored powders and is especially useful when collecting composite samples over a long period. Recovery by this method of the total solids in solution his high, with the highest proportional loss being of volatile acids and ammonia; the recovery of the total solids in raw sewage losses are considered to be largely manipulative and nonselective. Sewage and biological effluents have also been concentrated at about 54°C under reduced pressure in rotary and cyclone evaporators designed to minimize loss of heat-labile substances.

Since volatile compounds are preferentially lost by the evaporative methods, special means must be used to determine such compounds in the original sample. Volatile acids and bases have been separated by appropriate steam distillation, whereas some of the neutral volatile compounds have been extracted by solvents. Solvent extraction is also extensively used to separate fats, greases, and detergents from solids using petroleum ether, chloroform, or alkaline methanol; to separate nonvolatile acids and pesticides from solution; and to separate carbohydrates of differing degrees of complexity by successive extraction with aqueous ethanol, perchloric acid, 0.25N HCl, and 72% H_2SO_4. Examples of methods for removing interfering substances prior to analysis are the use of ion exchange resins to remove salts and amino acids before separating individual sugars by paper chromatography and the use of electrodialysis to eliminate nitrite and nitrate from concentrated effluents prior to the determination of sugars by the anthrone method.

Analytical Procedures

Methods are used to determine such parameters as, for example, BOD, COD, and SS. Hitherto most analyses for organic constituents have been made by conventional titrimetric and calorimetric procedures and, after suitable concentration and separation, by long-established chromatographic methods. Examples are amino acids by reaction with higher fatty acids by reversed phase column chromatography. More use is being made of gas-liquid chromatography of aqueous solutions for volatile fatty acids and also of infrared spectroscopy for degradation products of alkylbenzene sulfonates (ABS) and for poly-β-hydroxybutyric acid in sewage microorganisms. The *wet* combustion method for organic carbon has been replaced by high-temperature catalytic oxidation and determination of the carbon dioxide formed by a nondespersive infrared method. Bacterial numbers have been determined by most probable numbers (MPN) or plate counts, and protozoa have been counted by direct microscopic observation.

Sewage Characteristics

Fresh sewage is normally turbid and appears gray to yellow-brown depending on the time of day collected. If industrial wastes are discharged into the sewer, the sewage sometimes takes on the color of the waste. When viewed in ultraviolet light, a colored fluorescence is often seen which is probably due to minor constituents of packaged detergents. Sewage when fresh has a musty, but not offensive, odor; on standing, however, putrefaction sets in and objectionable odors are produced. Occasionally the odor of an industrial waste is evident or identifiable.

The temperature of sewage is normally a degree or two above that of the water supply; in winter in moderate climates, the temperature range is 8–12°C and in summer 17–20°C. When hot discharges are made to the sewer, higher temperatures are observed, and similarly, infiltration of storm or surface waters can be expected to cause decreases in temperature. Solids suspended in sewage range from colloids up to recognizable matter.

Changes occur in the flow, strength, and composition of sewage hourly, daily, and seasonally; of these, hourly changes are usually greatest. Variations in flow are normally larger, the smaller the community served; the hourly variation is usually 50–200% of the average and can be as wide as 20–300%. The strength and composition also vary considerably during a day, and a fairly regular pattern is followed.

The polluting strength of sewage is assessed by such parameters as 5-day BOD, COD, SS, and ammonia content. The strength varies widely and depends on such factors as the quantity of water used per head of population, the amount of groundwater and surface water entering the sewer, and local habits.

Constituents

Inorganic content of sewage depends on the nature of the water supply from which it is derived as well as on the nature of the polluting material. The major groups present in solution in sewage are sugars, free and bound amino acids, volatile and nonvolatile acids, anionic detergents, and unspecified, ether-soluble neutral compounds, whereas minor groups included bases, amphoterics, phenols, sterols, and various nitrogen-containing substances.

In all fractions of suspended solids, fats, carbohydrates, and proteins are the main identified constituents and together account for 60–80% of the organic matter present.

There are wide variations of fats and greases. For example, total fat and grease is usually 40–100 mg/L, but values as low as 16 mg/L and as high as 1480 mg/L have been reported for sewage containing industrial wastes. Free fatty acids reported include all the saturated ones from C_8 (caprylic) to C_{14} (myristic), including those whose odd numbers of carbon atoms, the saturated acids C_{16}, C_{18}, and C_{20}, and the unsaturated acids C_{16}-2H, C_{18}-2H, and C_{18}-4H. A number of these acids are bound as esters; free acids as a proportion of total free plus ester acid varies. The major acids are palmitic, stearic, and oleic, which together form the majority—over two-thirds or even as much as 90%; myristic, lauric, and linoleic acids are present in relatively low proportions.

Proteins and amino acids comprise the largest single nitrogen-containing group in sewage, with the proportion of total organic nitrogen present in this form varying.

A number of N-containing compounds other than amino acids have been found in sewage. Of those derived from urine, urea is the most abundant in fresh material; concentrations as high as 55 mg N/L, but more usual values were in the range 2–16 mg N/L. Changes in urea content are fairly rapid, aerobically and anaerobically, leading to higher ammonia content and lower organic carbon content. Uric acid is present at a fairly constant level of 0.2–1.0 mg/L; hippuric acid has also been detected.

Miscellaneous Constituents

Sterols as a group were thought to be normally present in sewage at 0.1–0.2 mg/L and seldom in excess of 1–2 mg/L. Cholesterol has been reported at 0.04–0.26 and 0.03–0.05 mg/L, whereas the concentration of coprostanol ranges from 0.096 to 0.75 mg/L.

Various pigments derived from plants, e.g., chlorophyll and lycopene, from urine, e.g., urochrome, and from feces, e.g., stercobilin, have been detected in very low concentrations in sewage.

The increasing use of synthetic detergents is reflected in their increasing concentration in sewage.

Organisms present in sewage originate from feces, soil, and water and range from viruses through bacteria and fungi to protozoa and worms. The identity and concentration of such organisms is imprecisely known and effort has been concentrated on organisms pathogenic to humans and the degree to which they are removed by treatment. Many species of the various types of organisms have been described, and in some cases the numbers of individuals of the species have been reported, but nothing like a comprehensive analysis of the total number of organisms present, even of bacteria, is available.

INDUSTRIAL WASTEWATERS

The total polluting load of industrial wastewaters has been estimated to be at least as great as that of domestic sewage. The volume and strength of industrial wastes vary considerably from industry to industry and even within each industry there are wide variations. General properties of wastes from a given industry are usually similar. For many installations there are diurnal variations associated with batch production, a weekly pattern with decreased flow at weekends, or seasonal variations associated with availability of raw materials. Many industrial wastes, especially from the food industry, are similar to domestic sewage and can be purified alone or in admixture with sewage by the usually biological processes. Other wastes are characteristically different. Others cannot be purified by the usual methods either alone or in admixture with sewage. Some wastes are therefore discharged, with or without treatment, to sewers and others are discharged to rivers or the sea, again with or without prior treatment.

Any given industrial discharge must be examined in detail, not only by analysis of suitable samples but also by examining in detail the processes that produce each of the waste waters that together form the discharge. Table 8 summarizes common impurities found in water, their chemical formula, difficulties caused, and means of treatment.

GROUND WATER CHARACTERISTICS

Ground water technology involves the art and science of investigating, developing, and managing ground water. The technology involves specialized fields of soil science, hydraulics, hydrology, drainage, geophysics, geology, mathematics, agronomy, metallurgy, bacteriology, and electrical, mechanical, and chemical engineering. Ground water engineering is an important source of water. Ground water recovery for water supply—ground water engineering—is important in problems concerning seepage from surface reservoirs and canals, the effects of bank storage, stability of slopes, recharging of ground water reservoirs, controlling of saltwater intrusion, dewatering of excavations, subsurface drainage, and construction, land subsidence, waste disposal, and contamination control.

Ground waters involve the determination of aquifer properties and characteristics and the application of hydraulic principles to ground water behavior for the solution of engineering problems. Determination of aquifer characteristics and the application of data by appropriate methods are essential to the solution of complex problems in which ground water is a factor. The extent to which the determination of aquifer properties and characteristics must be made depends on the complexity of the problem involved. A required investigation may range from cursory to detailed and may entail study or consideration of all or only one or two aquifer properties and hydraulic principles. Conditions often may be so complex as to preclude the determination of finite values and the application of available theory to the solution of some problems. In such cases, solutions may be largely subjective and their reliability dependent on the experience and judgment of the ground water specialist.

Shallow, hand-dug wells and crude water-lifting devices marked the early exploitation of ground water. The introduction of well-drilling machinery and motor-driven pumps made possible the recovery of ground water in large amounts and at increased depths. The benefits of ground water development have become increasingly important. The use of water for domestic purposes usually has the highest priority, followed by industrial requirements, and then agricultural usage (irrigation). Development of the ground water resources has been increasing in

Table 8 Common Impurities Found in Water

Constituent	Chemical formula	Difficulties caused	Means of treatment	Minimum effluent—guarantee
Turbidity	None—expressed in analysis as units	Imparts unsightly appearance to water. Deposits in water lines, process equipment, boilers, etc. Interferes with most process uses.	Coagulation, settling, and filtration	Coagulation—average 5 ppm; filtration—2 ppm
Color	None—expressed in analysis as "units" of color on arbitrary scale	May cause foaming in boilers. Hinders precipitation methods such as iron removal, hot phosphate softening. Can stain product in process use.	Coagulation and filtration, chlorination, adsorption by activated carbon	Coagulation filtration—5 on cobalt scale
Hardness	Calcium and magnesium salts expressed as $CaCO_2$	Chief source of scale in heat exchange equipment, boilers, pipe lines, etc. Forms curds with soap, interferes with dyeing, etc.	Softening, distillation, internal boiler water treatment, surface-active agents	Cold lime soda—16 ppm (railway process); hot lime soda—10 ppm; ion exchange—zero
Alkalinity	Bicarbonate (HCO_3), carbonate (CO_3), and hydrate (OH), expressed as $CaCO_3$	Foaming and carryover of solids with steam. Embrittlement of boiler steel. Bicarbonate and carbonate produce CO_2 in steam, a source of corrosion.	Lime and lime soda softening, acid treatment, hydrogen zeolite softening, demineralization, dealkalization by anion exchange, distillation, split-stream ion exchange	Cold lime soda—35 ppm; hot lime soda—17 ppm; H_2Z ion exchange—zero; anion exchange—10 ppm; split-stream ion exchange—zero
Free mineral acid	H_2SO_4, HCl, etc., expressed as $CaCO_3$	Corrosion	Neutralization with alkalies, anion exchange	Anion exchange—zero

Table 8 *(Continued)*

Constituent	Chemical formula	Difficulties caused	Means of treatment	Minimum effluent—guarantee
Carbon dioxide	CO_2	Corrosion in water lines and particularly steam and condensate lines.	Aeration, deaeration, neutralization with alkalies, filming and neutralizing amines	Decarbonator—15 ppm; degasifier—5 ppm; vacuum deaerator—2 ppm; deaerator—zero
pH	Hydrogen ion concentration defined as: $pH = \log \cdot \frac{1}{(H^+)}$	pH varies according to acidic or alkaline solids in water. Most natural waters have a pH of 6–8	pH can be increased by alkalies and decreased by acids	
Sulfate	$(SO_4)^{2+}$	Adds to solids content of water, but in itself is not usually significant. Combines with calcium to form calcium sulfate scale.	Demineralization, distillation	Demineralization—zero
Chloride	Cl^-	Adds to solids content and increases corrosive character of water.	Demineralization. Distillation.	Demineralization—zero
Nitrate	$(NO_3)^-$	Adds to solids content, but is not usually significant industrially. High concentrations cause methemoglobinemia in infants. Useful for control of boiler metal embrittlement.	Demineralization. Distillation.	Demineralization—zero

Fluoride	F^-	Cause of mottled enamel in teeth. Also used for control of dental decay. Not usually significant industrially.	Adsorption with magnesium hydroxide, calcium phosphate, or bone black. Alum coagulation. Anion exchange.	
Silica	$S10_2$	Scale in boilers and cooling water systems. Insoluble turbine blade deposits due to silica vaporization.	Hot process removal with magnesium salts. Adsorption by highly basic anion exchange resins, in conjunction with demineralization. Distillation.	Cold Lime Soda—15 ppm Hot Lime Soda—1 ppm Na_2Z + anion exchange—0.5 ppm Demineralization—0.01 ppm
Iron	Fe^{2+} (ferrous) Fe^{3+} (ferric)	Discolors water on precipitation. Source of deposits in water lines, boilers, etc. Interferes with dyeing, tanning, paper mfr., etc.	Aeration. Coagulation and filtration. Lime softening. Cation exchange. Contact filtration. Surface active agents for iron retention.	Aeration-Filtration—0.3 ppm Manganese Zeolite—0.1 ppm Cation Exchange—zero
Manganese	Mn^{2+}	Same as iron.	Same as iron.	Same as iron.
Oil	Expressed as oil or chloroform extractible matter.	Scale, sludge and foaming in boilers. Impedes heat exchange. Undesirable in most processes.	Raffle separators. Strainers. Coagulation and filtration. Diatomaceous earth filtration.	Coagulation Filtration—Clear bright and free from opalescent turbidity as visible to the naked eye in a clean quart glass bottle. Filtration with preformed floc-Diatomaceous earth.
Oxygen	O_2	Corrosion of water lines, heat exchange equipment, boilers, return lines, etc.	Deaeration, Sodium sulfite. Corrosion inhibitors.	Vacuum Deaeration—0.1 ppm Open Heater—0.3 ml/L Deaerator—0.005 ml/L

Table 8 *(Continued)*

Constituent	Chemical formula	Difficulties caused	Means of treatment	Minimum effluent—guarantee
Hydrogen sulfide	H_2S	Cause of "rotten egg" odor. Corrosion.	Aeration. Chlorination. Highly basic anion exchange.	Aeration Anion Exchange—zero
Ammonia	NH_3	Corrosion of copper and zinc alloys by formation of complex soluble ion.	Cation exchange with hydrogen zeolite. Chlorination. Deaeration.	Cation Exchange—zero Aeration
Conductivity	Expressed as micromhos $\mu\Omega$ specific conductance	Conductivity is the result of ionizable solids in solution. High conductivity can increase the corrosive characteristics of a water.	Any process which decreases dissolved solids content will decrease conductivity. Examples are demineralization, lime softening.	Multibed Demineralization—1 mW Mixed Bed Demineralization—0.1 microhm
Dissolved solids	None	"Dissolved solids" is a measure of total amount of dissolved matter, determined by evaporation. High concentrations of dissolved solids are objectionable because of process interference and as a cause of foaming in boilers.	Various softening processes, such as lime softening and cation exchange by hydrogen zeolite, will reduce dissolved solids. Demineralization. Distillation.	

Suspended solids	None	"Suspended Solids" is the measure of undissolved matter, determined gravimetrically Suspended solids plug lines, cause deposits in heat exchange equipment, boilers, etc.	Subsidence. Filtration, usually preceded by coagulation and settling.
Total solids	None	"Total Solids" is the sum of dissolved and suspended solids, determined gravimetrically.	See "Dissolved Solids" and "Suspended Solids" above.

Note: The above minimum effluent guarantees are not applicable to all plants. The effluent water depends on its final end use and the economics of treatment.

recent years as development of surface water sources approaches the point of full potential use.

Precipitation, storage, runoff, and evaporation of the earth's water follow an unending sequence known as the hydrological cycle and are the source of groundwater. During this cycle, the total amount of water in the atmosphere and in or on the earth remains constant even though its form may change. Although minor quantities of magmatic water or water from other deep-seated sources may find its way to the surface, all water is part of the hydrological cycle.

The movement of water within the hydrological cycle consists of water vapor in the atmosphere condensed into ice crystals or water droplets that fall to the earth as rain or snow. A portion evaporates and returns to the atmosphere. Another portion flows across the ground surface until it reaches a stream and flows to the ocean. The remaining portion infiltrates directly into the ground and seeps downward. Some of this portion may be transpired by the roots of plants or moved back to the ground surface by capillarity and evaporated. The remainder seeps downward to join the ground water body.

Ground water returns to the ground surface through springs and seepage to streams where it is subject to evaporation or is directly evaporated from the ground surface or transpired by vegetation. Thus, the hydrological cycle is completed. When sufficient water vapor again gathers in the atmosphere, the cycle repeats. The elements of the hydrological cycle for any area can be quantified.

A basic ground water equation which permits an approach to a quantitative estimate of ground water availability can be established for an area to account for those factors of the hydrological cycle which have a direct effect on flow and storage of ground water. The equation is

$$\Delta S_{gw} = \text{recharge} - \text{discharge}$$

where ΔS_{gw} is the change in ground water storage during the period of study. Theoretically, under natural conditions and over a long period of time, which include both wet and dry cycles, ΔS_{gw}, will be zero and inflow (recharge) will equal outflow (discharge).

The natural recharge to the ground water body includes deep percolation from precipitation, seepage from streams and lakes, and subsurface underflow. Artificial recharge includes deep percolation from irrigation and water spreading, seepage from canals and reservoirs, and recharge from recharge wells. The natural discharge or outflow from the ground water body consists of seepage to streams, flow from springs, subsurface underflow, transpiration, and evaporation. Artificial discharge occurs by wells or drains. If ground water storage in an area is less at the end of the selected period of time than at the beginning, discharge is indicated as having exceeded recharge. Conversely, recharge may exceed discharge.

Recharge from natural sources include the following:

- Deep percolation from precipitation: Deep percolation of precipitation is one of the most important sources of ground water recharge. The amount of recharge in a particular area is influenced by vegetative cover, topography, and nature of soils as well as the type, intensity, and frequency of precipitation.
- Seepage from streams and lakes: Seepage from streams, lakes, and other water bodies is another important source of recharge. In humid and subhumid areas where ground water levels may be high, the influence of seepage may be limited in extent and may be seasonal. However, in arid regions where the entire flow of streams may be lost to an aquifer, seepage may be of major significance.
- Underflow from another aquifer: An aquifer may be recharged by underflow from a nearby, hydraulically connected aquifer.

The amount of this recharge depends on the head differential, the nature of the connection, and the hydraulic properties of the aquifers.

- Artificial recharge: Artificial recharge to the ground water may be achieved through planned systems or may be unforeseen or unintentional. Planned major contributions to the groundwater reservoir may be through spreading grounds, infiltration ponds, and recharge wells. Irrigation applications, sewage effluent spreading grounds, septic tank seepage fields, and other activities have a similar but usually unintentional effect. Seepage from reservoirs, canals, drainage ditches, ponds, and similar water impounding and conveyance structures may be local sources of major ground water recharge. Recharge from such sources can completely change the ground water regimen over a considerable area.

Ground water discharges—losses from the ground water reservoir—occur in the following ways:

- Seepage to streams: In certain reaches of streams and in certain seasons of the year, ground water may discharge into streams and maintain their base flows. This condition is more prevalent in humid areas than in arid or semiarid areas.
- Flow from springs and seeps: Springs and seeps exist where the water table intersects the land surface or a confined aquifer outlets to the surface.
- Evaporation and transpiration: Ground water may be lost by evaporation if the water table is near enough to the land surface to maintain flow by capillary rise. Also, plants may transpire ground water from the capillary fringe or the saturated zone.
- Artificial discharge: Wells and drains are imposed artificial withdrawals on

ground water storage and in some areas are responsible for the major depletion. Ground water moves in response to a hydraulic gradient in the same manner as water flowing in an open channel or pipe. However, the flow of ground water is appreciably restricted by friction with the porous medium through which it flows. This results in low velocities and high head losses as compared with open channel or pipe flow.

An aquifer is a water-bearing bed or stratum of earth, gravel, or porous stone. Some strata are good aquifers, others are poor. The most important requirement is that the stratum must have interconnected openings or pores through which water can move. The nature of each aquifer depends on the material of which it is composed, its origin, the relationship of the constituent grains or particles and associated pores, its relative position in the Earth's surface, its exposure to a recharge source, and other factors.

Rocks—used here to denote all material of the earth, whether consolidated and firm or unconsolidated and loose or soft—are generally classified as sedimentary, igneous, and metamorphic. The geological structure, lithology, and stratigraphy of rocks in an area provide general knowledge of their potentials as aquifers.

In general, the best aquifers are the coarse-grained, saturated portions of the unconsolidated, granular sedimentary mantle which cover the consolidated rocks over much of the surface of the Earth. The widespread presence of unconsolidated sediments is more common at lower elevations in proximity to streams. These sediments consist of stream alluvium, glacial outwash, wind-deposited sand, alluvial fans, and similar water- or wind-deposited coarse-grained, granular materials. In addition, some residual materials resulting from the weathering in place of consolidated rock are good aquifers.

The coarser grained consolidated rocks such as conglomerates and sandstones are also often good aquifers but are usually found below the unconsolidated granular sedimentary mantle. Their value as aquifers depends to a large extent on the degree of cementation and fracturing to which they have been subjected. In addition, some massive sedimentary rocks such as limestone, dolomite, and gypsum may also be good aquifers. These rocks are relatively soluble and over the years solution along fractures or partings may form voids which range in size from a fraction of an inch to several hundred feet. Some of the best-known and most productive aquifers are cavernous limestones.

The value of igneous and metamorphic rocks as aquifers depends greatly on the amount of stress and weathering to which they have been subjected after their initial formation. In general, the crystalline igneous rocks are very poor aquifers if they remain undisturbed. However, mechanical and other stresses cause fractures and faults in these rocks in which ground water may occur. Such openings may range from hairline cracks to voids several inches wide. In general, these openings disappear with depth and do not yield significant quantities of water below depths of several hundred to a thousand feet.

In coarse-grained crystalline igneous rocks, where inplace weathering has occurred, a thin permeable zone may be found in the transition zone between the sound rock and the thoroughly weathered, usually relatively impermeable, overlying residual material. Some lavas, especially those of viscous basaltic composition, may contain good to excellent aquifers in the zones between successive flows. The scoriaceous upper and lower surfaces of flows are usually porous and permeable, and cooling fractures may be present in a zone extending into the flow from the upper and lower surfaces. Furthermore, coarse-grained sedimentary material may also be present between flows.

An unconfined aquifer is one that does not have a confining layer overlying it. It is often referred to as a free or "water-table" aquifer or as being under "water-table conditions." Water infiltrating into the ground surface percolates downward through air-filled interstices of the material above the saturated zone and joins the ground water body. The water table, or upper surface of the saturated ground water body, is in direct contact with the atmosphere through the open pores of the material above and is everywhere in balance with atmospheric pressure. Movement of the ground water is in direct response to gravity.

A confined or artesian aquifer, as shown in Figure 4, has an overlying, confining layer of lower permeability than the aquifer and has only an indirect or distant connection with the atmosphere. Water in an artesian aquifer is under pressure and when the aquifer is penetrated by a tightly cased well or piezometer,

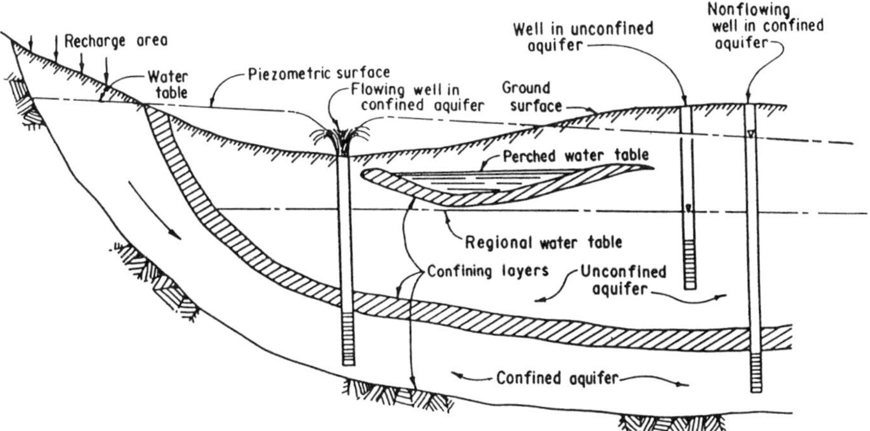

Figure 4 Types of aquifers.

the water will rise above the bottom of the confining bed to an elevation at which it is in balance with the atmospheric pressure and which reflects the pressure in the aquifer at the point of penetration. If this elevation is greater than that of the land surface at the well, water will flow from the well. The imaginary surface, conforming to the elevations to which water will rise in wells penetrating an artesian aquifer, is known as the potentiometric, or piezometric, surface.

Beds of clay or silt, unfractured consolidated rock, or other material with relatively lower permeability than the surrounding materials may be present in some areas above the regional water table. Downward percolating water may be intercepted and a saturated zone of limited areal extent formed. The result is a perched aquifer with a perched water table. An unsaturated zone is present between the bottom of the perching bed and the regional water table. A perched aquifer is a special case of an unconfined aquifer. Depending on climatic conditions or overlying land use, a perched water table may be a permanent phenomenon or one which varies seasonally.

Water may occur in several recognizable subsurface zones under different conditions, as shown in Table 9. The thickness of each zone above the zone of rock flowage varies according to the area and with time. During a period of recharge, the zone of saturation thickens at the expense of the zone of aeration. When discharge exceeds recharge, the zone of saturation thins and the zone of aeration thickens. During periods of recharge, a temporary downward migrating saturated lense may move through the zone of aeration.

Table 9 Status of Water in Various Soil Zones

Zone	Horizon	Condition of water	Condition of soil
Aeration (above water table)	Soil water	Under tension	Unsaturated
	Intermediate	Under tension	Unsaturated
	Capillary fringe	Under tension	Saturated and unsaturated
Saturation (below water table)	Unconfined groundwater	Under pressure but upper surface at atmospheric pressure	Saturated
	Confined or artesian ground water	Under pressure but upper surface above atmospheric pressure	Saturated
Rock flowage		Combined—no free water	Dry

Precipitation typically contains minute amounts of silica and other minerals and dissolved gases such as carbon dioxide, sulfur dioxide, nitrogen, and oxygen which are present in the air and are entrapped as precipitation occurs. As a result, the pH value of most precipitation is below 7 (acidic) and the water is corrosive. On reaching the surface, the water may pick up organic acids from humus and similar materials which increase its corrosive characteristics. While the water is percolating through rock, minerals are attacked by the acid waters and dissolved forming salts which are taken into solution. The amount and character of the salts depend on the chemical composition of the water, the mineralogical and physical structure of the rocks encountered, and the temperature, pressure, and duration of contact.

Nearly all elements may be present in groundwater, and its mineral content varies from aquifer to aquifer and from place to place within an aquifer. Commonly encountered elements and compounds are listed in Table 10.

Less common constituents which are also important because of their effects in use of water are boron (B), manganese (Mn), lead (Pb), arsenic (As), selenium (Se), barium (Ba), copper (Cu), zinc (Zn), hydrogen sulfide (H_2S), methane (CH_4), oxygen (O_2) carbon dioxide (CO_2), and nitrite (NO_2). The mineral content of water is so variable and the acceptable quality for various uses has such a large range that it is not practical to discuss them other than generalities.

Water quality is also important because of its influence on the efficiency and life of equipment. Acidic water with a pH value of less than 7 is usually corrosive; whereas alkaline water ($pH > 7$) is usually less corrosive. However, alkaline water is likely to form deposits on well screens and on pipes. Hard water with a pH value over 7 may be corrosive and may form deposits if it contains relatively large amounts of sulfate, bicarbonate, and chloride radicals. Gases such as hydrogen sulfide, carbon dioxide, methane, and oxygen may be damaging by both corrosion and cavitation.

Contaminated or polluted water contains organisms and substances that make it unsuitable or unfit for use. Ground water may become contaminated as

Table 10 Chemical Constituents Commonly Found in Groundwater

Cations	Anions
Calcium, Ca	Bicarbonate, HCO_3
Magnesium, Mg	Sulfate, SO_4
Sodium, Na	Chloride, Cl
Potassium, K	Nitrate, NO_3
Iron, Fe	Fluoride, F
	Silica, SiO_2

a result of leakage from septic tanks, sewage effluent spreading grounds, garbage dumps, or similar features for the disposal of vegetable and animal waste. Other sources of contamination which are concern are improperly sealed wells, other subsurface structures and excavations, and those areas created by the disposal of oil field brines and industrial wastes through evaporation ponds, spreading fields, and disposal wells.

The distance that organisms may migrate in ground water varies. In general, it should be assumed that in crevassed, fissured, and cavernous rock and in coarse, clean gravels, organically contaminated water may travel as far as several miles. In finer-grained materials, natural filtering action and adsorption may remove such organisms in less than a 100-foot distance. Chemical contaminants may persist indefinitely in ground water. Accordingly, no water should be considered suitable until both chemical and bacterial analyses have shown it to be so.

Differences and changes in the chemical content of water may be useful in determining the source or sources of recharge, direction of flow, and presence of boundaries. The age of water determined by tritium content, carbon 14 dating, and similar analysis may be useful in estimating the age, recharge conditions, or potential direction of flow of ground water or other aspects of ground water hydrology. Also, quality data may be essential in determining the compatibility of water intended for artificial recharge.

Ground water in humid areas maintains the base flow of streams by seepage into stream channels. However, the headwater reaches of some streams may be above the water table and therefore are dry during seasons of low precipitation. In such reaches, seepage from the streambed may charge an underlying aquifer. Some reaches of a stream may be replenished by ground water, whereas others lose water to the ground water reservoir.

In many arid drainage basins, the perennial master streams receive seepage from the ground water reservoir, whereas other streams may be above the water table and streamflow occurs only during periods of high surface runoff. Where the water table is below the streambed, practically all the streamflow may be lost by seepage to the ground water reservoir. Beneath many such streambeds, considerable underflow may be present in the channel fill, although the channel is dry.

It is in the semiarid to arid areas, where irrigation is usually practiced, that water losses from canals and deep percolation from irrigation applications frequently alter natural ground water conditions. Such alterations include water table rise and waterlogging and salination of soils. Artificial drainage by open or buried pipe drains, wells, or other means is often required to lower the water table, maintain a salt balance, and permit the continued production of crops.

Interest has developed in recharging ground water reservoirs with excess surface water. Such recharge is intended to maintain ground water levels, store water for use during droughts, control salt water intrusion, dispose of treated sewage effluent, or for other purposes. In addition, pollutants such as oil field brines and toxic and radioactive industrial wastes are often disposed of by storing them in deep isolated aquifers.

Because suitable surface water reservoir sites are becoming scarce, interest has increased in the underground storage of water. Although underground reservoirs are not as obvious or as readily recognized as surface reservoirs, they offer a possible alternative in many areas where conventional storage would be costly or otherwise undesirable. As is true of all alternative solutions, each type of reservoir offers advantages and disadvantages.

The major application of ground water has been and probably always will be the provision of a water supply by means of wells and infiltration galleries. Facilities range from isolated individual small wells yielding less than a gallon a minute for domestic and stock purposes to well fields consisting of a number of irrigation, municipal, or industrial water supply wells with individual discharges of more than 5,000 gallons per minute. The small individual well seldom presents a problem if it is designed according to proper engineering practice. Larger installations, particularly those with numerous wells, require evaluation of the aquifer characteristics, estimates of well spacing, drawdowns, quality water, and possibly recharge-discharge relationships. Wells should be designed and pumps selected for economical, long, and trouble-free operation within the capabilities of the aquifer, with the consideration of any possible corrosion and encrustation problems which may be present.

Some aquifers have little measurable recharge or discharge but contain large quantities of water in storage which have accumulated over long periods. Estimates can be made of the desirability of mining the water and the probable economic life of such aquifers under various degrees of development.

2

Dissolved Matter

INORGANIC MATTER

Natural water supplies, either in their raw state or after treatment, contain dissolved mineral matter. Mineral constituents vary greatly in amounts and relative proportions present in various water supplies. The most abundant are the bicarbonates, sulfates, and chlorides of calcium, magnesium, and sodium that are present as mixtures of their respective anions and cations. Solubilities are given in terms of $CaCO_3$ (calcium carbonate) equivalents.

Calcium Bicarbonate. Calcium bicarbonate, $Ca(HCO_3)_2$, exists only in solution and is formed by the action of water containing carbon dioxide on limestone, marble, chalk, calcite, dolomite, and other minerals containing calcium carbonate according to the following reaction:

$CaCO_3$	+	CO_2	+	H_2O	=	$Ca(HCO_3)_2$
Calcium carbonate		Carbon dioxide		Water		Calcium bicarbonate

Water at 32°F, saturated with carbon dioxide at atmospheric pressure, the solubility of calcium bicarbonate is 1620 ppm. This is the maximum solubility and most natural water supplies do not have calcium bicarbonate contents which even approach this figure. When the temperature is raised, the solubility decreases because of increasing reversal of the above reaction. In a steam boiler, for example, the reversal is complete and the entire content of calcium bicarbonate breaks down into carbon dioxide, water, and calcium carbonate, which has very low solubility (15 ppm at 3°F and 13 ppm at 212°F).

Magnesium Bicarbonate. Magnesium bicarbonate, $Mg(HCO_3)_2$, exists only in solution and is formed by the action of water containing free carbon dioxide on magnesite, dolomite, dolomitic limestone, and other minerals containing magnesium carbonate according to

$MgCO_3$	+	CO_2	+	H_2O	=	$Mg(HCO_3)_2$
Magnesium carbonate		Carbon dioxide		Water		Magnesium bicarbonate

Water at 3°F, saturated with carbon dioxide at atmospheric pressure, the solubility of magnesium bicarbonate is 37,100 ppm. Although the maximum solubility of magnesium bicarbonate is high, the amounts found in natural water supplies are usually under 75 ppm and seldom reach twice this figure. As the temperature rises, solubility decreases because of an increasing reversal of the above reaction forming first carbon dioxide, water, and magnesium carbonate, which in pure water has a solubility of 101 ppm at 32°F and 75 ppm at 212°F. At higher temperatures, as in a steam boiler, a further change takes place as the magnesium carbonate reacts with water to liberate carbon dioxide and form magnesium hydroxide, which has a very low solubility (17 ppm at 32°F and 8 ppm at 212°F).

Sodium Bicarbonate. Sodium bicarbonate, $NaHCO_3$, is commonly known as bicarbonate of soda and baking soda. Solubility in water at 32°F is 38,700 ppm and increases as the temperature rises, but above 100°F it begins to lose carbon dioxide and it completely breaks down into carbon dioxide, water, and highly soluble sodium carbonate at 212°F. In a steam boiler, it reacts with water, so that most of it is converted into carbon dioxide and sodium hydroxide (caustic soda).

Calcium Sulfate. Calcium sulfate, $CaSO_4$, occurs as the dihydrate ($CaSO_4 \cdot 2H_2O$) in such minerals as gypsum, alabaster, and selenite; as the anhydrous form ($CaSO_4$) in the mineral anhydride; and as the hemihydrate ($CaSO_4 \cdot 1/2H_2O$) in the material known as plaster of Paris, which is made by the partial dehydration of gypsum.

Magnesium Sulfate. Magnesium sulfate, $MgSO_4$, occurs as the heptahydrate ($MgSO_4 \cdot 7H_2O$) that is commonly known as Epsom salt or epsomite when found, e.g., in salt beds, mines, and caverns. Another mineral form is kierserite, which is a monohydrate ($MgSO_4 \cdot H_2O$). It also occurs in other minerals, such as kainite, picromerite, and loweite, as double salts with potassium chloride, potassium sulfate, sodium sulfate, etc. Unlike calcium sulfate, magnesium sulfate is a very soluble salt, having a solubility of 170,000 ppm at 32°F and 356,000 ppm at 212°F. In steam boilers it is corrosive.

Sodium Sulfate. Sodium sulfate, Na_2SO_4, occurs as the decahydrate ($Na_2SO_4 \cdot 10H_2O$) which is an efflorescent salt that is commonly known as Glauber's salt. It is also known as the mineral mirabilite when found deposited, e.g., in salt lakes, salt beds, and caverns. The anhydrous sodium sulfate of commerce, a by-product of the manufacture of hydrochloric acid from common salt and sulfuric acid, is commonly known as salt cake. Like the other sodium salts found in natural and treated waters, sodium sulfate is very soluble and is not a scale former.

Calcium Chloride. Calcium chloride, $CaCl_2$, occurs in natural brines, salt beds, and elsewhere and is obtained as a by-product in the chemical industry. Two mineral forms are tachhydrite, a hydrated double salt with magnesium chloride ($CaCl_2\ 2MgCl_2 \cdot 12H_2O$) and hydrophilite, an anhydrous form ($CaCl_2$). It forms several hydrates such as monohydrate ($CaCl_2 \cdot H_2O$), the dihydrate ($CaCl_2 \cdot 2H_2O$), and the hexahydrate ($CaCl_2 \cdot 6H_2O$). All at 32°F and 554,000 ppm at 212°F. In steam boilers, it is corrosive.

Magnesium Chloride. Magnesium chloride, $MgCl_2$, occurs in sea water, natural brines, salt beds, and elsewhere. Four mineral forms are the anhydrous chloromagnesite ($MgCl_2$), the hexahydrate bischofite ($MgCl_2 \cdot 6H_2O$), the double salt with potassium chloride carnallite ($MgCl_2 \cdot KCl\ 6H_2O$), and tachhydrite (see above paragraph). Magnesium chloride is deliquescent and highly soluble—362,000 ppm at 32°F and 443,000 ppm at 212°C. It is very corrosive in steam boilers, as it reacts with water at such temperatures to form hydrochloric acid and magnesium hydroxide.

Sodium Chloride. Sodium chloride, NaCl, is most widely known as salt without any qualifying adjective. It is also known as common salt, and the mineralogical name for sodium chloride is halite. It is the chief mineral constituent of sea water (2.7% NaCl) and occurs in salt beds, salt lakes, connate waters, other natural brines, and elsewhere. It is anhydrous in composition, but the crystals often enclose some water so that it often decrepitates on heating. Like the other sodium salts found in natural and treated waters, it has a high solubility and is not a scale former.

Other mineral matter which may be found in natural water supplies may include silica, iron, manganese, nitrates, nitrites, potassium, and mineral acidity. Silica, SiO_2, occurs in nearly pure form in quartz and many sands. It is also a major constituent of granite, feldspar, clay, and a host of other minerals. It is found in practically all natural waters and may be present in amounts ranging from as little as 1 ppm or less to over 100 ppm.

Iron is present in most water supplies in small amounts, and if it is present to the extent of 0.1 ppm or less, it may be considered as negligible for most industrial uses. If iron is present in greater amounts than 0.2 or 0.3 ppm, it is usually very objectionable. Manganese is a rarer constituent but is even more objectionable, and tolerances for special uses may be even lower than the tolerances for iron.

Nitrates are usually absent or present only as traces, but there are exceptions, and if appreciable amounts are present, nitrates are determined. Nitrites are either absent or present in such minute amounts that they are rarely determined in industrial water analyses. Ammonia, which possibly might more appropriately be considered under gases, is also usually such a minor constituent that it is seldom determined in industrial water analyses. There are exceptional cases, however, where heavily contaminated waters have given off enough ammonia in the boilers to have an effect on brass valves and fittings.

Potassium, because it is usually present in such small amounts and is so much like sodium in most of its properties, is usually grouped with the sodiums. Fluorides, which may occur in the waters in certain localities in amounts of less than 1 to over 8 ppm, are of importance from a health aspect but are probably of little significance for most industrial uses. A few of the waters in certain western areas contain borates in sufficient amounts to be of importance when used for irrigation. Small amounts of alumina are also commonly found in water supplies, but usually they are of little significance.

Small quantities of carbonate alkalinity may be found, at times, in both natural surface waters and well waters, and objectionable amounts may be present in the so-called alkali waters. Caustic alkalinity is practically never found except in treated waters.

Mineral acidity is found in many waters and some surface waters which have been contaminated by industrial wastes or seepage from mines.

For industrial uses and the usual run of natural and treated water supplies, the mineral constituents shown in Table 1 are the only ones usually considered. As to the amounts of mineral matter dissolved and carried away in solution by natural water supplies. Industrial contamination and contamination from man-made sources require a detailed and full analysis beyond this list.

Alkalinity

Alkalinity can be determined by titration with a standard acid solution, using phenolphthalein and methyl orange as indicators. Results of titration with the methyl orange indicator are expressed as methyl orange alkalinity or total alkalinity. Results of titration with phenolphthalein indicator are expressed as phenolphthalein alkalinity; most natural water supplies contain some free carbon

Table 1 Mineral Constituents Typically Determined in Water Analyses

Name	Formula	Expressed as	Commonly known as
Calcium	Ca	$CaCO_3$	Calcium hardness
Magnesium	Mg	$CaCO_3$	Magnesium hardness (Total hardness = Calcium hardness + Magnesium hardness)
Sodium	Na	$CaCO_3$	—
Bicarbonate	HCO_3	$CaCO_3$	Bicarbonate alkalinity
Carbonate	CO_3	$CaCO_3$	Carbonate alkalinity
Hydroxide	OH	$CaCO_3$	Caustic alkalinity
Chloride	Cl	$CaCO_3$	—
Sulfate	SO_4	$CaCO_3$	—
Nitrate	NO_3	$CaCO_3$	Mineral acidity
Mineral acid	—	F	—
Fluoride	F	F	—
Silica	SiO_2	$SiSO_2$	—
Iron	Fe	Fe	—
Manganese	Mn	Mn	—

dioxide and show no phenolphthalein alkalinity. If no phenolphthalein alkalinity is present, all of the alkalinity is assumed to be bicarbonate alkalinity. If phenolphthalein alkalinity is present, twice the phenolphthalein alkalinity, if less than or equal to the methyl orange alkalinity, is assumed to be carbonate alkalinity. If twice the phenolphthalein figure exceeds the methyl orange alkalinity, the excess is presumed to be caustic or hydroxide alkalinity.

In solutions as dilute as freshwater supplies are, the salts are not present as such but instead are practically dissociated into the corresponding anions and cations. However, it is often convenient and simpler to visualize some of the salts as if they were undissociated, so it is common practice to refer to the calcium bicarbonate or calcium sulfate content of a water as if these substances were present as undissociated salts instead of as ions. In other cases, it may be simpler to consider the ions that are present.

Hardness

The hardness of a water is due to its calcium and magnesium content. Hardness is expressed in terms of calcium carbonate. Calcium hardness is that hardness due to soluble calcium salts; magnesium hardness is that due to soluble magnesium salts. Total hardness is the sum of the calcium hardness plus the magnesium hardness.

Carbonate hardness (the preferred term), *bicarbonate hardness*, and *temporary hardness* are terms for that hardness attributed to the bicarbonates of calcium and/or magnesium. *Noncarbonate hardness* (the preferred term), *sulfate hardness*, and *permanent hardness* are terms for that hardness due to the sulfates chlorides and/or nitrates of calcium and/or magnesium.

The amounts of carbonate and noncarbonate hardness present are determined as follows:

- If the methyl orange alkalinity of the water equals or exceeds the total hardness, all of the hardness is present as carbonate hardness.
- If the methyl orange alkalinity of the water is less than the total hardness, the carbonate hardness equals the alkalinity.
- The noncarbonate hardness, under the conditions above, is equal to the total hardness minus the methyl orange alkalinity.

Calcium Carbonate Scale

Calcium carbonate is much less soluble than magnesium carbonate and both of these are much less soluble than sodium carbonate. It is convenient to visualize the alkalinity as bound first to the calcium, second to the magnesium, and third to the sodium. When a water containing the bicarbonates of calcium, magnesium, and sodium is heated, the calcium carbonate deposits first magnesium carbonate, next, while sodium carbonate is so extremely soluble that no separation in the solid form would occur until the solution was greatly concentrated by evaporation and crystallization takes place. The solubilities of these three carbonates at 212°F are shown in Table 2.

Calcium Carbonate and Magnesium Hydroxide Scale

At the temperatures prevailing in steam boilers, both calcium bicarbonate and magnesium bicarbonate break down to form scale, whereas the calcium scale consists of calcium carbonate, the magnesium deposits as magnesium hydroxide. This is due to the magnesium bicarbonate decomposing first to magnesium carbonate and reacting with the water to form magnesium hydroxide and free carbon dioxide. Sodium bicarbonate undergoes a similar decomposition in a

Table 2 Carbonates: Solubilities of the Carbonates of Calcium, Magnesium, and Sodium at 212°F

Name	Formula	Solubility as $CaCO_3$ ppm
Calcium carbonate	$CaCO_3$	13
Magnesium carbonate	$MgCo_3$	75
Sodium carbonate	Na_2CO_3	289,000

boiler, but since caustic soda is extremely soluble, it remains in solution. These reactions are shown as follows:

$Ca(HCO_3)_2$ Calcium bicarbonate	=	$CaCO_3$ Calcium carbonate	+	H_2O Water	+	CO_2 Carbon dioxide
$Mg(HCO_3)_2$ Magnesium bicarbonate	=	$MgCO_3$ Magnesium carbonate	+	H_2O Water	+	CO_2 Carbon dioxide
$MgCO_3$ Magnesium carbonate	+	H_2O Water	=	$Mg(OH)_2$ Magnesium hydroxide	+	CO_2 Carbon dioxide
$NaHCO_3$ Sodium bicarbonate	=	Na_2CO_3 Sodium carbonate	+	H_2O Water	+	CO_2 Carbon dioxide
Na_2CO_3 Sodium carbonate	+	H_2O Water hydroxide	=	2NaOH Sodium dioxide	+	CO_2 Carbon

Carbon dioxide formed leaves the boiler with the steam. In the case of the sodium carbonate, the conversion to sodium hydroxide often exceeds 80%. The magnesium hydroxide in the scale may lose part of its water content by baking on the tubes and appear partly as magnesium oxide. The solubilities of the hydroxides of calcium, magnesium, and sodium, expressed as parts per million of $CaCO_3$, at 212°F are shown in Table 3.

Under higher temperatures and pressures, the solubilities of both calcium carbonate and magnesium hydroxide decrease. At about 210 psi (392°F), the solubility of calcium carbonate is a little less than 5 ppm and that of magnesium hydroxide is only slightly over 1 ppm, expressed as $CaCO_3$. Calcium hydroxide also has a solubility that decreases with rise in temperature. At 32°F, its solubility is 2390 ppm, at 212°, 888 ppm, and at 210 psi (392°F), 134 ppm, expressed as

Table 3 Hydroxides: Solubilities of the Hydroxides of Calcium, Magnesium, and Sodium at 212°F

Name	Formula	Solubility as $CaCO_3$ (ppm)
Calcium carbonate	$Ca(OH)_2$	888
Magnesium carbonate	$Mg(OH)_2$	8
Sodium carbonate	NaOH	975000

$CaCO_3$. It is not found in a boiler scale unless boiler feed water was grossly overtreated with lime.

Calcium Sulfate Scale

Calcium sulfate is the only scale-forming salt in the noncarbonate hardness for magnesium sulfate, and the chlorides and nitrates of both calcium and magnesium have solubilities exceeding 150,000 ppm by weight expressed as $CaCO_3$, even at 32°F and over 356,000 ppm at 12°F. Calcium sulfate, in the form of gypsum, has a solubility curve which, on elevations of temperature, shows a rise up to about 104°F after which it falls until, at 212°F, it is slightly below the solubility at 32°F. At boiler temperatures the curve falls rapidly, until at 322 psi it is only some 40 ppm, expressed as calcium carbonate. Table 4 shows calcium sulfate solubilities.

Tolerance for calcium sulfate in a cooling water should be large—over 1200 ppm expressed as $CaCO_3$. This is taken into account in the acid treatment of cooling waters. The blowdown on a cooling pond is regulated so as to maintain the calcium sulfate content below some 100 ppm, expressed as $CaCO_3$. Owing to low solubility at the temperatures prevailing in steam boilers and because it forms a very hard and adherent scale, no amount of calcium sulfate can be tolerated in steam boilers.

Thermal Conductivities of Scales

The thermal conductivities (K) of adherent boiler scales, expressed in Btu/ft^2/ft/hr/°F (British thermal units per square foot of area per foot of thickness per hour per 1°F temperature difference) have been found to lie between 0.66 and 2.06 with a mean value of approximately 1.5. The degree of porosity of the scale affects the thermal conductivity, since these pores may be filled with steam instead of water under boiler operating conditions, thus reducing the thermal conductivity in much the same manner as the air cells in the commonly used

Table 4 Calcium Sulfate: Solubilities at 32–428°F

	Solubility as $CaCO_2$	
Temperature (°F)	(ppm)	(gpg)
32	1293	75.5
104	1551	90.5
212	1246	72.7
338 (100 psi)	103	6.0
392 (210 psi)	56	3.3
428 (322 psi)	40	2.3

heat-insulating materials. There is some evidence that porous scales may have thermal conductivities of only 0.2 Btu/ft^2/ft/hr/°F.

Thermal conductivity of firebrick is about 0.75 and steel about 26. Hard-water scales therefore have thermal conductivities of only some 3–8% (average about 5%) of that of steel or about the same as that of firebrick. Hardwater scales are practically as good heat-insulating materials as firebrick. Unfortunately, scale forms in the wrong places, which accounts for its being called misplaced insulation.

Heat-insulating properties of hard-water scales are objectionable not only in steam boilers and water heaters and water-cooled equipment such as condensers, internal combustion engines, and other water-jacketed equipment. Furthermore, these scale deposits do not form a layer which is of even thickness and uniformly distributed over the entire heat transfer area. Instead scale forms most rapidly and consequently is thickest at points where the rate of heat transfer is greatest.

Sodium Salts

Sodium salts that may occur in various natural water supplies are sulfate, chloride, nitrate, bicarbonate, and rarely carbonate. Quantities present in different fresh-water supplies vary over a wide range. For instance in the analyses of the various river lake, spring, and well waters, sodium salts may range from as little as 2% to as much as 98% of the total salts present.

Sodium salts are extremely soluble in either cold or hot water, as is shown in Table 5. Because of these high solubilities, the sodium salts do not form scales, either on heating or in evaporating in the steam boiler, unless the evaporation is carried out to extreme lengths. Also, they do not waste soaps, since soaps that are most widely used are sodium salts of certain of the higher fatty acids. There is no reaction between the sodium salts in water and soap. Strong solutions of sodium salts will throw soap out of solution; an effect is known as salting out. The composition of the soap is unaltered by salting out

Table 5 Sodium Salts and Their Solubilities

		Solubility as ppm of $CaCO_3$	
Name	Formula	(at 32° F)	(at 212°F)
Sodium bicarbonate	$NaHCO_3$	38400	Decomposes
Sodium carbonate	Na_2CO_3	62600	289000
Sodium chloride	NaCl	225000	241000
Sodium hydroxide	NaOH	370000	965000
Sodium nitrate	$NaNO_3$	248000	378000
Sodium sulfate	Na_2SO_4	33200	398000

and it can be redissolved either in fresh water or by diluting the saline solution.

Therefore, for many industrial uses moderate amounts of sodium salts in the water supply are not significant. They do increase the amount of blowdown on the steam boiler, and if much sodium bicarbonate is present, it may be advisable to remove it or to neutralize most of it with sulfuric acid and then to aerate or degasify the acid-treated water before feeding it to the boiler in order to reduce the free carbon dioxide content of the steam.

Sodium salts are objectionable in the processing of manufacture of certain cellulose products, dielectrics, fine drugs and chemicals, synthetic rubber, plastics, photographic materials, silver-plated ware, and other materials. High sodium alkalinity waters are objectionable in high-pressure steam boilers, laundries, and textile plants. Sodium bicarbonate may be removed by a hydrogen cation exchanger and the sulfate, chloride, or nitrate by this treatment followed by treatment with an anion exchanger.

Silica

Silica, in amounts ranging from less than 1 to over 100 ppm, is found in all natural water supplies. In rain, hail, and snow, silica contents range from 0.1 to 2.8 ppm. In the analyses of various surface and ground waters, silica contents range from 1 to 107 ppm. This refers to soluble silica content and not to the silica which may be present in the suspended matter. Suspended matter may be removed from a water supply by coagulation and filtration; such processes have little effect in reducing the soluble silica content.

Silica is particularly objectionable where it may have pronounced scale-forming tendencies. If calcium hardness is present in salines, the scale formed may be a calcium silicate; if soluble alumina is present, an aluminosilicate scale, such as analcite, may be formed; and under other conditions, the scale may consist almost entirely of silica. Silica scales are typically very hard, glassy, adherent, and difficult to remove. Thermal conductivities are usually very low and tube failures can occur with even very thin silica scales. Silica is frequently carried over with steam, forming scale in superheater tubes and on turbine blades. Although this action is usually ascribed to mechanical carryover, silica is soluble to a certain extent in high-pressure steam.

In low- and moderate-pressure steam boilers, silica scale can usually be avoided by maintaining a small excess of phosphate and a ratio of alkalinity to silica of at least 1:1 in the boiler salines. In boilers operated at over 600 psi, treatment of the make-up water to reduce its silica content is frequently necessary. This not only prevents scale in the boilers but also prevents silica deposits on the turbine blades, which may occur even when no serious silica deposits are formed in boilers.

With proper treatment, the silica content of water may be reduced to unobjectionable amounts. If the water to be treated is a surface water which requires coagulation, the use of ferric sulfate as a coagulant will assist in reducing the silica content. Dolomitic lime or activated magnesia, especially in the hot lime soda process, is effective in treatment. The same materials are also used in the cold lime soda process. In ion-exchange demineralizing processes, silica may be removed by direct anion exchange in a strongly basic anion exchanger which has been regenerated by caustic soda (NaOH).

Iron

Iron is present in practically all water supplies but if amounts present are 0.1 ppm or less, they may be considered as negligible for most industrial uses. The iron in various water supplies may be present in several forms.

In deep well waters, iron, if present in amounts of over 0.1 ppm, is almost always present as the soluble, colorless ferrous bicarbonate. Such waters are usually clear and colorless when first drawn, but on standing in contact with the atmosphere, they may slowly cloud and finally deposit a yellowish to reddish brown precipitate of ferric hydroxide. The amounts of iron present in the great majority of such waters will be found in the range below 5 ppm, with a few in the range of 5–15 ppm, and very few above the latter figure. Iron in the form of ferrous bicarbonate may also be found in carbon dioxide–containing, but oxygen-free, corrosive waters which effect iron pickup from the water mains.

So-called red waters, which are caused either by the action of corrosive waters containing dissolved oxygen on iron piping or by the aeration of a water containing ferrous bicarbonate, have suspended iron in the form of a more or less hydrated ferric oxide.

In some surface waters, iron in amounts of over 0.1 ppm may be present in an organic (chelated) form. Usually such waters will show appreciable color. In acid mine waters or in acid surface waters contaminated with acid mine waters or acid industrial wastes, part or even all of the iron may be present as ferrous sulfate. Such waters may also contain manganese sulfate, aluminum sulfate, free sulfuric acid, and suspended ferric hydroxide. Water supplies containing over 0.2 ppm of iron can be objectionable for industrial uses, and for many uses the tolerance should not exceed 0.1 ppm.

Iron-bearing waters also favor the growth of iron bacteria, called iron crenothrix. Such growths form abundantly in water mains, recirculating systems, and other places and can exert a clogging action and cut down the flow rates. Also, these growths can break loose in large, clogging masses.

Iron present as ferrous bicarbonate may be removed by

- Aeration and filtration
- Cation exchange
- Filtration through manganese zeolite

Iron present as suspended ferric hydroxide may be removed by filtration, which may be preceded by settling. Iron present in organic or colloidal form may be removed by coagulation and filtration. Iron present as ferrous sulfate may be removed by neutralization, aeration, and filtration, which may be preceded by settling.

Manganese

Manganese, rarer in occurrence than iron, can be even more troublesome. In clear, deep well waters, it usually occurs as manganous bicarbonate. Such waters may contain appreciable amounts of ferrous bicarbonate as well as manganous bicarbonate. When first drawn, such waters are clear and colorless, but on standing in contact with air, cloud and then usually deposit the yellowish or reddish brown ferric hydroxide first. This is because the iron oxidizes more readily and at lower pH values than manganese. For air to oxidize manganous bicarbonate rapidly, it is usually necessary to have a pH value of above 9.0. The oxidation is catalyzed if previously formed higher oxides of manganese are present as a contact medium.

When not masked by the color of admixed ferric hydroxide, the oxidized manganese forms gray to black deposits and stains. The color of the manganese crenothrix (manganese bacteria) is also black, and these, like iron bacteria, form clogging growths in pipe lines and recirculating systems, which may break loose in the form of large masses. Very small amounts of manganese—0.2–0.3 ppm—may form heavy incrustations in piping, whereas even smaller amounts may form noticeable black deposits.

In acid mine waters, manganous sulfate is frequently found with the ferrous sulfate. In colored surface waters, manganese, like iron, may be present in an organic or colloidal form. Manganese may also be found as manganous bicarbonate in surface waters, especially in rather quiescent waters such as lakes, ponds, or reservoirs. In such cases, the processes of decay may, especially at the bottom, use up the dissolved oxygen, generate free carbon dioxide, reduce manganese to the manganous, state and dissolve it as manganous bicarbonate.

Manganese present as manganous bicarbonate may be removed by

- Aeration and filtration
- Cation exchange
- Filtration through manganese zeolite

Manganese present as manganous sulfate in acid waters may be removed by neutralization, aeration, and filtration. Manganese present in surface waters in colloidal form may be removed by coagulation, settling, and filtration, but such waters may also contain manganous manganese, which requires oxidation.

Aluminum

Small amounts of aluminum are present in most natural water supplies. Since these are usually of little or no importance for most industrial uses, aluminum is determined separately only in exceptional cases.

Aluminum content of a water may be reduced for special uses by

- Coagulation at a favorable pH value, settling, and filtration
- Passing through a hydrogen cation exchanger
- By the ion-exchanger demineralizing process
- By distillation

The addition of an alum coagulant, adjustment of pH values to a favorable point, settling, and filtration will usually reduce the total alumina content of the water to 0.6–1.5 ppm as Al_2O_3.

Fluoride

In certain areas, the water supplies and especially, but not exclusively, the ground water supplies show presence of natural fluoride. Communities are now fluoridating their water supplies as a dental health measure. Supplies which are being fluoridated, the amounts added are small ranging from 0.7 to 1.1 ppm, expressed as F (fluorine). Waters which contain a natural fluoride content, in general, most of these fluoride waters will contain from a fraction of a ppm of fluoride, expressed as fluorine, up to about 3 ppm; a few will range from this to 8 ppm; and only in exceptional cases will the content be above this. The solubility of fluorite—calcium fluoride, CaF_2—is some 16–17 ppm at temperatures of about 65–80°F. This corresponds to about 8 ppm of fluorine. The magnesium salt is over four times as soluble, with a solubility, expressed as MgF_2, of about 76 ppm at 65°F and the sodium salt is quite soluble—about 4% at 60°F.

Owing to the low solubility of calcium fluoride, it can form scale, but this rarely occurs. For most industrial uses, it is of little or no consequence if the water has a small content of fluoride. Some exceptions include where the material is evaporated with a water containing only 1 ppm of fluoride; this would result in a higher concentration in the finished product.

Mineral Acidity

Water supplies that contain mineral acidity are relatively small in number and are usually confined to

- Mine waters
- Ground waters in the vicinity of mines or contaminated by acid trade wastes
- Surface waters contaminated by mine waters or acid trade wastes

The acid present is almost invariably sulfuric acid. The water may contain sulfates of iron, aluminum, manganese, calcium, magnesium, and sodium. Also, the free carbon dioxide content is usually high and, in the case of acid ground waters, ferrous bicarbonate is frequently present. In testing such waters, the mineral acidity is determined by titration with a standard solution of sodium hydroxide and using methyl orange as the indicator. The result includes free acid plus any of the acidic substances mentioned above, is noted as free mineral acidity, and is expressed as the calcium carbonate equivalent.

In mines, the free sulfuric acid and metallic sulfates have formed by the oxidation of sulfur-containing materials, notably pyrites. The most common of these is marcasite, or brassy iron pyrites, which yields sulfuric acid and ferrous sulfate on oxidation in the presence of moisture. Pumps, drains, and piping used in removing such acid waters from mines must be made of special, acid-resistant materials or their useful life will be very short. Not all mine waters are acid; many are alkaline with calcium and magnesium bicarbonates. Even in the same mine, waters from some of the levels may be alkaline and from others acid. Some of these acid mine waters seep through strata to emerge later as springs or to be tapped by wells. In passing through carbonate-containing strata, as the acid or iron sulfate is neutralized to a varying extent by the carbonates, corresponding amounts of carbon dioxide are liberated. If the extent of this neutralization is sufficient to neutralize the free sulfuric acid and part of the ferrous sulfate, the resultant water will contain a large part of its iron in the form of ferrous bicarbonate. In certain cases, this neutralization may proceed so far that the originally acid water becomes alkaline.

DISSOLVED GASES

Dissolved gases that may occur in various water supplies are carbon dioxide (CO_2), oxygen (O_2), nitrogen (N_2), hydrogen sulfide, and methane (CH_4).

Carbon Dioxide

Free carbon dioxide is found in varying amounts, in most natural water supplies, and that picked up by rain water from the atmosphere is very small, usually ranges from about 0.5 to 2.0 ppm on freely analyzed samples. Most surface waters also have low contents of free carbon dioxide; ranging from zero to about 5 ppm. In the case of lakes, surface samples will usually range from zero to about 2 ppm, whereas depth samples may show much higher contents.

In most rivers, small amounts of carbon dioxide, ranging from slightly above zero to about 5 ppm, are usually present most of the time. However, since the oxidation of organic matter furnishes carbon dioxide, rivers containing considerable organic matter may at times show much higher quantities of carbon dioxide and contents of 50 ppm or over are occasionally found. Rivers receiving acid mine waters or acid wastes may also show high contents of carbon dioxide. Table 6 shows carbon dioxide solubilities in water at various temperatures.

In addition to the carbon dioxide formed by the decaying organic matter,

Table 6 Carbon Dioxide: Solubilities in Water at Various Temperatures

Temperature (°C)	Temperature (°F)	Milliliters per liter (ml/L)	Parts per million (ppm)
0	32	1690	3350
5	41	1400	2770
10	50	1170	2310
15	59	996	1970
20	68	855	1690
25	77	733	1450
30	86	637	1260
40	104	491	970
50	122	384	760
60	140	293	580

carbon dioxide may also be picked up from strata or fissures in the Earth's crust. The amount of carbon dioxide contributed to natural surface and ground waters from the atmosphere is practically negligible compared with that contributed by decaying organic matter.

Although most natural waters, freshly sampled, usually have some content of free carbon dioxide, some will show, at times, an appreciable phenolphthalein alkalinity. This is due to photosynthesis; that is, plants, whether large or microscopic, under the influence of sunlight breathe in carbon dioxide and breathe out oxygen. This action does not necessarily stop with the exhaustion of the free carbon dioxide supply but may continue on part of the half-bound carbon dioxide content of the bicarbonates, thus forming the normal carbonate and imparting a phenolphthalein alkalinity to the water.

When carbon dioxide dissolves in water, it forms, to a certain extent, a weakly dissociated acid—carbonic acid. (Ionization constant for the first hydrogen, at 18°C, is 3×10^{-7}; and for the second hydrogen, at 25°C, is 6×10^{-11}.) If the water is free from all traces of alkali, and is saturated with carbon dioxide—about 1450 ppm at 77°F—the pH value is approximately 3.8. This low pH would not be found in natural waters; except for waters containing free mineral acidity or exactly neutral waters, there is some bicarbonate alkalinity always present. Pure distilled water in equilibrium with the carbon dioxide content of the atmosphere will have a pH of about 5.7. Trace of alkalinity will raise this figure, so that most distilled waters in glass vessels will have pH values of around 6.4.

If bicarbonate alkalinity is also present, then the pH value is not dependent on free carbon dioxide alone but instead on the ratio of free carbon dioxide to the methyl orange alkalinity of the water.

Carbon dioxide is corrosive; an example is the attack of oxygen-free condensate on return piping in a boiler system. This attack may be severe owing to the low pH value of condensate which contains carbon dioxide. Natural water supplies, which contain free carbon dioxide but not dissolved oxygen, when passed through pipe, dissolve iron as ferrous bicarbonate. The ferric or ferroso-ferric hydrated oxides react with the iron of the pipe, being first reduced to the ferrous state and then uniting with the free carbon dioxide to form ferrous bicarbonate. This is known as iron pickup and may be stopped by aeration or partial aeration of the water.

Carbon dioxide is an accelerating factor in dissolved-oxygen corrosion. Water which contains in addition to dissolved oxygen a high content of carbon dioxide in relation to its alkalinity will be much more corrosive than a water which contains a low content of carbon dioxide in relation to its alkalinity. Water having a dissolved oxygen content is much more corrosive if it has a low pH than if it has a high pH. If the calcium alkalinity of the water and its pH value also are such that it is saturated with a calcium carbonate, the corrosion will be greatly reduced, for a very thin film of $CaCO_3$ will protect the metal. Sodium silicate and caustic soda treatment of water to prevent corrosion are more fully discussed later.

Carbon dioxide may be reduced to certain limits or removed in cold temperatures by means of an aerator, a degasifier, or vacuum deaerator or in hot temperatures by means of a deaerator. Carbon dioxide may also be neutralized by the addition of lime or an alkali such as caustic soda, but these procedures are usually employed only for raw or treated waters which contain relatively small amounts of carbon dioxide. The reactions for forming either the carbonates or bicarbonates are

$$\underset{\text{Carbon dioxide}}{2CO_2} + \underset{\text{Hydrated lime}}{Ca(OH)_2} = \underset{\text{Calcium bicarbonate}}{Ca(HCO3)_2}$$

$$\underset{\text{Carbon dioxide}}{CO_2} + \underset{\text{Hydrated lime}}{Ca(OH)_2} = \underset{\text{Calcium carbonate}}{CaCO_3} + \underset{\text{Water}}{H_2O}$$

$$\underset{\text{Carbon dioxide}}{CO_2} + \underset{\text{Caustic soda}}{NaOH} = \underset{\text{Sodium bicarbonate}}{NaHCO_3}$$

$$\underset{\text{Carbon dioxide}}{CO_2} + \underset{\text{Caustic soda}}{2NaOH} = \underset{\text{Sodium carbonate}}{Na_2CO_3} + \underset{\text{Water}}{H_2O}$$

Carbon dioxide may also be partially removed from water by filtration through a neutralizing filter which employs a bed of calcite granules. Some of the calcite dissolves in the water forming calcium bicarbonate. In practice it will be found that this automatically brings the pH up to 7.3. or 7.3. Such filters are widely used in the household field of water treatment, in the filtration of coagulated swimming pool waters, and to a certain extent in industry. The reaction is

$CaCO_3$	+	CO_2	+	H_2O	=	$Ca(HCO_3)_2$
Calcite		Carbon dioxide		Water		Calcium bicarbonate

Solubilities of carbon dioxide from the atmosphere in pure water, at temperatures ranging from 32 to 104°F, are shown in Table 7. In equilibrium with the carbon dioxide content of the atmosphere, pure water can contain only very small amounts of carbon dioxide—from approximately 0.3 to 1.0 ppm. If instead of pure water, water containing amounts of bicarbonate alkalinity is aerated until equilibrium is obtained, it will be found that some of the half-bound carbon dioxide will be given up to the atmosphere, thus converting some of the bicarbonates to carbonates. Removal of free carbon dioxide by aeration is usually not complete enough to establish equilibria. Free carbon dioxide residuals, varying from 3 to over 15 ppm, are usually obtained. Free carbon dioxide removal together with dissolved air may be accomplished by boiling the water and venting off the noncondensible gases. In the usual form of deaerator used for deaerating boiler feed waters, the boiling takes place at or slightly above atmospheric

Table 7 Solubilities of Air, Oxygen, and Nitrogen in Dissolved Air at Atmospheric Pressure and at 0–100°C

Temperature		Milliliters per liter (ml/L)		
°C	°F	air	=	oxygen + nitrogen
0	32	28.64	=	10.19 + 18.45
5	41	25.21	=	8.91 + 16.30
10	50	22.37	=	7.87 + 14.50
15	59	20.11	=	7.04 + 13.07
20	68	18.26	=	6.35 + 11.91
25	77	16.71	=	5.75 + 10.96
30	86	15.39	=	5.24 + 10:15
40	104	13.15	=	4.48 + 8.67
50	122	11.40	=	3.85 + 7.55
60	140	9.78	=	3.28 + 6.50
80	176	6.00	=	1.97 + 4.03
100	212	0.00	=	0.00 + 0.00

pressure. Under vacuum, deaeration is effected by boiling cold water under reduced pressures.

In the vacuum type of deaerator, the degree of deaeration required is usually less than in other deaerators, and since these operate at lower temperatures, decomposition of the bicarbonates is less. Depending on the operating conditions and the content of bicarbonate alkalinity, carbonate alkalinity may or may not develop.

Oxygen and Nitrogen

The solubilities of pure oxygen and pure nitrogen in water, at 32°F and atmospheric pressure, are 48.89 ml/L for oxygen and 23.54 ml/L for nitrogen; the oxygen therefore being a little more than twice as soluble as nitrogen. When air is dissolved in water, the two main components exist in different proportions in solution than they do in the atmosphere. Oxygen constitutes 21% by volume of the atmosphere, on a moisture-free basis, whereas as shown in Table 7, the percentage of oxygen in dissolved air ranges from about 33 to 35%.

The total volume of dissolved air diminishes rapidly with rising temperatures. The solubility of air at 140°F is only about one-third of its solubility at 32°F and at 176°F only about one-fifth, whereas at 211°F its solubility is zero. A method of deaerating water is to boil it and vent off the dissolved gases. According to Henry's law, the solubility of a gas is proportional to the absolute pressure. Therefore, if the pressure is increased, the amount of air that can be held in solution at a given temperature is proportionately increased. The water from a pneumatic tank or pressure aerator may contain much more dissolved air than water saturated with air at atmospheric pressure.

Nitrogen gas is practically never determined in analyzing water, since it is inert and relatively unimportant as far as water treatment is concerned. Certain bacteria, such as found on the roots of clover, peas, and beans, have the ability to take nitrogen out of the air and build compounds. Also, electric discharges through the atmosphere cause some of the nitrogen to combine with oxygen, so that rain water may contain nitrates, which are of value as a plant food. Nitrogen is a rather inert material which has no corrosive effects on meals and, except for air binding, it is of small interest whether or not it is present in a water supply. It is present in surface waters and aerated waters. Also, nitrogen has been found in spring waters and well waters.

Oxygen, however, is an active element which readily combines with many materials. A solution of oxygen in water is very corrosive to those metals—iron, steel, galvanized iron, and brass—which are widely used for containing and transporting water. Low pH values accelerate the rate of dissolved oxygen corrosion and high pH values tend to retard it. With waters containing sufficient calcium, building up the pH value to the calcium carbonate saturation point is usually effective in reducing the rate of corrosion.

Temperature elevations greatly accelerate the rate of corrosion. If all of the dissolved oxygen stayed in solution, if protective films did not form, and if we assumed the reaction followed the general rule of roughly doubling in speed for each 18°F rise in temperature, corrosion would be some 500 times as fast at 194°F than at 39°F. The factor which works in the opposite direction is the diminishing solubility of oxygen with rising temperatures. The greatest rate of corrosion in water heaters takes place at 160–180°F. For corrosion to take place, it is necessary to have liquid water present. Dry steam which contains oxygen is not corrosive, but the condensate formed from such steam is. In condensate return lines, it also can be found that the greatest attack is where liquid water is pocketed, so that proper pitching of the lines so they drain rapidly greatly checks the rates of corrosion.

Dissolved oxygen corrosion can be severe because of air leaking into a system at periods when the steam is off. The pH value of the condensate may also be low owing to the absence or extremely low content of alkalinity in it; even rather small amounts of carbon dioxide will greatly depress the pH value. At low pH values, the carbon dioxide of itself will attack the metal and will greatly accelerate the rate of the dissolved oxygen corrosion when oxygen is present; it is important to keep the carbon dioxide content of the steam low.

As the solubility of air increases directly with the absolute pressure, the content of dissolved oxygen in the water in pneumatic tanks may be much higher than in water saturated with air at atmospheric pressure. The rates of dissolved oxygen corrosion in vessels in contact with such waters may be very rapid.

Dissolved oxygen attacks iron and steel piping with formation of tubercles under each of which is a pit. If a tubercle is rapidly scraped off from a freshly removed piece of pipe, a flash of the green ferrous hydroxide can be seen. The main body of the tubercle consists of a black material, which represents either a hydrated ferroso-ferric oxide or an intimate mixture of ferrous and ferric hydroxides. The outside of the tubercle shows the typical yellowish or reddish brown color of the hydrated ferric oxide.

If tuberculated pipe which has been carrying a water containing dissolved oxygen is later used to carry an oxygen-free water which contains free carbon dioxide, an "iron pickup" frequently occurs and soluble ferrous bicarbonate is found in the water. This is because of reduction by the iron pipe of the ferric and intermediate compounds to the ferrous state and solution by the free carbon dioxide content of the water. Such iron pickups are frequently encountered when a deep well supply is substituted for a surface supply. It can be stopped by aerating the new water supply.

Galvanized iron and brass pipe are attacked by dissolved oxygen just about as rapidly as black iron. In the atmosphere, galvanized iron resists weathering well because of the formation of an adherent basic zinc carbonate. When water containing dissolved oxygen is passed through an internally galvanized pipe, the adherent basic zinc carbonate does not form. Instead the zinc rapidly oxidizes and washes away. Yellow brass pipe also corrodes rapidly owing to the attack of dissolved oxygen on the zinc of the brass. The result is that the zinc is eaten away leaving a spongy and structurally weak skeleton of copper. As the action is usually localized, perforations and leaks occur long before the main body of the brass has been dezincified. Red brass is less subject to this attack than yellow brass, but it and even copper may be attacked by aggressive waters.

Although this decay of organic matter is using up dissolved oxygen, the plant life of the water (e.g., algae, water plants, certain diatoms) by photosynthesis takes in carbon dioxide and gives out oxygen.

Inhibition of reduction of dissolved oxygen corrosion methods of reducing or inhibiting dissolved oxygen corrosion have been mentioned. The cathodic system of corrosion prevention is used both for tanks and pipe lines. It depends for its action on a film of hydrogen on the surface of the cathode (in these cases, the pipe or tank) protecting the metal against the attack of dissolved oxygen. The current is feeble and power consumption is low, but care in installation, placing of anodes, and painting such as in tanks or of dry surfaces above the water line must be taken.

Hydrogen Sulfide

Waters which contain sulfides are commonly known as "sulfur waters." The prominent properties characteristic of hydrogen sulfide gas are an offensive, rotten-egg odor and corrosiveness. The odor is noticeable even in the cold when hydrogen sulfide is present in a water to the extent of 0.5 ppm, and when it is present to the extent of 10 ppm or more, the odor is very offensive. If the water has a high pH value, the odor may be slight; in which a case much of the sulfur may be present as an alkaline sulfide instead of as hydrogen sulfide.

Much of the corrosion deposits from sulfur waters is ferrous sulfide. Corrosion deposits may also consist of FeS_2 instead of FeS. Aeration is widely used for the reduction of the sulfide content of sulfur waters, but the reduction especially in waters of rather high alkalinity is usually only partial. Both hydrogen sulfide and carbon dioxide when dissolved in water are very weakly ionized, and it is possible to displace either one of these substances from its alkaline or alkaline earth salt by blowing a stream of the other gas through the solution. In treating sulfur waters, the carbon dioxide comes out much more easily than the more soluble hydrogen sulfide. Also, as it comes out, the pH value rises, and this upsets the equilibrium between alkaline sulfides and hydrogen sulfide so that the reaction proceeds in the wrong direction for the removal as hydrogen sulfide. Solubilities of hydrogen sulfide are shown in Table 8.

Oxidation of sulfides by dissolved oxygen is apparently a slow process. The oxidation of the hydrogen of the hydrogen sulfide follows:

Table 8 Hydrogen Sulfide: Solubilities at 760 mm and at 0–100° C

Temperature		Milliliters per liter (ml/L)	Parts per million (ppm)
°C	°F		
0	32	4590	7070
5	41	3900	6000
10	50	3320	5110
15	59	2870	4410
20	68	2500	3850
25	77	2190	3380
30	81	1940	2980
40	104	1530	2360
50	122	1220	1880
60	140	962	1480
80	176	497	765
100	212	0	0

H_2S	+	$1/2O_2$	=	H_2O	+	S
Hydrogen sulfide		Oxygen		Water		Sulfur

Presumably, then, some of this finely divided sulfur oxidizes further with a sulfate as the final endproduct.

Chlorine may be used to oxidize sulfides, but this process is rather expensive on raw sulfur waters; depending on the pH of the water and the amount of sulfide present, it may take up to eight atoms of chlorine to oxidize one molecule of hydrogen sulfide instead of the two atoms theoretically required to oxidize the hydrogen of the hydrogen sulfide to water and liberate the sulfur. This is because the sulfides are oxidized to sulfates and this oxidation takes place to a great extent even when a deficiency of chlorine is added.

The amount of chlorine required to oxidize 1 ppm of H_2S to the sulfate is 8.32 ppm and the reaction is

H_2S	+	$4Cl_2$	+	$4H_2O$	=	H_2SO_4	+	$8HCl$
Hydrogen sulfide		Chlorine		Water		Sulfuric acid		Hydrochloric acid

The amount of chlorine required to oxidize 1 ppm of H_2S to water and sulfur is 2.08 ppm and the reaction is

H_2S	+	Cl_2	=	S	+	H_2SO_4
Hydrogen		Chlorine		Sulfur		Water

Chlorination of waters containing high contents of sulfides would be expensive. Chlorination is of value in eliminating the small residuals from the effluents of other sulfide removal processes.

Methane is commonly given off in stagnant, marshy waters where putrefactive processes are taking place—hence its common name, marsh gas. However, well waters can contain methane. The amounts of methane in these waters ranges from 0.1 to 10 cubic feet per 1000 gallons. This is equivalent roughly to 0.8–87 ml of methane per liter of water. As the solubility of methane at 60°F at atmospheric pressure is only a little over 36 ml/L, a water containing 87 ml/L would effervesce at ordinary temperatures when pressure is released.

DRINKING WATER AND WATER SUPPLY

The municipal water supply should be safe to drink. Many natural water supplies, especially most deep ground water supplies, are perfectly safe as drawn and need nothing more than to be guarded against subsequent contamination. Most surface supplies though, especially in highly populated and industrial areas, and also some ground water supplies are not safe and consequently need treatment to render them safe before they are fed into the distribution systems. There are also some waters which have become so polluted that they are beyond redemption.

Municipal water supplies are generally safe to drink owing to the quality standards and to the strict supervision exercised by the regulatory authorities. This is a most important, since it eliminates municipal water supplies as potential carriers of such waterborne diseases such as cholera and typhoid fever.

In addition to being safe, a drinking water should also be attractive in appearance and palatable, since people do not like to drink a water which has a marked turbidity or a high color, nor do they relish a water which smells like rotten eggs.

Turbidity and Color Removal

Water supply can come from ground water supply, surface water supply, or part of it from a ground water supply and the rest from a surface water supply. Uncontaminated ground water supplies from deep ground are usually clear and practically colorless, but some contain iron and/or manganese and some hydrogen sulfide. Although some surface waters are clear and practically colorless, most of them contain appreciable amounts of turbidity and/or color.

Turbidity in slow sand filtration practice may often be removed without coagulation but has little or no effect in reducing color. In rapid sand filter practice, both turbidity and color are removed by coagulation, settling, and filtration. If the raw water contains large amounts of coarse sediment, then sedimentation may precede these steps.

Sedimentation is effected without the aid of a coagulant, whereas *settling* refers to the settling of a coagulated water. If raw water contains large amounts of suspended solids, an appreciable percentage of which is coarse enough to be easily removed by sedimentation, then sedimentation preceding coagulation, settling, and filtration may be worthwhile. Impounding of surface water supplies is quite common, and the reservoir formed usually accomplishes a satisfactory degree of sedimentation. Sedimentation basins, with detention periods measured in hours instead of in weeks, months, or even years, may be used if the raw water has a high content of relatively coarse sediment.

Coagulation and Settling

The most widely used coagulant in municipal filtration plants is aluminum sulfate, which is also commonly known as filter alum. The formula is generally given as $Al_2(SO_4)_3 \cdot 18H_2O$, but the commercial product usually has a lower water content—about 13 or 14 moles instead of 18 moles. It is also available in solution and is shipped in tank cars or tank trucks. Other coagulants used in municipal filtration plants are ferric sulfate, chlorinated copperas (ferrous sulfate), and sodium aluminate.

In raw waters which are high in color and low in turbidity and dissolved mineral matter, a coagulant aid, such as clay, activated silica, or one of the polyelectrolyte coagulant aids may be required. When clay is used, it should be from a tested source, because different clays vary widely in their properties as a coagulant aid. Activated silica is made from sodium silicate by several different processes.

Coagulation and settling may be carried out in various designs of settling basins, tanks and floc-formers, but the suspended solids contact or sludge blanket units have been widely used. In these, coagulation and the development of the floc are carried out while the water is flowing downwardly through an inner section. It then rises and filters upwardly through a suspended blanket of previously formed precipitates in an outer section. Emerging from the upper surface of the blanket, it flows upwardly to the drawoffs from which it passes to the rapid sand filters, which may be divided into two main types: gravity-type and pressure-type filters. Gravity-type rapid sand filters are used in filtering municipal water supplies. The filtration rate is usually 2 gpm/ft^2 of 2.5–3.0 gpm, but higher rates have been employed. In industrial practice, the standard rate of 3.0 gpm/ft^2 is used. For the removal of iron and/or manganese from ground waters, pressure-type rapid sand filters can be employed.

The shells of the gravity type rapid sand filters may be made of either concrete or steel. Concrete shells are usually rectangular in shape; the smaller sizes of units are square and the larger ones are oblong. When steel shells are employed, they are cylindrical in shape. Filter medium may be sand or may be crushed and graded anthracite. If the filter medium is sand, the supporting layers under the fine filter sand are coarse sand and several layers of graded gravel. If the filter medium is anthracite, then the supporting layers may be either graded anthracite or coarse anthracite and graded gravel.

During filtration, the accumulations of strained out material gradually clog the filter bed and build up an increasing resistance to the flow of water through it. When this reaches a pressure drop ranging from about 8–12 feet loss of head in different gravity type filters, the automatic valveless gravity-type rapid sand filters use a 4–5 feet loss of head, the filter unit is taken out of service, cleaned by backwashing, filtered to waste for a few minutes, and then returned to service. Backwash rates for gravity-type sand filters are usually 20 gpm/ft^2 if not equipped with surface washers and 15 gpm/ft^2 if equipped with surface washers. Backwash rates for gravity type anthracite filters are usually 15 gpm/ft^2 if not equipped with surface washers and 12 gpm/ft^2 if equipped with surface washers. Gravity filters are usually installed in batteries of two or more units.

A separate rate of flow controller is usually employed with each filter unit so as to maintain a uniform rate of flow in spite of the variations in back pressure which occur during the filtration run. For backwashing, usually only one wash rate of flow controller, installed in a common backwash leader, is required, since only one filter unit is backwashed at a time.

Pressure-type rapid sand filters are vertical and horizontal. Vertical pressure filters range in size from 30 to 120 inches in diameter with flow rates of 10–250 gpm per unit at 3 gpm/ft^2. Horizontal filters are typically 8 feet in diameter and from 10 feet 6 inches to 25 feet in length with capacities ranging from 200 to 510 gpm at 3 gpm/ft^2. Backwash rates are usually 10 gpm/ft^2.

In municipal practice, pressure-type filters are used mostly in connection with the treatment of ground waters such as in the removal of iron or manganese from aerated and settled iron-bearing and/or manganese-bearing waters. In industrial applications, pressure-type filters are also employed. In operation, the filter run is continued until the pressure loss rises to some 5–8 psi, when the filter is cut out of service, backwashed, filtered to waste for a few minutes, and then returned to service.

Other filter media employed for special uses include neutralizing filters that have crushed and graded calcite as the filter media for industrial uses to neutralize an aggressive water and automatically raise the pH value to 7.2–7.3 or to stabilize a treated water. Activated carbon filters using granular activated carbon as the filter medium are used for taste and odor removal, chiefly for industrial uses but also to a limited extent for removing tastes and odors from relatively small municipal water supplies. Manganese zeolite filters use granular manganese zeolite as the filter medium and are used both in industrial and municipal practice for removing small amounts (1 ppm or less) of iron and/or manganese from waters containing them as the soluble, divalent bicarbonates.

Disinfection

Disinfection may be accomplished by the use of

- Chlorine
- Chlorine dioxide
- Chloramides
- Hypochlorites
- Ozone

Chlorine, the most widely used disinfectant, is a very strong oxidizing agent. It is available as the liquefied gas in pressure cylinders holding 100, 150, or 2000 lb. It may be applied before other forms of treatment (prechlorination) and/or after other forms of treatment (postchlorination). It may be applied at one or more points in the distribution system (rechlorination).

Chlorine dioxide is a very strong oxidizing agent which, in municipal practice, is made in solution by reacting chlorine with sodium chlorite. It is used for oxidizing disagreeable taste- and odor-producing organic matter as well as for its disinfectant action.

Chloramines are made by the reaction of chlorine with ammonia. The chloramines are not as strong or as rapid oxidizing agents as chlorine or the hypochlorites but are of value in maintaining chlorine residuals.

Hypochlorites are often used in place of liquefied chlorine in treating small water supplies. They are available as "bleaching powder," which is equivalent in oxidizing power to 35% of chlorine or as high-test hypochlorites equivalent to 70% of chlorine.

Ozone is an unstable allotropic form of oxygen made by passing a high-voltage silent electric discharge through a stream of air. It is of value in oxidizing taste- and odor-producing organic matter and has a high bactericidal action but decomposes so rapidly that no residual oxidizing action can be maintained.

Taste and Odor Removal

Objectionable tastes and odors in water are usually due to various organic compounds, except the rotten egg odor of sulfur waters which is due to hydrogen sulfide). As these compounds vary greatly in composition and character in different waters and also from time to time in the same water, it is advisable to make preliminary tests to find the best method of removal.

Many of these compounds have a certain degree of volatility, so that aeration is often of value as a preliminary step. Chlorination may remove some of the odors but with others may intensify them, whereas heavier dosages or the use of chlorine dioxide or ozone may completely remove them. Activated carbon is an excellent taste and odor removal agent. Pulverized activated carbon has been with longer and more intimate contact.

Iron and Manganese Removal

For iron-bearing, deep well waters which have a bicarbonate alkalinity, the iron is present as colorless, soluble ferrous bicarbonate—$Fe(HCO_3)_2$. Iron in this form may be removed by oxidation to insoluble ferric hydroxide, settling, and filtration. The oxidation is usually effected by aeration, but it may also be effected with such oxidizing agents as chlorine, hypochlorites, or chlorine dioxide; by the sodium cation exchanger (zeolite) water-softening process in which it is simultaneously removed with the hardness; by filtration through a manganese zeolite filter (this process is usually limited to waters containing not much more than 1 ppm of iron); or for industrial uses, by the hydrogen cation exchange process.

In acid waters, iron and/or manganese may be partly or wholly present as divalent sulfates ($FeSO_4$ or $MnSO_4$) and may be removed by aeration followed by neutralization and a build-up of the pH value, settling, and filtration. Iron may be present as insoluble ferric hydroxide—$Fe(OH)_3$—and may be removed by settling and filtration or by coagulation, settling, and filtration, which often may well be preceded by aeration. Organic (chelated) iron and/or manganese may be present in high-color waters and may be removed by coagulation, settling, and filtration, usually preceded by aeration.

Water Softening

Municipal water supply is suitable for drinking purposes, but this does not necessarily mean that it is even fairly suitable for other uses such as heating, cooking, washing, bathing, laundering, or dish washing. Excessive hardness, for example, may scale up water heaters and hot water piping causing impacted flows and eventually serious clogging and burnouts for which expensive repairs and replacements are required. Hardness can waste large proportions of the soap used in laundering and deposit insoluble calcium and magnesium soap curds in the laundered materials. These not only prevent thorough cleansing, resulting in dingy laundry, but develop rancid odors on standing and embrittle fibers and thus greatly shortening the useful life of laundered materials. Washing, bathing, shampooing, and shaving become neither satisfying nor pleasurable. Owing to these disadvantages, many municipalities soften their water supplies. Usually in municipal practice, this softening is not carried out to zero hardness but instead to an acceptable level which furnishes much better and more usable wastes than the very hard raw waters. Municipal water softening benefits all of its users, including those who have household water softeners because they get so much greater capacities from their softeners.

There are two basic methods of softening municipal water supplies plus a third that is a two-step combination of the first two. These are:

- Cold lime (or lime soda) water-softening process
- Sodium cation exchanger (zeolite) water-softening process
- Two-step cold lime and sodium cation exchanger water-softening process (lime zeolite process)

Cold Lime Soda Processes

In these cold precipitation processes, hydrated lime—$Ca(OH)_2$—alone may be used to reduce only the bicarbonate hardness, commonly known as carbonate hardness or hydrated lime plus soda ash—Na_2CO_3—may be used to reduce both the bicarbonate hardness and the noncarbonate hardness. If hydrated lime alone is used, the process is properly known as the cold lime process, whereas if both hydrated lime and soda ash are used, the process is known as the cold lime soda process. However, the term *cold lime soda process* is commonly used collectively to recover both processes.

In addition to these primary chemicals, hydrated lime or hydrated lime and soda ash, a small amount of either an aluminum coagulant, such as aluminum sulfate, or an iron coagulant is also added because the precipitates which are produced are very finely divided and coagulation is necessary to effect good settling and filtration.

Hydrated Lime and Chemical Lime

Instead of using hydrated lime, chemical lime (quicklime)—CaO—may be purchased and hydrated (slaked) before use. The reaction is as follows:

$$\underset{\text{Chemical lime}}{CaO} + \underset{\text{Water}}{H_2O} = \underset{\text{Hydrated lime}}{Ca(OH)_2}$$

The use of chemical lime instead of hydrated lime is generally cheaper in treatment costs, but chemical lime is harder to handle and contact with water, except in proper slaking equipment, must be avoided owing to the large amount of heat evolved in the reaction.

Precipitates Produced—Calcium Carbonate and Magnesium Hydroxide

The calcium hardness which is removed in these cold water-softening processes, the free carbon dioxide and the hydrated lime which was added to the water are precipitated as calcium carbonate—$CaCO_3$. The magnesium hardness which is removed is precipitated as magnesium hydroxide—$Mg(OH)_2$. As these precipitates are slightly soluble, there will be some residual hardness in the clear, filtered effluent. If the process is carried out to reduce the total calcium and magnesium hardness as much as possible, without the use of excess chemicals, this residual will be about 68 ppm.

The amounts of chemicals required to effect various degrees of hardness removal with various types of wastes, together with the reactions involved, are briefly: one equivalent of hydrated lime is required to remove one unit of calcium bicarbonate hardness; two equivalents of hydrated lime to remove one unit of magnesium bicarbonate hardness; one equivalent of soda ash to remove one unit of calcium noncarbonate hardness; and both one equivalent of soda ash and one equivalent of hydrated lime to remove one unit of magnesium noncarbonate hardness.

Although hydrated lime has a higher equivalent weight and higher cost than chemical lime, both of them are low-priced chemicals. Soda ash is higher in price and has a higher equivalent weight than either chemical lime or hydrated lime, so the costs for removing noncarbonate hardness are higher than the costs for removing bicarbonate hardness.

The cold lime soda water softening consists of

- Chemical feeders
- Softener unit(s), in which the softening and coagulation reactions take place and the bulk of the precipitates are settled out
- Filters which remove the last traces of turbidity
- A clear well from which the water is pumped to service

In addition, if the water contains iron and/or manganese or is high in free carbon dioxide or is a sulfur water, aeration is employed ahead of the softener unit, with the aerator or degasifier being so mounted that a gravity flow to the softener unit may be obtained. In some plants, recarbonation is practiced, in which the flue gases from a stack or from a coke, oil, or gas burner are first blown through a scrubber and then, entering through a grid, are passed upwardly through the treated and settled water in a recarbonation tank, which is situated ahead of the filters.

The oldest type of cold lime soda water-softening processes is the batch type. In this process, a tank was filled with water, a weighed dose of chemicals was stirred in, the batch was allowed to settle for 4 hr or more, and the treated and settled water was then drawn off from the sludge and filtered. The batch process is practically obsolete, having been replaced by the continuous processes in which the water is softened as it is flowing through the plant.

Iron and Manganese Removal in the Cold Lime Soda Water Softening

Iron and/or manganese present as soluble bicarbonates may be removed in the cold lime soda water softener by aeration of the water before it enters the softener. The aerator, which may be of the coke tray, wood slat tray, forced draft or other type, is usually mounted above the softener so as to secure a gravity flow to it. Owing to the relatively high pH in the softener, the soluble iron and/or manganese compounds are then oxidized to and precipitated as their insoluble, hydrated higher oxides (ferric hydroxide and manganic hydroxide). Acid waters in which these metals are present, either partially or wholly, as sulfates may be freed from them in the same way. Aeration is also practiced with waters high in free carbon dioxide to reduce the content to below some 10 ppm, which is then removed by precipitation as calcium carbonate in the softening process.

Sodium Cation Exchange Process

In softening municipal water supplies by the sodium cation exchange (zeolite) process, one portion of the hard water is completely softened by passing it through the softeners and this flow then mixes with a sufficient flow of the hard water to give a mixed effluent of the desired composition.

Softener units used may either be of the pressure type or of the gravity type, and they are usually installed in batteries of two or more units so that when one of the units is being regenerated, the other unit or units will carry the load and thus ensure an uninterrupted soft-water service. Units operate on a staggered time basis so that no two units require regeneration at the same time. Whichever type of unit is employed, it contains a bed of a granular or bead-type sodium cation exchanger (zeolite) supported by several layers of graded gravel. Depending on the hardness of the water to be softened, the type of sodium cation exchanger used and the desired length of the softening run, the bed depths of the exchanger may range from 2–7 feet.

To soften a hard water by this process, all that is necessary is to flow the water downwardly through the softener. As the hard water comes in contact with the insoluble sodium cation exchanger, the calcium and magnesium cations, which constitute the hardness, are taken out and held by the exchanger, which gives in exchange an equivalent amount of sodium cations to the water. At the end of the softening run, the unit is cut out of service and regenerated.

The regeneration consists of three steps: backwashing, salting, and rinsing. Backwashing is accomplished by passing a strong, upward flow of water through the bed to loosen, cleanse, and hydraulically regrade it. Salting is carried out by flowing a predetermined volume of a solution of sodium chloride through the bed. When the salt solution comes in contact with the cation exchanger, it removes the accumulated calcium and magnesium from it in the form of their soluble chlorides and simultaneously restores the cation exchanger to its former sodium state. Rinsing consists of washing the chlorides of calcium and magnesium plus any excess of salt out of the softener. After rinsing, the softener unit is returned to service ready to soften a further equal amount of the hard water.

Iron and or manganese as soluble bicarbonates may be removed in the sodium cation exchange water-softening process simultaneously with the removal of the hardness. Furthermore, on regenerating the spent exchanger with salt, these metals are removed from the cation exchanger bed simultaneously with the removal of calcium and magnesium and in the same form; that is, as their soluble chlorides.

In municipal practice, many hard, iron-bearing waters if free from manganese are freed from iron by aeration, settling, and filtration of all of the water before softening by the sodium cation exchanger process. Then one portion of the hard, iron-free water is completely softened by the sodium cation exchanger units and this is then mixed with another portion of the hard, iron-free water, with the flows being adjusted so as to obtain the desired residual.

The great majority of iron-bearing hard waters do not contain manganese and this process is applicable with such waters. Manganous bicarbonate requires much higher pH values for rapid oxidation than ferrous bicarbonate. Therefore, although manganese may be removed from all but acid waters by either sodium cation exchanger water softeners or manganese filters without any necessity of raising the pH value, it is only partially removed by aeration, settling, and filtration unless the pH value is also raised.

Two-Step Cold Lime and Sodium Cation Exchange Process (Cold Lime Zeolite Process)

In this process, either the calcium bicarbonate hardness or the total bicarbonate hardness is reduced by treatment with hydrated lime and a coagulant, settled, and filtered in the first step. If necessary, recarbonation or a slight dosage of acid may be applied before filtration. Then, in the second step, one portion of the lime treated and filtered effluent is completely softened by passing it through sodium cation exchanger water softener and this is mixed with another portion of the lime-treated and filtered effluent, with the flows being so adjusted as to produce a mixed effluent with the desired residual.

Operating costs for removing noncarbonate hardness by the sodium cation exchanger process are lower than the operating costs for removing it by the cold lime soda process. On the other hand, the operating costs for removing bicarbonate hardness (especially calcium bicarbonate hardness) by the cold lime process are usually somewhat lower than the costs for removing it by the sodium cation exchanger process.

In waters which are high in both bicarbonate and noncarbonate hardness, this two-step cold lime and sodium cation exchanger process usually has lower operating costs than either the cold lime soda process or the sodium cation exchanger process. Also, there is a reduction of total solids corresponding to the reduction in bicarbonate hardness effected by the lime treatment in the first step.

So-called "red waters" are caused by the action of corrosive waters containing dissolved air on ferrous metals with the production of pits, tubercles, and reddish brown ferric hydroxide. Corrosive, low pH waters containing dissolved air may also attack copper tubing with the production of greenish stains on porcelain fixtures, but the attack is usually slow. As for yellow brass, this may be very badly attacked with spotted pitting and leaks occurring through the spongy copper formed by the dezincification of the brass.

Two widely used corrosion-inhibition methods are the production of a very thin protective coating of calcium carbonate and the use of a small amount of sodium silicate (0.1 lb per 1,000 gal plus usually a little caustic soda to build the pH up to about 8.3) which forms a very thin protective film on the metal surfaces.

SOLUTIONS AND THEIR PROPERTIES

A careful distinction should be made between solutions, compounds, and mixtures. In a mechanical *mixture*, no matter how finely the mixture may be ground, it is possible to detect nonhomogeneity, because the properties of the component substances are unchanged. In a *compound*, the simple substances of which it was formed have lost their characteristics during the change and can no longer be identified as such. For example, in carbon dioxide we find none of the characteristics of the elements carbon and oxygen. In the formation of a *solution*, the properties of a substance are not greatly altered; the change is usually physical involving dispersion into molecular state. However, there are instances when the substances undergo a chemical change during the process of solution resulting in the formation of a new material which dissolves in the liquid.

Suspensions

When coarse insoluble material, such as sand, is mixed with water and the mixture is allowed to stand, the sand settles to the bottom. If the coarse material is less dense than water, such as sawdust, it floats to the top on the surface. If the sand is first ground to a fine dust and then shaken with water, it will settle more slowly. Such mixtures can be readily separated by filtration; they are not homogeneous and are called suspensions.

A difference between a solution and a suspension is the relative size of particles. In a solution, the dissolved particles are of molecular size, either individual molecules or a small multiple of molecules, whereas in a suspension, the particles consist of very large groups of molecules. Between the two limits there is a wide range. Solutions can be prepared in which the size of the particles are within these limits. Such is the case with mixtures in which the solid is so finely divided that the particles pass through all common filters, do not settle out on standing for a long time, and are invisible even under view of a microscope. When a special type of microscope (electron microscope) is employed, it is often possible to see not the individual particles, but the light refracted by the particles. Such solutions (not true solutions) are known as colloidal dispersions, which are sometimes called colloidal solutions. Colloidal dispersions may be as clear and transparent as true solutions, but when a beam of light is passed through the liquid, its path becomes illuminated owing to reaction or refraction of part of the light by the colloidal particles. This phenomenon is called the Tyndall effect.

When the suspended material of a suspension is a liquid like oil, and when the fine droplets of the oil are so small that they pass through a filter and do not settle out readily, the mixture is called an emulsion. An emulsion may be white in appearance owing to the reflection of light from the droplets of oily matter. On standing for a very long time the emulsion breaks, or separates, and the oil, usually lighter than water, floats on top.

Solutions may exist in the liquid or solid state, although they may involve substances which were originally solids, liquids, or gases. Since all gases mix in all proportions with no alteration in their properties except when chemical reaction occurs, these are not considered a mixture of gases to be a solution. The following types of solutions are possible:

Liquid State	*Example*
• Gas in a liquid	Carbon dioxide dissolved in water
• Liquid in a liquid	Alcohol dissolved in water or ether
• Solid in a liquid	Salt dissolved in water

Solid State	*Example*
• Gas in a solid	Gas dissolved (occluded) in a metal
• Liquid in a solid	Mercury dissolved in copper
• Solid in a solid	Alloy of one metal in another

Solute and Solvent

The terms *solute* and *solvent* refer to the components of the solution. When a solution is made by dissolving a solid substance in a liquid, the solid substance is usually termed the solute and the liquid is referred to as the solvent. When two liquids are mixed and dissolve in one another, the one present in the larger proportion is often termed the solvent. *Solvent* carries with it the idea of dispersion medium; *solute*, that of the dispersed substance. However, solute and the solvent are terms of convenience and should not be used when they lead to ambiguity.

When the solution contains a relatively small amount of solute, it is called a dilute solution. If the relative amount of solute is large, the solution is said to be concentrated. It is obvious that concentrated solutions are possible only when the solute is very soluble. When an excess of the solute, preferably in a finely divided state, is mixed with the solvent for a sufficient length of time, a solution which is in equilibrium with the excess undissolved solute results. Such a solution is saturated with the solute and is called a saturated solution.

The concentration of a solution may be expressed in physical or in chemical units of weight. When *physical* units are employed, one may state the number of grams of solute in a fixed quantity of the solvent; thus, a solution of sodium sulfate, made by dissolving 15 g of Na_2SO_4 in 100 g of water. We may also express the concentration of this solution as 15 g of solute in 115 g of solution, 15/115 or 13.05% by weight of sodium sulfate. If the density (weight of 1 ml) of the solution is known, it is a simple matter to calculate the weight of solute in a given volume of the solution.

Very often concentrations are expressed as the weight of solute per unit volume of solution; as 100 g of sodium chloride per liter of solution, or 1 g of potassium sulfate per 100 ml of solution. To make the latter solution, 1g of potassium sulfate is dissolved in a small quantity of the solvent and then more solvent is added until the total volume of the solution is 100 ml.

Employing *chemical* units, two methods of expressing concentration are used; in one the gram-equivalent weight of the solute is taken as the unit, and in the other, the gram-molecular weight. The gram-equivalent weight of a substance is that weight which contains or will interact with 1 gram-equivalent weight of an element. Thus the molecular or formula weight of hydrochloric acid, HCl, is 36.465 g, which is also the equivalent weight, for it contains one equivalent of hydrogen, and this weight of hydrogen is displaced by one equivalent weight of a metal. The molecular weight in grams of sulfuric acid (98.08 g) contains 2 gram-equivalents of hydrogen and 1 mol of phosphoric acid H_3PO_4 (98.051 g) contains three equivalents. In the case of mixed compounds such as $KAl(SO_4)_2$ or $MgNH_4PO_4$, it becomes necessary to refer to the element or group which takes part in the reaction. The equivalent weight of $KAl(SO_4)_2$ with reference to potassium is that weight of the salt which contains one equivalent of that element; the equivalent weight is the same as the formula weight. One mole of the same salt contains 3 gram-equivalents of aluminum, so that the equivalent weight with reference to aluminum is one-third of the formula weight, and similarly, the equivalent weight with reference to the SO_4 group is one-fourth of the formula weight. A normal solution contains 1 gram-equivalent of the solute in 1 L of solution. The normality is the number of gram-equivalent weights per liter of solution. We frequently use solutions which are one-tenth normal or decinormal ($N/10$ or $0.1N$), semi-normal ($N/2$), four times normal ($4N$), and so forth.

A molar solution is one which contains 1 mol of solute in 1 L of solution. A liter of molar (1 M) sulfuric acid solution contains 1 gram-molecular weight of H_2SO_4 (98.08 g), and such a solution is twice normal ($2N$). When the molecular weight of a solute is the same as the equivalent, as in the case of HCl, the molar solution is also normal.

When a crystal of hydrated cupric sulfate is placed in water, it gradually becomes smaller and ultimately disappears, and the liquid in the immediate vicinity becomes colored blue. Molecules of the solid leave the surface of the crystal and move about in the solution as liquid particles. Although the motion of the individual molecules is rapid, the rate of dispersion of the dissolved material throughout the entire liquid is impeded by the close packing of the molecules of the solvent. The blue color slowly rises as the molecules move against the force of gravity into the upper layers of the water. This phenomenon, the intermingling of the solute, is called diffusion. Eventually, the solute is uniformly distributed. Stirring the mixture or suspending the solid to be dissolved in the solvent will hasten the distribution of the dissolved material.

If more of the solid is added, the rate at which it dissolves decreases. The rate at which molecules leave each unit of surface of the solid is constant for each temperature. But as the solution becomes more concentrated, the rate at which they return to the solid increases, so that the overall rate of solution grows smaller until finally no more dissolves. The solution is then in equilibrium with the excess solid. Such a solution is called a saturated solution. A condition is reached where the number of molecules which come to the surface of the undissolved solid and deposit is just equal to the number of molecules which leave the surface. A condition of equilibrium exists between the dissolved and the solid states of the solute.

Rate of Solution

Factors affecting the rate of solution are evaporation where solution takes place at a surface. The greater the surface exposed to the solvent, the greater is the speed of solution of a given amount of solute. Finely divided material dissolves more rapidly than the same weight of the solid in the form of one large crystal. Stirring or suspending the solid in the solvent is analogous to exposing the surface of an evaporating liquid to air currents. The stirring of the solute with the solvent results in a constant removal of the saturated solution from the surface of the solid, thus exposing it to fresh or unsaturated solvent. When the material is suspended, the solution, which is denser than the solvent, sinks through the solvent, fresh solvent, rises, and a circulation is started. At an elevated temperature, the molecular motion is increased, so that the rate at which the solute molecules move away from the solid is increased. With this increased molecular motion, diffusion increases, thereby tending to maintain a condition of unsaturation around the solid. In preparing solutions where time is an important factor, one should bear in mind the preceding conditions and their effect on the rate of solution.

Solubility

The amount of a substance required to saturate 100 g of water at a given temperature is called the solubility of that substance at that temperature. Changes in temperature always change the solubility of a solute. As a rule, solids are more soluble in hot than in cold water, although occasionally the reverse is true. Some compounds of calcium, such as calcium hydroxide, $Ca(OH)_2$, and calcium chromate, $CaCrO_4$, belong to the latter group. The solubility of sodium chloride is only slightly affected by a change in temperature.

A saturated solution in equilibrium with excess solute behaves in accord with Le Chatelier's principle. As the temperature is increased, a change takes place which will absorb heat. Most solutes (solids) absorb heat during the process of dissolving. The dissolving of a solid to form a liquid solution involves a change from the solid to the liquid state, an endothermic process. Hence if a saturated solution with excess solid present is heated, more solid will dissolve. In some cases, an exothermic chemical reaction accompanies the solution of a solute and the heat evolved by the chemical change may be greater than the *heat of solution*. In such cases, the temperature will rise as the material goes into solution, even though the process of solution itself is endothermic. This is the case for sodium hydroxide and for potassium hydroxide. It is found, however, that when sodium or potassium hydroxide dissolves in an almost saturated solution, the process is endothermic. As Le Chatelier's law applies to systems at or near equilibrium, we should expect and actually find that sodium and potassium hydroxides are more soluble at higher temperatures.

The most useful way of representing solubility data is by means of solubility curves. The graph in Figure 1 shows the curves for several common substances. The ordinates represent the solubility, expressed in grams of the *anhydrous compound* in 100 g of water, and the abscissae represent the temperatures.

The solubility curve for calcium chloride is not continuous. There are two breaks: one at 29.8°C and the other at 45.3°C. Whenever such breaks occur, it is found that some chemical change takes place at the temperature of the break. In the case of calcium chloride, at 29.8°C, the hexahydrate, $CaCl_2 \cdot 6H_2O$, loses water and becomes the tetrahydrate $CaCl_2 \cdot 4H_2O$

$$CaCl_2 \cdot 6H_2O \rightarrow CaCl_2 \cdot 4H_2O + 2H_2O$$

At 45.3°C, the tetrahydrate loses water and becomes the dihydrate $CaCl_2 \cdot 2H_2O$

$$CaCl_2 \cdot 4H_2O \rightarrow CaCl_2 \cdot 2H_2O + 2H_2O$$

The solubility curve for sodium sulfate (Fig. 2) shows a break at 32.4°C, at which temperature the decahydrate $Na_2SO_4 \cdot 10H_2O$ changes to the anhydrous salt Na_2SO_4. Such solubility curves are really made up of separate solubility curves for the different substances; that is,

$$CaCl_2 \cdot 6H_2O;\ CaCl_2 \cdot 4H_2O;\ \text{and}\ CaCl_2 \cdot 2H_2O$$

and

$$Na_2SO_4 \cdot 10H_2O\ \text{and}\ Na_2SO_4$$

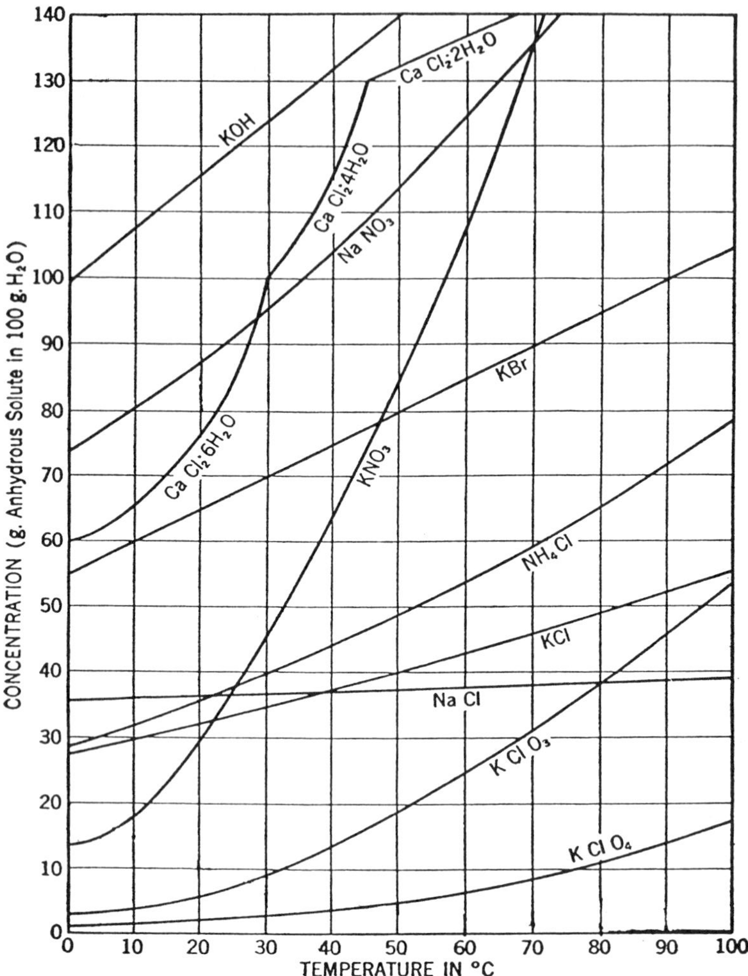

Figure 1 Solubility for various substances in water.

respectively, and these separate curves intersect. The temperature at which the change takes place, for example, 32.4°C on the sodium sulfate curve, is called the transition point between the hydrate and the anhydrous salt or between two hydrates. At the transition point, the two materials coexist in equilibrium with each other and with a saturated solution. Above this temperature, the decahydrate is unstable and loses water, whereas below it the decahydrate is stable and the anhydrous salt combines with water to form the decahydrate.

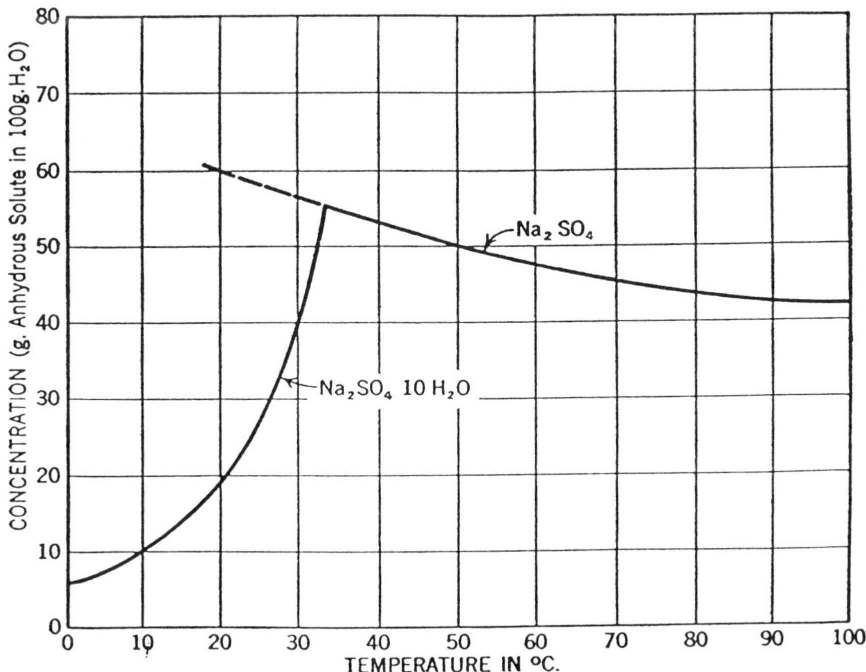

Figure 2 Sodium sulfate solubility curve.

Supersaturation

Generally, solids are more soluble at higher temperatures. If a solution is saturated at one temperature and is then cooled, the excess of solute usually deposits from the solution. If there is no excess of solid solute and the solution is cooled slowly, there is often a delay before crystals of the solute appear. During this time, the solution is supersaturated. This delay before the excess is deposited seems to be indefinite. A supersaturated solution of sodium sulfate hexahydrate or of sodium thiosulfate, if protected from dust and shaking, will remain in this condition for long periods. If, however, a minute crystal of the substance be added, crystallization of the excess of solute commences immediately, with the added crystal acting as a nucleus for the formation of the crystals. The addition of a crystal to start the formation of more crystals is known as *inoculation*.

A supersaturated solution may be obtained in a slightly different way. Inspect the curve in Figure 2. The unbroken portion of the curve represents the solubility at the various temperatures when the solution is in equilibrium with the solid indicated on the curve. At 20°C, the solution contains 19.5 g of Na_2SO_4 in 100 g of water when in equilibrium with the solid decahydrate. It is possible to prepare another saturated solution at 20°C by dissolving anhydrous sodium sulfate, and analysis of the liquid shows that 60 g of Na_2SO_4 is dissolved in 100 g of water. The solubility of the anhydrous salt at temperatures below 32.4°C, *in the absence of solid decahydrate*, is indicated by the dotted portion of the curve. Saturated solutions of the anhydrous sodium sulfate exist at temperatures below the transition point. Such a solution contains more dissolved salt than is present when the solution is in equilibrium with solid decahydrate and is supersaturated with respect to the hydrated salt. When a crystal of the decahydrate is introduced into such a solution, crystallization starts. In the presence of the decahydrate crystal, the solution becomes supersaturated with the hydrated variety, the excess separates out, and the resulting solution is then saturated with the decahydrate; the crystals which separate are the decahydrate, $Na_2SO_4 \cdot 10H_2O$.

Supersaturated solutions may be prepared with other salts, such as sodium thiosulfate, $Na_2S_2O_3 \cdot 5H_2O$; manganese sulfate, $MnSO_4 \cdot 7H_2O$; sodium acetate, $NaC_2H_3O_2 \cdot 3H_2O$; and so forth.

Liquids in Liquids

When two different liquids are mixed, they may dissolve, one in the other, in all proportions, as alcohol and water. Such pairs of liquids are said to be miscible in all proportions. Other pairs may dissolve in one another to a limited extent only, forming two layers; the less dense floating on the more dense, as oil or organic solvents and water. Such pairs are immiscible. The upper layer is either saturated with water and the lower layer is water saturated with organic or oil. The solubility in water decreases with rising temperature, whereas the solubility of water increases.

Law of Partition

When a solute is added to two immiscible liquids in contact with each other, thoroughly shaken and mixed, and then allowed to stand until the two immiscible liquids separate, the solute is distributed between the two solvents. If sufficient time is allowed for the system to come to equilibrium, the ratio of the concentrations of the solute in each of the two solvents is equal to the ratio of the solubilities of the solute in each solvent.

Solutions of Gases

Solutions of gases in liquids may be obtained by bubbling the gas through the liquid or by confining the liquid in a vessel containing the gas. The quantity of gas which will dissolve in a given volume of liquid depends upon four factors:

- Nature of the liquid
- Nature of the gas
- Temperature
- Pressure

There are no general rules by which it is possible to calculate the solubility of any given gas in a given solvent. Some gases are very soluble in water and others are only slightly soluble; e.g., at 0°C and 760 mm, one volume of water dissolves 1200 volumes of ammonia, 69 volumes of sulfur dioxide, 4.7 volumes of hydrogen sulfide, 0.04 volume of oxygen, and 0.02 volume of hydrogen. The solubility of a gas in a liquid decreases with rise in temperature. Table 9 shows the solubilities of several gases at various temperatures; the figures are the volumes of gas in one volume of water.

When a gas dissolves in a liquid, the process involves a change of a substance from the gaseous to the liquid condition, an exothermic process, which according to Le Chatelier's law explains why the solubility of gases decreases as the temperature rises.

Kinetic theory of gases furnishes a picture of the process of solution of a gas in a liquid. When a liquid is introduced in a vessel containing a gas, such as oxygen, the molecules of the gas strike the surface of the liquid and some are dissolved. The dissolved molecules move about in the solution in all directions, and some come back to the surface and leave the liquid. As the number of molecules in the solution increases, the number leaving the solution increases until the number leaving just equals the number entering the solution. The system is then in dynamic equilibrium.

The greater the pressure of the gas above a liquid, the greater will be the number of molecules entering the liquid per unit of time and therefore the greater will be the concentration of the gaseous substance when equilibrium is reached. Or the greater will be the solubility of the gas. W. Henry (1803) discovered that the volume of the gas absorbed by a liquid is directly proportional to the pressure of the gas. This relationship is known as Henry's law.

Those gases which are very soluble in water, such as ammonia and sulfur dioxide, usually have high critical temperatures and are strongly polar. The

Table 9 Solubilities of Gases

Temperature (°C)	Carbon dioxide	Nitrous oxide	Oxygen	Hydrogen	Nitrogen
0	1.79	1.30	0.041	0.021	0.020
5	1.44	1.09	0.036	0.020	0.018
10	1.18	0.92	0.032	0.019	0.016
20	0.90	0.67	0.028	0.018	0.014

molecules exert cohesive influence upon each other, and in many cases chemical combination takes place between the dissolved gas and the solvent. For example, sulfur dioxide combines with water to form sulfurous acid (H_2SO_3).

Mixed Gases

When a mixture of two gases is exposed to the action of a solvent, the quantity of each gas which remains in solution when equilibrium is reached depends on the frequency with which its own molecules strike the surface of the liquid and is independent of the presence of the other gas. The solubility of each gas is the same as if it were present alone at a pressure equal to its partial pressure in the mixture.

The solution of air in water illustrates this. Since air is approximately 1/5 oxygen and 4/5 nitrogen by volume, the partial pressures of these gases are 1/5 and 4/5 of an atmosphere 1.3, respectively. The solubilities of oxygen and nitrogen in a unit volume of water at 0°C and 760 mm are 0.04 and 0.02 volume, respectively. When, therefore, air is bubbled through water at 0°C, the quantity of oxygen which dissolves should be the product of its partial pressure (1/5 atm) by its solubility at 1 atmosphere pressure, $1/5 \times 0.04$ or 0.008 volume, and that of nitrogen, $4/5 \times 0.02$ or 0.016 volume.

Solution Properties

Properties of solutions has shown that the more concentrated the solution, the more will the physical properties of the solution differ from those of the pure solvent. The observed changes in properties may be divided into two classes.

The magnitude of the change varies with the nature of the solute as well as with the concentration of the solute. Some of these changes are difficult to explain. For example, when sodium chloride is dissolved in water, there is a decrease in the total volume (the total volume of the solution is less than the volume of the salt plus the volume of the solvent), whereas when ammonium chloride is dissolved, there is an expansion in the total volume. In the case of sugar and water, there is almost no change in volume. The electrical conductivity of a solution is particularly dependent on the nature of the solute. The change in the property varies with the number of dissolved molecules in a given quantity of solvent. Such properties as vapor pressure, osmotic pressure, boiling point, and freezing point show this.

The vapor pressure of a solution is lower than the vapor pressure of the pure solvent when the solute is nonvolatile. F. M. Raoult (1881) formulated the following generalization: In dilute solutions, the depression of the vapor pressure is proportional to the number of moles of the solute in a given weight of solvent.

Solutions containing the same weight of a given solvent and 1 mol of a different nonvolatile solute in each have the same vapor pressure. When the solvent is benzene (1000 g) in each case, and the solutes are camphor (1 mol) and naphthalene (1 mol), the two solutions have the same vapor pressure at any given temperature. The concentration of such solutions, containing 1 mol of solute in 1000 g of solvent, are called molal. When water is the solvent, organic substances, such as glucose ($C_6H_{12}O_6$) and urea (N_2H_4CO), yield solutions with vapor pressures in accordance with Raoult's law, whereas most inorganic solutes, such as acids and salts, yield solutions whose vapor pressures are lower than those of sugar solutions of the same molal concentrations. This abnormal depression of the vapor pressure is due to the fact that there are a greater number of solute particles due to ionic dissociation.

The molecular hypothesis serves to explain Raoult's law. The vapor pressure is dependent on the number of molecules leaving the surface of the liquid per unit of time. In a pure liquid, the molecules at the surface are all alike, whereas in a solution, some of the molecules at the surface are not volatile. The concentration of the solvent molecules at the surface in the pure solvent is greater than the concentration of the solvent molecules in the solution. Hence, the number leaving the surface of the solution will be less than in the case of a pure solvent. For example, in a solution where the number of solute molecules is one-fifth of the total, the number of solvent molecules leaving the surface will be reduced by one-fifth. At equilibrium, the concentration of the solvent molecules in the space above the liquid (partial pressure) will be such that the number returning to the liquid will just equal the number leaving the surface of the solution. But the number leaving the surface has been reduced one-fifth by the presence of the solute, and the number returning is also one-fifth less than that in the case of the pure solvent; hence the vapor pressure of the pure solvent has been reduced one-fifth. The depression of the vapor pressure of a solvent is equal to the product of the vapor pressure of the pure solvent by the mole fraction of nonvolatile solute in the solution:

$$\text{Depression} = \underset{\text{(pure solvent)}}{\text{Vapor Pressure}} \times \frac{\text{No. of moles of solute}}{\underset{\text{(solute + solvent)}}{\text{Total no. of moles}}}$$

Mole fraction of solute refers to the number of moles of solute divided by the total number of moles of solute plus moles of solvent. Since the number of molecules of a substance is directly proportional to the number of moles (1 mol contains 6.03×10^{23} molecules), the fraction of solute molecules is

$$\frac{\text{Number of solute molecules}}{\text{Total no. of molecules}} = \text{Mole fraction}$$

The depression of the vapor pressure is independent of the nature of the solute but depends on the number of molecules of solute in solution.

Deliquescence

A solution of a very soluble substance may have a vapor pressure which is less than the pressure of the water vapor in the atmosphere surrounding it. Such a solution will not evaporate. The solution will take up moisture from the air until its vapor pressure equals the partial pressure of the water vapor in the air. Equilibrium will then exist. All solids tend to adsorb or condense on their surfaces small quantities of water from the air. This adsorbed moisture forms a film of saturated solution. If the solid is very soluble, this film of solution will have a low vapor pressure. The solution absorbs more water from the air in an effort to reach equilibrium; more of the solid dissolves as the solution is diluted by the absorbed moisture. This process goes on until the entire bulk of the material has dissolved in the water extracted from the air. The process stops when the vapor pressure of the solution equals the partial pressure of water vapor in the air. This behavior of a substance which in fact is due to the low vapor pressure of a solution is known as deliquescence.

Osmotic Pressure

A substance in solution tends to distribute itself uniformly throughout the entire volume of the solvent. This tendency becomes apparent when an attempt is made to prevent this diffusion by separating the solution from the pure solvent by a membrane or partition, which offers no obstacle to the free circulation of the solvent molecules but resists the passage of the solute molecules. Such a membrane is said to be semipermeable. Various substances may serve as semipermeable membranes. The reader is referred to Chapter 13 for further discussion on membrane properties.

When a solution and the pure solvent (water) are separated by such a membrane, water flows through the partition from the pure solvent into the solution. This phenomenon is called osmosis. The water will continue to flow unless a pressure sufficient to prevent it is exerted on the solution or until the solution rises and builds up a sufficient hydrostatic pressure. The pressure needed to prevent the flow of solvent from the pure solvent through a semipermeable membrane into a solution is known as the *osmotic pressure of the solution*.

Osmotic pressure of a solution is independent of the nature of the solute molecules but is proportional to the number of solute molecules in the solution (mole fraction of the solute). Solutions of different substances having the same molal concentrations have the same osmotic pressure at the same temperature, with the exception of solutions of acids and salts. A solution containing 1 mol of solute (nonelectrolyte) dissolved in 1000 g of water produces an osmotic pressure of 22.4 atm at 0°C. Osmosis will occur not only from pure solvent to a solution, but also from a solution of lower to one of higher concentration.

Equilibrium is established only when the concentrations of the two solutions are equal.

Boiling and Freezing Point Changes

The boiling point of a liquid substance is the temperature at which its vapor pressure is 760 mm, and the addition of a nonvolatile solute lowers the vapor pressure; the addition of a nonvolatile solute will raise the boiling point. The solution must be heated to a temperature above the boiling point of the pure solvent in order for the vapor pressure of the solution to be 760 mm. The addition of 1 mol of glucose, $C_6H_{12}O_6$, (180 g) to 1000 g of water—a molal solution—raises the boiling point to 100.52°C; 2 mol raises it to 101.04°C. The addition of a nonvolatile solute lowers the vapor pressure in proportion to the molality of the solution.

Eutectic Mixtures

Practical use can be made of the fact that a solute lowers the freezing point of a solvent. To prevent the water in an automobile radiator from freezing, various substances can be added in concentrations which depend on the extent of the low temperatures anticipated. Water solutions of the following composition are safe to use if the temperature does not go below –23°C (–9.4°F): 33.8% alcohol; 50% glycerine, $C_3H_8O_3$; 38.5% ethylene glycol, $C_2H_6O_2$, by weight. Such *antifreeze* solutions may be changed in composition to withstand even lower temperatures; for example, 71.9% alcohol by weight will freeze at –51.3°C (–60.3°F).

When the pure solvent freezes out of a solution, the concentration of the solute increases and the freezing point of the remaining solution is decreased. Further cooling causes more solvent to crystallize, and as the process continues, the freezing point becomes progressively lower as the concentration of the solute increases. The solution may eventually become saturated with the solute, for example, salt and water solution, and at this point further cooling causes both solvent and solute to crystallize out together in the same proportion as they are in solution. There is then no further change in concentration and hence no change in freezing point. The composition of a solution of such concentration, in which both solvent and solute freeze out together at a constant temperature, is called a eutectic mixture, and its freezing point is called the eutectic point. The eutectic point of a sodium chloride solution is –21°C. Below –21°C, salt added to ice will not cause it to melt, but calcium chloride can be employed, since the eutectic point of calcium chloride solution is –54°C.

Fractional Distillation

When two miscible volatile components form a liquid solution, the vapor pressure of the product is always less than the sum of the vapor pressures of the two components of the mixture. Each particular mixture of the two components has its own boiling point and gives off a vapor which is in equilibrium with the mixture. The vapor pressure might be accurately calculated if each component exerted a partial vapor pressure which was proportional to the mole fraction of that component (Raoult's law). Very few solutions of volatile components act strictly in accord with Raoult's law. The deviations from the law are due to factors which involve attractive forces between the molecules in the solution. When benzene and toluene, two very similar substances, are mixed to form a solution, Raoult's law holds, and the solution is *ideal*. The reason for this ideal behavior of benzene (C_6H_6) and toluene (C_7H_8) is the similarity between the two components; the attraction between a molecule of one and a molecule of the other is about the same as the attraction between two molecules of the same component. Were this not the case, the solution would be nonideal. For all solutions of two volatile components, the composition of the vapor changes as the composition of the liquid varies.

By employing fractionating columns, devices in which the vapors are condensed and redistilled, separation of a mixture of two or more volatile components may be realized by one continuous distillation. Fractionating columns are employed in the separation of crude oil into products such as benzene, gasoline, and kerosene, and in the distillation of liquors and commercial alcohol.

Constant Boiling Mixtures

When a solution of a gas in a liquid is heated, it is the gas and not the liquid which is driven off. Thus water which has been boiled for some time becomes free from dissolved gases. However, some very soluble gases, such as HCl and HBr, do not exhibit this behavior. When a dilute solution of HCl (5% HCl) is boiled, the vapor which comes from the boiling liquid is mainly water and very little HCl. The concentration of the residual liquid increases in HCl and the boiling point of the mixture rises until the composition of the boiling liquid is 20.2% HCl. Further boiling does not change the composition of the liquid or the boiling point. These remain constant until the last drop of liquid evaporates. If the original solution contains more than 20.2% HCl, the vapor which comes off is richer in HCl, and as the boiling continues, the concentration of HCl in the residual liquid decreases until 20.2% HCl is present. At the same time, the temperature of the boiling liquid increases. The boiling point of 20.2% HCl solution is 110.0°C at 760 mm. This mixture of HCl and water (20.2% HCl) is called the constant boiling mixture at 760 mm. In the formation of constant boiling mixtures, it is not necessary that one of the components be gaseous at ordinary temperatures. Two liquids may form a constant boiling mixture. Water

Table 10 Binary Mixtures Having Maximum Boiling Points

Components	Composition	Boiling point (°C)
Water—HNO_3	68% HNO_3	120.5
Water—HCl	20.2% HCl	110.0
Water—HBr	47.5% HBr	126.0
Water—HI	57.0% HI	127.0
Water—H_2F_2	43.2% H_2F_2	111.0
Water—HCOOH (formic acid)	77.9% HCOOH	107.1
$CHCl_3$—CH_3COCH_2	80.0% $CHCl_3$	64.7

and ethyl alcohol form a constant boiling mixture of 95% alcohol; boiling point 78.2°C. A partial list of binary liquid mixtures having constant maximum boiling points under atmospheric pressure is given in Table 10. Mixtures of substances which form constant boiling mixtures cannot be completely separated by fractional distillation.

3

Preliminary Treatment

INTRODUCTION

The objectives of preliminary treatment are to selectively remove materials which could interfere with the physical operation of subsequent treatment processes and to precondition the wastewater in order to prevent odor problems and to improve the treatment efficiency of subsequent processes. The unit operations most commonly used in wastewater pretreatment include

Screening
Flow equalization
Comminution
Mixing
Flotation for oil and grease removal
Flocculation
Sedimentation

Pretreatment in the sewerage system is aimed at controlling conditions further downstream in the treatment process. Aerobic conditions may be ensured by providing adequate sewer venting, designing sewers with steep gradients, and by minimizing the total detention time of sewage in the collection system. The first stage in sewage treatment is usually the removal of large floating objects, such as rags, pieces of wood, and heavy mineral particles such as sand and grit. This is done in order to protect from damage the equipment used in the subsequent stages of treatment, for example, floating aerators in aerated lagoons or any pumps which may be used. This pretreatment comprises screening and grit removal. A common alternative to screening is comminution. The principal applications of primary treatment are summarized in Table 1.

FLOW EQUALIZATION

Flow equalization simply is the damping of flowrate variations so that a constant or nearly constant flowrate is achieved. The application of flow equalization in wastewater treatment is illustrated in the two flowsheets given in Figure 1. In the in-line arrangement, all of the flow passes through the equalization basin. It can be shown that this arrangement can be used to achieve a considerable amount of constituent concentration, uniformity, and flowrate damping. In the off-line arrangement, only the flow above the average daily flowrate is diverted into the equalization basin.

The principal benefits derived from the application of flow equalization are

- Wastewater treatability has been claimed to be enhanced after equalization.
- Biological treatment is enhanced, because shock loadings are eliminated or can be minimized, inhibiting substances can be diluted, and pH can be stabilized.
- The effluent quality and thickening performance of secondary sedimentation tanks following biological treatment is improved through constant solids loading.

Table 1 Primary Treatment Processes and Applications

Operation	Application
Screening	Removal of coarse and settleable solids by interception (surface straining)
Comminution	Grinding of coarse solids to a more or less uniform size
Flow Equalization	Equalization of flow and mass loadings of BOD and suspended solids
Mixing	Mixing of chemicals and gases with wastewater and maintaining solids in suspension
Flocculation	Promotes the aggregation of small particles into larger particles to enhance their removal by gravity sedimention
Sedimentation	Removal of settleable solids and thickening of sludges
Flotation	Removal of finely divided suspended solids and particles with densities close to that of water, and also thickens biological sludges

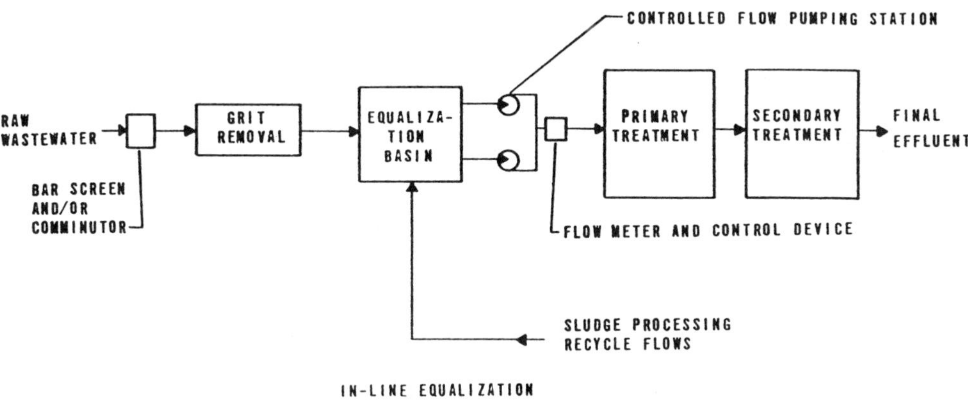

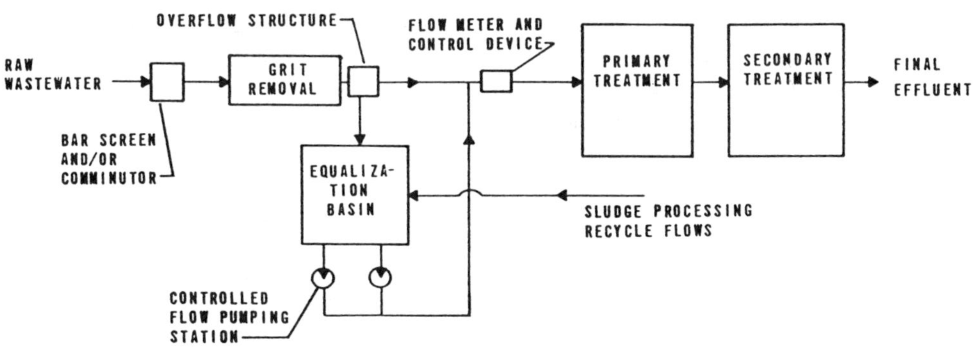

Figure 1 Flow diagrams of equalization.

- Effluent filtration surface area requirements are reduced, filter performance is improved, and more uniform filter-backwash cycles are possible.
- In chemical treatment, damping of mass loadings improves chemical feed control and process reliability.

In some cases, equalization after primary treatment and before biological treatment may be appropriate. Equalization after primary treatment causes fewer problems with sludge and scum. If flow-equalization systems are to be located ahead of primary settling and biological systems, the design must provide for sufficient mixing to prevent variations in solids deposition and concentration and aeration to prevent odor problems.

COMMINUTION

To improve the downstream operations and processes and eliminate problems caused by the varied sizes of solids that are present in wastewater, the solids are often reduced to a smaller, more uniform size. Devices that are used to cut up (comminute) the solids in wastewater are known as comminutors. Often comminuting devices are used to cut up the material retained on the screens so that it may be returned to the flow stream for removal in the subsequent downstream treatment operations and processes.

This device combines screening and disintegration. It consists of a slotted sheet–brass cylinder rotating on a vertical axis through which the whole of the flow passes. Solids arrested on the outside of the cylinder are shredded by fixed cutting edges as the cylinder rotates until they will pass through the slots with the flow. The comminutor avoids any handling of screenings, mechanical or otherwise, and requires very little power. For a small plant which is left unattended for long periods, the comminutor is a good choice provided an electrically supply can be provided without too much expense.

Wear on the cutting edges is usually reduced by installing comminutors downstream of grit-removal tanks. The head loss through a comminutor is greater than that through screens. This may mean that subsequent treatment units will have to be at lower levels than with a screen. In some cases, this results in increased costs. Comminutors are available in sizes up to 0.9 m diameter. They are used especially in smaller plants that are served by separate sanitary sewers carrying a minimum of grit.

GRIT CHAMBERS

Grit chambers are designed to remove grit, which consists of sand, gravel, cinders, or other heavy solid materials that have subsiding velocities or specific gravities substantially greater than those of the organic putrescible solids in wastewater. Grit may also include large organic particles, eggshells, bone chips, coffee grounds, and other food wastes typically found in municipal sewage. Grit chambers are used to protect moving mechanical equipment from abrasion and accompanying abnormal wear; to reduce formation of heavy deposits in pipelines, channels, and conduits; and to reduce the frequency of digester cleaning that may be required as a result of excessive accumulations of grit in such units. There are two general types of grit chambers: horizontal-flow and aerated.

Horizontal-Flow Grit Chambers

In the past, most grit chambers were of the horizontal-flow, velocity-controlled type. These chambers were designed to maintain a velocity as close to 0.3 m/s (1.0 ft/s) as practical. Such a velocity will carry most organic particles through the chamber and will tend to resuspend any that settle but will permit the heavier grit to settle out. The design of horizontal-flow grit chambers should be such that under the most adverse conditions the lightest particle of grit will reach the bed of the channel prior to its outlet end.

Aerated Grit Chambers

The discovery of grit accumulations in spiral-flow aeration tanks preceded by grit chambers led to the development of the aerated grit chamber. Aerated grit chambers are usually designed to provide detention periods of about 3 min at the maximum rate of flow. The cross section of the tank is similar to that provided for spiral circulation in activated-sludge aeration tanks, except that a grit hopper about 0.9 m (3 ft) deep with steeply sloping sides is located along one side of the tank under the air diffusers.

The velocity of roll or agitation governs the size of particles of a given specific gravity that will be removed. If the velocity is too great, grit will be carried out of the chamber; if it is too small, organic material will be removed with the grit. The quantity of air is easily adjusted. With proper adjustment, almost 100% removal will be obtained, and the grit will be well washed. Wastewater should be introduced in the direction of the roll.

Quantities of grit will vary greatly from one treatment plant to another depending on the type of sewer system or wastewater treatment, the characteristics of the drainage area, the condition of the sewers, the frequency of street sanding to counteract icing conditions, types of industrial wastes, the number of household garbage grinders on-line, and the proximity and use of sandy bathing beaches. Possibly the most common method of grit disposal is as fill, covered if necessary to prevent objectionable odors. Generally, the grit must be washed before removal.

OIL/WATER SEPARATION

Many industries discharge liquid wastes contaminated with hydrocarbon or oil-like pollutants. This type of pollution is most prevalent in processes where commercial oils are refined or used as well as some manufacturing operations. Common sources of such waste include refining and processing crude petroleum and petrochemicals, metal fabricating waste, utility operations, sanitary sewage, bilge and ballast wastes, and contaminated surface runoff. Oil-contaminated wastes are also a major problem in the food-processing industry, particularly in animal rendering.

Oils discharged into the environment typically have deleterious effects. Oily waste discharges may have objectionable odor, cause undesirable appearance, burn on the surface of the receiving water creating potential safety hazards, and consume dissolved oxygen necessary to forms of life in the water. Oils in drinking water sources cause objectionable taste and odors, turbidity and film, and make filtration treatment difficult. Bioassay data indicate that oil is toxic to fish. In subacute levels, oil contains fish and shellfish taste. In greater quantities, it limits oxygen transfer, hindering biological activity. Floating puddles of oily waste entrap waterfowl, damage watercraft, and wash up on recreational beaches.

Federal, state, and some local regulations have established standards for the discharge of wastewaters containing oily residues. These standards vary from finite quantities stated in milligrams per liter to qualitative standards requiring, for example, that the wastewater have no visible sheen. International regulations aimed at controlling at-sea discharge of oily wastes define objectionable oily discharges as containing greater than a specified quantity of grains per million. Although these standards, and the analytical methodology on which they are based, differ and pose many problems to the discharger trying to comply, their intent is clear: to control the discharge of oil residues into the environment.

Oil in Water

The difficulty in complying with the regulations is largely due to the complex chemical and mechanical interactions of water and oily substances. Five categories describe the ways in which oil can exist in water:

- Free oil is that which rises quickly to the water surface when given a short quiescent settling period.
- Mechanical dispersions are distributions of fine oil droplets ranging in size from microns to fractions of a millimeter and having stability due to electrical charges and other forces but not due to the presence of surface-active materials.
- Chemically stabilized emulsions are distributions of oil droplets similar to mechanical dispersions but which have additional stability due to chemical interactions typically caused by surface-active agents present at the oil/water interface.
- Dissolved oil is that which is truly dissolved in a chemical sense plus that oil dispersed in such fine droplets (often less than 5 micrometers) that removal by normal physical means (i.e., filtration, coalescence, or gravity settling) is impossible.
- The oil that adheres to the surface of particulate materials is referred to as oil-wet solids.

The degree of difficulty of an oil/water separation problem is, therefore, a function of the oil particulate size distribution (distinguishing free oils from dispersions) and the presence of surface-active agents, dissolved oils, and oil-wet (or non–oil wet) solids. Most separation problems also involve chemicals other than oil, which have a wide variety of effects on the treatment required. The most important factor that can be applied to a given oil/water separation task is, therefore, a common sense management approach. Oil spills, leaks, and points of contamination should be contained. Oil-laden wastes should be treated in their most concentrated state at their source. To limit the dispersion of entrained oil, centrifugal pumps and other equipment that has strong shearing forces should not be used. To limit chemical emulsification, wastes containing surface-active agents (typically used to wash down oil-coated equipment) should not be mixed with other oil-laden wastes. Good housekeeping practices limit the complexity of the separation and in many cases may totally eliminate the problem.

Treatment of Oil in Wastewater

Treatment of oily wastewaters is similar in many respects to the treatment of domestic sewage. In domestic sewage treatment, a primary level of treatment is employed to separate the easily settleable solids from the wastewater. In the treatment of oily wastewaters, a primary treatment is used to separate the floatable free oils from the dispersed emulsified and soluble fractions. Primary treatment is also used to remove oil-wet solids. Common primary separation devices utilize sedimentation, flotation, and centrifugation related techniques. Secondary treatment is then used to break oil/water emulsions and to remove dispersed oil. Technology typically consists of chemical treatment and filter coalescence. In certain applications, a tertiary level of treatment is applied to remove finely dispersed and soluble oil fractions. Tertiary treatment in this context includes ultrafiltration, biological treatment, and carbon adsorption.

Gravity Separation

Gravity separation is the primary and most common treatment utilized. It is based on the specific gravity difference between water and immiscible oil globules and is used to move free oil to the surface of a water body for subsequent skimming and removal. The American Petroleum Institute (API) has specified design criteria for simple gravity separators based on the removal of free oil globules larger than 0.015 cm (150 μ) in diameter. The rise rate of these oil globules is described by Stokes law:

$$V_R = \frac{g\, D^2\, (\rho_w - \rho_o)}{18\mu}$$

where

V_R = rise velocity
g = gravity constant
D = oil globule diameter
ρ_w = water density
ρ_o = oil density
μ = fluid viscosity

The effectiveness of a gravity separator depends on proper hydraulic design and the period of wastewater detention for a given rise velocity. Longer retention times generally increase separation efficiency. The effect of detention time on oil removal is shown in Figure 2. The effective removal of oil droplets with a given rise velocity is therefore a function of the system geometry. The liquid detention must be efficient to permit oil droplets rising at a given velocity to come to the fluid boundary where they can be removed by skimming. This is shown schematically in cross section in Figure 3, where

V_R = rise velocity vector
V_L = liquid velocity vector
V_E = effective path of removed oil

As in the case of conventional clarification equipment, which is sized on the basis of a solids settling velocity, the rise velocity, expressed in units of feet per minute, can be translated into a design overflow rate, expressed in units of gallons per day per square foot. The overflow rate term is quite useful in calculating the surface area required for a given clarification problem.

The design of a given gravity separator is sometimes controlled by the effect of water-in-oil emulsions, where oil appears to be the continuous phase. This is only true, however, when this settling velocity is smaller and therefore limiting.

API SEPARATOR

API separator design criteria control the velocities within the unit by specifying that

- The horizontal velocity through the separator may be up to 15 times the rise velocity of the critical (i.e., slowest rising) oil globule, up to a maximum 3 ft per minute. Above this limit, the effect of turbulence tends to redistribute oil droplets.
- The depth of flow in the separator should be within the limits 3–8 ft. This limits the height that must be traversed by a rising oil globule.

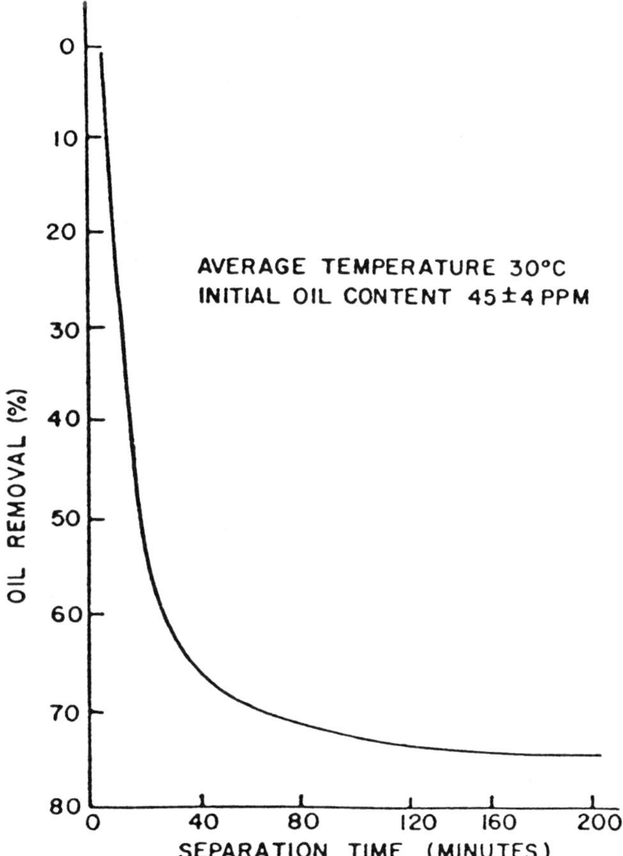

Figure 2 Effect of detention time on oil removal by gravity separation.

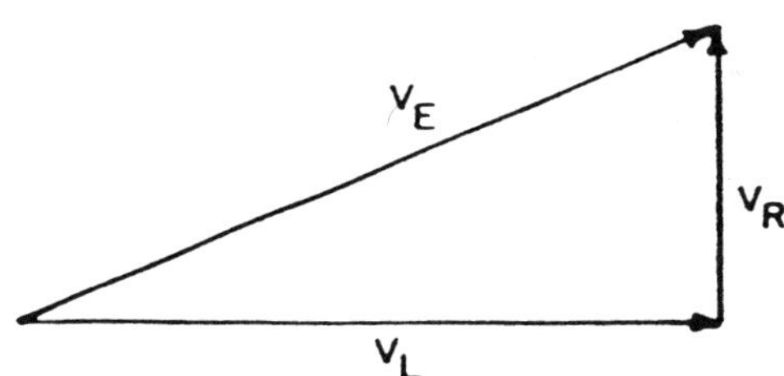

Figure 3 Vector diagram of velocities.

- The width of the separator should be between 6 and 20 ft.
- The depth-to-width ratio should be within the limits 0.3–0.5.
- An oil-retention baffle should be located no less than 12 inches downstream from a skimming device and should have a maximum submergence of 55% of water depth.

Gravity separation often includes provision for heating to lower the viscosity and extended plate surfaces to decrease the effective rise height that must be traversed by a rising oil globule. This latter effect is illustrated in Figure 4. Common separator designs using gravity differential as the primary force include so-called API (based on the above-described API design criteria), CPI (corrugated plate interceptor), and PPI (parallel plate interceptor) units. These units are shown schematically in Figures 5 and 6. Careful handling of flows in a gravity separator by these latter methods often permits separation of oil droplets finer than those of

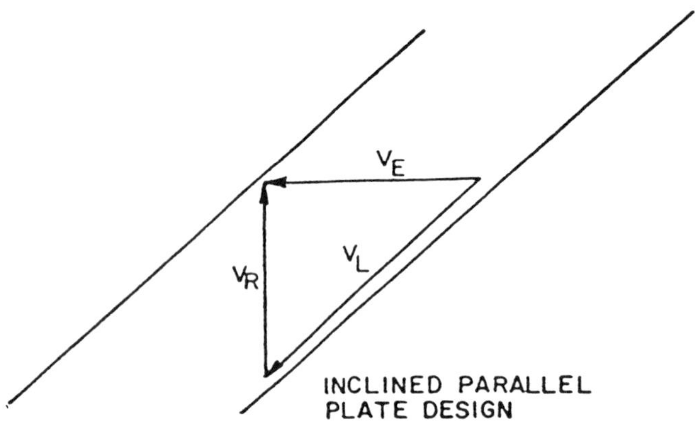

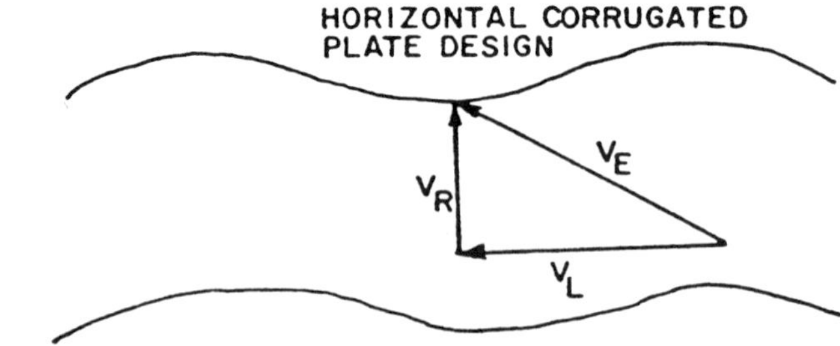

Figure 4 Gravity separation schemes.

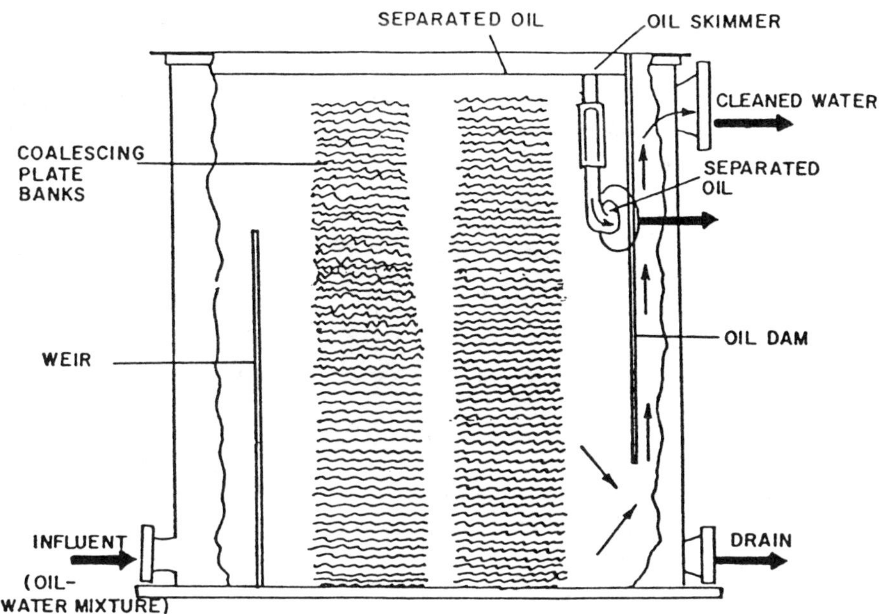

Figure 5 Corrugated plate interceptor.

free oil. The available literature indicates that gravity separation is effective in producing an effluent with an average of 20–100 ppm oil.

Flotation Devices

Flotation devices utilize the gravity separation concept but tend to be more effective than sedimentation devices in removing dispersed oil, since the buoyancy differential is increased by the attachment of small air bubbles to the slow-rising oil globules. One use of air flotation equipment is in the treatment of oil-wet solid laden wastes. Coagulant aids such as polyelectrolytes are commonly used to promote agglomeration of the oil-bearing matter into large flocs which are more easily removed. Air flotation devices are normally preceded by one of the gravity separation techniques described above to remove gross quantities of free oil and settleable solids. This reduces the required volume of dissolved air and flocculating chemicals to economical levels. Also, effective prior removal of the easily floatable oil is necessary for optimum water clarification in the air flotation unit. Air flotation type equipment is reported effective in producing an effluent with 1–20 ppm oil.

Two general methods are commercially used in forming the minute air bubbles. One method involves aerating the waste to saturate it with air at

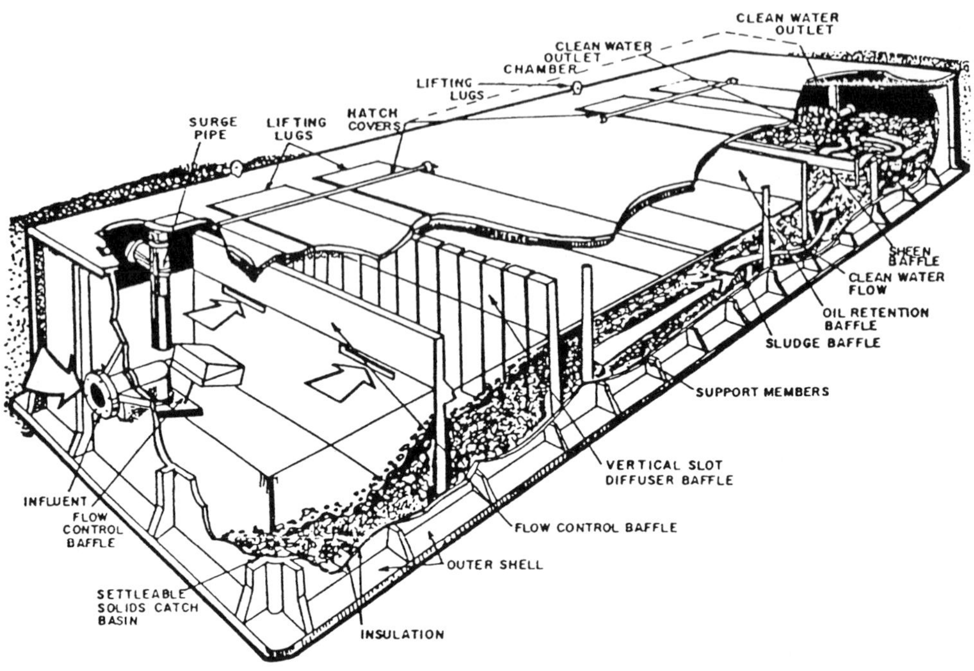

Figure 6 API separator.

atmospheric pressure, releasing the excess air, and then forming the small bubbles by applying a vacuum of approximately 9 inches of mercury. In the other method, air is dissolved into the waste under 2–3 atm of pressure and then the pressure is released forming the minute bubbles. The latter method is more common in the treatment of oily wastes. There are three variations of this latter method, full-flow, split-flow, and recycle operation.

In full-flow operation, shown schematically in Figure 7, the entire waste stream is saturated under pressure, followed by the subsequent release of pressure and bubble formation at the inlet to the flotation chamber. This scheme offers advantages as follows:

- It provides maximum gas solution at any particular pressure, thereby achieving maximum bubble formation and bubble contact.
- For equal flow rates, a small flotation chamber is required.

However, this orientation requires a pressurizing pump large enough to handle the full waste flow, and that the raw waste, which may be loaded with solids, must pass through the pressurizer.

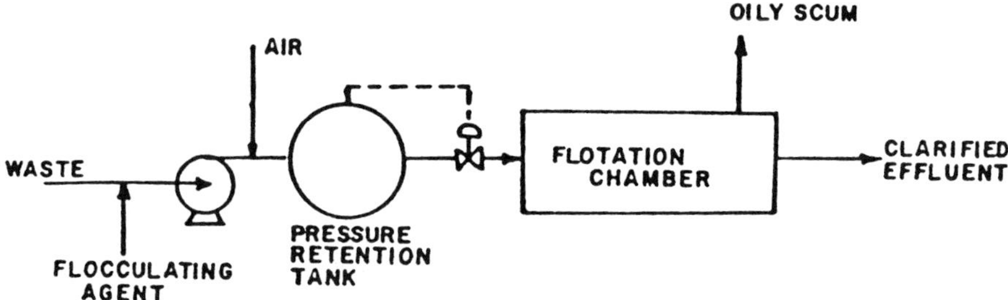

Figure 7 Full-flow flotation.

Split-flow operation, shown schematically in Figure 8, consists of pressurizing and dissolving air in only part of the waste flow and diverting the remainder directly into the flotation chamber, where it is mixed with the pressurized fraction. Split-flow operation

1. Uses a smaller pressurizing pump than full-flow operation
2. Reduces the amount of emulsion that might be formed by the pressurizing pump
3. Uses a small flotation chamber

Recycle operation, depicted in Figure 9, consists of pressurizing and dissolving air in a recycle stream of clarified effluent. The pressure is released and the bubble-containing recycle stream is mixed with the wastewater influent flow. The recycled stream usually is 20–50% of the influent flow. This system

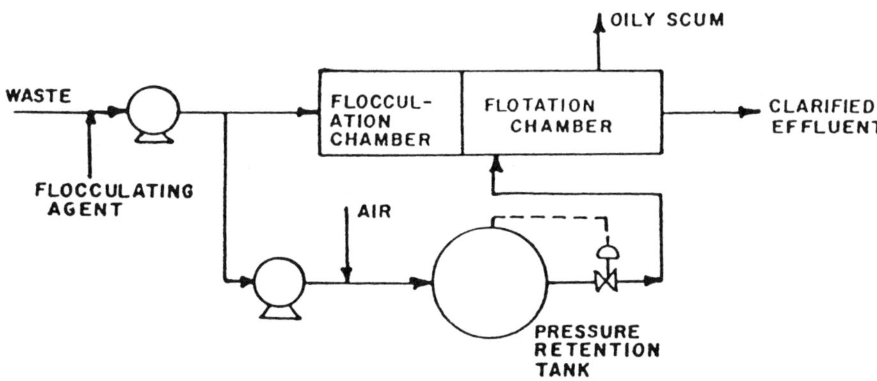

Figure 8 Split-flow air flotation.

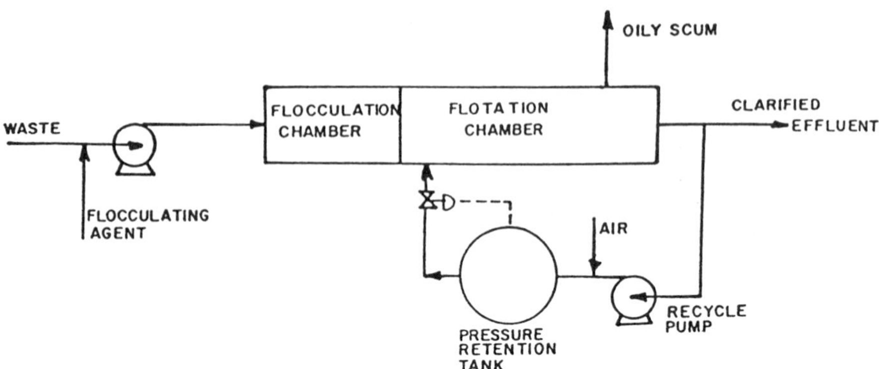

Figure 9 Recycle-flow air flotation.

uses the smallest pressurizing pump and operates the pressurization at a constant flow rate. It minimizes emulsion formation and plugging problems by pressurizing clear effluent. In systems incorporating flocculation, it does not tend to disintegrate the floc by the shearing action of the pressurization pump. This system, however, does require a large flotation chamber.

Several vendors of flotation equipment incorporate the use of interceptor plates in the flotation tank. As in the case of gravity separators, these plate surfaces tend to substantially reduce the height to which a rising droplet of oil or solid must traverse before being separated.

Centrifugal Separators

Centrifugation separators also take advantage of the difference in specific gravity between oil and water. In this technique, the more dense water phase is moved to the outer region of a rotating volume of fluid. The lighter oily materials collect near the vortex and are subsequently removed. This geometry requires that the oil-collecting mechanism must be designed to remove a small column of oil at the centerline if it is to be effective in oil-in-water emulsions. The maximum benefit of centrifugal forces is realized at the outer radial regions apart from the small column of separated oil. For these reasons, centrifugal separators have found limited use in the treatment of oil-in-water emulsions. They have, however, found widespread use in the treatment of water-in-oil emulsions. Both vertical and horizontal designs are available. Separators based on centrifugation principles are quite effective in the removal of oil-wet solids. These solids tend to bind or plug filter-coalescence media used to polish the wastewater effluent. A modification of the basic centrifuge, which incorporates parallel plates to provide laminar flow regions, has been found effective in treating dispersed oil. Effluent qualities averaging 50–70 ppm oil have been reported. Other devices which use centrifugal separation principles include hydroclones, swirl concentrators, vortex separators. The reader is referred to Chapter 9 for further details.

Demulsification

Several different processes are used to break or demulsify oil/water emulsions. These include chemical, electrical, and physical methods. Chemical methods are, by far, in widest use. The electrical process is directed toward emulsions containing mainly oil with small quantities of water. Physical emulsion breaking methods include heating, high-speed centrifugation, and precoat filtration. Chemical treatment of an emulsion is usually directed toward destabilizing the dispersed oil droplets or chemically binding or destroying any emulsifying agents present. Chemical demulsifying processes include

- Acidification
- Coagulation
- Salting out
- Demulgation with organic cleaving agents

Acids generally cleave emulsions more effectively than coagulant salts but are more expensive and the resultant wastewater must be neutralized after the separation. Salting out of an emulsifier is achieved by adding large quantities of an inorganic salt, thereby increasing the dissolved solids content of the water phase. Coagulation with aluminum or iron salts is generally effective, and this is in common use even though the resultant hydroxide sludges are difficult to dewater and limit the reusability of the recovered oil. These coagulants generally react as follows:

$$Fe_2(SO_4)_3 + 6H_2O \rightarrow 2Fe(OH)_3 + 3H_2SO_4$$
$$Al_2(SO_4)_3 + 6H_2O \rightarrow 2Al(OH)_3 + 3H_2SO_4$$

Organic demulgators are extremely effective demulsifying agents but are generally very expensive and specialized. The science of these demulgators is to a large extent a "black art." Two specific modes of operation are to add a specific chemical agent with sufficient properties to neutralize any other charges that are acting to inhibit coalescence or chemically to react with and/or break down the specific chemical species causing the emulsification. Chemical demulgators are normally considered only if effective in extremely low concentrations owing to their cost and, in many cases, their toxicity at higher concentrations. Demulsification is normally followed by sedimentation or flotation for removal of the destabilized oil.

Determination of the effectiveness of demulsifiers is normally carried out by jar testing. A representative waste sample is split into several aliquots, each of which is reacted with a known quantity of reagents. After each addition, the cells are examined to note any reaction. Breaking of the emulsion into separate surface-active properties tends to alter the surface wetting properties of the coalescing fibers, which usually lends to "poisoning" of the media. In addition, the effectiveness of the system depends on, among other things, the mechanical forces of the influent passing through the filter. If the volume and/or force of the pumping is too great, the oil droplets tend to be prematurely carried into the mainstream flow and are insufficient in size to gravity-separate from the effluent. Despite drawbacks, filter coalescers are quite effective. The effluent quality achievable with such devices is in the range of 1–50 ppm oil depending on such factors the surfactant content, loading conditions, and oil type.

Coalescing media used for oil separation vary in the materials used and the effective pore size. In some coalescing media, a fibrous material such as nylon or propylene is wound about a rigid spool to form a cartridge. The tightness of the wrap and the fiber diameter largely control the effective porosity for these devices. Other coalescing media incorporate the use of tightly woven or tightly wrapped sheets of fiberglass. Since coalescing media does tend to plug with particulates, less costly pleated paper–type elements are often used as coalescing media or as prefiltration media. More recently, reticulated polyurethane foams have come into use as coalescing media. These foams are natural sorbents, are light in weight, are relatively inexpensive, and can be molded in such a manner as to control readily effective pore size. In most cases, the separators incorporating coalescence are designed for the replacement of the media once it is poisoned, plugs, or otherwise fails. The cartridge-type media are favored in these designs. End caps are provided on the cartridges to ensure that the oily water, fed from the inside out, does not have the opportunity to "bypass" the media. Some separators incorporate means to back-wash the media as a means of removing entrained solids and any adhering surface-active agents. Steam, hot water, and/or solvents are at times prescribed for these situations.

The geometry and orientation of coalescing elements in separator devices varies from one design to another. Most manufacturers utilize long, relatively small-diameter cartridges of a standard size and stack these in parallel to handle the required throughput. Several stages of these groupings operating in series are often utilized to achieve a greater degree of removal and to act as a built-in backup. The media is normally installed vertically or horizontally depending on the design. Horizontal orientation normally decreases the effective oil droplet rise height. However, this design is normally more difficult and time consuming to service, since the entire vessel must be completely drained before opening. Figure 10 illustrates a typical setup.

Biotechnology

The treatment of dissolved oils and other types of chemically stabilized emulsions that cannot be destabilized by chemical additions can pose serious problems.

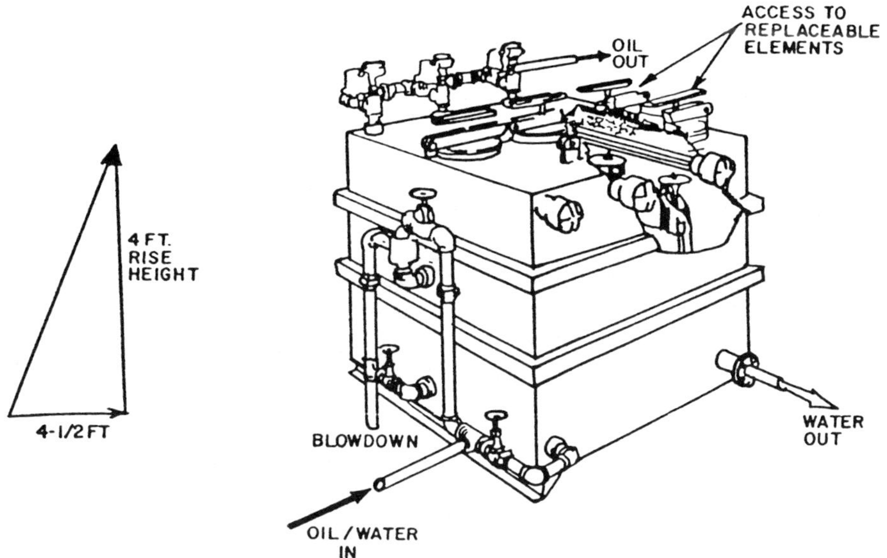

Figure 10 Rise height and maintenance access for horizontally and vertically oriented filter coalescing separators.

Biological treatment with acclimated microorganisms is generally effective in degrading much of this material and is commonly used in petroleum refineries and animal-rendering plants. However, the systems are only effective if suitable pretreatment and high dilution can be achieved. Too much oil is a problem in biological systems, because it is adsorbed by the microorganisms faster than it can be metabolized. In trickling filters, oil tends to coat the microbial surfaces and reduce the transfer of more readily oxidizable organics. In activated sludge systems, the adsorbed oil tends to impair sludge-settling characteristics. Resulting sludge losses may be so high as to reduce the microbial level in the system enough to cause reduced efficiency and possible system failure. The microbial metabolism of oil is limited by the low solubility of oil, the chemical configuration of oil molecules, and the microbial surfaces. Trickling filters can treat oil concentrations of up to 100 ppm with no effect. Activated sludge systems show no effect if the oil concentration is kept less than 25 ppm. Biologically treated effluents typically contain less than 10 ppm oil.

Selection of the optimum biological waste-treatment process for a particular waste is a complex problem. Many factors are involved, including economics, land availability, and effluent quality required. Bench-scale testing is also useful in determining the toxic limit of the oily contaminant. It should be realized, however, that the oxygen transfer rates and the sludge separation/return efficiencies in pilot systems are higher than the rates in a full-scale system and must be compensated for in scale-up.

Carbon Adsorption

Carbon adsorption has been addressed quite extensively as a means of removing trace quantities of oil. Treatment by this method requires a suitable means of regenerating the carbon. Methods that have been addressed include steam, hot water, organic solvents, and pyrolysis. Treatment by carbon adsorption generally requires a large capital investment for carbon inventory and regeneration equipment and has, therefore, not found widespread use in oil separation where high concentrations are involved.

Ultrafiltration

Ultrafiltration is based on the sieving action of a polymeric membrane controlling the flow of molecules larger than the membrane pores. Applied pressure is used to increase the flux of the liquid across the membrane. The membranes tend to foul with particulates and the flux therefore decreases. The fouling is normally removed by back flushing and/or detergent washing. Reverse osmosis is similar to ultrafiltration in that an applied pressure forces the water through the membrane against a concentration gradient while oil is retained owing to the small size of the membrane pores. However, in reverse osmosis, the pore sizes are smaller and the applied pressures are significantly higher. These treatments can be used to produce essentially oil-free effluents. However, these treatments require large capital investment and have high operating costs.

In selecting the appropriate separator equipment, the specifics of the oily waste problem should be carefully studied. The characteristics of the oily water mixture should be determined after all reasonable water-management techniques have been instituted. Typical characteristics that should be determined include the oil and bulk fluid densities, the oil rise velocity, the oil droplet size distribution, the presence of emulsifying agents, and the suspended solids content and distribution. Every effort should be made to treat the waste in its most concentrated form and to prevent contamination with particulates and chemical emulsifiers. Low-shear positive-displacement pumps should be used to prevent shear of the fluid. If the oil is present in quantities greater than roughly 1%, gravity separation or similar methods should be used to achieve bulk separation. If chemical emulsions are present, they should be treated before contacting coalescer media. Demulsifying agents should be evaluated by jar testing. Prefiltration should be used to increase coalescer life when applicable.

It may be necessary to engineer a separation system for a specific separation problem. In such cases, the available off-the-shelf components can be added together in series to meet the requirements of the specific situation. For example, a separation system may consist of gravity separation, emulsion breaking, flocculation-assisted air flotation, prefiltration, and filter-coalescence.

SCREENING

Water supply or wastewater effluents may contain coarse suspended and floating matter which may damage or interfere with the operation of pumps and treatment plant equipment. This material is usually removed by simple screening devices. Screens are sometimes provided at the inlet to sewage pumping stations where rags, paper, and other coarse materials could be the source of damage by fouling pump impellers, and on sewer overflow structures where it is desirable to prevent coarse materials from fouling the receiving process. The principal types of screens in use are described in Table 2.

Bar screens or racks are usually constructed as a series of metal bars arranged in one plane across a slightly expanded channel and inclinded upwards in the downstream direction. Less frequently the bars are arranged vertically, being spaced, for example, in a circle to form a basket grate. Bar screens fall into two groups: coarse with spacings wider than 38 mm and fine with openings between approximately 12 and 25 mm. The bars are commonly circular, otherwise rectangular, in section and spaced to suit the wastes being treated.

Screens serve mainly to withhold larger pieces of extraneous solids which might otherwise cause blockage in sludge lines and other process equipment. The slope is usually about 30 degrees but for mechanical cleaning say 75 degrees. The channel must not be unduly expanded or sand and silt might separate. A bypass overflow channel, usually fitted with vertical bars spaced 102 mm apart, is essential. Coarse bar screens normally remove less than 30 per capita yearly of solids from municipal sewage, fines from 31 to about 81. Screenings may be buried, incinerated, or digested. Incineration leads to odor nuisance, unless a temperature of about 1500 degrees or higher is reached. Figure 11 shows a manually raked bar screen and Figure 12 shows a mechanically cleaned bar rack.

Cleaning is performed by rakes the tines of which fit between the bars and allow the screenings to be drawn upwards out of the flowing waste. The material may then be transported by conveyors or wheeled buckets, usually after dewatering on simple perforated trays draining through small screens. Manual cleaning of coarse bar screens is usually sufficient if regular attention is given, otherwise it may be done mechanically, usually automatically when the upstream level builds up to a predetermined height.

Fine Screens

Fine screen openings vary from 2.3–6.0 mm (0.09–0.25 in.). Static screens with openings less than 2.3 mm have been used for pretreatment and/or primary

Table 2 Screening Devices Used in Wastewater Treatment

Type of screen	Screening surface			
	Size classification	Size range	Screen material	Application
Inclined				
Fixed	Medium	250–1500	Stainless steel wedge-wire screen	Primary treatment
Rotary	Coarse	0.8–2.4 mm × 50-mm slots	Milled bronze or copper plates	Pretreatment
	Medium	100–1000 m	Stainless steel wire cloth	Primary treatment
Drum (Rotary)	Coarse	0.8–2.4 mm × 50-mm slots	Milled bronze or copper plates, wire screen	Pretreatment
	Medium	250–1500mm	Stainless steel and polyester screen cloths	Removal of residual secondary suspended solids
	Fine	15–60 mm	Stainless steel or other noncorrosive material	Removal of residual secondary suspended solids
Traveling	Coarse to medium		Stainless steel or other noncorrosive material	Primary treatment
Centrifugal	Fine-medium	10–500 mm	Stainless steel, polyester, and various other fabric screen cloths	Primary treatment, secondary treatment with settling tank, and the removal of residual secondary suspended solids

Note: mm × 0.03937 = 1 inch.

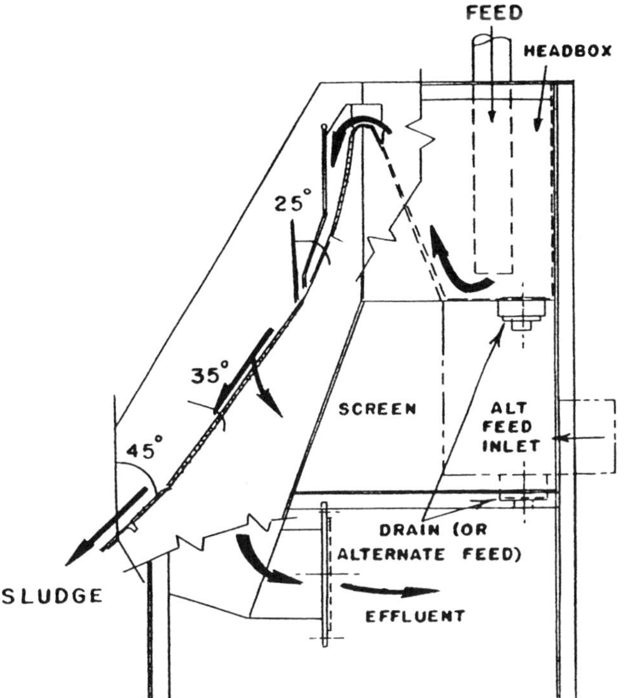

Figure 11 Inclined screen.

treatment. Revolving drum screens are being used for upgrading secondary treatment plants. Perforated plate and closely spaced bars normally have openings greater than 0.02 mm. Woven wire screens are used when finer screening is required.

Fixed Screens

The application of static screens to municipal wastewater treatment can result in BOD and suspended solids (SS) removal in the range of 20–35%. The use of fine screens may offer pretreatment and possibly even for providing primary treatment. This is particularly true in the upgrading of existing small treatment plants. The screening media employed may be perforated plate, woven wire cloth, or closely spaced bass.

Although fine screens could be expected to remove only 5–10% of the SS in raw wastewater, static screens with 0.8- to 1.5-mm (0.03- to 0.06-in.) openings can be expected to remove 25–35%, with some removals as high as 60%. Primary sedimentation tanks can be expected to remove 50–60% SS. In addition,

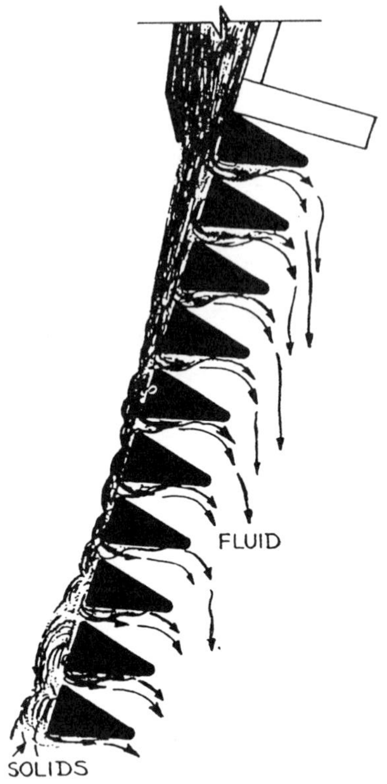

Figure 12 Screen detail.

the wastewater dissolved oxygen (DO) is raised to 2 mg/L or more, thus assisting in grease removal and subsequent biological treatment.

Moving Screens

Moving fine screens closely approximate continuously cleaned screens. The strainer consists of a rotating cylinder having a screen attached to the circumferential area of the drum. Different screens (stainless steel, nylon, polyester) can be employed, with openings commonly varying from 0.02 to 3.0 mm. Screens with pore sizes as small as 0.005 mm are now available from some commercial sources but are seldom used.

Fine screens have been used to polish secondary effluents as well as aerated lagoon effluents. They are also used to provide protection for close clearance equipment and may be used for the treatment of combined sewer flows. The performance of a fine screen device varies considerably depending on influent solids concentrations, operating hydraulic head, and the degree of biological conditioning of the solids. A fine screen will perform well over a range of influent suspended solids, although it is extremely sensitive to influent solids fluctuations. Suspended solids removals of 60–90% for 0.020-mm screens and 55–75% for 0.035-mm screens have been achieved when processing secondary effluents.

Coarse Screens

Modern coarse screens are of the disk or drum type, with stainless steel or nonferrous wire-mesh screen cloth. Typically, the openings vary from 6 to 29 mm or more. The disk type has a vertical circular screening surface that rotates on a horizontal shaft set slightly above the water surface. It is available in sizes from 4 to 18 ft in diameter. The drum type revolves at about 4 r/min around a horizontal axis and operates slightly less than half submerged. The wastewater flows in one end of the drum and outward through the screen cloth.

In both the disk and drum types, the solids are raised above the liquid level by rotation of the screen and are backflushed into receiving troughs by high-pressure jets. With the finer-mesh cloth, effluent may be used for spray water.

Physical straining processes as those processes which remove solids by virtue of physical restrictions on a media which has no appreciable thickness in the direction of liquid flow. Physical straining devices may be grouped according to the nature of their straining action and are summarized in Table 2.

A typical inclined screen is shown in Figure 11. These devices operate by gravity and function as an inclined drainage board with a screen of wedge wire construction having openings running transverse to the flow.

The screen consists of three sections with successively flatter slopes on the lower sections. The screen wires are triangular in cross section as shown in Figure 12 and usually spaced 0.06 in. apart for raw wastewater screening applications. In the unit in Figure 13, these wires bend in the plane of the screen as shown. They are straight and transverse to the flow. Inclined screening units are generally constructed entirely of stainless steel. Lighter units with a fiberglass housing and frame are also available.

Wastewater enters and overflows the headbox and onto the upper portion of the screen. On the screen's upper slope, most of the fluid is removed from the influent. The solids mass on the following slope, because it is flatter, and additional drainage occurs. On the screen's final slope, the solids stop momentarily, simple drainage occurs, and the solids are displaced from the screen by oncoming solids.

Inclined screens cannot remove suspended solids to the same extent as a sedimentation unit but do an excellent job of removing trashy materials which may foul subsequent treatment of sludge handling units. Their ability to remove fine grit is limited by size openings. Separate grit-removal equipment should be

Figure 13 Curved screen bars.

installed after the inclined screens if required. Incidental to the removal of suspended solids in this process is the aeration of the separated water.

Rotating Wedge Wire Screens

In a rotating wedge wire screen which backwashes itself, wastewater passes virtually downward from the outside to the inside of the drum by gravity. The screened wastewater then passes out through the lower half of the drum to a collection trough.

Solids are retained on the outside of the drum and are removed by a fixed scraper blade screen spacing of 0.06 in. is recommended for service on raw wastewater. In comparison with static screens, rotating units have been claimed to require less maintenance, lower operating head, and smaller space and produce dryer solids.

Microscreening

A microscreen unit typically consists of a motor-driven rotating drum mounted horizontally in a rectangular chamber. A fine screening media covers the periphery of the drum. Feedwater enters the drum interior through the open end and passes radially through the screen with accompanying deposition of solids on the inner surface of the screen. At the top of the drum, pressure jets of effluent water are directed onto the screen to remove the mat of deposited solids. The dislodged solids together with that portion of the backwash stream which penetrates the screen are captured in a waste hopper. Solids flushed from the unit are sent to sludge-handling systems or recycled to the head of the plant. Units may be equipped with ultraviolet lights to control biological growth on the screen media. Effluent passes from the chamber over control weirs oriented perpendicular to the drum axis.

Microscreening has been used for the removal of algae from uncoagulated lagoon effluents. However, many classes of algae, such as *Chlorella*, are too small to be removed, even on fine screens (23 μ) and excessive loadings (up to 2×10^6 algae/ml) makes this application a limited one. Parameters of mesh size, submergence, allowable head loss, and drum speed (rpm = peripheral speed/(π/4) [diameter]) are sufficient to determine the flow capacity of a microscreen with given suspended solids characteristics.

The basic screen support structure is a drum-shaped, suitably stiffened rigid frame supported on bearings to allow rotation. Designs using water-lubricated axial bearings or greased bearings located on the upper inside surface of the rotating drum allow submergence well above the central axis. Both plastic (polyester) and stainless steel are used for the microscreen media itself. Greater mechanical strength, especially at higher temperatures, are an advantage of stainless steel. Greater economy and chemical resistance are advantages of plastics. Microscreen fabrics normally are woven of stainless steel or plastic (polyester with polypropylene supporting grid) with openings in the range of 15–60 μ. Plastic fabric is less subject to chemical attack by strong chlorine or acid cleaning solutions. Stainless steel can better withstand temperatures encountered in steam cleaning.

Backwash jets are directed against the outside of the microscreen drum as it passes the highest point in its rotation. About half the flow penetrates the fabric dislodging the mat of solids formed on the inside. A hopper inside the drum receives the flushed-off solids. The hopper is positioned to compensate for the trajectory that the solids follow at normal drum peripheral velocities. Microscreen effluent is usually used for backwashing. Straining is required to avoid clogging of backwash nozzles. In-line strainers used for this purpose will require periodic cleaning; the frequency of cleaning will be determined by the quality of the backwash water.

SEDIMENTATION

Sedimentation is the separation from water by gravitational settling of suspended particles that are heavier than water. Sedimentation is used for grit removal, particulate matter removal in the primary settling basin, biological floc removal in the activated-sludge settling basin, and chemical floc removal when the chemical coagulation process is used. It is also used for solids concentration i[n] sludge thickeners. The primary purpose is to produce a clarified effluent, bu[t] also produces sludge with a solids concentration that should be easily handled and treated.

Intermediate and final settling basins are used to remove the settleable solids produced by biological treatment solids produced by biological treatment processes. Settling basins are also used to remove settleable solids produced as the result of tertiary treatment. Settling units can be used to accomplish a certain amount of thickening of the settled solids. However, the major purpose of the settling process to obtain maximum removal of SS is best achieved when the settled sludge is rapidly removed from the tank.

Settling basins can be either circular or rectangular or are designed to operate on a continuous flow-through basis. Circular settling basins frequently are called clarifiers. Sedimentation basins can be divided into the four zones: inlet, clarification, sludge, and outlet as shown in Figure 14. The suspension to be clarified is admitted to the basin through the inlet zone. Separation of the solids from the liquid takes place in the relatively quiescent clarification zone. The clarified liquid or effluent is then removed through the outlet zone. Separated solids are allowed to accumulate, compact, and are then withdrawn from the sludge zone. Figure 14 shows the flocculation process in settling basins.

Settling tanks have scrapers, ploughs, or suction devices for transporting sludge along the bottom into a sump or direct removal through pipes. The scrapers, ploughs, and suction devices are attached to rotating arms, traveling bridges, or endless chains. The tanks may be circular, square, or rectangular.

The efficiency of settling tanks varies from 45 to 70% suspended solids, from 85 to 99% settleable solids removal, and from 25 to 40% BOD reduction depending on the strength, nature, and age of the sewage; tanks design; detention period; and operation.

MIXING

Mixing is an important unit operation in many phases of wastewater treatment where one substance must be completely intermingled with another. The mixing of chemicals with wastewater, where chlorine of hypochlorite is mixed with the effluent from the secondary settling tanks are examples. Chemicals are also mixed with sludge to improve its dewatering characteristics before vacuum filtration. the digestion tank, mixing is used frequently to assure intimate contact between nd microorganisms. In the biological process tank, air must be mixed with ted sludge to provide the organisms with the oxygen required.

mixing can be carried out in different ways, including

ns in open channels

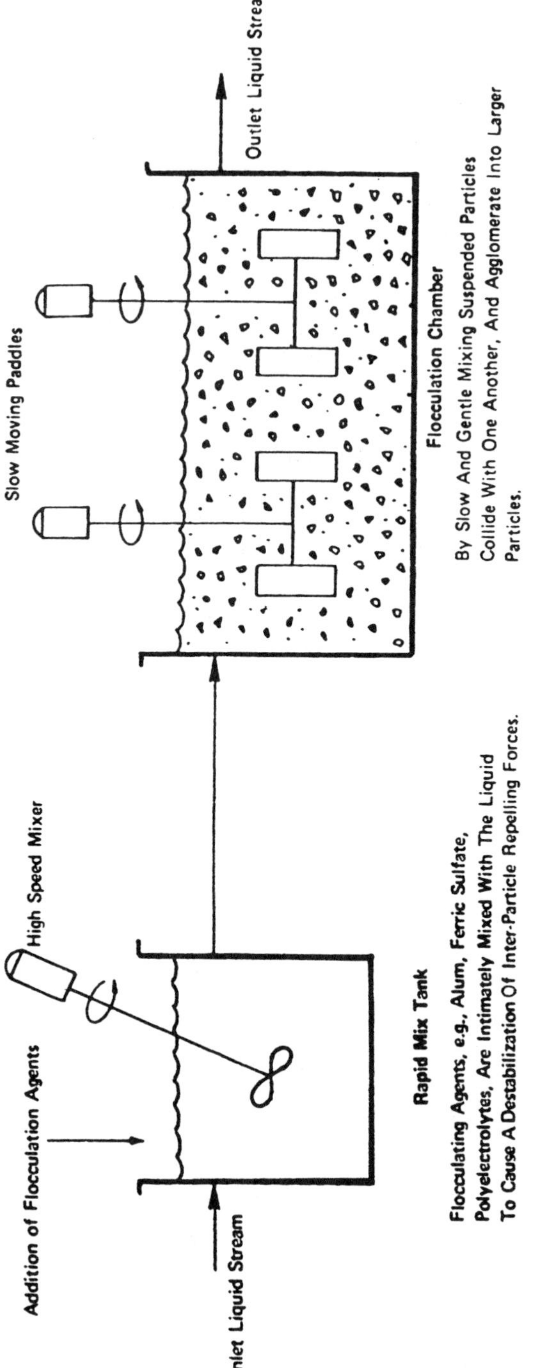

Figure 14 Flocculation process.

- Pumps
- Vessels with the aid of mechanical means

A typical mixer used in wastewater-treatment plants is shown in Figure 15. Paddles generally rotate slowly, as they have a large surface exposed to the liquid. They are used as flocculation devices when coagulants, such as aluminum or ferric sulfate, and coagulant aids, such as polyelectrolytes and lime, are added to wastewater or sludges. The production of a good floc usually requires a detention time of 15–30 min.

Turbine mixers are used in sludge-blending tanks. Large blenders are designed to rotate at moderate speed, about 25–100 r/minute. Propeller-type impellers have also been used for high-speed mixing with rotational speeds up to 2000 r/min. They are shaped like ship propellers and generate strong axial currents that rapidly mix chemicals or gases with liquids. For fairly small mixing vessels, vortexing can be prevented by mounting the impellers off center or at an angle. The usual method in both circular and rectangular tanks is to install four or more vertical baffles extending approximately one-tenth the diameter our from the wall. These effectively break up the mass rotary motion and promote vertical mixing. Concrete mixing tanks may be made square and baffles may be omitted.

COAGULATION

The removal of oxygen-demanding and turbidity-producing colloidal solids from wastewaters is often called intermediate treatment, since colloids are intermediate

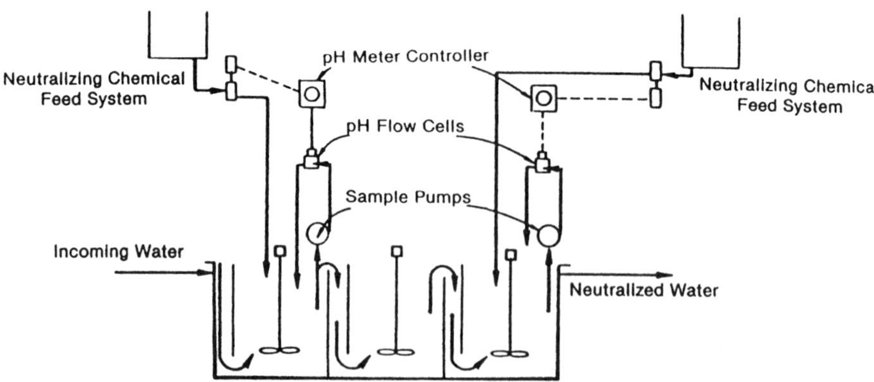

Figure 15 Continuous-flow neutralization.

in size between suspended and dissolved solids. The most common method of removing these solids is by chemical coagulation. This process consists of destabilizing colloids, aggregating them, and binding them together for ease of sedimentation. It involves the formation of chemical flocs that absorb entrap, or otherwise bring together suspended matter, more particularly suspended matter that is so finely divided as to be colloidal.

Chemicals most commonly used are

- Alum ($Al_2(SO_4)_3 \cdot 18H_2O$)
- Copperas ($FeSO_4 \cdot 7H_2O$)
- Ferric sulfate ($Fe[SO_4]_3$)
- Ferric chloride ($FeCI_4$)
- Chlorinated copperas, a mixture of ferric sulfate and chloride

The process of chemical coagulation involves complex equilibria among a number of variables, including colloids of dispersed matter, water or another dispersing medium, and coagulating chemicals. Driving forces such as the electrical phenomenon, surface effects, and viscous shear cause the interaction of these variables.

FLOTATION

Air flotation methods can be used in water treatment and wastewater treatment for separation of particles which are too light to settle effectively but too heavy to float unaided.

Flotation is a unit process by which solid particles in liquid suspension become attached to microscopic air bubbles giving the air-solids agglomerate buoyancy. Under the right conditions, the agglomerate will rise to the surface to join other particles and form a blanket that can be removed mechanically. Once the suspended solids have been floated to the surface, a simple skimming operation can be used to collect and dispose them.

Flotation can be incorporated with wastewater-treatment schemes in the following ways:

- As a pretreatment unit ahead of primary sedimentation, called "scalping." Grit removal may be incorporated if desired.
- As a primary treatment unit ahead of secondary treatment units.
- For pretreatment of industrial waste prior to treatment of combined industrial and domestic waste.
- As a unit process for removing specific suspended materials which are not readily removed by other processes.
- For sludge thickening.

Advantages include

- Grease, light solids, grit, and heavy solids are removed all in one units.
- High overflow rates and low detention periods mean smaller tank sizes resulting in less space requirements and possible savings in construction costs.
- Odor nuisance is minimized because of the short detention periods and the presence of dissolved oxygen in the effluent.
- In many cases, thicker scum and sludge are obtained from a flotation unit than by gravity settling and skimming.
- The recovered solids may be reusable as a source of fuel.

Disadvantages include

- The additional equipment required results in higher operating costs.
- Flotation units generally do not give as effective treatment as gravity-settling units, although the efficiency varies with the waste.
- The pressure type has high power requirements, which increase operating cost.
- The vacuum type requires as relatively expensive airtight structure capable of withstanding a pressure of 9 in. of mercury; any leakage to the atmosphere will adversely affect performance.
- More skilled maintenance is required for a flotation unit than for gravity-settling unit.

NEUTRALIZATION

Excessively acid or alkaline wastes should not be discharged without treatment into a receiving stream. A stream even in the lowest classification; that is, one classified for waste disposal and/or navigation is adversely affected by low or high pH values. This adverse condition is even more critical when sudden surges of acids or alkalis are imposed on receiving streams. There are many acceptable methods for neutralizing overacidity or overalkalinity of waste waters. Figure 16 shows a schematic for a continuous flow neutralization.

Mixing Wastes

Mixing of wastes can be accomplished within a single plant operation or between neighboring industrial plants. Acid and alkaline wastes may be produced individually within one plant and proper mixing of these wastes at appropriate times can accomplish neutralization, although this usually requires some storage of each waste to avoid slugs of either acid or alkali.

Passing acid wastes through beds of limestone was one of the original methods of neutralizing them proceeds chemically according to the following typical reaction:

$$CaCO_3 + H_2SO_4 \rightarrow CaSO_5 + H_2CO_3$$

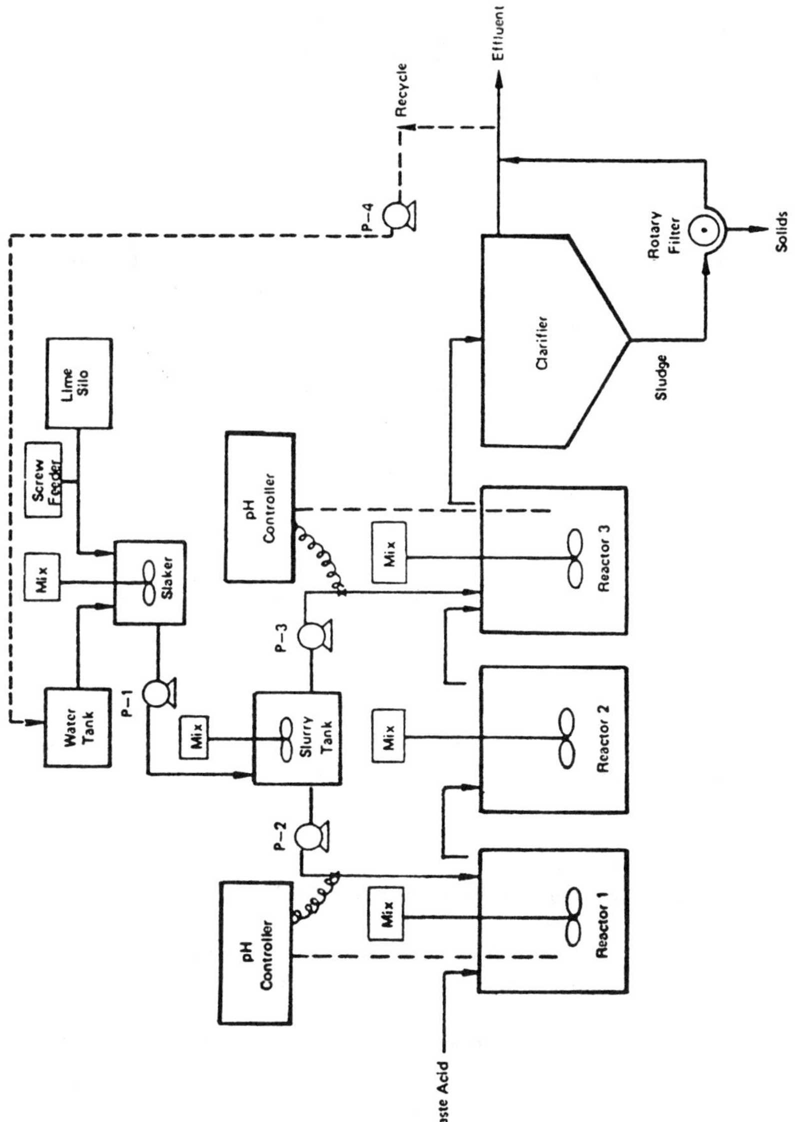

Figure 16 Lime neutralization of sulfuric acid waste.

The reaction will continue as long as excess limestone is available and in an active state. Disposing of the used limestone beds results in a waste problem in this method of neutralization.

Mixing acid wastes with lime slurries is an effective procedure for neutralization.

Caustic Soda Treatment for Acid Wastes

Adding concentrated solutions of caustic soda (NaOH) or sodium carbonate to acid wastes in the proper proportions results in faster, but more costly, neutralization. Smaller volumes of the agent are required, since these neutralizers are more powerful than lime or limestone.

When sodium hydroxide is used as a neutralizing agent for carbonic and sulfuric acid wastes, the following reactions take place:

$$Na_2CO_3 + CO_2 + H_2O \rightarrow 2NaCHO_3$$
$$2NaOH + CO_2 \rightarrow Na_2CO_4 + H_2O$$
$$NaOH + H_2SO_4 \rightarrow NaHSO_4 + HOH$$
$$NaHSO_4 + NaOH \rightarrow Na_2SO_4 + HOH$$

Both these neutralizations take place in two steps and the end products depend on the final pH desired.

Boiler-Flue Gas

Blowing waste boiler-flue gas through alkaline wastes has been proposed as a method for neutralizing them. CO_2 dissolved in wastewater will form carbonic acid (a weak acid), which in turn reacts with caustic wastes to neutralize the excess alkalinity.

Carbon Dioxide Treatment for Alkaline Wastes

Bottled CO_2 is applied to wastewaters in much the same way as compressed air is applied to activated sludge basins. It neutralizes alkaline wastes on the same principle as boiler-feed gases.

Another way to produce carbon dioxide is to burn gas under water. This process is called submerged combustion and has been used in the disposal of nylon wastes to neurtralize the waste prior to biological treatment.

Sulfuric Acid Treatment for Alkaline Wastes

The addition of sulfuric acid to alkaline wastes is a fairly common but an expensive means of neutralization. The neutralization reaction which occurs when it is added to wastewater is as follows:

$$2NaOH + H_2SO_4 \rightarrow Na_2SO_4 + 2H_2O$$

Acid Waste Utilization

It may be possible to use acid wastes to effect a desired result in industrial processing to wash, cool, or neutralize products. Mine waste water occurs in large quantities in the coal industry. These waters are usually acid and contain sulfates of iron and aluminum; if they are used to wash raw coal, neutralization results, since coal contains calcium and magnesium carbonates.

Prechlorination

Prechlorination of sewage by injection chlorine into rising mains is an established method of avoiding hydrogen sulfide (H_2S) odor and corrosion problems at the inlet and in primary phases of treatment works. Control of H_2S is achieved by its oxidation, usually to precipitating molecular sulfur. Each milligram per liter of sulfide present requires 2 mg/L of chlorine to be added. Therefore, a typical septic sewage might require 10–15 mg/L of chlorine for precipitation of sulfur from the sulfide. In practice, however, the chlorine demand by organic material in the sewage often requires chlorine to be dosed at up to 25 mg/L. With increasing concern over the possible formation of carcinogenic chloro-organic compounds as a result of chlorination, the use of prechlorination for H_2S control has declined in favor of aeration or oxygenation processes.

Preaeration

The objectives that are often given for aerating wastewater prior to primary sedimentation are to

- Provide grease separation
- Control odor
- Remove grit and produce flocculation
- Promote uniform distribution of suspended and floating solids to treatment units
- Increase BOD removals

Of these objectives, the promotion of a more uniform distribution of suspended and floating solids is probably the best application. Short periods of preaeration of 3–5 min do not significantly improve BOD or grease removal. When preaeration is used, it frequently consists of increasing the detention period in aerated grit chambers.

The amount of air required ranges between 0.02 and 0.05 m/linear ft min (2 and 5 ft/lin ft min) of channel. Aerated channels are used for distributing mixed liquor to activated-sludge final settling tanks.

FLOCCULATION

The purpose of wastewater flocculation is to form aggregates or flocs from the finely divided matter. Although not used routinely, the flocculation of wastewater by mechanical air agitation may be considered when it is desired to

- Increase the removal of suspended solids and BOD in primary settling facilities
- Condition wastewater containing certain industrial wastes
- Improve the performance of secondary settling tanks following biological treatment processes, especially the activated-sludge process

In both mechanical and air-agitation flocculation systems, it is common practice to taper the energy input so that the flocs initially formed will not be broken as they leave the flocculation facilities. Figure 17 illustrates a solids contact unit.

SEDIMENTATION

Sedimentation is the process of removing settleable materials. It is accomplished by decreasing the velocity of flow to lessen the ability of the water to hold particles in suspension. Many particles are removed in this way during storage of the water in ponds and reservoirs. In very turbid waters, pretreatment of the

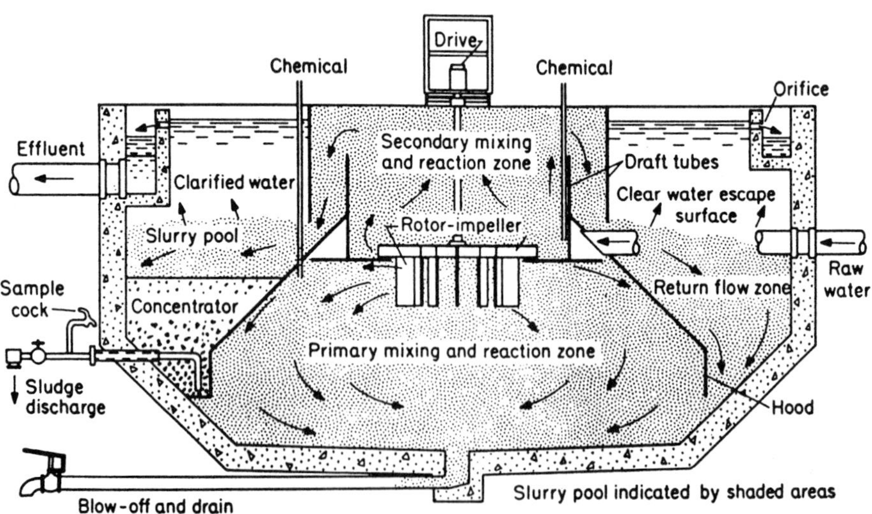

Figure 17 Solids contact unit.

water by sedimentation without the use of chemical coagulants is provided in specially built basins, in which the velocity may be controlled.

In a water-treatment plant, sedimentation usually follows mixing and flocculation for the removal of floc masses prior to filtration. With effective flocculation, a satisfactory filter effluent can be produced without the use of sedimentation, but the filters would clog very quickly and operating costs would be so high that the process would not be practical.

A sedimentation basin is a structure through which water flows at an extremely low velocity and suspended material will settle toward the bottom of the basin, with a relatively clear water passing out of the tank. Detention periods between 3 and 6 hr and horizontal velocities less than 3 ft/min are specified. Practice is to remove as much floc and other settleable material as possible by coagulation and sedimentation. This increases rates of filtration and the length of filter runs, without deterioration of finished water quality.

The most important features in the operation of a sedimentation tank are

- Introduction of water into the tank with a minimum turbulence
- Prevention of short circuiting or direct currents between inlet and outlet
- Removal of the effluent with a minimum disturbance so that settled material will not be carried out of the tank

A basin which is not functioning properly can be modified by making changes to the inlet and outlet devices or by installing stilling baffles to improve any or all of the important features mentioned above.

If the amount of suspended material is not too great, sedimentation tanks may be cleaned by first draining out the water then manually removing the material on the bottom and sides by means of squeegees and fire hoses. Many modern tanks are cleaned by mechanical scrapers which bring the settled material to one end or to the center in circular tanks. This material is removed by opening a drain valve. Mechanical removal has the following advantages:

- The basin need not be taken out of operation
- Water is not wasted by draining of the tank
- More frequent removal of the settled material is possible, thus minimizing bacterial action in the settled material which might cause taste and odors

Laboratory control of the combined coagulation and sedimentation process consist of making determinations of turbidity, color, alkalinity, and pH. Other determinations may be included if needed. Tests are made of the water before and after the various treatment process steps.

Turbidity, alkalinity, and pH of the raw water generally give a rough indication of the coagulant dosage and additional alkalinity that are required. Experience is gained in treating a specific water; the ranges of pH and alkalinity that will produce the best results are soon disclosed and can be used as a general guide in the control of the process. Turbidity and color determinations give an indication of the efficiency.

Coagulation can be facilitated by mixing a portion of the settled floc or sludge from a sedimentation tank with the raw water. This material forms the nuclei for subsequent floc formation. This is known as the coagulation-sedimentation technique commonly referred to as the *solids-contact process*. Units are used which are basically similar in that chemical mixing, flocculation, and sedimentation are carried on in a single tank. In all of them, the coagulated water is introduced at or near the bottom of the sedimentation compartment so that upward flow occurs. Raw water may enter near the top of the tank, in which case coagulation takes place in a central compartment having downward flow. As the rate of rise of water in the sedimentation compartment is lower than the rate of settling of the heavy floc, the floc remains in suspension to a depth of several feet and the rising water passes through it. The scrubbing action of the previously formed floc leads to the precipitation of insoluble compounds, as well as to the effective removal of fine suspended particles which would not be settled otherwise.

The volume of the tank for the complete process provides a detention period from 1 to 2 hr. This is considerably less than is provided with the more conventional units. Often coagulant aids are required for satisfactory operation with the low detention periods, especially when the water temperature is low. As these units are small, proper and careful control is even more essential than with the more conventional units.

Solids-contact units may be used either in the clarification or in the softening of a water supply. The precipitate that is formed in the softening process is much heavier than that formed in the coagulation process. Smaller units are generally provided when a water supply is softened. In either case, however, the water is subsequently filtered as the final step. Both softening and filtration are more fully discussed elsewhere.

4
Coagulation and Mixing

INTRODUCTION

In Chapter 3, it was mentioned that large, heavy particles suspended in water will settle out quite readily, but smaller lighter particles settle very slowly, or in some cases not at all. Chemical coagulation consists of treating the water with certain chemicals, to bring nonsettleable particles together into larger, heavier masses of solid material, which are then easier to remove. Masses are called "floc."

Coagulation includes several operations, and these operations are discussed because each of them involves different considerations. The first step in the process consists of feeding one or more chemicals to the water. Second, *mixing* is required so that the chemicals are distributed, and the water will be equally well treated. During mixing, or thereafter, certain chemical reactions occur. Some of the reaction products are insoluble and will begin to precipitate as solid particles. The term *coagulation* is sometimes used to refer specifically to these chemical reactions and the beginning of the formation of the floc. The word *coagulation* is also used to mean the total overall process up to sedimentation, and it must be used carefully to avoid confusion. The final step in the process is *flocculation*, which refers to gentle agitation of the treated water for a period of time. This favors the collision of the small floc particles with each other and with the other suspended particles in the water. This causes them to stick together, or agglomerate, and grow into large, readily settleable masses. Again, the word *flocculation* may have two slightly different meanings. Theoretically it may mean the growth of the floc particles. In practice, it more often refers to the gentle agitation of the water which brings about this growth. *Sedimentation* generally follows flocculation, but it is usually considered a separate process and not a part of chemical coagulation and refers to the removal of the floc from the water by settling in basins especially designed for the purpose. The terms *mixing*, *coagulation*, *flocculation*, and *sedimentation* are interrelated. Mixing refers to rapid, uniform distribution of a coagulant or other chemical in the water being treated before chemical reactions have occurred to any marked degree. Coagulation refers to the physical-chemical changes which take place between the soluble coagulant and the alkalinity in the water to form precipitated or incipient floc. Flocculation refers to the gentle agitation of a coagulant-treated water for an appreciable period of time to foster the coagulation reactions and to provide conditions favorable for the flocculent material to meet and adhere together into large floc masses. Sedimentation refers to the settling of floc in basins specifically designed for that purpose.

COAGULATION

Chemicals Used

Different chemicals may be used in coagulation depending on the characteristics of the water being treated. In some waters, a combination of two or more chemicals produces better results than any one chemical alone. It is usually necessary to perform coagulation tests in the laboratory to decide which chemical or combination should be used.

Aluminum Sulfate

Filter alum or merely alum is the most commonly used coagulant in water treatment. It may be purchased in solid form or as a solution called "liquid alum." "Activated alum" contains about 9% sodium silicate. This improves the coagulation in some waters. "Black alum" is alum containing activated carbon.

Sodium Aluminate

Sodium aluminate is a much more alkaline compound than alum. When added alkalinity is found to be necessary for the best coagulation, sodium aluminate may be substituted for all or part of the needed alum dose.

Iron Salts

The iron salts most commonly used as coagulants include ferric sulfate, ferric chloride, ferrous sulfate, also known as copperas, and chlorinated copperas.

These compounds often produce good coagulation when conditions are too acid for best results with alum. Sometimes the particles are best removed under acid conditions and iron compounds give better results.

Coagulant Aids

These materials do not produce coagulation when they alone are added to the water but improve the results when they are used with a coagulant. Several of the chemicals used in this way are either acid or alkaline compounds and are used to adjust the pH for best coagulation. These include lime; sodium carbonate, which is known as soda ash; sodium hydroxide, which is known as caustic soda or lye; and sulfuric acid.

Other coagulant aids have little or no effect on the pH but enter into the coagulation process. Some of them act by providing solid surfaces where the first formation of floc can begin more easily. Materials used for this purpose include certain clays such as kaolinite and bentonite, sodium silicate, and activated silica. These materials may also add weight to the floc and improve sedimentation.

Other coagulant aids may react with the fine particles in the water and assist in bringing them together into larger and heavier masses for settling. Coagulant aids of this type have molecules which consist of long chains of atoms, with many reactive groups scattered along the length of the chain. These are mostly artificial products. Those known as "synthetic polymers" or "synthetic polyelectrolyes" are entirely manmade. Others are manufactured from starch or cellulose. The molecules of starch and cellulose are already long molecular chains and in manufacturing the reactive groups are added.

Chemical Application

There are a number of chemical compounds which are used as water coagulants, but aluminum sulfate, commonly called alum, is most widely used. Alum is a granular powderlike material which is readily soluble in water and is easily applied as a solution or as a dry material. Reactions between alum and the natural constituent in various waters are influenced by many factors and it is good operation practice to determine the required amount of coagulant to be added by trial or experiment.

Normally best coagulation results are obtained with an alum dosage from 10 to 50 ppm, with the optimum dosage being obtained through the use of the so-called "jar test." This technique consists of adding known mounts of a coagulant into several jars of the water to be treated, gently stirring this mixture for a definite period of time, and observing the quality and settling characteristics of the floc. Sometimes a coagulant dosage is given in grains per gallon (gpg) or more conveniently in pounds per million gallons (lb/mg).

Coagulants are usually fed by a machine which can be set to deliver a known amount during a certain period of time. The equipment may be calibrated in convenient units, such as pounds per 24 hr. There are many types of available equipment or chemical feeders.

Mixing is necessary to provide rapid and uniform dispersion of the alum throughout the water being treated. It may be done either mechanically or hydraulically in special tanks, in sections of other tanks, or in piping systems. The fundamental principle is the violent agitation of the water with the applied chemical for a short period of time. This may be accomplished either by motor-driven propellers or baffled channels to create turbulent flow conditions. Contrary to popular belief, centrifugal pumps in themselves do not provide good mixing, as their efficient performance depends on a minimum amount of turbulence as the water passes through them. Valves, elbows, and other devices normally used in conjunction with pumping operations may cause adequate turbulence for satisfactory mixing.

Flocculation takes place following mixing, usually in a single tank with a detention period of between 15 and 45 min. During this period, the water is gently agitated to make floc particles. Bacteria and suspended solids meet and adhere together into large masses. The physical-chemical mechanism by which this is accomplished is complex. Only by gentle agitation will the coagulated particles become large enough to settle readily. Generally, soft waters of low mineral content coagulate best over a narrow range of pH values (between pH 5.8 and 6.4), whereas harder waters coagulate quite readily at pH values of 6.8–7.8. Under normal conditions, the reaction between alum and the alkaline materials usually found in a water supply are quite effective for the removal of turbidity, and this is accompanied by the absorption of a moderate amount of color. Water with high concentrations of color greater than about 30 ppm require that the coagulation reaction be in the acid zone of pH about 5.0–6.0. In this range, a more complex action takes place to form a color floc instead of the conventional aluminum hydroxide floc.

With sufficient alkalinity present in the water to complete the coagulation process. It may be necessary to add alkalinity in the form of lime or soda ash. If it is necessary to coagulate in the acid zone in order to remove color, an acid may be added for proper pH value. Such waters are corrosive and must be subsequently treated with an alkali to prevent corrosion from occurring in the distribution system.

The gentle agitation that is essential in flocculation may be accomplished either hydraulically or mechanically. A common hydraulic method is the well-known baffled basin, in which the water flows around the end or over and under baffles. Baffles are not readily adjustable, and the degree of agitation depends on the rate of velocity of the water. The optimum degree of agitation varies, and this lack of flexibility is undesirable. Flow velocities between baffles from 0.3 to 1.0 ft/s usually give satisfactory results. Figure 1 shows a typical baffle flocculation basin.

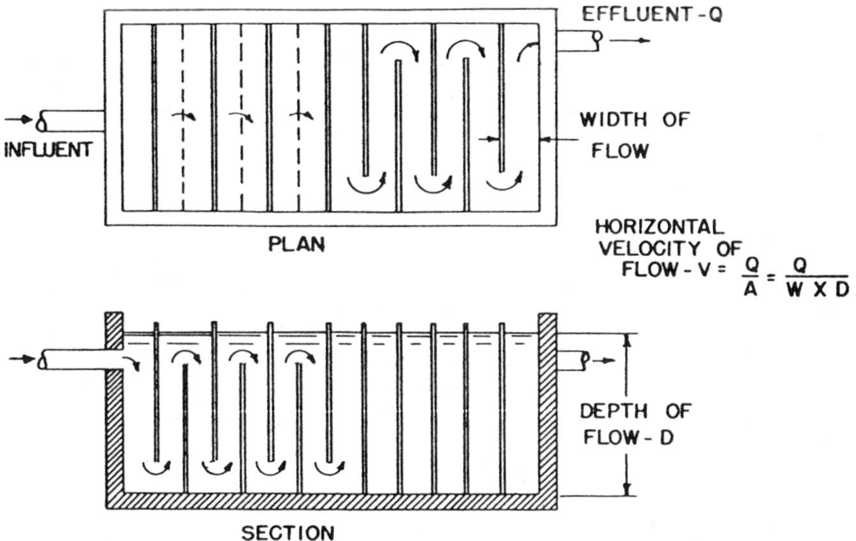

Figure 1 Baffle flocculation basin.

Mechanically driven equipment is used to provide the required agitation. The most common type of equipment consists of power-driven paddles or agitators so proportioned that appreciable but not undue agitation is produced. The speed of rotation can be controlled and the equipment will give the optimum degree of agitation regardless of the characteristics or quantity of water being treated or type of coagulant being used. Paddles may rotate on either a vertical or horizontal axis and either longitudinally or at right angles to the direction of flow. Sometimes, several paddles are operated in series and the first paddle is rotated at a greater speed than the succeeding ones. Modern practice suggests sudden violent agitation as the coagulant is added with a gradual decrease in turbulence in each successive step through the entire process of mixing, flocculation, and sedimentation. Figure 2 shows a mechanically driven paddle-type flocculation basin.

Control of flocculation basins consists of regulating the amount of coagulant that is added as well as the degree of agitation to secure the most effective floc formation with a minimum coagulant dosage. It may also be desirable to use coagulant aids in order to produce a floc which settles readily. The outlet end of the basins may be equipped with submerged lights to permit observation of the floc formed. Properly coagulated water should show visible floc in clear water, like snowflakes in clear air, as contrasted to cloudy water or foggy air.

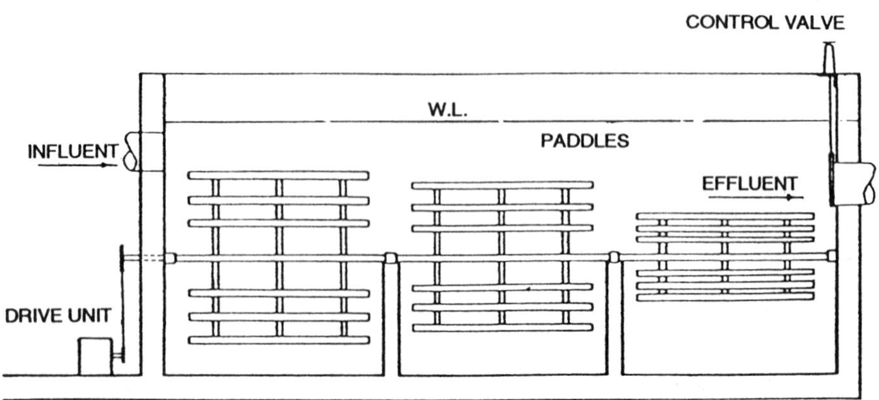

Figure 2 Mechanical flocculation basin.

Aluminum Compounds

Principal aluminum compounds that are commercially available and suitable for suspended solids removal are dry and liquid alum. Sodium aluminate has been used in activated sludge plants, but for phosphorus removal and its applicability for suspended solids removal is limited.

Commercial dry alum most often used in wastewater treatment is known as filter alum, and has the approximate chemical formula $Al_2(SO_4)_3 \cdot 14H_2O$ and a molecular weight of about 600. Alum is white to cream in color, and a 1% solution has a pH of about 3.5. The commercially available grades of alum and their corresponding bulk densities and angles of repose are

Grade	Angle of repose	Bulk Density (lb/ft^3)
Lump	—	62–68
Ground	43	60–71
Rice	38	57–71
Powdered	65	38–45

Each of these grades has a minimum aluminum content of 2%, expressed as Al_2O_3, and maximum Fe_2O_3 and soluble contents of 0.75 and 0.5%, respectively. Viscosity and solution crystallization temperatures are included in the subsequent section on liquid alum.

Since dry alum is only partially hydrated, it is slightly hygroscopic. However, it is relatively stable when stored under the extremes of temperature and humidity usually encountered.

The solubility of commercial dry alum at various temperatures is as follows:

Temperature (°F)	Solubility (lb/gal)
32	6.03
50	6.56
68	7.28
86	8.45
104	10.16

Dry alum is not corrosive unless it absorbs moisture from the air, such as during prolonged exposure to humid atmospheres. Precautions should be taken to ensure that the storage space is free of moisture. Alum is shipped in 100-lb bags or drums, or in bulk (minimum of 40,000 lb) by truck or rail. Bag shipments may be ordered on pallets if desired.

General design considerations of alum. Ground and rice alum are the grades commonly used by utilities because of their superior flow characteristics. These grades have less tendency to lump or arch in storage and therefore provide more consistent feeding qualities. Hopper agitation is seldom required for such grades.

Alum dust is present in the ground grade and will cause minor irritation of the eyes and nose on breathing. A respirator may be worn for protection against alum dust. Gloves may be worn to protect the hands. Because of minor irritation in handling and the possibility of alum dust causing rusting of adjacent machinery, dust-removal equipment is desirable. Alum dust should be thoroughly flushed from the eyes immediately and washed from the skin with water.

A typical storage feeding system for dry alum is shown in Figure 3. Bulk alum can be stored in mild steel or concrete bins with dust collector vents located in, above, or adjacent to the equipment room. Recommended storage capacity is about 30 days. Dry alum in bulk can be transferred with screw conveyors, pneumatic conveyors, or bucket elevators made of mild steel. Pneumatic conveyor elbows should have a reinforced backing, as the alum can contain abrasive impurities. Bags and drums of alum should be stored in a dry location to avoid caking. Bag- or drum-loaded hoppers should have a nominal storage capacity for 8 hrs at the nominal maximum feed rate so that personnel are not required to charge the hopper more than once per shift. Converging hopper sections should have a minimum slope of 60 degrees to prevent arching. Bulk storage hoppers should terminate at a bin gate so that the feeding equipment may be isolated for servicing. The bin gate should be followed by a flexible connection and a transition hopper chute or hopper which acts as a conditioning chamber over the feeder.

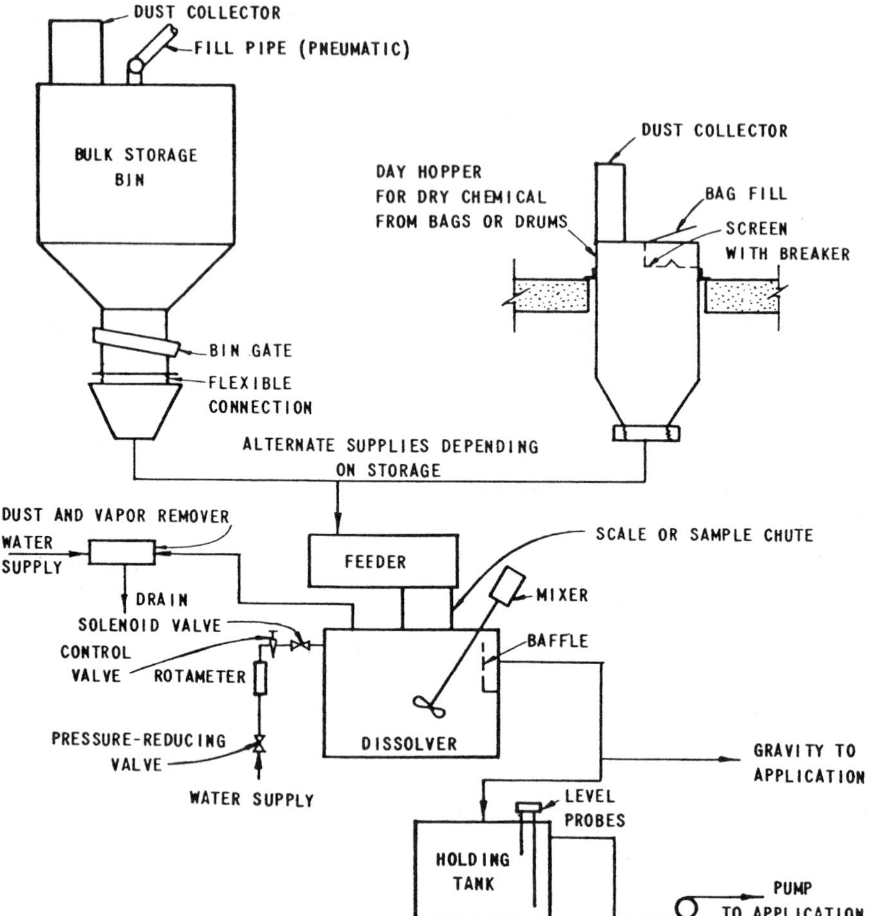

Figure 3 Typical dry feed system.

Feed systems. A feed system includes all of the components required for the proper preparation of the chemical solution. Capacities and assemblies should be selected to fulfill individual system requirements. Three basic types of chemical feed equipment are used: volumetric, belt gravimetric, and loss-in-weight gravimetric. Volumetric feeders are usually used where initial low cost and usually lower capacities are the basis of selection. Volumetric feeder mechanisms are usually exposed to the corrosive dissolving chamber vapors which can cause corrosion of discharge areas. Control of this problem is by use of an electric heater to keep the feeder housing dry or by using plastic components in the exposed areas.

Volumetric dry feeders in general use are screw type. Two designs of screw feed mechanisms are available. Both allow even withdrawal across the bottom of the feeder hopper to prevent hopper dead zones. One screw design is the variable pitch type, with the pitch expanding unevenly to the discharge point. The second screw design is the constant pitch type expanding evenly to the discharge point. This type of screw design is the constant pitch–reciprocating type. This type has each half of the screw turned in opposite directions so that the turning and reciprocating motion alternately fills one half of the screw while the other half of the screw is discharging. The variable pitch screw has one point of discharge, while the constant pitch–reciprocating screw has two points of discharge, one at each end of the screw. The accuracy of volumetric feeders is influenced by the character of the material being fed but ranges between ±1% for free-flowing materials and ±7% for cohesive materials.

Belt-type gravimetric feeders span the capacity ranges of volumetric and loss-in-weight feeders and can usually be sized for all applications encountered in wastewater-treatment applications. Initial expense is greater than for the volumetric feeder and slightly less than for the loss-in-weight feeder. Belt-type gravimetric feeders consist of a basic belt feeder incorporating a weighing and control system. Feed rates can be varied by changing either the weight per foot of belt, or the belt speed, or both. Controllers in general use are mechanical, pneumatic, electric, and mechanical-vibrating. Accuracy specified for belt-type gravimetric feeders should be within ±1% of set rate. Materials of construction of feed equipment normally include mild steel hoppers, stainless steel mechanism components, and rubber-surfaced feed belts.

Because alum solution is corrosive, dissolving or solution chambers should be constructed of type 316 stainless steel, fiberglass-reinforced plastic (FRP), or plastics. Dissolvers should be sized for preparation of the desired solution strength. The solution strength usually recommended is 0.5 lb of alum to 1 gal of water, or a 6% solution. The dissolving chamber is designed for a minimum detention time of 5 min at the maximum feed rate. Because excessive dilution may be detrimental to coagulation, educators or float valves that would ordinarily be used ahead of centrifugal pumps are not recommended. Dissolvers should be equipped with water meters and mechanical mixers so that the water to alum ratio may be properly established and controlled.

Piping and accessories. FRP, plastics (polyvinyl chloride, polyethylene, polypropylene, and other similar materials), and rubber are in general use and are recommended for alum solutions. Care must be taken to provide adequate support for these piping systems, with close attention given to spans between supports so that objectionable deflection will not be experienced. The alum solution should be injected into a zone of rapid mixing or turbulent flow.

Solution flow by gravity to the point of discharge is desirable. When gravity flow is not possible, transfer components should be selected that require little or no dilution. When metering pumps or proportioning weir tanks are used, return of excess flow to a holding tank should be considered. Metering pumps are discussed further in the section on liquid alum. Valves used in solution lines should be plastic, type 316 stainless steel or rubber-lined iron or steel.

Liquid alum is shipped in rubber-lined or stainless steel, insulated tank cars or trucks. Alum shipped during the winter is heated prior to shipment so that crystallization will not occur during transit. Liquid alum is shipped at a solution strength of about 8.3% as Al_2O_3 or about 49% as $Al_2(SO_4)_3 \cdot 14H_2O$. The latter solution weighs about 11 lb/gal at 60°F and contains about 5.4 lb dry alum (17% Al_2O_3) per gallon of liquid. This solution will begin to crystallize at 30°F and freezes at about 18°F.

Liquid alum is stored without dilution at the shipping concentration. Storage tanks may be open if indoors but must be closed and vented if outdoors. Outdoor tanks should also be heated, if necessary, to keep the temperature about 45°F to prevent crystallization. Storage tanks should be constructed of type 316 stainless steel; FRP; or steel lined with rubber, polyvinyl chloride, or lead. Liquid alum can be stored indefinitely without deterioration.

Storage tanks should be sized according to maximum feed rate, shipping time required, and quantity of shipment. Tanks should generally be sized for 1.5 times the quantity of shipments. A 10-day to 2-week supply should be provided to allow for unforeseen shipping delays.

Feeding equipment. Various types of gravity of pressure feeding and metering units are available. Figures 4 and 5 show commonly used feed systems.

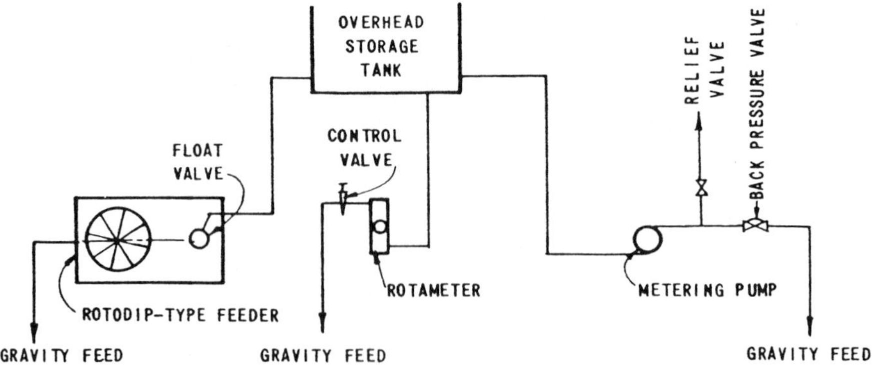

Figure 4 Liquid feed systems for overhead storage.

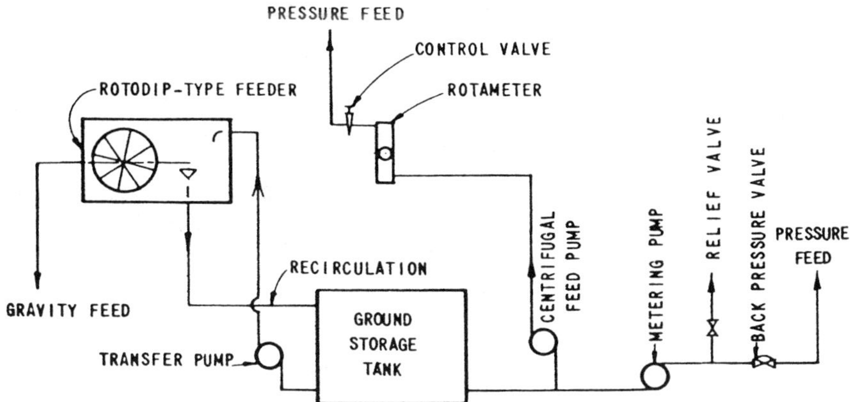

Figure 5 Liquid feed systems for ground storage.

The rotodip-type feeder or rotameter is often used for gravity feed and the metering pump for pressure feed systems.

The pressure or head available at the point of application frequently determines the feeding system to be used. The rotodip feeder can be supplied from overhead storage by gravity with the use of an internal level control valve, as shown by Figure 4. It may also be supplied by a centrifugal pump. The latter arrangement requires an excess flow return line to the storage tank, as shown by Figure 5. Centrifugal pumps should be direct-connected but not close-coupled because of possible leakage into the motor and should be constructed of type 316 stainless steel, FRP, and plastics.

Metering pumps allow a wide range of capacity compared with the rotodip and rotameter systems. Hydraulic diaphragm type pumps are preferable to other type pumps and should be protected with an internal or external relief valve. A back pressure valve is usually required in the pump discharge to provide efficient check valve action. Materials of construction for feeding equipment should be as recommended by the manufacturer for the service, but depending on the type of system, will generally include type 316 stainless steel, FRP, plastics, and rubber.

Aluminum sulfate reactions. Reactions between alum and the normal constituents of wastewaters are influenced by many factors. Theoretical reactions serve as a general guide, but in general the optimum dosage in each case must be determined by laboratory jar tests.

The reaction of Al^{3+} with OH^- ions made available by the ionization of water or by the alkalinity of the water.

Solution of alum in water produces

$$Al_2(SO_4)_3 = 2Al^{3+}3\ (SO_4)^{2-}$$

Hydroxyl ions become available from ionization of water:

$H_2O = H^+ + OH^-$

The aluminum ions (Al^{3+}) then react:

$2\ Al^{3+} + 6\ OH = 2\ Al\ (OH)_3$

Consumption of hydroxyl ions will result in a decrease in the alkalinity. Where the alkalinity of the wastewater is inadequate for the alum dosage, the pH must be increased by the addition of hydrated lime, soda ash, or caustic soda. The reactions of alum with the common alkaline reagents are as follows.

$Al_2\ (SO_4)_3 + 3Ca\ (HCO_3)_2 \rightarrow 2\ Al(OH)_3 \downarrow + 3CaSO_4 + 6CO_2 \uparrow$
$Al_2\ (SO_4)_3 + 3\ Na_2CO_3 + 3H_2O \rightarrow 2AL\ (OH)_3 \downarrow + 3CO_2 \uparrow$
$Al_2\ (SO_4)_3 + 3Ca\ (OH)_2 \rightarrow 2Al\ (OH)_3 \downarrow + 3CaSO_4$

In terms of quantities, the reactions can be expressed as follows:

1 mg/L of alum reacts with
0.50 mg/L alkalinity, expressed as $CaCO_3$
0.39 mg/L 95% hydrated lime as $Ca\ (OH)_2$
0.54 mg/L soda ash as Na_2CO_3

These approximate amounts of alkali when added to wastewater will maintain the alkalinity of the water unchanged when 1 mg/L of alum is added. For example, if no alkalinity is added, 1 mg/L of alum will reduce the alkalinity of 0.50 mg/L as $CaCO_3$, but alkalinity can be maintained unchanged in 0.39 mg/L of hydrated lime is added. This lowering of natural alkalinity is desirable in many cases to attain the pH range for optimum coagulation.

For each mg/L of alum dosage, the sulfate (SO_4) content of the water will be increased approximately 0.49 mg/L and the CO_2 content of the water will be increased approximately 0.44 mg/L.

Iron Compounds

Iron compounds have pH coagulation ranges and floc characteristics similar to aluminum sulfate. The cost of iron compounds may often be less than the cost of alum. However, the iron compounds are generally corrosive and often present difficulties in dissolving, and their use may result in high soluble iron concentrations in process effluents.

Liquid ferric chloride is a corrosive, dark brown oily-appearing solution having a weight as shipped and stored of 11.2–12.4 lb/gal (35–45% $FeCl_3$). The ferric chloride content of these solutions, as $FeCl_3$, is 3.95–5.58 lb/gal. Shipping concentrations vary from summer to winter because of the relatively high crystallization temperature of the more concentrated solutions. The pH of 1% solution is 2.0. The molecular weight of ferric chloride is 162.22.

Liquid ferric chloride is shipped in 3,000- to 4,000-gal bulk truckload lots, in 4,000- to 10,000-gal bulk carload lots, and in 5- and 13-gallon carboys.

Tank trucks and cars are usually unloaded pneumatically, and operating procedures should be closely followed to avoid spills and accidents. The safety vent cap and assembly should be removed prior to opening the unloading connection to depressurize the tank car or truck prior to unloading.

Ferric chloride solutions are corrosive to many materials and cause stains which are difficult to remove. Areas which are subject to staining should be protected with resistant paint or rubber mats.

Normal precautions should be employed when cleaning ferric chloride handling equipment. Workmen should wear rubber gloves, rubber apron, and goggles or a face shield. If ferric chloride comes in contact with the eyes or skin, flush with copious quantities of running water and call a physician. If ferric chloride is ingested, induce vomiting and call a physician.

Ferric chloride solution can be stored as shipped. Storage tanks should have a free vent or vacuum-relief valve. Tanks may be constructed of fiberglass-reinforced plastic (FRP), rubber-lined steel, or plastic-lined steel. Resin-impregnated carbon or graphite are also suitable materials for storage containers.

It may be necessary in most instances to house liquid ferric chloride tanks in heated areas or provide tank heaters and insulation to prevent crystallization. Ferric chloride can be stored for long periods of time without deterioration. The total storage capacity should be 1.5 times the largest anticipated shipment and should provide at least a 10-day to 2-week supply of the chemical at the design average dosage.

Feeding equipment and systems described for liquid alum generally apply to ferric chloride except for materials of construction and the use of glass tube rotameters. It may not be desirable to dilute the ferric chloride solution from its shipping concentration to a weaker feed solution because of possible hydrolysis. Ferric chloride solutions may be transferred from underground storage to day tanks with impervious graphite or rubber-lined self-priming centrifugal pumps having Teflon rotary and stationary seals. Because of the tendency for liquid ferric chloride to stain or deposit, glass tube rotameters should not be used for metering. Rotodip feeders and diaphragm metering pumps are often used for ferric chloride, and they should be constructed of materials such as rubber-lined steel and plastics. Materials for piping and transporting ferric chloride should be rubber or Saran-lined steel, hard rubber, FRP, or plastics. Valving should consist of rubber or resin-lined diaphragm valves, Saran-lined valves with Teflon diaphragms, rubber-sleeved pinch-type valves, or plastic ball valves. Gasket material for large openings such as manholes in storage tanks should be soft rubber; all other gaskets should be graphite-impregnated blue asbestos, Teflon, or vinyl.

Ferrous chloride, $FeCl_2$, as a liquid is available in the form of waste pickle liquor from steel processing. The liquor weighs between 9.9 and 10.4 lb/gal and contains 20–25% $FeCl_2$ or about 10% available Fe^{2+}. A 22% solution of $FeCl_2$ will crystallize at a temperature of –4°F. The molecular weight of $FeCl_2$ is 126.76. Free acid in waste pickle liquor can vary from 1 to 10% and usually averages about 1.5–2.0%. Ferrous chloride is slightly less corrosive than ferric chloride.

Waste pickle liquor is available in 4,000-gal truckload lots and a variety of carload lots. In most instances, the availability of waste pickle liquor will depend on the proximity to steel processing plants.

Ferric sulfate. Ferric sulfate is marketed as dry, partially hydrated granules with the formula $Fe_2(SO_4)_3 \cdot X\ H_2O$, where X is approximately 7. Typical properties are

Molecular Weight	526
Bulk Density	56–60 lb/ft^3
Water Soluble Iron Expressed as Fe	21.5%
Water Soluble Fe^{3+}	19.5%
Water Soluble Fe^{2+}	2.0%
Insolubles Total	4.0%
Free Acid	2.5%
Moisture at 105°C	2.0%

Ferric sulfate is shipped in car and truck load lots of 50 lb and 100 Abbe moistureproof paper bags and 200 Abbe and 400 Abbe fiber drums. Bulk carload shipments in box and closed hopper cars are available.

General precautions should be observed when handling ferric sulfate, such as wearing goggles and dust masks, and areas of the body that come in contact with the dust or vapor should be washed promptly.

Aeration of ferric sulfate should be held to a minimum because of the hygroscopic nature of the material, particularly in damp atmospheres. Mixing of ferric sulfate and quicklime in conveying and dust vent systems should be avoided, as caking and excessive heating can result. The presence of ferric sulfate and lime in combination has been known to destroy cloth bags in pneumatic unloading devices. Because ferric sulfate in the presence of moisture will stain, precautions similar to those discussed for ferric chloride should be observed.

Ferric sulfate is usually stored in the dry state either in the shipping bags or in bulk in concrete or steel bins. Bulk storage bins should be as tight as possible to avoid moisture absorption, but dust collector vents are permissible; and desirable. Hoppers on bulk storage bins should have a minimum slope of 36 degrees; however, a greater angle is preferred. Bins may be located inside or outside and the material transferred by bucket elevator, screw, or air conveyors. Ferric sulfate stored in bins usually absorbs some moisture and forms a thin protective crust which retards further absorption until the crust is broken.

Feed solutions are usually made up at a water to chemical ratio of 2:1 to

8:1 (on a weight basis), with the usual ratio being 4:1 with a 20-min detention time. Ferric sulfate solutions should not be diluted to less than 1% to prevent hydrolysis and deposition of ferric hydroxide. Ferric sulfate is actively corrosive in solution, and dissolving and transporting equipment should be fabricated of type 316 stainless steel, rubber, plastics, ceramics, or lead.

Dry feeding requirements are similar to those for dry alum except that belt-type feeders are rarely used because of open type of construction. Closed construction, as found in the volumetric and loss-in-weight–type feeders, generally exposes a minimum of operating components to the vapor and thereby minimizes maintenance. A water jet vapor remover should be provided at the dissolver to protect both the machinery and operator.

Ferrous sulfate. Ferrous sulfate, or copperas, is a by-product of pickling steel and is produced as granules, crystals, powder, and lumps. The most common commercial form of ferrous sulfate is $FeSO_4 \cdot 7H_2O$, with a molecular weight of 278, and containing 55–58% $FeSO_4$ and 20–21% Fe. The product has a bulk density of 62–66 lb/ft^3. When dissolved, ferrous sulfate is acidic. Composition of ferrous sulfate may be quite variable and should be established by consulting suppliers. Bulk, drum (400 lb), and bag (50 and 100 lb) shipments are available from producers.

Dry ferrous sulfate cakes at storage temperatures above 68°F, is efflorescent in dry air, and oxidizes and hydrates further in moist air. Precautions similar to those for ferric sulfate with respect to dust and handling acidic solutions should be observed when working with ferrous sulfate. Mixing quicklime and ferrous sulfate produces high temperatures and the possibility of fire.

The granular form of ferrous sulfate has the best feeding characteristics and gravimetric or volumetric feeding equipment may be used. The optimum chemical-to-water ratio for continuous dissolving is 0.5 lb/gal of 6 percent with a detention time of 5 min in the dissolver. Mechanical agitation should be provided in the dissolver to assure complete solution. Lead, rubber, iron, plastics, and type 304 stainless steel can be used as construction materials for handling solutions of ferrous sulfate. Storage, feeding, and transporting systems probably should be suitable for handling ferric sulfate as an alternative to ferrous sulfate.

Reaction of iron compounds. Ferrous sulfate and ferrous chloride react with the alkalinity of wastewater or with the added alkaline materials such as lime or soda ash. Reactions may be written to show precipitation of ferrous hydroxide, although in practice, as with alum, the reactions are more complicated. The reactions using ferric sulfate are shown as follows:

$$Fe_2(SO_4)_3 + 3Ca(HCO_3)_2 \rightarrow 2Fe(OH)_3\downarrow + 3CaSO_4 + 6CO_2\uparrow$$
$$Fe_2(SO_4)_3 + 3Na_2CO_3 + 3H_2O \rightarrow 2Fe(OH)_3\downarrow + 3Na_2SO_4 + 3CO_2\uparrow$$
$$Fe_2(SO_4)_3 + 3Ca(OH)_2 \rightarrow 2Fe(OH)_3\downarrow + 3CaSO_4$$

Ferric chloride can be substituted in these reactions.

In terms of useful quantities, the reactions can be expressed as follows:

1. 1 mg/L of $Fe_2(SO_4) \cdot 7H_2O$ reacts with
   0.57 mg/L alkalinity, expressed as $CaCO_3$
   0.44 mg/L 95% hydrated lime as $Ca(OH)_2$
   0.62 mg/L soda ash as Na_2CO_3
2. 1 mg/L of anhydrous $FeCL_3$ reacts with
   0.92 mg/L alkalinity expressed as $CaCO_3$
   0.72 mg/L 95% hydrated lime as $Ca(OH)_2$
   1.00 mg/L soda ash as Na_2CO_3

Ferrous sulfate and ferrous chloride react with the alkalinity of wastewater or with the added alkaline materials such as lime to precipitate ferrous hydroxide. The ferrous hydroxide is oxidized to ferric hydroxide by dissolved oxygen in wastewater. Typical reactions using ferrous sulfate are as follows:

$$FeSO_4 + Ca(HCO_3)_2 \rightarrow Fe(OH)_2 \downarrow + CaSO_4 + 2CO_2 \uparrow$$
$$FeSO_4 + Ca(OH)_2 \rightarrow Fe(OH)_2 \downarrow + CaSO_4$$
$$4Fe(OH)_2 + O_2 + 2H_2 \rightarrow 4Fe(OH)_3 \downarrow$$

Ferrous hydroxide is rather soluble and oxidation to the more insoluble ferric hydroxide is necessary if high iron residuals in effluents are to be avoided. Flocculation with ferrous iron is improved by addition of lime or caustic soda at a rate of 1–2 mg/mg Fe to serve as a floc-conditioning agent. Polymers are also generally required to produce a clear effluent.

Lime

The term *lime* applies to a variety of chemicals which are alkaline in nature and contain principally calcium, oxygen, and in some cases magnesium. Lime includes quicklime, dolomitic lime, hydrated lime, dolomitic hydrated lime, limestone, and dolomite.

Quicklime, CaO, has a density range of approximately 55–75 lb/ft^3, and a molecular weight of 56.08. A slurry for feeding, called milk of lime, can be prepared with up to 45% solids. Lime is only slightly soluble, and both lime dust and slurries are caustic in nature. A saturated solution of lime has a pH of about 12.4. Lime can be purchased in bulk in both car and truck load lots. It is also shipped in 80- and 100-lb multiwall moistureproof paper bags.

The CaO content of commercially available quicklime can vary quite widely over an approximate range of 70–96%. Content below 88% is generally considered less than standard in the municipal wastewater treatment. Purchase contracts are usually on 90% CaO content, with provisions for payment of a bonus for each 1% over and a penalty for each 1% under the standard. A CaO content less than 75% probably should be rejected because of excessive grit and difficulties in slaking.

Workers with lime should wear protective clothing and goggles to protect the skin and eyes, as lime dust and hot slurry can cause severe burns. Areas contacted by lime should be washed immediately. Lime should not be mixed with chemicals which have water of hydration. The lime will be slaked by the water of hydration causing excessive temperature rise and possibly explosive conditions. Conveyors and bins used for more than one chemical should be thoroughly cleaned before switching chemicals.

Pebble quicklime, all passing a 0.75-in. screen and not more than 5% passing a No. 100 screen, is normally specified because of easier handling and less dust. Hopper agitation is generally not required with the pebble form. Published slaker capacity ratings require "soft or normally burned" limes which provide fast slaking and temperature rise, but poorer grades of limes may also be satisfactorily slaked by selection of the appropriate slaker retention time and capacity.

Storage of bagged lime should be in a dry place and preferably elevated on pallets to avoid absorption of moisture. System capacities often make the use of bagged quicklime impractical. Maximum storage period is about 60 days. Bulk lime is stored in airtight concrete or steel bins having a 55- to 60-degree slope on the bin outlet. Bulk lime can be conveyed by conventional bucket elevators and screw, belt, apron, drag-chain, and bulk conveyors of mild steel construction. Pneumatic conveyors subject the lime to air slaking and particle sizes may be reduced by attrition. Dust collectors should be provided on manually and pneumatically filled bins.

Feeding equipment. A typical lime storage and feed system is illustrated in Figure 6. Quicklime feeders are usually limited to the belt or loss-in-weight gravimetric types because of the wide variation of the bulk density. Feed equipment should have an adjustable feed range of at least 20:1 to match the operating range of the associated slaker. Feeders should have an over-under feed rate alarm to warn immediately of operation beyond set limits of control. The feeder drive should be instrumented to be interrupted in the event of excessive temperature in the slaker compartment.

Lime slakers for wastewater treatment should be of the continuous type, and the major components should include one or more slaking compartments, a dilution compartment, a grit separation compartment, and a continuous grit remover. Commercial designs vary in regard to the combination of water to lime, slaking temperature, and slaking time in obtaining the "milk of lime" suspensions.

The paste-type slaker admits water as required to maintain a desired mixing viscosity. This viscosity therefore sets the operating retention time of the slaker. The paste slaker usually operates with a low water to lime ratio (approximately 2:1 by weight), elevated temperature, and 5-min slaking time at maximum capacity.

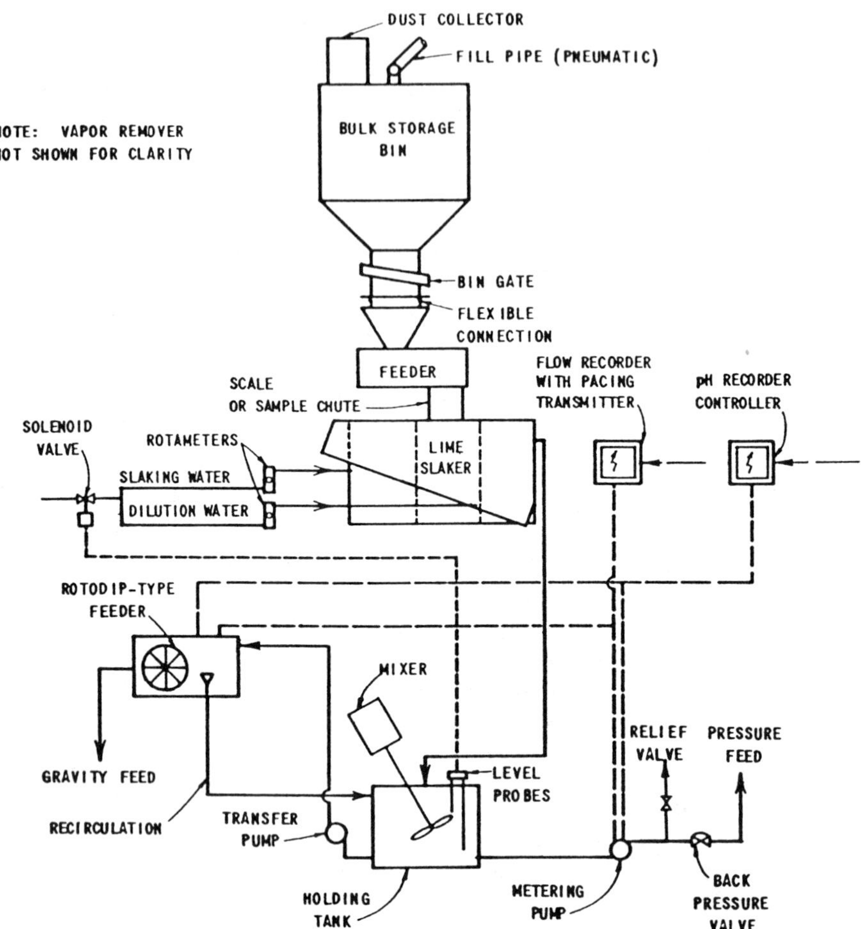

Figure 6 Typical lime feed system.

The detention type slaker admits water to maintain a desired ratio with the lime, and therefore the lime feed rate sets the retention time of the slaker. The detention slaker operates with a wide range of water to lime ratios (2.5:1.0 and 6:1), moderate temperature, and a 10-min slaking time at maximum capacity. A water to lime ratio of from 3.5:1.0 to 4:1 is most often used. The operating temperature in lime slakers is a function of the water to lime ratio, lime quality, heat transfer, and water temperature. Lime slaking evolves heat in hydrating the CaO to $Ca(OH)_2$ and, therefore, vapor removers are required for feeder protection.

Piping and accessories. Lime slurry should be transported by gravity in open channels wherever possible. Piping channels and accessories may be rubber, iron, steel, concrete, and plastics. Glass tubing, such as that in rotameters, will cloud rapidly and therefore should not be used. Any abrupt directional changes in piping should include plugged tees or crosses to allow rodding-out of deposits. Long sweep elbows should be provided to allow the piping to be cleaned by the use of a cleaning "pig"; daily cleaning is desirable.

Milk of lime transfer pumps should be of the open impeller centrifugal type. Pumps having an iron body and impeller with bronze trim are suitable for this purpose. Rubber-lined pumps with rubber-covered impellers are also frequently used. Make-up tanks are usually provided ahead of centrifugal pumps to ensure a flooded suction at all times. "Plating-out" of lime is minimized by the use of soft water in the make-up tank and slurry recirculation. Turbine pumps and educators should be avoided in transferring milk of lime because of scaling problems.

Lime slaker water proportioning is integrally controlled or paced from the feeder. Therefore, the feeder-slaker system will follow pacing controls applied to the feeder only. As discussed previously, gravimetric feeders are adaptable to receive most standard instrumentation pacing signals. Systems can be instrumented to allow remote pacing with telemetering of temperature and feed rate to a central panel for control purposes.

The lime feeding system may be controlled by an instrumentation system integrating both plant flow and pH of the wastewater after lime addition. However, it should be recognized that pH probes require daily maintenance in this application to monitor the pH accurately. Deposits tend to build up on the probe and necessitate frequent maintenance. The low pH lime treatment systems (pH 9.5–10.0) can be more readily adapted to this method of control than high-lime treatment systems (pH 11.0 or greater) because less maintenance of the pH equipment is required. In a closed-loop pH-flow control system, milk of lime is prepared on a batch basis and transferred to a holding tank with variable output feeders set by the flow and pH meters to proportion the feed rate. Figure 6 shows such a control system.

Hydrated lime. Hydrated lime, $Ca(OH)_2$, is usually a white powder, 200–400 mesh; has a bulk density of 20–50 lb/ft^3; contains 82–98% $Ca(OH)_2$; is slightly hygroscopic; tends to flood the feeder; and will arch in storage bins if packed. The molecular weight is 74.08. The dust and slurry of hydrated lime are caustic in nature.

Hydrated lime is slaked lime and needs only enough water added to form milk of lime. Wetting or dissolving chambers are usually designed to provide 5-min detention with a ratio of 0.5 lb/gal of water of 6% slurry at the maximum feed rate. Hydrated lime is usually used where maximum feed rates do not exceed

250 lb/hr; i.e., smaller plants. Hydrated lime and milk of lime will irritate the eyes, nose, and respiratory system and will dry the skin. Affected areas should be washed with water.

Storage of quicklime also applies to hydrated lime except that bin agitation must be provided. Bulk bin outlets should be provided with nonflooding rotary feeders. Hopper slopes vary from 60 to 66 degrees.

Volumetric or gravimetric feeders may be used, but volumetric feeders are usually selected only for installations where comparatively low feed rates are required. Dilution does not appear to be important; therefore, control of the amount of water used in the feeding operation is not considered necessary. Inexpensive hydraulic jet agitation may be furnished in the wetting chamber of the feeder as an alternative to mechanical agitation. The jets should be sized for the available water supply pressure to obtain proper mixing. Piping and accessories for quicklime are also appropriate for hydrated lime.

Reactions of lime. Lime is somewhat different from the hydrolyzing coagulants. When added to wastewater, it increases pH and reacts with the carbonate alkalinity to precipitate calcium carbonate. If sufficient lime is added to reach a high pH, approximately 10.5, magnesium hydroxide is also precipitated. This latter precipitation enhances clarification owing to the flocculent nature of the $Mg(OH)_2$. Excess calcium ions at high pH levels may be precipitated by the addition of soda ash. Reactions are shown as follows:

$$Ca(OH)_2 + Ca(HCO_3)_2 \rightarrow 2CaCO_3 \downarrow + 2H_2O$$
$$2Ca(OH)_2 + Mg(HCO_3)_2 \rightarrow 2CaCO_3 \downarrow + Mg(OH)_2 \downarrow + 2H_2O$$
$$Ca(OH)_2 + Na_2CO_3 \rightarrow CaCO_3 \downarrow + 2NaOH$$

Reduction of the resulting high pH levels may be accomplished in one or two stages. The first stage of the two-stage method results in the precipitation of calcium carbonate through the addition of carbon dioxide according to the following reaction:

$$Ca(OH)_2 + CO_2 \rightarrow CaCO_3 \downarrow + H_2O$$

Single-stage pH reduction is generally accomplished by the addition of carbon dioxide, although acids have been employed. This reaction, which also represents the second stage of the two-stage method, is

$$Ca(OH)_2 + 2CO_2 \rightarrow Ca(HCO_3)_2$$

As noted for the other chemicals, the above reactions are merely approximations to the more complex interactions which actually occur in wastewaters.

The lime demand of a given wastewater is a function of the buffer capacity of alkalinity of the wastewater.

Other Inorganic Chemicals

Soda ash. In addition to aluminum and iron salts and lime, a number of other inorganic chemicals have been used in wastewater treatment. These include soda ash, caustic soda, and carbon dioxide, but others have been and will be employed. Mineral and other acids are examples.

Soda ash (Na_2CO_3) is available in two forms. Light soda ash has a bulk density range of 35–50 lb/ft^3 and a working density of 41 lb/ft^3. Dense soda ash has a density range of 60–76 lb/ft^3 and a working density of 63 lb/ft^3. The pH of a 1% solution of soda ash is 11.2. It is used for pH control and in lime treatment.

The molecular weight of soda ash is 106. Commercial purity ranges from 98 to greater than 90% Na_2CO_3. Soda ash is available in bulk by truck, box car, and hopper car and in 100-lb bags.

Dense soda ash is used in municipal applications because of superior handling characteristics. It has little dust, good flow characteristics, and will not arch in the bin or flood the feeder. It is relatively hard to dissolve and ample dissolver capacity must be provided. Normal practice calls for 0.5 lb of dense soda ash per gallon of water or a 6% solution retained for 20 min in the dissolver. The dust and solution are irritating to the eyes, nose, lungs, and skin, and therefore general precautions should be observed and the affected areas should be washed promptly with water.

Soda ash is usually stored in steel bins and where pneumatic filling equipment is used. Bins should be provided with dust collectors. Bulk and bagged soda ash tend to absorb atmospheric CO_2 and water to form the less active sodium bicarbonate ($NaHCO_3$). Material recommended for unloading facilities is steel.

Feed equipment as described for dry alum is suitable for soda ash. Dissolving of soda ash may be hastened by the use of warm dissolving water. Mechanical or hydraulic jet mixing should be provided in the dissolver. Materials of construction for piping and accessories should be iron, steel, rubber, and plastics.

Caustic soda. Anhydrous caustic soda (NaOH) is available but its use is generally not considered practical in water- and wastewater-treatment applications. Consequently, only liquid caustic soda is discussed. Liquid caustic soda is shipped at two concentrations—50 and 73% NaOH. The densities of the solutions as shipped are 12.76 lb/gal for the 50% solution and 14.18 lb/gal for the 73% solution. These solutions contain 6.38 lb/gal NaOH and 10.34 lb/gal NaOH, respectively. The crystallization temperature is 53°F for the 50% solution and 165°F for the 73% solution.

Truck load lots of 1,000–4,000 gallons are available in the 50% concentration only. Both shipping concentrations can be obtained in 8,000-, 10,000-, and 16,000-gal car load lots. Tank cars can be unloaded through the dome eduction pipe using air pressure or through the bottom valve by gravity or by using air pressure or a pump. Trucks are usually unloaded by gravity or with air pressure or a truck mounted pump.

Liquid caustic soda is received in bulk shipments, transferred to storage, and diluted as necessary for feeding to the points of application. Caustic soda is poisonous and dangerous to handle. U.S. Department of Transportation Regulations must be observed. However, if handled properly, caustic soda poses no particular industrial hazard. To avoid accidental spills, all pumps, valves, and lines should be checked regularly for leaks. Workers should be thoroughly instructed in the precautions related to the handling of caustic soda. The eyes should be protected by goggles at all times when exposure to mist or splashing is possible. Other parts of the body should be protected as necessary to prevent alkali burns. Areas exposed to caustic soda should be washed with copious amounts of water for 15 min to 2 hr. A physician should be called when exposure is severe. Caustic soda taken internally should be diluted with water or milk and then neutralized with dilute vinegar or fruit juice. Vomiting may occur spontaneously but should not be induced except on the advice of a physician.

Liquid caustic soda may be stored at the 50% concentration. However, at this solution strength, it crystallizes at 53°F. Therefore, storage tanks must be located indoors or provided with heating and suitable insulation if outdoors. Because of its relatively high crystallization temperature, liquid caustic soda is often diluted to a concentration of about 20% NaOH for storage. A 20% solution of NaOH has a crystallization temperature of about –20°F. Recommendations for dilution of both 73 and 50% solutions should be obtained from the manufacturer, because special considerations are necessary.

Storage tanks for liquid caustic soda should be provided with an air vent for gravity flow. The storage capacity should be equal to 1.5 times the largest expected delivery, with an allowance for dilution water, if used, or a 2-week supply at the anticipated feed rate, whichever is greater. Tanks for storing 50% solution at a temperature between 75 and 140°F may be constructed of mild steel. Storage temperatures above 140°F require more elaborate materials selection and are not recommended. Caustic soda will tend to pick up iron when stored in steel vessels for extended periods. Subject to temperature and solution strength limitations, rubber, 316 stainless steel, nickel, nickel alloys, or plastics may be used when iron contamination must be avoided.

Further dilution of liquid caustic soda below the storage strength may be desirable for feeding by volumetric feeders. Feeding systems as described for liquid alum generally apply to caustic soda with appropriate selection of materials of construction. A typical system schematic is shown in Figure 7. Feeders will

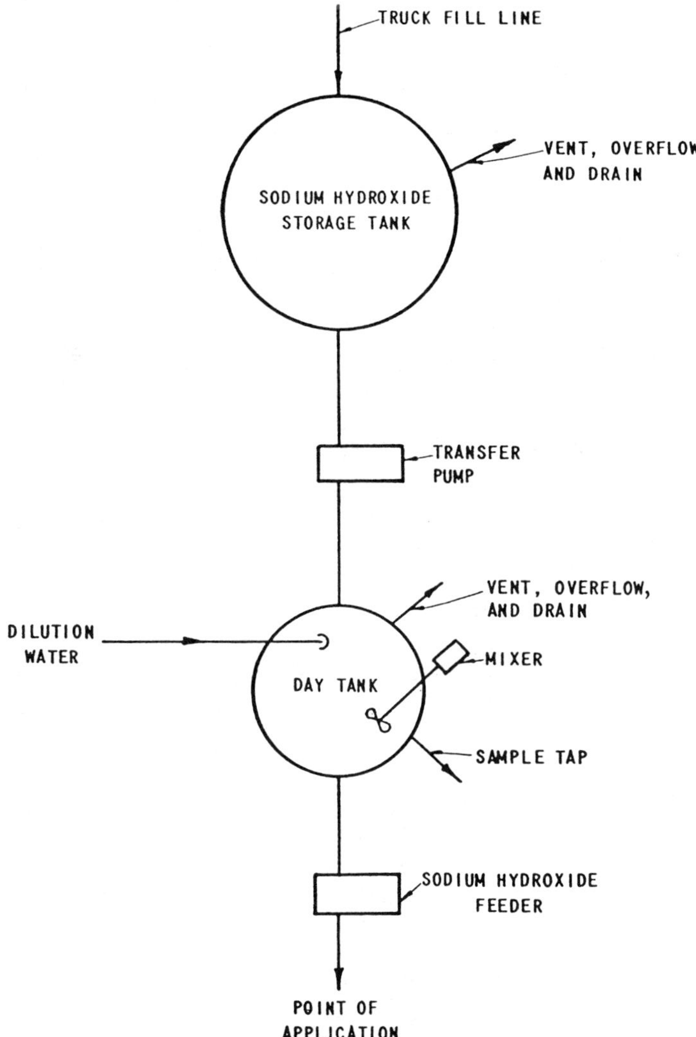

Figure 7 Typical sodium hydroxide feed system.

usually include materials such as ductile iron, stainless steels, rubber, and plastics.

Transfer lines from the shipping unit to the storage tank should be spiral wire–bound Neoprene or rubber hose, solid steel pipe with swivel joints, or steel hose. Because caustic soda attach glass, use of glass materials should be avoided.

Carbon dioxide. Carbon dioxide, CO_2, is available for use in wastewater-treatment plants in gas and liquid form. The molecular weight of CO_2 is 44. Dry CO_2 is not chemically active at normal temperatures and is a nontoxic safe chemical; however, the gas displaces oxygen and adequate ventilation of closed areas should be provided. Solutions of CO_2 in water are reactive chemically and form carbonic acid. Saturated solutions of CO_2 have a pH of 4.0 at 68°F.

The gas form may be produced on the treatment plant site by scrubbing and compressing the combustion product of lime recalcining furnaces, sludge furnaces, or generators used principally for the production of CO_2 gas only. These generators are usually fired with combustible gases, fuel oil, or coke and have yield CO_2. The gas forms, as generated at the plant site, usually have a CO_2 content of between 6 and 18% depending on the source and efficiency of the producing system. The liquid form is available from commercial suppliers in 20- to 50-lb cylinders, 10- to 20-ton trucks, and 30- to 50-ton rail cars. The commercial liquid form has a minimum CO_2 of 99.5%.

Recovery of CO_2 from recalcining furnaces or incinerators is the least expensive source, but maintenance of stack gas systems is likely to be extensive because of the corrosive nature of the wet gas and the presence of particulate matter. Scrubber systems are required to clean the stack gas, and specially designed gas compressors are necessary to provide the process injection pressure.

Pressure generators and submerged burners require less maintenance, because the system pressure is established by compressors or blowers handling dry air or gas. On-site generating units have a limited range of CO_2 production as compared with the liquid storage and feed system and therefore may require multiple units. The liquid CO_2 storage and feed system generally includes a temperature-pressure controlled, bulk storage tank, an evaporation unit, and a gas feeder to meter the gas. Solution feeders, similar in construction to chlorinators, may also be used to feed carbon dioxide. Table 1 is a summary of chemicals used in water treatment.

Polymers. Polymeric flocculants are high molecular weight organic chains with ionic or other functional groups incorporated at intervals along the chains. Because these compounds have characteristics of both polymers and electrolytes, they are frequently called polyelectrolytes. They may be of natural or synthetic origin. Synthetic polyelectrolytes can be classified on the basis of the type of charge on the polymer chain. Thus polymers possessing negative charges are called anionic, whereas those carrying positive charges are cationic. Certain compounds carry no electrical charge and are called nonionic polyelectrolytes.

Because of the great variety of monomers available as starting material and the additional variety that can be obtained by varying the molecular weight,

Table 1 Chemicals Used in Water Treatment

Chemical	Treatment
Activated Carbon	Color removal, dechlorination, and taste and odor control
Activated Silica	Coagulation
Aluminum Ammonium Sulfate	Coagulation and taste and odor control
Aluminum Chloride Solution	Coagulation
Aluminum Potassium Sulfate	Coagulation
Aluminum Sulfate	Coagulation, iron and manganese removal, and activation of silica
Amines, Neutralization	Boiler water treatment and pH control
Ammonia, Anhydrous	Taste and odor control
Ammonia, Aqua	Taste and odor control
Ammonium Silicofluoride	Fluroidation
Ammonium Sulfate	Taste and odor control
Barium Carbonate	Boiler water treatment
Bentonite	Coagulation and color removal
Bromine	Algae control and swimming pool disinfection
Calcium Fluoride	Fluoridation
Calcium Hydroxide	Coagulation, color removal, iron and manganese removal, pH control and softening
Calcium Hypochlorite	Disinfection and slime control
Calcium Oxide	Coagulation, color removal, iron and manganese removal, pH control and softening
Carbon Dioxide	pH control, recarbonation in water softening, and activation of silica
Chlorinated Copperas	Coagulation
Chlorinated Lime	Disinfection and slime control
Chlorine	Color removal, disinfection, iron and manganese, taste and odor control, activation of silica, slime control, and to produce chlorine dioxide
Chlorine Dioxide	Color removal, disinfection, iron and manganese removal, and taste and odor control
Cooper Sulfate	Algae control and taste and odor control
Diatomaceous Earth	Color removal, disinfection, iron and manganese removal, and taste and odor control
Disodium Phosphate	Boiler water treatment and pH control and softening
Dolomitic Lime	Silica removal in boiler water, pH control, softening by silica removal, and fluoride removal
Ferric Chloride	Coagulation
Ferric Sulfate	Coagulation
Ferrous Sulfate	Fluoridation
Hydrazine	Boiler water treatment, corrosion control, and O_2 scavenger
Hydrochloric Acid	Softening by regeneration of ion exchange materials and activation of silica
Hydrofluoric Acid	Fluoridation

Table 1 *(Continued)*

Chemical	Treatment
Iodine	Swimming pool disinfection
Magnesium Hydroxide	Fluoride removal
Magnesium Oxide	Fluoride removal
Otadecylamine	Boiler water treatment
Ozone	Color removal, disinfection, and taste and odor control
Potassium Iodine	Swimming pool disinfection
Potassium Permanganate	Iron and manganese removal and taste and odor control
Sodium Aluminate	Boiler water treatment, coagulation, and color and silica removal
Sodium Bicarbonate	pH control and activation of silica
Sodium Bisulfite	Dechlorination and O_2 scavenger
Sodium Carbonate	Corrosion and scale control and pH control and softening
Sodium Chloride	Softening by regeneration of ion exchange materials
Sodium Chlorite	Color removal, disinfection, iron and manganese removal, taste and odor control, and to produce chlorine dioxide
Sodium Chromate	Corrosion control
Sodium Dichromate	Corrosion control
Sodium Fluoride	Fluoridation
Sodium Hexametaphosphate	Corrosion and scale control and sequester ions
Sodium Hydroxide	Boiler water treatment, coagulation, corrosion and scale control, and pH control and softening
Sodium Hypochlorite	Disinfection, taste and odor control, and slime control
Sodium Nitrate	Boiler water treatment
Sodium Pentachlorophenate	Algae control and slime control
Sodium Phosphate Monoanhydrous	Boiler water treatment and pH control
Sodium Phosphate Monohydrous	Boiler water treatment and pH control
Sodium Silicate	Activated silica coagulation aid and corrosion and scale control
Sodium Silicofluoride	Fluoridation and activation of silica
Sodium Sulfite	Boiler water treatment, corrosion control, dechlorination, and O_2 scavenger
Sodium Thiosulfate	Dechlorination
Sulfamic Acid	Corrosion and scale removal and pH control
Sulfur Dioxide	Dechlorination and activation of silica
Sulfuric Acid	pH control and activation of silica
Terra-Sodium Pyrophosphate	Boiler water treatment, corrosion and scale control, and sequester ions
Tricalcium Phosphate	Fluoride removal
Tri-Sodium	Boiler water treatment, corrosion and scale control pH control, and softening

charge density, and ionizable groups, it is not surprising that a great assortment of polyelectrolytes are available.

Extensive use of any specific polymer as a flocculent is determined by the size, density, and ionic charge of the colloids to be coagulated. As other factors need to be considered, e.g., coagulants used, pH of the system, techniques and equipment for dissolution of the polyelectrolyte, it is mandatory that extensive jar testing be performed to determine the specific polymer that will perform its function most efficiently. These results should be verified by plant-scale testing.

Types of polymers vary widely in characteristics. Manufacturers should be consulted for properties, availability, and cost of the polymer being considered.

Dry polymer and water must be blended and mixed to obtain a recommended solution for efficient action. Solution concentrations vary from fractions of a percent up. Preparation of the stock solution involves wetting of the dry material and usually an aging period prior to application. Solutions can be very viscous, and close attention should be paid to piping size and length and pump selections. Metered solution is usually diluted just prior to injection to the process to obtain better dispersion at the point of application. General practice for storage of bagged dry chemicals should be observed. The bags should be stored in a dry, cool, low-humidity area and used in proper rotation; i.e., first in, first out. Solutions are generally stored in type 316 stainless steel, FRP, or plastic-lined tanks.

Two types of systems are frequently combined to feed polymers. The solution preparation system includes a manual or automatic blending system with the polymer dispensed by hand or by a dry feeder to a wetting jet and then to a mixing-aging tank at a controlled ratio. The aged polymer is transported to a holding tank where metering pumps or rotodip feeders disuse the polymer to the process. A schematic of such a system is shown in Figures 8 and 9. It is generally advisable to keep the holding or storage time of polymer solutions to a minimum (1–3 days or less) to prevent deterioration.

Chemical Feeders

Chemical feed systems must be flexibly designed to provide for a high degree of reliability in light of the many contingencies which may affect their operation. Thorough waste characterization in terms of flow extremes and chemical requirements should precede the design of the chemical feed system. The design of the chemical feed system must take into account the form of each chemical desired for feeding, the particular physical and chemical characteristics of the chemical, maximum waste flows, and the reliability of the feeding devices.

In suspended and colloidal solids removal from wastewaters, the chemicals employed are generally in liquid or solid form. Those in solid form are generally converted to solution or slurry form prior to introduction to the wastewater stream; however, some chemicals are fed in a dry form. In any case, some type of solids

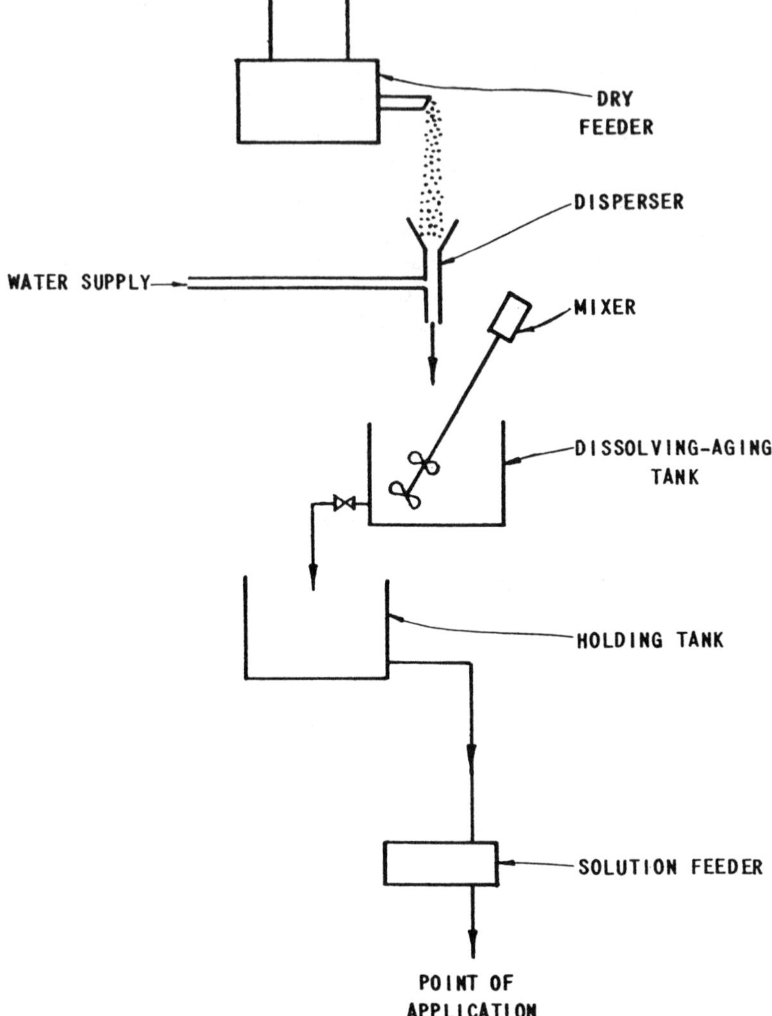

Figure 8 Schematic of a dry polymer feed system.

feeder is usually required. This type of feeder has numerous different forms owing to wide ranges in chemical characteristics, feed rates, and degree of accuracy required. Liquid feeding is somewhat more restrictive, depending mainly on liquid volume and viscosity.

The capacity of a chemical feed system is an important consideration in both storage and feeding. Storage capacity design must take into account the

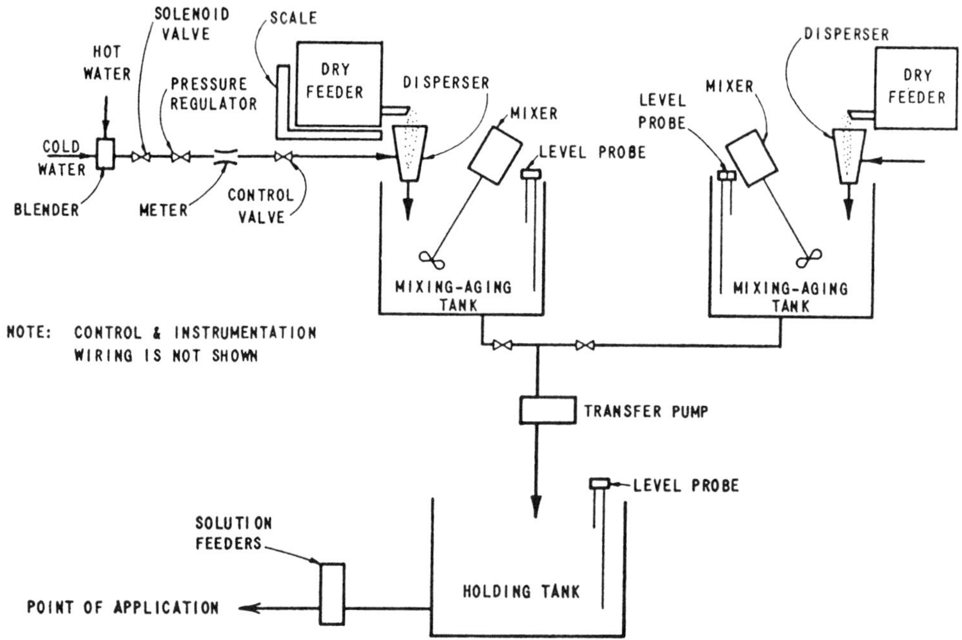

Figure 9 Automatic dry polymer feed system.

advantage of quantity purchase versus the disadvantage of construction cost and chemical deterioration with time. Potential delivery delays and chemical use rates are necessary factors in the total picture. Storage tanks or bins for solid chemicals must be designed with proper consideration of the angle of repose of the chemical and its necessary environmental requirements, such as temperature and humidity. Size and slope of feeding lines are important along with their materials of construction with respect to the corrosiveness of the chemicals.

Chemical feeders must accommodate the minimum and maximum feeding rates required. Chemical feeder control can be manual, automatically proportioned to flow, dependent on some form of process feedback, or a combination of any two of these. More sophisticated control systems are feasible if proper sensors are available. If manual control systems are used with the possibility of future automation, the feeders selected should be amenable to this conversion with a minimum of expense. Standby or backup units should be included for each type of feeder used. Reliability calculations will be necessary in larger plants with a greater multiplicity of these units. Points of chemical addition and piping should be capable of handling all possible changes in dosing patterns in order to have proper flexibility of operation. Designed flexibility in hoppers, tanks, chemical feeders, and solution lines is the key to maximum benefits at least cost.

Liquid feeders are generally in the form of metering pumps or orifices. Usually these metering pumps are of the positive displacement variety, plunger, or diaphragm type. The choice of liquid feeder is highly dependent on the viscosity, corrosivity, solubility, suction and discharge heads, and internal pressure–relief requirements.

Solids characteristics vary to a great degree and the choice of feeder must be considered carefully, particularly in the smaller-sized facility where a single feeder may be used for more than one chemical. Chemicals should be kept cool and dry. Dryness is important, since hygroscopic (water-absorbing) chemicals may become lumpy, viscous, or even rock hard; other chemicals with less affinity for water may become sticky from moisture on the particulate surfaces causing increased arching in hoppers. In either case, moisture will affect the density of the chemical and may result in underfeed. Dust-removal equipment should be used at shoveling locations, bucket elevators, hoppers, and feeders for neatness, corrosion prevention, and safety reasons. Collected chemical dust may often be used. Figures 10 and 11 show two types of feeders.

The simplest method for feeding solid chemicals is by hand. Chemicals

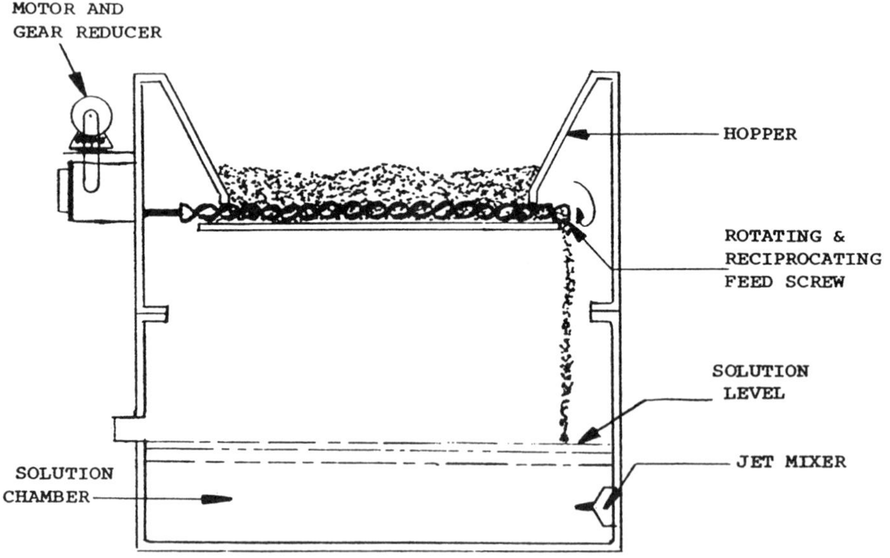

Figure 10 Screw feeder.

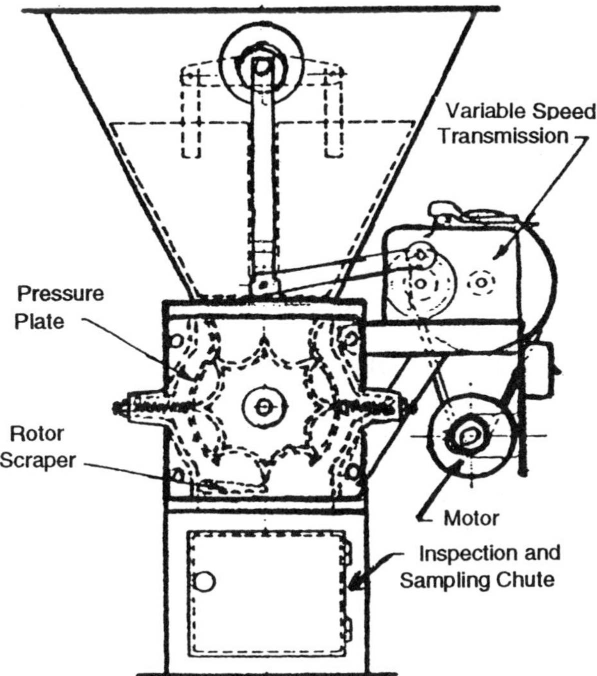

Figure 11 Rotary positive displacement solid feeder.

may be preweighed or simply shoveled or poured by the bagful into a dissolving tank. This method is of economic necessity limited to very small operations or to chemicals used in very weak solutions. Many factors, such as moisture content, different grades, and compressibility, which can affect chemical density (weight to volume ratio), volumetric feeding of solids is normally restricted to smaller plants, specific types of chemicals which are reliably constant in composition, and low rates of feed. Within these restrictions, several volumetric types are available.

MIXING

Mixing is a unit operation that is practiced widely to meet a variety of process and treatment requirements. The specific mixing system design, operating arrangement, and power requirements depend largely on the desired form of the intermediate or final products. Mixing is applied to achieve specified results in the following situations:

- Creating a suspension of solid particles
- Blending miscible liquids
- Dispersing gases through liquids
- Blending or dispersing immiscible liquids in each other
- Promoting heat transfer between a fluid (liquid) and the coil or jacket of a heat exchanging device

Operating characteristics and design configuration of a mixing system are established on the basis of the required energy expenditure to create or approximate a homogeneous fluid system. For example, in producing an emulsion one must supply sufficient energy to "break up" the dispersed phase. In doing so, high shear stresses, which depend on velocity gradients, are developed in the mixing medium. In the zones in which the velocity gradient approaches a maximum, an intensive breaking up of the dispersed phase occurs.

Mixing reduces concentration and temperature gradients in the processed system, thus exerting a favorable effect on the overall rates of mass and heat transfers. This applies in particular to dissolving applications, electrolysis, crystallization, absorption, extraction, heating or cooling, and heterogeneous chemical reactions, which proceed for the greater part in a liquid medium.

Increased turbulence of the fluid system caused by mixing leads to a decrease in the fluid's boundary layer thickness. This is derived from a continuous renewal of the surface contact area resulting in a pronounced rate of increase in heat and mass transfer mechanisms. Regardless of which medium is mixed with the liquid, i.e., gas, liquid, or solid particles, two basic methods are employed. These are mechanical mixers, which utilize different types of impellers, and pneumatic mixers, which utilize air or an inert gas to effect mixing. In addition to these designs, mixing also is achieved in normal fluid handling operations, such as in pumps and jet flows.

Two major characteristics of all mixing devices that provide a basis for comparative evaluations are

- Efficiency of a mixing device
- Intensity of mixing

The *efficiency* of a mixing device characterizes the quality of the process to be treated and may be expressed differently depending on the mixing purpose. For example, in producing suspensions mixing efficiency is characterized by the uniform distribution of the solid phase in the volume of equipment. For the intensification of thermal and diffusion processes, it is characterized by the ratio of mixed and unmixed heat and mass transfer coefficients, respectively. Mixing efficiency depends not only on the equipment design, but also on the amount of energy introduced in the liquid being agitated.

The *intensity* of mixing is determined by the time required to achieve a desired technological result or by the mixer rotations per minute at fixed process conditions (for mechanical mixers). From an economical standpoint, it is beneficial to achieve the required mixing effect in the shortest possible time. In evaluating the energy required for a mixing operation, one must account for the total energy consumption during the time needed to achieve a specified mixing result.

Mechanical mixers, which for the most part comprise rotating devices, are employed for liquids almost exclusively. Because of the widespread practice, most of the discussions in this chapter will revolve around fundamentals pertaining to such applications. Because these mechanical devices involve the flow of fluids around immersed bodies, such as impellers, mixing constitutes a unit operation in the *external problems of hydrodynamics*. The governing laws of hydrodynamics therefore are applicable to mixing.

The slow motion of a body immersed in a viscous fluid produces a laminar boundary layer which is influenced by the presence of a solid boundary. The shape and thickness of the layer depend on the geometry and size of the body. Boundary layer separation occurs whenever a change in fluid velocity, either in magnitude or direction, is so large that the fluid cannot adhere to the solid surface.

The maximum speed of an agitated fluid occurs at the periphery of the mixer impeller because it is proportional to the mixer diameter. As follows from Bernoulli's equation, a zone of reduced pressure develops in this region. Resulting flow patterns, including radial flows due to centrifugal forces, result in intensive mixing of materials contained in the reacting or process vessel.

Flow around bodies responsible for performing mixing may be described through application of the Navier-Stokes equations and continuity. However, their exact solution to mixing problems is complicated and, from an engineering standpoint, often applicable to a few specific cases only. Hence, analyses are usually based largely on similarity theory, which although less rigorous, affords a greater degree of flexibility in developing design-oriented formulations.

Mechanical Mixing Devices

Mechanical mixing devices consist of three basic parts: an impeller, a shaft, and a speed-reducing gearbox. The impeller constitutes the working element of the device, mounted on a vertical, horizontal, or inclined shaft. The drive may be connected directly to an electric motor (for high-speed mixers) or through a gearbox.

The multitude of impeller configurations can be grouped into five distinct categories of which only the first four are of commercial importance and are discussed concerning only three most widely used types: propeller, turbine, and paddle mixers.

Propeller Mixers

The two basic propeller mixer configurations are fixed to a rotating vertical, horizontal, or inclined shaft. The first is similar to an aircraft propeller, whereas the second resembles a marine propeller. Depending on the height of liquid layer, one shaft may carry one to three propellers.

Owing to their more streamlined shape, a propeller mixer's power requirements are less than the other types of mixers at the same Reynolds' number. Their transition in the self-modeling region is observed at relatively low values of Reynolds' number. $Re_M \simeq 10^4$. They are capable of high-speed operation without the use of a gearbox and, hence, provide a more cost-effective operation, because there are no mechanical losses in transmission. Propeller mixers produce an axial flow, which has a great pumping effect and provides short mixing times.

Disadvantages of a propeller mixer compared with paddle and turbine mixers are its high cost, the sensitivity of operation to the vessel geometry, and its location within the tank. As a general rule, propeller mixers are installed with convex-bottom vessels and should not be used in square tanks or in vessels with flat or concave bottoms.

A rotating propeller traces out a helix in the fluid, from which a full revolution moves the liquid longitudinally to a fixed distance depending on its pitch; i.e., the ratio of this distance to the propeller diameter. Pitch may be computed from the following formula:

$$s = 2\pi r \tan\psi \tag{1}$$

where r is the propeller blade radius and, therefore, also the radius of the cylinder created in the liquid as derived from the movement of the impeller, and ψ is the angle of tilt of the blade. Pumping and mixing efficiencies increase with pitch, as achieved by the axial flow of the liquid from the impeller. This flow results from the delivery head of the propeller and the helical turbulent flow of the entire contents of the vessel, which is caused by the radial velocity gradients in the liquid strata at different distances from the impeller. At high rotational speeds, the entire fluid mass swirls despite the axial flow, and a central vortex begins to form around the shaft.

Draft tubes are employed to improve the mixing of large quantities of liquids by directing the motion of the liquid. Figure 12 shows such an arrangement, which is favorable for large ratios of liquid depth to mixer diameter. In such applications, a high pumping capacity of the mixer is utilized, especially where mixtures of low viscosity are concerned. The draft tube directs the flow to the regions of the vessel that otherwise would not be agitated by the liquid stream. In the absence of draft tubes and at high rotational velocities of the

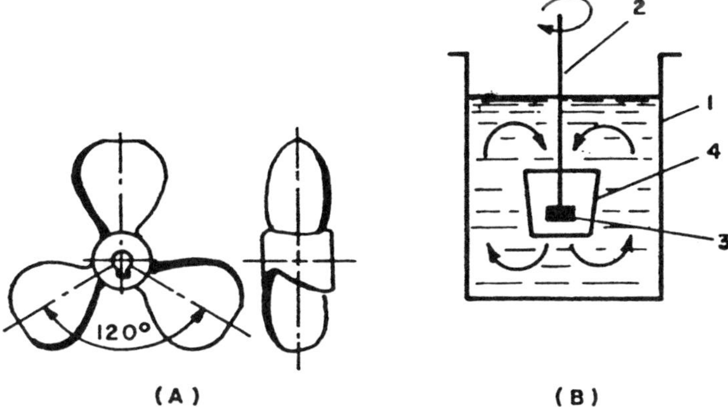

Figure 12 (A) A typical propeller-type mixer. (B) propeller mixer with a draft tube: (1) tank; (2) shaft; (3) propeller; (4) draft tube.

propeller, baffles generally are located at various points in the vessel. Baffles minimize vortex formation and divide it into a number of local eddies increasing the total turbulence of the tank.

Depending on the application, multiple impellers may be mounted on a single revolving shaft and more than one shaft may be employed in a given tank. In some applications, it is desirable to have two adjacent impellers rotating in opposite directions forming a beater. Sometimes the impellers actually touch the walls of the tank, giving a positive scraping action, which is desirable when thick layers of material tend to stick to the wall.

Propeller mixers are used for mixing liquids with viscosities up to 2000 centipoise (cp). They are suitable for the formation of low-viscosity emulsions, for dissolving applications, and for liquid-phase chemical reactions. For suspensions, the upper limit of particle size is 0.1–0.5 mm, with a maximum dry residue of 10%. Propeller mixers are unsuitable for suspending rapid-settling substances and for the absorption of gases. Propellers are designed on the basis of data obtained from properly executed modeling experiments.

Turbine Mixers

A turbine mixer is an impeller with essentially constant blade angle with respect to a vertical plane, over its entire length or over finite sections, having blades either vertical or set at an angle less than 90° with the vertical. Blades may be curved or flat, as shown by the various configurations in Figure 13. A brief description of each type impeller is given in Table 2.

Turbine mixer operation is analogous to that of a centrifugal pump working

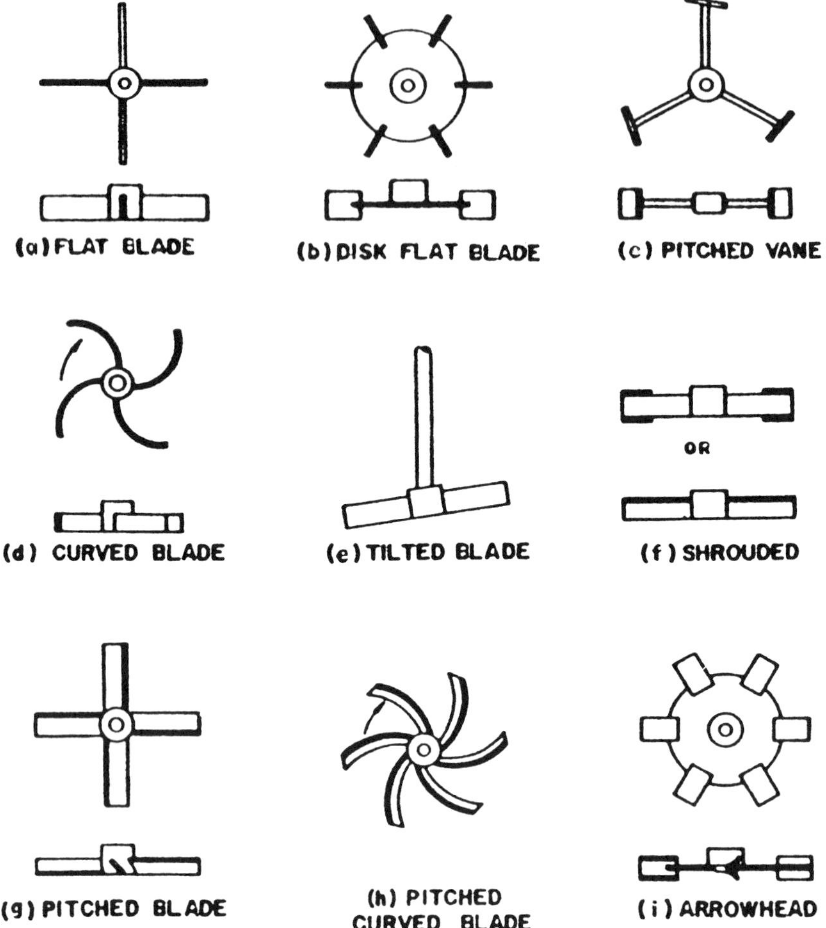

Figure 13 Various turbine impeller designs.

in a vessel against negligible back pressure. The mixing action is accomplished by the turbine blades, which entrain and discharge the liquid. The predominantly radial flow from the impeller impinges onto the vessel walls, where it splits into two streams. These streams cause mixing by their energy.

Open-tank agitators usually are supplied with mounting rails for bolting to mounting beams spanning the vessel top. Mounting rails often are designed to be removed and the drive adapts easily to a variety of special mounting requirements.

When turbine mixers are operated at sufficiently high rotational speeds,

Table 2 Description of Turbine Impeller Configurations Shown in Figure 13

Flat blade. Also termed a "straight-blade turbine," this impeller discharges radially, deriving suction from both top and bottom. Customary operation is in a peripheral speed range of 600–900 fpm. Blade widths generally are one-fifth to one-eighth of the diameter.

Disk flat blade. This turbine is widely used industrially and has been employed in many laboratory investigations. Although it has essentially the same performance characteristic as the flat-blade turbine, the difference in power consumption is significant.

Pitched vane. Also called a "radial propeller," it is an adaptation of the disk type, with the area reduced by pitching the blades to the vertical plane. Its advantage is its ability to obtain a high ratio of d/D and a high speed (for drive economy) without high power consumption.

Curved blade. Also called the "backswept," or "retreating blade," turbine, the blades curve away from the direction of rotation. This modification of the flat-blade style currently is thought to reduce mechanical shear effect at the impeller periphery. Industrial usage in suspensions of friable solids is widespread.

Tilted blade. In comparison to the curved-blade style, it has the effect of increasing the depth of the flow pattern and generally improving performance without an increase in power.

Shrouded. Addition of a plate, full or partial, to the top or bottom planes of a radial flow turbine is made to control the suction and discharge pattern. The upper unit usually has annular rings on the top and bottom. The lower design is fully shrouded on top to restrict suction to the lower side. The same impeller used in as dispersion (the latter style) is referred to as a "vaned disk." A full shroud on the lower surface of an impeller, which is located near the liquid surface, will increase the vortex considerably; e.g., for as reentrainment.

Pitched blade. This impeller has a constant blade angle over its entire blade length. Its flow characteristic is primarily axial, but a radial component exists and can predominate if the impeller is located close to the tank bottom. The blade slope can be anywhere from 0 to 90 degrees, but 45 degrees is the commercial standard. This impeller also is known as a "fan type."

Pitched curved blade. Sloping the blades of a curved-blade style is possible and has been practiced occasionally. No performance or power data arc available, and the high cost of construction of this impeller limits it to only a few special applications.

Arrowhead. This is a mixed-flow (axial and radial) impeller. It is a laboratory design with no commercial application.

both radial and tangential flows become pronounced along with vortex formation. This flow situation warrants the installation of baffles to ensure a more uniform flow distribution throughout the mixing vessel.

To concentrate the entraining action near the tank bottom, guide rings sometimes are mounted under the turbine. This is especially important with suspensions when the solids settle directly in the center of the vessel beneath the mixer. A stator ring is employed to eliminate the central vortex. Such an arrangement provides near-perfect radial flow from the turbine.

The optimum blade configuration depends on the properties of the materials to be mixed and the intended product state. For example, with mobile liquid mixtures, straight flat blades are suitable; if it is desirable to increase the pumping effect, inclined blades are recommended. For viscous liquids, inclining the blades in the opposite direction of rotation is advantageous. Profiled and curved blades are recommended because their starting moment is smaller and they facilitate the transmission of energy from the impeller to the liquid.

Paddle Mixers

Paddle mixers are devices consisting of two or more blades mounted on a vertical or inclined shaft. The basic paddle impeller configurations are shown in Figure 14, and a description of each is given in Table 3. The main advantages of paddle mixers are their simplicity and low cost. A disadvantage is their small pumping capacity (a slow axial flow), which does not provide a thorough mixing of the tank volume. Perfect mixing is attained only in a relatively thin stratum of liquid in the immediate vicinity of the blades. The turbulence spreads outward very slowly and imperfectly into the entire contents of the tanks; hence, circulation of the liquid is slow. Therefore, paddle mixers are used for liquids with viscosities only up to about 1000 cp. Because of a concentration gradient that often is created in the liquid when these type mixers are used, they are unsuitable for continuous operation. This can be remedied by tilting the paddle blades 30–45 degrees to the axis of the shaft, which results in an increase in axial flow and, consequently, a decrease in concentration gradients. Such a mixer can maintain particles suspended provided settling velocities are not high. Mixers with tilted blades are used for processing slow chemical reactions, which are not limited by diffusion.

To increase the turbulence of the medium in tanks with a large height to diameter ratio, a configuration is employed that consists of several paddles mounted one above the other on a single shaft. The separation between individual paddles lies in the range of 0.3–0.8 d (where d is the diameter of the paddle) and is selected according to the viscosity of the mixture. For mixing liquids with viscosities up to 1000 cp, as well as for heated tanks in cases in which sedimentation can occur, anchor or gate paddle mixers are employed (see Fig. 14d and f). Paddle diameters are almost as large as the inside diameter of the tank in such applications, so that the outer and bottom edges of the paddle scrape (or clean) the walls and bottom.

Leaf-shaped (broad-blade) paddle mixers provide a predominant tangential flow of liquid, but there is also turbulence at the upper and lower edges of the blade. Leaf-type blades are used for mixing low-viscosity liquids, intensifying heat-transfer processes, promoting chemical reactions in a reactor vessel, and for dissolution. For dissolving applications, leaf-blades generally are perforated.

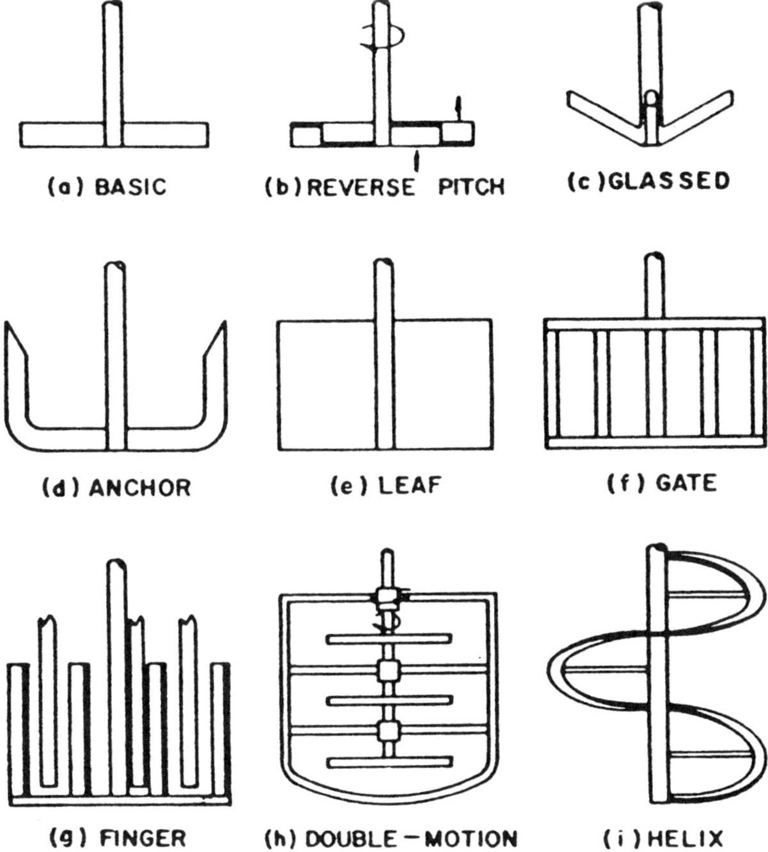

Figure 14 Major paddle impeller configurations.

During the mixer rotation, jets are formed at the exits from the holes and promote the dissolution of solid materials.

The rotational velocity of paddle mixers is in the range of 15–45 rpm. Under these conditions, pumping action is small and there is no danger of vortex formation. Therefore, paddle mixers are most often used without baffles. However, for broad-blade paddles, which operate at speeds up to 120 rpm, baffles are incorporated into the design to minimize vortex formation.

CHEMICAL MIXING

Chemical mixing and flocculation or solids contact are important mechanical steps in the overall coagulation process as described in Chapter 3. Application

Table 3 Description of Paddle Impeller Configurations Shown in Figure 14

Basic paddle. The simplest design is a single horizontal flat beam. The ratio of impeller diameter to tank diameter usually is 0.5–0.9. The peripheral speed range is generally 250–450 fpm. Paddles used in the United States generally have had a width/vessel diameter (b/D) ratio of one-sixth to one-twelfth; European practice is generally one-fourth to one-sixth.

Reverse pitch. To improve the top-to-bottom turnover characteristic of a simple paddle, the reverse pitch design in Figure 14b is used. The 45 degree blade angle is reversed at a diameter of approximately three-fourths of the impeller diameter, and rotation generally is set to produce upflow in the outer section.

Glassed steel. This impeller is a three-blade style common in glass-coated vessel applications. Blade form is either a pitch of about 30 degrees from the vertical or curved. Usually, d/D is 0.55–0.65. For low-viscosity fluids, its operation is definitely in the turbine category, but the difficulty of achieving a fully baffled condition in a glass-lined vessel causes it to perform like a paddle in many cases.

Anchor. Contouring a simple paddle to the shape of a tank bottom gives the anchor or "horseshoe" style. Extent of the blade may be limited to the lower vessel tangent line or the blades may continue upward along the straight side. Clearance between blade and vessel shell may be from 0.125–3.0 in., depending on tank diameter and heat transfer requirements.

Leaf. This has a extreme b/D ratio for a paddle. This geometry is used frequently in European practice.

Gate. A multiple-arm paddle with connecting vertical members, this design often is adopted for structural reasons in large tanks.

Finger. Also known as a "paste mixer," this combination of vertical blades meshing with stationary baffles has been in use for many years, but no data on power or performance have been published. The application is restricted to small batch sizes (less than 1000 gal) because of structural design difficulties.

Double-motion. This design combines a gate and anchor and a multiple-pitched paddle. Rotation of the two assemblies is countercurrent. As a special drive with concentric counterrotating output shafts is required, use is restricted to a few applications requiring intensive mixing of very viscous or non-Newtonian fluids.

Helix. This configuration least resembles the basic paddle. It does operate in the laminar range, but at normally high d/D ratios, and is an important member of the paddle group. One traditional use of a helix or screw is in a vertical-draft tubel.

of the processes to wastewater generally follows standard practices and employs basic equipment used for years in the water-treatment field. Chemical mixing thoroughly disperses coagulants or their hydrolysis products so the maximum possible portion of influent colloidal and fine supracolloidal solids are absorbed and destabilized. Flocculation or solids contact processes increases the natural rate of contacts between particles. This makes possible, within reasonable detention periods, for destabilized colloidal and fine supracolloidal solids to aggregate into particles large enough for effective separation by gravity processes or media filtration.

Most mixing processes in water depend on fluid shear for coagulant dispersal and for promoting particle contacts. Shear is most commonly introduced by mechanical mixing equipment. In certain solids-contact processes, shear results from fluid passage upward through a blanket of previously settled particles. Some designs utilize shear resulting from energy losses in pumps or at ports and baffles.

All theoretical approaches recognize the importance of time (t in seconds) and a velocity gradient (G, a measure of shear intensity in fps/ft) as controlling parameters in determining performance of mixing and flocculation processes. It should be noted that chemical mixing and flocculation differ only in intensity and duration and that some aggregation takes place in the chemical mixing stage.

In addition to velocity gradient and time, expressions for the rate of aggregation in flocculation or solids-contact processes involve parameters reflecting the total volume and the size and number of floc particles. When destabilized, particles in the fine colloidal range rapidly aggregate under natural conditions to form small floc of fine supracolloidal size, about 1 micrometer in diameter, often termed *primary* sized particles. In developing mathematical relations, this is generally the assumed initial size of particles to be further aggregated.

The rate of aggregation is usually taken as a function of the dimensionless product GCt, where C is the ratio of the volume of floc to total volume of suspension and G and t are as defined above.

The floc volume concentration resulting from a given coagulant dosage depends, among other things, on the amount of water entrained in the floc. More water is entrained and higher floc volumes result when flocculation takes place at lower values of G. The value of C may be increased greatly by recirculation of settled solids. This is used in certain types of solids contact reactors.

Design of rapid mix and flocculation units generally involves the choice of detention and G value and selection of configurations, of mixing equipment, tanks, piping and so forth. Unless the designer provides for direct control of floc volume concentration through solids recirculation, operating values of this parameter are determined indirectly through the chemical dosage and choice of G value and detention. Special attention should be given to avoiding excessive localized shear and reducing short circuiting. Pretreatment assures that wastewater is free of debris (e.g., rags, sticks) which could damage mixing equipment.

G represents the root mean square velocity gradient (fps/ft) over the mixing basin. For mechanically stirred basins, it can be calculated from the relation:

$$G = \left(\frac{P}{V\mu}\right)^{1/2}$$

where

P = power applied to stirring (ft-lb/s = HP × 550)
V = reactor volume (ft^3)
μ = viscosity of fluid ($lb\text{-}s/ft^2$)
G = root mean square velocity gradient (fps/ft)

Viscosity varies with temperature as follows:

T (°C)	μ ($lb\text{-}s/ft^2$)
1	0.361×10^{-4}
5	0.316×10^{-4}
10	0.273×10^{-4}
15	0.239×10^{-4}
20	0.210×10^{-4}
25	0.187×10^{-4}
30	0.167×10^{-4}

Chemical mixing should be designed to provide a thorough and complete dispersal of chemical throughout the wastewater being treated to ensure uniform exposure to pollutants which are to be removed. The intensity and duration of mixing of coagulants with wastewater must be controlled to avoid overmixing or undermixing. Over excessively disperses newly formed floc and may break up existing wastewater solids. Excessive floc dispersal retards effective flocculation and may significantly increase the flocculation period needed to obtain good settling properties. The breakdown of incoming wastewater solids may result in less efficient removals of pollutants associated with those solids. Undermixing inadequately disperses coagulants resulting in uneven dosing which may reduce efficiency of solids removal while requiring unnecessarily high coagulant dosages.

In water treatment several types of chemical mixing units have been used. These include high-speed mixers, in-line blenders and pumps, and baffled mixing compartments or static in-line mixers (baffled piping sections). Designs usually call for a 10- to 30-s detention time and approximately 300 fps/ft velocity gradient. Variable-speed mixers allow for varying requirements for optimum mixing. In solids-contact reactors, the G values in the immediate mixing zone approximate those for high-speed mixing.

High-speed mixers designed on the above basis should be satisfactory for wastewater applications. Where flows must be pumped just prior to coagulation, addition of chemicals at the pumps is feasible. The pump selection should take into account effects on organic solids of shear in centrifugal units. If problems are anticipated, lower-speed units such as screw pumps should be used. Baffled compartments or in-line static mixing devices are limited in their effectiveness as chemical mixing devices whether in water or wastewater treatment because

- Head losses of up to 3 ft are required.
- G cannot be changed to meet requirements but rather is a function of flow rate through the units.

In mineral addition to biological wastewater-treatment systems, coagulants may be added directly to mixed biological reactors such as aeration tanks or rotating biological contactors.

Based on typical power inputs per unit tank volume, mechanical and diffused aeration equipment and rotating fixed-film biological contactors produce average shear intensities generally in the range suitable for chemical mixing.

Flocculation and flocculating chemicals have been discussed in Chapter 3. Flocculation effectiveness is the performance of subsequent solids separation units in terms of both effluent quality and operating requirements, such as filter backwash frequency. Effluent quality depends greatly on the reduction of residual primary size particles during flocculation, whereas operating requirements relate more to the floc volume applied to separation units.

Tapered flocculation in which the flow is exposed to decreasing G values as it passes through the unit can provide a rapid build up of small dense floc with subsequent agglomeration at lower G into larger but still dense particles. The use of multicompartment flocculators permits tappered flocculation and reduces the high short-circuiting associated with single-compartment units. A wide variety of physical layouts are possible to achieve series flow through multiple compartments. Flocculation units should have multiple compartments and should be equipped with adjustable-speed mechanical stirring devices to permit meeting changed conditions. In spite of simplicity and low maintenance, nonmechanical, baffled basins are undesirable because of inflexibility, high head losses, and large space requirements.

Mechanical flocculators may consist of rotary, horizontal-shaft reel units as shown in Figure 15, rotary vertical shaft turbine units as shown in Figure 16, or other rotary or reciprocating equipment.

Tapered flocculation may be obtained by varying reel or paddle size on horizontal common shaft units or by varying speed on units with separate shafts and drives. A typical series of G values for successive compartments would be 100, 50, and 20 fps/ft. In most cases, equipment should provide overall Gt values up to 20×10^5 at maximum drive speed. Speed variation over a range of 1:3 or 1:4 should be possible. G values are determined from the hp actually transmitted to the fluid (water hp). This should be distinguished from the total input hp which

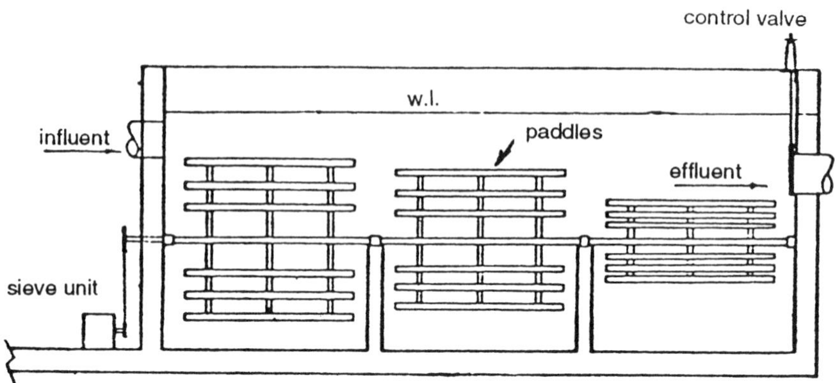

Figure 15 Mechanical flocculation basin horizontal shaft—reel type.

includes losses in, e.g., motors, drives, and bearings. It should be noted that G is a mean value for the entire flocculator volume. Practical limits are set to localized high values at the flocculator blades or paddles by specifying peripheral speeds below about 2 fps.

In applications other than coagulation with alum or iron salts, flocculation parameters may be quite different. Lime pates are granular and benefit little from prolonged flocculation or very low terminal G values. At Lake Tahoe, a detention of 4.5 min proved adequate. A minimum of 5 min, but as much as 10 min, may be needed to assure complete dissolution and reaction of CaO. As in water softening practice, G values of 100 or more are desirable. Polymers which already have a long chain structure may provide a good floc at low Gt values. Often the turbulence and detention in the clarifier inlet distribution is adequate. Settling and effluent clarity in the activated sludge process can frequently be improved by controlled flocculation between the aeration tank and clarifier.

Solids-contact processes combine chemical mixing, flocculation, and clarification in a single unit designed so that a large volume of previously formed floc is retained in the system. The floc volume may be as much as 100 times that in a "flow-through" system. This greatly increases the rate of agglomeration from particle contacts and may also speed up chemical destabilization reactions. Solids-contact units are of two general types: slurry-recirculation and sludge-blanket. In the former, the high floc volume concentration is maintained by recirculation from the clarification to the flocculation zone. In the latter, the floc solids are maintained in a fluidized blanket through which the wastewater under treatment flows upward after leaving the mechanically stirred flocculating compartment.

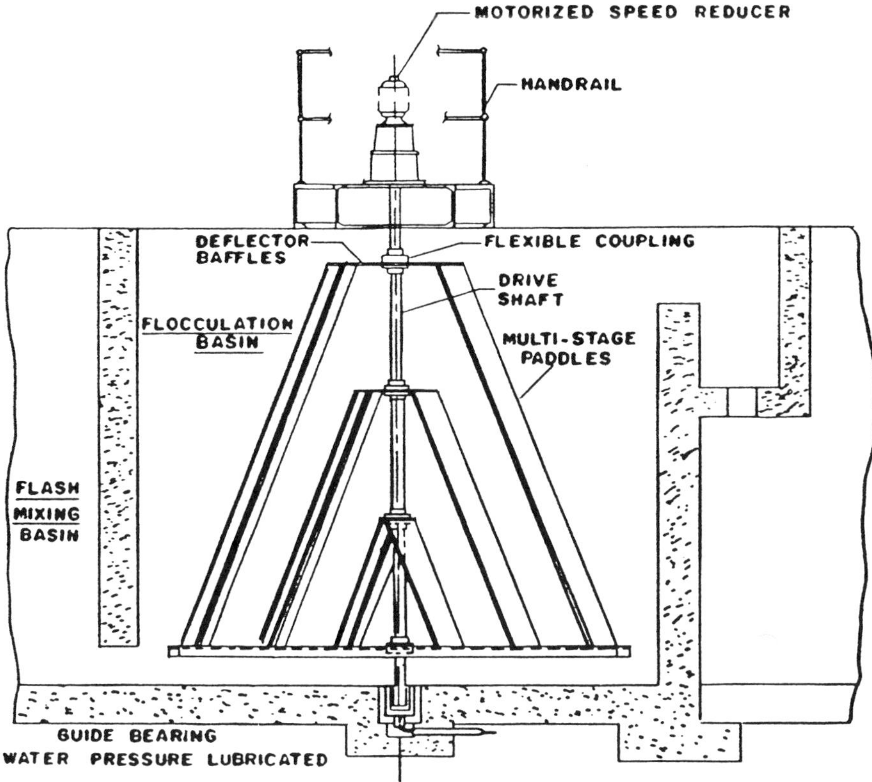

Figure 16 Mechanical flocculator vertical shaft—paddle type.

Solids-contact units have the following advantages:

- Reduced size and lower cost result because flocculation proceeds rapidly at high floc volume concentration.
- Single-compartment flocculation is practical because high reaction rates and the slurry effects overcome short circuiting.
- Units are available as compact single packages eliminating separate units.
- Even distribution of inlet flow and the vertical flow pattern in the clarifier improve clarifier performance.

PROBLEMS AND TREATMENT

The odor threshold recognition level is defined as the greatest dilution of the pure substance which can be detected by 100% of an odor panel. It may be expressed as parts per million (ppm) or in odor units (ASTM Standard Method D-1391-57), which are the dilution factors required to reach the level of detectability. Unfortunately, calculations of emissions based on these dilutions and threshold values do not always prove out in the field.

Generally, sewage-treatment facilities have been located to meet the requirements of effluent discharges to diluting bodies of water (receiving waters) and where flow in the sewers to the plant site from the population to be served would be accomplished by gravity. Consequently, many selected sites were located at low elevations relative to the population. These low-lying areas have a built-in potential for odor problems because of meterological effects associated with valleys.

Odor measurements show a wide degree of variability both between and within human subjects. Further, the objectionability of odors is a highly subjective evaluation, and often odors of low intensity (approaching threshold detection limits) emanating from sewage treatment facilities are characterized as objectionable because of subjective associations.

In the studies conducted by and for treatment facilities, odor strength (quantity) and offensiveness (quality) are evaluated by many different means. Methods vary from the elaborate (transporting samples to off-site panels), through chemical analysis of hydrogen sulfide concentration, to merely noting the numbers of complaints.

Chemical Origin of Odor

Fresh (aerobic) sewage has only a slight musty odor. As long as aerobic conditions are maintained, odors should be minimal.

Under anaerobic conditions, biological activity produces the classes of malodorous compounds listed in Table 4 with their threshold recognition values.

Hydrogen sulfide gas is the most prevalent odor associated with sewage. In the absence of oxygen, the sulfur in the sulfate ion acts as the oxidizing agent (electron acceptor) as follows:

Table 4 Threshold Recognition Values of Odors Associated with Anaerobic Sewage Decomposition

Compound	Threshold recognition values (ppm)
Hydrogen sulfide (H_2S)	0.0005
Mercaptans (R-SH)O	0.04–0.00005
Thioethers (R_1 - S - R_2)	0.001–0.00005
Pyridine	0.004
Skatole and indole	0.001
Ammonia (NH_2)	47

$$SO_4^{=} + \text{Organic Matter} \xrightarrow[\text{Bacteria}]{\text{Anaerobic}} S^{=} + H_2O + CO_2^{+}$$

Under acidic conditions hydrogen sulfide is produced:

$$2H^+ + S^{2-} \rightarrow H_2S \uparrow$$

The actual percent of the sulfide ion which is in the form of hydrogen sulfide is related to the pH of the solution.

The organic sulfur and nitrogen compounds (mercaptans, thioethers, pyridine, skatole, and indole) are formed by the degradation or incomplete oxidation of proteins. Proteins are made up of α-amino acids, all of which, by definition, have the general formula NH_2–CHR–COOH. In addition, tryptophan has in the R radical the group which comprises indole and skatole. Other amino acids with N in a five-membered ring are proline, hydroxyproline, and histidine.

Some of the amino acids also contain sulfur. Cysteine contains the sulfhydryl group R–SH characteristic of mercaptans, and methionine contains the R_1–S–R_2 thioether configuration (sometimes called sulfides).

The endproduct of protein metabolism in human beings is urea, CO–$(NH_2)_2$, which is readily converted to CO_2 and NH_3 and is the primary source of ammonia in domestic sewage.

Depending upon the constituents, the introduction of industrial wastes into the waste stream could pose unique odor problems to be solved on an individual basis.

The generation of odor-causing compounds in collection systems and treatment facilities are varied.

Collection System

The production of H_2S in collection systems not only causes odor problems in the system and at the treatment facility, but also is responsible for crown corrosion of the pipes in the system. Bacteria capable of the aerobic oxidation of sulfide to sulfuric acid are normally present on the walls and crown of sewers. The sulfuric acid that is produced attacks the concrete.

The production of H_2S in sewers is affected by several factors, including

- Temperature
- Sewage strength
- Velocity of sewage flow
- pH
- Sulfate concentration
- Aerobic conditions
- Detention time

H_2S production is enhanced by high temperatures, strong sewage, high sulfate concentration, long detention time, low pH, low velocities, and anaerobic conditions. Flatgraded sewers in the southern climates as well as those which require pumping stations that may retain sewage for longer periods of time experience the most serious crown corrosion problems.

Detention time alone does not account for the H_2S production; studies have shown that H_2S is not formed in fresh wastewater for 2–3 days. However, low velocities and pumping basins permit sedimentation in the collection system. In free-flowing sewers, sulfides are produced only by slimes on the submerged surface and by deposited sludge. Sludge deposits and slime growths provide sites where the microbial environment is suitable for H_2S generation.

Anaerobic conditions in sewers may also result in the production of methane (CH_4), an odorless but highly combustible gas which can be explosive.

Treatment Facilities

Odor-generation sources at treatment facilities may be

- Raw sewage entering the plant may have developed odors.
- Screening device may be a source of odors if they are not cleaned regularly and if the solids are not disposed of quickly. Accumulated grease and solids are often organic material or coated with organic material that can quickly give off an odor.
- Because of the organic coating on the particles collected, grit chambers can be another source of odor.
- The effluent weir of the primary settling tank can be a source of odor when it allows the sewage to be dropped into overflow troughs and then into collection troughs facilitating the release of odorous gases.
- Scum on the primary settling tanks will concentrate odorous materials on the air-water interface of the tank. This can be a significant source of odor generation where flotation clarifiers are used.
- The injection of air into an activated sludge reactor can promote the formation of aerosols, small bubbles on the order of 1,000–1,500 μ, which may contain odorous materials in the form of partially degraded organics as described elsewhere in this chapter. The actual distance traveled by the aerosol droplets before the water is evaporated varies with wind speed, relative humidity, and temperature, but it can be as much as 200 m. The odorous gases are released into the air when the water is evaporated.
- Odors from aeration tanks have been traced to poor hydraulic design of the tanks. Odor problems from extended aeration basins operating at only 60% of design capacity and operating under aerobic conditions are known. Basins

with poor mixing characteristics and a low turbulence near the recycle sludge inlets permit the build-up of sludge banks in the corners of the basins.

- Trickling filters can be a source of odor problems, particularly when they are overloaded. The spraying of the sewage onto the media may release odorous gases. Further, the inner slime layer coating the media becomes anaerobic and, on sloughing, can release odorous organics and H_2S.
- Sludge handling and storage processes may cause odor problems, since the sludge is usually anaerobic. Transfer of the sludge by plunger-type pumps to tank trucks and holding tanks of ships may allow odorous gases to escape as may the vents of holding tanks or transport vessels.
- The sludge-digestion process itself is not usually a source of odors, since the volatile organics and H_2S are usually collected and flared. However, incomplete combustion of the sludge gas will result in an odor problem.
- Dewatering or drying on beds of incompletely digested sludge is often a source of odors. Lime is often added to the sludge before dewatering. This raises the pH so that it is unfavorable for the production of H_2, but at the same time promotes the evolution of ammonia gas as follows:

  $$NH_4^+ + CO_3^= \rightarrow NH_3 + HCO_3^-$$

  making ammonia the predominant odor.
- Lagoons, stabilization ponds, and oxidation ponds can be an odor source. The organisms, algae, bacteria, and actinomycetes oxidize the organic matter for food.

Odors may arise from the decay of mats of algae that have been blown to a bank or corner. On decomposition, oxides of sulfur and nitrogen as well as H_2S, indole, sulfides, skatole, mercaptans, cadaverine, and amines are produced.

During periods of high water temperatures in shallow pond, sludge mats may rise from the bottom. These masses of organic debris usually become covered with blue-green algae resulting in odors from the decaying matter.

When algae are overproduced because of the overloading of the pond, they spread over the pond surface reducing the penetration of light to the remainder of the pond.

The principal odorous substances produced by the actinomycetes have been isolated and called geosmin ($C_{12}H_{22}O$) and mucidone ($C_{12}H_{18}O$).

Sewage odors and their control are covered in more detail in Chapter 15.

5
Gravity Separation/Sedimentation

GRAVITY SEPARATION

Gravity separation refers to the removal of suspended solids whose specific gravity difference from that of water causes them to settle or rise during passage through a tank or basin under quiescent conditions. Separation by settling is termed sedimentation; separation by rising is termed flotation. The size of particles determines the fluid drag retarding this separation. For a given specific gravity, smaller particles having greater surface area encounter more drag and hence are more difficult to separate.

Tanks or basins in which sedimentation is carried out, also called clarifiers, may be classified as horizontal flow or vertical flow according to the predominant direction of the flow path from inlet to outlet. It should be noted that, depending on placement of inlets and outlets, certain designs—particularly small radial flow tanks—may have a flow path with significant components in both horizontal and vertical directions.

Vertical-flow units may be annular or rectangular, and they are generally narrower at the bottom than at the top. In annular designs, the flow is distributed at the bottom along the circumference of the tank and rises to peripheral or radial effluent weirs or launders. Flow in rectangular tanks is distributed at the bottom along the length of the tank and rises to longitudinal or transverse effluent weirs or launders. Annular units have been built with outside diameters to 150 ft, but the width from inner wall to outer wall is much less. Horizontal flow units, both rectangular and circular, are most often used for sedimentation.

Flow through rectangular tanks enters at one end, passes a baffle arrangement, and traverses the length of the tank to effluent weirs. In narrow tanks, longitudinal collectors scrape sludge to single or multiple hoppers at one end (Fig. 1). In tanks with multiple wide bays, the longitudinal collectors scrape sludge to a cross collector which then moves the sludge to a central hopper. Circular designs employ three inlet/outlet configurations with corresponding flow paths as shown in Figure 2. In Figures 2a and 2c, sludge is removed by

(A)

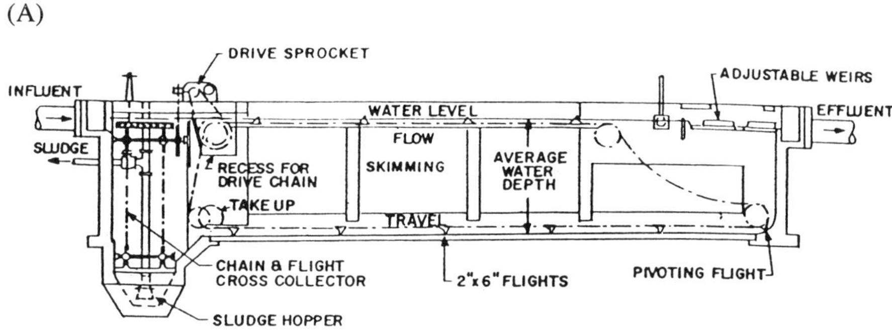

(B)

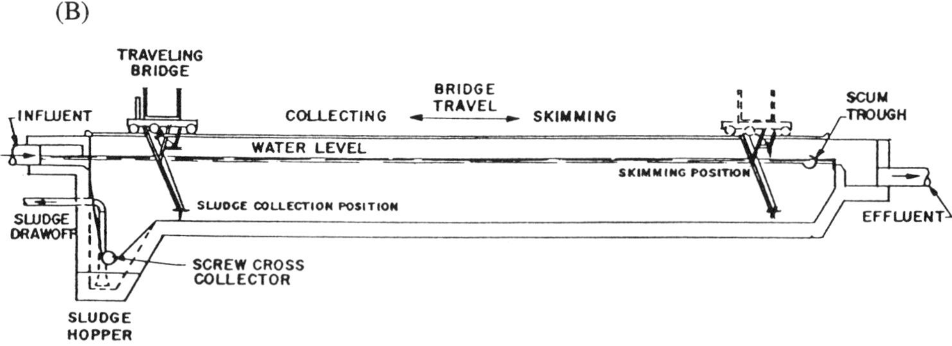

Figure 1 Rectangular sedimentation tanks: (A) with chain and flight collector; (B) with travelling bridge collector.

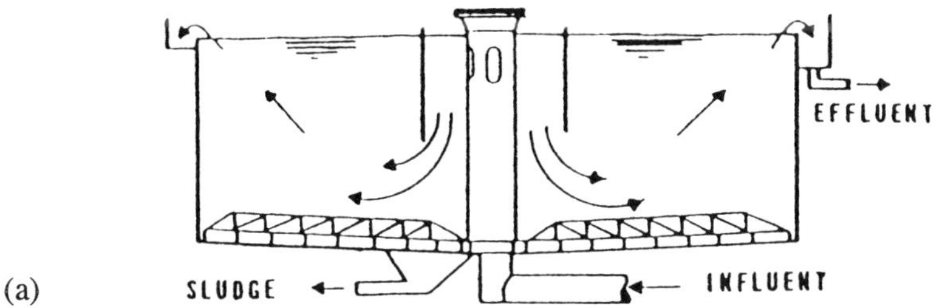

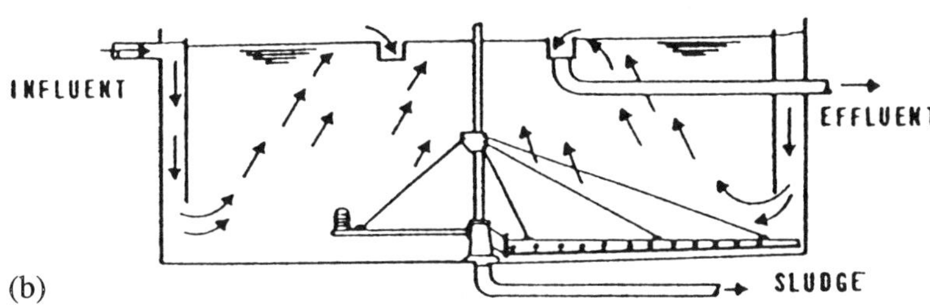

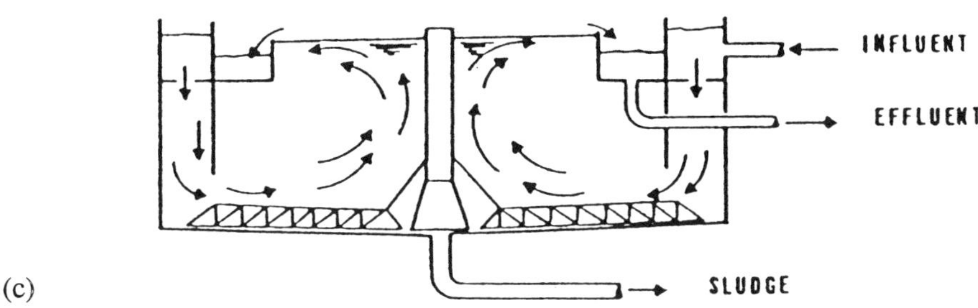

Figure 2 Typical clarifier configurations.

mechanical scraping to a central hopper or draw-off. In Figure 2b, a hydraulic suction sludge-removal system is employed.

Hydraulic Loading

The basic parameter to which settling tank performance is related is the surface hydraulic loading (Q/A). This is the inflow (Q) divided by the surface area (A) of the basin and is commonly expressed in units of gpd/ft^2 (gallons per day per square foot). Under these conditions, all particles whose settling velocity (V_s) exceeds Q/A are removed. In addition, in horizontal flow tanks particles of lower settling velocities are partially removed in proportion $V_s/(Q/A)$.

Short circuiting can greatly reduce the removal efficiency of a settling tank. Effects are most critical for flocculent suspensions whose removal is affected by detention time, but depend on current pattern, removal of discrete particles may also be affected. Short circuiting is caused by high inlet velocities, high outlet weir rates, close placement of inlets and outlets, exposure of tank surface to strong winds, uneven heating of tank contents by sunlight, and density differences between inflow and tank contents. Density-induced short circuiting can be a significant factor in secondary settling tanks handling activated sludge mixed liquor. Inlet and outlet conditions, tank geometry, and density differences due to influent suspended solids concentrations produce steady short circuiting, whereas effects of other factors are generally intermittent and unpredictable. The degree of short circuiting can be measured using tracer studies.

Turbulence levels in a settling basin are normally difficult to estimate. Good design minimizes other sources of turbulence such as inlet, outlet, and wind and density currents. These sources produce unpredictable levels of turbulence and may increase short circuiting. Even where the degree of turbulence during sedimentation can be definitely measured, the effect on removal of flocculent particles is not easily predicted because agglomeration induced by turbulence can alter particle sizes and localized settling velocities.

Particle Agglomeration

For the flocculent suspensions handled in wastewater treatment, particle contact and agglomeration continues during sedimentation. Mechanisms that produce particle contacts are velocity gradients within the settling tank and differential settling rates; each of which permits faster moving particles to overtake slower ones. Based on the nature of the influent suspension, either mechanism can significantly affect both the size and settling velocity of floc and the fraction of fine, with unsettleable particles remaining in suspension. Regardless of surface loading on a settling tank, attachment of smaller, unsettleable particles onto larger ones of separable size is essential in attaining high suspended solids removal efficiencies. In any case, these larger particles must have the opportunity to agglomerate to sizes which will be removed at the maximum surface loadings applied to the tank. Otherwise massive failure of the separation process will occur with significant loss of suspended solids in the effluent.

Design Considerations

The following should be provided in clarifier selection:

- Provide for even inlet flow distribution which minimizes inlet velocities and short circuiting.
- Minimize outlet currents and their effects by limiting weir loadings.
- Sufficient sludge storage depths to permit desired thickening of sludge.
- Sufficient wall height to give a minimum of 18 in. of freeboard.
- Reduce wind effects on open tanks by providing wind screens and by limiting fetch of wind on tank surface with baffles, weirs, or launders.
- Maintain equal flow to parallel units.

Sludge interfaces and sludge mass are shown in Figures 3 and 4.

SEDIMENTATION

Sedimentation is the separation of suspended particles that are heavier than water from water by gravitational settling. It is a widely used unit operations in wastewater treatment. This operation is used for grit removal, particulate-matter removal in the primary settling basins, biological floc removal in the activated-

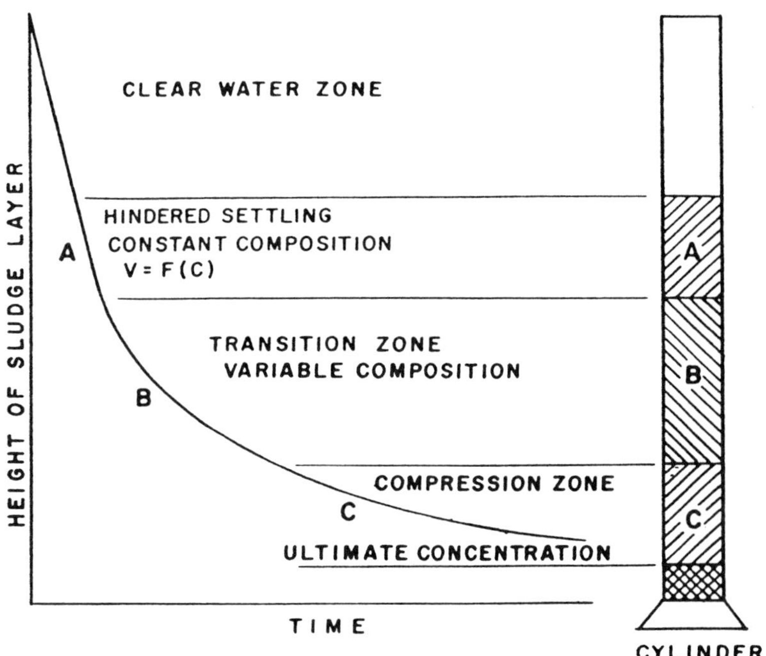

Figure 3 Schematic representation of settling zones.

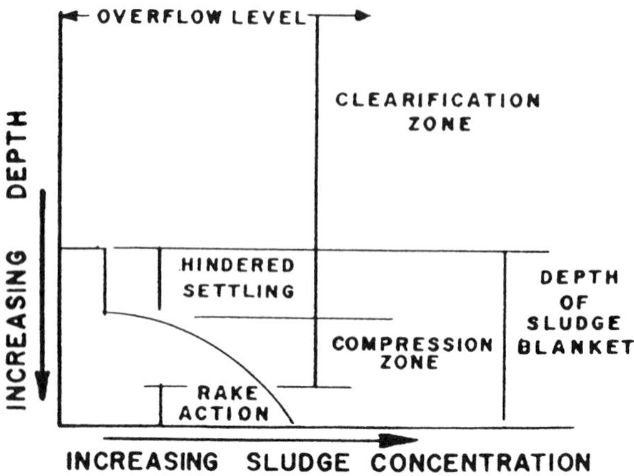

Figure 4 Sedimentation in a secondary settling tank.

sludge settling process, chemical floc removal when the chemical coagulation process is used, and for solids concentration in sludge thickeners. The primary purpose is to produce a clarified effluent, but it is also necessary to produce sludge with a solids concentration that can be easily handled and treated. In other processes, such as sludge thickening, the principal purpose is to produce a concentrated sludge that can be treated or transported more economically. In the design of sedimentation basins, consideration is for production of a clarified effluent and a concentrated sludge.

On the basis of the concentration and the tendency of the particles to interact, four general classifications in which particles settle can be made. More than one type of settling takes place at a given time during sedimentation, and it is possible to have all four occurring simultaneously.

Type 1 settling refers to the sedimentation of discrete particles in a suspension of low-solids concentration. Particles settle as individual entities, and there is no significant interaction with neighboring particles. An example is a dilute suspension of grit or sand particles. This type of settling is also called free settling.

Type 2 settling refers to a rather dilute suspension of particles that coalesce, or flocculate, during sedimentation. By coalescing, the particles increase in mass and settle at a faster rate.

Type 3 settling occurs in suspensions of intermediate concentration where interparticle forces are sufficient to hinder the settling of neighboring particles. The particles tend to remain in fixed positions with respect to each other and the mass of particles settles as a unit. A distinct solids-liquid interface develops at the top of the settling sludge mass and this is generally called zone settling.

Type 4 settling occurs when the particles are of such concentration that a structure is formed and further settling can occur only by compression of the structure. Compression takes place due to the weight of the particles, which are constantly being added to the structure by sedimentation from the supernatant liquor. This type of settling is known as compression settling and occurs usually in the lower layers of deep sludge masses.

Settling of discrete, nonflocculating particles can be described by means of the classic laws of sedimentation of Newton and Stokes. Newton's law yields the terminal particle velocity by equating the effective weight of the particle to the frictional resistance, or drag. The effective weight is simply

$$W = (\rho_s - \rho)gV \tag{1}$$

where

ρ_s = density of particle
ρ = density of fluid
g = acceleration due to gravity
V = volume of particle

The drag per unit area depends on the particle velocity, fluid density, fluid viscosity, and particle diameter. The drag coefficient C_D (dimensionless) is shown in

$$C_D = \frac{F_d/(\rho v^2)}{2} A \tag{2}$$

where

F_d = drag force
v = particle velocity
A = cross-sectional or projected area of particle at right angles to v

Equating the drag force to the effective weight of the particle, for spherical particles, yields Newton's law:

$$V_c = \left[\frac{4}{3}\frac{g(\rho_s - \rho)d}{C_{D_\rho}}\right]^{1/2} \tag{3}$$

where

V_c = terminal velocity of particle
d = diameter of particle

The drag coefficient takes on different values depending on whether the flow regimen surrounding the particle is laminar or turbulent. The drag coefficient is a function of the Reynolds' number. Although particle shape affects the value of the drag coefficient for spherical particles, the curve is approximated by the following equation (upper limit of $N_R = 10^4$).

$$C_D = \frac{24}{N_R} + \frac{3}{\sqrt{N_R}} + 0.34 \tag{4}$$

For Reynolds' numbers less than 0.3, the first term in Eq. (4) is the most important and substitution of this drag term into Eq. (3) gives us Stokes' law:

$$V_c = \frac{g(\rho_s - \rho)d^2}{18\mu} \tag{5}$$

For laminar flow conditions, Stokes' law finds the drag force to be

$$F_d = 3\pi\mu v d \tag{6}$$

This force related to the effective particle weight yields Eq. (5).

In the design of sedimentation basins, the usual procedure is to select a particle with a terminal velocity V_s and to design the basin so that all particles that have a terminal velocity equal to or greater than V_s will be removed. The rate at which clarified water is produced is then

$$Q = AV_s \tag{7}$$

where A is the surface area of the sedimentation basin. Eq. (7) becomes

$$V_c = \frac{Q}{A} = \text{overflow rate, gpd/ft}^2 \tag{8}$$

which shows that the overflow rate or surface loading rate, a common basis of design, is equivalent to the settling velocity. Eq. (7) indicates that for type 1 settling, flow capacity is independent of the depth.

For continuous-flow sedimentation, the length of the basin and the time a unit volume of water is in the basin, or detention time and should be such particles with the design velocity V_s will settle to the bottom of the tank. The design velocity, detection time, and basin depth are related as follows:

$$V_s = \frac{\text{Depth}}{\text{Detention Time}} \tag{9}$$

Design factors have to be included to allow for the effects of inlet and outlet turbulence, short circuiting, sludge storage, and velocity gradients due to the operation of sludge-removal equipment.

Thickeners and Clarifiers

A large number of waste effluents and water comprise suspensions of solid particles in liquids. The application of filtration techniques for the separation of heterogeneous systems can be very costly. If, however, the discrete phase of the suspension largely contains settleable particles, the separation can be effected by the operation of *sedimentation*. Sedimentation involves the removal of suspended solid particles from a liquid stream by gravitational settling. This unit operation is divided into *thickening*; i.e., increasing the concentration of the feed stream and *clarification*, which is removal of solids from a relatively dilute stream.

A thickener is a sedimentation unit that operates according to the principle of gravity settling. Compared to other types of liquid/solid separation devices, a thickener's principal advantages are

- Simplicity of design and economy of operation
- Its capacity to handle extremely large flow volumes
- Versatility, as it can operate equally well as a concentrator or as a clarifier

In a batch-operating mode, a thickener normally consists of a standard vessel filled with a suspension. After settling, the clear liquid is decanted and the sediment removed periodically.

The operation of a continuous thickener is also relatively simple. Figure 5 illustrates a cross-sectional view of a standard thickener. A drive mechanism powers a rotating rake mechanism. Feed enters the apparatus through a feed well designed to dissipate the velocity and stabilize the density currents of the incoming stream. Separation occurs when the heavy particles settle to the bottom of the tank. Some processes add flocculants to the feed stream to enhance particle agglomeration to promote faster or more effective settling. The clarified liquid overflows the tank and is sent to the next stage of a process. The underflow solids are withdrawn from an underflow cone by gravity discharge or pumping. Figure 6 provides descriptions and typical size ranges of thickeners of different configurations. Major industries and typical applications uses of thickeners are listed in Table 1.

Thickeners can be operated in a countercurrent fashion. Applications are aimed at the recovery of soluble material from settleable solids by means of continuous-countercurrent decantation (CCD). The basic scheme involves streams of liquid and thickened sludge moving countercurrently through a series of thickeners. The thickened stream of solids is depleted of soluble constituents as the solution becomes enriched. In each successive stage, a concentrated slurry is mixed with a solution containing fewer solubles than the liquor in the slurry and then is fed to the thickener. As the solids settle, they are removed and sent to the next stage. The overflow solution, which is richer in the soluble constituent, is sent to the preceding unit. Solids are charged to the system in the first-stage thickener, from which the final concentrated solution is withdrawn. Wash water or virgin solution is added to the last stage, and washed solids are removed in the underflow of this thickener.

The flow scheme for a three-stage continuous countercurrent decantation system is shown in Figure 7. The feed stream, F, is mixed with overflow O_2 (from thickener 2) before entering stage 1. The overflow of concentrated solution, O_1, is withdrawn from the first stage. The underflow from the first stage, U_1, is

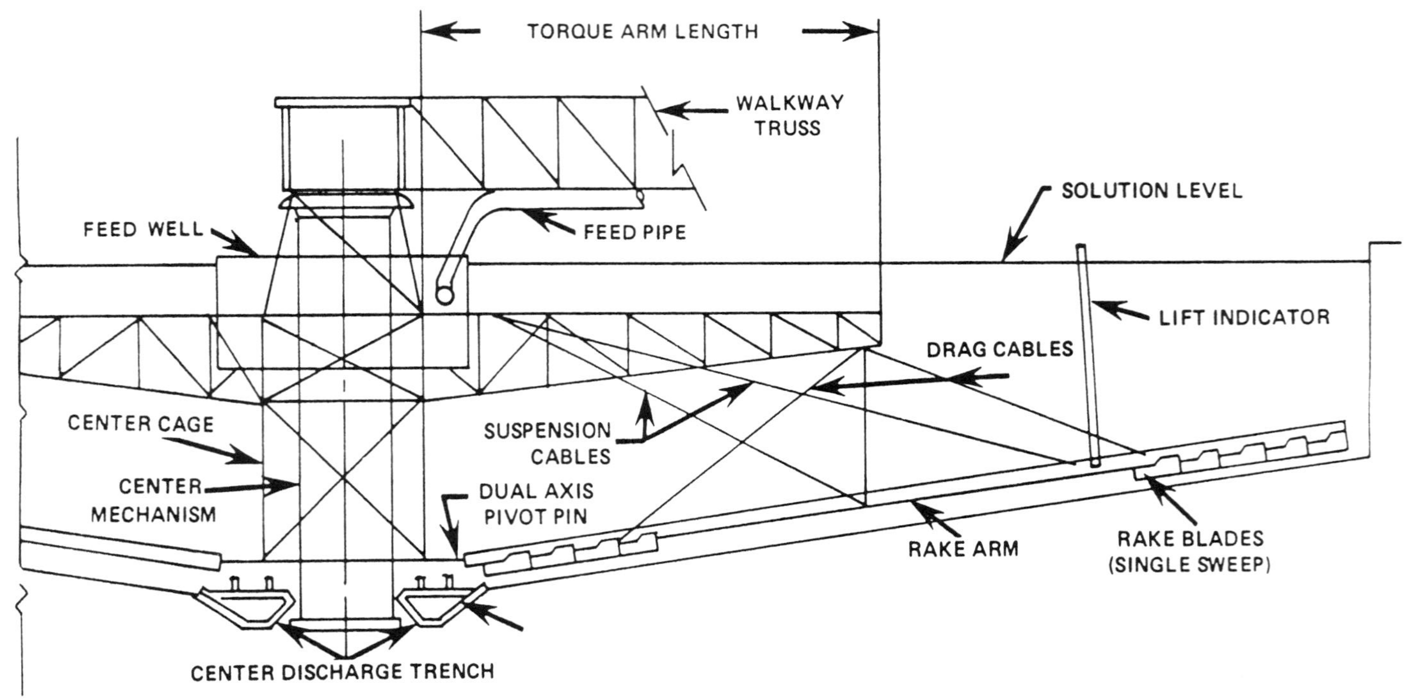

Figure 5 Cross-sectional view of a thickener.

Type		Drive	Diameter	Operating Torque	Description
Standard Bridge-Support Thickener (Type A)		Type A drive head	To 150 ft	To 750,000 ft-lb	"Bridge"–or "beam" or "truss"–spans diameter of tank and supports drive and rake mechanisms. Underflow is removed from discharge cone at bottom center.
Standard Center-Pier Thickener (Type S)	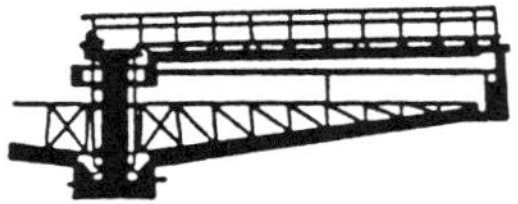	Type S center mechanism	To 400 ft	To 2,400,000 ft-lb	Stationary center pier supports drive and rake mechanisms. Truss extending from center pier to tank periphery supports walkway, power lines and feed launder.
Caisson Center-Pier Thickener (superthickener)	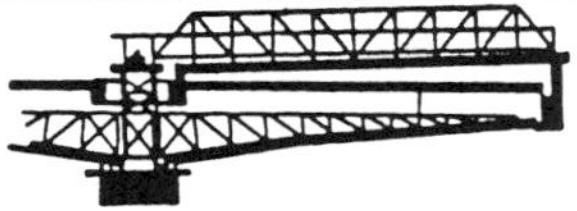	Type S center mechanism	To 600 ft	To 4,000,000 ft-lb	Center pier has been enlarged to form a control and pumping station, as well as a support for the rake assembly. Underflow tunnel is eliminated as underflow is pumped *up* through the caisson.
Traction Thickener	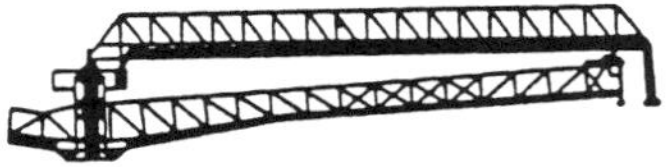	Electric motor	To 400 ft	To 1,300,000 ft-lb	Stationary center pier partially supports rake mechanism and serves as pivot about which rake rotates. Power is supplied by electric motor mounted on two traction wheels running around the periphery of the tank.
Hi-Rate Thickener	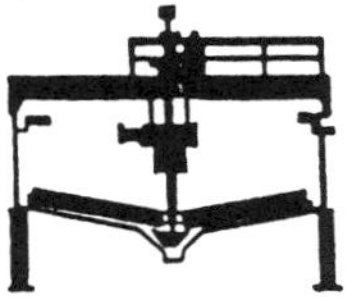	Type A or Type S drive unit	To 140 ft	To 300,000 ft-lb for Type A to 600,000 ft-lb for Type S	Mechanically similar to Standard Type A or Type S units, except for special flocculating feed well and necessary support equipment. Designed to provide roughly 15X the throughput of the conventional machine of similar size.

Figure 6 Descriptions and sizes of different configurations of industrial thickeners.

Table 1 Major Industries Served by Thickeners

Industry	Application
Alumina	Red mud and hydrate
Iron ore	Hydroseparation, concentrate, tailings
Copper/molybdenum ore	Concentrate, tailing
Nickel/cobalt	Concentrate, tailings
Steel	Mill wastes
Potash	Tailings
Phosphates	Slimes, phosphoric acid clarification
Flue gas desulfurization	Scrubber water
Coal	Refuse treatment
Cement	Acid-back residue
Uranium	Yellow cake
Zinc, gold, lead	Tailings
Sand	Slimes
Magnesia	Concentration
Chemical processing	Waste treatment
Industrial processing	Waste treatment

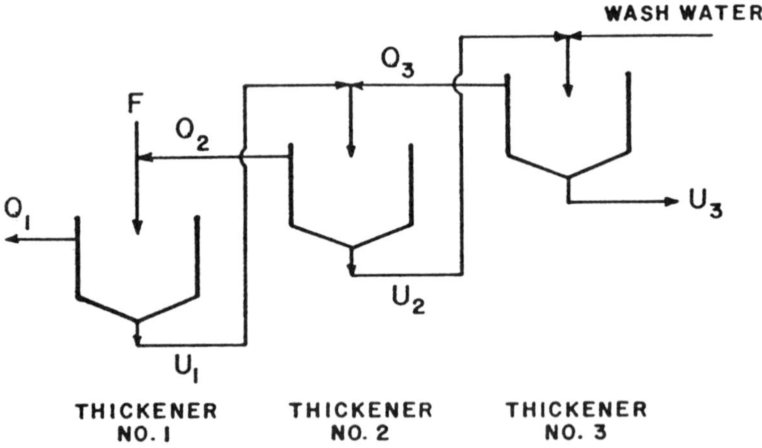

Figure 7 Flow scheme for a three-stage continuous countercurrent decantation. F, feed stream; O, overflow of concentrated solution; U, underflow of concentrated solution.

mixed with third-stage overflow, O_3, and fed to the second stage. Similarly, the second-stage underflow, U_2, is mixed with wash water and fed to thickener 3. The washed solids are removed from the third stage as the final underflow, U_3.

Continuous clarifiers handle a variety of process wastes, domestic sewage, and other dilute suspensions. They resemble thickeners in that they are sedimentation tanks or basins whose sludge removal is controlled by a mechanical sludge-raking mechanism. They differ from thickeners in that the amount of solids and weight of thickened sludge are considerably lower. Figure 8 shows one of the various types of cylindrical clarifiers. In this type of sedimentation machine, the feed enters up through the hollow central column or shaft, referred to as a *siphon feed system*. The feed enters the central feed well through slots or ports located near the top of the hollow shaft. Siphon feed arrangements greatly reduce the feed stream velocity as it enters the basin proper. This tends to minimize undesirable cross currents in the settling region of the vessel. Most cylindrical units are equipped with peripheral weirs; however, some designs include radial weirs to reduce the exit velocity and minimize weir loadings. Units can also be equipped with adjustable rotating overflow pipes (Fig. 9).

Sedimentation Principles

To examine sedimentation in greater detail, let us examine the events occurring in a small-scale experiment conducted batchwise, as depicted by Figure 9. Particles in a narrow size range will settle with about the same velocity. When this occurs, a demarcation line is observed between the supernatant clear liquid

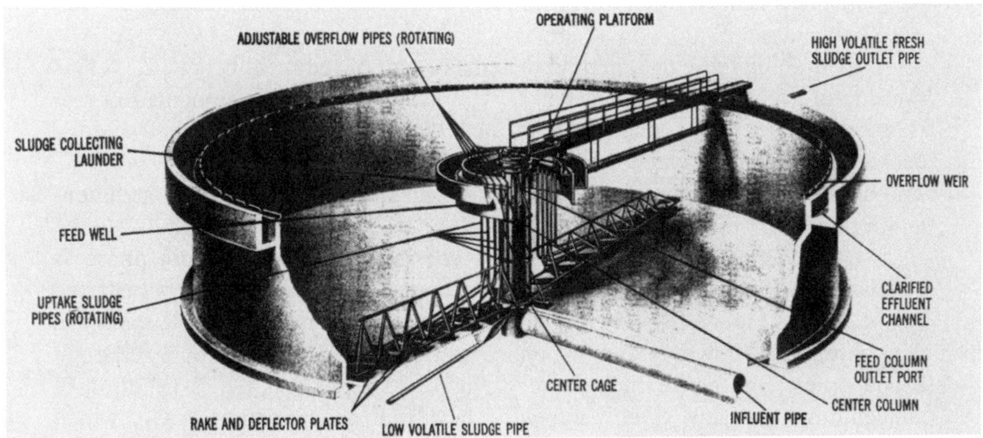

Figure 8 Cylindrical clarifier.

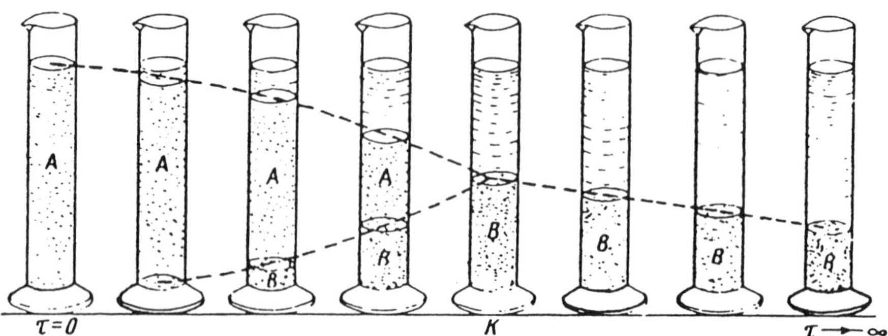

Figure 9 Batch sedimentation occurring in a glass cylinder.

(zone A) and the slurry (zone B) as the process continues. The velocity at which this demarcation line descends through the column indicates the progress of the sedimentation process. The particles near the bottom of the cylinder pile up and form a concentrated sludge (zone D) whose height increases as the particles settle from zone B. As the upper interface approaches the sludge build-up on the bottom of the container, the slurry appears more uniform as a heavy sludge (zone D); the settling zone B disappears; and the process from then on consists only of the continuation of the slow compaction of the solids in zone D.

By measuring the interface height and solids concentrations in the dilute and concentrated suspensions, a graphic representation of the sedimentation rates can be prepared as shown in Figure 10. The plot shows the difference in interface height plotted against time, which is proportional to the rate of settling as well as to concentration.

Examining these data in more detail as a plot of sediment height, Z, versus time, t, in Figure 11, we note that $Z \propto t$, meaning that the sedimentation rate is, and continues to be, constant. Then the sedimentation rate of the heavy sludge decreases with time, which corresponds to the curve on the graph after point K. The higher the concentration of the initial suspension, the slower the sedimentation process.

Observations show that the solids concentration in the dilute phase is constant up to the point of complete disappearance of phase A. This is illustrated by the plot in Figure 12 and corresponds to a constant rate of sedimentation in the phase. Note, however, that the concentration in phase B changes with height Z and time t, and hence each curve in Figure 12 represents the distribution of concentrations at any given moment. The initial concentration is C_1, which remains in the dilute phase during the process. After a sufficient period of time, the concentration increases to C_2 but in zone D. Obviously, if the concentration

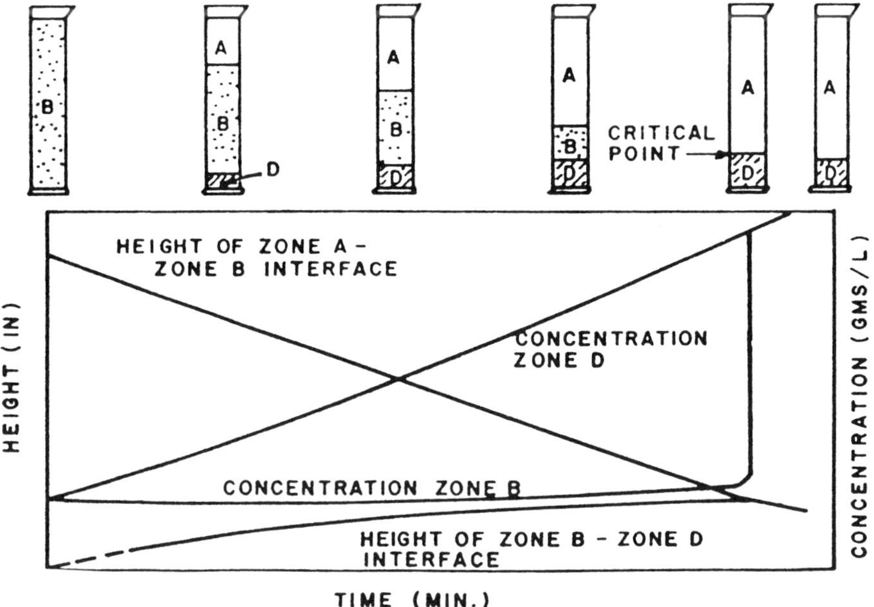

Figure 10 Plots of interface height and solids concentration versus time for the batch sedimentation.

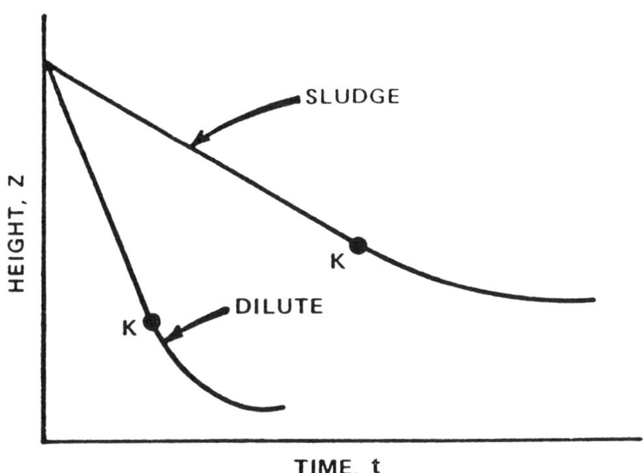

Figure 11 Graph describing the kinetics of sedimentation.

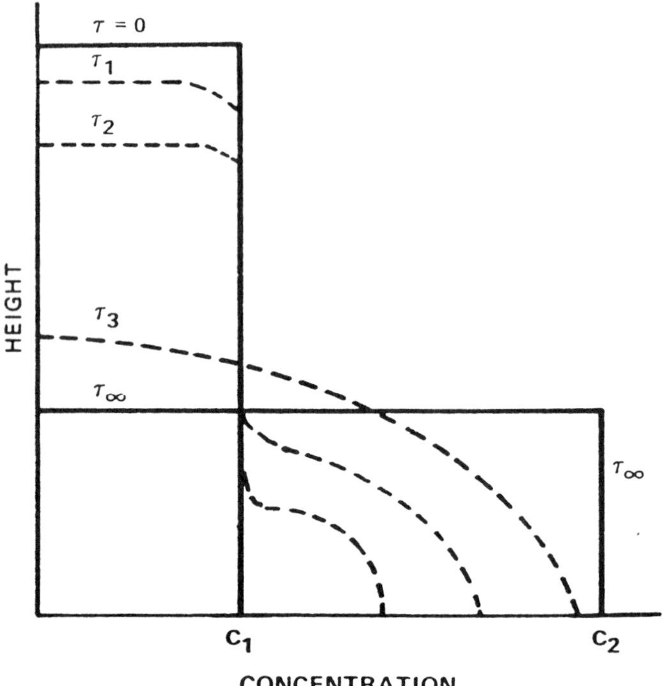

Figure 12 Plot of concentration versus interface height, denoting sedimentation rates.

of the feed suspension is too high, no dilute phase will exist, even during the initial period of sedimentation. Hence, there is no constant sedimentation rate. In this case, concentration, not height, will change with time only.

Chemical Sedimentation

Sedimentation of chemically coagulated or precipitated wastewaters is similar to sedimentation of wastewaters without chemicals. Design can proceed on essentially the same basis provided special consideration is given to the effects of chemical treatment on settling characteristics, sludge quantities, resistance of the sludge to movement by collecting and pumping equipment, and the special maintenance problems encountered with lime coagulation.

Actual surface loadings vary considerably from one application to another. This wide variation emphasizes the importance of testing and pilot work in designing chemical precipitation facilities. In the absence of testing indicating higher figures to be satisfactory, the following typical surface loading rates have been used for sizing tanks:

Chemical	Peak Surface Loading (gpd/ft^2)
Alum	500–600
Iron	700–800
Lime	1400–1600

In general, these design rates may be used for primary, secondary, or tertiary applications. Sludge quantities from chemical precipitation can be estimated from the suspended solids removal and the stoichiometry of chemical reactions involved. Volumes depend on sludge concentrations which are highly variable (1–15%) and are best determined by actual testing. Equipment suppliers should be consulted about the strength and power of collector equipment to handle the dense sludges expected from lime precipitation. Extra smooth piping glass-lined or PVC should be used for lime sludges. Average sludge productions determined from raw wastewater coagulation by lime, iron, and alum are 6,500, 1740, and 1120 lb/mg.

Advantages which might favor the use of flotation in special applications include

- Higher surface loadings and hence smaller tanks sizes (important where space is critical
- Ability to handle peak seasonal loads or storm flows (in some designs flotation may be used intermittently to increase capacity of settling tanks)
- Effectiveness in removing solids which are difficult to settle

Dissolved-Air Flotation

Dissolved-air flotation (DAF) units commonly employ rectangular tanks with separate chain-and-flight scum and sludge collectors (Figure 13). Circular units are also commercially available. The widest application for these units has been as thickeners for waste-activated sludge. Units used for suspended solids separation are similar, but design parameter values vary according to the application. To avoid fouling of pressurizing and pressure-regulating equipment and excessive shearing of influent solids, a stream of recycled effluent is usually pressurized. On pressure release, this stream is blended with the inflow to be treated. Other methods include pressurizing all or part of the influent stream.

Design of dissolved air flotation units involves selection of values for a number of parameters, including percentage recycle flow; operating pressure; pressurization retention time; air flow; and surface hydraulic loading, solids loading (area basis), and float detention period. Variables reflecting influent characteristics include flow, solids loading, liquid temperature, and type and quality of influent solids. Investigators have attempted to relate flotation performance to the air to solids ratio and a number of other variables with a

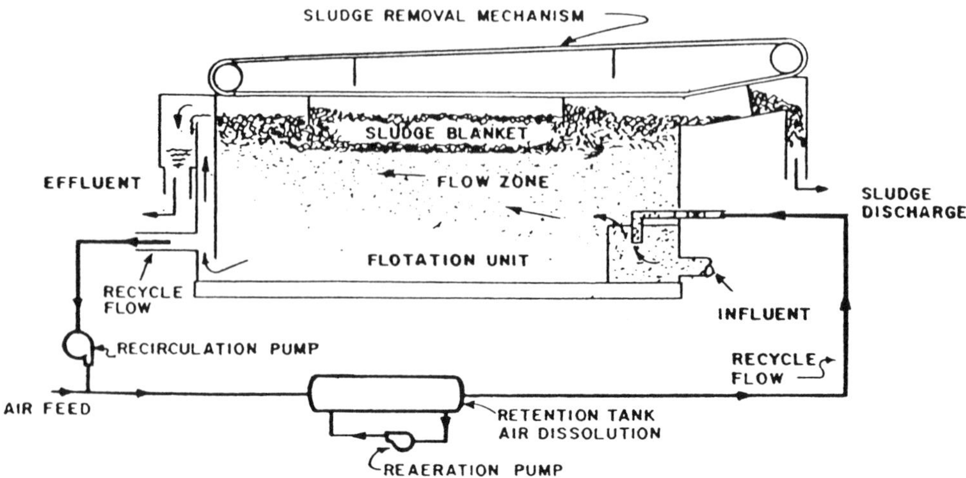

Figure 13 Dissolved-air flotation unit schematic.

limited amount of success. Values of specific parameters used in actual applications vary widely. Typical ranges include

Parameter	Range
Pressure, psig	25–70
Air to solids ratio, lb/lb	0.01–0.1
Float detention, min	20–60
Max. 24 hr	
Surface hydraulic loading, gpd/sq ft	500–4000
Recycle, %	5–120

In flotation equipment special attention must be given to the inlet, outlet, and collector mechanism configurations. The flotation tank must permit aggregate rise with a minimum of interference in the form of turbulence or obstructions and provide for removal of floated froth, settled sludge, and treated effluent. Effluent ports must be sufficiently submerged to prevent interference with the froth on the surface. The inlet conditions of the flotation tank are critical to proper performance. Baffles, walls, and other obstructive energy-dissipating devices tend to destroy aggregate bonding with resulting loss in flotation efficiency. Also turbulence in the region of the froth will result in losses of floated solids.

Shallow Settling

The potential advantages of multiple tray shallow settling devices have long been recognized, but early prototypes of such equipment were unsuccessful owing to practical problems of flow distribution and sludge removal. Shallow settling devices of improved design, such as tube settlers, have been applied to water and wastewater treatment. Tube settlers consist of bundles of small plastic tubes with hydraulic radii ranging from 1 in. upward and lengths of 2 ft or more depending on the particular application. Square tube sections are most common, but hexagonal and other shapes have been used by various manufacturers.

Tubes are commonly inclined steeply (60 degrees) to horizontal and fabricated in modules, as shown in Figure 14. These modules have beam strength which permits their installation in settling tanks, as shown in Figures 15 and 16. Clarifier influent is introduced beneath the tube modules. The flow passes upward through the modules with the solids moving countercurrently by gravity (Fig. 17) and falling from the tube bottoms into the sludge-collection zone beneath. The clarified effluent is collected above the tube modules. Free-standing package units with tubes only slightly inclined (5 degrees) have found some application in small chemical clarification/filtration systems for tertiary wastewater treatment.

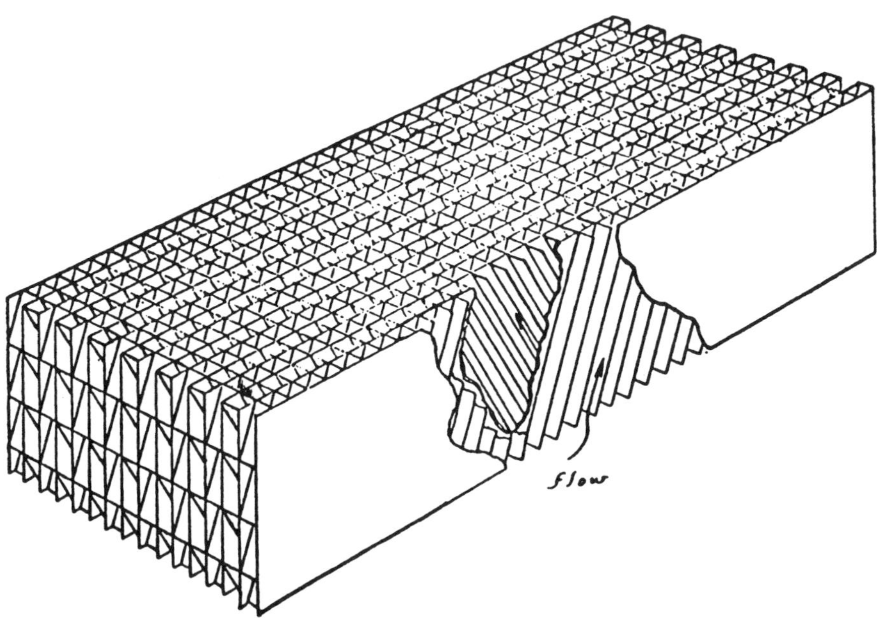

Figure 14 Module of steeply inclined tubes.

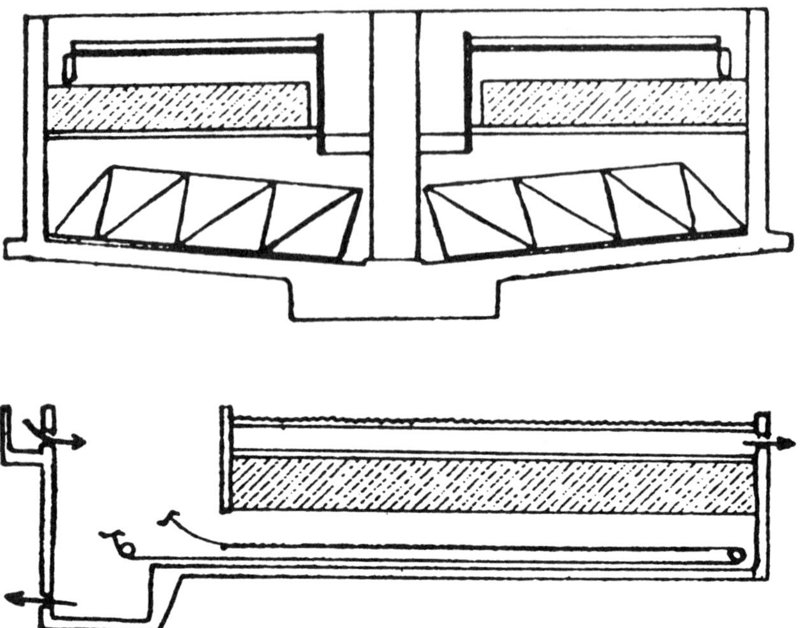

Figure 15 Tube settlers in existing clarifier.

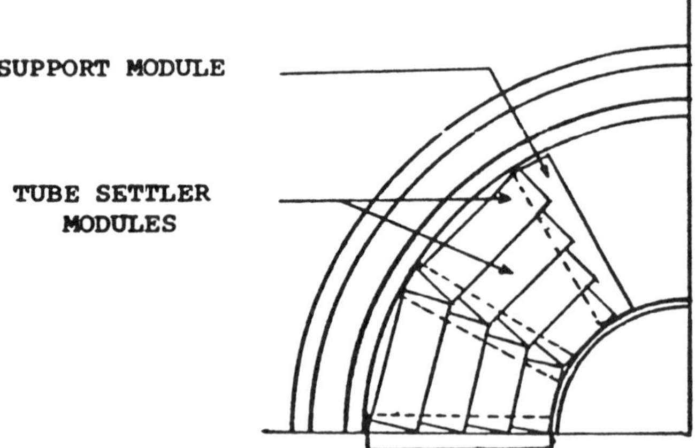

Figure 16 Plan view of modified clarifier.

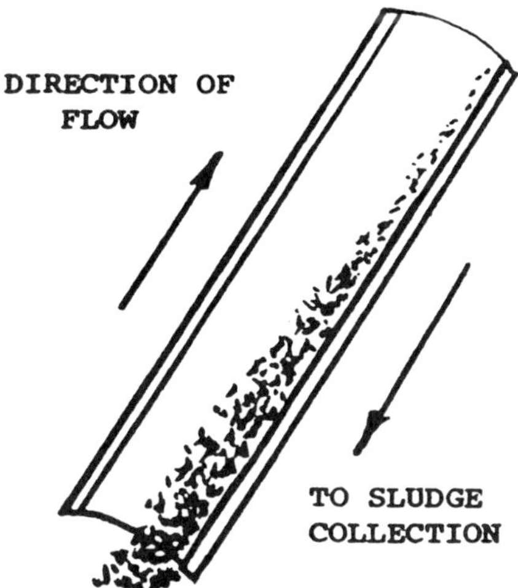

Figure 17 Tube settlers—flow pattern.

Tube settlers promote sedimentation in three ways:

- The multiple tubes stacked one above another provide an effective settling area several times that of the projection in plan of the modules.
- The small hydraulic radius of the tubes maintains laminar flow and promotes uniform flow distribution.
- In steeply inclined tubes, the movement of sludge against the direction of flow favors particle contact and agglomeration.

This additional flocculation offsets the reduction in their horizontal projected area caused by inclining the tubes.

6

Biological Treatment Systems

INTRODUCTION

The most efficient method for eliminating or removing organic material from wastewater is by utilizing biological treatment systems. There are three basic systems most often utilized for this removal: activated sludge, trickling filters, and aerobic oxidation lagoons. Although these three systems may differ somewhat regarding the detention times, oxygen requirements, and mode of utilization of biological slimes, the essential biochemistry that occurs within each system is identical.

The purpose of the biological treatment unit is to remove organic material either by oxidation to carbon dioxide, water, and other derivatives or by conversion of the organic material into a settleable form which can be removed by gravity sedimentation. Production of carbon dioxide, water, and ammonia is referred to as the respiration stage, whereas the conversion of the organic material into new bacterial cells which can be settled out is referred to as synthesis.

Since these are aerobic biological treatment processes, the final hydrogen acceptor for the oxidation of organic material is oxygen. During this hydrogen transfer, the liberation of energy from the organic molecule is used for synthesis and for energy requirements of cellular substances.

The quantity of oxygen that is necessary to oxidize the organic material depends mainly on the biological oxygen demand (BOD) satisfied during the biological treatment process. Various treatment methods have been estimated as satisfying the following percentages of applied BOD: conventional activated sludge, 90–95%; high-rate trickling filters, 65–85%; and low-rate trickling filters, 80–90%. The removal of organic material is accompanied by oxidation and synthesis of cells. The amount of new cell material produced per pound BOD added varies with the chemical composition of the substrate.

Variabilities among yields can be explained by visualizing that if the organic loading to the unit is quickly assimilated by the microorganisms, then the organisms metabolize themselves, oxidizing cellular material normally contributed to sludge yield. Energy yields of different compounds are not the same; consequently, more or less of a particular substrate may be used to satisfy the energy requirements of the system. Removal of organic material from liquid waste is achieved by complete destruction (oxidation), which yields energy, and by synthesis, which uses the energy produced during the oxidation of organic matter.

PRIMARY TREATMENT

Before undergoing secondary biological treatment, a wastewater is usually subjected to a preliminary form of treatment, known as primary in order to remove suspended and other insoluble matter. These systems usually include rough screens or racks, constant-velocity grit-removal tanks, and primary settling tanks and have been discussed in earlier chapters.

Typical microbial metabolic reactions of importance to biological treatment include

- Anaerobic nonphotosynthetic reactions:

  Nitrate reduction (denitrification)

  $5CH_5COOH + 8NO_3^- \rightarrow 10CO_2 + 4N_2 + 6H_2O + 8OH^-$

  $5S + 6NO_3^- + H_2O \rightarrow 5SO_4^= + 3N_2 + 4H^+$

  Sulfate reduction

  $2CH_3CHOHCOOH + SO_4^= \rightarrow 2CH_3COOH + H_2S + 2OH^-$

  $4H_2 + SO_2^= \rightarrow 2H_2O + H_2S + 20H^-$

  Organic carbon reduction (fermentation)

  $CH_3COOH \rightarrow CH_4Y + CO_2$

  $4CH_3OH \rightarrow 3CH_4 + CO_2\ 2H_2O$

  $C_6H_{12}O_6$ bacteria $\rightarrow 3CH_3COOH$

  $C_6H_{12}O_6$ yeast $\rightarrow 2CH_3CH_2OH + 2CO_2$

  Carbon dioxide reduction

  $2CH_3CH_2OH + CO_2 \rightarrow 2CH_3COOH + CH_4$

  $4H_2 + 2CO_2 \rightarrow CH_3COOH + 2H_2O$

- Anaerobic nonphotosynthetic bacterial reactions:
  - Oxygen-limited reactions
    - $CH_3CH_2OH + O_2 \rightarrow CH_3COOH + H_2O$
    - $2\ CH_3CHO + O_2 \rightarrow 2CH_3COOH$
    - $2CH_3CHOHCH_3O_2 \rightarrow 2\ CH_3COOH + 2\ H_2O$
  - Complete oxidation
    - $CH_3COOH + 2O_2 \rightarrow 2CO_2 + 2H_2O$
    - $2H_2 + O_2 \rightarrow 2H_2O$
  - Nitrification
    - $2NH_3 + 3O_2 \rightarrow 2NO_2^- + 2H^+ + 2H_2O$
    - $2NO_2^- + O_2 \rightarrow 2NO_3^-$
  - Sulfur oxidation
    - $2\ H_2S + O_2 \rightarrow 2S + 2H_2O$
    - $2S + 2H_2O + 3O_2 \rightarrow 2SO_4^= + 4H^+$
    - $S_2O_3^= + H_2O + 2O_2 \rightarrow SO_4^= + 2H^+$
  - Nitrogen fixation
    - $N_2 \overline{>}$ Nitrogenous organics
- Photosynthetic reactions:
  - $CO_2 + 2H_2S$ light $\rightarrow (CH_2O) + H_2O + 2S$
  - $3CO_2 + 2S + 5H_2O$ light $\rightarrow 3\ (CH_2O) + 4H^+ + 2SO_4^=$
  - $CO_2 + 2H_2O$ light, algae $\rightarrow (CH_2O) + H_2O + O_2$
  - $9CH_3COOH$ light $\rightarrow 2CO_2 + 4(C_4H_6O_2) + 6H_2O$
  - $CO_2 + 2H_2$ light $\rightarrow (CH_2O) + H_2O$
  - $2\ CH_3COOH + H_2$ light $\rightarrow (C_4H_6O_2) + 2H_2O$

Primary Settling Tanks

Primary tanks are utilized in all trickling filter plants and in some activated-sludge plants. Their purpose is to reduce the content of settleable solids so as to reduce the load on subsequent treatment systems and to prevent the formation of sludge deposits. They may be either manually or mechanically cleaned and typically have detention times of 2 hr or less, and they are responsible for as much as 35% of the BOD removal and 60% of the suspended solids removal.

Removal of the primary sludge removed (approximately 400 ft^3/million gal at 4–6%) and its subsequent effect on secondary treatment make it an attractive addition to a large system. On the other hand, the cost of disposing of the primary sludge removed (approximately 400 ft^3/L million gal at 4–6% solids) by anaerobic systems has caused some concern. It is necessary to use primary treatment prior to trickling filtration systems to avoid clogging of filter media.

ACTIVATED SLUDGE

The activated-sludge process of wastewater purification is a common process for the secondary treatment of wastes. The activated sludge consists of a gelatinous matrix in which filamentous and unicellular bacteria are imbedded and on which protozoa attach and feed. The bacterial genera which predominate depends on the characteristics of organic matter in the waste water, such as *Pseudomonas* for hydrocarbon and carbohydrate wastes and *Alcaligenes*, *Bacillus*, and *Flavobacterium* for proteinaceous wastes. The process consists of mixing activated sludge, recirculated from a final settling tank, with incoming raw or primary sewage to form a mixed liquor, which is subsequently aerated and from which activated sludge is later settled. When a plant is first started, it can be seeded with an activated sludge from a currently operating plant. If no seed sludge is available, then one can be built up over a short period of time (4–6 weeks) by simply continually aerating, settling, and returning the residue of the sewage.

Conventional Process

A flow diagram of a conventional activated-sludge plant is shown in Figure 1.

In the process, primary treated sewage is mixed with a portion of returned activated sludge and aerated for 4–6 hr. The process can consists of the following steps:

- Mixing in the activated sludge with the sewage to be treated
- Aeration and agitation of this mixed liquor for the required length of time
- Separation of the activated sludge from the effluent and the subsequent return of a portion of the settled sludge to be mixed with the incoming sewage

Theory

The mechanism of removal of organic material from sewage by activated sludge can be generalized in the following way:

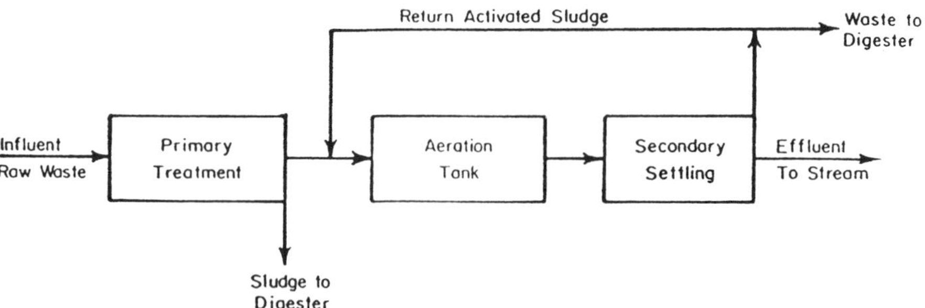

Figure 1 Flow chart of a conventional activated-sludge plant.

$$\text{Organic Material} + \text{Bacteria} \xrightarrow[O_2]{\text{Sufficient}} CO_2 + H_2O + NH_3 + \text{Energy} \quad (1)$$

The release of energy results in the formation of new cellular material; this formation is commonly referred to as synthesis, whereas the production of carbon dioxide, water, and ammonia is referred to as respiration. Thus, we can then say that the oxidation of organic material results in respiration and synthesis.

Bacteria necessary to utilize the organic material are present in the activated-sludge flocs. These flocs are gelatinous matrices in which unicellular and filamentous bacteria are present. Protozoa and metazoa are usually present on the surface of the flocs.

When sewage is first mixed with activated sludge, a portion of material is stored away (adsorbed) until the bacteria find it necessary to use it as food. The remaining portion of organic material is then oxidized and results in synthesis and respiration. The total removal of organic material is accomplished in two major parts—adsorption and oxidation. Adsorption occurs within the first section of the aeration tank during the first hour of aeration. Total removal of organic material by the activated-sludge process is from 90 to 95%.

Organic Loading

The main treatment units involved in the activated-sludge process is shown in Figure 1. They include the aeration tank and the sludge separation or secondary settling tank. Primary effluent, or the untreated wastewater containing colloidal and soluble organic material, enters the aeration tank where it is attacked and stabilized by the mixed flora and fauna known as activated sludge. When this activated sludge is combined with the influent wastewater, is known collectively as mixed liquor, and the sludge solids are designated as either the mixed liquor suspended solids (MLSS) or mixed liquor volatile suspended solids (MLVSS). The MLVSS is the volatile portion of the MLSS and is usually related by MLVSS = 0.80 × MLSS. The use of MLVSS may more closely approximate the total active (biological) mass in the system. The overflow from the secondary settling tank leaves the process as effluent. Since there is usually a net production of biological cellular material by the aerobic treatment process, some sludge must be removed from the system. It is either removed intentionally by separation of sludge or occurs unintentionally by the loss of solids in the process effluent.

The nutrient substrate level, or food value of a wastewater, is measured in terms of the biochemical oxygen demand (BOD) of the constituents present in the waste. The relationship existing between the quantity of substrate and the BOD removal efficiency of the process has led to the development of many concepts proposed for use as loading parameters.

Loadings have been expressed in terms of pounds of BOD/1000 ft^3 of aeration tank volume. However, constant BOD removal efficiencies (87–90%) have been reported at loadings of 30–120 lb/1000 ft^3, indicating the ineffectiveness of this ratio as a general loading parameter. Parameters based on tank volumes do not take into account the amount of solids under aeration or specify the length of the aeration period.

Sludge age is a measure of the quantity of organic material entering the process. This term is defined as the average length of time a particle of suspended solids remains under aeration. It was expressed as the ratio of the pounds of mixed liquor solids under aeration to the pounds of suspended solids entering the system day and thus had a unit of days.

In dealing with sewage, sludge age is adequate, since the suspended solids concentration in milligrams per liter usually is about equal to the BOD concentration in milligrams per liter. The parameter is modified for relatively soluble organic wastes by substitution of the pounds of BOD entering the system per day in place of the pounds of suspended solids. This substitution makes the sludge age equivalent to the ratio of the pounds of mixed liquor solids under aeration per pound of BOD entering the process per day.

A loading parameter combining the three major factors applicable to the activated sludge process are the BOD of the applied waste, the quantity of sludge present in the aeration tank, and the period of aeration. Ample evidence demonstrates that the effectiveness of the activated sludge process and the amount of aid required for its operation are primarily dependent on the daily BOD input/mixed liquor volatile solids ratio.

In the activated sludge process, it is difficult to obtain an exact measure of the amount of active cell material in a system. However, because of the simplicity of the test procedures involved, it has become standard practice to measure the weight of suspended solids (MLSS) contained in the mixed liquor and assume that there is a relationship between the amount of solids in the mixed liquor and the number of active organisms present.

A loading parameter can be expressed as pounds of BOD applied daily per 1000 lb of suspended solids in the aeration tank.

The sludge loading ratio (SLR) is the loading parameter to be used for activated sludge. The SLR is expressed as pounds of BOD applied per day per pound of MLVSS and thus amounts to a food to organisms ratio. The symbol F is also used for SLR, as in Figures 2–4 and 6. The concentration of suspended solids in the mixed liquor depends upon the waste being treated and the aeration capacity of the plant. Normally, the solids concentration is between 1000 and 2000 mg/L. Calculations of the required solids concentration necessary to obtain efficient treatment under specific conditions can be made by utilizing the SLR. The equation for the calculation of the SLR is

$$\text{SLR} = \frac{24\ \text{La}}{\text{Sa}\ t(1 + R)} \tag{2}$$

where La is the BOD in milligrams per liter, t is the detention time in hours, Sa is the concentration of MLVSS in milligrams per liter, and R is the recycle ratio.

It is usually expected that the sludge loading ratio should never exceed a value of 0.3/day in normal sewage plant operation utilizing conventional activated sludge.

If we recognize that the area of contact surface and the opportunity for contact are two of the most important factors in the activated sludge process, the sludge loading ratio can be readily accepted as a general parameter of loading intensity. It can be considered as (1) the weight of removable substrate applied in a unit of time to (2) a unit of contact surface for (3) a unit of contact time. the first factor is readily determined by analytical procedures (BOD) and measurements of sewage flow. The second factor can be evaluated only indirectly, by the volatile suspended solids concentration in the mixed liquor, whereas the third factor is usually represented as a statistical average for activated sludge plants. Situations are encountered in plant practice that indicate the ineffectiveness of this ratio.

The quality of the material making up the BOD is extremely important in plant performance and also for design purposes. BOD is not a simple entity, but must be evaluated in light of the composition of the substrate contributing the BOD. Some domestic sewage treatment plants experience process trouble at one SLR but other plants at the same SLR operate extremely well.

Sludge Volume Index

Knowledge of the volume of sludge present or of the solids (on a dry basis) in the aeration tank is not sufficient for good process control. A combination of these two is, however, an essential feature of good process control. This combination, called the sludge volume index (SVI), is the volume in milliliters occupied by 1 g (dry weight) of sludge after 30 min of settling, and it is sometimes referred to as the Mohlman index. The test is performed by settling a 1-L sample of mixed liquor for 30 min in a 1000-ml graduated cylinder. The volume occupied by the sludge is reported as percentage or in milliliters; the suspended solids are determined and reported in percentage by weight or in milligrams per liter. The SVI can be expressed as

$$\text{SVI} = \frac{\%\ \text{Volume of Sludge Settled in 30 min}}{\%\ \text{Suspended Solids}} \tag{3}$$

Although the 30-min period has been adopted as a standard, variations in sludge settling rates taking place in this time due to the effect of different mixed liquor solids concentrations have been reported. A modification of this test which somewhat eliminates these variations and is usually used for all the index calculations in experimental work. This modification calls for an adjustment of the mixed liquor solids concentration to between 1000 and 1500 mg/L before the SVI determination is made.

A well-settling sludge may have a Mohlman index between 50 and 100, but an index of 200 is indicative of a sludge with poor settling characteristics. Knowledge of the sludge volume index and the SLR is necessary for good process control, since there is a critical value of sludge volume index below which the volume of settled sludge in the final tanks will exceed the return sludge rate. If the SVI rises to 200, then the return sludge rate must be increased to maintain a constant solids concentration under aeration. If the concentration of mixed solids decreases, the sludge loading ratio will increase, thereby increasing the bulking tendency of the sludge and compounding the problem.

Sludge Density Index

The sludge density index (SDI) rather than the SVI can be used. This can be defined as the weight of a specific volume of sludge after it has settled for 30 min. It is calculated as follows:

$$\text{SDI} = \frac{100}{\text{SVI}} \tag{4}$$

SDI values are more amenable to graphical expression than are SVI values. Another reason for their use has been to encourage standardization of this nomenclature for usage in the field. This means that a sludge with good settling characteristics has an SDI of between 2.0 and 1.0, whereas an SDI of 0.5 indicates a bulky or nonsettleable sludge.

Sludge Bulking

During normal operation, mixed liquor flows into the final settling tanks from the aeration tanks. Activated sludge forms flocs, settles, and the effluent flows over the weirs of the final tank. At times, the activated sludge does not settle well; its volume becomes greater in comparison with its density and the return sludge pumps cannot keep up with the large volumes of light sludge settling in the final tanks. If this condition persists, the sludge in the final tanks will spill over the weirs and the BOD of the final effluent will increase. This phenomenon, known as sludge bulking, is a major problem of the activated-sludge process. Frequently, bulking occurs unexpectedly when the plant seems to be operating at its peak efficiency and producing an excellent effluent.

Theories have been advanced as to the cause of bulking, with none of them completely satisfactory. It was observed by many early investigators that the organism *Sphaerotilus natans* was often present in large numbers when bulking occurred. *S. natans* is a sheathed, filamentous bacterium and it was reasoned that its growth in excessive numbers caused the sludge to be less dense, and hence caused it to bulk. Bulking produced by carbohydrates is a direct response of *Sphaerotilus* to a relatively long contact with an available energy food; on the other hand, these organisms may be a result, not a cause, of bulking.

Bulking has been associated with characteristics of the raw waste such as septicity, heavy organic load, trade wastes, mineral oil, and excessive carbonaceous content. In the treatment plant, overaeration, underaeration, poor mixing, short circuiting, and too high or too low mixed liquor solids have been listed as causes of bulking. Other causes within the plant have been listed as septic return sludge and excessive detention periods.

Although the characteristics of the raw waste are associated with the causes of bulking, in-plant causes of bulking are more significant than those attributed to raw waste characteristics.

The many different measures used to control bulking reflect the incomplete knowledge of its causes. Reducing the suspended solids in the aeration tank, increasing the quantity of air, increasing the solids in the aeration tanks, use of inert materials, use of iron compounds, chlorination of the return sludge, reaeration of return sludge, and addition of lime to the mixed liquor are some of the methods which have been recommended at various times. Since most treatment operations are concerned with maintaining a constant SLR or sludge age, it becomes apparent that perhaps there are inconsistencies contained within the formulation of the parameter.

Active Mass and Loading

The food to organisms ratio, designated as the sludge loading ratio (SLR), is the parameter used in wastewater-treatment practice. The SLR corresponds to the symbol F, which is sometimes used in mathematical formulations of biological treatment processes and utilized herein in Figures 2–4 and 6. This representation of a biological system depends on an understanding of fundamental biological concepts within the activated-sludge process. It is easy to represent an analysis that is mathematically correct but biologically incorrect.

The constant assumption made in most mathematical formulations is that the suspended solids or volatile suspended solids contained in the mixed liquor represent the total number of bacteria present and are related to the total enzymatic activity of the system. This is not entirely correct. For any given system it may be possible to use some solids parameter as an index of active mass, but it must be used with extreme caution. In an unchanging ecological system the solids concentration may be adequate, but as a general estimation of active mass it is sorely lacking. The weight of nitrogen in the sludge may be a better estimation of the quantity of bacteria and hence the optimum loading parameter for activated sludge systems and can be designated loadings as BOD per milligram sludge nitrogen.

Anaerobic Digestion

When primary treatment is employed prior to secondary systems, a sludge is produced which is usually disposed of by anaerobic digestion. Excess activated sludge may also be treated in such fashion. The main purpose of sludge digestion is to produce an innocuous residue which can be easily disposed of, and also to reduce the volume of sludge which must ultimately be removed.

Usually the decomposition of complex organic matter is accompanied by production of intermediate products and endproducts. Such compounds as methane, hydrogen, organic acids, and alcohol are the main products of decomposition of carbonaceous organic materials under anaerobic conditions. Similarly, degradation of proteins will result in compounds such as ammonia, amino acids, amides, peptones, hydrogen sulfide, indole, skatole, and mercaptans.

Sewage solids subjected to anaerobic decomposition pass through three stages:

- A period of intensive acid production (acidification)
- A period of acid digestion (liquefaction)
- A period of intensive digestion and stabilization (gasification)

Each step is represented by the production of typical intermediate and endproducts. Under normal operating conditions, all three stages occur simultaneously.

The term *liquefaction* as applied to digestion denotes transformation of large solid particles into either a soluble or finely dissolved form. This process is brought about by hydrolysis utilizing extracellular enzymes. During this period, intermediate products of fermentation accumulate and gasification are at a minimum. When sludge is digested without a seed source (a sludge that has been digested under similar environmental conditions), this condition is greatly exaggerated. With seeded sludge, liquefaction is in balance with gasification and there is usually no undue accumulation of intermediate products. Although the terms *liquefaction* and *hydrolysis* are used interchangeably they are not strictly synonymous. Hydrolysis is a well-defined chemical term designating the addition of water to the molecule to break down complex substances into simpler ones. Liquefaction does not have such an exact connotation and refers merely to the transfer of substances from a solid sludge stage to a liquid phase.

Primary gases produced during the gasification stage are methane and carbon dioxide. These two gases normally form more than 95% of the gas evolved. The average heat value of the gas is approximately 700 BTU/ft^3. A good indicator of the degree of digestion is the percentage of methane gas contained in the total digester gas. Usually, when the methane production is low (under 65%) the digestion is poor. The maximum volume of gas that is generated from a heated anaerobic digester is approximately 10 ft^3/lb of volatile solids added to the tank, or approximately 0.7 ft^3/capita/day.

Methane production results from the breakdown of many compounds by numerous interdependent and interaction reactions which take place in an orderly and integrated fashion. Methane organisms which produce methane do not utilize such substances as cellulose, glucose, proteins, amino acids, or fats, but they do utilize a restricted group of simple compounds consisting of lower fatty acids (e.g., formic, acetic, propionic, and n-butyric acids). The transformation of complex organic materials contained in sludge to methane and carbon dioxide is brought about in two stages by two different groups of bacteria. Complex organics are converted by a variety of common bacteria to volatile acids and alcohols without the production of methane (acid production). These products are then converted to methane (methane fermentation stage) by a restricted and specialized group of bacteria among which are *Methanobacterium omelianskii*, which utilize primary and secondary alcohols; *M. suboxydans*, which partially utilizes butyrate, valeric acid, and other four- and six-carbon fatty acids to produce acetic and propionic acid; and *M. propionicum*, *M. mayei*, and *M. barkerii*, which utilize the simpler organic acids and alcohols and produce methane and carbon dioxide.

Organisms responsible for active and thorough digestion of waste solids require an environment in which the pH is about 7. The optimum pH value for digestion varies slightly with the characteristics of water supply and the types of waste present. Insufficient seed or low temperature results in retardation of gasification and accumulation of acidic intermediate decomposition products.

In the normal digestion of domestic sewage sludge small amounts of carbon monoxide, as well as hydrogen sulfide, may be present. Oxygen should not be present in the digester gas. Its presence is indicative of an air leak in the digester or an error in the gas-sampling procedures. Although some claims of improved digestion with aeration have been made, these can usually be attributed to increased mixing of the tank contents with the air. The introduction of large quantities of air on a regular basis will adversely affect methane fermentation.

Some of the most important considerations affecting sludge digestion are food supply, which is influenced by the type of primary sludge generated in the primary treatment system; time of digestion; utilization of seed sludge; temperature; mixing; pH; the volatile acids to alkalinity ratio; and the quantity and amount of chemicals added to the digestion system. The primary indices of digestive action are:

- Gas production: A rule of thumb is that 10 ft^3 of gas should be produced per pound of volatile solids added to the tank. The gas produced should be approximately 70% methane and 30% carbon dioxide with only traces of miscellaneous gases.
- Volatile solids: Sludge produced should have a solids content of approximately 5% with a volatile content of roughly 50%.
- Volatile acids: Volatile acid content of the digested sludge should be approximately 500 mg/L or less. Levels of higher acidity can be corrected by utilizing chemicals such as lime. Care should be taken, however, not to utilize great amounts of chemicals but rather to change environmental conditions to secure better digestion.
- Sludge characteristics: The digested sludge should be black in color, easily dewatered, and have a "tarry" odor that is not repellent.

In high-rate digestion, rates of digestion can be substantially increased by

- Thorough mixing of the tank contents either mechanically or by gas recirculation
- Optimum loading of the digesters by regulation of the solids in the feed sludge so that as dense a sludge as possible is fed to the digesters

Detention times can be decreased to as little as 5 days when adequate heating is provided.

The rate of activities of the organisms responsible for digestion is greatly influenced by temperature. The time required for digestion is indicated by the quantity of gas produced and the increased amount of volatile matter destroyed. The total amount of gas is not appreciably different at the end of digestion but is produced in a shorter time at higher temperatures. Increased gas production follows an increase in the amount of volatile matter destroyed. The optimal temperature for mesophilic digestion is about 85°C. Temperatures up to 95°F increase the rate of digestion slightly but may be more difficult to maintain throughout the year and may result in problems. The volume of gas may be greater in the presence of some organic industrial wastes, whereas other wastes may reduce the amount of gas. The carbon dioxide in the gas ranges from 15 to 35% and is affected by the degree of digestion and the types of trade waste present.

Process Modifications

Tapered aeration. In an aeration tank having a definite pattern of longitudinal flow, the impact of the high BOD of the influent entering the head end of the tank will create a relatively high oxygen demand in this point in the mixed liquor. As the oxygen demand of the waste is gradually decreased, the demand for oxygen becomes less and less. If one can envision a plug flow system

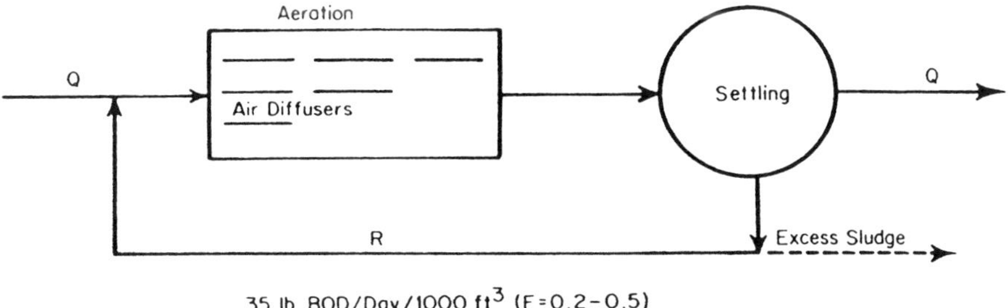

Figure 2 Tapered aeration. Q, R, and F refer to flow through plant, recirculated effluent, and sludge loading ratio, respectively.

passing down the length of the tank, then the tapered aeration process (Fig. 2) can be seen to take the gradually decreasing oxygen demand of the mixed liquor into account by making more oxygen available at the head end of the tank by the use of more diffusers. As with the conventional activated-sludge process, tapered aeration has a volumetric loading of about 35 lb BOD/day/1000 ft^3 of aeration tank capacity and an F value of 0.2–0.5, which is in the same order of magnitude as conventional activated sludge.

Step aeration. Another activated-sludge modification which is capable of handling shock loadings as well as evening out the oxygen demand in the mixed liquor entails the introduction of the waste flow at intervals throughout the length of the tank. This process is termed step aeration. This system, shown in Figure 3, has a volumetric loading of greater than 50 lb/day/1000 ft^3 of aeration tank capacity. The F ratio, however, is the same magnitude as that used in conventional activated-sludge plant operation.

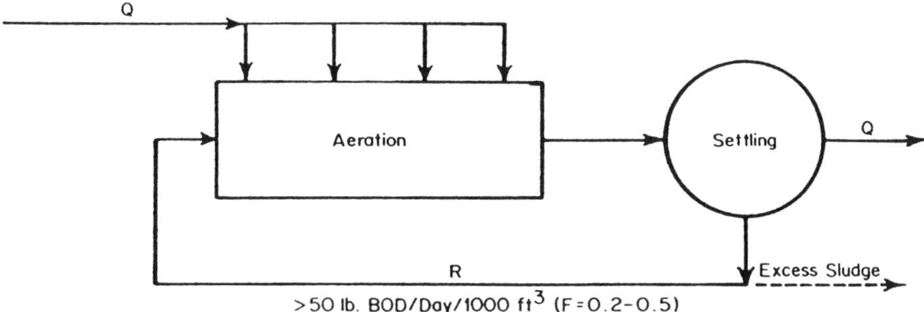

Figure 3 Step aeration (step loading). Q, R, and F refer to flow through plant, recirculated effluent, and sludge loading ratio, respectively.

Contact stabilization. Contact stabilization is another modified process that permits up to twice the volumetric loading of the conventional process. In contact stabilization, the volumetric loading for the system is in the order of 70 lb BOD/day/1000 ft^3 of aeration tank capacity. This system is shown in Figure 4. The mixed liquor, displaced to the settling unit, is settled out and pumped to the reaeration unit where it is aerated without further wastewater addition. The organisms thus enter a declining phase and, when finally discharged into the aeration tank, they have the capability of removing large amounts of substrate BOD by the anabolic processes involved in the assimilation and storage of substrate previously discussed. The actual contact time between waste water and mixed liquor is maintained between 30 min and 1 hr by the design of the system. Unlike the step aeration process, in which this contact time can be varied by the operator regulating the point(s) of waste entry, the contact stabilization process does not give extremely great flexibility in load assimilation. It is primarily used in the design of "package" treatment systems as well as in industrial waste applications.

Hatfield process. The Hatfield process, shown in Figure 5, is a system not widely used in the United States. This process differs from contact stabilization in that anaerobic digester supernatant or, in some cases, digested sludge is fed to the reaeration tank. Users of this process feel that in a waste flow containing large amounts of highly carbonaceous industrial material, the supplying of anaerobic digester effluent to the reaeration unit fortifies the active sludge solids with amino acids and other nitrogenous substances. Additionally, there is some thought that the addition of a heavier type of solid to the aeration tank would act to prevent the nonsettleability or bulking of the activated sludge.

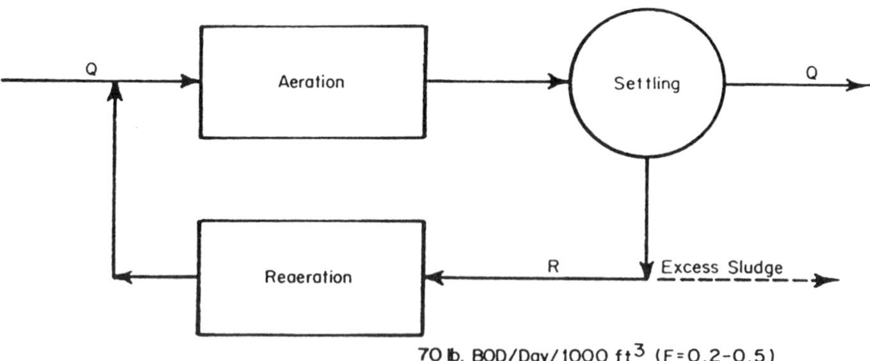

Figure 4 Contact stabilization. Q, R, and F refer to flow through plant, recirculated effluent, and sludge loading ratio, respectively.

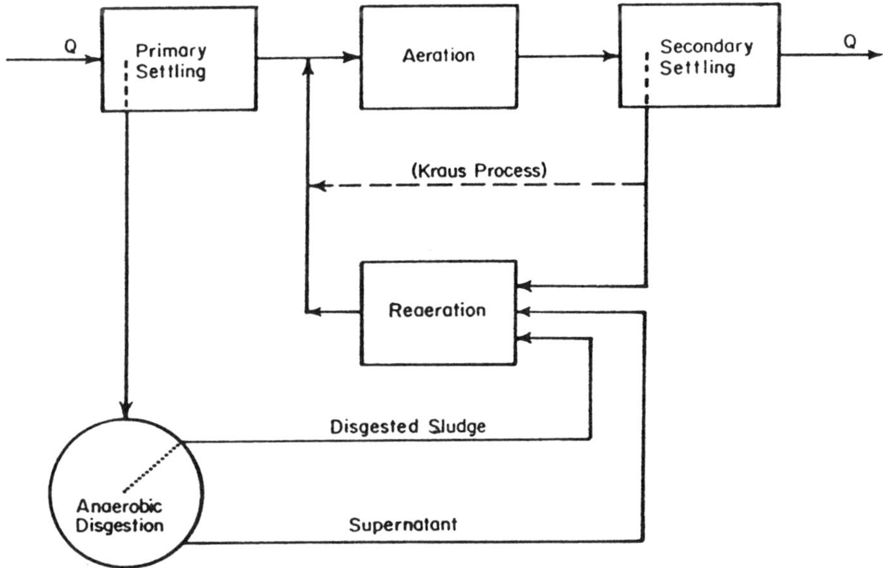

Figure 5 Hatfield process. Q refers to flow through plant.

As in the contact stabilization process, the Hatfield process has the advantage of being able to maintain a large weight of organisms under aeration in a relatively small aeration system.

Kraus process. Also shown in Figure 5 is the Kraus process modification. In this process, some of the return sludge bypasses the reaeration unit and is delivered directly to the mixed liquor aeration tank. In certain respects, the Kraus process is a hybrid between the conventional activated-sludge process and the Hatfield system.

Short-term or modified aeration. The short-term aeration processes have extremely high loading factors varying from about 0.5 up to 5 lb BOD/day/lb MLVSS. Modified aeration has a loading factor range of about 2–5 lb BOD/day/lb MLVSS. The volumetric loading for modified aeration is about 100 lb BOD/day/1000 ft^3 of aeration tank capacity.

The short-term aeration processes offer considerable economy of construction due to the very small aeration tank capacities that are required. However, it will be noted that aeration systems are able to contain only relatively low organism weights and that the effluent quality will suffer accordingly, since the lower the organism weight in a system, the greater will be the amount of unused BOD in the process effluent. However, this effect is not directly proportional to the weight of organisms contained in the system, and effluent quality reduction becomes serious only when the weight of organisms is extremely small, as is the case for the modified aeration and superactivation systems.

Short-term aeration systems produce a relatively large amount of net growth of MLVSS. Sludge disposal becomes a major problem if the sludge is intentionally removed from the system. If sludge is allowed to remain within the system, the BOD removal efficiency would range between 50 and 75%, since excess sludge would pass out in the process effluent. Removing the sludge intentionally, although creating a sludge handling problem, would increase the BOD removal efficiency to as high as 90%.

Extended aeration. The extended aeration process (Fig. 6) is characterized by a low loading spectrum. The objective of this process is to oxidize the biological solids produced by synthesis from the removal of BOD. Extended aeration typically operates at loading factors ranging from about 0.05 to 0.2 lb BOD/day/lb MLVSS, and the volumetric loading is generally about 20 lb BOD/day/1000 ft^3 of aeration tank capacity. One unique feature of the extended aeration process is the system's ability, because of its relatively large aeration tank volume, to contain a relatively large weight of volatile sludge. The low volumetric loading rate, together with the large weight of organisms, combine to give the typically low loading factors.

As usually operated, the extended aeration process has no intentional wasting of sludge from the system because theoretically no excess activated sludge is produced. Practically, however, net growth is produced and a large amount of it is wasted from the system unintentionally in the effluent. This escape of solids results in lowered efficiencies in the range of 75–85%. If sludge is removed intentionally, there are indications that removal efficiencies can be improved to those of conventional activated sludge.

Because of the relatively large concentration of microorganisms carried

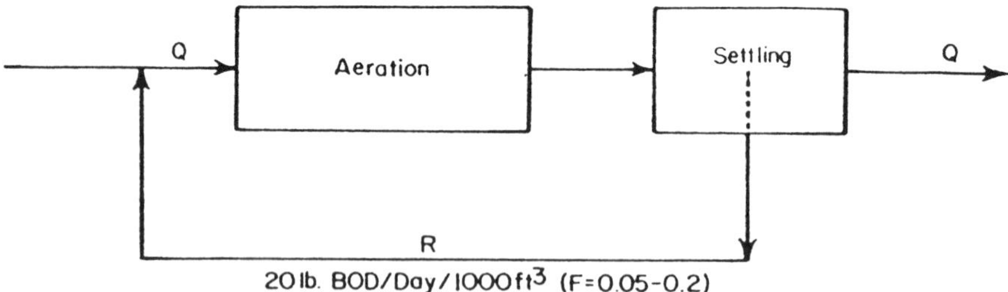

Figure 6 Extended aeration. Q, R, and F refer to flow through plant, recirculated effluent, and sludge loading ratio, respectively.

within the aeration tank and the extended length of aeration time, endogenous respiration plays a major role in sludge quality. The volatile portion of the sludge remaining is not degraded at the same rate as normal activated sludge and thereby results in a lower BOD exertion per unit weight of solids. Effluents from this process, therefore, contain higher suspended solids contents but relatively less BOD than conventional effluents containing equivalent solids concentrations. For this reason, effluents from extended aeration systems often meet regulatory agency requirements for BOD levels but contain unsatisfactory levels of suspended solids.

TRICKLING FILTERS

A second major biological treatment system is the trickling filter. The name is a misnomer since the biological unit neither filters nor trickles. The major difference between the trickling filter and the activated-sludge system is that the trickling filter utilizes a succession of biological communities established at different levels within the trickling filter and associated with correspondingly different degrees of purification. The activated-sludge system, however, has the same biological community within the floc at any one time, which is associated with the raw untreated waste entering the basin and with the purified effluent.

In the trickling filter the interrelationship and activities of the different members of the biological community are similar to those outlined for the activated sludge system, although they are limited to the extent that stratification occurs. In general, there seems to be an underlying consensus of opinion in the field of pollution control stating that trickling filters are more easily adapted or utilized to treat shock loadings of waste source. The concept of shock loadings and their relationship to both trickling filters and activated sludges is discussed later. There are, however, basic physical differences as well as operational differences between the two biological treatment systems.

A trickling filter is a bed of coarse, rough, hard, impervious material over which the sewage is sprayed or otherwise distributed through the air. A biological slime which grows on the filter packing is responsible for the biological reactions. The sewage then flows downward through the filter in contact with the air. The filter is usually 3–12 ft deep and is provided with an underdrainage system to remove the filter effluent and provide ventilation.

The main function of a trickling filter is to remove unstable, organic, pollutional materials in the form of dissolved and finely divided organic solids and to oxidize these solids biologically to form more stable materials. There are many variations in flow pattern for trickling filtration plants. The most common patterns are given in Figures 7a–d. No universally applicable flow system exists, and it is not unusual for one plant to operate on many different patterns during the course of a year.

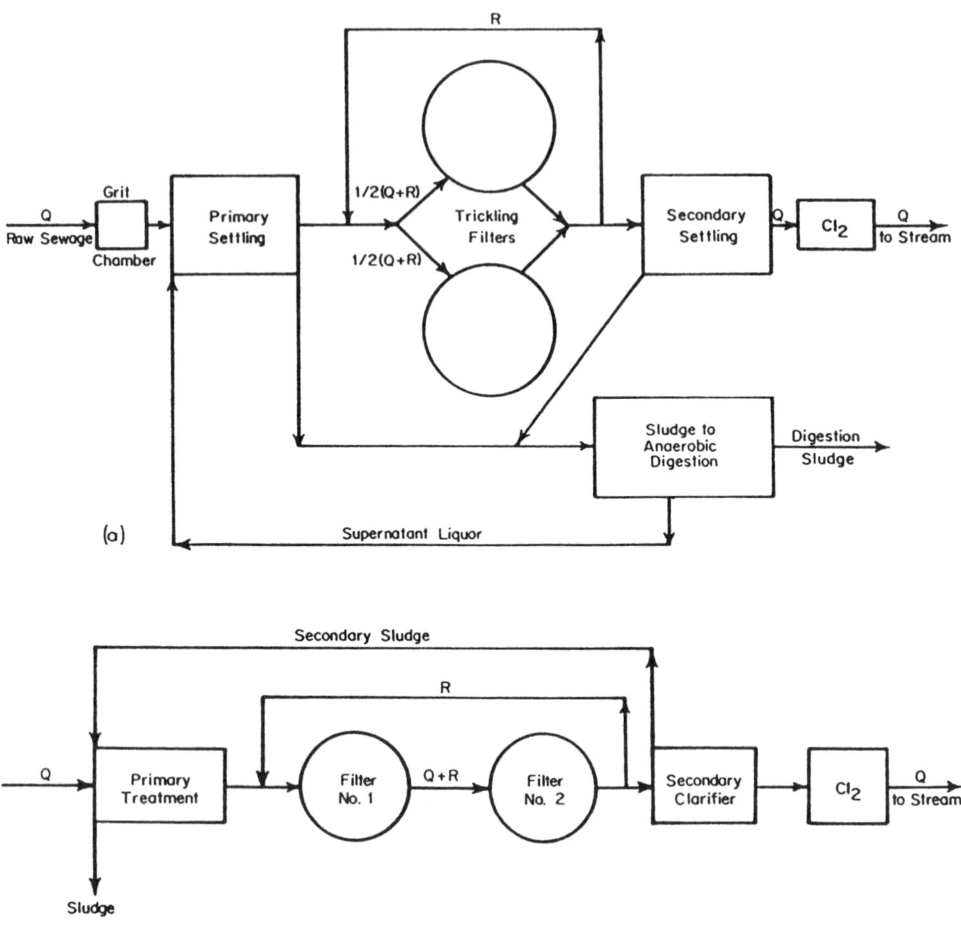

Figure 7 (a) Trickling filters in parallel with direct recirculation. (b) Two-stage plant with direct recirculation to primary stage. (c) Two-stage plant with direct recirculation within each stage. (d) Two-stage plant with recirculation clarification between stages. Q and R refer to flow through plant and recirculated effluent, respectively.

Theory of Operation

The trickling filter process depends upon the biochemical oxidation of complex organic material in sewage. Soon after a filter is placed in operation, the surface of the filter medium becomes covered with zoogleal slime, a viscous, jellylike substance containing bacteria and other biota. Under favorable environmental

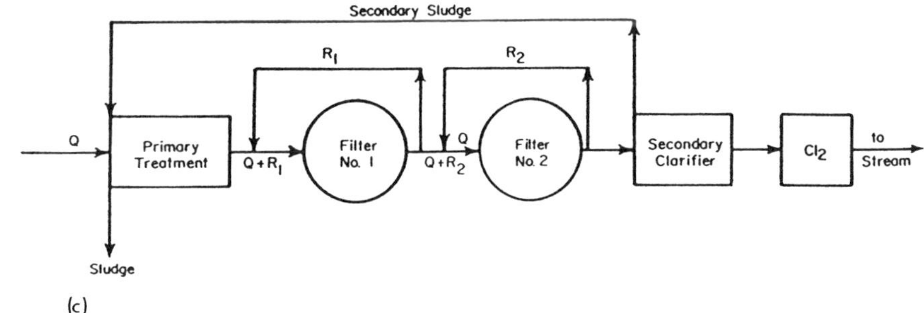

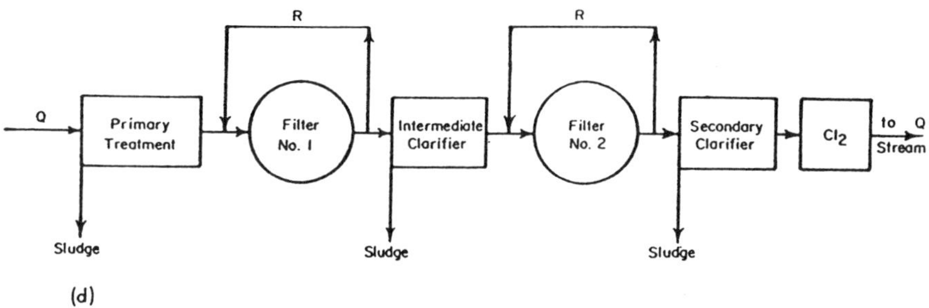

conditions, the slime adsorbs and utilizes suspended, colloidal, and dissolved organic matter from the sewage, which passes in a relatively thin film over the slime's surface. Eventually a population equilibrium is reached. As biota die, they are discharged from the filter together with the more or less partly decomposed organic matter. This "sloughing off" of material may occur periodically or continuously depending on the type of filter. A low- or standard-rate filter sloughs periodically, whereas a high-rate filter sloughs continuously. Generally, secondary settling is provided to retain the settleable solids sloughed from the filter.

The essential features necessary to the process are (1) surface area must be provided for biological growth, and the surface area/volume ratio must be large; (2) free oxygen must be available so that aerobic conditions exist and (3) waste must be amenable to biological treatment; sewage is no problem, but industrial wastes may be.

BOD and Suspended Solids Removal

Trickling filters are capable of providing adequate treatment of wastes susceptible to aerobic biologic processes where the production of a plant effluent of 20–30 mg/L BOD is acceptable (Table 1).

Table 1 Comparison of BOD and Suspended Solids (SS) Removal for Various Processes

Type of plant	Expected BOD removal (%)	Expected SS removal (%)
Primary treatment	15–35	45–65
Primary and trickling filtration treatment	80–90	80–95

BOD removals are affected by

- Quantity and quality of waste: A definite ratio exists between the quantity of a waste applied to a trickling filter and the BOD removal efficiency. In general, the greater the quantity of waste applied, the lower the BOD removal efficiency. In order to obtain a reasonable degree of treatment, the waste should be free from constituents that are toxic to the filter organisms, such as cyanide, copper, chromium, and other heavy metals.
- Temperature: Greater BOD removals should be expected during the summer months because of the increased activity of the microorganisms.

Construction

The shape of a filter is related to the type of distributor used. Most plants constructed since 1935 have rotary distributors and the filters are usually round. Plants utilizing fixed-nozzle distributors usually have rectangular filters. Some filters have been built without retaining walls for the media. However, such construction is seldom economical. The majority of filters have circumferentially reinforced concrete walls, usually 8–12 in. thick.

Choice of the filter medium is often governed by the material locally available and the cost of transporting it. Field stone, gravel, blast furnace slag, redwood blocks, and synthetic inert materials have all been used. The medium should be hard, clean, free of dust, insoluble in sewage, and approximately cubical in shape to obtain a large surface area/volume ratio. Ninety-five percent or more of the medium should pass a 42-in. square screen but be retained on a 2-in. square screen.

The medium must serve two primary purposes: it must provide surface area for slime growth, and it must leave sufficient voids for free circulation of air.

The underdrainage system in the filter serves two primary purposes: to carry sewage effluent passing through the filter away for further treatment, and to provide ventilation and maintenance of aerobic conditions.

The entire underdrainage system should be designed to permit free passage of air, and provisions should be made for flushing out of the lateral channels. The inlet openings should have an unsubmerged gross combined area equal to at least 5% of the surface area of the filter. The floor drainage system, the filter medium, and the water will load if the filter is to be flooded.

Sewage is applied to the filters by either fixed nozzles or rotary arm distributors. Fixed-nozzle filters are usually rectangular in shape and utilize deflector plates along with the orifices, as shown in Figure 8.

Rotary distributors are the most commonly employed. The arms are rotated by the reaction of the sewage discharge or, in some cases, are motor driven. The arms are usually set 6 in. above the filter bed.

Loadings

Trickling filters are classified according to applied hydraulic and organic loadings. The hydraulic load is the total volume of sewage (including recirculation) applied to the filter per day per square feet of surface area (gal/day [gpd] per ft^2 of million gpd/acre). The organic load is the pounds of 5-day BOD applied to the filter per day per cubic foot of filter medium (lb BOD/day/ft^3 medium) or (lb/day/1000 ft^3). Table 2 shows a comparison between standard-rate and high-rate filters.

Some filters are called "roughing filters." These are usually high-rate filters receiving high organic loadings. Although they may give a high pound per unit volume of organic load removal, their settled effluent still contains substantial

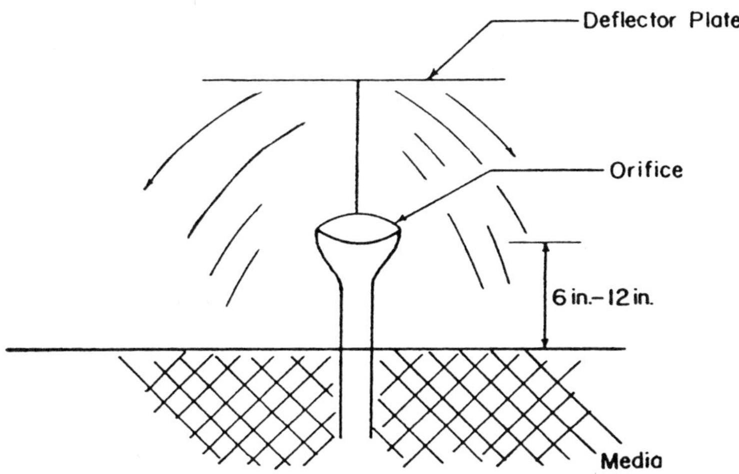

Figure 8 Fixed-nozzle distributor.

Table 2 Classification of Trickling Filters

Type of filter	Hydraulic load (gpd/ft^2)	Organic load (1b BOD/day/1000 ft^3)	Application	Sloughing	BOD removal through filter (%)
Standard rate	25–100	5–25	Intermittent	Largely periodic	80–85
High rate	200–1000	25–300	Continuous	Continuous	65–80
High rate	200–1000	25–300	Continuous	Continuous	65–80

BOD. They are used to provide intermediate treatment or as the first of a multistage biological treatment.

Most high-rate filters utilize recirculation, which is the recycling of filter effluent through the filter. In such cases, the ratio of recycled flow to sewage flow is known as the recirculation ratio. Among the useful purposes of recirculation are

- Reducing "out-of-service" periods (due to shock loadings)
- Minimizing by adjusting recirculation to influent flow
- Keeping self-propelled distributors turning by adjusting recirculation to influent flow
- Lowering film thickness and fly breeding by film sloughing
- Improving the quality of the effluent at constant efficiency of treatment by reducing the concentration of applied sewage

Biofilter

The biofilter (Fig. 9) employs recirculation and a high rate of application to a shallow trickling filter. The recirculation in this case involves bringing the effluent of the filter or the secondary settling tank back to the primary sedimentation basin. This requires designing the primary settling tank to

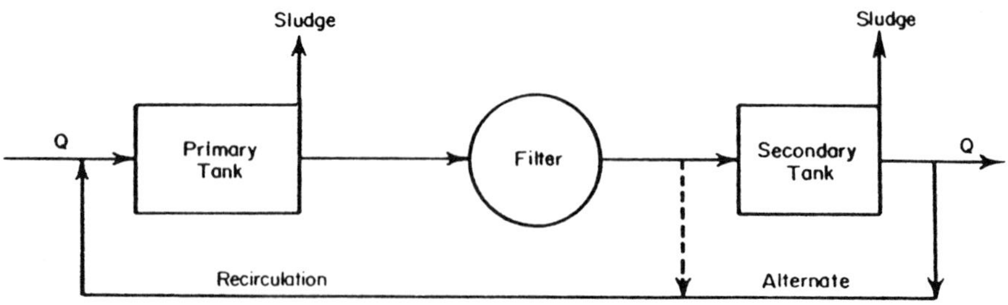

Figure 9 Biofilter. Q refers to flow through plant.

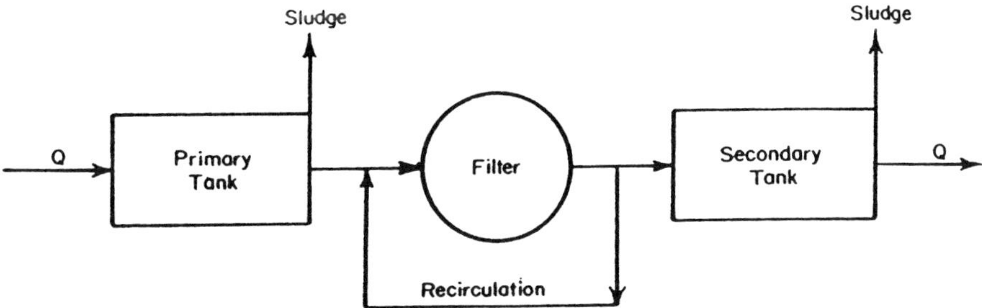

Figure 10 Accelofilter. Q refers to flow through plant.

accommodate not only the average daily sewage flow but also the recirculated flow, which may be as much as 10 times the average flow.

Accelofilter

This filter (Fig. 10) employs recirculation of the filter effluent directly back to the filter.

Aerofilter

The aerofilter (Fig. 11) is a filter over which the sewage is distributed by maintaining a continuous rainlike application of the sewage over the filter bed. A disk distributor revolving at speeds as high as 380 rpm and set 20 in. above the surface of the filter is sometimes used on small beds.

For large beds 10 or more, revolving distributor arms are used to obtain the uniform rainlike application. Aerofilters operate at hydraulic loadings greater than 300 gpd/ft^2 of surface area.

For certain industrial applications, sectional filters utilizing controlled quantities of forced air within each section can be utilized. Many states now

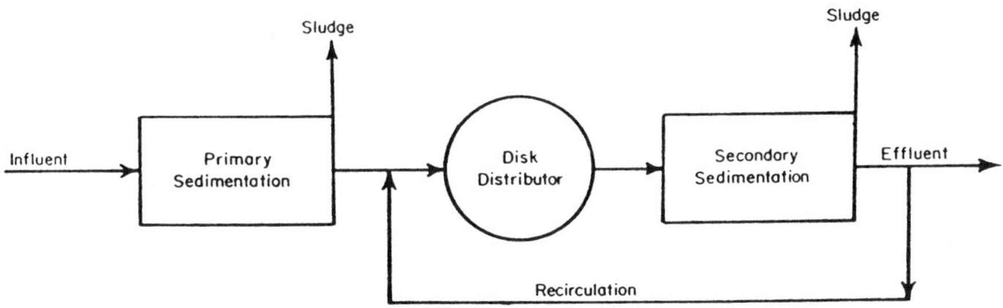

Figure 11 Aerofilter.

make provisions for the use of synthetic media which allows greater flexibility in construction. These filter are often referred to as "tower" filters and can receive greater organic loads at relatively high rates of removal efficiency.

Among the advantages of a trickling filter include

- Relatively high nitrifying effect; i.e., biological oxidation of amino nitrogen to nitrates
- Dependability to give a good effluent under wide variations in influent quality
- Relatively low operating and maintenance cost
- Ability to function under extreme weather conditions

Disadvantages include

- High head loss through the filter
- Odor and fly nuisance (low rate filters)
- Large surface area required
- Relatively high construction costs

Shock Loadings

An important consideration in the design of biological sewage treatment plants in the ability of the system to handle immediate or rapidly occurring changes in the chemical or physical environment of the biological population. Such changes may be broadly categorized as shock loads; the effect of which may vary from slight malfunctioning of the system to a complete cessation of metabolic activity resulting in the shutdown of the system.

Shock loading of systems may take any of several forms. The most common type is the quantitative shock, which involves a change in the concentration of the waste or the organic loading of the system. A rapid decrease in organic loading, termed a hydraulic shock, might result from an influx of storm waters which dilute the influent waste. In this same category is the rapid increase in loading which might result from the removal of blockage in the sewer system.

The chemical nature and structure of the waste are unchanged from those normally handled by the system, and the existing population requires no acclimation period for this type of shock.

A toxic shock may be the result of materials which damage or inhibit the existing metabolic pathways or disrupt the physiological condition of the microbial population. The occurrence of heavy metals, cyanide compounds, or materials producing rapid pH changes would be considered toxic shocks. The treatability of such a waste depends on a prolonged acclimation period and even then adaptation of the microorganisms to the new substrate is not guaranteed.

Qualitative shock loads are associated with a change in the structural configuration of the carbon source and do not of themselves entail any change in the organic loading of the system. This type of shock could occur when a new waste is introduced into the system, such as cannery or meat-packing wastes.

Trickling filters and the activated sludge process are the most commonly encountered biological treatment systems. Since quantitative shock loads involve an increase in the hydraulic or organic loading of the system, the apparent method of treatment would be to increase the capacity of the plant by having additional sedimentation tanks, filters, or aeration tanks on a standby basis. It is immediately apparent that this is an extremely uneconomical approach to the problem.

Inherently, a high-rate trickling filter has the potential to cope with quantitative shocks, since it employs recirculation to maintain the hydraulic loading on the system. Recirculation may be increased to dilute high organic loads or decreased to accommodate hydraulic shocks. By these adjustments the trickling filter effluent can be maintained at a fairly stable level irrespective of large variations in the influent.

The activated-sludge process does not provide sufficient latitude in operation to handle large quantitative shocks. Sludge bulking and inadequate treatment times caused by the shock result in the release of an unsatisfactory effluent. This inability of the activated sludge process to cope with shock loads has led to numerous modifications of the system in an attempt to alleviate this problem.

The step aeration system provides for the introduction of the primary effluent at several points along the course of the mixed liquor. All of the return sludge is introduced at the head of the plant. By varying the feed to the various passes, the BOD/suspended solids ratio may be completely controlled, thus evening out the oxygen demand in the mixed liquor. Step aeration provides great flexibility in the handling of quantitative shocks but BOD removal is sacrificed.

Other variation of the activated-sludge process is commonly known by the names sludge reaeration, contact stabilization, or biosorption (Fig. 12). In separating the adsorption and oxidation phases, double the volumetric loading of step aeration is facilitated. This ability to handle larger loadings than the standard process is the property which provides for adequate treatment of shock loads, but again, it is at the expense of BOD removal. It should also be noted that this system is not as flexible as step aeration.

A somewhat different approach to the problem of quantitative shock loading has been the use of multistage biological treatment. To this end, systems have been defined with high-rate biological filters in parallel with the activated-sludge process or trickling filters in series with the activated-sludge process. Systems such as these take advantage of the trickling filter's ability to handle shock loads and the ability of activated sludge to produce an effluent of high quality and in this fashion eliminate some of the shortcomings of each.

The activated-sludge aeration system is a combined approach to shock loading of a quantitative type. It employs a step aeration system run in parallel

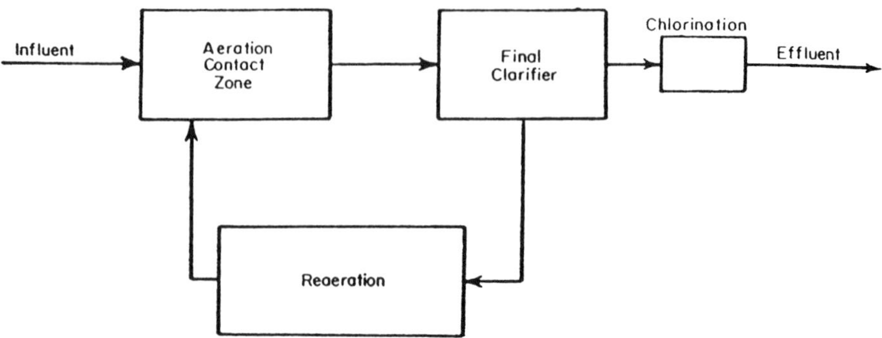

Figure 12 Sludge reaeration or biosorption.

with a conventional process, with the step aeration system serving as a source of return sludge for both legs of the circuit. This combination provides the flexibility of step areation and high-quality effluent.

Quantitative shock loads may be successfully handled, in general, by processes designed for increased volumetric and/or organic loading without provision for duplication of system components. This ability has been provided by the use of recirculation, separation of the adsorption and oxidation phases of treatment, or by multistage biological treatment, in general, at the expense of BOD removal.

The treatment of qualitative or toxic shock loads poses very different problems, since an adaptation phase is required for the biological population to adjust to the new substrate. The utility of trickling filters is open to question when nonquantitative shocks are considered. The utility of trickling filters is also questioned on the grounds that it is more time consuming and costly to clean filter material and renew biological mass in the activated sludge process. Since the successful response of a system to a qualitative shock must be initiated and completed during the time of substrate-biological solids contact, activated sludge is better than a trickling filter from a biochemical standpoint. The treatability of a toxic shock depends on a prolonged acclimation period which is very much greater than the delay time in the system, and hence neither method can adequately treat such a shock.

The sludge reaeration or biosorption system as shown in Figure 12 recovers from wide variations in pH of short duration in a short period of time and that long-term shocks (36 hr) require recovery times of 5 days. This is the system's ability to resist a toxic shock and has nothing to do with its ability to treat it. Similarly, it has been found that biological treatment systems can be acclimated to specific toxic shocks not exceeding some maximum concentration, but this is for industrial waste treatment such as the removal of cyanide and chromium from an acid medium. The best way to handle a toxic shock is prevention at the source before the material enters the biological treatment system.

TRICKLING FILTER/ACTIVATED SLUDGE COMBINED

The combination trickling filter and activated-sludge process is similar to the trickling process described except an aeration tank is added. The primary effluent passes into the aeration tank where a portion of the aeration tank contents are continually recirculated through the trickling filter. Solids from the secondary clarifier are returned to the primary holding tank. Some of the trickling filter effluent can also go to the aeration tank and the rest is recycled through the filter. Some solids from the secondary clarifier can be returned to the trickling filter unit.

BIO-DISC SYSTEM

The bio-disc system consists of a series of flat, parallel discs which are rotated while partially immersed in the waste being treated. Biological slime covers the surface of the discs and adsorbs and absorbs colloidal and dissolved organic matter present in the wastewater. Excess slime generated by synthesis of the waste materials is sloughed off gradually as necessary for aerobic biological activity of the slime. Disc rotation also provides contact between the slime and the wastewater. The rotating discs provide

- Mechanical support for a captive, microbial population
- A mechanism of aeration, the rate of which can be adjusted by changing the rotational speed
- Contact between the biological slime and the waste water, the intensity of which can be varied by changing the rotational speed

The use of closely spaced parallel discs achieves a high concentration of active biological surface area. Secondary treatment of domestic sewage is accomplished with a retention time of 1 hr or less, and 90% BOD reduction is obtained at a loading of 5 lb BOD/day/1000 ft.

The large microbial population achieves a high degree of treatment in a relatively short retention time. The turning action of the rotating contractors carries the media with attached organisms into the air where oxygen and organic materials from the wastewater. Shearing forces exerted on the biomass by the rotating banks of discs strip excess material from the media helping to maintain a uniform biological film. Mixing action within the rotating biological contactor (RBC) tank keeps the stripped solids in suspension until the flow of treated wastewater carries them out of the final pair of clarifiers.

The high turbulence and effective aeration of this process have produced

͜val rates two or three times higher than normally achieved with static fixed ˌılm systems such as trickling filters. This process is well suited to cold-weather operations. Variable speed is obtained through control of the air feed to the diffusers. Compared with an activated sludge system, this process is much easier to operate, monitor, and maintain and occupies less space.

Rotating biological contractors generally are more reliable than other fixed film processes because of the large amount of biological mass present (low operating F/M). This also permits them to withstand hydraulic and organic surges more effectively. The effect of staging in this plug-flow system eliminates short circuiting and damps shock loadings.

Rotating biological contacts are relatively new to secondary treatment in the United States, but many municipal and industrial facilities have installed these systems.

A checklist that summarizes the factors that should be considered in the design, construction, and operation of the facilities using the RBC process follows. Besides the introduction of corrugated media, other developments in the RBC process have included

- Less expensive, flat-bottom concrete basins, including removable baffles to permit optimum staging
- Fiberglass covers to limit wastewater hear loss inexpensively during winter operation, whereas preventing direct exposure of the plastic media to sunlight
- Increased surface density media to obtain low load application and nitrification
- Process application of RBCs to achieve BOD removal, nitrification, and denitrification efficiently in a continuous process train
- Innovative application in existing primary treatment plants to accomplish low-cost upgrading to secondary treatment levels
- Upgrading of existing activated sludge plants by combined use of RBCs and activated sludge on existing aeration basins
- Use of supplemental aeration, employed at the tank bottom, to achieve improved performance and better control of biofilm development and of nuisance organisms
- Air-driven RBCs which utilize air for both process improvement and control, whereas also efficiently rotating the equipment

PHYSICAL/CHEMICAL TREATMENT

Physical/chemical treatment consists of the following series: comminutor aerated flow equalization tank, flash mixer, flocculation tank, tube settlers (clarifiers with many small vertical tubes at the lower depths to enhance settling), filter, and optional chlorination.

Coagulants and powdered activated carbon are added to the sewage at the flash mixer to aid in the removal of suspended solids, BOD, and phosphates. Most of the removal occurs in the tube settlers where the precipitates, activated carbon, and flocculated solids are separated from the liquid. The resultant sludge is removed to the sludge holding tank. The effluent from the tube settlers can be chlorinated before it passes through the filter to prevent slime growths on the filter media. The filter removes most of the remaining suspended solids. Additional chlorine can be applied to the filter effluent for disinfection purposes. When the filter is backwashed, the waste backwash water goes to a holding tank where the solids settle and are pumped to a sludge holding tank. The supernatant goes to the head of the process. The sludge in the holding tank must be pumped out periodically.

Advantages of physical-chemical treatment over normal biological treatment are

- High suspended solids and BOD removals can be achieved (90–95% BOD and suspended solids).
- Phosphates can be removed (up to 90%).
- The process tolerates cold temperatures, toxic materials, and intermittent flows.
- It requires less space.

The disadvantages are that capital and operation and maintenance costs are significantly greater, and the process produces more waste sludge.

BIOLOGICAL/CHEMICAL TREATMENT

A combination of biological and physical-chemical treatment consists of the following: comminutor, aeration tank, clarifier, flash mixer, flocculator, tube settlers, filter, and chlorination. Also included are a sludge-holding tank and waste backwash waster-holding tank. The operation of the physical-chemical treatment scheme alternative; that is, coagulants and powdered activated carbon can be added, for example, at the flash mixer. Removals from this combination plant are 98% for suspended solids and 95% for BOD. Phosphates are significantly reduced. If nitrification is desired, an additional or second aeration tank and clarifier can be added. The combination biological and physical-chemical treatment process is expensive and should only be used where the strictest effluent quality requirements are needed.

BIOCHEMICAL OXYGEN DEMAND

Biochemical oxygen demand (BOD) is used as a measure of the presence in aqueous solution of organic materials which can support the growth of microorganisms. By simple definitions the interest is in oxygen utilization by microorganisms. BOD refers primarily to the results of a "standard" laboratory procedure. The procedure has been developed on the premise that biological reactions of interest can be "bottled" in the laboratory and that kinetics and extent of reaction can be observed empirically. Existing standard laboratory procedures for observation of BOD are not definable either as analytical or as empirical bioassay procedures.

Measurement of Water Quality

There is a relationship between microorganisms, organic materials, and utilization of oxygen in the waters of streams and lakes. Measurement of dissolved oxygen concentrations is not difficult, but measurement of the concentration of organic materials which are biologically stabilized is somewhat complex. The only approach applied to any significant extent is the measurement of oxygen utilization over extended periods, with correlation of the results to materials which existed at the start of the time period. This technique provides information as a historical array displaying past water quality. This examination record of information is useful, but as direct input to water quality control, it is questionable.

Two approaches to the development of BOD information free of the time delay disadvantages are development of a BOD technique which would provide information in a few minutes instead of 5 or more days or capability for quick identification and analysis of all the specific organic compounds present and calculation of biochemical oxidation results from available BOD-organic compound correlations.

BOD as Pollution Parameter

Measurements of quantities of materials discharged to surface water from a source are required for detection and control of pollution. The value of BOD is that it reflects the presence of materials which will be oxidized biologically in the receiving water. The accuracy is questionable.

BOD as measured by "standard methods" is a meaningful measurement within itself, but its usefulness as a parameter is impaired by misinterpretation, misuse, and misunderstanding which begins when the closed system in the BOD bottle is prepared. The assumption is that the closed system is parametrically related to the system from which the sample was obtained.

Another second assumption is that consumption of dissolved oxygen is an absolute and complete parameter of biological reactions occurring in the BOD bottle. This is reasonably accurate if the overall picture of oxygen consumption is developed but can be highly inaccurate when only the accumulated oxygen consumption for 5 days is observed. The accuracy may or may not be improved by daily observations of oxygen consumption for 5 to 7 days and projection of observations to an ultimate value by the fit of a first-order reaction curve. The oxygen consumptions which we observe are meaningfully related to reactions in the BOD bottle through the incubation period and at the time when the incubation is terminated. The results may not be readily correctable with conditions in the actual biological system being investigated.

Definition of water quality and the implementation of corrective measures are related to oxygen consumption through biological oxidation reactions. A form of BOD measurement must be utilized as the basis for decisions and action. The "standard" BOD measurement is not a truly quantitative entity in terms of water quality because the "analytical" technique as normally applied does not qualify as a quantitative measurement, and translation of the "analytical" information to the receiving waters is not always a valid interpretation.

Theory

The basic difference between the BOD bottle and a stream is that the BOD bottle is a static system which is controlled to provide aerobic conditions, whereas streams are a dynamic system which may include anaerobic areas. The BOD bottle is a batch operation in which the following operations take place. A portion of the organic (and inorganic) material in the system is oxidized in chemical reactions: organic carbon to carbon dioxide; organic hydrogen to water; sulfide and organic sulfur to sulfates; organic nitrogen and ammonia to nitrites or nitrates, referred to as nitrification; and other reactions. The purpose of the oxidation reactions is to provide energy, and the oxygen consumed is referred to as energy oxygen. Energy oxygen is defined as that amount of oxygen required in energy reactions which support the synthesis of organic materials into new biological cells or biologically stable organic materials. Energy oxygen will be a fraction of that required for complete biological stabilization. For example, the energy oxygen for sucrose is approximately 10% of the total requirement. Most of the energy produced does not appear as heat, but approximately 60–70% is utilized in coupled reactions which synthesize the remaining organic material into new microorganisms or convert it into stable compounds. The synthesis reactions generally do not consume oxygen. Thus, at the point where the original organic material no longer exists, oxygen consumption corresponds to energy oxygen and is equivalent to 10–40% of the chemical oxygen demand (COD) of the original organic material. Elapsed incubation time is in the order of 1–2 days. When related to organic materials, COD refers to the oxidation of carbon and hydrogen to carbon dioxide and water. Ammonia or organic nitrogen is not oxidized. As the synthesized microorganisms age, lysis of the cell wall occurs, and the cell contents are released. The remains of the cell wall are relatively stable materials, such as polysaccharides, which are not readily susceptible to further stabilization by biological reactions. Among the contents of the cells are biodegradable materials which are utilized in energy, synthesis, and system maintenance reactions. Therefore, a cyclic degradation of the biological mass progresses through lysis-synthesis-lysis (endogenous reactions). The organic materials in the BOD bottle are eventually altered to the point where the materials remaining are chiefly the stable endproducts of the cell wall. Oxygen consumption at this point is normally calculated to be the long-term or first-stage carbonaceous BOD and is in the range of 60–80% of the COD of the original organic material because of the remaining stable materials from the cell wall. Elapsed time is generally 10–20 days. During the period of cell stabilization, degradation of amino acids and other nitrogenous compounds will release ammonia. The ammonia may or may not be oxidized to nitrates during this period through the action of nitrifying bacteria (nitrification). If activity in the BOD bottle is allowed to continue for an extended period of time (beyond 100 days), the ultimate BOD development will represent essentially total oxidation of all the organic carbon, nitrogen, and hydrogen originally introduced into the system.

BOD theory has been based on the assumption that BOD development in the BOD bottle is a first-order progression and that the kinetics can be represented by a first-order rate constant and an ultimate BOD value. This assumption has not been substantiated, and as a result the BOD determination has not factually filled the quantitative need. Application of the technique has been oversimplified by utilization of a fixed incubation period.

In order to establish the use of a BOD procedure quantitatively, it is necessary for theoretical considerations to get into step with the observed facts, which are energy oxygen, synthesis, and endogenous reactions.

Surface Stream

Since prime interest is related to biological oxidation and to the resultant consumption of oxygen in rivers and lakes, a look at occurrences in surface waters is necessary. Development of the oxygen demand along a flowing stream below the point of an organic discharge depends to a great extent upon the physical configuration of the stream.

Deep Body of Water

In a relatively deep stream, the initial biological activity may be homogeneously dispersed if the organic materials in the discharge are soluble or colloidal. As microorganisms are synthesized, some flocculation may occur, and the flocculated microorganisms and other material may settle to form a bottom deposit where anaerobic stabilization takes place. Since the situation in the stream is dynamic in that the organic discharge, aerobic synthesis, and sludge deposition are continuous, the deposition of sludge can become significant.

The rate at which biological synthesis occurs is a function of the food to organism ratio. Therefore, the rate of stabilization of organic material discharged to the stream is very much related to upstream activity. If an established population of organisms already exists in the stream, an accelerated rate of synthesis and oxygen consumption can be anticipated. A BOD bottle is conveniently related to occurrences in the stream, and a BOD bottle does not involve anaerobic stabilization, it does not involve the same degrees of agitation, it does not involve the upstream input of organic materials and organisms, and it is static rather than dynamic.

Shallow Body of Water

Physical geometry of a shallow stream provides excellent food to organism potential. Shallow configurations provides a large bottom surface area/stream volume ratio, which is enhanced if the stream bottom is rocky. Additionally, a shallow stream normally has a high velocity, which provides turbulence in the flowing water. These characteristics of the shallow stream provide a high rate of stream reaeration and a suitable environment for growth of attached organisms. Immediately below an organic discharge attached organisms can grow. In effect, the stream becomes a trickling filter, and the rate of BOD removal as related to time of streamflow is accelerated compared with a deep stream. In the zone of the stream where the attached growths are functioning, oxygen demand may not correlate with BOD removal in the same manner as for a deep stream, but it may be more related to energy oxygen. Attached growths may grow to a size where they become detached and are carried downstream to be deposited in a quiescent zone. Anaerobic stabilization in the sludge deposit will reduce the quantity of material which will eventually be aerobically stabilized, and the net result is that oxygen required for stabilization of organic material can be less than the BOD of the initial quantity of organic material discharged to the stream.

BOD results obtained in a BOD bottle have a limited quantitative correlation with the oxygen demands which appear in a shallow stream. The statement is particularly true at this point in time because a definitive and fundamental evaluation has not been made of biological kinetics in a shallow stream.

Standard BOD Methods

Standard methods for the BOD determinations are basically similar and are directed toward the following:

- A standardized laboratory technique, covering dilution procedure, incubation condition, and calculation of results
- A standardized dilution water that does not contain toxic materials, is essentially free of oxidizable materials, and contains an adequate level of nutrients
- A viable culture of seed organisms that is capable of stabilizing the organic materials involved
- Acclimation of the biological culture, which may or may not be necessary

Once the procedure conforms to these elements of standardization, the question of how to interpret the observed oxygen consumptions remains.

The BOD curve of glucose as observed in three dilutions is shown in Figure 13. Oxygen consumption progressed in approximately 8 days to a point where synthesis and endogenous reactions had reduced biological cells and other materials in the system to relatively stable organic materials. Ammonia was not oxidized to nitrates.

The ultimate carbonaceous BOD is defined as oxygen required for complete conversion of organic carbon and hydrogen to carbon dioxide and water. This is not the same ultimate BOD that is predicted by fitting a first-order curve through the BOD data points. When successful, the ultimate value of the

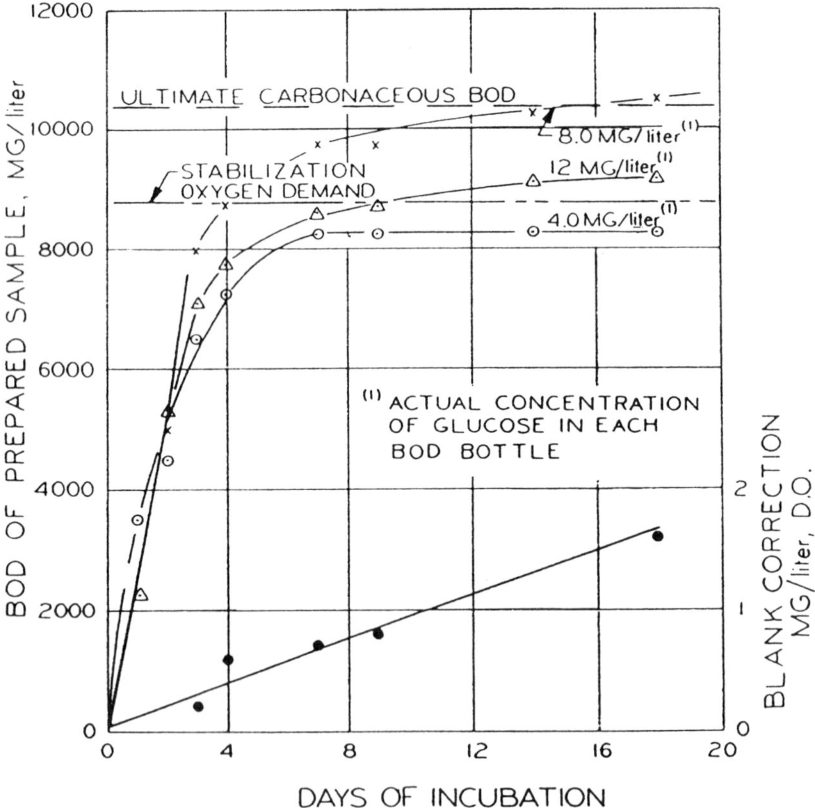

Figure 13 Biochemical oxygen demand, glucose, 10,000 mg/L prepared sample.

first-order projection is equivalent to the oxygen demand when relatively stable endproducts of cell walls of microorganisms remain.

The ultimate carbonaceous BOD for a pure compound can be arrived in two ways. For pure compounds, such as glucose, the maximum oxygen demand when all organic carbon and hydrogen are converted to carbon dioxide and water corresponds to the chemical oxygen demand. A second method, using glucose as an example, follows the biological reactions in the BOD bottle. Energy oxygen required in the synthesis of microorganisms from glucose is 0.34 g of oxygen per gram of glucose. Based on theoretical COD, the energy oxygen for glucose corresponds to oxidation of 32% of the material present and conversion of 68% into cell material or stable organic compounds.

The approximate molecular formula for cell material is $C_5H_7N_2O$, and the experimentally determined COD is approximately 1.42 g/g of cell material. On the basis of 1 g of glucose, the energy oxygen is 0.34 g which oxidizes 0.32 g of glucose. The remaining 0.68 g of glucose is synthesized into 0.50 g of cell material based on a carbon balance. Total oxidation of the cell material requires 0.72 g of oxygen, and total oxygen consumption is 1.06 g/g of glucose, which is essentially equivalent to the COD and the ultimate carbonaceous oxygen demand of glucose.

Calculation of oxygen requirements can be adjusted to show oxygen requirements for any particular level of stabilization of cell material. Degradation of cell material to stable end products represents approximately 75% reduction in organic material. For glucose, the BOD in the bottle should approach energy oxygen of 0.34 plus 0.75 × 0.72 for oxidation of cell material, or 0.88 g of oxygen per gram of glucose. The corresponding oxygen consumption as shown in Figure 9 as stabilization oxygen demand. The term stabilization is appropriate, since the cell material has been reduced to relatively stable endproducts. Progression of oxygen demand from stabilization to ultimate is a slow process.

OPERATING PROCESS SYSTEMS

Aerated Lagoons

Aerated lagoons are medium-depth basins designed for continuous biological treatment of wastewater. In contrast to stabilization ponds, which obtain oxygen from photosynthesis and surface reaeration, these employ aeration devices which supply supplemental oxygen to the system. The aeration devices may be mechanical such as surface aerator or diffused air systems. Surface aerators are divided into two types: cage aerators and the more widely used turbine and vertical shaft aerators. Diffused-air systems utilized in lagoons consist of plastic pipes supported near the bottom of the cells with regularly spaced sparger holes drilled in the tops of the pipes. Because aerated lagoons are normally designed to achieve partial mixing, aerobic-anaerobic stratification will occur, and a large fraction of the incoming solids and a large fraction of the biological solids produced from waste conversion settle to the bottom of the lagoon cells. As solids begin to build up, a portion will undergo anaerobic decomposition. Volatile toxics can potentially be removed by the aeration process, and incidental removal of other toxics can be expected to be similar to an activated sludge system. Several smaller aerated lagoon cells in series an have been used rather than one large cell. Tapering aeration intensity downward in the direction of flow promotes settling out of solids in the last cell. A nonaerated polishing cell following the last aerated cell is an optional, but recommended, design technique to enhance suspended solids removal prior to discharge.

Lagoons may be lined with concrete or an impervious flexible lining depending on soil conditions and environmental regulations. The use of various types of aeration are applicable. When high-intensity aeration produces completely mixed (all aerobic) conditions, a final settling tank is required. Solids are recycled to maintain about 800 mg/L MLVSS in this mode. Applications include domestic and industrial wastewater of low and medium strength, which are commonly used where land is inexpensive and costs and operational control are to be minimized. It is relatively simple to upgrade existing oxidation ponds, lagoons, and natural bodies of water to this type of treatment. Aeration increases the oxidation capacity of the pond and is useful in overloaded ponds and generate odors. This is useful when supplemental oxygen requirements are high or when the requirements are either seasonal or intermittent.

Limitations are in very cold climates aerated lagoons may experience reduced biological activity and treatment efficiency and the formation of ice.

Settled solids on pond bottom may require clean-out every 10–20 years, or possibly more often if a polishing pond is used behind the aerated pond. Design criteria include

Operation: One or more aerated cells, followed by a settling (unaerated) cell
Detention time: 3–10 days
Depth: 6–20 ft
pH: 6.5–8.0
Water temperature range: 0–40°C
Optimum water temperature: 20°C
Oxygen requirement: 0.7–1.4 times the amount of BOD_5 removed
Organic loading: 10–300 lb BOD_5/acre/day

The service life of a lagoon is estimated at 30 years or more. The reliability of equipment and the process is high and little operator expertise is required. Environmental impacts include volatile organic material and pathogens in aerated lagoons to enter air as with any aerated wastewater treatment process. This depends on air/water contact afforded by the aeration system. There is potential for seepage of wastewater into ground water unless a lagoon is lined. Compared with other secondary treatment processes, aerated lagoons generate less solid residue. Volatile toxics will be removed, and incidental removal of other toxics can be expected to be similar to an activated sludge system.

Anaerobic Lagoons

Anaerobic lagoons are relatively deep (up to 20 ft) ponds with steep sidewalls in which anaerobic conditions exist by keeping loading so high that complete deoxygenation is prevalent. Although some oxygenation is possible in a shallow surface zone, once greases form an impervious surface layer, complete anaerobic conditions develop. Treatment or stabilization results from thermophilic anaerobic digestion of organic wastes. The treatment process is analogous to that occurring in single-stage untreated anaerobic digestion of sludge in which acid-forming bacteria break down organics. Resultant acids are then converted to carbon dioxide, methane, cells, and other endproducts.

In the typical anaerobic lagoon, raw wastewater enters near the bottom of the pond (often at the center) and mixes with the active microbial mass in the sludge blanket, which is usually about 6 ft deep. The discharge is located near one of the sides of the pond, and submerged below the liquid surface. Excess undigested grease floats to the top forming a heat-retaining and relatively airtight cover. Wastewater flow equalization and heating are generally not practiced. Excess sludge is washed out with the effluent. Recirculation of waste sludge is not required. Anaerobic lagoons are capable of providing treatment of high-strength wastewaters and are resistant to shock loads.

Anaerobic lagoons are customarily contained within earthen dikes. Depending on soil characteristics, lining with various impervious materials such as rubber, plastic, or clay may be necessary. Pond geometry may vary, but surface area to volume ratios led to enhance heat retention. Anaerobic processes are common for sludge digestion. Anaerobic lagoons for wastewater treatment have found only limited application. The process is well established for stabilization of highly concentrated organic wastes.

Typically they are used in series with aerobic or facultative lagoons. Anaerobic lagoons are effective as roughing units prior to aerobic treatment of high-strength wastes.

Among the disadvantages of anaerobic lagoons are the requirement of relatively large land areas and the creation of unpleasant odors. For more efficient operation, water temperatures should be maintained above 75°F. There is also potential for seepage of wastewater into groundwater unless the lagoon is lined. BOD_5 removals of 50–70% are achievable depending on loading and temperature conditions. Total SS concentrations may increase, especially if the influent BOD_5 is primarily dissolved. Generally this procedure does not produce an effluent suitable for direct discharge to receiving waters; excess sludge is usually washed out in the effluent. Since anaerobic lagoons are often used for preliminary treatment, recirculation or removal of sludge is not generally required.

Nutrients are added as needed to make up deficiencies in raw wastewater; no other chemicals are required.

Typical operating conditions include

Operation: Parallel or series
Detention time: 20–50 days
Depth: 8–20 ft
pH: 6.8–7.2
Water temperature range: 35–120°F
Optimum Water Temperature: 86°F
Organic loading: 200–2200 lb BOD_5/acre/day

An anaerobic lagoon is valuable as a preliminary treatment process for combined industrial and municipal wastes containing high concentrations of organic materials. It can be used preceding most standard biological treatment processes. It is generally resistant to upsets. An anaerobic lagoon is highly reliable if pH in the relatively narrow optimum range is maintained.

Typical schematic for an anaerobic lagoon is shown in Figure 14.

Facultative Lagoons

Facultative lagoons are intermediate-depth (3–8 feet) ponds in which the wastewater is stratified into three zones. Zones consist of an anaerobic bottom layer, an aerobic surface layer, and an intermediate zone. Stratification is a result of solids settling and temperature-water density variations. Oxygen in the surface stabilization zone is provided by reaeration and photosynthesis. This is in contrast to aerated lagoons in which mechanical aeration is used to create aerobic surface conditions. In general, the aerobic surface layer serves to reduce odors while providing treatment of soluble organic by-products of the anaerobic processes operating at the bottom.

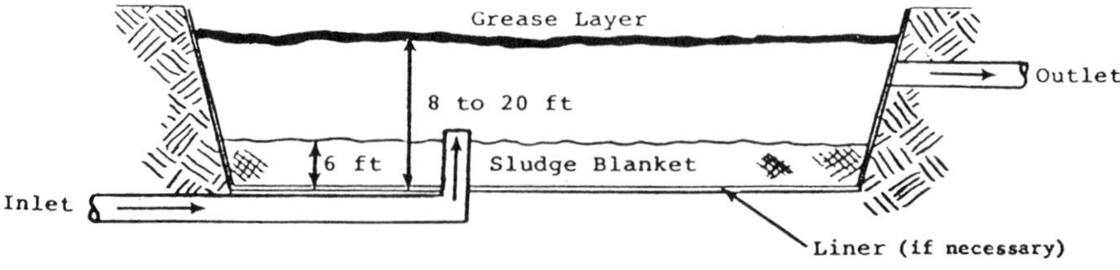

Figure 14 Anaerobic lagoon.

Sludge at the bottom of facultative lagoons undergoes anaerobic digestion producing carbon dioxide, methane, and cells. The photosynthetic activity at the lagoon surface produces oxygen diurnally, increasing the dissolved oxygen during daylight hours, whereas surface oxygen is depleted at night. Facultative lagoons are often operated in series. When three or more cells are linked, the effluent from either the second or third cell may be recirculated to the first. Recirculation rates of 0.5–2.0 times the plant flow have been used to improve overall performance.

Facultative lagoons are customarily contained within earthen dikes. Depending on soil characteristics, lining with various impervious materials such as rubber, plastic, or clay may be necessary. The use of supplemental top layer aeration can improve overall treatment capacity, particularly in northern climates where icing over of facultative lagoons is common in the winter. They are used especially for treatment of relatively weak municipal wastewater in areas where real estate costs are not a restricting factor and for treating raw, screened, or primary settled domestic wastewaters and weak biodegradable industrial wastewaters. Facultative lagoons are most applicable when land costs are low and operation and maintenance costs are to be minimized. In very cold climates, facultative lagoons may experience reduced biological activity and treatment efficiency. Ice formation can also hamper operations. In overloading situations, odors can be a problem.

BOD_5 reductions of 75–95% have been reported. Effluent suspended solids concentrations of 20–150 mg/L can be expected depending on the degree of algae separation achieved in the last cell. Efficiencies are strongly related to pond depth, detention time, and temperature. If wastewater is nutrient deficient, a source of supplemental nitrogen or phosphorus may be needed. No other chemicals are required. Settled solids may require clean out and removal once every 10–20 years.

Operation conditions typically include

Operation: At least three cells in series. Parallel trains of cells may be used for larger systems.
Detention time: 20 to 180 days.
Depth: 3 to 8 ft, although a portion of the anaerobic zone of the first cell may be up to 12 ft deep to accommodate large initial solids deposition.
pH: 6.5–9.0
Water temperature range: 35–90°F for municipal applications
Optimum water temperature: 68°F.
Organic loading: 10–100 lb BOD_5/acre/day.

The service life of the lagoon is estimated to be 50 years. Little operator expertise is required. Overall, the system is highly reliable. There is potential for seepage of wastewater into ground water unless the lagoon is lined. Compared with other secondary processes, relatively small quantities of sludge are produced. Figure 15 shows schematic for a facultative lagoon.

Oxidation Ditch

The oxidation ditch is an activated-sludge biological treatment process that is commonly operated in the extended aeration mode, although conventional activated-sludge treatment is also possible. Typical oxidation ditch treatment systems consist of a single- or closed-loop channel 4–6 ft deep with 45-degree sloping sidewalls.

Some form of preliminary treatment such as screening, comminution, or grit removal normally precedes the process. After pretreatment (primary clarification is usually not practiced), the wastewater is aerated in the ditch using mechanical aerators which are mounted across the channel. Horizontal brush, cage, or disc-type aerators especially designed for oxidation ditch applications are normally used. The aerators provide mixing and circulation in the ditch, as well as sufficient oxygen transfer. Mixing in the channels is uniform, but zones of low dissolved oxygen concentration can develop. Aerators operate in the 60–110 rpm range and provide sufficient velocity to maintain solids in suspension. A high degree of nitrification may occur in the process without special modification because of the long detention times and high solid retention times (10–50 days) utilized. Secondary settling of the aeration ditch effluent is provided in a separate clarifier.

Ditches may be constructed of various materials, including concrete, gunite, asphalt, or impervious membranes. Concrete is the most common

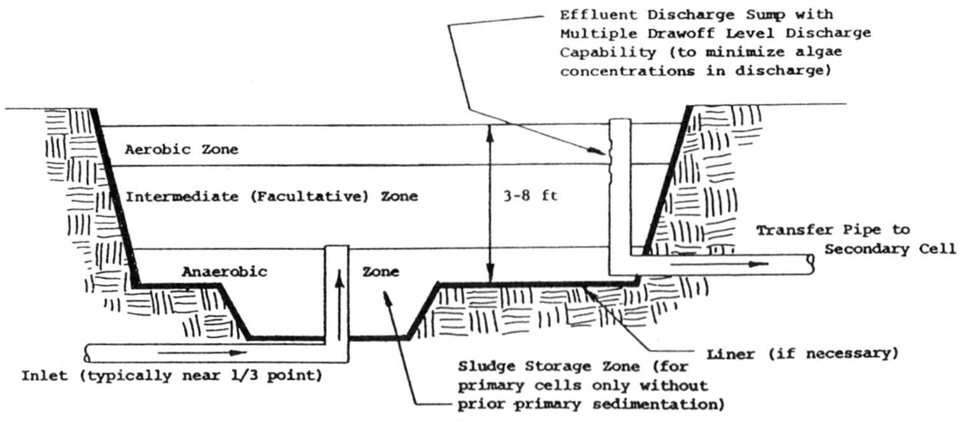

Figure 15 Facultative lagoon.

construction material. Ditch loops may be oval or circular in shape. "Ell" and "horseshoe" configurations have been constructed to maximize land usage. Conventional activated-sludge treatment, in contrast to extended aeration, may be practiced. Oxidation ditch systems with depths of 10 ft or more with vertical sidewalls and vertical shaft aerators may also be used.

Oxidation ditch technology is applicable in any situation where activated sludge treatment (conventional or extended aeration) is appropriate. The process cost of treatment is generally less than other biological processes in the range of wastewater flows between 0.1 and 10 Mgal/day. Oxidation ditches offer an added measure of reliability and performance over other biological processes but are subject to some of the same limitations that other activated-sludge treatment processes face. No primary sludge is generated. Sludge produced is less volatile owing to higher oxidation efficiency and increased solids retention times. Solid waste, odor, and air pollution impacts are similar to those encountered with standard activated-sludge processes. The same potential for sludge contamination, upsets, and pass through of toxic pollutants exists for oxidation ditch plants as standard activated sludge processes. Figure 16 shows flow diagrams for an oxidation ditch.

Biological/Chemical Precipitation

The combined biological/chemical precipitation process is based on the use of activated-sludge microorganisms to transfer phosphorus from incoming wastewater to a concentrated substream for precipitation. The activated sludge is subjected to anoxic conditions to induce phosphorus release into the substream and to provide phosphorous uptake capacity when the sludge is returned to the aeration tank. Settled wastewater is mixed with return activated sludge in the aeration tank. Under aeration, sludge microorganisms can be induced to take up dissolved phosphorus in excess of the amount required for growth. The mixed

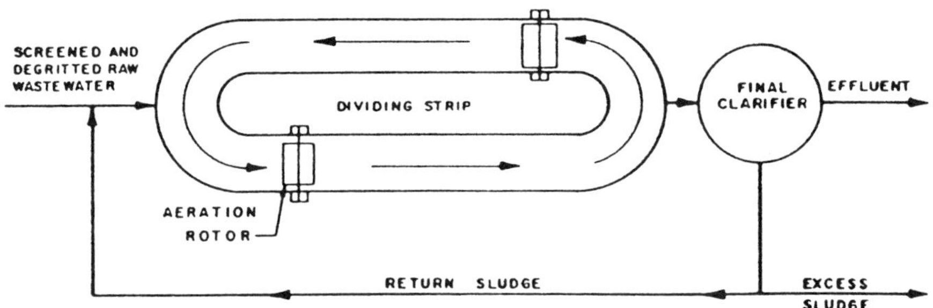

Figure 16 Oxidation ditch flow.

liquor then flows to the secondary clarifier where liquid effluent, now largely free of phosphorus, is separated from the sludge and discharged. A portion of the phosphorus-rich sludge is transferred from the bottom of the clarifier to a thickener-type holding tank: the phosphate stripper. The settling sludge quickly becomes anoxic and, thereupon, the organisms surrender phosphorus, which is mixed into the supernatant. The phosphorus-rich supernatant, a low-volume, high-concentration substream, is removed from the stripper and treated with lime for phosphorus precipitation. The thickened sludge, now depleted in phosphorus, is returned to the aeration tank for a new cycle.

The process has demonstrated a compatibility with the conventional activated-sludge process and appears to be compatible with modifications of it. The process can operate in various flow schemes, including full or split flow of return activated-sludge through the phosphate stripper, use of an elutriate to aid in the release of phosphorus from the anoxic zone of the stripper, or returning lime-treated stripper supernatant to the primary clarifier for removal of chemical sludge. The equipment for this process includes phosphate stripper tanks, chemical feeders, mixers, and precipitator tanks.

This method involves a modification of the activated sludge process and can be used in removing phosphorus from municipal wastewaters to comply with most effluent standards. Direct chemical treatment is simple and reliable, but it has the two disadvantages of significant sludge production and high operating costs. A photostrip system reduces the volume of the substream to be treated, thereby reducing the chemical dosage required, the amount of chemical sludge produced, and associated costs. Lime is used to remove phosphorous from the stripper supernatant at lower pH levels (8.5–9.0) than normally required. The cycling of sludge through an anoxic phase may also assist in the control of bulking by the destruction of filamentous organisms to which bulking is generally attributed.

More equipment and automation, along with a greater capital investment, are normally required than for conventional chemical addition systems. Since this method relies on activated-sludge microorganisms for phosphorus removal, any biological upset that hinders uptake ability will also affect effluent concentrations. It has been found that sludge in the stripper tank is very sensitive to the presence of oxygen. Anoxic conditions must be maintained for phosphorus release to occur.

Chemicals required for this process include lime (CaO). Chemical sludge containing hydroxides is formed from lime treatment. The fraction of the total sludge flow which must be processed through the stripper tank is determined by the phosphorus concentration in the influent wastewater to the treatment plant and the level required in the treated effluent. Required detention time in the stripper tank ranges from 5–15 hr. Typical phosphorus concentrations produced in the stripper are in the range of 40–70 mg/L. The volume of the phosphorus-rich supernatant stream to be lime treated is 10–20% of the total flow. Figure 17 shows a process flow diagram.

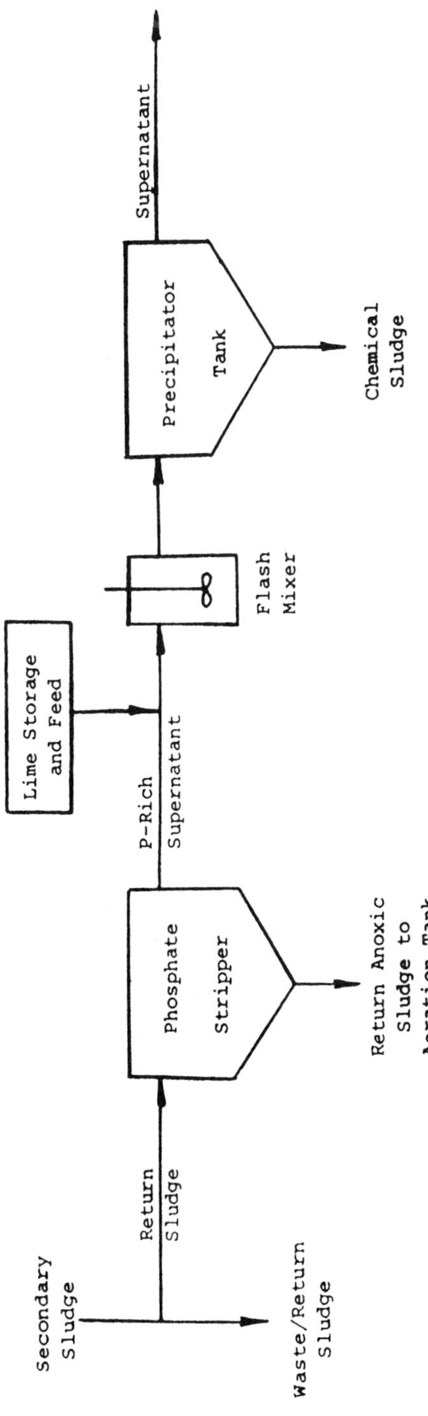

Figure 17 Biological/chemical precipitation.

Biological Contactors

The process is a fixed film biological reactor consisting of plastic media mounted on a horizontal shaft and placed in a tank. Common media forms are a disc type made of styrofoam and a denser lattice type made of polyethylene. While wastewater flows through the sump-storage tank, the media are slowly rotated, about 40 immersed, for contact with the wastewater for removal of organic matter by the biological film that develops on the media. Rotation results in exposure of the film to the atmosphere as a means of aeration. Excess biomass on the media is stripped off by rotational shear forces and the stripped solids are maintained in suspension by the mixing action of the rotating media. Multiple staging of RBCs increases treatment efficiency and could aid in achieving nitrification year round. A complete system could consist of two or more parallel trains with each train consisting of multiple stages in series.

Multiple staging; use of dense media for latter stages in the treatment train; use of molded covers or housing of units; various methods of pretreatment and after treatment of wastewater; use in combination with trickling filter or activated sludge processes; use of air driven system in lieu of mechanically driven system; addition of air to the tanks; addition of chemicals for pH control; and sludge recycling all enhance nitrification.

This process has only been in use in the United States since 1969. Because of its characteristic modular construction, low hydraulic head loss, and shallow excavation, which make it adaptable to new or existing treatment facilities, its use has been growing. It is applicable to the treatment of domestic and compatible industrial wastewater amenable to aerobic biological treatment in conjunction with suitable pre- and posttreatment; and it can be used for nitrification, roughing, secondary treatment, and polishing.

This process can be vulnerable to climatic changes and low temperatures if not housed or covered. Performance may diminish significantly at temperatures below 55°F. Enclosed units can result in considerable wintertime condensation if heat is not added to the enclosure. High organic loadings can result in first-stage septicity and supplemental aeration may be required. Use of dense media for early stages can result in media clogging. Alkalinity deficit can result from nitrification; a supplemental alkalinity source may be required.

Criteria for operation include

Organic loading: Without nitrification, 30–60 lb BOD_5/day/1000 ft^3 media; with nitrification, 15–20 lb BOD_5/day/1000 ft area

Hydraulic loading: Without nitrification, 0.75–1.5 gal/day/ft^2 of media surface area; with nitrification, 0.3–0.6 gal/day/ft^2 of media surface area

Number of stages per train: One to four depending upon treatment objectives

Rotational velocity: Peripheral velocity = 60 ft/min for mechanically driven; 30–60 ft/min for air driven

Typical media surface area: Disc type, 20–25 ft^2/ft^3; standard lattice type, 30–40 ft^2/ft^3; high-density lattice, 50–60 ft^2/ft^3
Percent media submerged: 40%
Tank volume: 0.12 gal/ft^2 of disc area
Detention time based on 0.12 gal/ft^2: Without nitrification, 40–120 min; with nitrification, 90–250 min
Secondary clarifier overflow rate: 500–800 $gal/day/ft^2$
HP: 3.0–5.0 consumed/25-ft shaft; 5.0–7.5 connected/25-ft shaft

Nitrification

The process by which ammonia is converted to nitrate in wastewater is referred to as nitrification. In the process, *Nitrosomonas* and *Nitrobacter* species act sequentially to oxidize ammonia (and nitrite) to nitrate. The biological reactions involved in these conversions may take place during activated-sludge treatment or as a separate stage following removal of carbonaceous materials. Separate stage nitrification may be accomplished via suspended growth or attached growth unit processes. In either case, the nitrification step is preceded by a pretreatment sequence to reduce the carbonaceous demand. Possible pretreatment schemes include activated sludge, trickling filter, roughing filter, primary treatment with chemical addition, and physical-chemical treatment. In general, if the pretreatment effluent has a BOD_5/total nitrogen ratio of less than 3.0, sufficient carbonaceous removal has occurred such that the following nitrification process may be classified as a separate stage. Low BOD is required to assure a high concentration of nitrifiers in the nitrification biomass.

The most common separate stage nitrification process is the plug-flow suspended growth configuration with clarification. In this process, pretreatment effluent is pH adjusted (as required) and aerated in a plug-flow mode. Because the carbonaceous demand is low, nitrifiers predominate. A clarifier follows aeration, and nitrification sludge is returned to the aeration tank. A possible modification is the use of pure oxygen in place of conventional aeration during the plug-flow operation.

Less prevalent are attached growth separate stage nitrification processes. These processes may be operated analogously to trickling filter, packed bed, or rotating biological disc systems. Since the biomass is attached to the reactor surface and solids synthesis is low, a clarifier may not be required. Final filtration is sometimes practiced to reduce effluent suspended solids, although this is often not required.

Nitrification is a well-known phenomenon in biological treatment processes. Separate stage nitrification has been well demonstrated throughout the United States and England in numerous pilot studies and several full-scale designs. Separate stage suspended growth systems outnumber separate stage attached growth systems in these applications by about four to one. These are applicable for conversion of ammonia to nitrate, particularly as a preliminary step prior to denitrification. Commonly used as an add-on process after secondary treatment.

A limitation of this process is that it is sensitive to toxicant upset. The design should compensate for reduced efficiency at low temperature. It only oxidizes ammonia to nitrate. It cannot remove nitrogen effectively and does not significantly treat organic nitrogen. Conversions of ammonia (and nitrite) to nitrate of up to 98% are achievable. Properly designed systems have effluent ammonia in the 1–3 mg/L range. BOD_5 reductions are generally 70–80% (influent BOD_5 assumed as approximately 50 mg/L).

A separate nitrification sludge is generated as a result of separate stage suspended growth systems. Attached growth systems may generate a filter backwash wastewater. Under controlled pH, temperature, loading and toxicant conditions, high levels of reliability are achievable. Nitrification sludge by itself is relatively difficult to dewater. However, it is usually combined with other sludges, resulting in a very small impact on overall dewaterability. Figure 18 shows a flow schematic of a separate stage nitrification system.

Typical operating parameters are

Suspended growth systems		Attached growth systems (trickling filters)	
Flow scheme	Plug-flow (preferable, but not mandatory)	Media area	3,000–10,000 ft^2/lb NH_4-N oxidized per day
Optimum pH	8.2–8.6	Recirculation rate	Up to 100% (variable)
MLVSS	1,200–2,400 mg/L		
Min. aeration tank DO	2 mg/L		
Clarifier surface loading rate	400–600 gal/day/ft^2		
Solids loading	20–30 lb/day/ft^2		
Return sludge rate	50–100%		
Detention time	0.5–3 hr		
Mean cell residence time	10–20 days		

Denitrification Filter (Coarse Media)

In denitrification, nitrates and nitrites are reduced to nitrogen gas through the action of facultative heterotrophic bacteria. Coarse media denitrification filters are attached growth biological processes in which nitrified wastewater is passed through submerged beds containing natural (gravel or stone) or synthetic (plastic) media. The systems may be pressure or gravity. Minimum media diameter is

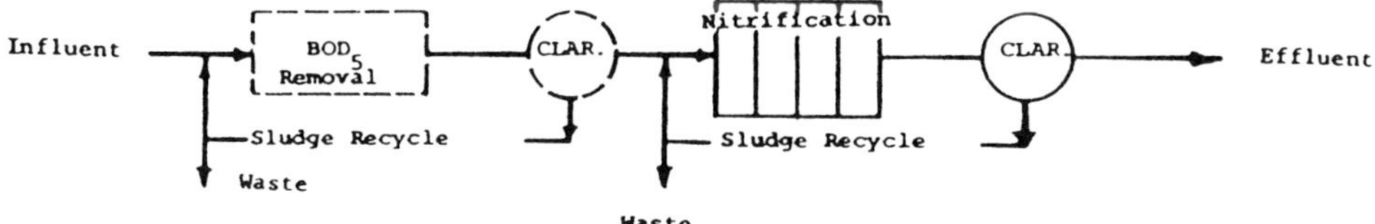

Figure 18 Nitrification, separate stage.

about 15 mm. Anaerobic or near anaerobic conditions are maintained in the submerged bed, and since the nitrified wastewater is usually deficient in carbonaceous materials, a supplemental carbon source (usually methanol) is required to maintain the attached denitrifying slime. Because of the high void percent and low specific surface characteristic of high-porosity coarse denitrification filters, biomass (attached slime) continuously sloughs off. As a result, the coarse media column effluent is usually moderately high in suspended solids (20–40 mg/L) requiring a final polishing step.

A wide variety of media types are used as long as high void volume and low specific volume are maintained. Both dumped plastic media and corrugated sheet media have been used. Backwashing is infrequent and is usually done to control effluent suspended solids rather than pressure drop. Alternate energy sources such as sugars, volatile acids, ethanol, or other organic compounds, as well as nitrogen deficient materials such as brewery wastes may be used. Nitrogen gas–filled coarse media denitrification filters are a possible modification. This process is used almost exclusively to dentrify municipal wastewater that has undergone carbon oxidation and nitrification. It may also be used to reduce nitrate in industrial wastewater. It specifically acts on nitrate and nitrite, and will not affect other forms of nitrogen. This process is capable of converting nearly all nitrate in a nitrified secondary effluent to gaseous nitrogen. Overall nitrogen removals of 70–90% are achievable. Figure 19 shows a flow schematic of a denitrification filter.

Denitrification Filter (Fine Media)

The fine media denitrification filter is an attached growth biological process in which nitrified wastewater is passed through a pressurized submerged bed of sand or other fine filter media (up to about 15 mm in diameter) in which anoxic conditions are maintained. The nitrified wastewater contains very little carbonaceous material and consequently requires a supplemental energy source, usually methanol to maintain the attached denitrifying slime. Because of the relatively fine media used, physical filtration analogous to that occurring in a pressure filter takes place. As a result, a clear effluent is produced eliminating the need for final clarification. Backwashing is required to maintain an acceptable pressure

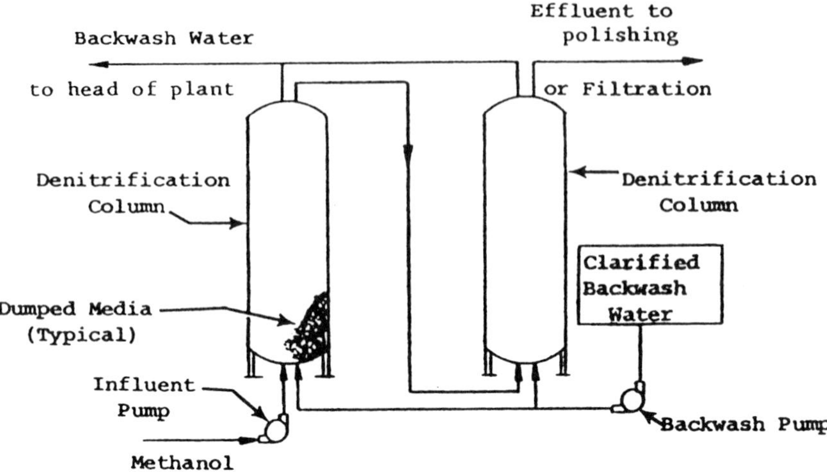

Figure 19 Schematic of a denitrification filter.

drop. Surface loading rates may be somewhat lower than those common for pressure filtration. The development of the denitrifying slime and consequent denitrification efficiency are a function of the specific surface area of the filter, and in practice fine media denitrification filters convert nitrates to nitrogen gas at a much higher rate than suspended growth systems. The coarser the media, the less frequent the backwashing, although the effluent may be more turbid.

Process variations include the use of various media such as garnet sand, silica sand, or anthracite coal with varying size distributions. Multimedia systems have also been used. Alternate energy sources such as sugars, volatile acids, ethanol, or other organic compounds, as well as nitrogen deficient materials such as brewery wastewater. An air scour may be incorporated into the backwashing cycle; however, temporary inhibition of denitrification may result. Various types of underdrains may be used. A bumping procedure (short periodic flow reversals) has been used to remove entrapped nitrogen gas bubbles produced during denitrification. Denitrification may be combined with refractory organic removal. Upflow systems utilizing fine media (sand or activated carbon) have been operated as fluidized bed reactors.

This process is used almost exclusively to denitrify municipal wastewaters that have undergone carbon oxidation and nitrification. It may also be used to reduce nitrate in industrial wastewater. It specifically acts on nitrate and nitrite and will not affect other forms of nitrogen. This process is capable of converting nearly all nitrate and nitrite in a nitrified secondary effluent to gaseous nitrogen. Overall nitrogen removals of 75–90% are achievable. Suspended solids removals of up to 93% have been achieved. Figure 20 shows a schematic flow diagram of a denitrification filter.

NITROGEN IN THE ENVIRONMENT

Nitrogen compounds exist in many forms and are abundant in the environment. Concern is with nitrogen as nitrogen gas (N_2), nitrate ions (NO_3^-), nitrite ions (NO_2^-), ammonia (NH_3) as ammonium ions (NH_4^+), and organic nitrogen. These are the compounds related to plants, animals, and nitrification bacteria.

Important reactions involving nitrogen include fixation, where bacteria or blue-green algae make use of nitrogen gas in the atmosphere; ammonification, whereby organic (dead animal and plant tissue or fecal matter) is converted by microorganisms to ammonia or the ammonium ion; assimilation, in which a nitrogen compound is used by an organism; nitrification, whereby ammonia is converted to nitrate; and denitrification, whereby nitrate is converted to nitrogen gas.

Nitrogen compounds often move within the environment as they change forms. Most of the problems caused by nitrogen compounds occur when they enter ground or surface water bodies. Nitrogen reaches fresh surface water through precipitation, dustfall, surface runoff, subsurface groundwater entry, and the discharge of wastewater effluents. There also are blue-green algae and some bacteria which are able to fix nitrogen from nitrogen gas in the atmosphere.

Discharges of conventionally treated domestic and industrial wastewater effluents high in nutrients are the main sources of nitrogen pollution in the form of ammonia-nitrogen and sometimes organic nitrogen. Nitrogen levels in

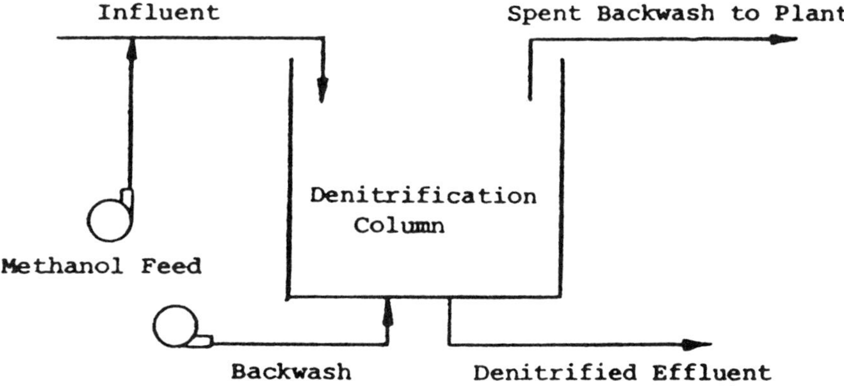

Figure 20 Schematic of a denitrification filter.

industrial effluents vary. Some industries posing significant nitrogen pollution problems include meat processing plants, milk processing plants, petroleum refineries, ice plants, fertilizer manufacturers, synthetic fibers facilities, and ammonia scouring and cleansing operations.

With increasing uses of fertilizers and concentrated animal growth farms (feedlots), storm water runoff is becoming a significant source of nitrogen pollution in agricultural areas. Subsurface drainage of agricultural lands carries excess nutrients to the receiving body of water (approximately 19 mg/L). Runoff from feedlots can contain up to 300 mg/L of ammonium owing to the hydrolysis of urea and up to 600 mg/L or organic nitrogen compared with fertilizer alone, which can contribute anywhere from 1–9 mg/L total nitrogen.

Storm water runoff in urban areas also contributes nitrogen to surface waters. Combined sewer (storm and sanitary) overflows during rainstorms can add a significant occasional nitrogen load.

Nitrogen enters soil through precipitation, dustfall, application of wastewater or fertilizers, plant residues, composting, or through fixation by bacteria directly from nitrogen gas in the atmosphere. Precipitation and dustfall contain nitrogen compounds owing to the combustion of fossil fuels. The planting of legume crops (peas and beans) increases the bacterial fixation of nitrogen gas and can account for 25% of the total nitrogen source in such areas.

Ninety percent of nitrogen found in soil is in the organic form, since it is derived from the decomposition residues of plants and animals. Most of the remaining 10% of nitrogen is as ammonium ion (NH_4^+) and tightly bound to soil particles. Septic tanks in heavily populated areas can contribute high nitrate levels to the soil at rates faster than can be assimilated.

When nitrogen is supplied to soil at a faster rate than it can be assimilated by the soil's bacteria, it will be filtered down to ground water pockets below.

Marine environments receive nitrogen compounds mostly from land drainage. It is the nitrate form which is of concern in such environments.

Algae are the basic link in the conversion of inorganic compounds in water into organic matter. They perform this function through a mechanism known as photosynthesis in which inorganic compounds and carbon dioxide are converted to a carbohydrate energy source plus oxygen in the presence of light. The inorganic compounds required include those with hydrogen, nitrogen, phosphorous, sulfur, potassium, magnesium, calcium, and trace elements such as zinc, copper, iron, and molybdenum. Since all of these chemicals except for phosphorus and nitrogen are usually present, phosphorus and nitrogen normally are considered to be the limiting nutrients.

Nitrogen reaching surface waters via the various pathways discussed above serve to trigger and/or sustain the growth of algae. Although some algal growth is desirable, since by day photosynthesis removes carbon dioxide and produces oxygen, dense growths can create problems for the environment.

When the natural balance of nutrients is thrown off by an excess supply of nitrogen and other necessary compounds, algae will grow in large numbers creating a dense mat on the water surface known as an algal bloom or scum. Depending on the type of algae present, the water will take on a green, yellow, red, black, or turbid appearance ruining the water for recreational purposes. The odors produced by the bacterial decomposition of algae further curb recreational use.

Similarly, growth of other water plants and some diatomes (freshwater and saltwater algae types which form silica shells) also is stimulated by an abundance of nutrients.

Algae growth in water supplies requires taste and odor removal through the use of filtration followed by carbon adsorption. In addition, the water supply itself must be chemically treated to kill algal growths.

Algae produce oxygen by photosynthesis during the daylight, however carbon dioxide is released at night during the algae's respiration process. Although more oxygen is produced than respired, the presence of large quantities of algae can produce wide swings in the water's oxygen content which can be harmful to other oxygen-dependent organisms living there.

Similar swings (diurnal) can be found in pH levels and alkalinity of the water as a result of CO_2 level changes due to algal photosynthesis and respiration. During the night, respiration increases dissolved carbon dioxide levels thus lowering the pH. Waters with high calcium contents are able to buffer the pH effects of carbon dioxide reduction by the precipitation of calcium carbonate.

Dense algal growths limit photosynthesis to the top layer of the water. Depths of greater than 3 ft are shaded from the sunlight. The algae below die and are decomposed by bacteria, which in turn deplete the available oxygen.

In extreme cases, algal production leads to eutrophication, which is defined as a process of enrichment in which a water body is so fertilized with nutrients that aquatic productivity is greater than the decay rate. Eutrophication is more of a problem in relatively slow-moving waters such as surface streams, lakes, and reservoirs where nutrients have time to build up.

Oxygen depletion in atrophying water bodies has a snowball effect as other organisms die as a result creating a further oxygen depletion problem by their own decay. Once eutrophication starts, it cannot be stopped.

Changes in organism distribution patterns occur in atrophying water bodies. Predator-prey relationships change, and trash fish tend to flourish while desirable sport fish disappear, thus further altering the value of the water body.

Besides potentially destroying the esthetic and recreational value of water bodies, there are some species of algae which can cause gastric disturbances in humans if the water is drunk or accidentally swallowed. Some people are allergic to planktonic algae.

Ammonia (NH_3) is a significant pollutant in raw water. It reacts with chlorine to form chloroamines which reduces chlorine disinfection properties.

Ammonia can be toxic to fish at certain pHs. It exists as ammonia NH_3 at pHs above 7 and as an ammonium ion (NH_4^+) at pHs below 7. It is the ammonia molecule that is toxic to fish. A maximum concentration of 0.02 ppm has been set as a water quality standard for freshwater aquatic wildlife. Acute toxicity at a given pH will increase with corresponding increases in dissolved oxygen, carbon dioxide, temperature, or bicarbonate alkalinity levels within a range of 0.01–2.0 mg/L ammonia. At levels up to 25 mg/L, ammonia toxicity can affect all aquatic life. Diurnal pH fluctuations due to photosynthesis therefore can play a significant role in creating toxic conditions.

Nitrogen and algal growth can also interfere with industrial and water-treatment operations. Nitrogen in the form of ammonia is corrosive to certain metals. Plankton and filamentous algae can clog sand filters in water-treatment plants, cause foam when heated, corrode metal, and create undesirable tastes and odors or oily substances which interfere with their use. Algal growth is a problem in cooling towers.

As with untreated carbonaceous matter, ammonia-nitrogen exerts an oxygen demand on receiving water bodies as it slowly oxidizes to nitrite and then nitrate.

In the early days of water treatment, nitrification of wastewaters was the primary concern. However, it was neglected with the advent of the biochemical oxygen demand (BOD_5) test when emphasis turned to control of carbonaceous matter. In the past 10–20 years, concern for control of nitrogen discharges in the environment has returned and efforts have been directed toward determining design considerations for upgrading existing carbonaceous oxidation plants to accomplish nitrification as well and toward developing independent nitrification or nitrification denitrification systems.

NITRIFICATION

Nitrification is the first of two stages in biological nitrogen removal in which ammonia nitrogen is biologically oxidized to nitrate, a less objectable form which does not exert an oxygen demand on the receiving water.

In the first step of nitrification, ammonium ions are oxidized to nitrite ions according to the following reaction in which 58–84 kcal/mol of ammonium is released: $NH_4^+ + 1.5O_2 \rightarrow H^+ + H_2O + NO_2^-$. The bacteria responsible for this oxidation are usually *Nitrosomonas* species, although sometimes *Nitrosococcus* species can be involved. These bacteria are aerobic autotrophs. Autotrophs, unlike heterotrophs which obtain their energy from the oxidation of carbonaceous (organic) matter, get their energy for growth from the oxidation of inorganic nitrogenous matter and use inorganic carbon rather than organic carbon, as heterotrophs do, for cell synthesis. Being aerobic, these autotrophs require the presence of oxygen to convert the nitrogen into a useable form.

In the second step of nitrification, the nitrite ion is further oxidized by *Nitrobacter* bacteria to nitrate releasing only 15–21 kcal/mol of nitrile oxidized as follows: $NO_2^- + 0.05_2 \rightarrow NO_3^-$. *Nitrobacter* organisms also are aerobic autotrophs.

The energy freed by the nitrification reactions is used by the bacteria for growth. Since the *Nitrosomonas* bacteria obtain more energy than the *Nitrobacter* bacteria per mole of nitrogen oxidized, their mass in any nitrification system is greater. The *Nitrobacter* bacteria require three times the substrate needed by nitrosomonas to get the same energy and therefore their population is one-third that of the *Nitrosomonas* bacteria.

Nitrifiers grow at a much slower rate than heterotrophs. This growth rate difference can be measured by the BOD test. BOD_5, which represents biochemical oxygen uptake after 5 days, indicates the oxidation of carbon by heterotrophs. BOD_{20} similarly represents the final oxygen uptake after 20 days by bacteria through nitrification.

Although nitrification reactions appear very simple, there are various intermediates and enzymes involved. The enzymes which control the rates of reactions conducted by the bacteria are sensitive to pH, temperature, and substrate concentration. The conditions necessary for proper functioning of these enzymes are reflected in the overall cell preference. The enzymes are substrate specific.

Therefore, there can be many enzyme reactions involved in the bacteria cell synthesis. The enzymes must convert the nitrogen compounds into an amino acid form before they can be used directly by the bacteria.

As discussed above, the growth of *Nitrosomonas* is limited by the ammonia nitrogen content of the wastewater, which in turn limits the growth of *Nitrobacter*. When the food (substrate) supply is plentiful and other conditions are favorable, growth will increase unchecked. As the population begins to exceed the available food supply, the growth rate will decline. As the food supply becomes scarce, the bacteria begin to obtain their nutrition from the dead bacteria through a method known as lysis.

Growth rates also increase with rises in temperature. The growth of *Nitrobacter* is more greatly influenced by temperature than that of *Nitrosomonas*. Increases are exponential in nature, reaching a maximum at some optimum temperature and then quickly falling to zero once the optimum temperature is passed.

The alkalinity and pH of a system also play an important role. The nitrification reaction releases carbon dioxide and free acid (H^+) during the oxidation of 1 mg ammonia to nitrate which destroys about 7 mg of alkalinity as calcium carbonate ($CaCO_3$). Depending on the alkalinity available, the reduction in $CaCO_3$ can have a depressing effect on the pH.

When the pH goes below 7, a considerable decrease in the nitrification rate will result. This is true for both acclimated and unacclimated systems, although the short-term effect on an acclimated system is less significant. It has been demonstrated that pH drops from 7.2 to 6.4 have no immediate adverse effects. Drops in the pH to 5.8, however, create significant reductions in the nitrification rate. An abrupt pH change from 5.8 back to 7.2 will cause an immediate rise in the nitrification rate. Therefore, pH has an inhibitory rather than toxic effect. Nitrifiers have been known to adapt to a pH range of 5.5–6.0. Many wastewaters do not have a sufficient alkalinity buffer and alkalinity maintenance becomes very important.

In order to obtain complete nitrification of a waste, 4.6 mg of oxygen is required for every milligram of ammonia present. Generally, the nitrification rate will increase with an increase in dissolved oxygen of the system if other conditions are favorable. Studies have shown that rates are 10–50% lower at dissolved oxygen levels of 1 or 2 mg/L than at 4 to 7 mg/L.

Nitrification therefore is an important factor in stabilizing the oxygen demand of the waste. Controlled biological treatment is necessary to obtain nitrification, since the population of nitrifying organisms is minimal in surface waters.

Another advantage of nitrification is that a highly nitrified effluent is immune to petrification, helping to preserve the esthetic quality of the receiving body of water.

Nitrification can be required when standards or limitations have been set on the receiving waters or effluent or where the reduction of the residual oxygen demand from ammonia is specifically required. The overall transformation of ammonia to nitrate will depend on how much organic nitrogen has been transformed to ammonia prior to the nitrification process.

When the total nitrogenous content of the effluent must be reduced because of regulatory limitations and/or prevent or reduce the growth of algae in the receiving water, denitrification is required.

DENITRIFICATION

What is denitrification? Denitrification is the second and final stage in the biological removal of nitrogen. With denitrification, nitrates are reduced to nitrogen gas. When methanol is used as a source of carbonaceous matter, the reaction for denitrification is $5CH_3OH + 6H^+ + 6NO_3^- \rightarrow 5CO_2 + 3N_2 + 13H_2O$. It also is possible for nitrites to be converted directly to nitrogen gas. The bacteria responsible for this transformation are heterotrophs which derive their energy from organic chemicals through the reduction of nitrate or nitrite. These bacteria include those of the genera *Pseudomonas*, *Achromobacter*, *Bacillus*, and *Micrococcus*, which are facultative, meaning they can survive with or without the presence of oxygen. These bacteria prefer oxidizing organic matter with oxygen rather than by reduction of nitrite or nitrate. Therefore anaerobic (no oxygen) conditions must be maintained in a denitrification system.

With both nitrification and denitrification of wastes, nitrogen removals of 70–90% can be obtained. Although such removal will serve to reduce or prevent most algal growth and eutrophication, many blue-green algal blooms cannot be affected, since these algae can fix nitrogen gas for their synthesis from the atmosphere.

BIOLOGICAL NITRIFICATION METHODS

Biological nitrification can be achieved by several means. With the renewed interest in nitrogen control, efforts have been made to determine various add-on treatment and upgrading methods for new and existing treatment systems. The first working nitrification system in the United States during recent times was started up in 1969 at Lake Tahoe.

Domestic waste effluent is one of the main contributors of nitrogenous compounds to our environment. Nitrogen in such wastes is usually as organic nitrogen or as free ammonia. The nitrate and nitrite concentrations are generally small in raw wastes in relation to the other forms. Typical values of nitrogen concentrations of raw domestic waste are given below.

	Waste Effluents (mg/L)		
Nitrogen Form	Strong	Moderate	Weak
Nitrogen as total nitrogen	85	40	20
Organic nitrogen	35	15	8
Free ammonia	50	25	12
Nitrites	*	*	*
Nitrates	*	*	*

*Negligible or nonexistent.

The most important of the above compounds is ammonia, since it can lower the dissolved oxygen of a receiving stream by nitrification. The ammonia content is derived from urea and to a lesser extent proteins. The organic nitrogens are in the form of purines, pyrimidines, proteins, urea, and amino acids. Much of the organic nitrogen is transformed to ammonia through hydrolysis before it reaches the wastewater-treatment plant. Conversion of organic nitrogen to ammonia continues to occur within a conventional treatment plant due to the actions of heterotrophic bacteria. The organic nitrogen compounds usually are in a soluble form, whereas most of that ammonia is particulate. Primary sedimentation removes some of the particulate nitrogen forms, which is usually less than 20% of the total nitrogen. Secondary sedimentation (clarification) removes another 10–20% of the nitrogen. A domestic waste, with predominantly the nitrate nitrogen form, has been stabilized with respect to its oxygen demand and is considered to be an old waste.

After passing through conventional biological treatment, the secondary sanitary effluent will have a typical nitrogen content of 20–50 mg/L indicating that nitrogen just passes through such systems. However, biological nitrification does not remove nitrogen any better than conventional treatment.

Biological nitrification can be achieved in conventional carbonaceous removal systems which have also been modified to combine nitrification by addition of a separate tertiary system. Generally if the BOD_5/total Kjeldahl nitrogen (TKN) ratio is less than 3, a separate nitrification system must be added. If the BOD_5 to TKN ratio is greater than 5, a combined system should work. There is no special recommendation for wastes with ratios between 3 and 5 at this time. There are varied opinions as to the merits of both methods.

There are two basic concepts of biological treatment available. These are suspended growth, where the bacterial masses are suspended in a mixed liquor and separated via clarification, and attached growth, whereby the bulk of the bacterial growth occurs on a plastic or stone media and solids separation is not necessary. The types of suspended growth systems available include various activated sludge setups. Attached growth systems include trickling filters, bio-discs, and fluidized beds.

Activated Sludge

Most of the studies done related to biological nitrification have been in the operation of activated-sludge treatment systems which is one of the major recognized effective methods. The important variables which have been studied extensively include the organic loading and sludge age, pH, dissolved oxygen, temperature, and the presence of inhibitory substances which are discussed below.

Organic Loading and Sludge Age

The organic loading to a system is the single most important factor in nitrification. High organic loadings favor the growth of heterotrophic bacteria which then overrun the system. These bacteria have a faster rate of substrate oxidation than autotrophs. The heterotrophs' faster growth rate is reflected in the oxygen uptake and sludge production and (bacterial) sludge is wasted faster than the nitrifiers can multiply. Therefore, with high organic (BOD_5) loadings, little nitrification will occur, and at lower BOD_5 loadings (approximately 10 mg/L) higher nitrification rates can occur. Increases in the organic loading of a waste can be compensated for by increasing the retention time of the activated sludge to prevent washouts of nitrifier populations before they can become established. Sufficient oxygen supplies also must be carefully monitored in such cases to supply both the carbonaceous and nitrogenous demands. These are important concerns in combined treatment systems. However, most of the carbonaceous oxygen demand already is removed when nitrification serves as tertiary treatment. A shock load of organics to a treatment plant would not have a significant impact on nitrification in a tertiary system.

Retention time is not an important parameter by itself but may be used to moderate effects of changes in other parameters, such as organic loading, sludge age, dissolved oxygen, and temperature. The time factor does have a directly proportional relationship to the amount of nitrifiers which will be present. The average sludge retention time for a conventional activated sludge system is 3.5 days. Six to 10 days would be needed in a combined nitrification-carbon removal system to prevent washout of nitrifying populations.

Dissolved Oxygen

As mentioned previously, dissolved oxygen (DO) has a significant effect on nitrification. The stoichiometry of nitrification reaction shows that four atoms oxygen are needed to oxidize one molecule of ammonia to nitrate. This translates to a 50% greater oxygen requirement for good nitrification of a typical domestic waste than is required for carbonaceous removal. Pilot plant studies have shown that nitrification is possible at DO levels of 1 mg/L and may not occur at all at DO levels of 7 mg/L if other important factors are not favorable. Generally though barring unfavorable conditions, higher DO levels will increase the rate of nitrification.

Temperature

Temperature affects bacterial metabolic activities, gas transfer rates (available DO), and settling characteristics of waste effluents. At temperatures above 40°C and below 5°C, nitrification rates are very slow. Studies have indicated that the optimum temperature for nitrifying bacteria is 22 and 30°C. Because the temperatures in summer and warm climates are within this optimum range, treatment plants can be operated at less favorable pHs and lower substrate levels that would be required during colder conditions to achieve the same degree of nitrification. In order to make up for the temperature difference, in winter up to five times the summer detention time (capacity) may be required. Temperature deficiencies also may be made up by increasing the mixed liquor suspended solids level of the system and or pH adjustment.

pH

Nitrification is most rapid when the pH is maintained at or slightly above neutral. Results have shown that the optimum pH for nitrification is 8.4. With all other conditions favorable, 90% nitrification can be obtained at pHs of 7.8–8.9 but less than 50% below 7.0 and above 9.8. Further reductions in nitrification rapidly occur below a pH of 6.0 and nitrification may cease entirely below pH

5.0. Nitrifiers have a low tolerance to the hydrogen ion concentration. Breakdowns of sludge flocs also have occurred when the pH drops below 7.0.

Inorganic Loadings

To a lesser extent, the ammonia-nitrogen level of the waste plays a role in affecting nitrification rate. Studies have shown that at a given organic loading, increases in the influent ammonia concentration increase nitrate production levels. These increases are not proportional in nature. There is a point where the nitrate production will be limited.

The ammonia levels found in domestic wastewater are not sufficient to inhibit the rate of nitrification of such effluents.

Toxicity

The following compounds are toxic to nitrifying bacteria.

Organics	Inorganics
Thiourea	Zn
Allyl-thiourea	OCN^{-1}
8-hydroxyquinoline	ClO_4^{-1}
Salicyladoxine	Cu
Histidine	Hg
Amino acids	Cr
Mercaptobenzthiazole	Ni
Perchloroethylene	Ag
Trichloroethylene	
Abietec acid	

Only inhibitory effects may be felt from heavy metals concentrations of 10–20 mg/L provided that the pH is 7.5–8.0. Precipitated metals in the sludge can redissolve if the pH goes down resulting in a system upset.

Industrial discharges which are unusually high in ammonia or nitrite can exert a temporary effect on the system too.

To screen for a toxicity problem, batch oxygen uptake tests may be used, and batch nitrification jar tests may then be run in order to determine the best pretreatment.

Pretreatment can afford some protection. Heavy metals may be removed by lime additions, carbon adsorption can be used for organics, and two-stage systems where the organic toxics are biodegradable.

However, perchloroethylene and trichloroethylene are not biodegradable and are toxic to nitrifiers.

For toxics that come and go, breakpoint chlorination may be used at the end of the system for added safety in ammonia removal.

Activated Sludge—General

In activated-sludge systems, waste is biologically oxidized under aerobic conditions. Large and easily settleable solids are removed prior to entry of the waste effluent into a reactor. Air (oxygen) is supplied to the reactor by diffusion of mechanical aeration. After oxidation has occurred, the mass of bacteria which has grown is separated from the liquid in a settling tank or clarifier. Some of the solids are returned to the reactor while the rest are wasted. There are many variations of this method.

Combined Carbon and Nitrogen Removal Systems

The first nitrification processes developed were in combined systems by modifying extended aeration systems. Combining operations is advantageous in terms of cost for existing carbonaceous systems which can be upgraded to include nitrification.

Combined carbon and nitrogen removal systems have at high proportion of influent organic loading relative to the ammonia-nitrogen concentration. As a result, the population of nitrifiers is small compared with heterotrophs. In addition, the conditions required for the carbon oxidizing heterotrophs and nitrifying autotrophs are different and therefore operating parameters must be carefully controlled in combined systems.

The design approach of combined systems therefore must be based on the sludge growth rate or solids retention time. This generally means an additional oxygen supply, longer mean cell residence times (about 10 days), and operation temperatures of 21–22°C.

Contact stabilization systems (Fig. 21), where sludge is reaerated prior to being recycled with the influent, will not provide complete nitrification. Even though the solids can be retained for a long time, there is an insufficient mass developed in the reactor.

The influent entering the last pass in step aeration systems (Fig. 22) may

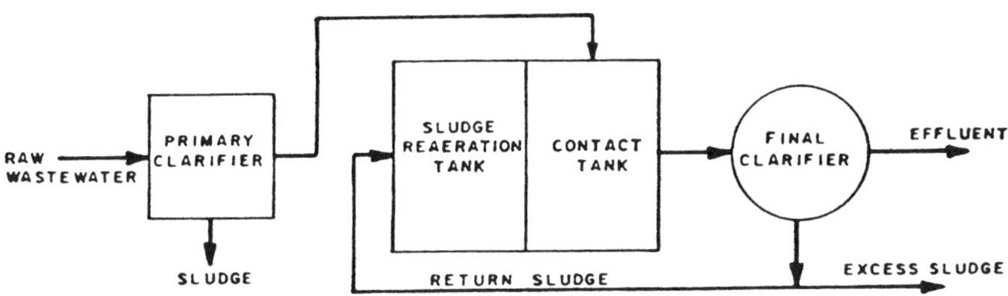

Figure 21 Contact stabilization.

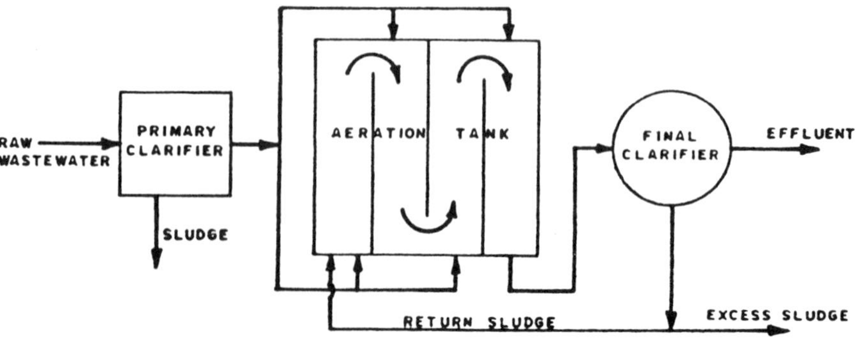

Figure 22 Step aeration.

not have enough time for hydrolyzation of the organic nitrogen to ammonia to permit nitrification. This problem can be somewhat alleviated by setting up artificial sludge reaeration zones in the first pass by not feeding influent to that section. However, back mixing is not prohibited. Neither is short circuiting and it is possible for ammonia bleed-through to occur.

Extended aeration plants normally are operated at such long retention times that except during cold temperatures of 5–10°C and below, nitrification usually is obtained if the plant is operated properly. These systems are similar to complete mix systems except that the hydraulic retention times are 24–48 hr rather than 2–8 hr. Complete mix systems (Fig. 23) can provide complete nitrification at typical domestic waste concentrations. In such systems waste is distributed uniformly to all points within the aeration tank.

Conventional (plug-flow) activated-sludge plants (Fig. 24) can be designed

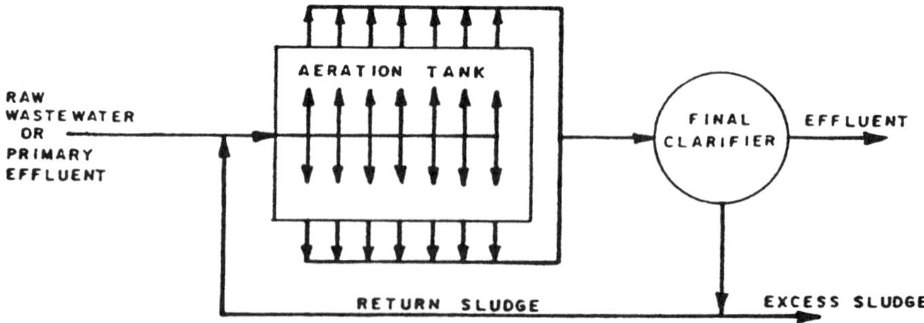

Figure 23 Complete mix system.

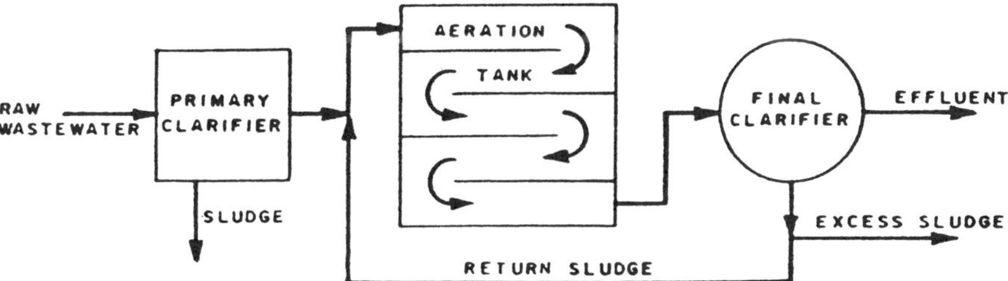

Figure 24 Conventional activated sludge system.

to prevent back mixing to the head of the tank by the addition of weirs, since the first portion of the tank may be ineffective for nitrification. Theoretically, plug-flow can be more efficient or require less tank volume for the same strength waste than complete mix or extended aeration plants. However, unless a diffused air system is installed, the carbonaceous oxygen demand can overpower the nitrifiers' needs. If lime is used for flocculation before the plug-flow reactor, carbon dioxide should be added to avoid pH toxicity.

High-purity oxygen (UNOX) systems have been experimented with for nitrification. Since the cover prevents the escape of carbon dioxide, a build-up in the system occurs. pH levels of 6.0 are not uncommon. This has a depressing effect on the nitrification rate and even longer solids retention times are required. If pH is carefully maintained in the system, UNOX plants are no different in the degree of nitrification achievable than conventional aeration plants. The choice must be based on economic and social (odor) considerations.

Separate, Two-Stage Carbon- and Nitrogen-Removal Systems

It has been demonstrated that the physical separation of carbon- and nitrogen-removal functions can improve the control and efficiency of nitrification in certain cases. By reducing significantly the BOD's load in the first stage to the influent ammonia concentration, more nitrifiers can be established.

The value of separating the heterotrophic and autotrophic populations is realized in the reduced residence time required: 6 days total versus 10 days in combined systems.

Plug-flow systems are the favored activated-sludge method for obtaining nitrification. Lower effluent ammonia concentrations can be achieved than in complete mix units.

Figure 25 is an example of a typical two-stage activated-sludge system.

Comparison of Combined and Two-Stage Systems

Although both types of systems can be operated to achieve complete nitrification, there are advantages and disadvantages to each type.

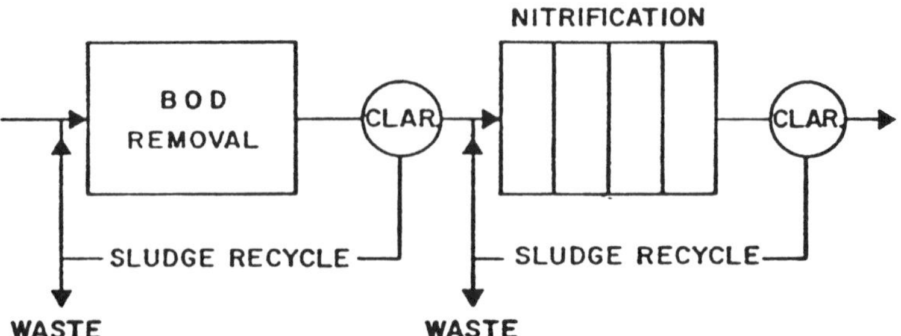

Figure 25 Two-stage activated sludge system.

Results of modifying activated-sludge systems for nitrification have been inconsistent. Problems sometimes occur with rising sludge because of the long retention times required where denitrification begins to take place owing to lack of oxygen. Nitrogen gas bubbles cause the sludge mass to rise.

Combined systems receiving a primary effluent with a weak (BOD_5) waste can be operated for nitrification satisfactorily with high mixed liquor suspended solids down to temperatures of 10°C. However, two-stage systems are necessary in northern climates when watewater temperature often goes below 18°C. It has been demonstrated that two-stage systems can handle seasonal load variations where combined systems cannot. Another advantage of combined rather than single-stage systems is the lower quantity of sludge (about half the amount) which must be handled and the normally better settling characteristics of the sludge produced. Two-stage systems tend to have more control problems, since two systems are involved and the clarifier is the least stable component of a system. However, with careful monitoring, two-stage systems can be managed and a greater degree of control obtained over microbial processes.

Toxics may not be a problem to combined systems if the primary effluent is treated with a coagulant. Some people think that toxics can be reduced by two-stage systems, but others think that there is no advantage to sacrificing a carbonaceous system for a combined system. There are some substances though which may be toxic to nitrifiers, yet biodegradable by heterotrophs, but there are indications that the toxics' advantage may not matter with domestic sewage. The first reactor in a two-stage system also may serve to reduce the possibility of organic surges to the nitrifying bacteria thus preventing overpowering carbonaceous bacterial growth.

Combined systems require less land and capital expenditures than separate systems and have lower sludge disposal costs. However, the power costs for separate systems are less.

Trickling Filters

Trickling filter systems are the major type of attached growth systems used to perform nitrification. Owing to larger land requirements relative to activated sludge systems, trickling filters are often used for sanitary treatment in smaller cities of less than 10,000 people and less populated areas. Because of trickling filters' relative stability compared with other biological systems, they are used for treating high-strength industrial wastes, too.

Trickling filters are usually circular in form. Waste effluent is distributed by rotary sprays over the media and is collected underneath by an underdrain system.

The media to which the organisms attach can be made of rock, plastic, or redwood. The liquid waste percolates down through the media and the substrate, and inorganic and organic waste matter is assimilated by the organisms attached to the media. Aerobic degradation takes place on the outer portions of the biological film which develops on the media. As the mass of organisms becomes thicker, anaerobic conditions occur near the media surface. The surface organisms then die and are washed off periodically (sloughing).

Effluent peaks can be taken care of by designing the system with additional surface area and increasing recirculation rates during low flow periods to keep the media from drying out. Clarification normally is not needed following trickling filers, since the solids are maintained within the units.

Trickling filters harbor a varied assortment of organisms including aerobic and anaerobic, or facultative, bacteria; fungi (at low pH); algae; protozoans; and worms, insect larvae, and snails which in turn feed on the bacteria. Owing to the unstable characteristics of the slime, a kinetics theory for the biological activities has not yet been developed. Conclusions regarding nitrification in such systems therefore are based on empirical results.

Nitrification and carbonaceous oxidation can occur simultaneously in trickling filters. It is better to use a media with a lower specific surface and higher voids such as a maximum of 35 ft^2/ft^3, to prevent clogging in combined systems. When nitrification is separate from carbonaceous oxidation, plugging is less of a problem for the nitrification system and application of a media with a high specific surface (up to 67 ft^2/ft^3) is okay and reduces space requirements.

In two-stage systems where organic carbon and nitrogen activities occur separately, increases in nitrification have been proportional to increases in surface areas. The surface area requirements for two-stage systems increases greatly at temperatures of 7–11°C than at 13–19°C. Surface area requirements for nitrification also increase with the degree of ammonia-nitrogen reduction desired.

If ammonia removal must be below 2.5 mg/L, breakpoint chlorination may be added following the trickling filter, since the cost of removing such ammonia levels is much higher with trickling filters.

The important variables in operating a trickling filter for either combined carbonaceous oxidation and nitrification or only nitrification as tertiary treatment include the organic loading, temperature, pH, dissolved oxygen, and toxicants.

Organic Loading

The organic loading has a significant effect on the ammonia content of the effluent. If it becomes too large, the media will be dominated by heterotrophs and significant nitrification will not occur. For combined systems, the organic loading must be reduced for cold-weather operations which increases the nitrification costs beyond those of adding a separate biological nitrification system or use of physical-chemical treatment. It has been demonstrated that nitrification efficiency levels of 75–100% can be achieved with BOD_5 loadings of less than 10 lb/1000 ft^3/day. Efficiency diminishes at greater loadings.

Temperature

Greater nitrification can be achieved with higher temperatures. Temperatures within 15–30°C are preferable, with 30°C being the optimum. Nitrification can be achieved down to 7°C, but as mentioned previously, cost factors make it impractical below 13°C. However, attached growth systems can compensate for cold temperatures better than suspended growth systems by thickening of the slime.

pH

Nitrifying bacteria are limited by pH in attached growth systems as they are in suspended growth systems.

Dissolved Oxygen

The oxygen mass transfer limits within the bacterial slimes may limit the nitrification reaction. In order to prevent oxygen from being the limiting factor, the dissolved oxygen supply must be 2.7 times the ammonia-nitrogen concentration. This can be achieved by increasing the recirculation rate to dilute the ammonia or by adding a high purity oxygen to increase the oxygen transfer rate.

Toxicants

Nitrifying bacteria are subject to the same toxicant effects whether they live in an attached or suspended growth system. However, trickling filters can handle shock loads better than can activated sludge.

Hydraulic Loading

The hydraulic loading of the system can have a profound effect on the degree of nitrification attainable in a trickling filter. An increase from 10–30Mgad reduced nitrification from 72 to 52%.

Filter Depth

It has been shown that greater nitrification can be achieved at media depths of 6 ft than at less than 6 ft.

Trickling Filters Versus Activated-Sludge Systems

Since most of the work to date on nitrification has been done with activated-sludge systems, the theory of trickling filter operation is not as precise. However, biofilm models developed indicate that trickling filters can handle adverse loading and lower temperature conditions better than activated sludge systems. Trickling filters do not quickly show ammonia breakthrough with changes in loading rate.

Bio-Discs

Bio-discs are attached growth systems which consist of a series of large-diameter plastic discs rotating on a horizontal shaft which runs across the top section of a troughlike reactor as shown in Figure 26.

Only 40% of the discs are in the wastewater at any time. Biological films develop on the discs (the media). Such systems can be operated for carbonaceous oxidation or combined nitrification-carbonaceous oxidation. In combined systems, organic oxidation occurs on the first discs and nitrification on the last. Nitrification does not begin until most of the BOD has been removed.

Problems have arisen in application owing to diurnal variations or shock loads. Then these load changes create an increased organic loading, some or all of the nitrifying discs convert to heterotrophic colonies. Depending on the variations involved, this situation can be prevented by derating the disc loading

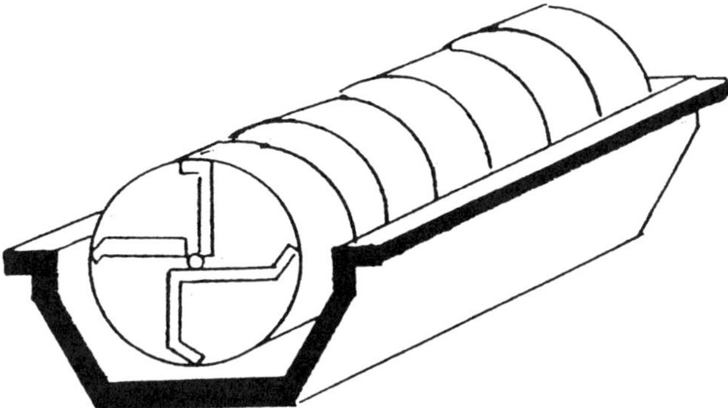

Figure 26 Cutaway view of a bio-disc unit.

or installing equalization ahead of the system. This is the disadvantage which has limited acceptance of bio-discs for nitrification.

Temperature shows no effect on nitrification rates above 13°C. The discs are normally housed to reduce the effects of external temperatures, to prevent algal growths, and to keep out rain or hail which can shear growths off the discs.

Fluidized Beds

Fluidized bed systems in large-scale applications have not been accepted on a wide basis. However, they are discussed here since they have been shown to promote nitrification.

Fluidized bed systems are purported to combine the best features of activated-sludge and trickling filter systems into one process. As in the trickling filters, the degradation organisms coat the media which for fluidized beds is sand grains in suspension.

Fluidized bed can handle shock loads and toxics as can trickling filters, but there is a minimum sloughing of growth. Secondary clarifiers are not needed.

In fluidized beds, water is passed up through a bed of sand at a velocity high enough to impart motion to or "fluidize" the sand. An enormous surface area is obtained using sand: greater than 1000 ft^2/ft^3 of reactor.

ALTERNATIVE MEANS OF AMMONIA-NITROGEN CONTROL

An effluent ammonia-nitrogen problem or concern can be eliminated by several other methods as briefly described in the following discussion.

Algal Ponds

These are shallow lagoons where intensive algal growths are cultivated under aerobic conditions. Algal ponds contain algae which reduce the nitrogen content of the effluent through photosynthesis as could occur in the receiving body of water if the effluent was discharged without such treatment. The algae is harvested from the pond along with the nitrogen which has been assimilated.

This form of nitrogen has its application in small cities with plenty of available land. It is a seasonal treatment method dependent on light and temperature. Ice and cold winter weather significantly reduce metabolic activities and ponds go anaerobic. During the spring which follows, hydrogen sulfide odors are released as ponds return to their aerobic states. As a result, algal ponds must be located as far as possible from existing and future residential communities. Although construction and operating costs are low, the land costs and requirements can make ponds prohibitive.

Ion Exchange

Ion exchange involves the removal of ionic species (in this case ammonium ions) from an aqueous phase. Nitrite, nitrate, and organic nitrogen cannot be removed by this method.

For ammonium removal, effluent is passed through a column of clinoptilolite, a naturally occurring zeolite (resin) with a high selectivity for ammonium. Organics can foul the resin and must be removed prior to this treatment. When all the available sites are taken up by the ammonia, breakthrough will occur and the resin must be regenerated. Ammonium removals between 90 and 97% have been obtained by this method.

There are some major disadvantages associated with ion exchange. First, the possibility of organic fouling. Second, high regeneration costs. Finally, there is no ultimate disposal of the ammonium ion, since it is contained in the waste brine from regeneration.

Land Disposal

Secondary effluent has been disposed of by spray irrigation providing less soil preparation and more crops for farmers or as a soil conditioner for marginal or drastically disturbed land. Nitrogen removals between 30 and 95% can be obtained.

Toxics management is important in land disposal because of the variability of soil capacity to filter, buffer, absorb, and chemically or biologically react with nitrogen. Land disposal can be a reliable method of nitrogen disposal if care is taken in utilization.

If the nitrogen-loading rate is too high, plants and soil bacteria will not be able to assimilate all of it and an increase in ground water nitrate will result. Other problems associated with land disposal include large land requirements, high management costs, climate dependence, potential health hazards from bacterial contamination in ground water, and accumulation of trace elements (toxicity).

Air Stripping for Ammonia Removal

Air stripping mainly is used for ammonia removal from raw wastewater or digester supernatant and is discussed elsewhere. Ammonia-nitrogen is achieved by aeration of wastewater in a stripping tower. Effluent is pumped to the top of the tower and as it falls to the bottom, fans force air countercurrent to the falling water. Ammonia is vaporized and discharged to the atmosphere. The ammonia must be in the molecular form of NH_3 not as an ammonia ion, NH_4^+. Up to 98% removal can be obtained; however, residual levels of less than 5 ppm cannot be removed by this method.

Nitrite, nitrate, and organic nitrogen levels are not affected by air stripping.

Air temperature and effluent pH also affect the amount of ammonia which can be stripped. The pH must be raised to 10 or 11 with lime and the tower must be shut down during freezing weather. Problems in efficiency and scaling occur with cold-weather operation.

The ammonia removed from the effluent is discharged to the air. Although there are no US air standards for ammonia at this time, rainfall washouts and nearby stormwater runoff can carry the ammonia to a receiving body of water. The net effect to the receiving water is not as bad though as it would be if the effluent were discharged directly.

Breakpoint Chlorination

In breakpoint chlorination or superchlorination, enough chlorine is added to oxidize ammonia-nitrogen to nitrogen gas. Approximately 10–20 mg/L of chlorine is needed to oxidize 1 mg/L of ammonia-nitrogen. With this method, ammonia-nitrogen levels can be brought down near zero. The effect of chlorine on organic nitrogen though is still uncertain. Nitrite and nitrate are not removed by this method, and therefore breakpoint chlorination becomes a possible follow-up to incomplete biological nitrification where low or negligible levels of ammonia are required.

Optimum breakpoint chlorination can be obtained at a pH of 10 and temperature of 30°C. Up to 90% ammonia removal can be obtained in 4–60 hr. With chlorine gas, ammonia, and to a lesser extent organic nitrogen removal can be obtained. Results using chlorine dioxide are not very good for ammonia and nonexistent for organic nitrogen.

The acidity produced by the chlorination must be compensated for by lime and caustic soda addition which increases the total dissolved solids of the effluent. Breakpoint chlorination can be expensive operation. Chlorine is rarely added to the actual breakpoint through.

Several different methods of ammonia and nitrogen removal have been presented herein. In selecting the method most suited to the specific situation, the following should be considered:

- Form and concentration of the influent nitrogen compounds
- Required effluent quality
- Other existing treatment processes
- Costs
- Degree of reliability required
- Flexibility of the system

AERATORS

There is a wide choice of equipment which can be used for effective aeration of waste. Three common classes of equipment which are most often considered are:

- Surface aerators
- Submerged turbine aerators
- Diffused aeration systems

Increasingly used in most new system designs is the surface aerator. Submerged turbine aeration has found application in plants where high-horsepower requirements and relatively short detention times are needed and where land is at a premium. This system is also be used to increase oxygen input where existing diffused air systems have reached capacity limits. Diffused air was the earliest aeration system available. Many plants are still in existence today with this type of equipment and more are now being installed, but the use of surface aerators and submerged turbine aerators has increased substantially.

Selection of mechanical aerators is usually based on

- Oxygen transfer efficiency
- Solids suspension and mixing
- Flexibility

Diffused air and submerged turbine devices accomplish oxygen transfer by bringing quantities of air into contact with the liquid; i.e., the air in this case is the transported or principal phase.

Surface aerators are different in that the waste is the transported or principal phase brought in contact with the air. Various submerged turbine devices operate between these two extremes where both air and waste are transported with nearly equal importance.

Mixing is also to be considered as a separate operation. Mixing requirements often become secondary to oxygen transfer when actually in reality there are applications where the mixing requirement may control the equipment selection. Mixing requirements can be met by the various classes of equipment but by two different operations. In the case of diffusers, mixing is accomplished as a secondary effect of air rising and at the other extreme it is a result of circulating substantial quantities of water from the basin bottom to the surface.

Diffused Air

Diffused aeration equipment can be classified into two general types depending on bubble size. General design techniques for these units are well established and considerable information is available on optimum tank geometry and sparge locations.

Fine bubble devices are generally porous media comprised of carborundum, nylon, or tightly wrapped Saran. Principal advantages over the large bubble diffusers is that a greater percentage absorption is obtained owing to the increased interfacial area of the relatively small bubbles. A problem often encountered with such devices is plugging resulting in high maintenance costs.

Large bubble devices offer the advantage of reduced maintenance costs over the fine bubble devices but a reduction in percentage absorption and transfer efficiency results. Efficiencies calculated indicate most diffused air systems fall in the low range of the devices. Fine bubble devices are consistently more efficient than large bubbled diffusers and are competitive with the low end of submerged turbine aeration.

Submerged Turbine

Operating goals of a good submerged turbine system are

- To provide, by mechanical and fluid action, sufficient shear to create a fine bubble distribution
- To provide, by a suitable hydraulic device, the proper fluid regimen to maximize the air hold-up in the system

There are several systems possible to achieve goals, the most usual system consists of a radial flow impeller located above a large orifice sparge ring or simply an open pipe. Gas rising from the pipe is dispersed by the impeller and distributed throughout the tank.

Balance between airflow and impeller flow is quite important in that at too high air rates the gas may completely overcome the pumping action of the impeller. This flooding results in loss of oxygen transfer efficiency. At all lower air rates, good dispersion results allowing a wide operating range of air rate while maintaining adequate mixing independently. That oxygen transfer can be varied independently of the mixing is a unique advantage of submerged turbine aerators and should be of benefit for plants having widely varying loads.

Typical transfer efficiencies for submerged turbine devices fall in an intermediate range of the devices tested. The ability to install submerged turbine aerators to increase the total horsepower and oxygen transfer of existing diffused air systems has offered many plants the chance to increase their oxygen transfer capability to suite present needs at a minimum of construction cost.

Surface Aeration

A surface aerator is a device which brings to the surface quantities of waste for contact with the atmosphere. Two general classes of equipment can be further defined as to whether the pumping device operates substantially below the surface or at the liquid surface.

Oxygen transfer occurs through two principal mechanisms: direct transfer to the waste while being sprayed through the air and air entrainment which can occur both at the impeller and throughout an area around the device as a result of splashing liquid impinging into the bulk liquid.

Surface aeration offers low operating costs for given oxygen requirements. Surface aerators provide efficiencies generally higher than competitive devices.

Ordinarily, adequate mixing is obtained, but in unusual tank or lagoon geometrics, additional precautions such as lower impellers for deep basins may be required to insure satisfactory fluid regimen.

Surface aerators are readily controllable for varying oxygen demand situations by using such techniques as speed control, cycle timers, and submergence adjustment. Combination units where a surface aerator is placed on a common shaft with a submerged turbine aerator have been successfully used in a number of installations. Such units are quite adaptable to existing diffused air systems for increasing oxygen transfer and are also for plants with considerable variance in the oxygen demand. Efficiencies cover a wide range depending on the combination split between submerged and surface aeration. A rotary biological contractor (RBC) consists of large-diameter circular media (disc) mounted on a horizontal shaft. The shaft slowly rotates the media as approximately 40% of its surface area is submerged in the wastewater. Microorganisms are attached to the media forming a biomass of approximately 50,000 mg/L suspended solids. This is an equivalent of 10,000–20,000 mg/L mixed concentrations in a suspended growth system. Wastewater retention times are shortened due to the large microbial population.

Microorganisms consume oxygen and organic material while the RBC rotates them through wastewater and air. The wastewater film on the media absorbs oxygen during exposure to the air. Organisms remove the oxygen by the rotation through the bulk of the wastewater in the tank. Unused dissolved oxygen in the film mixes with the mixed liquor dissolved oxygen concentration. Excess biomass formation is stripped off the media by shearing forces or a stripping blade into the mixed liquor. The treated wastewater is then forwarded out of the process. The solids which settled at the bottom are forwarded for drying and disposal.

RBC biomass is shaggy and projects outward into the adjacent film of wastewater. Unlike trickling filters, this provides much larger active biological surface area. Such outward protection of RBC biomass facilitates more efficient interaction with substrate and oxygen. The shaggy biogrowth increases in size until it can no longer carry its own weight against the forces exerted due to the rotation. The sheared biomass which is in the form of relatively large aggregates settles readily in the final clarifier.

The RBC process can be viable application for wastewater treatment with the following advantages:

- High performance for carbon and nitrogen removal.
- Short process retention time.
- Excellent shock and toxic load capabilities.
- High sludge settleability and filterability.
- Simplicity of the process control.

- Low costs for operation and maintenance.
- BOD removal and nitrification can be achieved in a single step of treatment with no change in operation.
- Ease of upgrading: Trickling filter and activated sludge plants can be conveniently and economically upgraded to RBC process due to its modular construction, shallow excavation, and low head loss.

However, RBC systems may suffer from following disadvantages:

- Susceptibility to harsh weather unless covered
- Possible odor problem due to lack of even oxygen transfer
- Mechanical failures of shaft and drive system

7
Filtration with Granular Media

INTRODUCTION

Filtration with granular media has long been applied in the treatment of municipal and industrial waters as well as for wastewater treatment both in upgrading existing conventional plants and in designs of new advanced treatment facilities. Next to gravity sedimentation, it is the most widely used process for separation of wastewater solids. The following applications have found wide use:

- Removal of residual biological floc in settled effluents from secondary treatment by trickling filters or activated sludge processes
- Removal of residual chemical-biological floc after alum, iron, or lime precipitation of phosphates in secondary settling tanks of biological treatment processes
- Removal of solids remaining after the chemical coagulation of wastewaters in tertiary or independent physical-chemical waste treatment

For these applications, filtration may serve as either an intermediate process to prepare wastewater for further treatment or as a final polishing step following other processes.

Granular media filtration involves passing the water through a bed of granular material with resulting deposition of solids. Eventually the pressure drop across the bed becomes excessive or the ability of the bed to remove suspended solids falls off. It is then necessary to clean to restore operating head and effluent quality to acceptable levels. Most filters operate on a batch basis, with the entire unit being removed from periodically for cleaning. Time in service between cleanings is termed the run length. Head loss at which filtration is interrupted for cleaning is called the terminal head loss.

The filter involves the following parameters:

- Filter configuration
- Media sizes and depths and materials
- Filtration rate (gallons per million [gpm])/ft^2
- Terminal head loss ft of water
- Flow control
- Backwashing

The most important characteristic in equipment costs is the filtration rate, which determines the filter size. Operating costs depend on filtration rate, terminal head loss, media characteristics, and backwash design. The first three filter characteristics determine the cost of power for operating head and the production of the filter per run. Backwash determines the cost per cleaning of operator attention, washwater pumping, air scouring (compressor operation), and treatment of dirty wash water. The cost of cleaning per unit volume treated (cost per cleaning divided by production per run) depends on all of these factors. Wherever possible, designs should be based on pilot filtration studies of the actual waste. Pilot studies can also determine effects of pretreatment variations or character *filterability* in terms of performance attainable with a specific filter design.

Filter units usually consist of a containing vessel, the granular media, structures to support or retain the media, distribution and collection devices for influent, effluent and wash water flows, supplemental cleaning devices, and necessary controls for flows, water levels, or pressures.

Most filter designs employ a static bed with vertical flow either downward or upward through the bed. Downflow designs traditionally used in potable water treatment (Figs. 1 and 2) are most common, but a number of installations have been designed for upward flow. Figure 2c uses both flow directions with the effluent withdrawn from the interior of the bed. Upflow washing is used regardless of the operating flow direction. Special filter designs have been employed horizontal radial flow through an annular bed. Media is cycled downward through the bed, withdrawn at the bottom, externally washed, and returned to the top.

Filters may be designed with closed vessels permitting influent pressures above atmospheric as shown in Figure 3 or with open vessels where only the hydrostatic pressure over the bed is available to overcome filter head losses. Pressure units are preferable where high terminal head losses are expected or where the additional head will permit flow to pass through downstream units

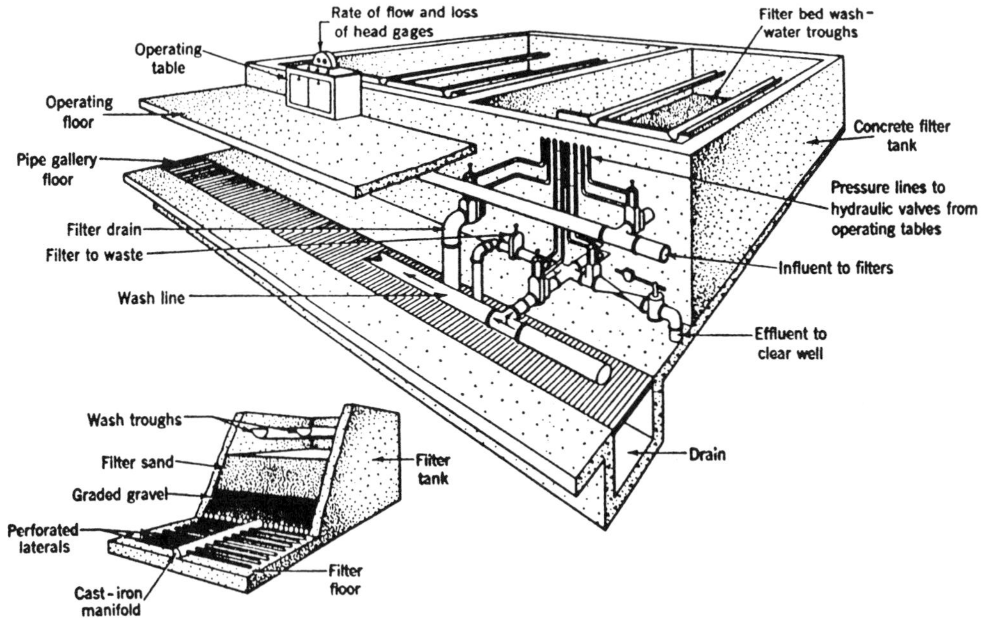

Figure 1 Typical rapid sand filter.

without repumping. They are most commonly used in small to medium sized treatment plants where steel-shell package units are economical.

Figure 2 illustrates schematically a number of different filter configurations using fixed bed media. The beds shown are all graded during upflow washing so that the finer material of a given specific gravity is on top. It should be noted that the conventional single-media filter used in potable water treatment (Fig. 2a) is generally unsatisfactory for wastewater treatment, because the wastewater solids cause a high-pressure drop build-up at the fine surface layer.

In upflow designs, flow passes first through the coarser media which for a given head loss build-up has greater capacity for retaining filtered solids. This is advantageous in lengthening filter runs and increasing output. Dual and multimedia (Fig. 2d and 2e) obtain the same effect under downflow operation by placing coarser layers of lighter material over finer denser material. An alternative downflow single-media configuration attempts to get the same advantage from use of beds of uniform-sized coarse media with depths of 60 in. or more.

Filters of this type are typically designed to operate on a batch basis with entire units taken out of service for cleaning as required. Special designs, however, can provide more or less continuous cleaning either externally with

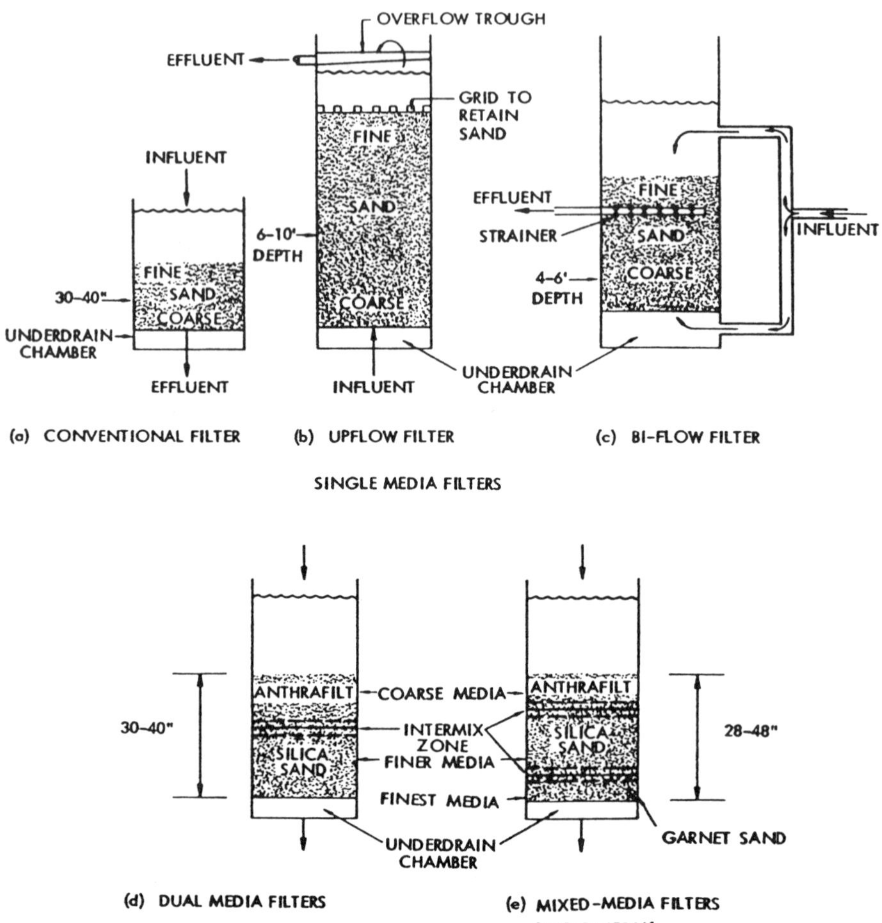

Figure 2 Filter configurations.

media cycled through the bed or in-place with techniques such as traveling backwash or air pulsing of the bed and air mixing of the liquid above.

PROCESS VARIABLES

The measures of filter performance are output quality and quantity. Variables which determine or limit performance are influent characteristics and physical characteristics of the filter. The latter include media characteristics, filtration

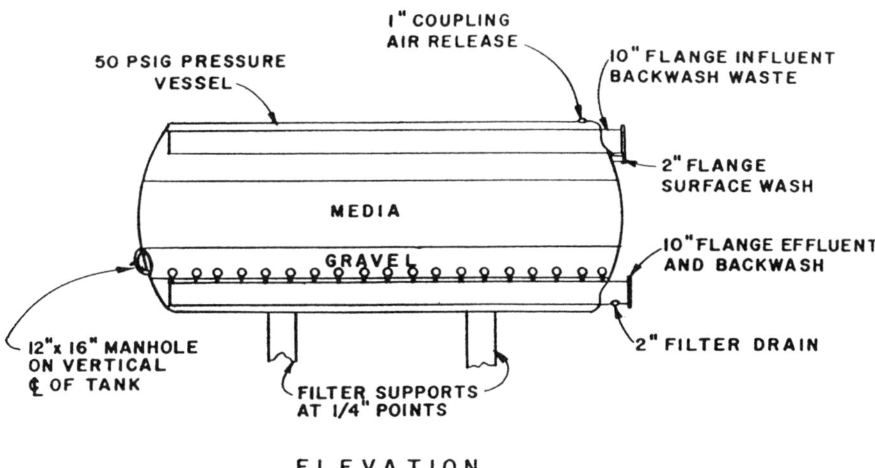

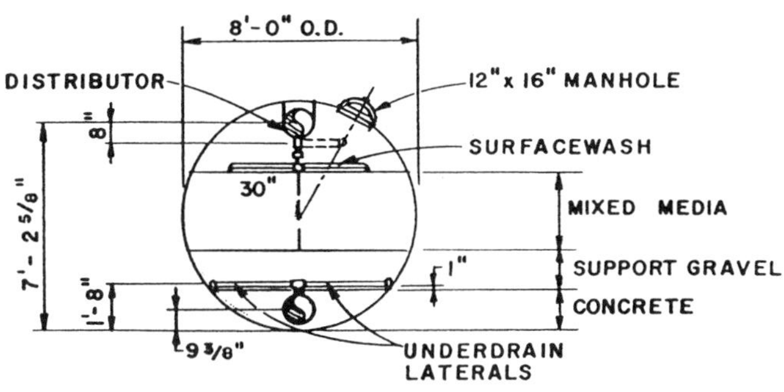

Figure 3 Typical pressure filter.

rate, available and applied operating head, and the design and operating parameters of the filter-cleaning system.

Limits on quality performance of importance is the influent solids: concentration, strength, size, and the physical-chemical properties governing adhesion of particles to each other or to the media surfaces. Commonly a number of filters with different physical characteristics can come close to the limiting quality performance for a given influent. Quality performance of given filters can vary widely for different solids characteristics. In determining output quantity from filters, the influent solids characteristics—especially floc strength and solids concentration—are again very important, but the physical characteristics of the filter become significant as well.

At run lengths of 24 hr or more, output depends almost totally on filter rate. As length of run becomes shorter, the effects of downtime and wash water recycle during cleaning become important. Wash water recycle volume depends on the backwash flow rates and the wash cycle duration needed for adequate cleaning. Factors governing the backwash system include

- Size distribution, depth, and specific gravity of media
- Nature of solids removed, principally their adhesion to the media and their tendency to compact in a dense layer at the media surface
- Type of supplementary cleaning provided

Run length may be limited either by available head or by deterioration of effluent quality as the filter bed becomes filled with solids (breakthrough). Variables affecting this include

- Influent solids characteristics (all those which affect quality performance)
- Flow rate
- Temperature and viscosity of the wastewater
- Media characteristics
- Amount of head available

Head loss in a clean bed varies directly with filter rate and inversely with grain size. In determining head loss build-up, the most significant media characteristic is the grain or pore size at the influent surface of the bed (or in some cases within finer denser layers of multimedia filters). In downflow filtration through a graded bed, influent solids particles larger than about 7% of the minimum grain size will be removed by straining provided their strength is sufficient to withstand the shear at the surface. Shear varies with filter rate and liquid viscosity.

In surface straining, head loss increases exponentially with time or solids accumulation. Where significant solids loads are removed mainly by surface straining, head loss build-up will be rapid, filter runs short, and backwash frequency high. In addition, the solids removed at the surface tend to be compressed into a dense mat which is difficult to remove in backwashing.

Removal of solids within the bed rather than just at the surface is termed depth filtration. Both surface and depth filtration are usually involved to some degree in any given application. In depth filtration head loss tends to build up linearly with time or with solids accumulation. Compression of the solids removed is limited by the granular structure of the bed. For downflow filtration within a single media, the farther solids penetrate into the bed, the slower will be the rate of head loss build-up, but the sooner solids will breakthrough into the effluent. Factors which determine breakthrough for a single media are the media size and depth, the flow rate, and the resistance of deposited materials to shear within the bed.

INFLUENT

Influent characteristics that determine filter performance are those of the solids to be removed. The only significant characteristic of the wastewater liquid—as opposed to the solids—is viscosity, which varies with temperature. Its effects on development of filter head loss are small in comparison to the effects of solids accumulations or filter rates.

Characteristics of wastewater solids which govern or limit filter performance are determined by the treatment processes ahead of filtration. In direct filtration of secondary biological effluent, the residual solids applied to the filter are predominantly biological floc grown in the treatment process. Filtration of effluent following tertiary coagulation for phosphate removal the residual solids are predominantly chemical flocs. Filtration of chemically precipitated raw wastewater or primary effluent, the solids consist of inorganic chemical floc with varying quantities of precipitated organics.

FLOC

Biological flocs tend to be significantly stronger or more resistant to shear than chemical flocs, at least those from alum or iron coagulants. In filtering biological flocs, surface straining is generally significant and runs are almost always terminated by excessive head loss with breakthrough being rarely observed. Alum and iron hydroxide flocs which have been shown to penetrate readily into filters and to breakthrough at relatively low heads ranging from 3 to 6 ft. In contrast to flocs from other coagulants, calcium carbonate precipitates are strongly removed at the filter surface where they may form a dense compressed layer that is hard to remove during washing.

Polymer filter aids may be added to the filter influent to strengthen weak chemical flocs, thereby permitting operation at higher rates without breakthrough. Polymers added as coagulant aids in upstream settling or flocculating units may similarly strengthen the residual floc applied to the filters. Head loss must be available to meet losses due to the tougher floc, and doses must be kept as low as possible to avoid excessive head loss. Floc particle sizes in settled biological effluent tend to be bimodally distributed. Mean sizes for the two modes in one study were 3–5 μm and 80–90 μm. About half of the weight was in each mode.

The filterability of residual solids from secondary settling varies with solids retention time and with liquid contact time in the biological process. For biological systems with higher solids retention times and longer liquid contact times, filtered effluents tend to have lower suspended solids. Solids in extended aeration effluents filter particularly well, in as much as they often settle poorly, leaving high concentrations in the secondary effluent.

HEAD LOSS BUILD-UP VERSUS SOLIDS CAPTURE

Although effluent quality reflects the solids which pass through the filters, head loss reflects the amount and location of solids which deposit in the bed. Both solids loading (solids concentration times flow rate) and filter efficiency are important in determining the build-up of head loss with increasing solids capture. Results generally vary widely in view of the wide range of solids characteristics, media characteristics, and filter rates, and the very different head loss patterns that result from surface and depth filtration.

FILTER CHARACTERISTICS

Most wastewater filter designs employ media configurations and loadings which minimize surface straining and promote depth filtration. Special designs with fine media are intended to remove solids primarily by surface filtration or straining. Such designs include provisions for overcoming the adverse effects of rapid headloss build-up. Where surface filtration predominates, the media characteristics have little effect on quality performance or head loss. In addition, removal of solids is independent of filter rate or influent solids concentration.

MEDIA CHARACTERISTICS

The most important media characteristic in determining performance is size. In a media graded from fine to coarse in the direction of flow, the highest solids concentration is applied to the layers with the greatest removal efficiency. This results in removal concentrated in a small depth with accompanying high head losses. In media graded from coarse to fine in the direction of flow, substantial penetration occurs but most of the solids are removed in the coarser media where less head loss build-up results. The finer layers, protected from heavy solids loadings, are available for polishing and to prevent breakthrough as the coarser layers become filled with solids. Media depth is most significant in coarse uniform beds. Because of the uniformity, the efficiency of removal (as a percent of the solids applied to each depth) is nearly constant for all layers of the filter. Penetration is substantial and extra depth is relied on for polishing and to retard breakthrough. Size and specific gravity of media together are significant in determining expansion during backwash and the degree of intermixing in multimedia beds.

The effect of filter rates on quality performance can vary widely depending on application. In filtering biological floc at reasonably low influent solids concentration, the effect on effluent quality of rates up to 10 gpm/ft^2 is not very significant. Sudden changes in filter rates may affect effluent quality more adversely than sustained higher rates. Higher filter rates tend to increase solids penetration. In cases where this significantly reduces surface removal, head loss build-up per unit volume filtered may actually be less at higher rates. It has also been suggested that the advantages of using a coarse top media layer may be lost if the filter rate is not high enough to force solids into the bed and limit surface straining.

Besides upflow washing, some form of auxiliary scouring of the media appears essential to clean adequately wastewater filters. If cleaning is not adequate, two serious problems can develop: filter bed cracking and mud ball formation. Cracks open in filter beds because of compression of excessively thick coatings on the filter grains. The resulting localized heavy penetration of solids may both lower effluent quality and contribute to mud ball formation. Mud balls are compressed masses of filtered solids large and dense enough to remain in the bed during backwashing. If conditions favoring their formation persist, mud balls tend to increase in size and to sink deeper in the bed. Their presence increases head loss and may lead to loss of effluent quality. Both air scrubbing and surface or internal water jets have been used for auxiliary scouring of the media. Air injected below the media produces shear as the bubbles rise through the bed. Water jets positioned at the top of the expanded bed produce high shear around the surface media, which is the most heavily loaded with solids. In multimedia beds, jets may be similarly provided at the expanded height of the media interface.

The main upflow wash and the auxiliary scouring systems should be controlled independently to permit use together or separately. Key parameters for design of the cleaning system are the upflow wash rate capacity and the air scour rate or surface wash rate capacity. Typically, upflow wash rates are about 20 gpm/ft^2. The maximum capacity is selected to provide the desired degree of fluidization and expansion of the media under critical high temperatures. Capacities for auxiliary scouring are generally established empirically. Air scour rates typically range from 3 to 5 (standard cubic feet per minute per square foot) and surface wash rates from 1 to 3 gpm/ft^2.

Given adequate information on performance, the filter rate and terminal head loss for a particular media design should be selected by making tradeoffs between filter size, operating head requirements, and run length, all within the limits dictated by effluent quality requirements. This section outlines procedures for such tradeoffs and provides an alternative basis for selection where specific performance information is lacking. Tradeoffs can be obtained only from pilot studies of the specific media application. Pilot studies should indicate the build-up of head loss with time for various filter rates and for average and peak influent solids concentrations. With this information it is possible to estimate the filter run length, the net production, and the capital and operating costs of the filter for the given influent solids concentrations and for different combinations of filter rate and terminal head loss.

The following factors should be considered:

- Maximum flows and solids loadings for various durations up to 24 hr. Use of equalization to limit maximum wastewater flows should be considered.
- Run length limits: Lower limit should be 6–8 hr to maintain reasonable net production. The upper limit should be 36–48 hr to avoid anaerobic decomposition of solids in the filter.
- Head loss limits: For gravity filters allowable head losses generally are below 10 ft. Use of heads much above this commits the design to pressure filters. Use of pressure filters would be favored where pressurized discharge to following facilities is needed. Gravity filters would be favored where the extra head for pressure filters would require intermediate pumping but head for the gravity units is available without such pumping.
- Backwash design and expected cost per cycle: Manpower costs should reflect whether the operation is to be automated. Backwashing costs should include costs of treating recycled backwash in units ahead of the filters, and the recycled flow should be deducted in determining the net production of the filters.
- Space limitations may force use of higher filter rates.
- Number of filter units: This should be tentatively selected to facilitate cost estimates but may be varied with little effect on the tradeoff calculations provided labor is not a major factor in the operating cost per backwash. For reliability and economy, a minimum of four to six units should generally be provided, with at least two in even the smallest installations. Above these minimums, the number of filter units depends on the actual size of individual units. The practical maximum size of gravity filters is about 800 ft^2.

In addition to the above, pilot testing may reveal upper limits on head loss or rate required to avoid solids breakthrough and effluent quality deterioration and an optimum filter rate for minimizing head loss build-up. No filter rates lower than the optimum should be considered.

Where it is impossible to test proposed filter media on the actual influent, guidance may be obtained from results with the same media treating similar influents. In the absence of specifically applicable test results, filter rates and head loss allowances should be very conservatively selected based on ample estimates of influent solids concentrations. To assure adequate capacity it is suggested that, as a minimum, sufficient filter area be provided to handle the 24-hr design flow at 4 gpm/ft^2 or the 4-hr maximum design flow at 6 gpm/ft^2, whichever is more stringent. For predominantly chemical floc, the surface media should be no finer than 1 mm and allowance should be made for a terminal head loss of 10 ft.

MEDIA FOR FILTRATION

Media commonly used in water and wastewater filtration include silica sand (specific gravity [sp gr] 2.65), anthracite coal (sp gr 1.4–1.6), and in special multimedia designs garnet (sp gr 4.2) or ilmenite (sp gr 4.5). Since they occur in nature, these materials are not of uniform size but instead typically have a varying grain size distribution. Natural grain size distributions frequently are close to geometrically normal; i.e., plot as a straight line on log probability paper. As shown in Figure 4 grain size distributions are often characterized by two points, the 10 and 60% size (d_{10} and d_{60}). These are sizes such that the weight of all smaller particles constitutes respectively 10 or 60% of the whole. Media are frequently specified in terms of effective size (d_{10}) and the uniformity coefficient (d_{60}/d_{10}).

It is possible to change the characteristics of a given media material by

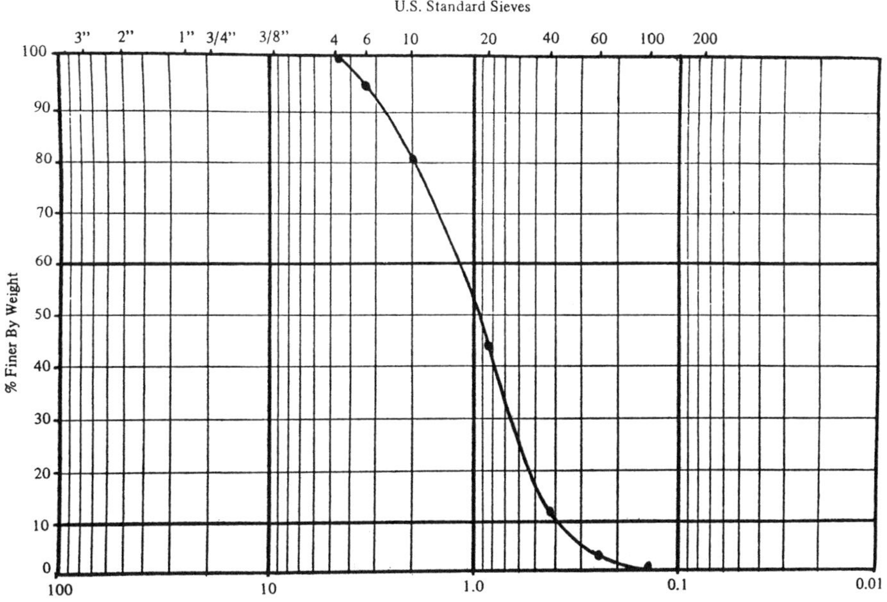

Figure 4 Grain size curve.

removing certain size fractions. Coarser fractions may be sieved out, whereas finer fractions may be removed by "scalping"—removing surface layers after hydraulically grading material during upflow washing. The most important modification for most media is to remove any very fine particles, say, those less than 80% of the effective size. Such fine material never constitutes more than a small fraction of the media volume, but if not removed, it may cause head losses far greater than would be expected for the given effective size.

With sufficient size separation, it is possible to produce almost uniform media. Such media are frequently used in experimental investigations, but usually the extra cost is not justified in full scale installations.

DUAL MEDIA AND MULTIMEDIA

Upflow washing stratifies a bed in accordance with the settling velocities of the media particles as determined by their size, shape, and specific gravity. In a dual or trimedia bed, although each media component is still graded fine to coarse in a downward direction, lighter coarser media can be maintained above finer denser media. This makes it possible to approximate a coarse-to-fine gradation in downflow filtration units. Another advantage of dual or multimedia over a single medium is that mud balls formed in the filter remain above the coal-sand interface where they are subject to auxiliary scrubbing action.

The maximum settling velocities of media particles also determine the minimum wash rate required for adequate fluidization of the bed during backwash. Hence, for a given media size at the top of the bed, lower wash rates can be used if each media component is more uniform and the top portion of the filtration is of anthracite rather than a heavier material.

Hydraulic behavior and filtration performance of any given media are more properly related to pore size than to grain size. For single-media component, pore size is directly proportional to grain size, and the porosity (percentage of volume represented by pores) is a constant depending only on media shape. Coal, which tends to be angular, has a porosity of almost 0.5, whereas sand porosity is closer to 0.4. In water-treatment applications, coal media, because of its greater porosity, has been found to give poorer removals but lower pressure losses than sand of the same grain size.

The pore size in multicomponent filter media depends on the degree of intermixing of the components. With no mixing, pore size distribution simply follows that of the components. With intermixing, the finer layers of the denser material below are dispersed into the voids of the coarser layers of lighter material above. Where such information is of interest, it may be obtained from test columns or from experience with specific combinations of components in other installations.

The single-medium configurations employ depths of 60 in. or more. In downflow filtration, this great depth is intended to improve efficiency, whereas in upflow units, it has an additional purpose of adding weight to restrain the bed from uplift due to differential pressures during operation. Where uplift exceeds the submerged weight of the media, it will either fluidize the bed or lift it in a "piston" effect (small-diameter filters).

Some of the theoretical advantage of upflow coarse-to-fine filtration is lost because minimum grain sizes must be coarse enough to avoid excessive uplift. Additional resistance to uplift is provided in many upflow designs by placing a restraining grid on top of the media. The spacing between bars of the grid must be large enough to prevent upward bed movement during filtration. Although these two requirements appear contradictory, arching of the grains takes place between the bars allowing a reasonably large spacing in the range of 100–150 times the diameter of the smallest grain size in the beds.

Pilot testing is required to provide the information necessary for meaningful comparison of different media designs or to assure the effluent quality performance of any media design selected. Without pilot testing, media should be selected which on the basis of experience with similar influents may be expected to provide good solids removal with low head loss build-up. Any such media would include an ample depth of coarse media followed by fine media in the size ranges indicated for dual media configurations. Pilot testing to guide media selection should define headloss development versus time for each media design under all test conditions.

Major filter functions requiring monitoring and/or control are

- Head loss
- Effluent quality
- Initiation of backwash where automatic
- Flow rate through the filters
- Backwash sequence, rate, and duration

Performance control is the automatic turbidimeter which can continuously monitor the filter feed and product. This allows determination of difficulties from changes in feed quality and rapidly remedy process failures. Such devices allow the operator to rapidly evaluate the effects of changes in process variables and provide a continuous record of plant performance. All turbidimeters operate on the principle of measurement of scattered or transmitted light and a variety of commercial instruments are available.

Types of flow control systems used for filters include

- Effluent rate control
- Influent flow splitting
- Variable declining rate control

Effluent rate control maintains a set flow for each filter by throttling the effluent (Fig. 5). The throttle valve may be controlled directly by mechanical

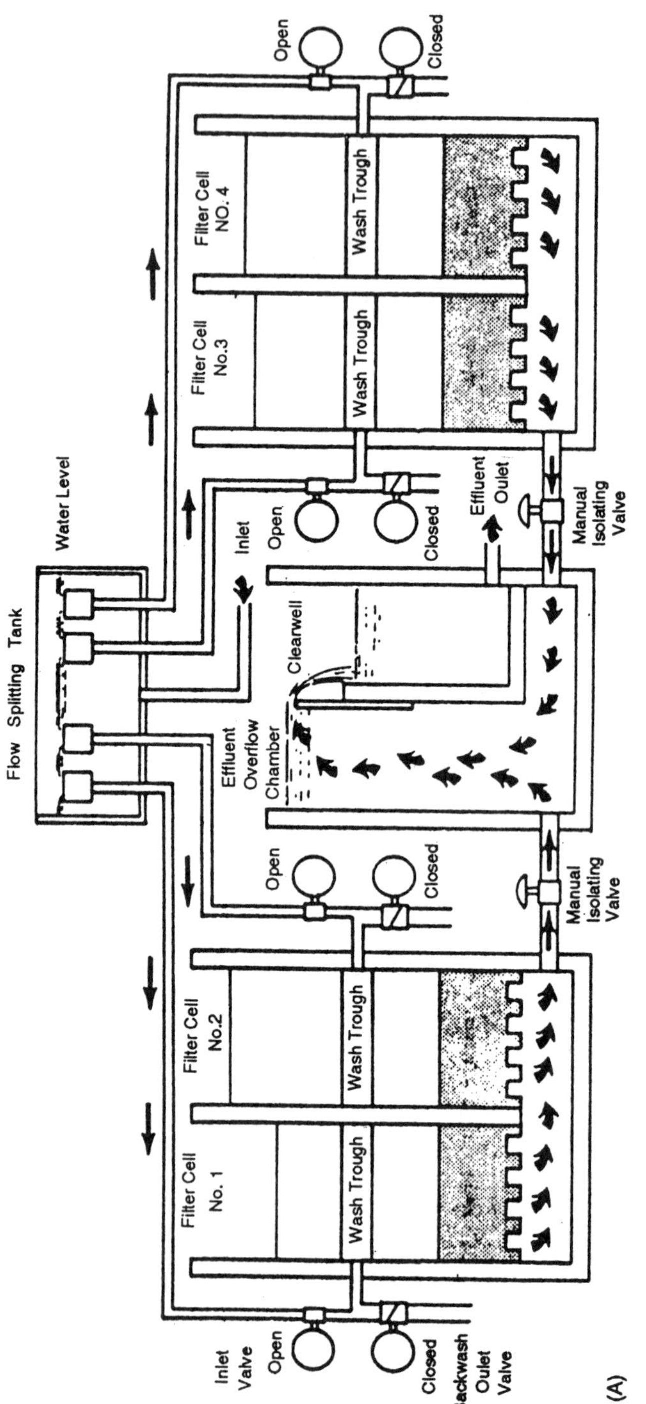
Inlet Valve
Open
Closed
Backwash Oulet Valve
Filter Cell No. 1
Wash Trough
Filter Cell No.2
Wash Trough
Open
Closed
Manual Isolating Valve
Flow Splitting Tank
Effluent Overflow Chamber
Clearwell
Inlet
Open
Closed
Water Level
Effluent Oulet
Manual Isolating Valve
Filter Cell No.3
Wash Trough
Filter Cell NO.4
Wash Trough
Open
Closed
(A)

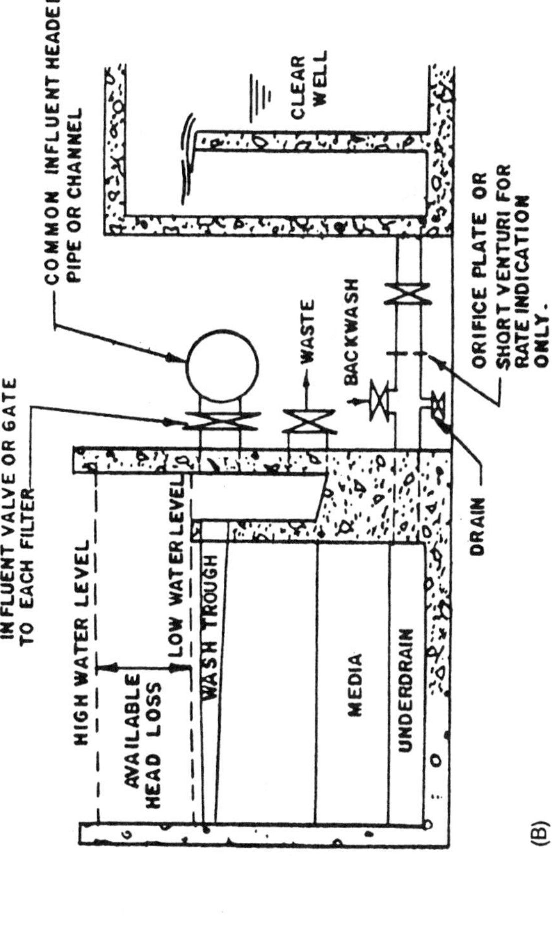

Figure 5 Flow control systems. (A) influent flow splitting; (B) variable cyclining rate filtration.

linkage to a venturi controller or indirectly by a set point controller linked to a pneumatic or hydraulic valve operator. The direct-acting system is unsuitable if flows to individual filters must vary over the day. The indirect system is complex and both may be troublesome in maintenance. The system is also wasteful of head, since available head not needed in a clean filter is lost in the controller. Control valves may produce high frequency surges in the filter bed with accompanying loss of efficiency.

Influent flow splitting system evenly divides among filters in a splitter box located at or above the level of the top of the filter boxes (Fig. 5A). The boxes themselves are made deep so that the water level in them can build up to provide the maximum operating head needed when the filter bed is dirty. A weir on the filter outlet maintains a constant back pressure or minimum water level to prevent accidental dewatering of the bed. Advantages include

- Rate controllers with attendant maintenance and surging problems are eliminated.
- Flow variations are distributed to filters automatically.
- Head loss may be read directly from water levels in filter boxes.
- Only a single master flow meter is needed.
- Changes in filter rate are gradual because of time required for head to build up in filter boxes.

Disadvantages include

- The head not needed for filter operation is lost in the drop between the splitter and the filters.
- Capital cost of filter box construction is increased by the greater depth.

Declining rate filtration requires multiple filters which operate under the same head but at different flow rates depending on the degree of clogging. Under constant head, the output from a single filter declines as the run progresses. The filter selected to be backwashed is always the one which has been on line the longest and is most clogged. Total output from all filters is controlled by varying the head applied. The head on the filters may be controlled by varying either the upstream or downstream water level. With downstream water level control, an equalizing chamber must be provided to limit the rate of change of head and hence of flow when filters are taken off line or restored to service. It is common to apply maximum design loadings to the filters as a group and to limit maximum rates on individual clean filters to from 20–40% above these design loadings.

Programmed backwash systems are widely used in designs. Such systems consist of interlocked controllers and timers programmed to open and close valves, make or break siphons, start and stop pumps and blowers, and limit backwash flows to control the rate, duration, and sequence of activities during backwash. Even where backwash is manually initiated, the rest of the control system may be entirely automatic. Proprietary systems with various features are available from different manufacturers.

FILTER CLEANING

Accumulated solids are removed from filters by a rapid upflow of washwater. The waste flow is then recycled to some prior treatment unit, usually primary settling. Wash water sources may include filter influent, filter effluent, or effluent from subsequent treatment units. Storage of wash water supply may be needed if rates required exceed the flow available. Recycled spent wash water flows should be equalized by storage so they do not disrupt prior treatment processes. Backwash rates for most effective cleaning vary with media size and density. Whatever wash rate and duration are expected, design of piping, valves, pumps, and storage tanks should provide extra capacity of at least 25% for bed expansion.

UNDERDRAINS

Underdrains should distribute wash water as uniformly as possible over the area of the filter. Excessive variation in wash water rate results in uneven and ineffective cleaning. Moreover, the accompanying excessive jet action can lead to lateral displacement of gravel and clogging of the underdrains with filter media. Underdrain systems developed for water filtration may also be used in wastewater applications.

FILTER MEDIA

Sand or other materials used as filter in granular media already discussed must possess good hydraulic qualities and have good filtration characteristics. It must be hard and durable, free of impurities, and insoluble in water. Fine sand is very effective in removing suspended matter but shortens the filter run. Coarse sand provides good hydraulic characteristics but reduces the removal of suspended materials. A thicker filter bed can somewhat offset the use of coarser sand. To adjust between these two important factors, filter sands are selected on the basis of *effective size* and *uniformity coefficient*. When these values are within certain limits, the sand will provide good filtering characteristics and allow an economical period of filtration between cleanings. The effective size is that size of sieve opening which will permit the passage of 10% by weight of the sand grains. The uniformity coefficient is the ratio of the sieve sizes which will pass 60 and 10% of the sample, respectively.

Depending on the desired rate of filtration, the effective size ranges from 0.25 to 0.55 mm and the uniformity coefficient varies from 1.5 to 3.0. The most commonly used medium for municipal water filtration is sand. However, it may be any inert material. Anthracite, diatomaceous earth, or a finely woven fabric or some similar material have been used. The use of some of these materials will be discussed under specific types of filters.

The mechanisms or processes by which a porous filter medium, such as sand, removes suspended particles smaller than the openings between the grains. Water filtration is based on five processes which occur in the sand. These are straining, sedimentation, adsorption, flocculation, and biological metabolism. The most important of these is straining, which occurs at the surface of the sand bed. Particles too large to pass through the openings are strained out. As filtration continues, smaller and smaller size particles are removed, and a mat is formed at the surface. When the water contains organic matter, part of the organic content is deposited within the mat. Bacterial growth develops within the mat when it is left for relatively long periods of time. The deposited material causes the pore space to become gradually filled. Resistance to flow and head loss are increased as filtration continues. When the head loss becomes excessive, the filter medium must be cleaned. The period between cleanings is known as a filter run.

The most widely used filters for water have sand or anthracite as a medium and may be either the gravity or pressure type. Two basic types of gravity filters are the slow and the rapid sand filter. Pressure filters may use sand or diatomaceous earth as a media. They are not usually employed in large plants. The use of the diatomaceous earth filters is increasing for small communities where a relatively clear source of water is available, at swimming pools, and at emergency installations.

SLOW SAND FILTRATION

Slow sand filters have been effective in providing a safe, potable water supply. There are limitations, however, in the use of this type of treatment. Generally, the average turbidity is limited to 10 ppm, with a maximum concentration of 30 ppm. Within these limits, effective bacterial and turbidity removals are realized accompanied by a color reduction of about 40%. It is evident that to remove such low turbidities, bacteria, and color, something in addition to straining action is necessary. Adsorption of substances on the surface of a particle undoubtedly is an important factor. Because of this, slow sand filters must be operated at the relatively slow rate of up to 5 mg/day/acre of sand bed. Clean sand is relatively ineffective until an adsorptive film is developed on the sand grains, so the filtered water from beds in which new sand has just been placed is normally wasted for several days until this film is developed. Usually the rate of filtration with cleaned beds starts out very low and is gradually increased until the desired rate is reached and the quality of the effluent is satisfactory. A slow sand filter is shown in Figure 6.

A slow sand filtration plant consists essentially of a covered concrete basin

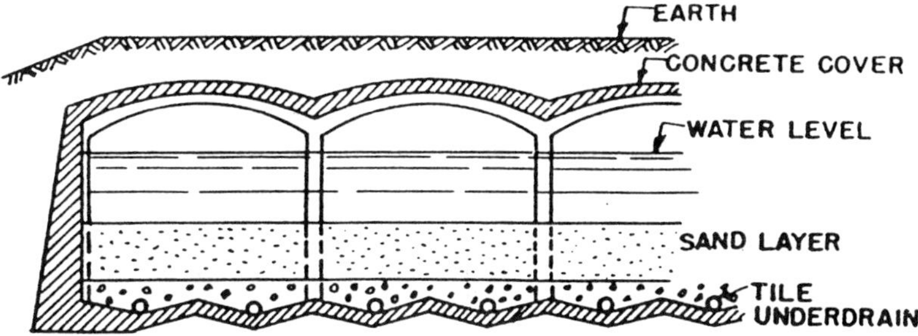

Figure 6 Slow sand filter.

about 10 or 12 ft deep. Open-jointed tile drains are placed on about 6-ft centers and lead to a central connecting pipe or main drain. These tile drains are covered with about 12–18 in. of graded gravel—the largest sizes on the bottom—which is in turn covered with about 3 ft of sand. The cover of the structure should be at least 6 ft above the surface of the sand in order to provide for an adequate depth of water over the sand and sufficient head room during the cleaning operation. The cover usually consists of a concrete slab supported by columns with several feet of soil on top to prevent freezing. Figure 6 is a typical cross section of such a filter.

During a filter run, the rate of filtration is normally held constant. Usually, both the influent and effluent lines are controlled by valves, either automatically or manually operated, to accomplish this condition. Figure 7 is a diagrammatic cross section to illustrate the operation of a filter.

With clean sand at the start of a filter run the head loss is only a few inches and the water is not effectively filtered. Effectiveness of filtration is dependent on the building up of an active surface mat. Since the turbidity of the raw water is generally low, it may take several days for a satisfactory mat to form; during this period, the effluent is wasted. As the sand becomes dirty, the rate of filtration will be reduced unless something is done. The proper rate of discharge can be restored by opening the effluent valve slightly, but the water level in the indicator tube will be lowered thereby. As the sand gets still dirtier, the rate of filtration will again decrease. The proper rate is restored by opening the valve further, and the water level in the indicator tube will drop still lower. Eventually, as the sand gets dirtier and the effluent valve is opened further, the water level in the indicator tube will drop to the level of the top of sand, at which point the filter should be taken out of service and cleaned, because the resistance to the passage of water through the sand has reached the available pressure or head of water on the sand. If the water level in the indicator tube is allowed to drop below this level, the pressure within the sand bed will be less than atmospheric, and

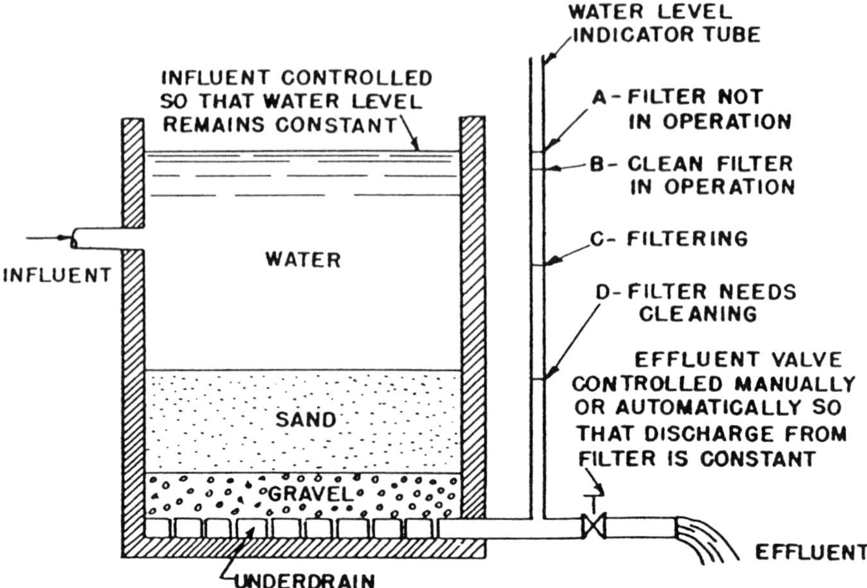

Figure 7 Diagrammatic section of sand filter.

operating difficulties or an unsatisfactory effluent may result. As the period between cleanings is usually a matter of weeks and the change in the rate of filtration is very slow, the manual operation of the effluent valve is reasonably satisfactory.

Filtering material of the proper quality and size characteristics is important. Slow sand filters should have sand with an "effective size" between 0.25 and 0.35 mm and a "uniformity coefficient" of between 2.5 and 3.5. The trend is toward the use of larger sizes of sand in order to provide better hydraulic characteristics if the quality of the water to be treated is satisfactory and permits the use of larger sand.

Quality of the gravel should be equal to that of the sand and the sizes should vary from about two inches on the bottom to one-eighth inch or less on the top. The gravel *is usually* placed in about six layers, each 2–3 in. thick and of gradually decreasing size, giving a total gravel depth between 12 and 18 in.

The first step taken in cleaning the filter sand is to close the influent valve and allow the water to drain from the filter. Most of the suspended material is removed at the surface of the sand bed and most of the head loss also occurs within the first few inches of the sand bed. The filter is cleaned by removing the first 3 or 4 in. of sand. The filter may be cleaned several times before any sand is replaced. However, the depth of sand on the filter must never be less than 24–30 in. The removed sand is washed and stored. Several mechanical devices are available to wash the top layer of sand without draining the filter.

RAPID SAND FILTRATION

The terms *rapid sand filters* and *mechanical sand filters* are synonymous. The first term is based on the fact that the rate of filtration is usually about 40 times the rate of filtration through slow sand filters, whereas the latter term comes from the fact that mechanical washing equipment is used to clean the beds.

Attempts to reduce filter areas by increasing the rate of filtration when treating turbid waters without subsurface clogging of beds occurring until coagulation was used. Coagulated material or floc as found to be sufficiently coarse to be retained by larger sand grains and the "mat" produced by the accumulated floc effectively removed bacteria and fine suspended solids not originally enmeshed as the floc was being formed.

A rapid sand filtration plant consists fundamentally of a clean bed of fairly coarse sand to remove previously coagulated solids remaining after sedimentation. The effective size of the sand is usually in the range from 0.35 to 0.55 mm as compared with 0.25 to 0.35 mm for slow sand filters. The depth of sand must be sufficient to prevent the penetration of the floc through the bed and facilities must be provided for washing the sand at periodic intervals so that it is maintained in a clean condition. The effectiveness of rapid sand filtration depends on the effectiveness of preliminary coagulation and sedimentation and on the condition of the sand. The depth of sand usually is between 24 and 30 in. and rests on from 9–18 in. of graded gravel or on some special type of patented filter bottom.

Besides the gravel, or in conjunction with a special filter bottom, an underdrainage system is provided which is capable not only of uniformly collecting the filtered water but also of uniformly distributing the relatively large flow of water when the filter is being cleaned or backwashed. The most common arrangement is the header and lateral system. The header or manifold is a large central pipe to which smaller pipes, or laterals, are connected at relatively close intervals on either side and extending horizontally into the gravel. Small holes are located at intervals of several inches along the laterals through which the water flows. Figure 8 is a cross-sectional view of a typical rapid sand filter showing the location of the underdrainage system, as well as the gravel and sand layers.

During the filtering process, the hydraulic principles of operation are identical with those of the slow sand filter, although the rate of filtration is 125 mg/day or 2 gpm/ft^2 of filter area. The opening through the effluent valve is always controlled by an automatic device, as the period between cleanings is

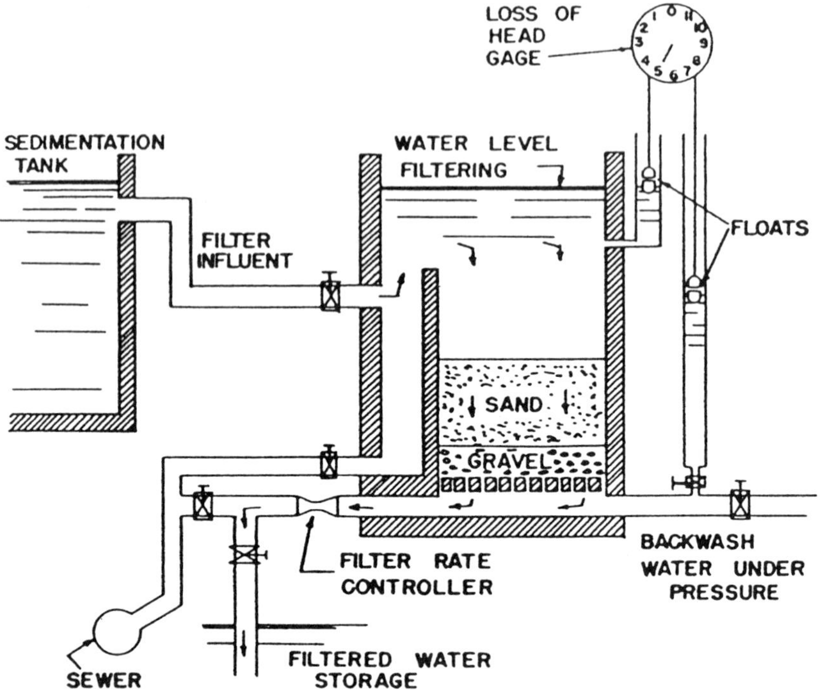

Figure 8 Rapid sand filter operation.

usually from several hours to a few days. Otherwise frequent adjustments would be required if the unit were manually operated. This valve is called a filter rate controller and may operate on any one of a number of principles.

The most common type of filter rate controller employs a Venturi tube principle in which the difference of pressures at the large and small section of the tube is a measure of the rate of flow through the tube.

Filter Washing

Washing of filters is one of the most important operations in a rapid sand filter. It should be done either when the loss of head equals the distance from the water surface on the filter to the bottom of the sand layer or when the effluent is no longer satisfactory. A filter is backwashed by first shutting of the influent line, opening the sewer valve, and then running clean water through the underdrainage system in a reverse direction at seven or eight times the rate of filtration. The dirty water coming from the top is collected by means of wash water troughs and discharged to the sewer.

After the filter is taken out of operation, the water surface should be lowered to the top of the wash water troughs. After the necessary valves have been closed and the drain valve opened, the wash water should be turned on gradually until the desired maximum rate of flow is reached. Otherwise "boils" may be created and the sand and gravel unduly disturbed in those areas where clogged sand is broken suddenly or moved as a mass. The required rate of flow for the backwashing operation is that which will expand the sand bed until the individual grains are not continuously in contact with each other but will vibrate back and forth and dislodge any dirty material adhering to the surface of the sand grains. The rate should also be great enough to raise the small dirt particles vertically and carry them out into the wash water troughs. Backwash rates may be expressed in terms of gallons per minute per square foot of bed, usually about 15 gpm/ft^2, or as the vertical rise of water per minute, usually 24 in./min:

$$\frac{15\text{ gal}}{\text{min} \times \text{sq ft}} \times \frac{1\text{ cu ft}}{7.5\text{ gal}} \times \frac{12\text{ in.}}{\text{ft}} = \frac{24\text{ in.}}{\text{min}}$$

Backwashing should be continued until the wash water rising through the filter is clear, at which time the wash water and sewer valves should be closed and the filter put back in operation.

Sand expansion is necessarily limited by the elevation of the wash water troughs. It follows, therefore, that these troughs should be located at an elevation above the sand high enough to preclude the possibility of any sand being washed into the troughs. The most effective sand washing occurs when the sand bed is expanded about 40%. Too great an expansion of sand is detrimental because the sand grains may be so far apart as not to impinge on each other to the degree required for effective scrubbing action. The rate of flow required to produce such an expansion is, of course, dependent on the size of sand grains. In addition, the temperature of the water is also an influencing factor and a greater backwash rate is required in the summer months than in the winter. Backwash rates as high as 22.5 gpm/ft^2 (36-in. rise per min) may be required in summer. Depending on the character and size characteristics of the filter media and the effectiveness of filter washing, the total wash water requirements for a well-operated rapid filter plant will vary from 2 to 5% of the water filtered.

In many plants total dependence is not placed on hydraulic backwashing alone; additional facilities are provided to assist in the cleaning operation. In addition to a possible increase in the quality of cleaning, such aids all have the advantage of permitting the use of somewhat lower wash water rates than would be necessary with hydraulic backwashing alone. The formation of mud balls is minimized if not prevented altogether. The three main types of backwashing aids are:

- Hydraulic surface agitators
- Mechanical rakes
- Compressed air

Hydraulic surface agitators are horizontal pipes located a short distance above the sand bed with small perforations or nozzles at fairly close intervals through which water is forced at high velocities. The jet action of these small high speed streams provide violent local scrubbing action in the zone of dirtiest sand. The horizontal pipes may be stationary or may rotate.

Circular filters are often equipped with mechanical rakes suspended above the sand and rotated horizontally about a vertical axis during backwashing operations. The tines of these rakes extend well into the sand beds.

Compressed air is sometimes introduced into the underdrainage system before or at the same time the wash water is turned on. Generally a separate piping system is provided for the air. The operating costs of an "air wash" system are usually higher than other forms of backwashing aids. This system is not used as much as in former years. Normal operation of a rapid filter is relatively simple provided none of the component parts of the filter or the various pretreatment processes fail.

As discussed, one of the requirements for effective operation is that floc should not pass through the filter. However, for the most efficient operation full penetration is desirable and may be considered to have been reached when the turbidity of the effluent is about one-fifth of a unit.

The rate of floc penetration is dependent on several factors. These are porosity of the filter media, head loss, and temperature. Size and toughness or strength of the floc is also a factor. Formation of floc may be observed by examining a sample taken from the flocculating basin and using transmitted light. A method commonly used is to lower a submergible electrical light bulb into the basin and observe the floc formation by reflected light. Floc may be detected in the clear water basin by this procedure.

The term *negative head* describes the existence of a pressure below atmosphere. Magnitude of the pressure deficiency is the intensity of the vacuum at that point. A negative head or partial vacuum occurs in a filter when the head loss at any point is greater than the pressure head at that point. This condition develops when a high loss of head occurs within a relatively small depth of the filter media, such as the first few inches of the sand bed.

The existence of excessive negative head in the filter is objectional because the reduction in pressure allows air which is dissolved in the water to escape. This results in air binding of the filter. Air binding of a filter is caused by the release of dissolved air from the water. The released air forms bubbles which may remain in the sand, break through the sand bed or accumulate as an air mass within the filter.

Air bubbles in the sand cause a decrease in its porosity and may result in a loss of filter capacity or cause unequal rates of filtration. If the bubbles break through the sand bed, channels are formed through which water may pass without proper filtration. Accumulated air masses may break through the sand at the beginning of the washing process and allow the wash water to develop high velocities at these points to displace the gravel and carry sand particles to the wash water troughs.

Dissolved air may be released from the water being filtered as follows:

- The reduction of pressure resulting from operation under a negative head
- An increase in the temperature of the water during filtration
- The release of oxygen by algae collected within the filter

All of these conditions are detrimental to efficient filter operation.

Materials carried over from the flocculating or sedimentation basin are deposited on the surface of the sand may form clumps varying from pea size up to 1 or 2 in. or more in diameter. These are known as mud balls and consist primarily of grains of sand and gelatinous materials. They may be distributed throughout the sand and at times become firmly attached to the gravel. Mud ball formation may be due to ineffective washing. This allows some gelatinous material to remain on the surfaces of the sand particles and results in the cracking or clogging of a filter bed.

Methods used for removal of mud balls are

- Breaking up by the use of rakes during washing.
- Removing by the use of dipper during washing.
- Breaking up by the use of water jets.
- Soaking the bed for a period of about 12 hr with solution of caustic soda. After use of caustic soda, the sand is thoroughly agitated and washed.

Formation of cracks and clogging of a filter may occur when a dense blanket of floc, organic matter, and mud is formed on the filter media by gelatinous coating on the sand is compressed and the sand settles. This causes the blanket to shrink and cracks are formed in the filter bed and at the sides of the filter. Cracks may extend for some distance into the sand and are soon filled with mud and floc. The rising wash water compacts these materials into clumps or balls during the washing process. As their density increases, they sink to the level of the gravel.

During later washing periods, these balls cause water to be unevenly distributed and mounds are formed by the movement of the gravel. Improper washing of the sand occurs and, if continued, a clogged mass may extend from the gravel layer to the top of the sand bed. Preventions are similar to those for controlling mud balls. In severe cases, it may be necessary to remove, clean and replace the sand.

Filter Cleaning

Cleaning the filter refers to the freeing of the sand of incrustations and mud ball formation due to calcium, magnesium, alum, or other deposits which accumulate on the surface of the sand particles. Beside contributing to mud ball formation, such deposits have the effect of enlarging the grain size and changing the filtering characteristics of the media.

Sand may be cleaned either in place or after removal from the filter. Methods used may be mechanical or chemical. Mechanical methods are similar to those given for the control of mud ball formation. When acid or other chemicals are to be applied to the filter, a test using a sample of the sand and the chemical should be conducted in the laboratory.

Caustic soda, soda ash, sulfuric or hydrochloric acid, sulfur dioxide, and chlorine have been used. A solution of the chemical is allowed to stand in contact with the sand for periods of 4 hr when sulfuric acid is used to as long as 48 hr for caustic soda and soda ash. After chemical treatment, the filter must be thoroughly washed before it is put back into service. A chlorine solution of about 50 ppm and a contact period of 4–24 hr or more is effective in removing material of biological origin. Alum and organic deposits may be removed with caustic soda. The acids are effective in removing calcium carbonate. Sulfur dioxide is used to remove iron, alum, and manganese. Copper sulfate may be used for the removal of algae. A patented process consisting of pumping a 2% aqueous solution of sulfur dioxide through the unit for about 24 hr has been used for the removal of iron and manganese. Treating filter sand with chemicals is hazardous and extreme care must be used. Mechanical or chemical cleaning processes provide only temporary relief. It is good practice to seek and correct the conditions which make the cleaning operation necessary.

Crushed anthracite may be used to replace the sand as a filter medium. Among the advantages claimed for its use are lower head loss, less wash water needed, a lower rate of application of wash water, a higher rate of filtration, and longer filter runs. One disadvantage is the loss of material due to abrasion. Anthrafilt has an effective size between 0.70 and 0.75 mm and a uniformity coefficient of 1.75. Graded anthracite varying from 1 5/8 in. at the bottom to 5/16 in. at the top is placed in four layers to replace the gravel. A layer of the fine material may also be placed as a cover for a sand bed. This combination of sand and anthrafilt is claimed to reduce mud ball formation and to increase the filter run and is called a "capped filter."

FILTER LOADING

To produce an effluent of highest quality by filtration through rapid filters, the importance of effective coagulation and sedimentation cannot be overstressed.

The ability of a rapid filter to remove turbidity, bacteria, and the materials causing taste and odor depends on the prefiltration processes. These filters will remove turbidity when coagulation and sedimentation are effective. Removal of algae, diatomes, and other large organisms, clogging of the filter may occur, taste and odor may be created in the water, and oxygen may be released by the organisms, which will tend to cause air binding. Control of the concentration of these organisms by treatment in the impounding reservoir is considered to be better practice than their removal by filtration.

Permissible bacterial (coliform) loading on a rapid filter is based on the overall efficiency of flocculation, sedimentation, filtration, and chlorination following filtration. It is generally agreed that the use of the above processes will be adequate for waters having coliform counts (MPN) no greater than 5,000/100 ml, when not more than 5% of the samples tested show that count. For counts from 5,000 to 20,000/100 ml, the use of prechlorination, presettling, and postchlorination is required. When the concentration is greater than 20,000, a long period of storage and possibly other precautions should be taken. Bacterial removal varying from 90 to 99% can be expected for concentrations in the lower ranges.

Taste and odor caused by algae, dissolved gases, and dissolved mineral matter can be removed by aeration, chlorination, the application of activated carbon, and other means and is discussed elsewhere.

DIATOMACEOUS EARTH FILTRATION

This type of filtration has been adapted for municipal water supplies. Diatomaceous earth filtration units were found during World War II to be effective in prevention of amebic dysentery by the removal of *Endomoebia histolytica* cysts. Such units are compact and may be made portable. Construction costs are generally lower than rapid sand filtration plants and operational costs can be comparable. This type of filtration is also discussed in Chapter 7.

Diatomaceous earth is basically the skeletal remains of algal forms known as diatomes. Deposits of the skeletons of these organisms which contain siliceous materials are found throughout the world. In the United States, the largest deposits exist at Lompoc, California. In the geological past when warm-water inland seas existed there, abundant numbers of diatomes thrived and their remains were deposited as sediment.

Silica, the basic material of the particles of diatomaceous earth is high in strength and the particles are rigid and abrasive. After processing, the final product is white to gray in color and weighs 9–13 lb/ft^3. When wet, the weight varies from 18 to 21 lb/ft^3, which indicates their high void capacity. The size of particles in any grade can vary from 1 to 200 μm. However, the variation in a particular grade generally ranges from 30 to 40 μm. A characteristic of diatomaceous earth is high permeability due to the high void capacity.

Diatomaceous earth filtration is radically different from either rapid or slow sand filtration. The greatest differences are in the depth and size of the media. In rapid sand filtration beds, the particle size of the sand ranges between 0.35 and 0.80 mm (350–800 μm). Diatomaceous earth particles vary from 0.01–0.2 mm (10–200 μm). Particles smaller than 10 μm and larger than 200 μm can be expected. Twenty-four to 42 in. of sand bed is provided in slow and rapid sand filtration units. The filtration media in a diatomaceous earth unit is 1/16 to 1/8 in. in depth. When the rate is low, the area must be increased and the need for pretreatment is lessened. The higher rates used in rapid sand filtration require pretreatment by coagulation, flocculation, and sedimentation. The penetration of suspended materials into the sand aries directly with the grain size. And deeper penetration can be expected when large particles are used. With diatomaceous earth filtration, most of the suspended material is retained on the surface of the filter media and media depth increases as filtration progresses.

There are four basic operations involved in the diatomite filter process:

- Precoat
- Body feed
- Filtration
- Backwash

In Figure 9, the essential elements of a pressure diatomite filter system are shown. The system consists of a pump, a filter tank containing the filter septums, and auxiliary body feed equipment. Operation of either pressure or vacuum-type units is essentially the same. Differences are in design to meet a particular hydraulic condition.

Precoat

When diatomaceous earth filter units are placed in operation, a slurry of diatomaceous earth is introduced at the tank influent line and is hydraulically deposited by the filtered water on each septum. It is desirable to recirculate by pump suction and to recycle until all the diatomaceous earth is deposited. This will ensure an even deposition on the filter septa. The minimum amount of diatomaceous earth used under recirculation conditions is 0.1 lb/ft^2 of filtering area. This provides approximately a 1/16-in. coating on the elements. When the precoat is applied without recirculation a minimum of 0.15 lb/ft^2 must be added. This additional amount is added to ensure an adequate and even precoat. Applying diatomaceous earth without recirculation can result in uneven deposition on the septa. The precoat formed on the septa is caused by the "bridging" of the diatomaceous earth particles across each opening on the septum. This is shown in Figure 9A. The opening is much larger than the particles which form the "bridging."

(A)

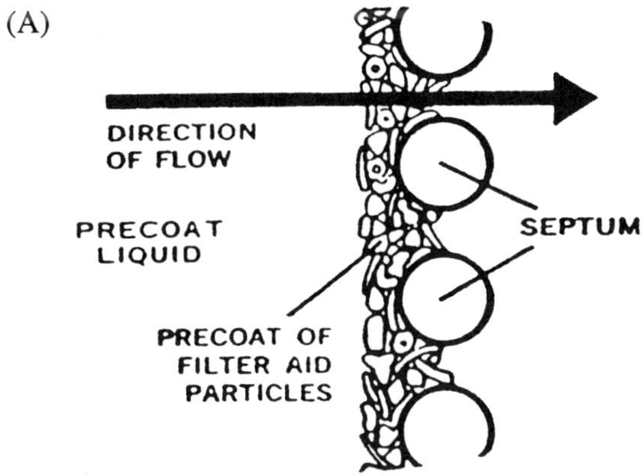

(B)

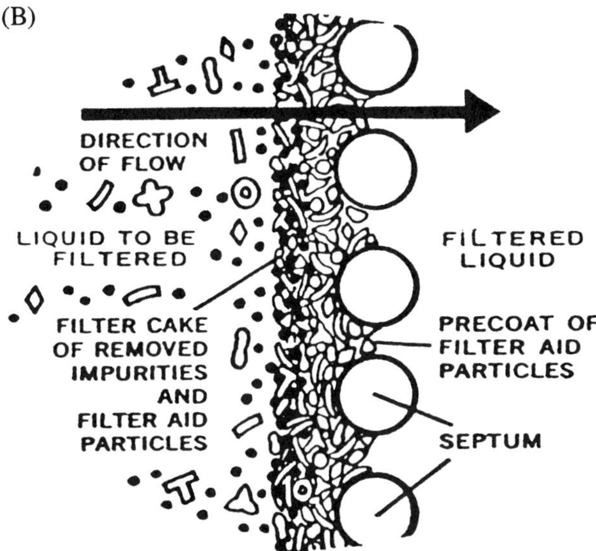

Figure 9 Precoat and filtration functions at the septum face.

Body Feed

Figure 10 shows a body feed tank from which an additional amount of slurry of diatomaceous earth can be mixed with filtered water and discharged into the water line to the filter. The addition of diatomaceous earth to the water to be filtered will lengthen the time between filter cleaning operations. Continuous addition allows the porous earth particles to mix with the suspended particles in the water. This mixing maintains the cake of diatomaceous earth on the filter septa in a more porous condition and results in a lower head losses. Figure 9A is a schematic cross section of one type of septum. Figure 9A shows the precoat of diatmaceous earth on a filter septum. Figure 9B shows the condition when diatomaceous earth is also added to the raw water during the filtering operation. The material is added by pumping reduction. When high levels of body feed are used, the accumulated precoat, body feed, and suspended material may be relatively deep on the filter just prior to backwashing. A minimum spacing of 1 in. between the outside of the septa is normally provided. If a cake depth of more than 0.5 in. accumulates on filter septa, problems in removal of the cake will result.

Filtration

After the precoat has been applied and the body feed operation started, the filtering process is initiated. A flow of water must be maintained through the filter tank in order to preserve the precoat and flow in the filter water line should

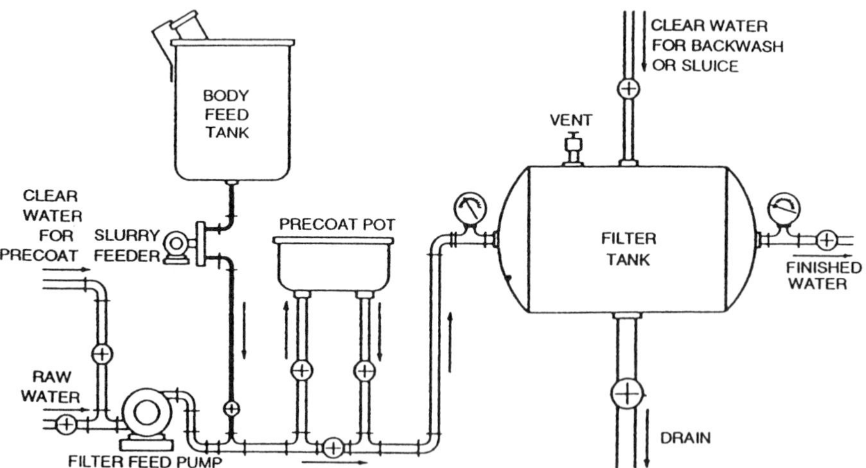

Figure 10 Pressure diatomite filter system.

be started before the precoat line is closed. The filter is operated until the desired head loss is reached. This is usually between 25 and 30 $lb/in.^2$ for pressure units and 15 in. of mercury vacuum for vacuum units.

Stopping the flow of water through a filter will cause the filter media to drop from the septa. In some cases, continuous operation of a filter is not required and the filter cake is kept on during interruptions. In order to keep the cake on the septa when a filter is operated intermittently, an auxiliary pump can be installed to recirculate water through the filter unit at a reduced rate of flow. This rate is usually not less than 0.1 $gal/min/ft^2$ of filter area.

Backwash

When the head loss through the filter has reached the desired maximum, the filter is cleaned by backwashing. Various methods are used to backwash filters. These are "air-bump," reverse flow, jetting, air gurgle, or any combination. High velocity of flow may be used in some filters for backwashing. Filters in which cloth or synthetic fabric is used may not withstand the high velocities used in wash rates. Poro-stone (silica bound aluminum oxide crystals) or monel wire mesh is generally used when the rate of backwash flow is high. The cleaning of a filter is usually completed in 5–8 min.

Removal of diatomaceous earth from septa during backwashing may be improved by using and air-bump procedure. In this method, the filter is backwashed and the tank is drained. After draining of the tank, the backwash line is closed and filtered water is introduced under pressure through the influent line. The air in the tank is compressed and 50–60 psi may be reached. When this pressure is released very quickly to atmosphere by the operation of a quick-opening valve on the backwash water line, a reverse surging action occurs through the filter septa.

Some units are designed with internal jets to aid in the removal of spent diatomite in the cleaning process. Vacuum diatomite filter systems use a manually operated hose to clean the filters and to flush material which has been removed to the waste line. Five to 20 min may be needed to backwash a vacuum filter.

Pretreatment Considerations

When a diatomaceous earth filter is used, the water to be filtered is not generally pretreated if the unit has been properly designed for the particular water. It has been found that when a filter of this type is used to treat water which contains appreciable amounts of clay and/or silt some operational problems can be expected. When turbidities are consistently less than 10 units and filtration rates are between 1–2 gpm/ft^2, little difficulty has been experienced when adequate body feed has been provided.

Microorganisms are often the major contributor to head loss during diatomaceous earth filtration. These organisms consist of various forms of algae, protozoa, and crustaceans and the products of their decomposition are also present. Microscopic organisms, products of their decomposition, and other amorphous organic material are removed by diatomaceous earth filtration. Removal of this material on the filter elements can reduce the length of time between filter cleaning operations. When a microscopic examination of water indicates that the enumeration of microorganisms is consistently at 4,000 standard units per milliliter or greater and when the turbidities are generally fewer than 10 units, pretreatment by microscreening to remove the microorganisms and their decomposition products may be indicated.

Diatomaceous earth filters have been used with ground waters which contain up to 4.0 mg/L of iron. Ferrous iron may be removed by mixing the water with magnesite (magnesium sulfate activated by heating) and discharging the treated water to a pressure-type unit. The iron is adsorbed by the magnesite and the magnesite particles are removed by the filter.

8
Mechanical Filtration

INTRODUCTION

Filtration can be described as a process in which relatively small quantities of solid matter are removed from liquids in which the solid is suspended. There are many types of mechanical filtering equipment, some of which may be classified as follows:

- Filter presses of various types
- Pressure or vacuum filters in which pressure or vacuum is used to force the liquid through filtering material such as sand or other media
- Leaf filters which were first developed in the metallurgical industry
- Rotary continuous vacuum filters
- Belt filters
- Hydro-extractors and centrifugals

Figure 1 shows a broad classification of filtering devices.

FILTER PRESSES

A filter press is a frame in which a number of loose slabs of filter surface may be clamped to form a series of hollow chambers capable of withstanding internal pressure. This arrangement gives a large filtering surface for a given volume of

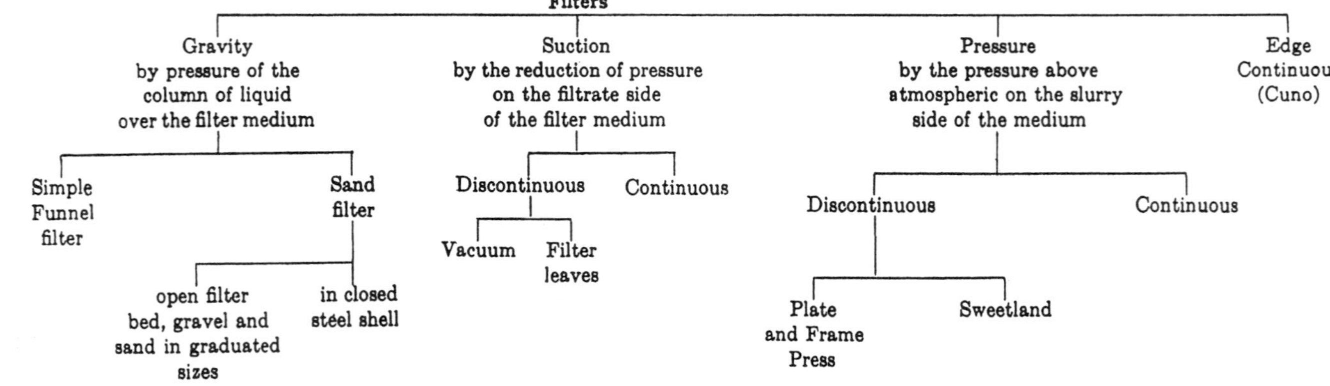

Figure 1 Classification of filters.

equipment and, when properly constructed of suitable materials, gives long service with a degree of reliability. Two forms of filter press in general use are the chamber and frame types.

Chamber Presses

In this type of filter press, as shown in Figures 2 and 3, the edges of the plates are raised so that when two are brought together a hollow chamber is formed. Filter cloths are laid over each side of the plate forming a tight joint when the plates are clamped together. The feed passage is taken through the body of the plates and through holes left in the filter cloths as indicated in the drawing. The

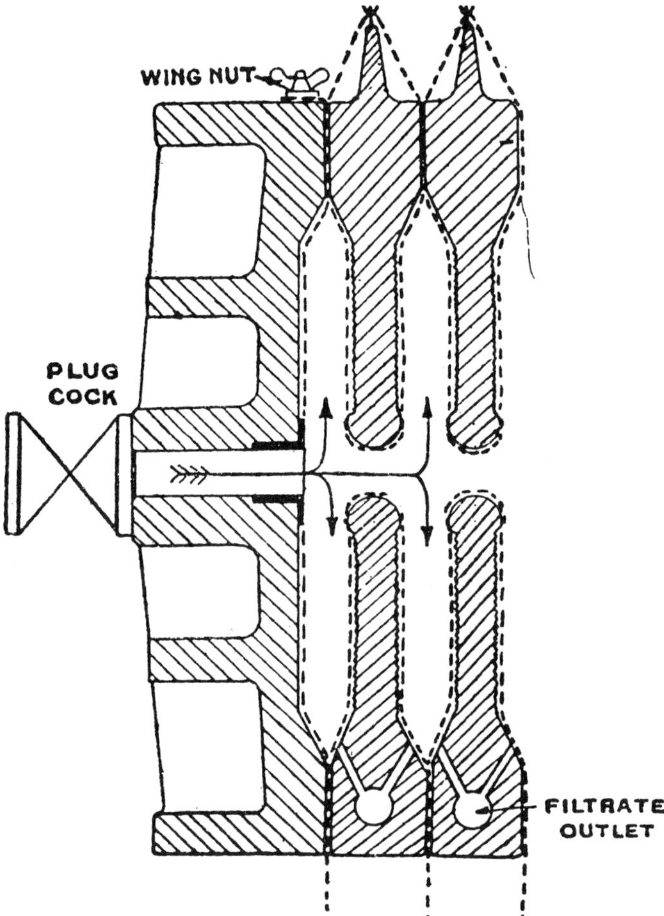

Figure 2 Chamber-type filter press.

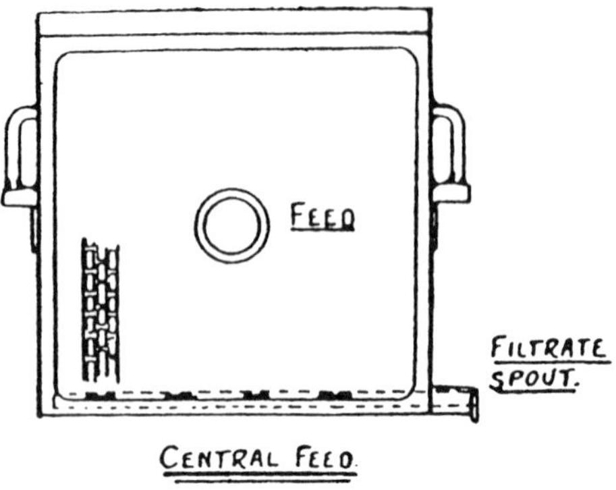

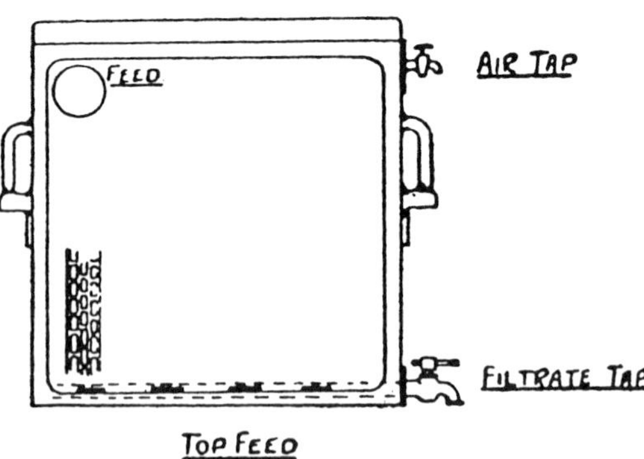

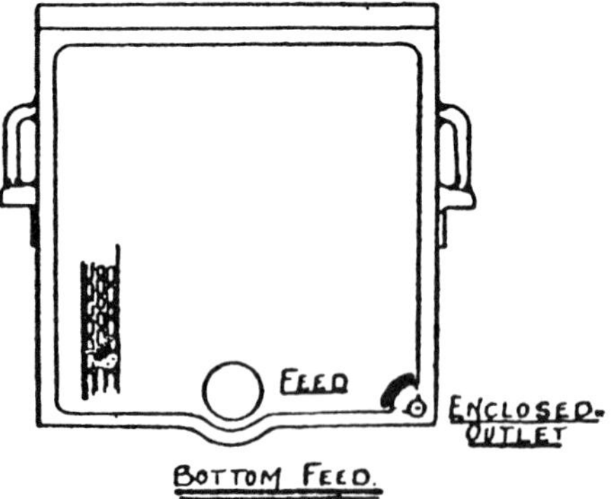

Figure 3 Methods of feed inlet and filtrate outlets in chamber filter press.

liquid enters the chambers under pressure and the solid portion is retained on the cloth. The clear liquid passes through the cloth and drains via corrugations on the chamber sides as indicated to the outlet at the bottom of the plates. Usually a tap is fitted to the outlet of each recessed plate so that any given chamber can be shut off if the filtrate is cloudy because of faulty filter cloth or other causes. If it is desired to raise the filtrate to a level higher than the press, this is done by a sealed outlet pipe into which all the enclosed chamber discharges are led.

Chamber or recessed plate presses are usually fitted with central feed, but the feed passage may be arranged in any other position if desired. For example, if a solid cake is not to be formed, bottom feed is convenient, as in this position the liquid contents of the chambers can be properly emptied before the press is dismantled for cleaning. In handling heavy suspensions, top feed is desirable, and sometimes in such cases two feed passages are arranged at different heights to obtain maximum density of cake.

Chamber presses are the least expensive type of filter press and require the least labor to operate and are particularly suitable for thin cakes, with the usual cake thicknesses of 1 in. for the smaller sizes and up to 2 in. for the largest chambers.

Plate and Frame Presses

This type of filter press is sometimes known as the flush plate type and is more versatile than the chamber pattern, but it is correspondingly more expensive to manufacture and requires more labor to handle. Figures 4 and 5 show a section of a group of such plates and frame units, and it will be noted that the plates are flat with the chamber formed by introducing a hollow frame behind successive plates. This type is suitable when thick cakes are desirable, and for cases where the conditions tend to rot the filter clothing, as the cloths lie quite flat on and are supported by the plate. The design permits the use of filter paper instead of cloth.

When handling materials which form a hard firm cake, an internal feed is desirable to avoid choking of the feed and uneven filtering pressure. In most cases, passages arranged externally to the cake are satisfactory, and the usual position for the passage is at the top center of the press; another favorite position is in the side of the press just a little above the center line. For sticky and gummy liquors, these presses are sometimes made with four passages, one at each corner of the chamber, as by this arrangement the liquor has a minimum distance to travel through the grooved surface to get to the outlet. Also, the upper passages permit of the free escape of air from the chamber.

Filter Press Materials

Plates are often made of cast iron for neutral or alkaline liquors, or in wood for acid or corrosive conditions, or of special metals and alloys to deal with special

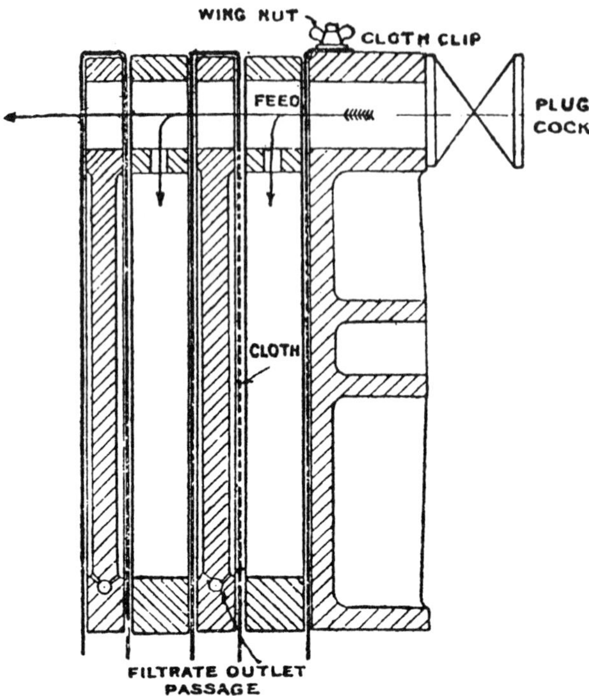

Figure 4 Plate and frame type of filter press.

problems. Sometimes vulcanite, rubber, or wood is used for handling strong hydrochloric acid liquors. Pitch pine is used for wooden plates, and for acetic acid liquors oak has been successfully used. Cypress wood is employed, and teak has been recommended for many acid liquors. Beech, maple, and sycamore woods have been used, but woods described are more reliable. Paraffin wax impregnation of wood frames before use has been recommended, and it is important when once in service that wooden plates should be kept wet to avoid shrinkage.

Filtering Surface

The surface of the plates upon which the filter cloth rests takes a variety of forms depending on the design of the plate, the position of the feed, and the material to be handled. It is important that the liquor channels be sufficiently deep and narrow to prevent complete closure of the passage by the filtering cloth.

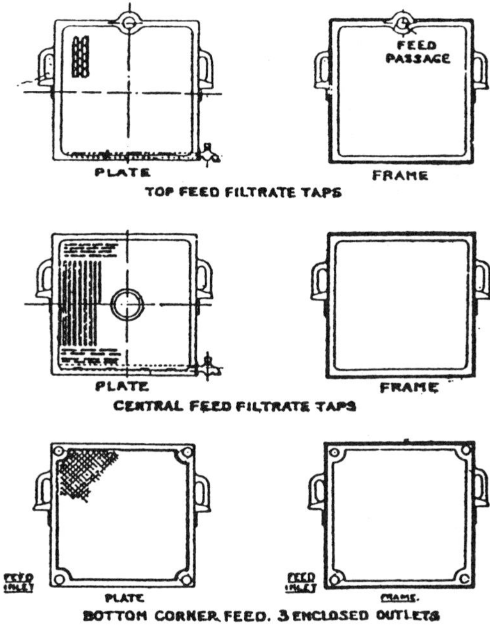

Figure 5 Various types of frames for plate and frame filter press.

Smoothness of surface and absence of sharp projections is most important, as otherwise the life of the cloths will be materially reduced, particularly if high filtering pressures are used. Coarse mesh gauze is sometimes used for difficult filtration, as also are perforated plates.

Plates and Dead Plates

A dead plate, or cutting off plate, is a desirable fitting if batches of varying bulk are to be handled. Such a plate has but one working surface, and is sufficiently robust to withstand the unbalanced end pressure, which, it must be noted, may be considerable at high filtering pressures. Dead plates can be placed anywhere in the assembly, thus permitting a selected number of chambers to be employed, with the remainder being isolated.

Closing Gear for Plates

The typical form small press is of the center screw pattern. It is not desirable to use this method for plates greater than 24 in.^{2}, in which case it will readily close and maintain tightness against 100 lb fluid pressure. Means should be provided for quick access to the plates for purposes of removal or cleaning, and this is

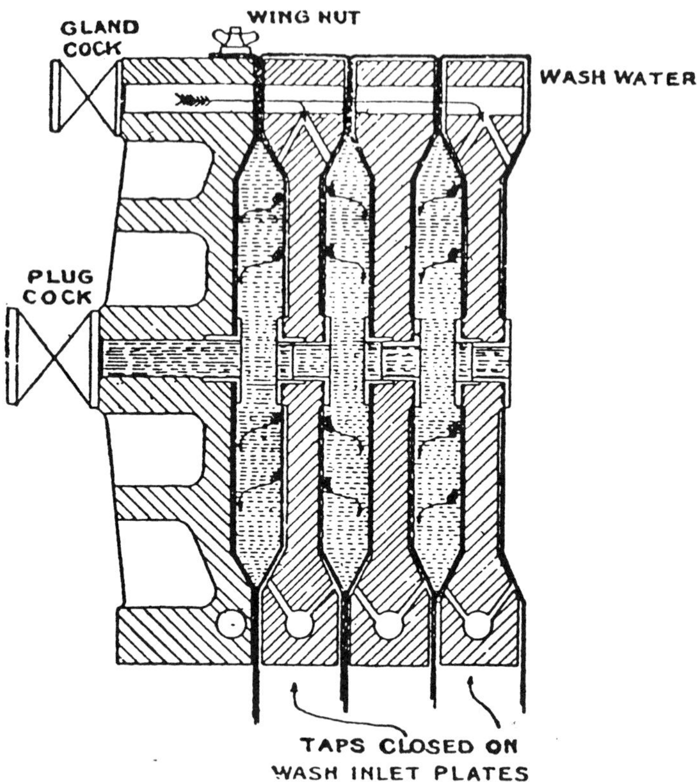

Figure 6 Recessed plate filter press arranged for washing.

achieved in designs by pivoting the loose head carrying the screw so that the head may be swung clear immediately the pressure is released. Hydraulic or pneumatic closure is adopted where speed of operation is essential and a pneumatic method of plate closure is used in sewage filtration.

The side rods or bars of the press act as supports for the plates as well as part of the press structure. Flat and round rods are equally suitable, and care must be taken to ensure that there is ample room for the plate handles and for filter cloth.

Filter Cloths

In chamber or recessed plate presses, double cloths are used, and in such cases the cloths are stitched together at the edges of the feed opening and one cloth is passed through the feed hole in the plate. Both halves are then spread on the filter surface and fastened at the top by means of clips, or they may be stitched. Another way is

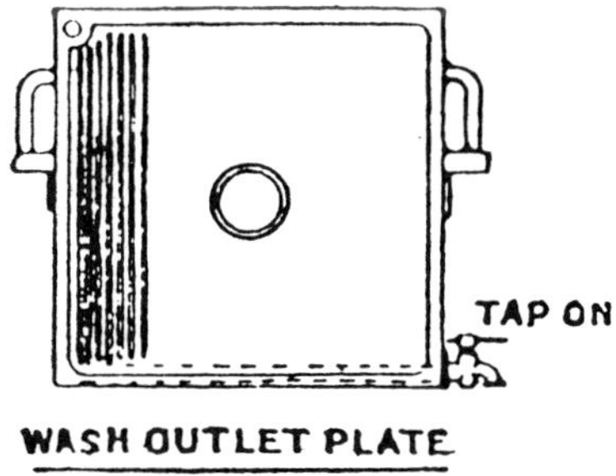

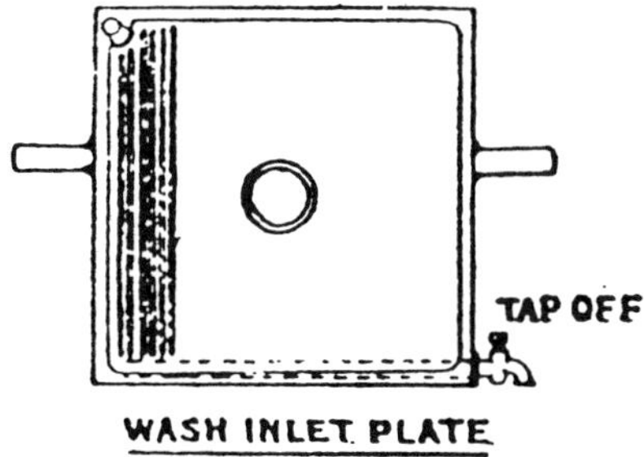

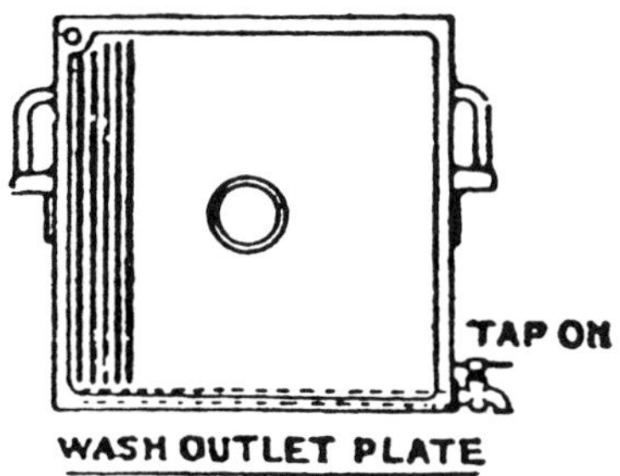

Figure 7 Plates used for recessed plate filter press.

to take a strip of cloth long enough to cover both sides of the plate with two holes cut to correspond with the position of the feed passage. This assembly is then hung over the plate and the joint around the feed passage made by clips of the screw pattern to make a fluid tight joint. In frame presses, cloths of double size may be simply hung over the plate without further attachment.

A common form of filtering material is cotton duck or twill jute, and hessian cloths have been used in place of cotton; these cloths are also used as undercloths to support a tightly woven twill. Synthetic filter cloths are also in use. Filter paper may be used, but it must be well supported and of the crinkled variety and backed by a suitable material such as hessian.

For strongly acid liquors, camel hair is reported as satisfactory for 30% sulfuric and 10% hydrochloric acid. Wool or felt cloths can resist up to 20% sulfuric acid, but are not recommended for hydrochloric acid.

Filter cloths should not be allowed to dry on the press saturated with possibly corrosive liquors and full of retained solids, but should be washed immediately after filtration is finished and kept moist until required again. Suitable washing machines are available for washing filter cloths under controlled conditions.

Operation

In a filter press, the liquor is fed to the press by gravity, by pump, or by some form of pressure egg, and in all cases the pressure of filtration should be carefully controlled. Generally, the best results are obtained with a relatively low filtering pressure with high pressures in the last stages of the run. In ordinary cases, pressures vary from 10 to 150 lb/in.^{2}; the higher pressure are only used for viscous liquors containing hard solids of an open granular nature.

Washing

Washing is carried out in the filter press with the simplest method to force water in through the feed passage. This is best done in the plate and frame type, and it is preferable that a solid cake should not be formed, but that a space should be left between the layers built up on each cloth. Here a bottom feed is advantageous, as the excess liquor can be drained from the chambers prior to washing, although in the ordinary type, it can generally be removed by air pressure. It is possible to wash this way in any plate and frame press or even in recessed presses, although they may not have been designed for the purpose.

The problem is that owing to sedimentation, the cake is apt to be thicker at the bottom, and in order to keep the two sides from touching, space may be wasted at the top. The press capacity is diminished and more labor is necessary in emptying. In some cases, however, solid cakes can be rough washed if they are sufficiently permeable. This method of washing is usually referred to as "simple" washing. Another method consists in forcing the water from one side of the cake to the other, and may be referred to as "through" washing. It is also called back washing, as the water is admitted to the back of the cloths.

The wash water inlet passage forms a continuous channel in an upper corner of the rims. It has inlet ports to each alternate plate on which the filtrate cocks must be closed during washing. The water enters behind the cloths and passes through these and the cakes to the back of the cloths on the intermediate plates, the cocks on which have been left open to permit it to escape. This is sometimes called tap washing, and it is very simple, as only one special passage is required. A disadvantage of an internal feed is that a soft core will permit short circuiting, whereas a hard one may not get washed properly. For best results, plate and frame presses should be employed because of the even thickness of the cake at the edge and the fact that any soft material in the feed passage is isolated. Provision should be made for the escape of any air which may have leaked into the space behind the filter cloths at the conclusion of filtration, as this may prevent the water flowing through the upper portions of the cakes. Even ordinary tap washing is distinctly improved by the provision of air cocks. Then, again, the washing should be taken out at the top of the press, so that there is exactly the same pressure tending to force the water through the cake, whether this be measured at the top or at the bottom. Except where provision has been made, time will be required under such conditions to force trapped air out of the press, and in addition, any unbalanced hydrostatic head due to unrestricted draining on the outlet plates will assist causing a disproportionate amount of wash water to percolate through the bottom of the cakes.

In the case of iron plates, wash passages are made either in the plate rims or can be in external lugs as shown in Figure 8, with advantage in certain cases, since this permits plain cloths without holes to be used, and there is no difficulty in fitting and keeping them in place. The joints on the lugs are often made by cloth sleeves, which are slipped over them. These must be changed with the cloth, and for that reason rubber joint rings are often preferred. These may be let into special grooves around the wash passages or may fit inside.

The cost of outside lugs is only justified where the filtering medium is subject to frequent washings or renewals. Normally, the cloth is kept in place to the plates once it has been used, so that even where passages are in the rims, no difficulty is experienced in keeping the holes in their right positions, especially if the cloths last a reasonable amount of time and proper arrangements are made for cutting the port holes.

Economical washing depends on even displacement, and good practice results when this ideal is approached as nearly as possible. If the cake is not complete, obviously there will be gaps through which water can pass. Further, if the deposit is soft and not properly compacted, such gaps may readily be formed during washing, whereas if this operation is delayed, some cakes tend to sag a little and do not fill the chamber up to the top. Frames may be fitted with an internal rib or fin to counteract the effect of this, but some users consider there is more difficulty in cleaning out the cake. If the grain is too fine, it will be difficult to get the wash water to penetrate without undue pressure, which may cause channeling and short circuiting. In such cases, adsorption effects are also likely to hold back soluble matter in spite of even the most perfect washing.

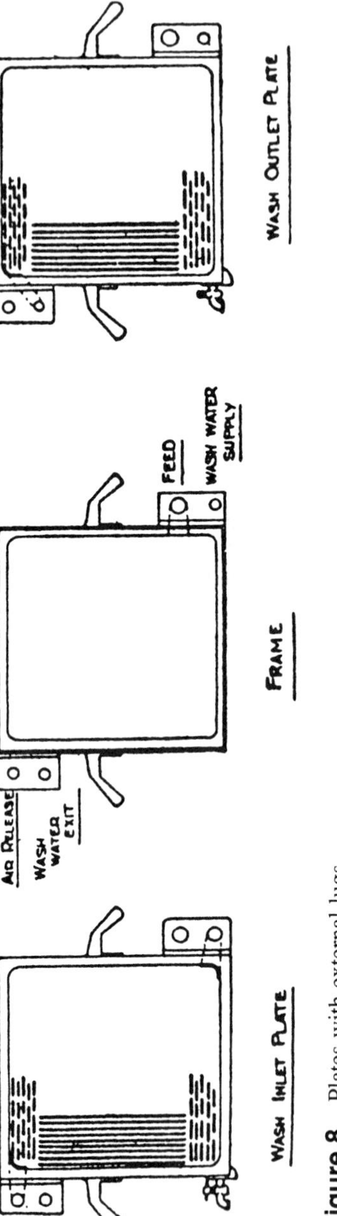

Figure 8 Plates with external lugs.

If the grains are reasonably coarse, they will be more permeable and will have less surface to retain strong liquor; truer and quicker displacement is then likely to result. If they are not of fairly uniform size, the coarser ones will probably have settled to the bottom of the chamber, and the wash water will pass in preference through them giving mixed or weak washing. If the particles themselves are porous, time will be needed to extract the absorbed soluble matter quite apart from surface effects.

High pressures should only be applied with caution where necessary, as this may cause channeling and short circuiting. The lowest pressure which will give a reasonable flow should therefore be used. Supply must be steady, and a gravity feed or a connection to the town water supply is usually preferable to the use of a pump where either can be employed. Although the best results are usually obtained at pressures much less than those in general use for consolidating the case, it should not be overlooked that the filtrate has only to penetrate half of its thickness, whereas the wash water must press through the whole depth of the consolidated mass. In order that washing not last too long, the best cake thickness is usually somewhat less than for plain filtration.

FILTER PRESS (DIAPHRAGM)

The diaphragm filter press is an extension of filter press technology to increase the throughput of press and provide a higher solids content in wastewater filter cake. The press makes use of a rubber diaphragm with pressurized water to provide a high pressure on the partially dewatered sludge cake in the press after conventional dewatering methods have been used. This diaphragm provides the support for the filter cloth on one side of the press cavity. Filtration of the sludge is accomplished by charging the chambers of the press with sludge under pump pressure in the conventional manner, but at a generally lower pressure, allowing a cake to develop. The time allotted to this cycle is dependent on the characteristics of the sludge but is scheduled to continue only as long as high filtration rates are in progress. This cycle usually is in the 10- to 30-min range. The pump pressurizing system is then turned off, and water pressure is applied inside the diaphragm. This pressure, in the 200-psi range, applies a uniform pressure over the cake and further reduces the water content. The squeezing cycle has been shown substantially to reduce the overall cycle time for the press, yet produces a low moisture–content cake. The filter cake produced is thinner than in the conventional press but has a uniform moisture content in contrast to the conventional press. The reduction in operating time for the sludge pumps is reported to substantially reduce wear and resulting required maintenance. The pressurized water for the diaphragm actuation is a closed recycle system; therefore, components operate under predictable conditions and no effluent is produced. Diaphragm presses are designed to make use of a number of automation features to reduce labor and recycle time.

Opening of the filter press allows simultaneous discharge of filter cake from all cavities. Rejection of the sludge cake is accomplished by physical movement of the filter cloth by vibration or actual movement of the cloth in a forced rejection mode by partial withdrawal of the filter cloth loop. Filter cloths should be washed at each cycle or as conditions dictate. Air purging of feed and filtrate lines should also be done between cycles. Full automation of press operation. The diaphragm filter press is Japanese or European technology. Several hundred presses were reported to be in operation in Japan's wastewater industry. In the United States the press is being demonstrated by the use of portable pilot units. Applications include dewatering of a wide variety of wastewater sludges including those with a high level of solids content. Production of an autocombustible filter cake is used where a large filtration area is required in a minimum of floor area. A relatively high operator skill is required. The life of filter cloths and diaphragms is limited. The moisture content of sludge is highly dependent upon proper sludge conditioning.

Pressure or Vacuum Filters

Application of pressure filters of the sand filter type is used in water purification. They give rapid filtration in a small space and are easily cleaned by backwashing combined with a little agitation.

Leaf Filters

There is a whole range of types and sizes of enclosed pressure filters, which consist of a vertical cylinder in which a number of leaves are placed. These leaves usually consist of coarse iron wire mesh, supported in a framework, and covered with cloth bags or with Monel metal filter cloth. The bottom edge of each leaf is fitted with a hollow conical projection which is at once a support and an outlet. These projections fit holes in a pipe (Fig. 9) placed horizontally across the filter a small distance above the bottom. The top of the filter is fitted with a removable cover, and the sump at the bottom has a discharge door. Liquid is admitted to the filter either at the side or at the bottom and passes through the leaves and out through the horizontal pipe. When sufficient cake is built up, the filter is drained and the cake scraped off the leaves into the sump, whence it is removed through the bottom door. In some clarifications, it is possible to clean the leaves quickly simply by removing the top cover and shaking them gently in the remaining liquid. The cake then falls to the bottom and filtration can continue until the leaves are again coated. This serves to shorten the cleaning time, as the filter only needs to be emptied after every second operation.

Forms are made abroad in which scrapers are placed between the leaves and are attached to an endless chain operable at any time from outside the filter.

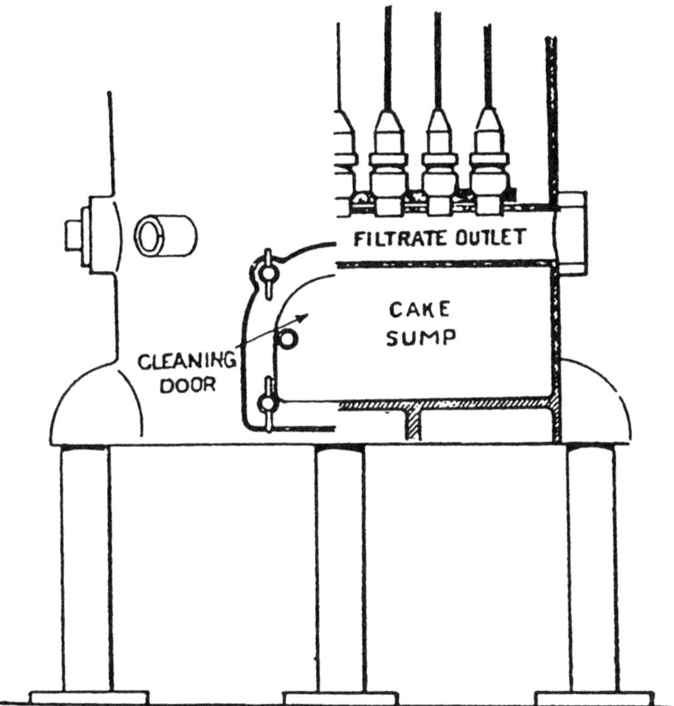

Figure 9 Arrangement of leaf filters.

This complication, however, is scarcely necessary owing to the ease and speed with which the cake can be made to drop to the leaves.

One of the best known of large pressure leaf filters is the *Sweetland filter*. This consists of a long cylindrical cast-iron body, split horizontally, the two parts being hinged to one another. The closing gear is a particularly neat arrangement. A shaft extending the whole length of the filter is carried in bearings on the upper stationary portion, and has attached to it eccentric cams supporting a number of bolts and nuts which engage with slotted brackets on the lower hinged portion of the filter. The action of the tightening gear is that the press is in the closed position with the bolts holding the parts tightly together. The first action on commencing to turn the tightening shaft is to revolve the cams, easing the pressure on the bolts, and allowing them to drop so that the heads are loose in the slotted brackets. Pins mounted on the shaft come into play and lift the bolts clear of the brackets. The lower part of the press can then be forced down or drops of its own accord, being balanced almost, but not quite, completely by a balance weight at the back. In very large presses, a large hand-wheel and spur gearing are employed to give additional pressure in operating the tightening gear, and hydraulic cylinders may be used to raise and lower the hinged portion of the body. Inside the press are mounted a number of circular leaves at suitable centers (from 2 up to 6 in.). These leaves each consist of a wire framework or drainage base which may be covered either with cloth bags or with Monel metal or other screens. The outlet, which is at the top, consists of a hollow shank which is pushed into a hole in the top of the press and secured by a nut. A hole in the side of the shank corresponds with a filtrate port which is led through a cock and a sight glass to the main filtrate outlet header. It is therefore possible to see the outlet from each individual leaf and to control it if it is not clear. The inlet to the press is at one end, where a vertical channel in the upper portion faces a similar channel in the lower portion, from which the feed is led along a passage extending the whole length of the press bottom. A deflector plate is placed above this so that the flow is directed sideways between the press casing and the leaves. A very smooth feed is thus obtained. The bottom of the press is provided with drain outlets so that any liquid remaining in the filter at the completion of filtration can be removed.

Often, in large presses, these filters are fitted with a spray pipe by which the cake may be washed off the leaves at the completion of filtration. This spray pipe has an opening between each pair of leaves, and there is an ingenious arrangement by which the pipe may be rocked so as to cover the whole surface of the leaves while it also moves a short distance backwards and forwards along the length of the press, so that there is a good chance of the spray of water reaching all portions of the cake. The cake, reduced to liquid mud, is run out of the bottom of the press, which is then ready to receive a further charge.

Washing in such presses may be carried out either by simply following up the feed with water or at the end of filtration the remaining liquid may be blown out after the press either through the leaves or by using the draining openings under compressed air. Wash water may then be introduced without mixing with the liquid to be filtered. It is important, however, during all these changes, always to maintain a positive pressure within the press across the leaf surface so that there is always a little liquid or air pressure in the leaves, holding the cake close to them. If this is not done, some of the cake may drop off causing difficulty during washing. Unevenness of flow due to broken or cracked cakes may be avoided by mixing with the wash water a small amount of already washed cake. This is carried to any points where excessive flow of wash water is occurring and helps to fill up the cracks and even out the resistance of the cake. It should be pointed out that providing the cake is held properly on the leaves and any cracks are filled up by the methods stated, enclosed pressure filters give good washing results. Since the cake is not subjected to any consolidating period, it follows that it is built up in an equally resisting state during filtration, and if this state does not disappear during the changes over incidental to the introduction of wash water, practically perfect theoretical washing should take place provided the cake is of reasonably equal thickness all over.

Leaves are obtainable with top drainage or with an internal tube giving bottom drainage, leaving a drier cake. Combinations of the two types are also available and, in another form, the stiffening rim is used as a draining channel.

Rotary Continuous Filters (Vacuum)

When large quantities of rapid cake-building products are to be handled, continuous vacuum filters of the rotary type can generally be employed. In these a large drum is arranged with filter elements around its periphery, as shown in Figure 10.

The surface of a rotary filter is divided into sections which are placed under suction while they revolve through a tank containing the material to be filtered. They are maintained under suction as they emerge from the tank, when they are usually subjected to washing by means of spray pipes. After this, air is drawn through the cake in order to dry it as far as possible, and finally each section comes opposite the knife, or "doctor," which scrapes the cake off. During this process, it is usual to subject each individual section to a blast of compressed air in order to free the cake from the cloth as shown in Figures 10 and 11. The heart of this arrangement is the rotary valve, by means of which the outlets from the sections are connected, first to vacuum through a port connecting to a

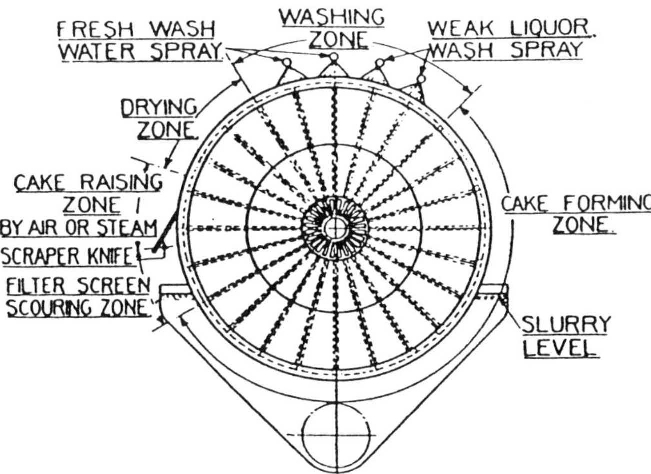

Figure 10 Operations during a cycle of continuous filtration.

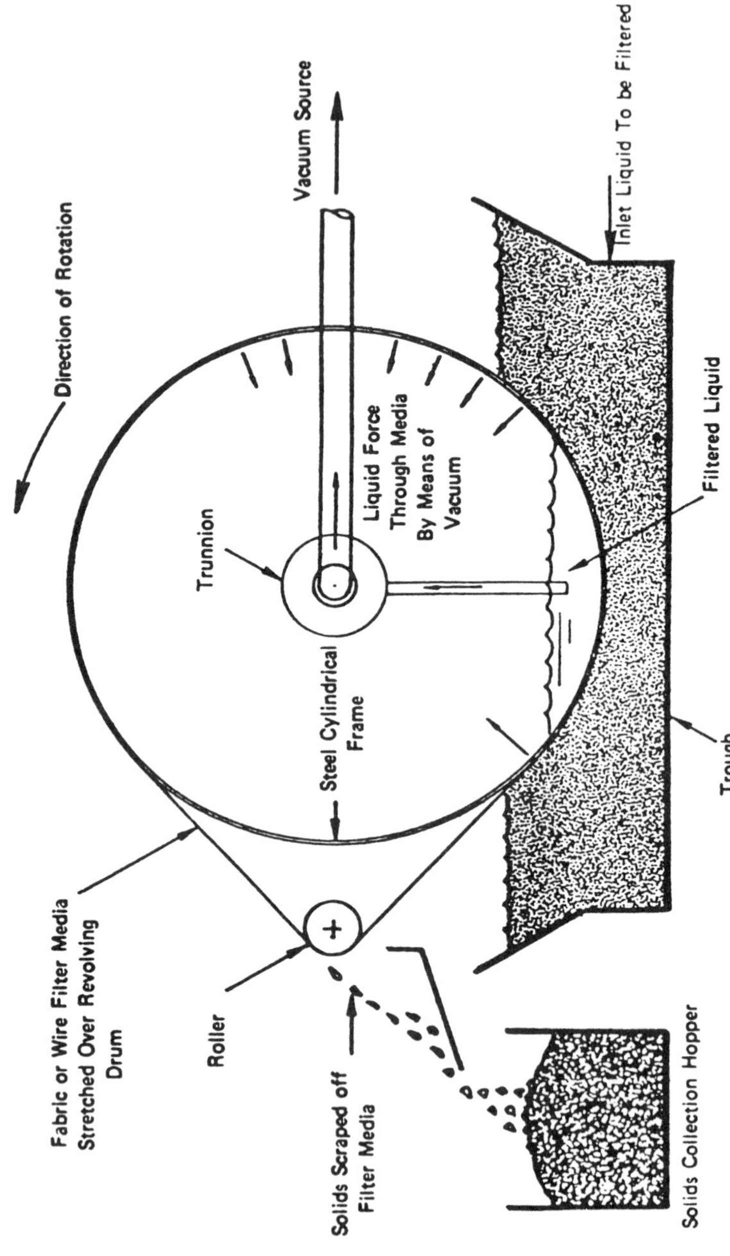

Figure 11 Rotary drum vacuum filter (cross-sectional side view).

filtrate-receiving tank, and then to vacuum through another port connecting to a washing receiving tank. There is also a port for compressed air.

The slurry tank is generally provided with an agitator, paddle or oscillating type, to keep the solids from settling. The filtrate and washings are drawn to separate receiving tanks connected to a dry vacuum pump. The filtrate or washings are withdrawn from the bottom of the tanks by centrifugal or rotary pumps. It is not unusual to arrange these tanks with a float or trip device so that if they become full the vacuum is broken in order to top liquid being carried out through the vacuum pump, which may be protected by a barometric tube. Washing arrangements in such filters are comparatively simple consisting of perforated pipes or sets of sprays arranged across the filter toward its top. The filter drum may have a cast iron or a wood periphery, divided into sectionalized chambers. In large filters, steel construction can be used. These chambers support the filter cloth. It is usually laid on and bound into place by wire turns. This wire also serves as support for the scrapers, preventing their coming into close contact with the cloth. A practice is to divide the cloth also into sections, each section being tamped into place, a device which improves the ease with which the filter can be reclothed.

A difficulty in rotary filters is the cracking of the cake during drying. This allows air to leak, reducing the vacuum and hindering drying. In order to prevent this it is usual to fit a number of rollers toward the top of the filter to squeeze the cake. A wide canvas band is sometimes stretched over these rollers in such a way that it presses on the upper surface of the filter cake, helping any cracks to close up. This is a help to improve drying in some cases, and use can be made of the band to keep the cake whole and uncracked while wash water is distributed over the band. Rollers for closing the cracks are sometimes made to reciprocate in order to increase the efficiency.

Another discharge arrangement consists of a number of endless cords which are passed over the filter drum, around another smaller roller, and through a kind of tooth-comb or spacer. These strings are placed fairly close together, and the cake clings to them and is lifted off the roller with them. As they pass over the small roller, the bending causes most of the cake to break up and drop off, with the remainder being scraped off the bands by the comb.

In a variation of this arrangement, an endless band, consisting of a variety of erosely meshed links, is passed round the filter. A cake is formed on this band and is lifted away from the roller with it. The band is taken into a drying chamber having hot-air circulation, where it travels in festoons to the outlet point, when it returns to the filter again. The dried material falls off just prior to the return journey owing to the great bending of the endless belt which occurs at that point.

Another form of rotary filter consists of a number of disclike leaves fixed to a single revolving shaft, each disc dipping into a trough. Wheels are built up of radial fanlike sections, each of which is independently clothed with a specially shaped bag. The end of each section has a pipe which fits into a recess in the shaft connecting to the vacuum main. The filter elements and cloths are secured in position by a segmental hoop device which encircles them. This arrangement is very helpful where large amounts of filter surface have to be got into a small space. It is best suited to problems in which no washing has to be done, since the face of the cake is vertical and, in consequence, does not hold wash water very readily.

Such filters can be sectionalized so that each disc in a multidisc filter can handle a separate product. Some slurries contain particles which settle rapidly that is is difficult to get them to adhere to the filter. Filter drums made with internal filter surfaces are intended to deal with this difficulty. Figure 12 shows a rotary filter process schematic.

For products which are too coarse or granular for treatment, a type of drum

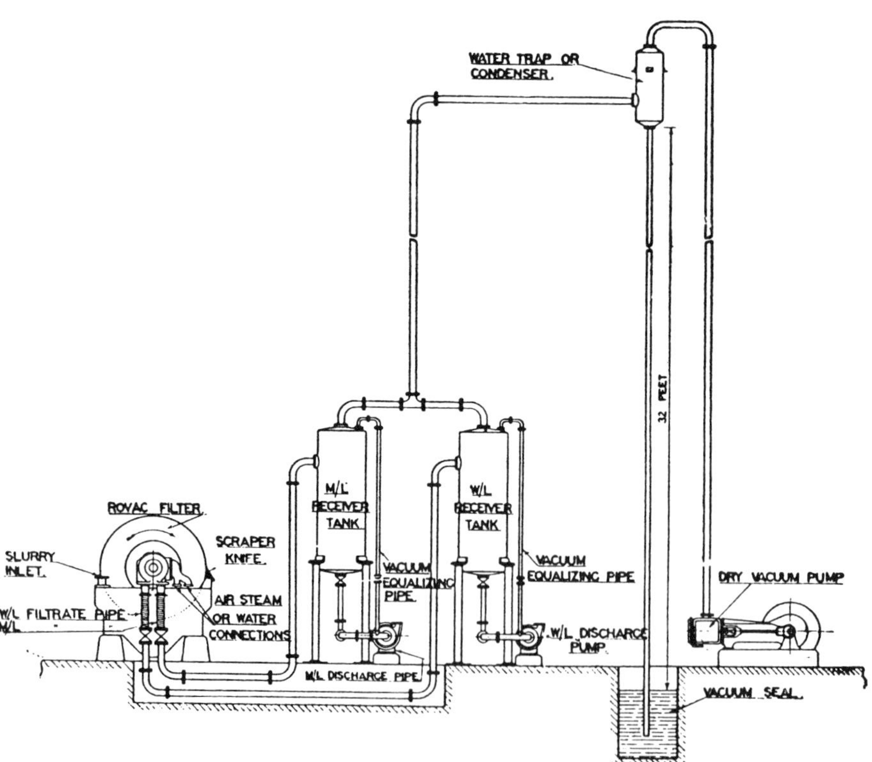

Figure 12 Rotary filter process schematic.

is used in which the periphery is divided into compartments into which the material is poured and subjected to suction. As the drum revolves, the compartments are turned over, and the drained product drops away. Occasionally a horizontal disc or table filter is used, but although this solves the problem of keeping difficult materials securely on the filter surface, it is limited by consideration of floor space, especially since the discs are necessarily one sided, and the filters are normally made with only one disc.

Vacuum filters have been used to dewater sludges so as to produce a cake having the physical handling characteristics and moisture contents required for subsequent processing. A rotary vacuum filter, as described, consists of a cylindrical drum rotating partially submerged in a vat or pan of conditioned sludge. The drum is divided radially into a number of sections, which are connected through internal piping to ports in a valve body (plate) at the hub. This plate rotates in contact with a fixed valve plate with similar ports, which are connected to a vacuum supply, a compressed air supply, and an atmospheric vent. As the drum rotates, each section is thus connected to the appropriate service. Various operating zones are encountered during a complete revolution of the drum. In the pick-up or form section, vacuum is applied to draw liquid through the filter covering (media) and form a cake of partially dewatered sludge. As the drum rotates, the cake emerges from the liquid sludge pool while suction is still maintained to promote further dewatering. A lower level of vacuum often exists in the cake drying zone. If the cake tends to adhere to the media, a scraper blade may be provided to assist removal.

The three principal types of rotary vacuum filters are the drum type, coil type, and the belt type. The filters differ primarily in the type of covering used and the cake discharge mechanism employed. Cloth media are used on drum and belt types, whereas stainless steel springs are used on the coil type. Infrequently, a metal media is used on belt types. The drum filter also differs from the other two in that the cloth covering does not leave the drum but is washed in place, when necessary. The design of the drum filter provides considerable latitude in the amount of cycle time devoted to cake formation, washing, and dewatering, whereas it minimizes inactive time.

A variation of the conventional drum filter is the top feed drum filter. In this case, sludge is fed to the vacuum filter through a hopper located above the filter. The potential advantages are that gravity aids in cake formation; capital costs may be lower, since the feed hopper is smaller and no sludge agitator and related drive equipment is required, and blinding of the media may be reduced.

The coil-type vacuum filter uses two layers of stainless steel coils arranged in corduroy fashion around the drum. After a dewatering cycle, the two layers of springs leave the drum and are separated from each other so that the cake is lifted off the lower layer of springs and discharged from the upper layer. Cake release is essentially free of problems. The coils are then washed and reapplied to the drum. The coil filter has been and is widely used for all types of sludge. However, sludge with particles that are both extremely fine and resistant to flocculation dewater poorly on coil filters.

Media on the belt-type filter leaves the drum surface at the end of the drying zone and passes over a small diameter discharge roll to facilitate cake discharge. Washing of the media next occurs before it returns to the drum and to the vat for another cycle. This type of filter normally has a small-diameter curved bar between the point where the belt leaves the drum and the discharge roll which aids in maintaining belt dimensional stability. In practice, it is frequently used to ensure adequate cake discharge.

A great many types of filter media are available for the belt and drum filters. There is some question whether increases in yield due to operating vacuums greater than 15 in. of mercury are justifiable. The cost of a greater filter area must be balanced against the higher power costs for higher vacuums. An increase from 15 to 20 in. of vacuum is reported to have provided about 10% greater yield in three full-scale installations.

Chemical conditioning is often employed to agglomerate a large number of small particles. It is almost universally applied with mixed sludges and is the most common method of mechanical sludge dewatering utilized in the United States.

Chemical conditioning is generally used in larger facilities where space is limited or when incineration is necessary for maximum volume reduction. A relatively high operating skill is required in this method. Operation is sensitive to type of sludge and conditioning procedures. As raw sludge ages (3–4 hr) after thickening, vacuum filter performance decreases. Poor release of the filter cake from the belt is occasionally encountered. Chemical conditioning costs can sometimes be extremely large if a sludge is hard to dewater. Solids capture ranges from 85.0 to 99.5% and cake moisture is usually 60–90% depending on feed type, solids concentration, chemical conditioning, machine operation, and management. Dewatered cake is suitable for landfill, heat drying, incineration or land spreading. $FeCl_3$ and/or lime or polymer dosing is a function of type of sludge and vacuum filter characteristics. Typical loadings in pounds dry solids/hr/ft^2 are 7–15 for raw primary sludges, 4–7 for digested primary sludges, and 3.5–5.0 for mixed digested sludges. Loading is a function of feed solids concentrations, subsequent processing requirements, and chemical preconditioning.

Diatomaceous Earth Filters

Diatomaceous earth filters have been applied to the clarification of secondary effluents at pilot scale. They produce a high-quality effluent but appear unable to handle the solids loadings normally expected in this application. Diatomaceous earth filters utilize a thin layer of precoat formed around a porous septum to strain out the suspended solids in the feedwater which passes through the filter cake and septum. The driving force can be imposed by vacuum from the product side or pressure from the feed side. As filtration proceeds, head loss through the cake increases owing to solids deposition until a maximum is reached. Cake and associated solids are then removed by flow reversal and the process is repeated. In the cases where secondary effluents have been treated by this process, a considerable amount of diatomaceous earth (body feed) has been required for continuous feeding with the influent in order to prevent rapid build-up of head losses. The diatomaceous earth filtration process is capable of excellent removal of suspended solids but not colloidal matter.

A wide variety of diatomaceous earth (diatomite) grades are available for use. The coarser grades have greater permeability and solids holding capacities than do the finer grades which will generally produce a better effluent. Some grades of diatomite are pretreated to change their characteristics for improved performance. A number of vessel configurations are available, with open-basin vacuum and vertical pressure designs being most common (see Fig. 11).

The filtration cycle can be divided into two phases, run time and downtime. Downtime includes the periods when the dirty cake is dislodged from the septum and removed from the filter and when the new precoat is formed. Run time commences when the feed is introduced to the filter and ends when a limiting head loss is reached. The single most important factor in secondary effluent filtration by diatomaceous filters is the amount of body feed required during the filtration or run time. The body feed rate is the largest operating cost factor and strongly affects the operating economics of the process. Similarly, it is related to cycle time between backwashing which determines the installed filtering area, hence economics.

BELT FILTERS

Belt filters consist of an endless filter belt that runs over a drive and guide roller at each end like a conveyor belt. The upper side of the filter belt is supported by several rollers. Above the filter belt is a press belt that runs in the same direction and at the same speed; its drive roller is coupled with the drive roller of the filter belt. The press belt can be pressed on the filter belt by means of a pressure roller system whose rollers can be individually adjusted horizontally and vertically. The sludge to be dewatered is fed on the upper face of the filter belt and is continuously dewatered between the filter and press belts. After having passed the pressure zone, further dewatering in a reasonable time cannot be achieved by only applying static pressures. However, a superimposition of shear forces can effect this further dewatering. The supporting rollers of the filter belt and the pressure rollers of the pressure belt are adjusted in such a way that the belts and the sludge between them describe an S-shaped curve. Thus, there is a parallel displacement of the belts relative to each other due to the differences in the radii. After further dewatering in the shear zone, the sludge is removed by a scraper.

Some units consist of two stages where the initial draining zone is on the top level followed by an additional lower section wherein pressing and shearing occur. A significant feature of the belt filter press is that it employs a coarse mesh, relatively open weave, metal medium fabric. This is feasible because of the rapid and complete cake formation obtainable when proper flocculation is achieved. Belt filters do not need vacuum systems and do not have the sludge pick-up problem occasionally experienced with rotary vacuum filters. The belt filter press system includes auxiliaries such as polymer solution preparation equipment and automatic process controls.

Belt filters may include the added feature of vacuum boxes in the free drainage zone. About 6 in of mercury vacuum is applied to obtain higher cake solids. A "second generation" of belt filters have extended shearing or pressure stages that produce substantial increases in cake solids but are more costly. Hard to dewater sludges can be handled more readily. Low cake moisture permits incineration of primary/secondary sludge combinations without auxiliary fuel. A large filtration area can be installed in a minimum of floor area. To avoid penetration of the filter belt by sludge, it is usually necessary to coagulate the sludge. Figure 13 shows a schematic of a belt filter.

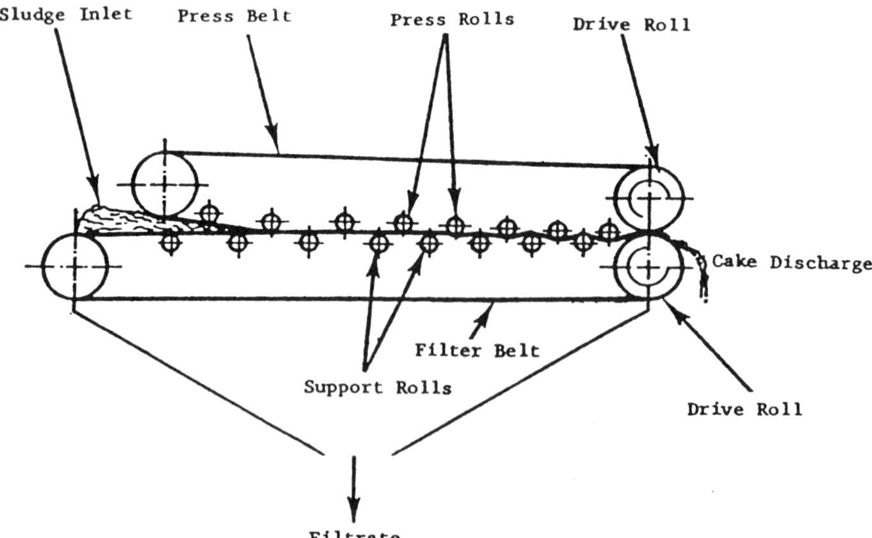

Figure 13 Filter belt flow diagram.

CENTRIFUGALS

Centrifugal machines are widely used in the chemical and allied industries for the removal of water and liquors from slurries or pastes or from fibrous materials. Some use has been found for them in wastewater treatment. It revolves at high speed round a vertical axis, and under the action of the centrifugal force developed, the liquid content of the substance contained in the drum is ejected through the perforations. To prevent the ejection of fine sediments, the perforations are covered by suitable fabric or filter cloth. Figure 14 illustrates the under-driven belt type, and the water or liquid removed from material in the drum leaves by the spout in the outer casing. A brake is provided to bring the basket quickly to rest.

Centrifuges can be conveniently classified into three principal types: fixed spindle machines, suspended machine, and self-balancing machines of either the underdriven or top-driven classes.

The fixed-spindle machine is shown in Figures 14 and 15. A problem with this machine is any unbalanced centrifugal force due to uneven loading must be taken by the bearings of the machine and therefore to the foundations. Vibration caused is a serious problem and can be avoided by the use of suspended machines, which, in the case of underdriven machines, are ordinary fixed-spindle machines hung by suspension rods from columns carried on an independent base ring.

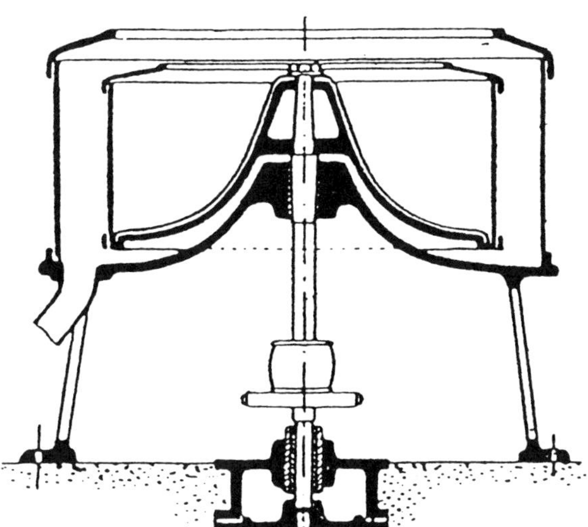

Figure 14 Perforated basket underslung centrifuge with fixed spindle.

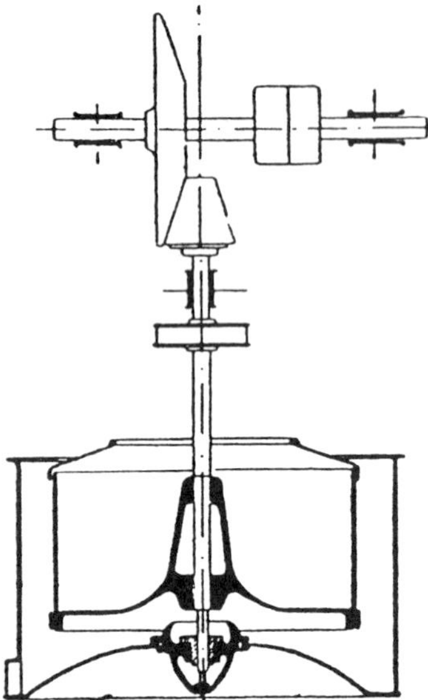

Figure 15 Fixed-spindle hydro-extractor.

These rods have hemispherical ends fitting in cups at top and bottom, and the machine is therefore free to vibrate without affecting its surroundings.

The self-balancing machine may be of either the underdriven or top-driven class. In this design, the spindle bearing is mounted in a box or casing to which a certain amount of free play is given under constraint from rubber buffers or springs. The top-driven type is shown in Figure 16. There is a single top bearing contained usually in a conical rubber buffer, although sometimes a cylindrical rubber buffer is used. The under-driven type is shown in Figure 17, which shows an electrically driven suspended machine of the self-balancing type.

The speeds at which centrifugal baskets can be run are of significance. A formula for centrifugal force, in pounds exerted by a mass of 1 lb, is $C = V^2/57{,}888D$, where V is the surface, or peripheral, speed in feet per minute and D is the diameter of the basket in feet. From this formula it will be noted that the force exerted for a given surface speed must vary inversely as the diameter. Hence, the problem is similar to that of a rotating pulley insofar as strength is concerned, and if the liquid load is proportional to the area of the basket wall, and to its thickness, bursting will always occur at the same surface speed whatever the size of the basket and whatever the wall thickness. *Surface speed* is a limiting factor insofar as strength is concerned, and it is to be noted that it is of no use to increase greatly the wall thickness, because the self-stress in the wall due to its own weight is a material factor (Fig. 17).

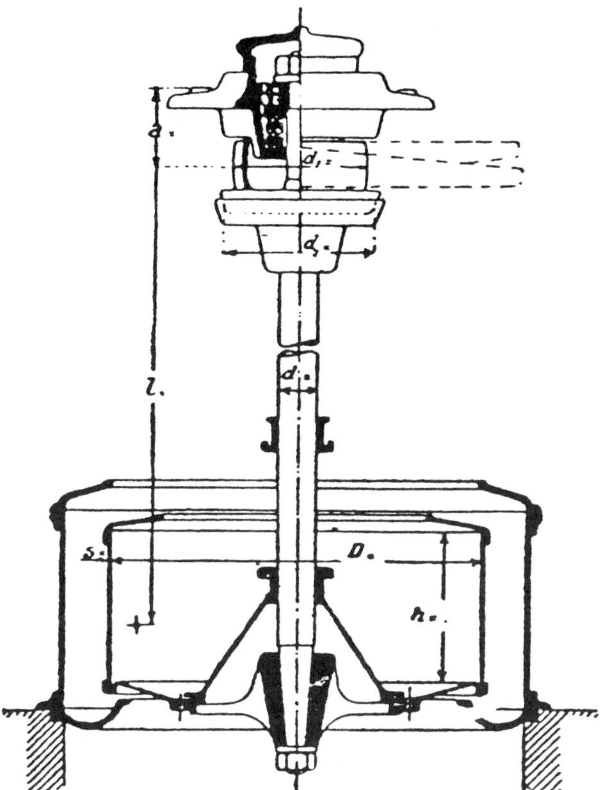

Figure 16 Top-driven, self-balancing hydro-extractor.

Speeds usually vary between 8000 and 12,000 ft/min, although smaller machines of the separator type on run as high as 18,000 ft/min. If the surface speed is kept constant, the centrifugal force would always be inversely proportioned to the diameter, but in practice the running speed of commercial machines is often an average between that for equal surface speed and that for equal effect. Although the surface speed increases in the case of the larger machines, the centrifugal force decreases.

Solid bowl centrifuges or separators are designed similar to perforated baskets but they have no holes in the baskets, which make them solid bowls. They can be used to separate from liquids solid particles that are so fine that would plug filter cloths, or for separating two liquids such as water and tar. The mixture is fed into a solid basket, which rotates at high speed, which precipitates the solids against the circumference, and the clarified liquid forms an interior

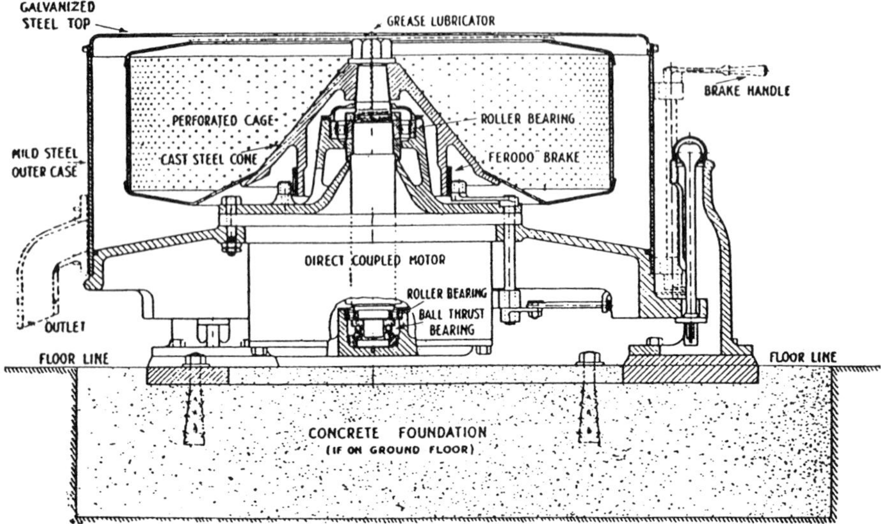

Figure 17 Direct coupled under-driven self-balancing hydro-extractor.

wall. The liquor is discharged by means of a specially shaped pipe having a knife-edge nozzle which is applied to the rotating liquor by means of a hand wheel. A flexible hose is attached to the end of the skimmer pipe and the surface speed of the liquor causes it to be discharged out of this pipe into storage vessels. When the cake is sufficiently thick, the machine is stopped and the cake removed.

9

Centrifugal Separation

INTRODUCTION

Equipment used for centrifugal separation includes cyclone separators for gas-solid separations, hydrocyclones for liquid-solid separations, and centrifuges for liquid-solid, liquid-liquid, and gas-gas separations. The theory and action is virtually the same. In cyclone separators and hydrocyclones, the heterogeneous suspension is subject to centrifugal force by its own rotation with respect to a stationary apparatus boundary by introducing it through a tangential inlet. In centrifuges, the fluid undergoes rotation with respect to a revolving apparatus boundary. Principles of operation and design methodology are presented in this chapter.

Centrifugal Dewatering

Centrifuges are used to dewater municipal sludges. Centrifugal force is used to increase the sedimentation rate of sludge solids. The three most common types of units are the solid-bowl type, the disc type, and the basket type.

The solid-bowl continuous centrifuge assembly consists of a bowl and conveyor joined through a planetary gear system designed to rotate the bowl and the conveyor at slightly different speeds. The solid cylindrical bowl, or shell, is supported between two sets of bearings and includes a conical section at one end. This section forms the dewatering beach over which the helical conveyor screw pushes the sludge solids to outlet ports and then to a sludge cake discharge hopper. The opposite end of the bowl is fitted with an adjustable outlet weir plate to regulate the level of the sludge pool in the bowl. The centrate flows through outlet ports either by gravity or by a centrate pump attached to the shaft at one end of the bowl. Sludge slurry enters the unit through a stationary feed pipe extending into the hollow shaft of the rotating bowl and passes to a baffled, abrasion-protected chamber for acceleration before discharge through the feed ports in the rotating conveyor hub into the sludge pool. Owing to the centrifugal forces, the sludge pool takes the form of a concentric annular ring on the inside of the bowl. Solids settle through this ring to the wall of the bowl where they are picked up by the conveyor scroll. Separate motor sheaves or a variable speed drive can be used for adjusting the bowl speed for optimum performance.

Bowls and conveyors can be constructed from a large variety of metals and alloys to suit specific applications. For dewatering of wastewater sludges, mild steel or stainless steel normally has been used. Because of the abrasive nature of many sludges, hard facing materials are applied to the leading edges and tips of the conveyor blades, the discharge ports, and other wearing surfaces. Such wearing surfaces may be replaced by welding when required. In the continuous concurrent solid-bowl centrifuge, incoming sludge is carried by the feed pipe to the end of the bowl opposite the discharge. Centrate is skimmed off and cake proceeds up the beach for removal. As a result, settled solids are not disturbed by incoming feed.

In the disc centrifuge, the incoming stream is distributed between a multitude of narrow channels formed by stacked conical discs. Suspended particles have only a short distance to settle, so that small and low-density particles are readily collected and discharged continuously through fairly small orifices in the bowl wall. The clarification capability and throughput range are high, but sludge concentration is limited by the necessity of discharging through 0.05–0.10-in. diameter orifices. Therefore, it is generally considered a thickener rather than a dewatering device.

In the basket centrifuge, flow enters the machine at the bottom and is directed toward the outer wall of the basket. Cake continually builds up within the basket until the centrate, which overflows a weir at the top of the unit, begins to increase in solids. At that point, feed to the unit is shut off, the machine decelerates, and a skimmer enters the bowl to remove the liquid layer remaining in the unit. A knife is then moved into the bowl to cut out the cake, which falls out the open bottom of the machine. The unit is a batch device with alternate charging of feed sludge and discharging of dewatered cake.

Solid-bowl and disc centrifuges are in widespread use. Basket centrifuges have been fully demonstrated for small plants, although not widely used. Solid-bowl and disc types are generally used for dewatering sludge in larger facilities where space is limited or where sludge incineration is required. The basket type is used primarily for partial dewatering at small plants. Disc centrifuges are more useful for thickening and clarification than dewatering.

Centrifugation requires sturdy foundations because of the vibration and noise that result from centrifuge operation. Adequate electric power must also be provided, since large motors are required. The major difficulty encountered in the operation of centrifuges has been the disposal of the centrate, which is relatively high in suspended, nonsettling solids. With disc-type units, the feed must be degritted and screened to prevent plugging of discharge orifices. Pluggage of discharge orifices is a problem on disc-type units if feed to the centrifuge is stopped, interrupted, or reduced below a minimum value. Wear is a serious problem with solid-bowl centrifuges. Centrate can be relatively high in suspended, nonsettling solids which, if returned to treatment units, could reduce effluent quality from primary settling system. Noise may require some control measures.

Solid-bowl centrifuge solids recovery is 50–75% without chemical addition and 80–95% with chemical addition. Solids concentration is 15–40% depending on type of sludge. For basket centrifuges, solids capture is 90–97% without chemical addition and cake solids concentrations are 9–14%. Disc centrifuges can dewater a 1% sludge to 6% solids concentration. Each installation is site specific and dependent on a manufacturer's product line. Maximum capacities of about 100 tons/hr of dry solids are available in solid bowl units with diameters up to 54 in. and power requirements up to 175 hp. Disc units are available with capacities up to 400 gal/min of concentrate.

CYCLONE SEPARATORS

A cyclone is a nonmechanical device that separates solid particulates from a relatively dry gas stream. Hydrocyclones are essentially the same device and effect solids separations from liquids. Both devices are the simplest and most economical separators (also called solids collectors or, simply, collectors). Operations, in which forces both of inertia and gravitation are taken advantage of, are identical and their primary advantages are high collection efficiency in certain applications, adaptability, and economy in power. The main disadvantage lies in their limitation to high collection efficiency of large-sized particles only. In general, cyclones are not capable of high emergencies when handling streams containing large concentrations of particulates less than 10 μm in size.

Cyclones generally are efficient handling devices for a wide range of particulate sizes. They can collect particles ranging in size from 10 to above 2,000 μm, with inlet loadings from less than 1 to greater than 100 gr/scfm (grains per standard cubic feet per minute). There are many design variations of the basic cyclone configuration. Because of the cyclone's simplicity and lack of moving parts, a wide variety of construction materials can be used to cover relatively wide operating temperatures—up to 2000°F for gases.

Cyclones are employed in the following applications: (1) collecting coarse dust particles; (2) handling high solids concentration gas streams between reactors; (3) classifying particulate sizes; (4) operations in which extremely high collection efficiency is not critical; and (5) precleaning devices in line with high-efficiency collectors for fine particles.

Operating Principles

Figure 1 shows the primary design features of a cyclone separator. The unit consists of a cylindrical barrel section (1) fitted with a conical base (2). The solids-laden gas stream enters through the tangential inlet (3) while collected solids fall through the outlet duct (5) and the cleaned fluid discharges through the exit duct (4).

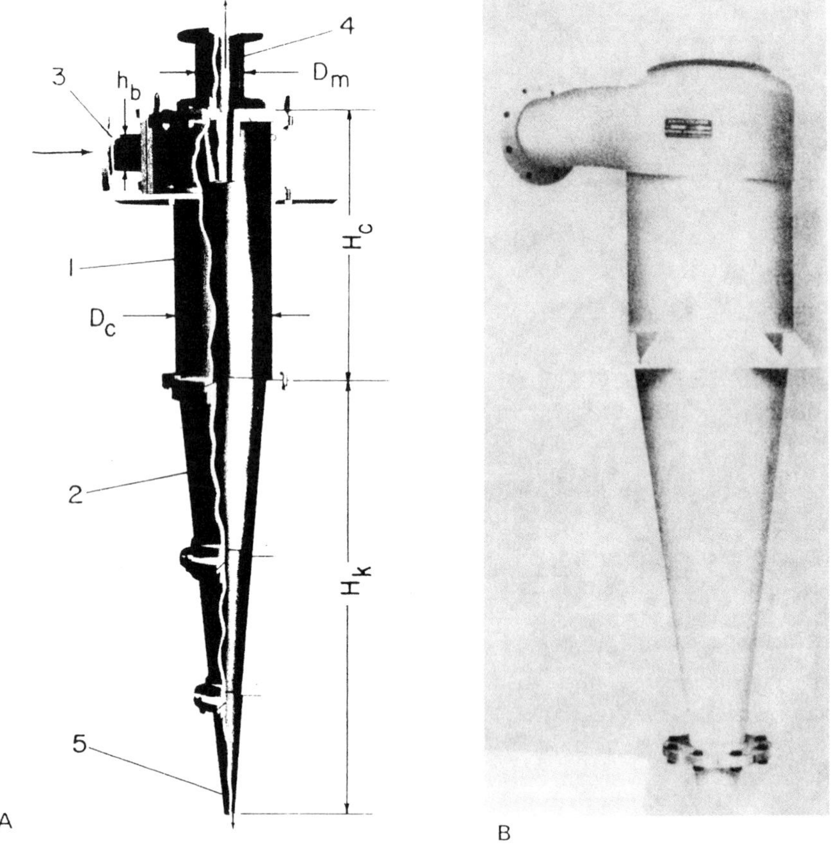

Figure 1 Cyclone separator. (A) Cross-sectional view: (1) cylindrical barrel; (2) conical base; (3) gas stream inlet; (4) fluid discharge outlet duct; (5) collected solids outlet duct. (B) Full view.

The flow pattern in a cyclone is complex. Three main flow patterns prevail in all cyclones:

- Descending spiral flow: This pattern carries the separated dust down the walls of the cyclone to a dust hopper.
- Ascending spiral flow: This rotates in the same direction as the descending spiral, but the cleaned gas is carried from the cyclone or the dust receptacle to the gas outlet.
- Radially inward flow: This feeds the gas from the descending to the ascending spiral.

The final solids separation occurs in the dust-collecting hopper located below the exit duct. In the duct leading to the hopper, the total gas flow reverses direction and transforms to an ascending spiral flow.

The flow patterns are generated by the creation of a double vortex, which centrifuges the dust particles to the walls. The two distinct vortices present in a cyclone are (1) a large-diameter descending helical current in the body and cone, and (2) an ascending helix of smaller diameter extending up from the dust outlet section through the gas outlet.

Particles at the walls are transported into the collecting hopper, which is isolated from the influence of the spinning gases. The gas spirals downward and upward through the inside of the cyclone. On entering the cyclone, the gas undergoes a redistribution of its velocity, so that the tangential velocity component exceeds the inlet gas velocity by several times. As the gas spins in a vortex in the cyclone body, the tangential velocities increase as the axis of the cyclone is approached at any horizontal plane. The tangential velocity at any radius appears to be relatively constant at all levels. However, tangential velocities at the extreme top of the cyclone are not included, because the proximity of the cyclone cover slows the spin. When the downward gas flow is smooth and unbroken, the dust particles flow spirally downward and pass the bottom dust outlet without reentrainment.

As depicted in Figure 2, the particle-laden fluid enters the tangential inlet and swirls through several revolutions in the body and cone while dropping its dust or particle load. The clean fluid is emitted through the axial cylindrical gas outlet. Particles, which were dispersed uniformly in the entering stream, tend to concentrate in the layer next to the cyclone wall under the influence of centrifugal force. The helical motion of the downward-moving main gas stream and the small quantity of fluid through the cyclone's outlet assist the separated solids into the discharge. Ideally, the fluid should be distributed to a cyclone in a radially thin layer so that the radial distance through which a particle must travel for separation is at a minimum. It is important that the inlet be exactly tangent to the body, because separation is affected adversely by flow abnormalities. The

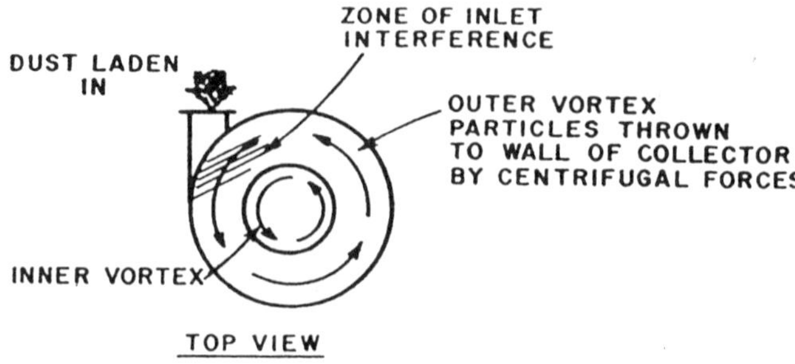

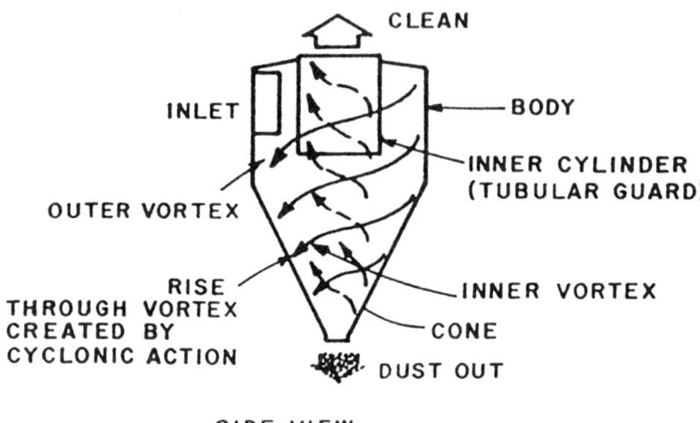

Figure 2 Details of internal flow patterns.

discharge through the vertical exist duct maintains the upward helical pattern for some distance until pipe bends and/or flow restrictions dampen out the spiral.

Three forces influence particle motion in a cyclone:

- Centrifugal due to the particle rotation within the flow
- Gravity force
- Archimedes' force

The last two forces are small in comparison with centrifugal force, because the density of the medium is hundreds of times smaller than that of the solid particles.

In principle, the separation of finely dispersed solid particles from gases or liquids is governed by

- Established centrifugal field
- Radial velocity pattern
- Residence time of the particles to be separated
- Turbulence that develops

Criteria for particle separation are based on the concept of "equilibrium orbits." This term refers to the regions of the flow comprising circular tracks having a radius such that the outwardly directed centrifugal force, acting on a particle in this orbit is in equilibrium with the inwardly directed drag according to Stokes' law due to the radial flow.

Collection Efficiency

Figure 3 shows typical tangential and axial velocity distributions in a cyclone. The maximum values of the tangential velocities are observed close to the cyclone axis (but not on the axis), as shown in Figure 3A. In tracing the tangential velocity component down the cyclone, the maximum shifts closer to the axis.

The maximum axial velocities fall exactly on the centerline axis at the lower elevations of the cyclone, as shown in Figure 3B. The axial components of velocity decrease the collection efficiency, because particles can be entrained into the discharge pipe. Owing to the counterbalancing effects of the velocity components, particle collection efficiency is not uniform in the different zones of cyclone, as shown in Figure 3C.

Overall collection efficiency is defined as:

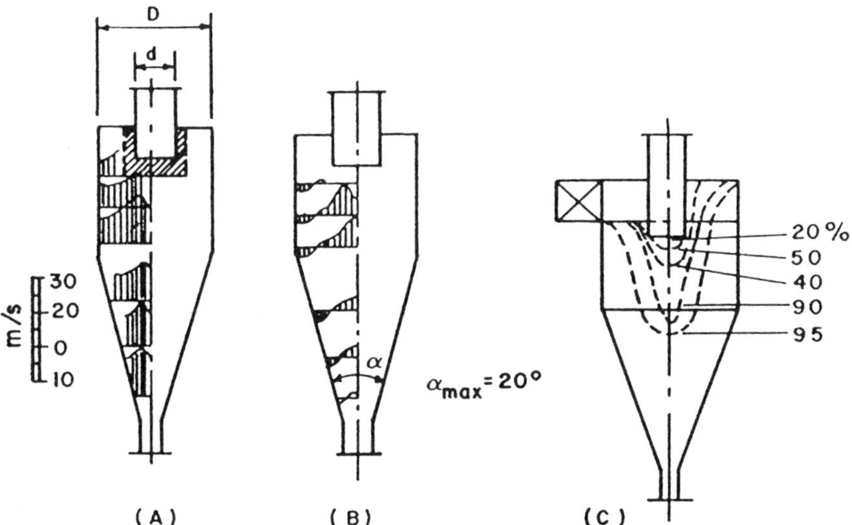

Figure 3 Variations of velocities in a cyclone. (A) tangential; (B) axial; (C) collection efficiency at different zones of cyclone.

$$h = \frac{C_1 - C_2}{C_1} \tag{1}$$

where

C_1 = particle concentration in the entering gas stream
C_2 = particle concentration in the exit gas

Efficiency depends on several factors, the primary ones being the physical properties of the heterogeneous system; i.e., fluid and particle densities, particle sizes, viscosity of the medium, the linear dimensions of a cyclone, solids loadings, and the gas inlet velocity.

Collection efficiency is a major parameter in the selection and design of a cyclone. Cyclones may be designed for any required efficiency; however, efficiency is both a function of the energy expended and the available space for the unit. Thus, proper optimization of an acceptable efficiency at moderate pressure drop within reasonable space requirements is necessary. Efficiency can be improved markedly at the expense of pressure drop without tampering with space requirements. The principal parameter in the prediction of collection efficiency is particle size. For each particular cyclone design there is a critical size particle at a given density on which the centrifugal and inward viscous forces are balanced; that is, at the equilibrium state the particle neither moves outward toward the walls nor inward toward the cyclone axis. Particles larger than this critical diameter (called the "theoretical cut," measured in microns) are collected, whereas all smaller particulates escape.

Efficiency is enhanced as the axis of the cyclone is approached until the edge of the core is reached. This occurs because the centrifugal force increases more rapidly than the inward drift up to the edge of the gas core. Therefore, particles rotate in orbits whose radii depend on the balance between the viscous forces due to the inward drift and outward centrifugal force. Particles then will be transferred to the outer orbits, where they pick up finer particles by collision or are captured by eddy currents. A knowledge of the theoretical cut has very little direct significance in the prediction of collection efficiency. It does, however, give a rough indication of cyclone performance. The factors effecting particle settling depend on the relationship of forces acting on the particle in the flow.

The most common method of estimating total collection efficiency is by plotting a fractional efficiency curve or grade efficiency curve. A fractional efficiency curve is a plot of particle size versus percentage collection. Figures 4 and 5 show the specific features of such a curve and a typical plot based on experimental data, respectively.

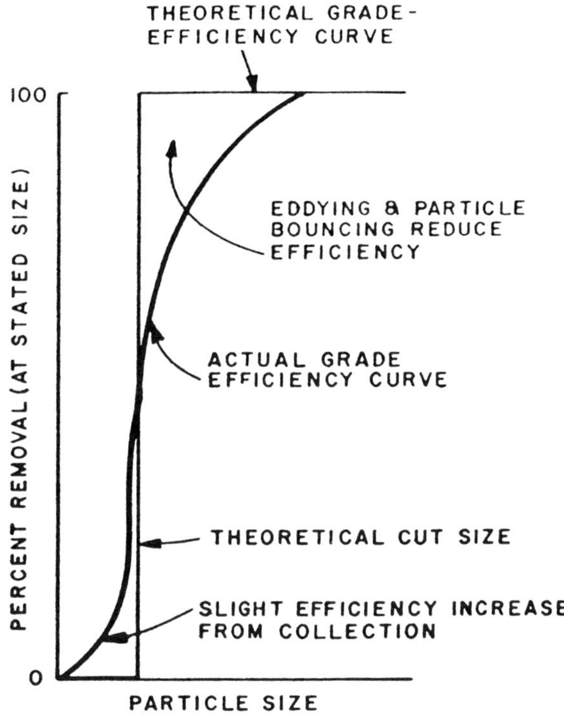

Figure 4 Typical grade efficiency curve.

Figure 4 shows that the cyclone has 0% efficiency for all particles smaller than the cut size and 100% for all larger sizes. Under actual operating conditions, however, a considerable amount of particulates smaller than the cut size are separated along with the coarser particles. This can be explained by collision between particulates or by particle aggregation. Furthermore, a portion of the particles larger than the cut size escape collection. They are carried into the inner vortex by eddies or by collisions. Poor success has been achieved in predicting efficiencies with particle sizes less than the cut size. The degree of efficiency in this range depends largely on the properties of the particles.

It is apparent that the curve in Figure 5 may be used accurately in predicting the total collection efficiency of a cyclone only if the particle size distribution of the suspension is known.

The specific operating parameters on which efficiency depends are

- Pressure drop
- Particle size distribution

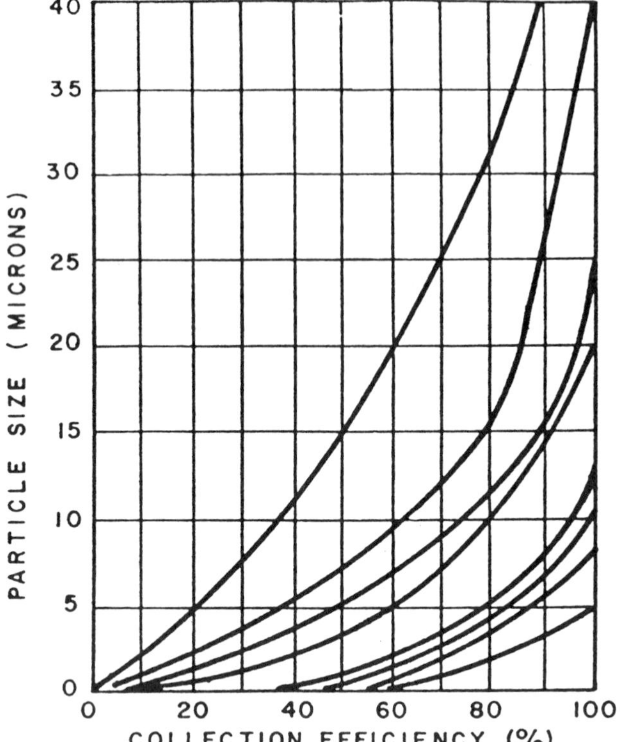

Figure 5 Typical collection efficiency curves. Data were obtained on a 12-in. diameter cycloner using cracking catalyst.

- Inlet particle loading
- Temperature of the inlet gas stream
- Specific gravities of gas and solids

Table 1 summarizes the relationships of these factors to efficiency. (Pressure drop is considered separately below.)

Several operating characteristics contribute to lower collection efficiencies for particles exceeding the cut size. One important characteristic is that the distribution, or drift, is highly nonuniform. As noted by the contour plots shown earlier, at certain points in the cyclone local velocities may exceed the mean inlet velocity by a factor of 2 or 3. This is particularly pronounced at the ends of the cyclone where additional surfaces induce precession currents. A doubling or tripling of the drift velocity at any point results in particles 40% larger than the theoretical cut in reaching the exit.

Table 1 Factors Influencing Efficiency

Condition	Effect on Efficiency
Temperature	Decreases as temperature increases due to gas viscosity changes
Velocity	Increases with velocity and falls off sharply below 25 fps
Specific gravity	Increases with higher specific gravity
Inlet loading	Increases with solids loading

Eddies in the vortex also are responsible for poor separation of coarse particles from the gas. The velocities of eddies normal to the main flow are generally one-fifth of the main flow. Thus, if the gases near the walls of the cyclone are spinning at about 50 fps, the eddies may add 10 fps or more to the inward drift velocity. The inward drift velocity should only be on the order of 1 fps. The eddies thus will cause particles of the magnitude of three times or more than the cut size to appear in the exit stream.

The generation of double eddies superimposed in the vertical plane on the main flow is also a problem. Figure 6 is a schematic view of the eddy currents in a cyclone. These assist the particle descent into the collection hopper but at the same time carry a portion of the collected particles back into the inner ascending vortex. Upswept particles have an opportunity to be separated while being transported to the clean gas exit because the gas spins very rapidly in the inner vortex. However, a large portion of particles does escape collection in this manner.

Base pick-up is a major cause of excessive emissions. Some cyclone systems are designed with a predisengaging hopper before the final collecting hopper. These are designed to provide disengagement space for extracting particles from the return; however, when they are undersized, additional vertical flow is induced and particulates become reentrained.

Pressure Drop

Pressure drop is an important factor affecting efficiency and design. Total pressure drop across a cyclone consists of separate losses in (1) the inlet pipe, (2) the vortex, and (3) the cyclone exit duct.

The pressure drop across a cyclone may be determined from Bernoulli's equation:

$$\frac{\Delta P}{\gamma} = \frac{{w_0}^2}{2g} - \frac{{w_1}^2}{2g} + Z \qquad (2)$$

where

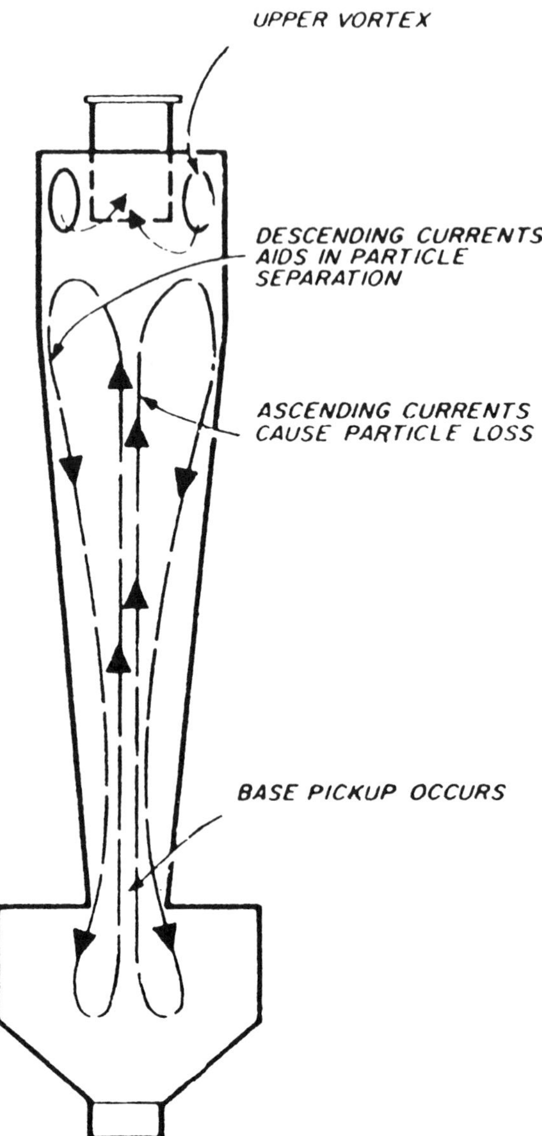

Figure 6 Eddying in a cyclone.

ΔP = total pressure drop
γ = specific gravity of gas
w_0 = superficial gas velocity at the inlet duct
w_1 = superficial gas velocity at the discharge duct
Z = total cyclone resistance consisting of the resistance inside the cyclone, Z_0, and at the discharge duct, Z_1

The gas resistance in the discharge duct may be obtained from Darcy's equation:

$$Z_1 = \lambda \frac{L_1}{d_1} \frac{w_1^{\ 2}}{2g} \tag{3}$$

As a first approximation, the resistance to gas flow in a cyclone will be proportional to the kinetic energy of the gas. As the gas velocity in the cyclone is practically equal to the initial velocity, w_0, we have

$$Z_0 = K_0 \frac{w_0^{\ 2}}{2g} \tag{4}$$

The resistance coefficient, K_0, for cyclones with a rectangular inlet duct (h, height; b, width) is

$$K_0 = \frac{16bh}{d_1^{\ 2}} \tag{5}$$

As the ratio w_0 / w_1 is equal to the ratio of cross sections F_1/F_0, Eq. (2) may be rewritten as follows:

$$\frac{\Delta P}{\gamma} = \frac{W_0^{\ 2}}{2g} \left[Z_0 + Z_1 - 1 + \left(\frac{F_0}{F_1} \right)^2 \right] \tag{6}$$

Hydroclones

Hydroclones are employed for the separation of solid particles from medium- to low-viscosity liquids. Like their cyclone counterparts, hydroclones are simple in design, and the degree of separation can be altered by either varying loading conditions or changing geometrical proportions.

Unlike other types of separating equipment, they are better suited for classifying than for clarifying, because high shearing stresses in a hydroclone promote the suspension of particles which oppose flocculation. However, by properly specifying dimensions and operating conditions, they can be used as thickeners in such a manner that the overflow contains mostly solid particles, whereas the clear overflow constitutes the largest portion of the liquid. The fluid vortices and flow patterns characteristic of gas cyclone operations are equally descriptive of liquid hydroclones. However, the density differences between particles and liquids are significantly smaller than for gas-solid systems. For example, the density of water is approximately 800 times greater than that of air. This means that high fluid-spinning velocities cannot be employed in hydroclones, as excessive pressure drop becomes a limitation. Obviously, the efficiency of hydroclones is low in comparison to gas cyclones.

The design features of a hydroclone are illustrated in Figure 7. It consists of an upper short cylindrical section (1) and an elongated conical bottom (2). The suspension is introduced into the cylindrical section (1) through the nozzle (3) tangentially, whence the fluid acquires an intensive rotary motion. The larger particles, under the action of centrifugal force, move toward the walls of the apparatus and concentrate on the outer layers of the rotating flow. Then they move spirally downward along the walls to the nozzle (4), through which the thickened slurry is evacuated. The largest portion of liquid containing small particles (clear liquid) moves in the internal spiral flow upward along the axis of the hydroclone. The cleared liquid is discharged through the nozzle (5) and fixed at the partition (6) and the nozzle (7). The actual flow pattern is more complicated than described because of radial and closed circulating flows. Because of peripheral flow velocities, the liquid column formed at the hydroclone axis has a pressure that is below atmospheric. The liquid bulk flow limits the upward flow of small particles from the internal side and has a significant influence on the separating effect. Hydroclones are used successfully for

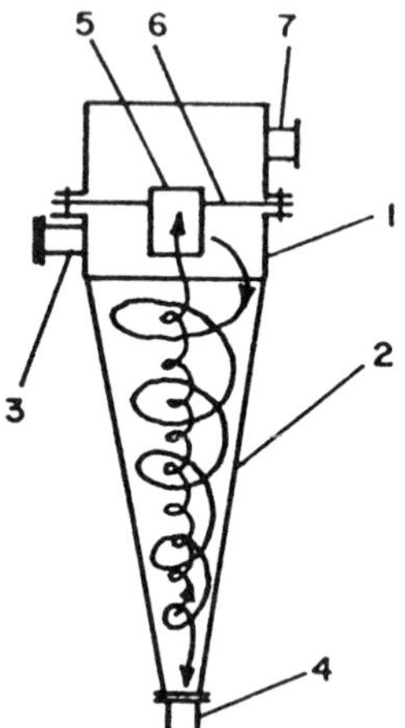

Figure 7 Design features of a hydroclone: (1) cylindrical section, (2) conical bottom, (3) feeding nozzle, (4) discharge nozzle, (5) nozzle, (6) partition, (7) nozzle for liquid discharge.

classification, clarification, and thickening of suspensions containing particles from 5 to 150 μm in size.

The smaller the hydroclone diameter, the greater the centrifugal forces developed and, consequently, the smaller the size particles that can be separated. The following are typical hydroclone diameters used for various general applications: for classification and degritting process streams, D = 300–350 mm; for thickening of suspensions, D = 100 mm; for clarification (where it is necessary to apply powerful centrifugal fields), D = 10–15 mm. In the last case, multiclones are employed. Figure 8 shows an example of a hydroclone being used in a degritting operation. Sand is accumulated in a grit chamber for intermittent blowdown. Such an operation could be used off of a cooling tower installation.

Good separation of suspensions is achieved, especially in thickening and clarification, when hydroclones have an elongated shape with the slope of the

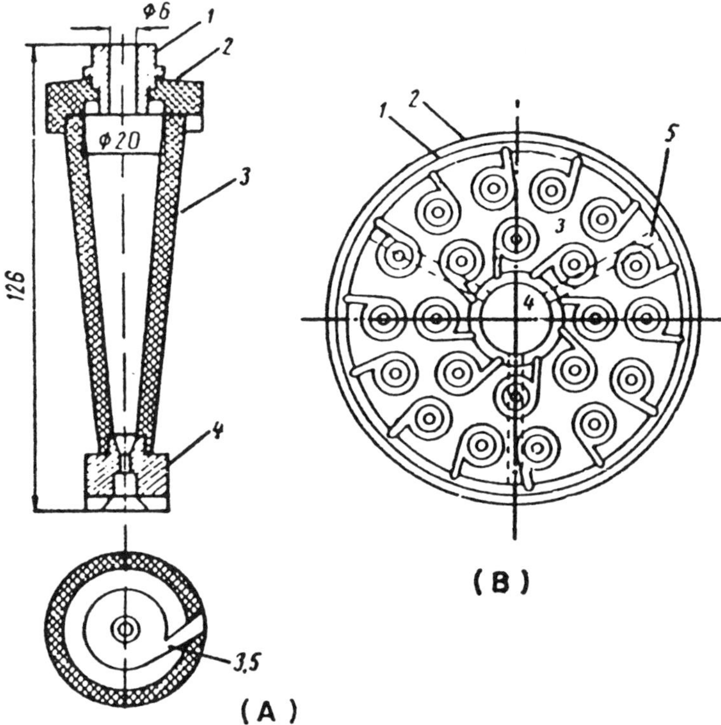

Figure 8 Multiclone arrangement. (A) Front view of single hydroclone: (1) bushing; (2) cover; (3) body; (4) endpiece. (B) Top view of a battery multihydroclone: (1) rubber blocks; (2) metal body; (3) multiclones; (4) central feeding pipes; (5) radial channels.

cone equal to approximately 10–15 degrees. At such a cone shape, the path of solid particles is increased as well as the residence time, which thus increases the separating efficiency. Design methodology for hydroclones and gas cyclones is not well developed. Current design principles are limited to determining capacity, approximate sizes of particles settled, and horsepower requirements. The flow rate of a suspension with density ρ through an inlet nozzle of diameter d_N at pressure drop ΔP may be calculated from the following formula:

$$V_{sec} = \tilde{\mu} \frac{\pi d_N^{\,2}}{4} \sqrt{2 \frac{\Delta P}{\rho}} \qquad (7)$$

where $\tilde{\mu}$ is a flow rate coefficient.

Introducing the hydroclone diameter, D, and the lower nozzle, d_{lN}, we obtain

$$V_{sec} = C'(Dd_{lN}) \sqrt{\frac{\Delta P}{\rho}} \qquad (8)$$

where factor C′ is

$$C' = \tilde{\mu} \frac{\pi \sqrt{2}}{4} \frac{d_N^{\,2}}{Dd_{o'N}} \qquad (9)$$

Coefficient C′ is constant for geometrically similar cyclones. Defining the coefficient

$$K = \frac{C'}{\sqrt{\rho}} \qquad (10)$$

and including K in Eq. (10), we obtain the design formula for calculating the suspension flow rate in the following form:

$$V_{sec} = KDd_{lN} \sqrt{\Delta P}, \ m^3/s \qquad (11)$$

If values in Eq. (11) are in S units, then from experimental data with D = 125–600 mm and cone angle 38°, the coefficient K is equal to 2.8×10^{-4}. The maximum size of particles in the cleared liquid is:

$$d = \frac{1.33 \times 10^{-2} d_N{}^2}{\Phi_x} \sqrt{\frac{\mu}{V_{sec} h \rho_f}}, \text{ m} \tag{12}$$

where h is the height of the central flow, which is assumed to be equal to one-third the cone height and Φ_x is the factor reflecting the change of liquid velocity. For hydroclones with a cone angle $\alpha = 38°$,

$$\Phi_x = 0.103 \frac{D}{d_{lN}} \tag{13}$$

The power required for a hydroclone operation is the horsepower of a pump providing the desired capacity V_{sec}, with an appropriate pressure.

The separating effect is influenced mostly by the ratio d_N/d_{lN} of nozzles, which may be assumed to be in the range of 0.37–0.40.

The diameter of the inlet nozzle, d_N, is assumed to be (0.14–0.3)D, and the diameter of discharge nozzle d_d = (0.2–0.167)D. The cone angle for hydroclones used as classifiers is 20°; for thickeners it is 10–15°.

The same operating arrangements used with gas cyclones are applicable to hydroclones, i.e., parallel and series operations. An example of a battery of multiclones is shown in Figure 8. The rubber block is enclosed in the metal body. On the head of the cyclone body are 24 multiclones. The suspension is introduced through a central pipe, which is connected with a block through three radial channels. The suspension enters into all cyclones at the same time from the feed pipe, and all products of separation are discharged to a common chamber.

SEDIMENTATION CENTRIFUGES

Another approach to the separation of solids from suspensions in a centrifugal field is through the use of sedimentation centrifuges. As in cyclones and hydrocyclones, particles of the heavier phase "fall" through the lighter phase away from the center of rotation. In centrifuges, liquid (or gas) and solids are acted on by two forces: gravity acting downward and centrifugal force acting horizontally. In commercial units, however, the centrifugal force component is normally so large that the gravitational component may be neglected.

The magnitude of the centrifugal force component is defined by the ratio $R_c/G{:}\omega^2 r/g$, which is referred to as the relative centrifugal force (RCF), or *centrifugal number* (N_c), where $R_c = m\omega^2/2$ or $R_c = mv^2/2$ is the centrifugal force and $F_g = mg$ is the gravitational force. The RCF typically varies from 200 times gravity for large-basket centrifuges to 360,000 for high-speed tubular-gas centrifuges and ultracentrifuges.

For liquid-solid separations, centrifugal force may be applied in sedimentation-type centrifuges, centrifugal filters or in a combination of both.

Sedimentation-type centrifuges also are used for size or density classification of solids, the separation of immiscible liquids of different densities, and for concentrating gases of different molecular weights. The principles of centrifugation are illustrated in Figure 9. In Figure 9A, a stationary cylindrical bowl contains a suspension of solid particles in which the particle density is greater than that of the liquid. Because the bowl is stationary, the free liquid surface is horizontal and the particles settle owing to the influence of gravity. In Figure 9B, the bowl is rotating about its vertical axis. Liquid and solid particles are

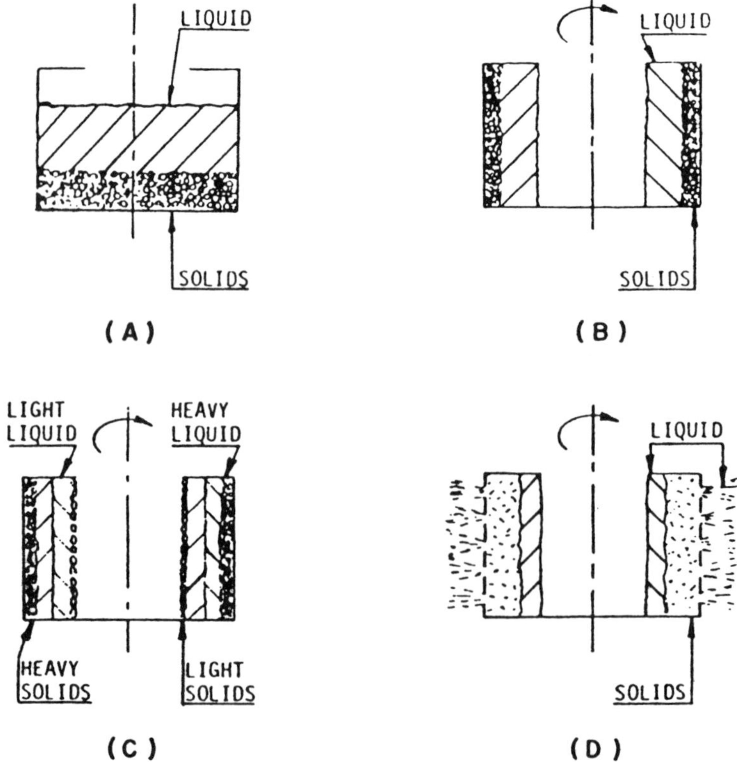

Figure 9 (A) Stationary cylindrical bowl (liquid plus heavy solid particles); (B) rotating cylindrical bowl (liquid plus heavy solid particles); (C) rotating cylindrical bowl (two liquids plus light and heavy solid particles); (D) rotating perforated bowl.

acted on by gravity and centrifugal forces resulting in the liquid assuming a position with an almost vertical inner surface (free interface).

If the suspension consists of several components, each with different densities, they will stratify with the lightest component nearest the axis of rotation and the heaviest adjacent to the solid bowl wall (Fig. 9C).

In Figure 9D, the bowl wall is perforated and lined with a permeable membrane, such as filter cloth or wire screen, which will support and retain the solid particles but allow the liquid to pass through because of the action of centrifugal force.

The main components of a centrifuge are

- Rotor or bowl in which the centrifugal force is applied to a heterogeneous system to be separated
- Means for feeding this system into the rotor
- Drive shaft
- Axial and thrust bearings
- Drive mechanism to rotate the shaft and bowl
- Casing or "covers" to contain the separated components
- Frame for support and alignment

There are three main types of centrifuges, which may be classified according to the centrifugal number and the range of throughputs or the solids concentration in suspension that can be handled. The first of these is the *tubular-bowl centrifuge*. This type has a centrifugal number in the range of 13,000 but is designed for low capacities (50–500 gallons per hour) and can handle only small concentrations of solids.

The second is the *solid-bowl centrifuge*, with maximum bowl diameters ranging from 4 to 54 in. The larger-diameter machines can handle up to 50 ton/hr of solids with a centrifugal number up to 3000. Similar centrifuges are manufactured with a perforated wall on the bowl. These machines operate exactly like filters with the filtrate draining through the cake and bowl wall into a surrounding collector.

The third type is the *disc-bowl centrifuge*, which is larger than the tubular-bowl centrifuge and rotates at slower speeds with a centrifugal number up to 14,000. These machines can handle as much as 30,000 gph of feed containing moderate quantities of solid particles.

The migration of particles in sedimentation centrifuges is radially toward or away from the axis of rotation depending on whether the density of the dispersed particles is greater or less than that of the continuous phase. There must be a measurable difference between the density of the continuous and dispersed phases to provide effective separation.

In commercial machines, the discharge of the liquid or separated liquid phases is performed almost always in a continuous fashion. The heavy solid phase deposited against the bowl wall is discharged and recovered intermittently, manually or by the action of an unloader knife or skimmer; continuously by the action of a differential screw conveyor; or intermittently or continuously with a portion of the continuous phase through openings in the wall of the bowl. Variations are summarized in Table 2. Manual solids removal units can operate continuously up to 1 hr and generally only require a few seconds for a fully automated intermittent operation. In systems in which solids have a lesser density than the continuous phase, particulates can be removed continuously from the surface of the liquid via a skimming tube or exit as an overflow from the bowl with a portion of the continuous phase.

Tubular-Bowl Centrifuges

Tubular-bowl centrifuges are used extensively for the purification of oils by separating suspended solids and free moisture from them; for removal of oversize particles from dye pastes, pigmented lacquers, and enamels; for "polishing" citrus and other aromatic oils; and for other small-scale separating applications.

The tubular-bowl clarifiers and separators comprise small-diameter cylinders (about 100 mm) which allow operation at very high velocities. Commercial machines typically work at 15,000–19,000 rpm, which corresponds to N_c = 13,000–18,000. For special applications (e.g., treatment of vaccines), the diameter of the tubular-bowl centrifuge is only several centimeters, with values of N_c as high as 50,000. Figure 10 shows the operating principle for this type of centrifuge. The tubular centrifuge rotor is suspended from its drive assembly on a spindle that has a built-in degree of flexibility. It essentially hangs freely with a sleeve bushing in a dampening assembly at the bottom. In some designs, a similar dampening assembly is included at the upper end of the rotor. This permits the rotor to determine its own mass axis after it exceeds its critical speed. The details of a specific design are illustrated by the cutaway view in Figure 11.

Table 2 Classification of Sedimentation Centrifuges

Flow Arrangement	Centrifuge Types
Liquid-batch	Analytical and clinical
Solid-batch	Ultra and preparatory miscellaneous batch
Liquid-continuous	Tubular bowl
Solids-batch	Multipass clarifier
	Disk (solid wall)
	Basket type (solid wall)
Liquid-continuous	Disk-valve discharge
Solids-intermittent	Disk-automatic opening
Liquid-continuous	Disk-peripheral nozzles
Solids-continuous	Disk-light solid-phase skimmer, continuous decanter

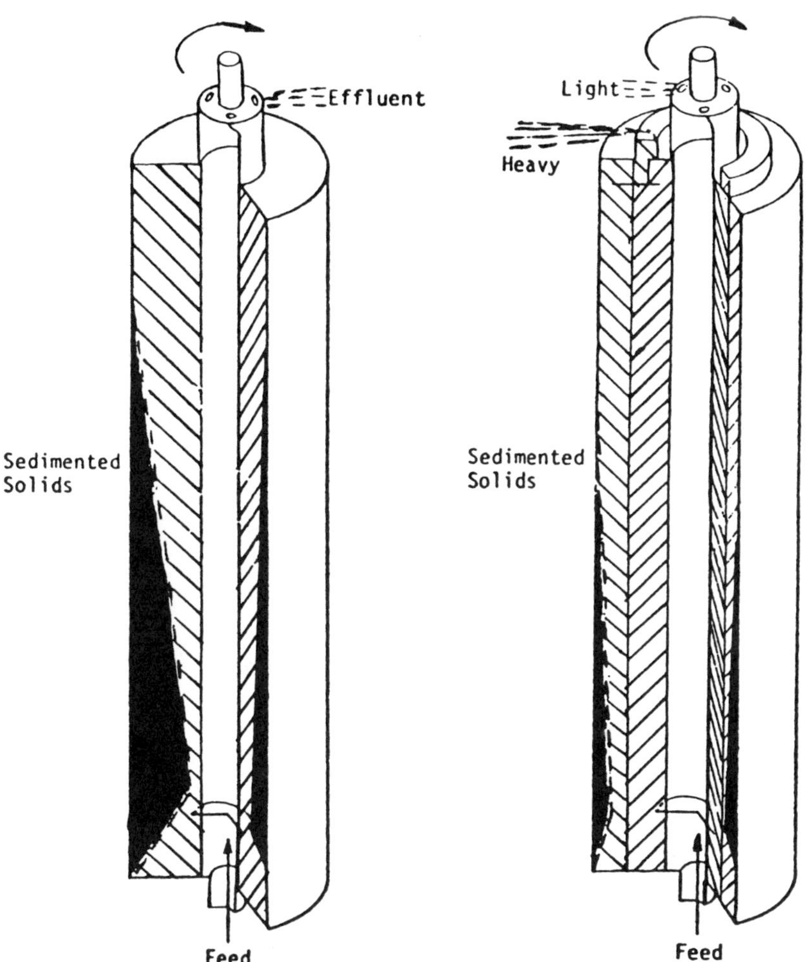

Figure 10 Operation of tubular-bowl centrifuges for clarification and separation.

The feed liquid is introduced to the rotor at the bottom through a stationary feed nozzle. The inlet feed is under sufficient pressure to create a standing jet, which ensures a clean entrance into the rotor. Often, an acceleration device is provided at the bottom of the rotor to bring the feed stream to the rotational speed of the bowl. The feed moves upward through the bowl as an annulus and

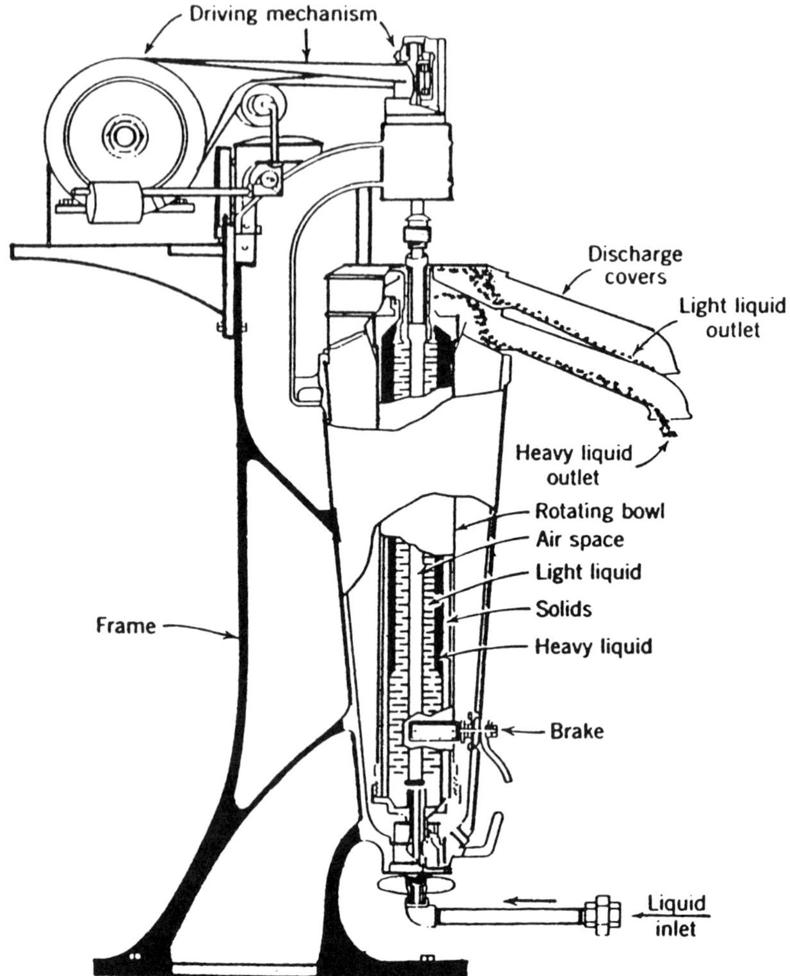

Figure 11 Cutaway view of a tubular-bowl centrifuge.

discharges at the top. To affect this, the radius of the discharge at the top must be larger than the opening at the bottom through which the feed enters. Solids move upward with the velocity of the annulus and simultaneously receive a radial velocity that is a function of their equivalent spherical diameter, d^2, their relative density, $\Delta\rho$, and the applied centrifugal force, ω^2 r. If the trajectory of a particle intersects the cylindrical bowl wall, it is removed from the liquid; if it does not, the particle flows out with the effluent overflow.

Multichamber (Multipass) Centrifuges

This machine combines the process principles of a tubular clarifier with mechanical drive and the bowl contour of a disk centrifuge. The suspension flows through a series of nested cylinders of progressively increasing diameter. The direction of the flow from the smallest to the largest cylinders is in parallel to the axis of rotation, as in the tubular bowl.

Such a rotor usually contains six annuli, so that the effective length of suspension travel is approximately six times the interior height of the bowl. The multipass bowl can be considered as a multistage classifier because the centrifugal force acting in the machine is greater in each subsequent annulus. Consequently, larger, heavier particles are deposited in the first annulus (the zone of least centrifugal force), whereas smaller, lighter particles are deposited in the last annulus (the zone of greatest centrifugal force). The radial distance particles must migrate to reach the cylinder wall is thus minimized. Multipass rotors typically have a total holding volume of up to 65 L for the largest size, of which about 50% is available for the retention of collected solids before the clarification process is impaired. As with the tubular type, the collected solids must be removed manually. These machines are applied to the clarification of fruit and vegetable juices, wine, and beer.

Solid-Bowl Centrifuges

The continuous solid-bowl centrifuge basically consists of a solid-wall rotor, which may be tubular or conical in shape, or a combination of the two. The rotor may rotate about a horizontal or a vertical axis because the centrifugal force is many times that of gravitational force (for many units N_c is more than 3,000).

A typical example of this kind of equipment is a continuous horizontal centrifuge, as shown in Figure 12. It consists of a cylindrical rotor with a truncated cone-shaped end and an internal screw conveyor rotating together. The

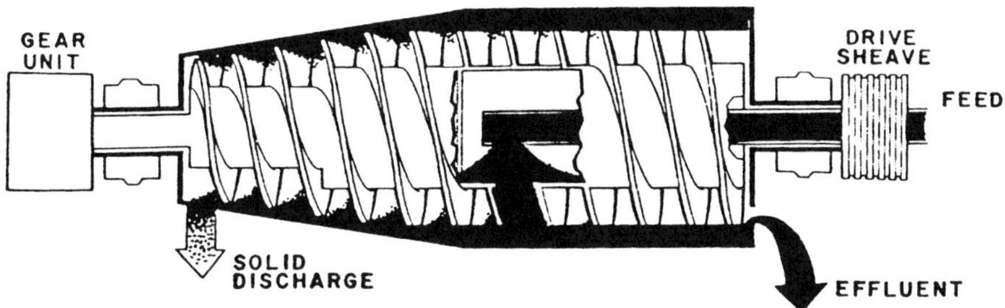

Figure 12 Continuous solid-bowl centrifuge.

screw conveyor often rotates at a rate of 1 or 2 rpm below the rotor's rate of rotation. The suspension enters the bowl axially through the feed tube to a feed accelerated zone and then passes through a feed port in the conveyor hub into the pond. The suspension is subjected to centrifugal force and thrown against the bowl wall where the solids are separated. The clarified suspension moves toward the broad part of the bowl to be discharged through a port.

The solid particles being scraped by the screw conveyor are carried in the opposite direction (to the small end of the bowl) across discharge ports through which they are ejected continuously by centrifugal force.

As in any sedimentation centrifuge, the separation takes place in two stages: settling (see Fig. 12, in the right part of the bowl) and thickening or pressing out of the sediment (left-hand side of the bowl). Because the radius of the solid discharge port is usually less than the radius of the liquid overflow at the broader end of the bowl, part of the settled solids is submerged in the pond. The remainder, closer to the center, is inside the free liquid interface, where they can drain before being discharged. The total length of the "settling" and "pressing out" zones depends on the dimensions of the rotor. Their relative length can be varied by changing the pond level through suitable adjustment of the liquid discharge radius. When the pond depth is lowered, the length of the pressing out zone increases with some sacrifice in the clarification effectiveness.

The critical point in the transport of solids to the bowl wall is their transition across the free liquid interface, where the buoyancy effect of the continuous phase is lost. At this point, soft amorphous solids tend to flow back into the pond instead of discharging. This tendency can be overcome by raising the pond level so that its radius is equal to or less than that of the solids discharge port. In reality, there are no dry settled solids. The solids form a dam, which prevents the liquid from overflowing. The transfer of solids becomes possible because of the difference between the rotational speed of the screw conveyor and that of the bowl shell. The flights of the screw move through the settled solids and cause the solids to advance. To achieve this motion, it is necessary to have a high circumferential coefficient of friction on the solid particles with respect to the bowl shell and a low coefficient axially with respect to the bowl shell and across the conveyor flights. These criteria may be achieved by constructing the shell with conical grooves or ribs and by polishing the conveyor flights. The conveyor or differential speed is normally in the range of 0.8% to five of the bowl's rotational speed.

The required differential is achieved by a two-stage planetary gear box. The gear box housing carrying two ring gears is fixed to, and rotates with, the bowl shell. The first-stage pinion is located on a shaft that projects outward from the housing. This arrangement provides a signal that is proportional to the torque imposed by the conveyor. If the shaft is held rotational (e.g., by a torque overload release device or a shear pin), the relative conveyor speed is equivalent to the bowl rotative speed divided by the gear box ratio. Variable differential speeds can be obtained by driving the pinion shaft with an auxiliary power supply or by allowing it to slip forward against a controlled breaking action. Both arrangements are employed when processing soft solids or when maximum retention times are needed on the pressing out zone. The solids-handling capacity of this type centrifuge is established by the diameter of the bowl, the conveyor's pitch, and its differential speed.

Feed ports should be located as far from the effluent discharge as possible to maximize the effective clarifying length. Note that the feed must be introduced into the pond to minimize disturbance and resuspension of the previously sedimented solids. As a general rule, the preferred feed location is near the intercept of the conical and cylindrical portions of the bowl shell. The angle of the sedimentation section with respect to the axis of rotation is typically in the range of 3–15°. A shallow angle provides a longer sedimentation area with a sacrifice in the effective length for clarification. In some designs, a portion of the conveyor flights in the sedimentation area is shrouded (as with a cone) to prevent intermixing of the sedimented solids with the free supernatant liquid in the pond through which they normally would pass.

In other designs, the clarified liquid is discharged from the front end via a centrifugal pump or an adjustable skimmer that sometimes is used to control the pond level in the bowl. Some displacement of the adhering virgin liquor can be accomplished by washing the solids retained on the settled layer, particularly if the solids have a high degree of permeability. Washing efficiency ranges up to 90% displacement of virgin liquor on coarse solids.

The longer section of a dry shallow layer provides more time for drainage of the washed solids. This system is especially effective for washing and dewatering such spherical particles as polystyrene. In either washing system, the wash liquid that is not carried out with the solids fraction returns to the pond and eventually discharges along with the effluent virgin liquor.

Disc-Bowl Centrifuges

Disc-bowl centrifuges are used widely for separating emulsions, clarifying fine suspensions, and separating immiscible liquid mixtures. More sophisticated designs can separate immiscible liquid mixtures of different specific gravities while simultaneously removing solids. Figure 13 illustrates the physical separation of two liquid components within a stack of discs. The light liquid phase builds up in the inner section, and the heavy phase concentrates in the outer section. The dividing line between the two is referred to as the "separating zone." For the most efficient separation this is located along the line of the rising channels, which are a series of holes in each disc, arranged so that the holes provide vertical channels through the entire disc set. These channels also provide access for the liquid mixture into the spaces between the discs. Centrifugal force

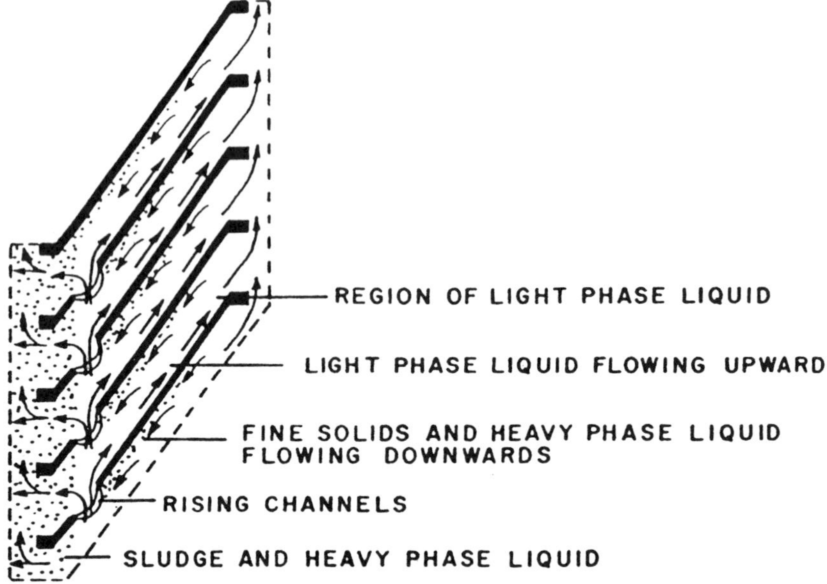

Figure 13 Separation using a stack of discs contained in a centrifuge.

causes the two liquids to separate, and the solids move outward to the sediment-holding space. The position of the separating zone is controlled by adjusting the back pressure of the discharged liquids or by means of exchangeable ring dams.

Figure 14 illustrates the main features of a disc-bowl centrifuge, which includes a bowl (2) with a bottom (13); a central tube (18), the lower part of which has a fixture (16) for discs; a stack of truncated cone discs (17), frequently flanged at the inside and outer diameters to add strength and rigidity; collectors (3 and 4) for the products of separation; and a feed tank (5) with a tube (6). The bowl is mounted to the tube (14) with a guide in the form of a horizontal pin. This arrangement allows the bowl to rotate along with the shaft.

The suspension is supplied from the feed tank (5) through the fixed tube (6) to the central tube (18), which rotates together with the bowl and allows the liquid to descend to the bottom. In the lower part of the bowl, the suspension is subjected to centrifugal force and thus directed toward the periphery of the bowl. The distance between adjacent discs is controlled by spacers that usually are radial bars welded to the upper surface of each disc. The suspension may enter the stack at its outside diameter or through a series of vertical channels cut through the discs, as described earlier.

The suspension is lifted up through vertical channels formed by the holes

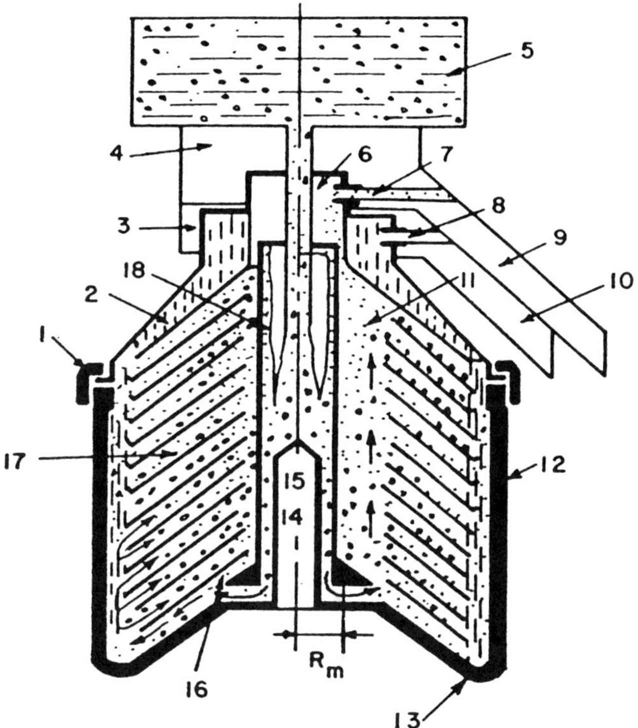

Figure 14 Schematic of a disc-bowl centrifuge: (1) ring; (2) bowl; (3,4) collectors for products of separation; (5) feed tank; (6) tube; (7,8) discharge nozzles; (9,10) funnels for collectors; (11) through channels; (12) bowl; (13) bottom; (14) thick-walled tube; (15) hole for guide; (16) disc fixator; (17) discs; (18) central tube.

in the discs and distributed simultaneously under the action of centrifugal force into the spacings between the discs (spacings are usually in the range of 0.5–3.0 mm).

Owing to a larger diameter, the disc bowl operates at a lower N_c than the tubular design. Its effectiveness depends on the shorter path of particle settling. The maximum distance a particle must travel is the thickness of the spacer divided by the cosine of the angle between the disc wall and the axis of rotation. Spacing between discs must be wide enough to accommodate the liquid flow without promoting turbulence and large enough to allow sedimented solids to slide outward to the grit-holding space without interfering with the flow of liquid in the opposite direction.

The disc angle of inclination (usually in the range of 35–50°) generally is small to permit the solid particles to slide along the discs and be directed to the solids-holding volume located outside of the stack. Dispersed particles transfer from one layer to the other; therefore, the concentration in the layers and their thickness are variables. The light component from the spacing near the central tube (18) falls under the disc, then it flows through the annular gap between the tube (18) and the cylindrical end of the dividing disc, where it is ejected through the port (7) into the circular collector (4) and farther via the funnel (9) on being discharged to the receiver. The heavier product is ejected to the bowl wall and raised upward. It enters the space between the outside surface of the dividing disc and the cone cover (2), then passes through the port (8), and is discharged into the collector (3). From there, the product is transferred to the funnel (10).

One variation of the disc-type bowl centrifuge is the nozzle centrifuge, so named because nozzles are arranged on the periphery or on the bottom of the bowl in a circle that is smaller in diameter than the bowl peripheral diameter. Figure 15 shows the conceptual operation of such a unit. The design is advantageous because it provides a high solids concentration in the discharge with nozzles of relatively large diameters. As centrifugal force is less in that area

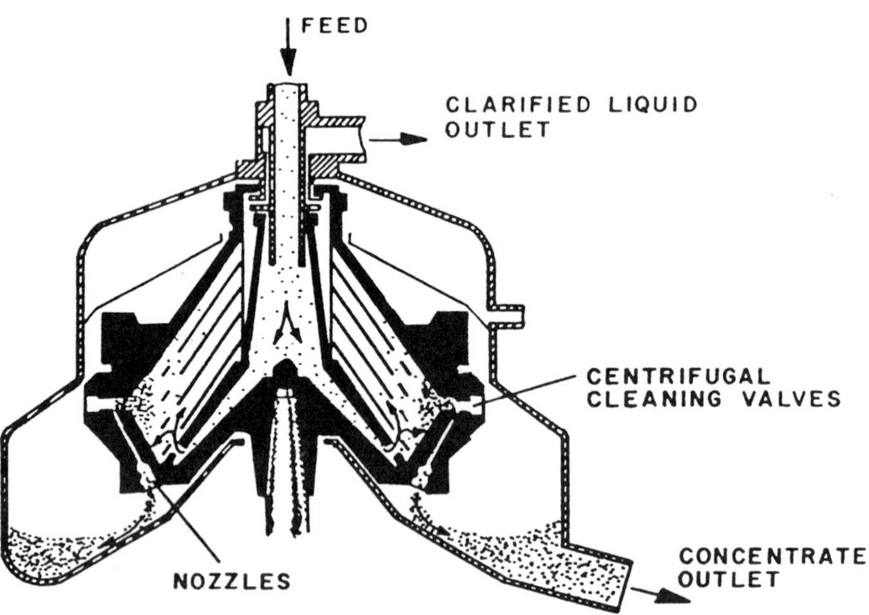

Figure 15 Cutaway view of a disc-bowl centrifuge fitted with nozzles on its underside for concentrate discharge.

than near the periphery of the bowl, the concentrated solids are ejected through the nozzles under a comparatively low pressure.

Figure 16 is a summary diagram of the various types of centrifuges discussed. Tables 3 and 4 show application ranges of centrifuges. Figure 17 is a schematic of an illustrative diagram of centrifuging a waste sludge.

EQUIPMENT SELECTION

Before selecting the type of equipment needed for a given separation problem, the sedimentation velocity of the discrete phase must be determined either experimentally or by the calculations. Based on this parameter and the volumetric flows to be handled, the residence time and capacity of the unit may be approximated.

To estimate sedimentation velocity, information on the physical properties of the system (particle or droplet sizes, densities, viscosity) and the equipment characteristics (e.g., revolutions per minute, characteristic size) must be known. It is always advisable to confirm estimates with laboratory data.

In all cases, after calculations and bench-scale experiments have been made, a check should be made on the actual sedimentation rate; hopefully in a geometrically similar machine. Such a confirmation is critical because it provides indicators of the negative influence of turbulence and the positive influence of internals (such as discs in a centrifuge).

For centrifuges that are to be applied to processing liquids containing solids and that must serve as clarifiers, the volume fraction of sediment in the liquid to be cleared must be determined. If this fraction approaches 1%, it generally is necessary to provide for pretreatment by a decanter or via hydroclones or by centrifugation with continuous withdrawal of the sediment. In this case, equipment selection depends on the relative cost of processing liquid and sediment, the sedimentation velocity necessary for obtaining an acceptable capacity and in the required form of the sediment, its properties, and the liquid capacity to be treated.

In practice, a clarifier must perform the separation within the same time as a separator. Often, it is necessary to obtain from the light liquid phase, the heavy phase, and the sediment, which generally is of little or no process value. If, for example, one must process a mixture of oil and water containing an amount of sediment having no value, a conical machine often is selected when the amount of sediment is large. If only small amounts of sediment are to be processed, it may be possible to achieve a preliminary treatment in a hydroclone or in a large-diameter centrifuge. A final clarification then can be performed in a more powerful centrifuge or in a disc centrifuge. The selection of the specific machine depends on the amount of liquid to be processed and the desired settling velocity (W_s).

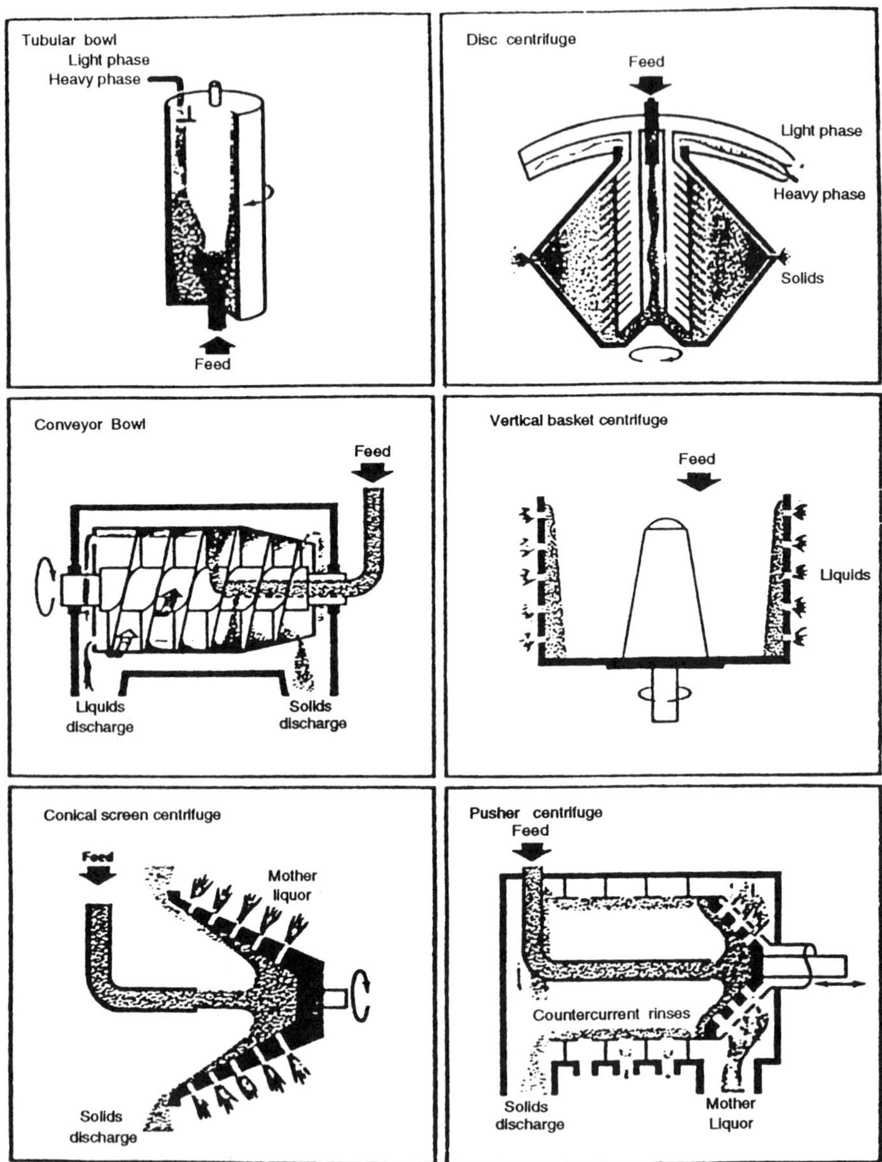

Figure 16 Types of centrifuges.

Table 3 Application Range of Sedimentation Type Centrifuges

Characteristics	Tubular	Disc (nozzle type)	Conveyor bowl
Minimum particle size (μ)	0.1	0.25	2
Minimum particle size (μ)	200	50	5,000
Minimum particle size (μ)	0.1	2–20	2–60
Allowable concentration of feed solids (%)			
Condition of cake	Pasty, firm	Fluid	Firm, pasty
Typical solids handling rate (lb/hr)	0.1–5.0	10–30,000	100–100,000

If the sediment is valuable and the liquid is a waste product, the equipment selected should produce a concentrated slurry, which may be processed further by filtration or direct drying. For preliminary sedimentation it is also possible to employ hydroclones. In some cases, an incomplete clarification is desirable, for example, in the treatment of wastewater in large-diameter centrifuges with lower port discharges or automatic discharge. Products such as this can be processed by a preliminary separating stage to remove the large-sized solids. Examples of precleaning devices are gravity settlers, slow-speed centrifuges of large diameters, hydroclones, or in some cases screening devices. In general, the efficiency of a centrifuge almost always depends on the concentration of the small particles that are settled, and all calculations made on the basis of gravitational sedimentation are valid for the final phase of the separation. An essential part of the operation of separators is the discharge level. In centrifuges designed for separating two liquids, the pureness of each fluid depends on this level.

For very stable emulsions with very fine globules, centrifugal separation

Table 4 Application Range of Filtering Centrifuges

Characteristics	Batch vertical	Conical basket	Pusher
Minimum particle size (μ)	10	250	40
Maximum particle size (μ)	1000	10,000	5000
Allowable concentration of feed solids (%)	2–10	40–80	15–75
Condition of cake	Pasty	Relatively dry	Relatively dry
Typical solids handling rate (1b/hr)	0.1–1.0	5–40	0.5–5.0

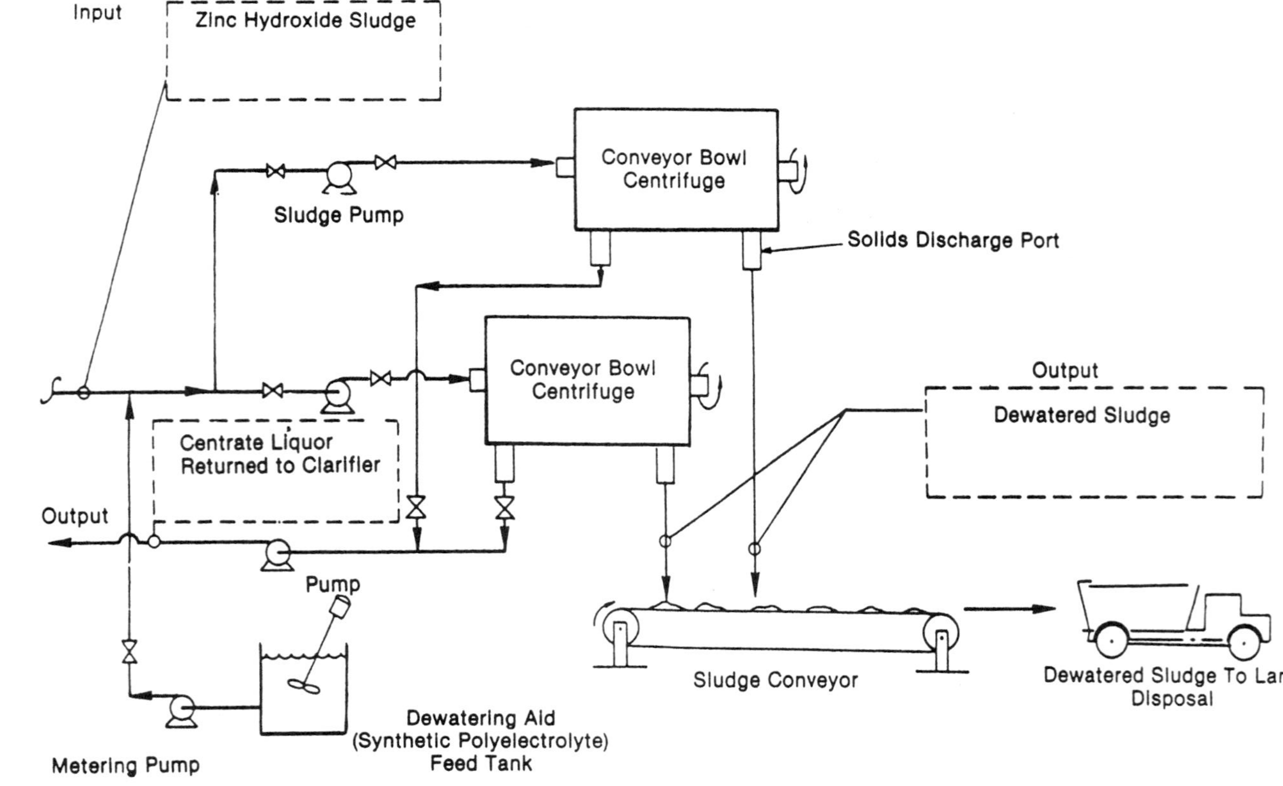

Figure 17 Centrifugation of waste sludge: an illustrative flowchart for zinc hydroxide sludge.

is often impossible. In this case, it is necessary to find a method of coalescence partially to break the emulsion or to make it less stable through moderate mixing, steam injection, neutralization of electrical charges, or inactivation with a demulsifier. Methods of breaking emulsions are very different and often empirical. For example, the injection of steam and agitation can have very different and sometimes opposite effects depending on the actual conditions. Emulsions first can be broken through pH changes or by adding certain surface-active agents. If the demulsifying agent is a product of a nonionic substance (e.g., an oxide of ethylene), it is possible to achieve separation by an increase in the emulsion temperature.

For the processing liquids that change in composition, centrifugation should be performed in hermetically sealed machines prior to pressurization. Hydroclones greatly increase the sedimentation velocity of particles up to 50 μm in size. They may be used also for classification in cases in which the velocities of gravitational decantation are almost the same.

10
Activated Carbon/Ion Exchange

ACTIVATED CARBON

Activated carbon has been used in numerous industrial products and processes for many years and has a long history of success. Activated carbon treatment of water and wastewater has been demonstrated in both municipal and industrial applications.

The use of granular activated carbon in wastewater treatment has been to use activated carbon as "tertiary" treatment following conventional primary and biological secondary treatment. Tertiary treatment processes involving carbon range from treatment of the secondary effluent with only activated carbon to systems with chemical clarification, nutrient removal, filtration, carbon adsorption, and disinfection. Another approach uses activated carbon in a physical/chemical treatment process in which raw wastewater is treated in a primary clarifier with chemicals prior to carbon adsorption. Filtration and disinfection may also be included in physical and chemical treatment but biological processes are not used. Flow diagrams for some alternate treatment schemes for such systems are shown in Figures 1 and 2.

If biological treatment and efficient filtration precede carbon treatment, there are several benefits: the applied loads of BOD (biochemical oxygen demand), COD (chemical oxygen demand), and other organics are reduced allowing either the

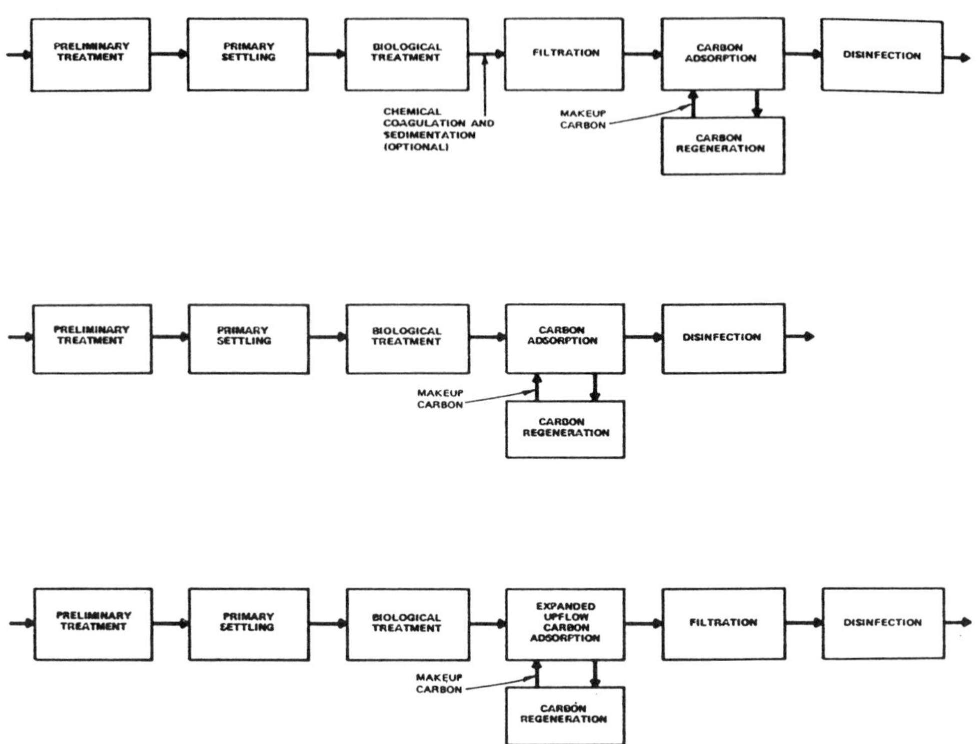

Figure 1 Typical treatment schemes utilizing carbon adsorption as a tertiary step.

production of a higher quality carbon column effluent at a given contact time or the production of equal water quality at a shorter contact period. The applied loads of suspended and colloidal solids are less, thus reducing headloss through the bed of carbon, which may aid in solving problems of physical plugging, ash build-up, and progressive loss of adsorptive capacity in the carbon particles after several cycles of regeneration, and problems of biological growth, septicity, and hydrogen sulfide production may be decreased by reducing the supply of bacterial food and oxygen-demanding substances applied to the carbon.

Maximum use of granular activated carbon by extending its function of removing refractory dissolved organics to adsorption of biodegradable organics as well and, in some cases, by using the granular bed of carbon as a filter to remove suspended and colloidal materials. With this application the carbon is loaded as heavily as possible within the limits of effluent water quality criteria. Physical-chemical treatment results in capital cost savings when compared with biological treatment followed by tertiary treatment. Capabilities must be evalu-

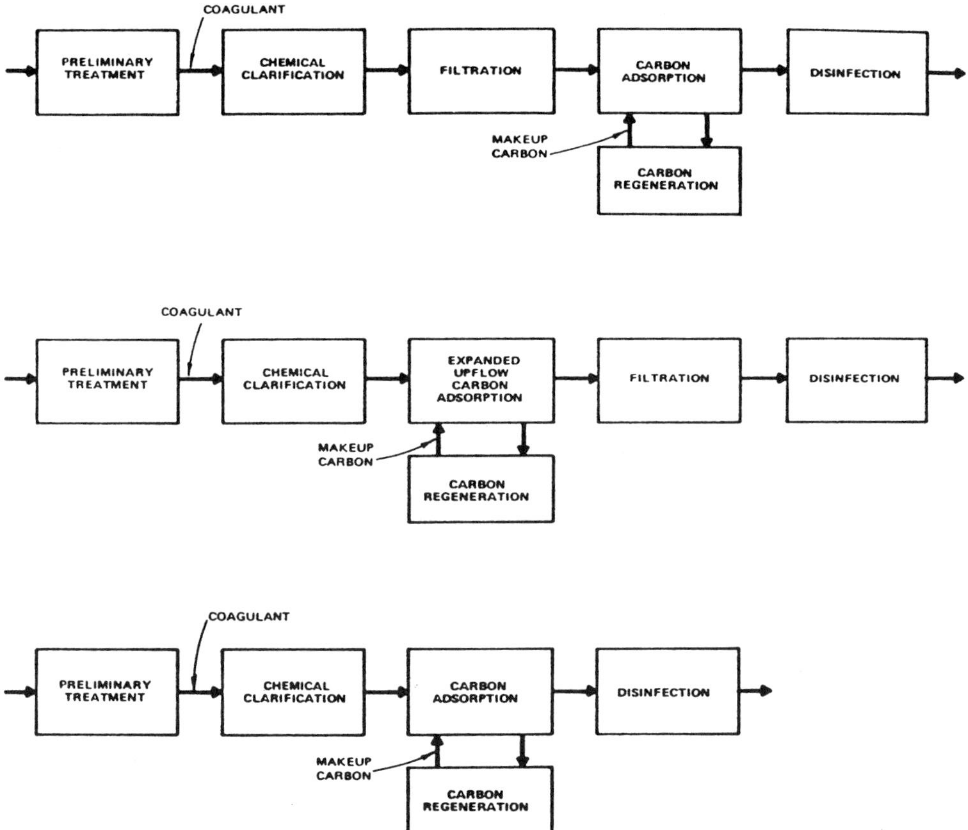

Figure 2 Typical physical-chemical treatment schemes.

ated in light of specific effluent quality requirements to determine its applicability to a given problem.

Carbon adsorption and carbon regeneration are basically simple processes, which are therefore quite reliable. Carbon is versatile and may be used in a variety of locations in wastewater-treatment plants. There is no single method of contacting or contractor design which is best for all conditions because of the placement of carbon adsorption at different points in the wastewater-treatment unit operation sequence and the varying qualities of wastewater which may be applied to the carbon. Design, equipment, and material selection may also be influenced by the total capacity of the plant under consideration. There are fewer design variables in carbon-regeneration systems, and facilities are therefore more uniform in their concept and design carbon transport, monitoring, and control systems are fairly well standardized.

Added capacity for removal of organics can be provided by adding extra spent carbon-storage capacity and by oversizing the regeneration facilities. It is often relatively inexpensive to oversize the regeneration equipment and to operate it initially on a part-time schedule. Application of more organics to the carbon will increase the quantity of carbon to be regenerated; however, there are some savings resulting from continuous rather than intermittent operation of the furnace. Thus, nonproductive operating costs, such as that associated with startup and shutdown, be reduced but will be offset by productive operating costs.

Carbon adsorbers readily conform to modular design concepts; treatment plant capacity may be increased by merely adding additional contactor vessels as required. The initial facility layout can anticipate future installation of additional contactors and associated apparatus in order to minimize construction costs and disruption of plant operations.

Streams of wastewater vary in their volume and chemical composition because of changes in the processes or in the events which generate these streams. Variations frequently exhibit clearly defined cycles. Municipal wastewater exhibits diurnal cycles corresponding to the life patterns of the population, length of sewer, and size of the town. Infiltration of ground water or storm water connections may have a marked effect on these cycles. Industrial wastes may be influenced by the working hours of the plants, shift changes, weekend shutdowns, summer holidays, or fluctuations in production rates caused by seasonal marketing patterns. Fluctuations occur in the chemical character of the wastewater as well as in its volume.

Wastewater Quality

Wastewater parameters of concern in activated carbon treatment systems include suspended solids, oxygen demand as measured by BOD, COD, or TOC (total organic carbon), other organics such as MBAS or phenol, and dissolved oxygen. If there are effluent requirements for any or all of these, then much of the design criteria must be established by laboratory and bench scale tests.

Types of Systems

Alternatives for carbon contacting systems include

- Downflow or upflow of the wastewater through the carbon bed
- Series or parallel operation (single or multistage)
- Pressure or gravity operation in downflow contactors
- Packed or expanded bed operation in upflow contactors
- Steel or concrete construction of carbon vessel, in either circular or rectangular cross section configuration

Upflow/Downflow Systems

Upflow beds have an advantage over downflow beds in carbon use efficiency because they more closely approach continuous countercurrent contact operation. Countercurrent operation results in the minimum use of carbon, or the lowest carbon dosage rate. Upflow beds can be designed to allow addition of fresh carbon and withdrawal of spent carbon while the column is in operation. When operations are conducted almost continuously, the bed may be referred to as a pulsed bed. A pulsed bed may be either an upflow packed bed or an upflow expanded bed. Upflow packed beds require a high clarity influent, usually a turbidity less than about 2.5 JTU (Jackson turbidity units), which may be considered a disadvantage. Upflow expanded beds have the advantages of being able to treat wastewater relatively high in suspended solids and to use finer carbon, which reduces the required contact time without excessive head losses. Whereas upflow packed beds typically use 8 × 30 mesh carbon, upflow expanded beds typically use 12 × 40 mesh.

The reason for using a downflow contactor is to use the carbon for two purposes: adsorption of organics and filtration of suspended materials. This dual use of granular carbon leads to some reduction in capital cost. This economical gain is offset, is not fully predictable, by loss of efficiency in both filtration and adsorption, and perhaps also by higher operating costs. There is a sacrifice in finished water quality resulting from combined adsorption-filtration by carbon.

Downflow beds may be operated in parallel or in series (multistage). Valves and piping are provided in series installations to permit each bed to be operated in any position in the series sequence, thus giving a countercurrent operation.

Provision is made to backwash downflow beds periodically and thoroughly to relieve the pressure drop associated with the accumulation of suspended solids. Continued operation of a downflow bed without backwashing may compact or foul the bed sufficiently to make it more difficult to expand the bed during backwash without the use of an excessive quantity of backwash water. Upflow beds may be flushed through a simple well screen inlet-outlet system. Downflow beds require a false bottom support system, backwash facilities, and controls similar to those used in waterworks practice for sand filters.

Equipment available for automatic operation of filters is available, reliable, and offers adequate process control. Upflow countercurrent carbon columns, following efficient filters, is best accomplished by simple, manual controls. Except for occasional flow reversal, valves serving separate carbon columns are usually operated only during withdrawal and replacement of carbon for regeneration; perhaps once every 4–6 weeks. Valve operation is best done manually with operator attendance and observation.

The use of downflow carbon contactors for the purposes of adsorption and filtration provides capital cost savings. The carbon filter contactor is basically a surface-type filter and is subject to the shortcomings of surface filters. Upsets in pretreatment which produce sudden increases in suspended solids or turbidity can completely blind the surface of a single media bed, which requires backwashing before it can be restored. If the upset in applied water quality continues, then the supply of high-quality water necessary for backwashing carbon filter contactors may be exhausted. Separate filters can be backwashed and placed back on line. Backwashing a carbon filter contactor may require longer and a proportionally larger volume of water of higher quality to avoid plugging the bottom of a deep bed and saturating it with adsorbed organics. Whether using upflow or downflow contactors, filters protect carbon columns from pretreatment upsets and increase the overall plant reliability.

The underdrain system used for downflow carbon beds in similar to that used for conventional water filters. Figure 3 shows a two-bed series downflow system for carbon contact. As indicated, water is first passed down through column A and then down through column B. When the carbon in column A is exhausted, the carbon in column B is only partially spent. At this time, all carbon in column A is removed for regeneration and is replaced with fresh carbon. Column B then becomes the lead column in the series. When the carbon in column B is exhausted, the carbon is removed for regeneration and is replaced with fresh carbon. With upflow columns, no spare contactors are needed, because carbon can be withdrawn for regeneration while the column remains in service.

The use of pressure vessels for carbon contactors will increase the flexibility of operation, since it will allow the system to be operated at higher pressure losses. This may allow the carbon contact system to operate during upsets or variations in the wastewater flow. Gravity contactors may be more economical, since concrete and common wall construction may be utilized.

The aspect ratio of a carbon column is the bed depth to diameter ratio. With good design of the flow distribution and collection systems, the aspect ratio is not a crucial factor, and ratios of even less than 1:1 are satisfactory. However, if the granular bed itself is to be used as a means of flow distribution, then high aspect ratios (greater than 4:1) are desirable to minimize short circuiting and dead spots in the bed. Designs provide inlet and outlet arrangements which distribute and collect the influent and effluent water very uniformly across the entire cross-section of the bed, in which case the aspect ratio is not important.

Contactors Required

There should be an adequate number of contactors (or pairs of contactors if series units are used) to provide adequate treatment even with one contactor or pair of contactors are out of service. In large plants, this is not a constraint, since the physical limitations on individual contactor's size dictate a large number of columns. Shop-assembled pressure vessels cannot be greater than 12 ft in diameter and 60 ft overall length and still be transportable. At 30-min contact,

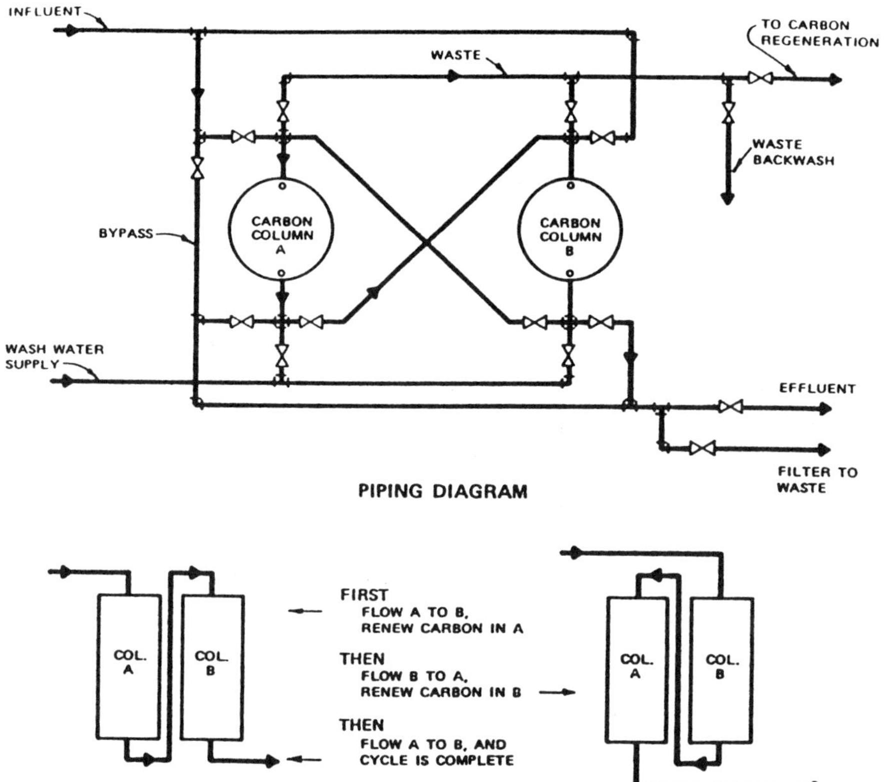

Figure 3 Downflow carbon beds in series.

the maximum capacity per shop-fabricated vessel is about 2 mg/day with more typical designs corresponding to about 1 mg/day per column (Fig. 4). Other upflow arrangements are shown in Figures 5 and 6.

Most open concrete contactors will have a maximum capacity of 2–4 mg/day per contactor at 30-min contact, as the maximum area consistent with good flow distribution is about 1,000 ft^2 for plants of 5 mg/day and above.

For small plants, it may be necessary to use two or more columns to provide flexibility of operation even though one larger column capable of handling the entire flow may be technically feasible. Some added carbon usage may be achieved by blending of the effluent from two columns.

The flow rate and bed depth necessary for optimum performance will depend on the rate of adsorption of impurities from wastewater by the carbon. General range of flow rates or hydraulic loadings 2–10 gpm/ft^2 of cross-sectional

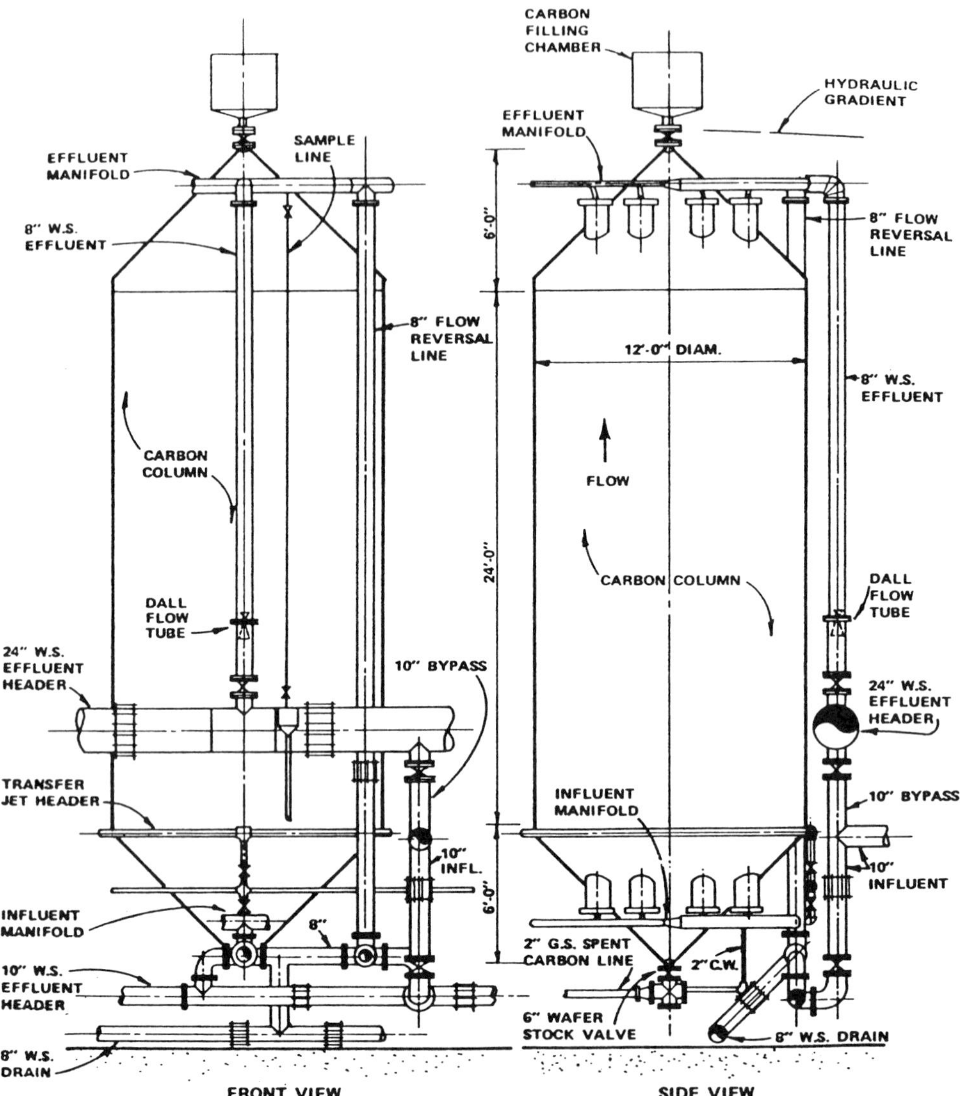

Figure 4 Upflow countercurrent carbon column.

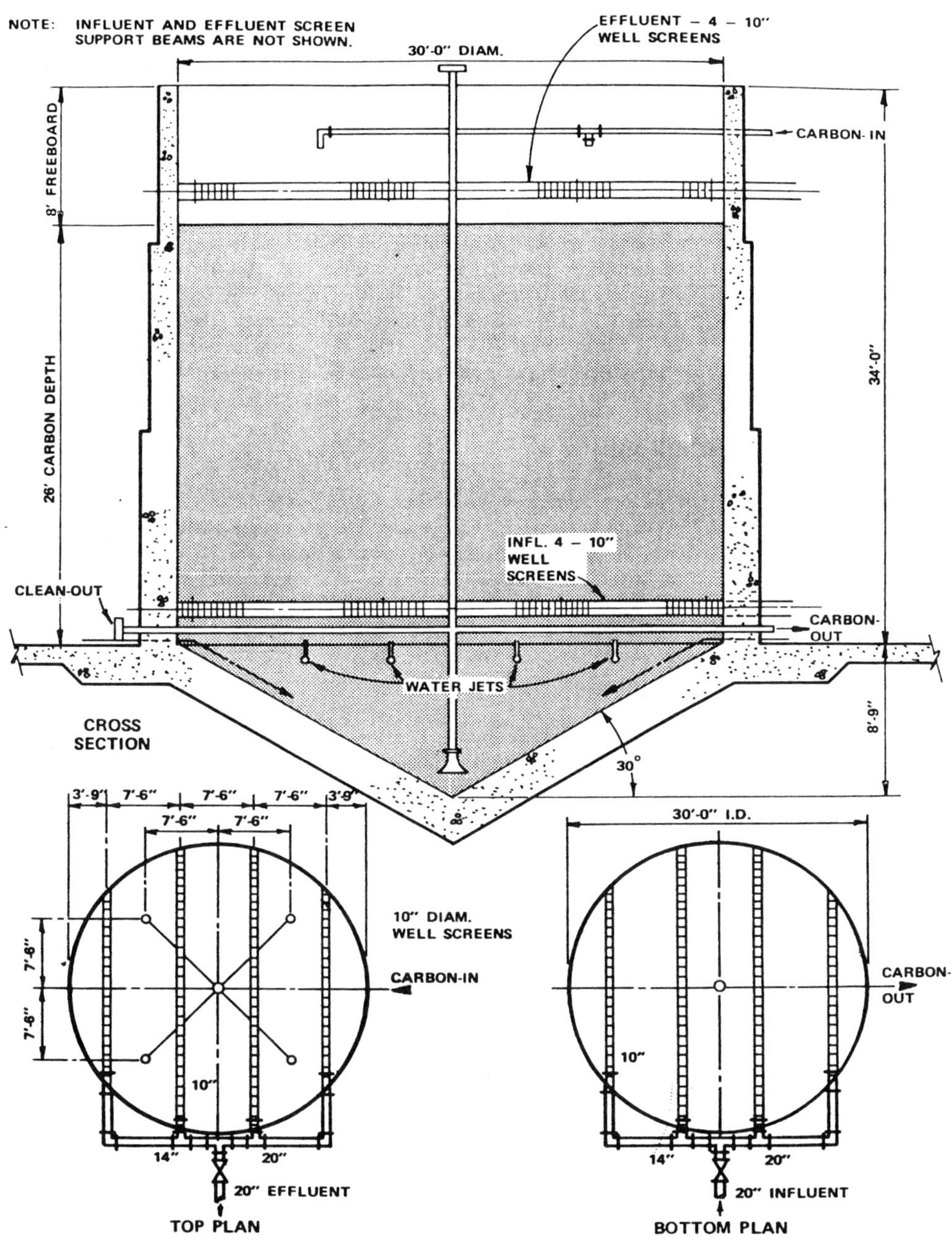

Figure 5 Upflow open-packed bed contactor.

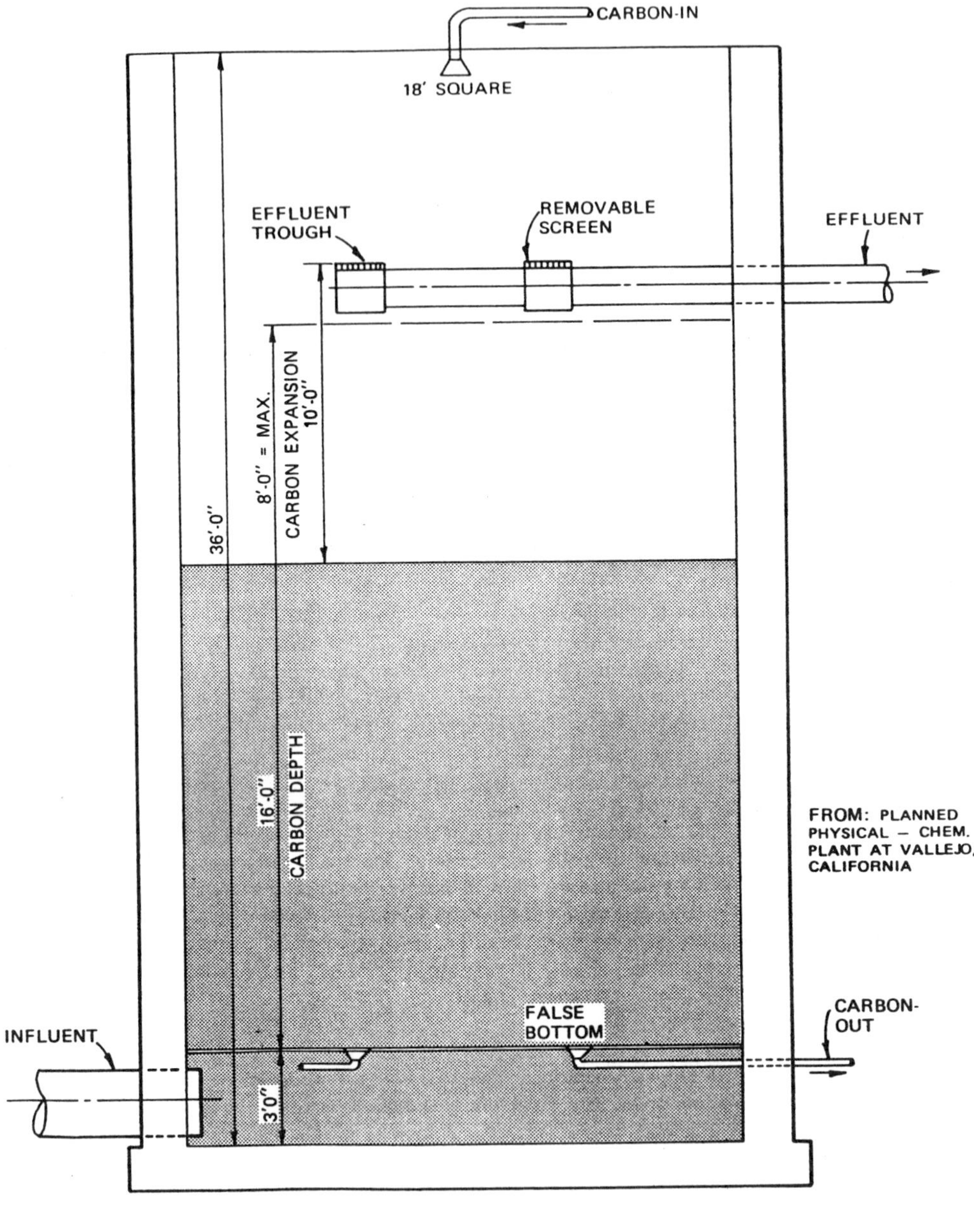

Figure 6 Upflow open-expanded bed contactor.

area. Bed depths are usually 10–30 ft. There are critical velocities for liquids passing through porous beds which change the nature of the resistance to diffusion. At low velocities, the solute content of the stagnant film surrounding the adsorbent particles may become depleted more rapidly than the solute can be replaced by diffusion from the main body of the liquid. The diffusional resistance across the film is controlling. As the velocity is increased, the point will be reached where the controlling effect will be the inability of the adsorbent material to remove the solute from solution as rapidly as it is transported to the surface from the main body of the stream. Within the range of loadings of 2–10 gpm/ft^2, several studies have found that velocity is not a limiting factor.

Hydraulic loading has an additional effect on carbon column operation. Increasing flow rates through the carbon will cause increasing head losses (ΔP). Head loss is dependent on the flow rate and carbon particle size. This relationship for clean water passing through a bed of clean carbon is expressed in the formula

$$\Delta P = \frac{K \upsilon V L_c}{D_p^2 D_c}$$

where

P = Pressure drop, inches
K = Constant
υ = Viscosity, centipoise
V = Flow rate, gallons per minute
L_c = Bed depth, feet
D_p = Mean particle diameter, millimeters
D_c = Column diameter, inches

Hydraulic head loss is related directly to flow rate and inversely related to particle size. Because of the more favorable headloss charactersitics, 8 $\times$ 30 mesh carbon is often preferred for downflow beds, whereas 12 $\times$ 40 mesh carbon may be preferable for upflow beds, because a lower upflow velocity is required for expansion. The head loss for a given hydraulic loading with wastewater feed must be determined by pilot testing. Because head loss development is such an important consideration in the design of a carbon bed, hydraulic loading cannot be discussed in isolation from several other design factors. If an excessive rate of head loss development (due to a high hydraulic loading) is anticipated, an upflow bed should be considered. Choice of gravity versus pressurized flow may be influenced by the anticipated rate of head loss development. Very high hydraulic loadings are practical only in pressurized systems. Gravity flow in downflow beds is considered practical only at hydraulic loadings less than about 4 gpm/ft^2.

Upflow expanded beds should be considered when high head loss is expected. At low flow rates, the particles are undisturbed and the bed remains fixed. As the flow rate is increased, however, a point is reached where all particles no longer remain in contact with one another and the carbon bed is expanded in depth. The flow rate required for initial expansion of the bed is accompanied by a sizable increase in head loss. As the flow rate is increased, there is further expansion of the bed. Flow rates required for further expansion of the bed are accompanied by lesser increases in head losses. Figure 5 illustrates the sharp increase in ΔP for initial bed expansion and the lower rate of increase for further expansion.

Backwash

Backwashing reduces the resistance to flow caused by solids that have accumulated in the bed. The rate and frequency of backwash depends on the hydraulic loading, the nature and concentration of the suspended solids in the wastewater, the carbon particle size, and the method of contacting, upflow, and downflow.

Backwash frequency may be determined by build-up of head loss, deterioration of effluent turbidity, or at regular predetermined intervals of time. For operational reasons arbitrarily backwash of beds at regular intervals without regard for headloss or turbidity may be practiced. Removal of solids trapped in a packed upflow bed may require two steps: first, the bottom surface plugging may have to be relieved by temporarily operating the bed in a downflow mode, and second, the suspended solids entrapped in the middle of the bed may have to be flushed out by bed expansion.

Backwashing a downflow contactor normally requires a bed expansion of 10–50%. Provisions for backwash flow rates of 12–20 gpm/ft^2 should be made with the granular carbons of either 8 × 30 mesh or 12 × 40 mesh. Effective removal of the solids accumulated on the carbon surface in downflow contactors requires

- Surface wash equipment utilizing rotating or stationary nozzles for directing high-pressure streams of water at the surface of the bed
- Air wash
- Combination air-water wash

A surface wash or air wash system is normally operated only during the first few minutes of a backwash of 10–15 min. When backwashing is supplemented by this scouring type of wash, the total amount of water to achieve a given degree of bed cleaning may be reduced. Also, surface wash or air wash overcomes bed plugging that may not be alleviated by normal backwash velocities. The total amount of backwash water required should not exceed 5% of the average plant flow.

Backwash water may be effectively disposed of by recirculating it into primary sedimentation or elsewhere near the inlet of the wastewater treatment plant. A return flow equalization tank may be advisable in order to reduce shock hydraulic loads on the plant from waste washwater.

Biological Activity in Carbon

Under certain conditions, granular carbon beds provide favorable conditions for the production of hydrogen sulfide (H_2S) gas and which may contribute to corrosion of metals and damage concrete. Hydrogen sulfide is produced by sulfate-reducing bacteria under anaerobic conditions. Conditions promoting or accelerating the production of hydrogen sulfide in carbon contactors include

- Low concentrations or absence of dissolved oxygen and nitrate in the carbon contactor influent
- High concentrations of BOD and sulfates
- Long detention times
- Low flowthrough velocities

It may be possible to prevent or correct problems of hydrogen sulfide generation by eliminating one or more of the conditions necessary to sustain growth of the sulfate-reducing bacteria. Most of the preventive measures must be provided in the design of the carbon contacting system, but there are also some corrective measures which can be taken in plant operation. The following measures may be taken to provide flexibility for dealing with problems of hydrogen sulfide production:

- Satisfy as much of the oxygen demand of the wastewater as possible by providing biological treatment prior to carbon treatment.
- Provide biological treatment and efficient filtration to reduce the load of suspended and dissolved organics thus permitting the use of higher flowthrough velocities and reduced detention times in the carbon columns.
- In packed beds of carbon, provide facilities for application of chlorine to the influent. In addition, in upflow expanded beds it may be desirable to provide for introduction of air, oxygen, or sodium nitrate (as a source of oxygen). Because of the mass of cell growth produced, it may be less desirable to introduce air or oxygen ahead of packed beds because of potential physical plugging of the beds. These growths are flushed through expanded upflow beds but may be removed in sections of the plant which follow such as filters or clarifiers.

Remedial measures available in the operation of carbon facilities are

- Columns may be backwashed more frequently or backwashed more violently by use of air scour or surface wash.
- Detention time reduced by taking some carbon contactor units off the line,

provided that the reduced carbon contact time is still sufficient to obtain the desired removal of organics and that head losses in the carbon columns remaining on the line do not become excessive.
- Chlorination or oxygen addition.

ION EXCHANGE

Ion exchange is a versatile separation process with the potential for broad application in the metal finishing industry, both for raw material recovery and reuse and for water pollution control. Three major areas have been demonstrated: wastewater purification and recycle, end-of-pipe pollution control, and chemical recovery.

The ion exchange process has been commercially available for many years. An impetus for the interest in ion exchange technology is the broad range of resins manufactured today. With proper resin selection, ion exchange can provide an effective and economical solution to pollution control requirements. As a further stimulus to the use of the process, the metal-bearing sludge generated by hydroxide treatment systems is considered a hazardous waste material and banned from landfills. Consequently, an alternative means of disposal in an environmentally safe manner must be found. The ion exchange process can concentrate the heavy metals in a dilute wastewater into a concentrated metal solution that is more amenable to metal recovery than is a sludge.

Water containing dissolved electrolytes undergoes changes when in contact with certain solids. Some of the ions present in the water can exchange for other ions on the surface or even in the interior of the solid ion exchange resin. When this exchange occurs, the total number of charges leaving the solution must equal the total number entering. In this order, the law of electroneutrality is obeyed. This process, in essence, is the phenomenon of ion exchange.

There are many solids, both natural and synthetic, that make ion exchange possible. Soils, humus, wool, cotton, and bacterial cells, which are *not* generally regarded as ion exchange materials, are but a few of the naturally occurring substances that are capable of exchanging ions when in contact with a solution containing electrolytes. The degree or extent of exchange that occurs depends on the size and valence (charge) of the ions entering into the exchange, the concentration of ions in the water or solution, the nature (both physical and chemical) of the ion exchange substance, and the temperature. Because of the limitations of natural ion exchange materials for specific commercial and research purposes, inorganic and organic ion exchangers have been used throughout the world.

WATER PURIFICATION

In the first area of application, mixed rinse solutions are deionized to permit reuse of the treated water. The contaminants in the rinses are concentrated in the small volume purge streams and are thereby made more economical to treat. Because ion exchange is efficient in removing dissolved solids from normally dilute spent rinse waters, it is well suited for use in water purification and recycle. Most plating chemicals, acids, and bases used in metal finishing are ionized in water solutions and can be removed by ion exchange. The ion exchange process is effective for this application because

- Ion exchange can economically separate dilute concentrations of ionic compounds from water solutions.
- The process can consistently provide high purity water over a broad range of loading conditions.
- Resins used for separation are durable in severe chemical environments.

Application of the ion exchange process in a wastewater purification and recycle system significantly reduces water consumption and the volume of wastewater discharged, thus reducing water use and sewer fees and the size and cost of the pollution control system.

End-of-Pipe Pollution Control

Toxic heavy metals and metal cyanide complexes can be removed selectively from combined waste streams before discharge. The key to this application is that the ion exchange resins remove only the toxic compounds and allow the nontoxic dissolved ionic solids to remain in solution.

The ion exchange process can be used in two different forms for end-of-pipe pollution control. It has been demonstrated as a means of polishing the effluent from conventional hydroxide precipitation to lower the metal concentration in the discharge, and it has been used as a means of directly treating wastewaters to remove heavy metals and metal cyanide pollutants.

Most plating shops can remove sufficient metal to comply with wastewater discharge regulations using the conventional hydroxide precipitation process. However, as the regulations get more stringent for effluent metal concentrations, or where the metals are complexed with chemical constituents that interfere with their precipitation as metal hydroxides, conventional treatment may not be reliable with the discharge limits. Ion exchange can be used in such cases to polish the effluent from the conventional treatment and reduce the metal concentration further. In this application, the process can provide a relatively inexpensive means of upgrading system performance for compliance with the discharge regulations.

Ion exchange has been used to a limited extent to remove toxic pollutants selectively from an untreated wastewater while allowing most of the nontoxic ions to pass through. Approaches employed to facilitate this application include using:

- Weak acid cation resin in an application of the wastewater softening type to remove heavy metals and other divalent cations from a wastewater solution with a high concentration of sodium ions
- Heavy-metal selective weak acid or chelating cation resin to remove only the heavy metal ions while allowing sodium, magnesium, and calcium ions to pass through
- A stratified bed of resin containing strong and weak acid cation and strong base anion resins to remove heavy metal and metal cyanide complex ions from solution while allowing most of the wastewater ionic constituents to pass through

In each of these approaches, wastewater pretreatment requires pH adjustment to ensure that the pH is within the operating range of the resin and filtration to remove suspended solids that would foul the resin bed. The pollutants removed from the wastewater are concentrated in the ion exchange regenerant solutions. The regenerants can be treated in a small batch treatment system using conventional processes. Firms with access to a centralized treatment system to dispose of the regenerant solutions resulting from treatment would not need to install chemical destruct systems. In neither case would it be necessary to invest in sophisticated pH control systems, flocculent feed systems, clarifiers, and other process equipment associated with conventional metal precipitation systems. And, as a further advantage, ion exchange units are compact and easy to automate compared with conventional precipitation systems.

Chemical Recovery

In the chemical recovery applications segregated plating rinse waters are treated to concentrate the plating chemicals for recycling to the plating bath. The purified rinse water is also recycled.

Ion exchange, evaporation, reverse osmosis, and electrodialysis have all been used in the plating industry to recover chemicals from rinse waters. These processes have the ability to separate specific compounds from a water solution, yielding a concentrate of those compounds and relatively pure water. The concentrate is recycled to the plating bath and the purified water is reused for rinsing. Determination of the separation process best suited for a particular chemical recovery application requires evaluating both general and site-specific factors:

- General factors include rinse water concentration, volume, and corrosivity among others.
- Site-specific factors include, for example, availability of floor space and utilities (e.g., steam, chemical reagents, electricity) and the degree of concentration needed to recycle the chemical to the bath.

Usually, ion exchange systems are suitable for chemical recovery applications where the rinse water feed has a relatively dilute concentration of plating chemicals and a relatively low degree of concentration is required for recycle of concentrate. Ion exchange is well suited for processing corrosive solutions. Ion exchange has been demonstrated commercially for recovery of plating chemicals from acid-copper, acid-zinc, nickel, tin, cobalt, and chromium plating baths. The process has also been used to recover spent acid solutions and to purify plating solutions for longer service life.

End-of-Pipe Systems

As discussed, most plating shops can achieve sufficient metals removal to comply with discharge regulations by employing the conventional hydroxide precipitation process. Conventional treatment may not be reliable, however, in achieving compliance with discharge limits in certain cases, including where

- Unusually strict limits are placed on the effluent metal concentration
- Metals are complexed with chemical constituents that interfere with their precipitation as metal hydroxides

In such cases, the use of ion exchange to polish the effluent can provide relatively inexpensive upgrading of system performance for compliance with the regulations.

The development of special chelating resins have ion exchange feasible for selective removal of trace heavy metals from a water solution containing a high concentration of similarity charged, nontoxic ions. These resins exhibit a strong selectivity, or preference, for heavy metal ions over sodium, calcium, or magnesium ions. Weak acid cation resins also display a significant preference for heavy metal ions, and in some applications they are superior to the chelating resins in performance characteristics. In a polishing application, both resins can remove the heavy metal ions from the wastewater while leaving most of the nontoxic ions in solution. The preference for heavy metal ions allows a large volume of water to be treated per unit of resin volume before the resin must be regenerated. The regenerant solution, which contains a high concentration of metal ions, is treated upstream in the conventional process (Fig. 7a).

Ion exchange has received limited commercial application for selective removal of heavy metal and metal cyanide pollutants from an untreated wastewater while allowing most of the nontoxic ions to pass through. In each of the three approaches discussed earlier for selective removal, wastewater pretreatment requirements consist of pH adjustment to assure operation within the pH operating range of the resin and filtration to remove suspended solids which could foul the resin bed (Fig. 7b). The pollutants removed from the wastewater are concentrated in the ion exchange regenerant solutions. The regenerant can be treated in a small batch treatment system using conventional processes or metals can be filtered out using reverse osmosis and then electrodeposited for disposal.

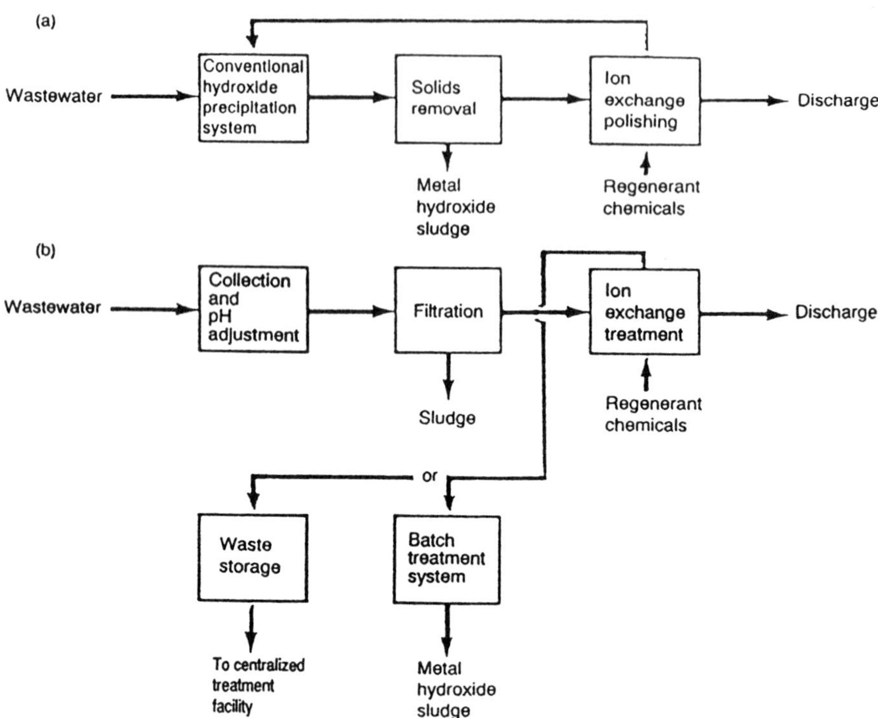

Figure 7 Ion exchange system: (a) polishing and (b) end-of-pipe treatment.

Ion exchange is a reversible chemical reaction wherein an ion (an atom or molecule that has lost or gained an electron and thus acquired an electrical charge) from solution is exchanged for a similarly charged ion attached to an immobile solid particle. As discussed previously, these solid ion exchange particles are either naturally occurring in organic zeolite or synthetically produced organic resins. Each resin has a distinct number of mobile ion sites that set the maximum quantity of exchanges per unit of resin.

Most plating process water is used to cleanse the surface of the parts after each process bath. To maintain quality standards, the level of dissolved solids in the rinse water caused by "drag-out" from the plating tank must be regulated. Fresh water added to the rinse tanks accomplish this purpose, and the overflow water is treated to remove pollutants and then discharged. As the metal salts, acids, and bases used in metal finishing are primarily inorganic compounds, they are ionized in water and could be removed by contact with ion exchange resins. In a water deionization process, the resins exchange hydrogen ions for the positively charged ions like nickel, copper, and sodium and hydroxyl ions for negatively charged sulfates, chromates, and chlorides. Because the quantity of H^+ and OH^- ions is balanced, the result of the ion exchange treatment is relatively pure, neutral water.

RESIN TYPES

Ion exchange resins are classified as cation exchangers that have positively charged mobile ions available for exchange and as anion exchangers whose exchangeable ions are negatively charged.

Wastewater-treatment systems employing ion exchange include the following components:

- Wastewater collection tank or sump
- Wastewater pretreatment (e.g., filtration, flocculation, pH adjustment)
- Ion exchange columns
- Ion exchange columns regeneration systems
- Batch treatment for regenerants (or waste storage if regenerants are shipped off site)

Both anion and cation resins are produced from the same basic organic polymers. They differ in the ionizable group attached to the hydrocarbon network. It is this functional group that determines the chemical behavior of the resin. Resins can be broadly classified as strong or weak acid cation or strong or weak base anion exchangers. Table 1 indicates the selectivity of ion exchange resins in order of decreasing preference. Table 2 lists some commercially demonstrated resin systems along with their application.

Table 1 Selectivity of Ion Exchange Resins in Order of Decreasing Preference

Strong acid cation exchanger	Strong base anion exchanger
Barium	Iodide
Lead	Nitrate
Calcium	Bisulfite
Nickel	Chloride
Cadmium	Cyanide
Copper	Bicarbonate
Zinc	Hydroxide
Magnesium	Fluoride
Potassium	Sulfate
Ammonia	
Sodium	
Hydrogen	

Table 2 Commercially Demonstrated Resin Systems for Wastewater Treatment

Resin system	Application
Weak acid	Selective removal of heavy metals from untreated wastes
Chelating cation	Selective removal of heavy metals from untreated wastes
	Removal of trace concentrations of heavy metals from solutions with high background cation concentrations
	Selective removal of heavy metals from solutions containing metal complexing compounds
Stratified bed (strong base, weak acid, strong acid)	Selective removal of metal cyanide, anionic metal complexes, and multivalent cations from solution

Wastewater Treatment

The conventional practice of converting the heavy metal pollutants in metal-finishing wastewater to a hydroxide sludge was thought to be a means of eliminating any environmental hazard the metals might pose. In fact, although the volume is much smaller than that of the wastewater, a solid waste stream is generated that requires further controls to ensure that the disposal of the metal residue is environmentally acceptable. Ion exchange represents an alternative approach of concentrating the pollutants. It usually presents a less costly means of complying with pollution control regulations. The key to using ion exchange for waste treatment is to remove only the toxic ions while allowing most of the nontoxic ions to remain. Ion exchange has proven successful in selectively removing many of the pollutants encountered in metal-finishing wastes. Proper application of the process requires selecting the appropriate resin and regeneration sequence and usually some pretreatment of the wastewater before ion exchange.

Ion exchange is a cost-effective and versatile separation process for removing ionic contamination (heavy metal) from wastewater. It is nondestructive and allows for recycling of the metals. Application of the ion exchange process in a wastewater purification and recycle systems will significantly reduce water consumption and the volume of wastewater discharged, thus reducing water use and sewer fees and the size and cost of the pollution control system. The ion exchange process can concentrate heavy metals in a dilute wastewater into a concentrated metal solution which is more amenable to metal recovery. With proper resin solution, ion exchange can provide an effective and economical solution to pollution control requirements. Figures 8–12 are schematics of various ion exchange systems.

IMPORTANCE OF HIGH-QUALITY WATER

Water problems in cooling, heating, steam generation, and manufacturing are caused in large measure from the kinds and concentrations of dissolved solids, dissolved gases, and suspended matter in the makeup water supplied. Table 3 lists the major objectionable ionic constituents present in many water supplies that can be removed by demineralization.

Prevention of scale and other deposits in cooling and boiling waters is best accomplished by removal of dissolved solids. Whereas in municipal water purification such removal is limited to the partial reduction of hardness and the removal of iron and manganese, in industrial water treatment, it is often carried much further and may include the complete removal of hardness, the reduction or removal of alkalinity, the removal of silica, or even the complete removal of all dissolved solids.

The two most frequently encountered water problems—scale formation and corrosion—are common to cooling, heating, and steam-generating systems.

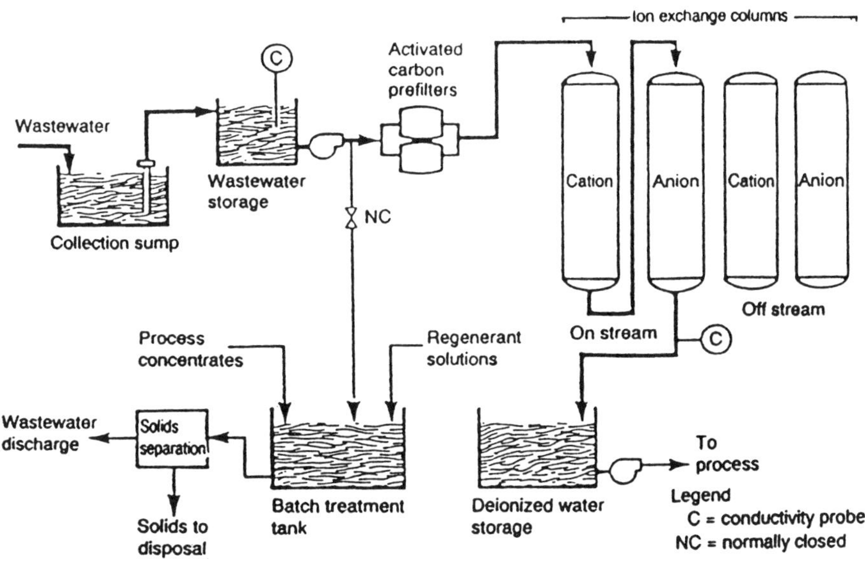

Figure 8 Ion exchange wastewater purification and recycle system.

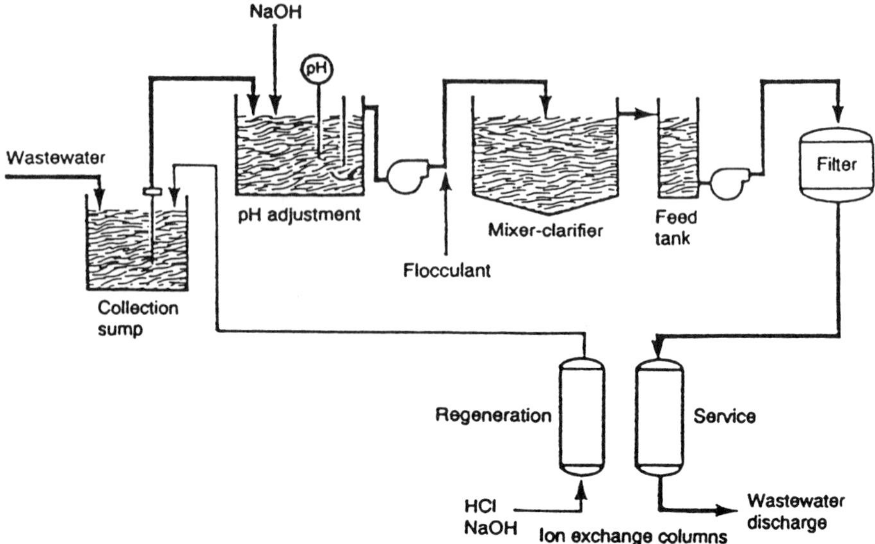

Figure 9 Conventional treatment system with ion exchange polishing.

Hardness (calcium and magnesium), alkalinity, sulfate, and silica all form the main source of scaling in heat-exchange equipment, boilers, and pipes. Scales or deposits formed in boilers and other exchange equipment act as insulation, preventing efficient heat transfer and causing boiler tube failures through overheating of the metal. Free mineral acids (sulfates and chlorides) cause rapid corrosion of boilers, heaters, and other metal containers and piping. Alkalinity causes embrittlement of boiler steel and carbon dioxide and oxygen cause corrosion, primarily in steam and condensate lines.

Low-quality steam can produce undesirable deposits of salts and alkali on the blades of steam turbines; much more difficult to remove are silica deposits which can form on turbine blades even when steam is satisfactory by ordinary standards. At steam pressures above 600 psi, silica from the boiler water actually dissolves in the gaseous steam and then reprecipitates on the turbine blades at their lower-pressure end.

In the operation of every cooling, heating, and steam-generating system, the water changes temperature. Higher temperatures, of course, increase both corrosion rates and scale-forming tendency. Evaporation in process steam boilers and in evaporative cooling equipment increases the dissolved-solids concentration of the water, compounding the problem.

In addition to the formation of scale or corrosion of metal within boilers, auxiliary equipment is also susceptible to similar damage. Attempts to prevent

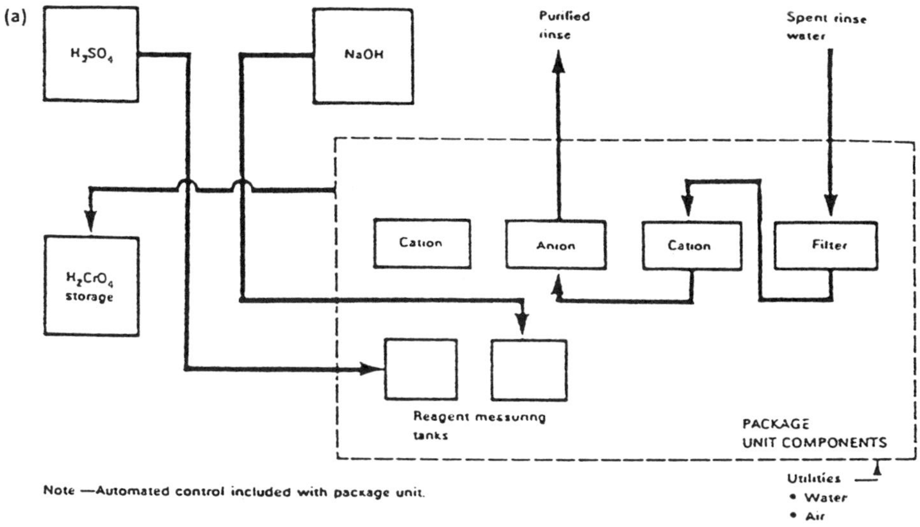

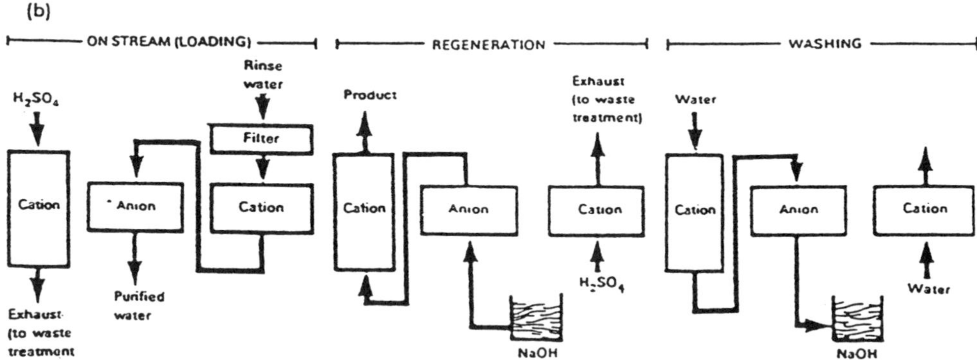

Figure 10 Chromic acid recovery system: (a) hardware components and (b) operating cycle.

scale formation within a boiler can lead to makeup line deposits if the treatment chemicals are improperly chosen. Thus, the addition of normal phosphates to an unsoftened feed water can cause a dangerous condition by clogging the makeup line with precipitated calcium phosphate. Deposits in the form of calcium or magnesium stearate deposits, otherwise known as "bathtub ring," can be readily seen and are caused by the combination of calcium or magnesium with negative ions of soap stearates. Table 3 shows some common ionic constituents in water.

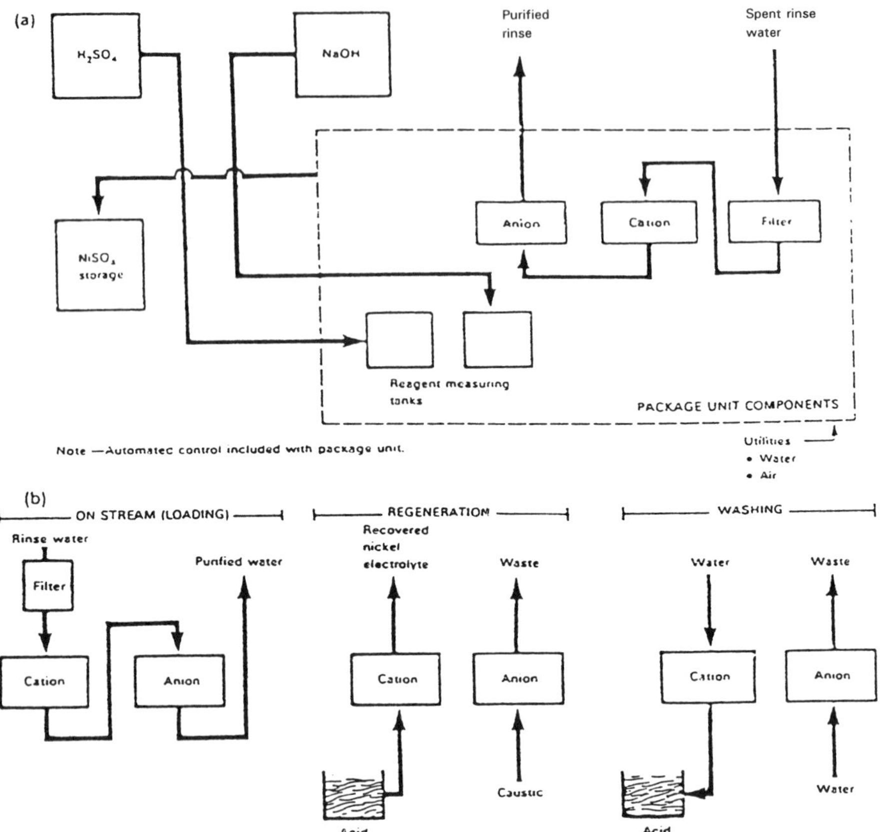

Figure 11 Metal salt recovery system: (a) hardware components and (b) operating cycle.

Theoretical Background

Ion exchangers are materials that can exchange one ion for another, hold it temporarily, and release it to a regenerant solution. In a typical demineralizer this is accomplished in the following manner. The influent water is usually passed through a hydrogen cation exchange resin which converts the influent salt (say, sodium sulfate) to the corresponding acid (say, sulfuric acid) by exchanging an equivalent number of hydrogen (H^+) ions for the metallic cations (Ca^{2+}, Mg^{2+}, Na^+). These acids are then removed by passing the effluent through an alkali regenerated anion exchange resin which replaces the anions in solution (Cl^-, SO_4^{2-}, NO_3^-) with an equivalent number of hydroxide ions. The hydrogen ions and hydroxide ions neutralize each other to form an

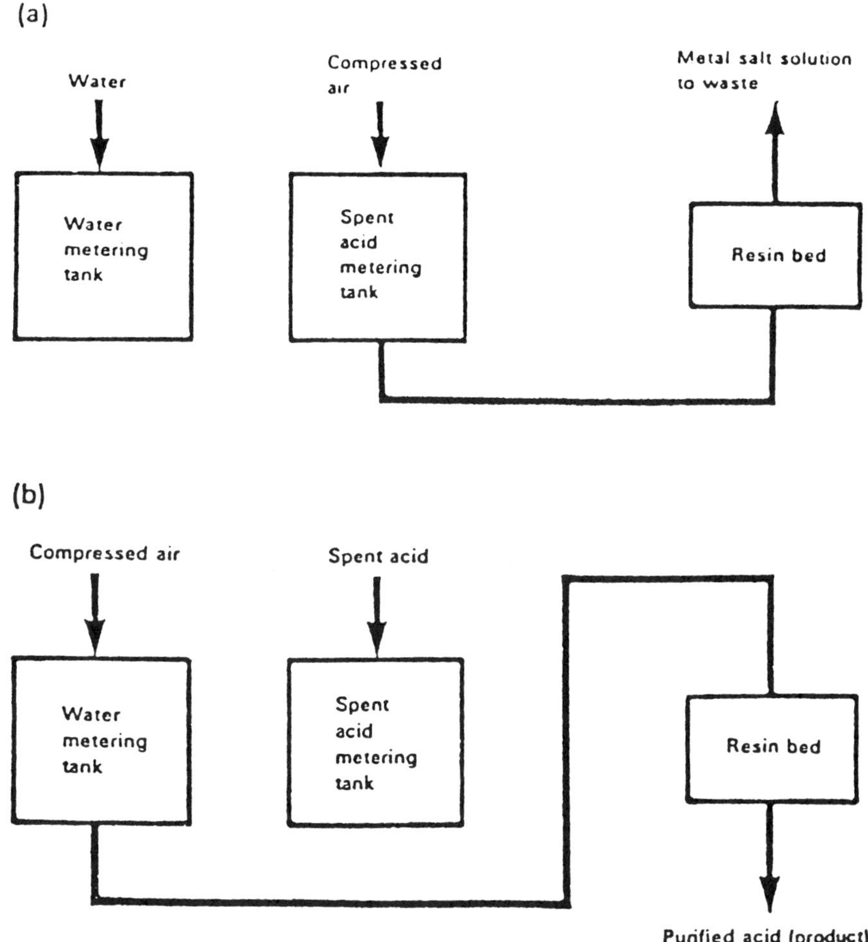

Figure 12 Acid recovery system operation: (a) upflow and (b) downflow.

equivalent amount of pure water. During regeneration, the reverse reaction takes place. The cation resin is regenerated with either sulfuric or hydrochloric acid and the anion resin is regenerated with sodium hydroxide. This is schematically shown in Figure 13.

There is a variety of possible arrangements and equipment but in all cases, except in mixed-bed demineralization, the water should first pass through a cation exchanger. In mixed demineralization, the two exchange materials (the cation exchange resin and anion exchange resin) are placed in one column instead of two separate columns. In operation, the two types of exchange materials are

Table 3 Ionic Constituents in Water

Constituent	Chemical designation	Problems caused
Hardness	Calcium and magnesium salts expressed as $CaCO_2$, Ca, Mg.	Main source of scaling in heat-exchange equipment, boilers, pipe lines, and so on. Forms curds with soap, interferes with dyeing.
Alkalinity	Bicarbonate (HCO_3), carbonate (CO_3), and hydrate (OH), expressed as $CaCO_3$.	Cause of foaming and carryover of solids with steam. Embrittlement of boiler steel. Bicarbonate and carbonate produce CO_2 in steam, a source of corrosion.
Free mineral acidity	H_2SO_4, HCl, and so on, expressed as $CaCO_3$.	Rapid corrosion and deterioration.
Chloride	Cl^-	Interferes with silvering processes and increases TDs.
Sulfate	$(SO_4)^-$	Calcium sulfate scale is formed.
Iron and manganese	Fe^{2+} (ferrous) Fe^{3+} (ferric) Mn^{2+}	Discolors water, deposits in water lines, boilers, and so on. Interferes with dying, tanning, paper manufacture, and process work.
Carbon dioxide	CO_2	Corrodes water lines, particularly steam and condensate lines.
Silica	SiO_2	Scale in boilers and cooling-water systems. Insoluble scale on turbine blades due to silica vaporization in high-pressure boilers (over 600 psi).

thoroughly mixed so that we have, in effect, a number of multiple demineralizers in series. Higher quality water is obtained from a mixed bed unit than from a two-bed system as shown in Figure 14. Operation of cation and anion exchangers is shown in Figures 15 (for fundamental processes) and 16.

Functional Groups

The molecular structure of the resin is such that it must contain a macroreticular tissue with acid or basic radicals. These radicals are the basis of classifying ion exchangers into two general groups:

- Cation exchangers, in which the molecule contains acid radicals of the HSO_3 or HCO type able to fix mineral or organic cations and exchange with the hydrogen ion H^+
- Anion exchangers, containing basic radicals (e.g., amine functions of the type

Theory

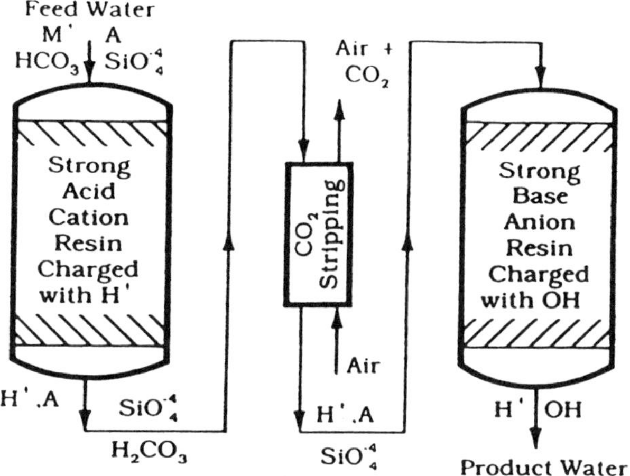

Figure 13 Ion exchange mineralization.

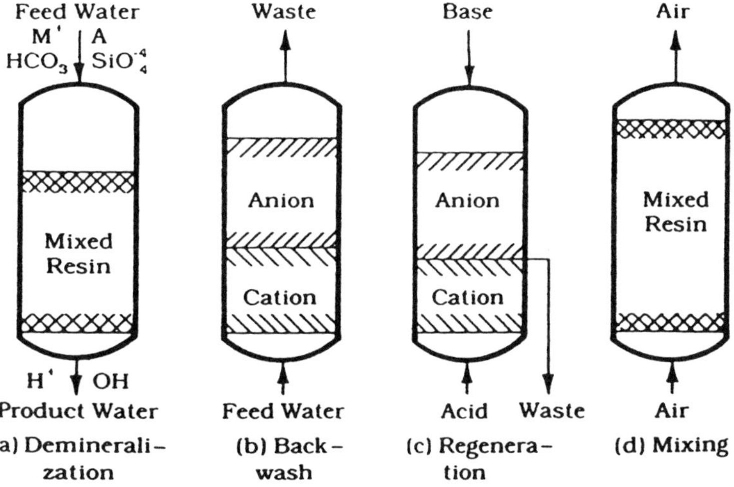

Figure 14 Mixed resin demineralization.

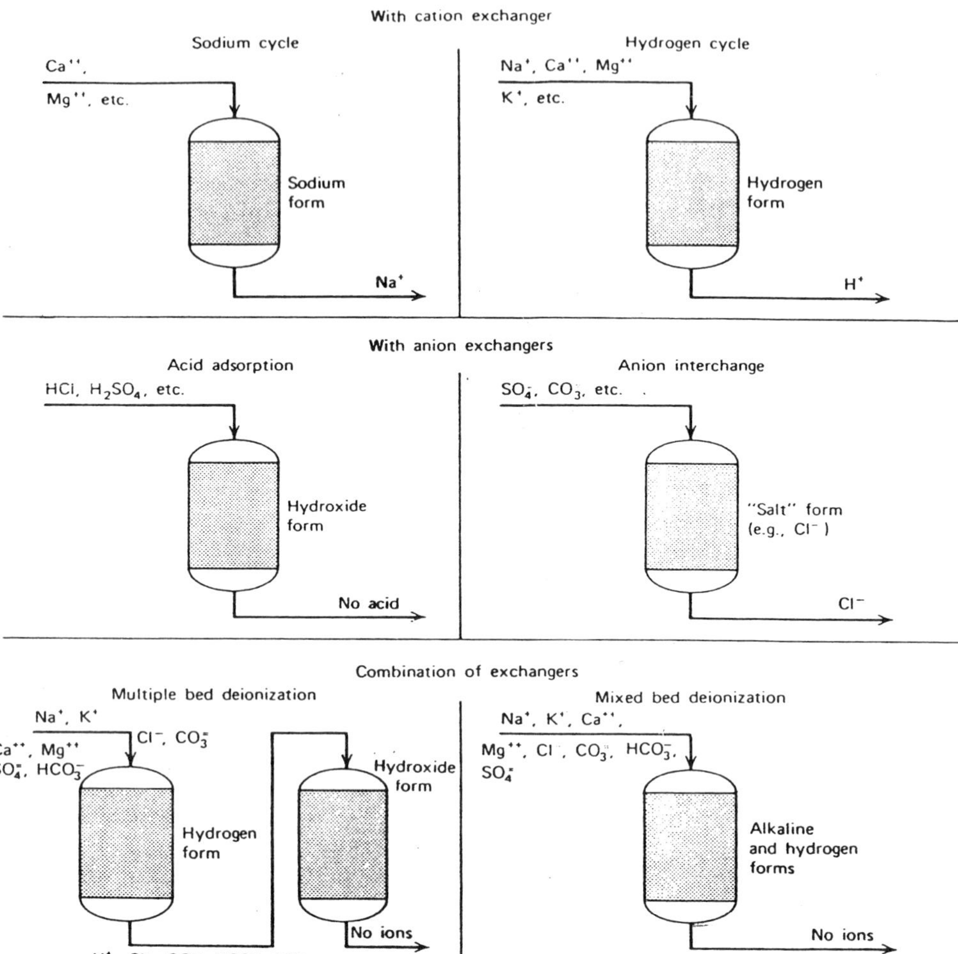

Figure 15 Fundamentals of ion exchange process.

NH_2) able to fix mineral or organic anions and exchange them with the hydroxyl ion OH^- coordinate to their dative bonds

The presence of these radicals enable a cation exchanger to be assimilated to an acid of form H–R and an anion exchanger to be a base of form OH–R when regenerated.

These radicals act as immobile ion exchange sites to which are attached the mobile cations or anions. For example, a typical sulfonic acid cation

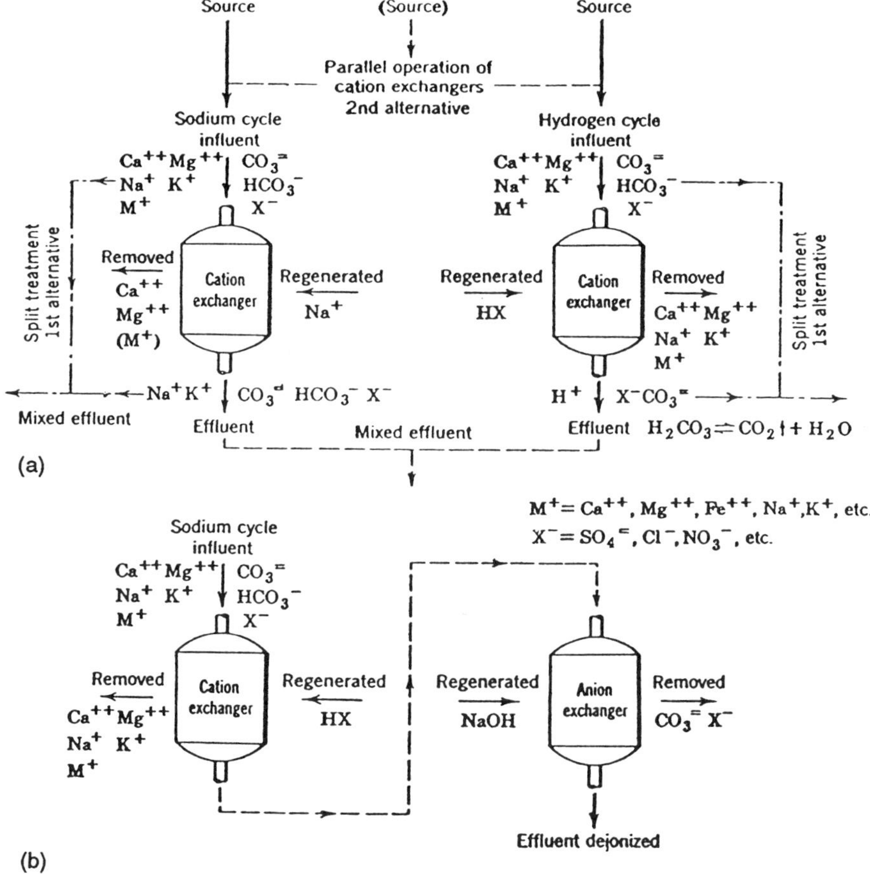

Figure 16 Operation of ion exchangers: (a) cation exchangers and (b) cation and anion exchangers.

exchanger has immobile ion exchange sites consisting of the anionic radicals SO to which are attached the mobile cations, such as H^+ or Na^+. An anion exchanger similarly has immobile cationic sites to which are attached mobile (exchangeable) hydroxide anions OH. The radicals attached to the molecular nucleus further determine the nature of the acid or base, whether it will be weak or strong.

Exchangers are divided into four specific classifications depending on the kind of radical, or functional group, attached: strong acid, strong base, weak acid, or weak base. Each of these four types of ion exchangers is described in detail later.

The ion exchange substance must be insoluble under normal conditions of use. All ion-exchange resins in current use are high molecular weight polyacids or polybases which are virtually insoluble in most aqueous and nonaqueous media. This is no longer true of some resins once a certain temperature has been reached. For example, some anion exchange resins are limited to a maximum temperature of 105°F. Tables 4–7 indicate the exchange of cations and anions with resin groups.

There also exists a type of resin with no functional groups attached. This resin offers no capacity to the system but increases regeneration efficiency in mixed-bed exchangers. These inert resins are of a density between cation and anion resins and when present in mixed-bed vessels help to separate cation and anion resins during backwash.

A steady falloff of resin-exchange capacity is a matter of concern to the operator and is due to several conditions.

Improper Backwash

Blowoff of resin from the vessel during the backwash step can occur if too high a backwash flow rate is used. This flow rate is temperature dependent and must be regulated accordingly. Also, adequate time must be allotted for backwashing to ensure a clean bed prior to chemical injection.

Channeling

Cleavage and furrowing of the resin bed can be caused by faulty operational procedures or a clogged bed or underdrain. This can mean that the solution being treated follows the path of least resistance, runs through these furrows, and fails to contact active groups in other parts of the bed.

Incorrect Chemical Application

Resin capacities can suffer when the regenerant is applied in a concentration that is too high or low. Another important parameter to be considered during chemical application is the location of the regenerant distributor. Excessive dilution of the regenerant chemical can occur in the vessel if the distributor is located too high above the resin bed. A recommended height is 3 in. above the bed level.

Mechanical Strain

When broken beads and fines migrate to the top of the resin bed during service, mechanical strain is caused which results in channeling, increased pressure drop, or premature breakthrough. The combination of these resulting conditions leads to a drop in capacity.

Resin Fouling

In addition to the physical causes of capacity losses listed previously, there are a number of chemically caused problems that merit attention, specifically the several forms of resin fouling that may be found.

Table 4 Sodium Cation-Exchanger Reactions

Softening

$$Na_2Z + \left.\begin{matrix} Ca \\ Mg \end{matrix}\right\} \left\{\begin{matrix} (HCO_3)_2 \\ SO_4 \\ Cl_2 \end{matrix}\right. = \left.\begin{matrix} Ca \\ Mg \end{matrix}\right\} Z + \left\{\begin{matrix} 2NaHCO_3 \\ Na_2SO_4 \\ 2NaCl \end{matrix}\right.$$

Sodium cation exchanger (insoluble) + Calcium and/or magnesium {Bicarbonates, sulfates, and/or chlorides (soluble) = Calcium and/or magnesiium} Cation exchanger (insoluble) + {Sodium bicarbonate, sodium sulfate, and/or sodium chloride (soluble)

Regeneration

$$\left.\begin{matrix} Ca \\ Mg \end{matrix}\right\} Z + 2\ NaCl = Na_2Z + \left.\begin{matrix} Ca \\ Mg \end{matrix}\right\} Cl_2$$

Calcium and/or magnesium} Cation exchanger (insoluble) + Sodium chloride (soluble) = Sodium cation exchanger (insoluble) + Calcium and/or magnesium} Chlorides (soluble)

The symbol Z represents sodium cation exchanger.

Compensated hardness: The hardness of a water for softening by the zeolite process should be compensated when:

1. The total hardness (TH) is over 400 ppm as $CaCO_3$, or
2. The sodium salts (Na) are over 100 ppm as $CaCO_3$

Calculated compensated hardness as follows:

$$\text{Compensated hardness (ppm)} = \text{total hardness (ppm)} \times \frac{9{,}000}{9{,}000 - \text{total cations (ppm)}}$$

Express compensated hardness as:

1. Next higher tenth of a grain up to 5 grains per gallon.
2. Next higher half of a grain from 5–10 grains per gallon.
3. Next higher grain above 10 grains per gallon.

Salt consumption: The salt consumption with the sodium cation-exchange water softener ranges between 0.275 and 0.533 lb of salt per 1,000 grains of hardness, expressed as calcium carbonate, removed. This range is due to two factors: (1) the composition of the water and (2) the operating exchange value at which the exchange resin is to be worked. The lower salt consumptions may be attained with waters that are not excessively hard nor high in sodium salts and where the exchange resin is not worked at its maximum capacity.

Table 5 Hydrogen Cation-Exchanger Resins

Reactions with Bicarbonates								
$\left.\begin{array}{r}Ca\\Mg\\Na_2\end{array}\right\}(HCO_3)_2$	+	H_2Z	=	$\left.\begin{array}{r}Ca\\Mg\\Na_2\end{array}\right\}Z$	+	$2H_2O$	+	$2CO_2$
Calcium, magnesium and/or sodium } Bicarbonate	+	Hydrogen cation exchanger	=	Calcium, magnesium, and/or sodium } Cation exchanger	+	Water	+	Carbon dioxide
(soluble)		(insoluble)		(insoluble)				(soluble gas)
Reactions with Sulfates or Chlorides								
$\left.\begin{array}{r}Ca\\Mg\\Na_2\end{array}\right\}\left\{\begin{array}{l}SO4\\Cl2\end{array}\right.$	+	H_2Z	=	$\left.\begin{array}{r}Ca\\Mg\\Na_2\end{array}\right\}Z$	+	H_2SO_4 / $2HCl$		
Calcium magnesium and/or sodium } { Sulfates and/or chlorides	+	Hydrogen Cation Exchanger	=	Calcium, magnesium, and/or sodium } Cation exchanger	+	Sulfuric and/or hydrochloric acids		
(soluble)		(insoluble)		(insoluble)		(soluble)		
Regeneration Reactions								
$\left.\begin{array}{r}Ca\\Mg\\Na2\end{array}\right\}Z$	+	H_2SO_4	=	H_2Z	+	$\left.\begin{array}{r}Ca\\Mg\\Na_2\end{array}\right\}SO_4$		
Calcium, magnesium and/or sodium } Cation exchanger	+	Sulfuric acid	=	Hydrogen cation exchanger	+	Calcium, magnesium, and/or sodium } Sulfates		
(insoluble)		(soluble)		(insoluble)		(soluble)		

The symbol Z represents hydrogen cation-exchanger radical.

Table 6 Anion-Exchanger Reactions (Weakly Basic and Strongly Basic Exchanges)

Reaction with Strongly Ionized Acids										
R_3N	+	H_2SO_4 / 2HCL	=	$R_3N\cdot$	H_2SO_4 / 2HCL					
weakly basic anion exchanger	+	Sulfuric and/or hydrochloric acids	=	Anion exchanger	Sulfate and/or hydrochloride					
(insoluble)		(soluble)		(insoluble)						
Regeneration Reaction										
$R_3N\cdot H_2SO_4$	+	Na_2CO_3	=	R_3N	+	Na_2SO_4	+	CO_2	+	H_2O
Anion exchanger hydrosulfate	+	Soda ash	=	Weakly basic anion exchanger	+	Sodium sulfate	+	Carbon dioxide	+	Water
(insoluble)		(soluble)		(insoluble)		(soluble)		(soluble gas)		
Reaction with Weakly Ionized Acids										
R_4NOH	+	H_2SiO_3	=	$R_4N\cdot HSiO_3$	+	H_2O				
Strongly basic anion exchanger	+	Silicic acid	=	Anion exchanger silicate	+	Water				
(insoluble)		(soluble)		(insoluble)						
Regeneration Reaction										
$(R_4N)_2SO_4$	+	2NaOH	=	$2R_4NOH$	+	Na_2SO_4				
Anion exchanger sulfate	+	Caustic soda	=	Strongly basic anion exchanger	+	Sodium sulfate				
(insoluble)		(soluble)		(insoluble)		(soluble)				

The symbol R_3N represents the complex weakly basic anion-exchanger radical.
The symbol R_4N represents the complex strongly basic anion-exchanger radical.

Organic fouling occurs on anion resins when organics precipitate onto basic exchange sites. Regeneration efficiency is then lowered, thereby reducing the exchange capacity of the resin. Causes of organic fouling are fulvic, humic, or tannic acids or degradation products of divinylbenzene (DVB) cross-linkage material of cation resins. The DVB is degraded through oxidation and causes irreversible fouling of downstream anion resins.

Iron fouling is caused by both forms of iron ions; the insoluble form will coat the resin bead surface and the soluble form can exchange and attach to exchange sites on the resin bead. These exchanged ions can be oxidized by subsequent cycles and precipitate ferric oxide within the bead interior.

Silica fouling is the accumulation of insoluble silica on anion resins. It is caused by improper regeneration which allows the silicate (ionic form) to hydrolyze to soluble silicic acid which in turn polymerizes to form colloidal silicic acid with the beads. Silica fouling occurs in weak-base anion resins when

Table 7 Removal and Recovery—Ion Exchange Reactions

Rinse Water Recovery										
(a)	$H_2Cr_2O_7$	+	$Al_2(Cr_2O_7)_3$	+	6RH	→	$2R_3Al$	+	$4H_2Cr_2O_7$	
	Chromic acid	+	Aluminum dichromate	+	Hydrogen cation exchanger		Aluminum cation exchanger	+	Chromic acid	
(b)	$H_2Cr_2O_7$	+	$2R_4NOH$	→	$(R_4N)_2Cr_2O_7$	+	$2H_2O$			
	Chromic acid	+	Hydroxide strongly basic anion exchanger		Dichromate strongly basic anion exchanger	+	Water			
Regeneration of Strongly Basic Anion Exchanger										
(a)	4NaOH	+	$(R_4N)2Cr2O7$	→	$2R_4NOH$	+	$2Na_2CrO_4$	+	H_2O	
	Sodium hydroxide	+	Chromate strongly basic anion exchanger		Hydroxide strongly basic anion exchanger	+	Sodium chromate	+	Water	
(b)	$2Na_2CrO_4$	+	4RH	→	4RNa	+	$H_2Cr_2O_7$	+	H_2O	
	Sodium chromate	+	Hydrogen cation exchanger		Sodium cation exchanger	+	Chromic acid	+	Water	
Plating or Anodizing Bath Recovery										
(a)	$H_2Cr_2O_7$	+	$Al_2(Cr_2O_7)_3$	+	6RH	→	$2R_3Al$	+	$4H_2Cr_2O_7$	
	Chromic acid	+	Aluminum dichromate	+	Hydrogen cation exchanger		Aluminum cation exchanger	+	Chromic acid	
(b)	$2R_3Al$	+	$3H_2SO_4$	→	6RH	+	$Al_2(SO_4)_3$			
	Aluminum cation exchanger	+	Sulfuric acid		Hydrogen cation exchanger	+	Aluminum sulfate			

The symbol R represents the cation-exchanger radical.
The symbol R_4N represents the complex strongly basic anion-exchanger radical.

they are regenerated with silica-laden waste caustic from the strong-base anion resin unless intermediate partial dumping is done.

Microbiological fouling (MB) becomes a potential problem when microbic growth is supported by organic compounds, ammonia, nitrates, and so on, which are concentrated on the resin. Signs of MB fouling are increased pressure drops, plugged distributor laterals, and highly contaminated treated water.

Calcium sulfate fouling occurs when sulfuric acid is used to regenerate a cation exchanger after exhaustion by a water high in calcium. The precipitate of calcium sulfate (gypsum) that forms can cause calcium and sulfate leakage during subsequent service runs. Given a sufficient calcium input in the water to treat, calcium sulfate fouling is especially prevalent when the percent solution of regenerant is greater than 5%, or the temperature is greater than 100°F, or when the flow rate is less than 1 gpm/ft^3. Stepwise injection of sulfuric acid during regeneration can help prevent fouling.

Aluminum fouling of resins can appear when aluminum floc from alum or other coagulants in pretreatment are encountered by the resin bead. This floc coats the resin bead and in the ionic form will be exchanged. However, these ions are not efficiently removed during regeneration so the available exchange sites continuously decrease in number.

Copper fouling is found primarily in condensate polishing applications. Capacity loss is due to copper oxides coating the resin beads.

Oil fouling does not cause chemical degradation but gives loss of capacity due to filming on the resin beads and the reduction of their active surface. Agglomeration of beads also occurs causing increased pressure drop, channeling, and premature breakthrough. The oil fouling problem can be alleviated by the use of surfactants.

Terminology Used in Ion Exchange

Acidity: An expression of the concentration of hydrogen ions present in a solution.

Adsorbent: A synthetic resin possessing the ability to attract and to hold charged particles.

Adsorption: The attachment of charged particles to the chemically active group on the surface and in the pores of an ion exchanger.

Alkalinity: An expression of the total basic anions (hydroxyl groups) present in a solution. It also represents, particularly in water analysis, the bicarbonate, carbonate, and occasionally the borate, silicate, and phosphate salts which will react with water to produce the hydroxyl groups.

Anion: A negatively charged particle or ion.

Anion interchange: The displacement of one negatively charged particle by another on an anion exchange material.

Attrition: The rubbing of one particle against another in a resin bed; frictional wear that will affect the site of resin particles.

Backwash: The countercurrent flow of water through a resin bed (i.e., in at the bottom of the exchange unit, out at the top) to clean and regenerate the bed after exhaustion.

Base exchange: The property of the trading of cations shown by certain insoluble naturally occurring materials (zeolites) and developed to a high degree of specificity and efficiency in synthetic resin adsorbents.

Batch operation: The utilization of ion exchange resins to treat a solution in a container wherein the removal of ions is accomplished by agitation of the solution and subsequent decanting of the treated liquid.

Bed: A mass of ion exchange resin particles contained in a column.

Bed depth: The height of the resinous material in the column after the exchanger has been properly conditioned for effective operation.

Bed expansion: The effect produced during backwashing when the resin particles become separated and rise in the column. The expansion of the bed due to the increase in the space between resin particles may be controlled by regulating backwash flow.

Bicarbonate alkalinity: The presence in a solution of hydroxyl (OH^-) ions resulting from the hydrolysis of carbonates or bicarbonates. When these salts react with water, a strong base and a weak acid are produced, and the solution is alkaline.

Breakthrough: The first appearance in the solution flowing from an ion exchange unit of unabsorbed ions similar to those which are depleting the activity of the resin bed. Breakthrough is an indication that regeneration of the resin is necessary.

Capacity: The adsorption activity possessed in varying degrees by ion exchange materials. This quality may be expressed as kilograins per cubic foot, gram-milliequivalents per gram, pound-equivalents per pound, gram-milliequivalents per milliliter, and so on, where the numerators of these ratios represent the weight of the ions adsorbed and the denominators represent the weight or volume of the adsorbent.

Carbonaceous exchangers: Ion exchange materials of limited capacity prepared by the sulfonation of coal, lignite, peat, and so on.

Carboxylic: A term describing a specific acidic group (COOH) that contributes cation exchange ability to some resins.

Cation: A positively charged particle or ion.

Channeling: Cleavage and furrowing of the bed due to faulty operational procedure in which the solution being treated follows the path of least resistance, runs through these furrows, and fails to contact active groups in other parts of the bed.

Chemical stability: Resistance to chemical change which ion-exchange resins must possess despite contact with aggressive solutions.

Color-throw: Discoloration of the liquid passing through an ion exchange material; the flushing from the resin interstices of traces of colored organic reaction intermediates.

Column operation: Conventional utilization of ion exchange resins in columns through which pass, either upflow or downflow, the solution to be treated.

Cycle: A complete course of ion exchange operation. For instance, a complete cycle of cation exchange would involve regeneration of the resin with acid, rinse to remove excess acid, exhaustion, backwash, and finally regeneration.

Deashing: The removal from solution of inorganic salts by means of adsorption by ion exchange resins of both the cations and the anions that comprise the salts. See *Deionization*.

Deionization: Deionization, a more general term than *deashing*, embraces the removal of all charged constituents or ionizable salts (both inorganic and organic) from solution.

Demineralizing: See *Deashing*.

Density: The weight of a given volume of exchange material backwashed and in place in the column.

Dissociation: Ionization.

Downflow: Conventional direction of solutions to be processed in ion exchange column operation, that is, in at the top, out at the bottom of the column.

Dynamic system: An ion exchange operation wherein a flow of the solution to be treated is involved.

Efficiency: The effectiveness of the operational performance of an ion exchanger. Efficiency in the adsorption of ions is expressed as the quantity of regenerant required to effect the removal of a specified unit weight of adsorbed material; for example, pounds of acid per kilogram of salt removed.

Effluent: The solution which emerges from an ion exchange column.

Electrolyte: A chemical compound which dissociates or ionizes in water to produce a solution which will conduct an electric current; an acid, base, or salt.

Elution: The stripping of adsorbed ions from an ion exchange material by the use of solutions containing other ions in concentrations higher than those of the ions to be stripped.

Equilibrium reactions: The interaction of ionizable compounds in which the products obtained tend to revert to the substance from which they were formed until a balance is reached in which both reactants and pacts are present in definite ratios.

Equivalent weight: The molecular weight of any element or radical expressed as grams, pounds, and so on divided by the valence.

Exchange velocity: The rate with which one ion is displaced from an exchanger in favor of another.

Exhaustion: The state in which the adsorbent is no longer capable of useful ion exchange; the depletion of the exchanger's supply of available ions. The exhaustion point is determined arbitrarily in terms of (1) a value in parts per million of ions in the effluent solution and

(2) the reduction in quality of the effluent water determined by a conductivity bridge which measures the resistance of the water to the flow of an electric current.

Fines: Extremely small particles of ion-exchange materials.

Flowrate: The volume of solution which passes through a given quantity of resin within a given time. Flowrate is usually expressed in terms of feet per minute per cubic foot of resin or as milliliters per minute per milliliter of resin.

Freeboard: The space provided above the resin bed in an ion exchange column to allow for expansion of the bed during backwashing.

Grain: A unit of weight: 0.0648 grams or 0.000143 pounds.

Grains per gallon: An expression of concentration of material in solution. One grain per gallon is equivalent to 17.1 parts per million.

Gram: A unit of weight: 15.432 grains or 0.0022 pounds.

Gram-milliquivalents: The equivalent weight in grams divided by 1,000.

Greensands: Naturally occurring materials composed primarily of complex silicates which possess ion exchange properties.

Hardness: The scale-forming and lather-inhibiting qualities which water high in calcium and magnesium ions possesses.

Hardness as calcium carbonate: The expression ascribed to the value obtained when the hardness-forming salts are calculated in terms of equivalent quantities of calcium carbonate; a convenient method of reducing all salts to a common basic for comparison.

Head loss: The reduction in liquid pressure associated with the passage of a solution through a bed of exchange material; a measure of the resistance of a resin bed to the flow of the liquid passing through it.

Hydraulic classification: The rearrangement of resin particles in an ion exchange unit. As the backwash water flows up through the resin bed, the particles are placed in a mobile condition wherein the larger particles settle and the smaller particles rise to the top of the bed.

Hydrogen cycle: A complete course of cation exchange operation in which the adsorbent is employed in the hydrogen or free acid form.

Hydroxyl: The term used to describe the anionic radical (OH^-) which is responsible for the alkalinity of a solution.

Influent: The solution which enters an ion exchange unit.

Ion: Any particle of less than colloidal size possessing either a positive or a negative electric charge.

Ionization: The dissociation of molecules into charged particles.

Ionization constant: An expression in absolute units of the extent of dissociation into ions of a chemical compound in solution.

Kilograin: A unit of weight: 1,000 grains.

Leakage: The phenomenon in which some of the influent ions are not adsorbed and appear in the effluent when a solution is passed through an underregenerated exchange resin bed.

Negative charge: The electrical potential which an atom acquires when it gains one or more electrons; a characteristic of an anion.

pH: An expression of the acidity of a solution; the negative logarithm of the hydrogen-ion concentration (pH 1, very acidic; pH 14, very basic; pH 7, neutral).

pK: An expression of the extent of dissociation of an electrolyte; the negative logarithm of the ionization constant of a compound.

pOH: An expression of the alkalinity of a solution; the negative logarithm of the hydroxyl-ion concentration.

Physical stability: The quality which an ion exchange resin must possess to resist changes that might be caused by attrition, high temperatures, and other physical conditions.

Positive charge: The electrical potential acquired by an atom which has lost one or more electrons; a characteristic of a cation.

Raw water: Untreated water from wells or from surface sources.

Regenerant: The solution used to restore the activity of an ion exchanger. Acids are employed to restore a cation exchanger to its hydrogen form; brine solutions may be used to convert the cation exchanger to the sodium form. The anion exchanger may be rejuvenated by treatment with an alkaline solution.

Regeneration: Restoration of the activity of an ion exchanger by replacing the ions adsorbed from the treated solution by ions that were adsorbed initially on the resin.

Rejuvenation: See *Regeneration*.

Reverse deionization: The use of an anion exchange unit and a cation exchange unit—in that order—to remove all ions from solution.

Rinse: The operation which follows regeneration; a flushing out of excess regenerant solution.

Siliceous gel zeolite: A synthetic, inorganic exchanger produced by the aqueous reaction of alkali with aluminum salts.

Static system: The batchwise employment of ion exchange resins wherein (since ion exchange is an equilibrium reaction) a definite endpoint is reached in which a finite quantity of all the ions involved is present. Opposed to a dynamic, column-type operation.

Sulfonic: A specific acidic group (SO_3H) on which depends the exchange activity of certain cation adsorbents.

Swelling: The expansion of an ion exchange which occurs when the reactive groups on the resin are converted from one form to another.

Throughput volume: The amount of solution passed through an exchange before exhaustion of the resin is reached.

Upflow: The operation of an ion exchange unit in which solutions are passed in at the bottom and out at the top of the container.

Voids: The space between the resinous particles in an ion exchange bed.

Zeolite: Naturally occurring hydrous silicates exhibiting limited base exchange.

11
Heavy Metals/Cyanide Treatment

HEAVY METALS

Treatment practices employed for heavy metals removal and cyanide destruction include the following methods:

- Precipitation
- Reduction
- Adsorption/selection
- Stabilization/fixation technologies
- Cyanide oxidation/removal

Precipitation

Precipitation is the treatment process most often employed to remove heavy metals from metal-finishing wastestreams. This method is used by a vast majority of the electroplating facilities which treat aqueous metal bearing wastes. The precipitation process is governed by the law of mass action which affects the equilibrium that exists between crystals of a metal salt compound in the solid state and its ions in solution by the following expression:

$$M(OH)n = M^{n+} + n(OH)^-$$

$$[M^{n+}]\,[OH^-]^n = Ksp$$

where n = 2, 3, 4, and so forth, and Ksp is the solubility product. The brackets represent the use of the major concentration of the ions. The primary mechanism driving the precipitation reaction involves the alteration of the ionic equilibrium of the metallic compounds to produce a salt and an insoluble metal complex.

Although most precipitation applications use alkali metal hydroxides or alkaline earth metal oxides as precipitating agents, other metal salts such as $Al_2(SO_4)_3$, $CaCl_2$, or $FeSO_4$ can be used to coagulate colloidal particles to form precipitates. A common hindrance to effective precipitation results when a diverse ion such as metal complexes and chelating agents or charge interaction decreases the effective concentration of the slightly soluble metal ions; hence shifting the equilibrium to put more into solution. pH has an effect on the precipitation of metal salts where control of pH is essential for the precipitation of solutions containing multimetals.

In general, chemical precipitation depends upon several variables:

- Maintenance of an alkaline pH through the precipitation reaction and subsequent settling
- Addition of a sufficient excess of treatment ions to drive the precipitation reaction to completion
- Addition of an adequate supply of sacrificial ions (such as iron or aluminum) to ensure precipitation and removal of specific target ions
- Effective removal of precipitated solids

Precipitation techniques practiced on aqueous metal wastes include hydroxide (e.g., lime), sulfide (soluble and insoluble), and carbonate precipitation processes. Hydroxide precipitation is effective in removing arsenic, cadmium, chromium III, copper, iron, manganese, nickel, lead, and zinc. Sulfide precipitation is highly effective in the removal of cadmium, cobalt, copper, iron, mercury, manganese, nickel, silver, tin, and zinc. Carbonate precipitation is effective in removing cadmium, nickel, and lead at a slightly lower pH than either hydroxide or sulfide techniques. Figure 1 shows a precipitation process schematic.

Useful precipitation reactions with the insoluble product underlined are summarized as follows:

- Fluoride, from glass manufacture and cupola washwater:

$$2F^- + Ca(OH)_2 \rightleftharpoons \underline{CaF_2} + 2\,(OH)^-$$

- Aluminate, from aluminum etching and anodizing:

$$2\,H_2O + Na_2Al_2O_4 + H_2SO_4 \rightleftharpoons Na_2SO_4 + \underline{2Al(OH)_3}$$

- Iron, from steel pickling:

$$FeSO_4 + Ca(OH)_2 \rightleftharpoons \underline{Fe(OH)_2} + CaSO_4$$

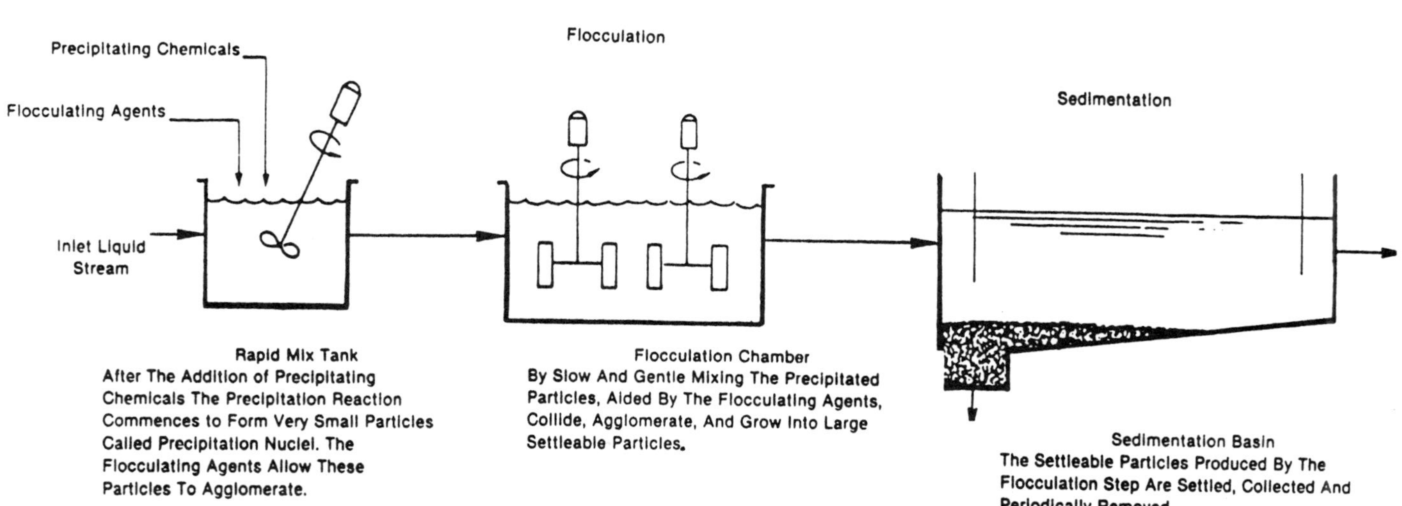

Figure 1 Configuration using precipitation flocculation and sedimentation.

$$FeCl_2 + Ca(OH)_2 \rightleftharpoons \underline{Fe(OH)_2} + CaCl_2$$

$$6FeCl_2 + 6\ Ca(OH)_2 + O_2 \rightleftharpoons \underline{2\ Fe_3O_4} + 6\ CaCl_2 + 6\ H_2O$$

- Sulfide, from refinery overhead systems and chemical operations:

$$S^{2-} + CuSO_4 \rightleftharpoons \underline{Cu} + SO_4^{2-}$$

- Phosphate, from municipal sewage and fertilizer plants:

$$2\ PO_4^{2-} + Al_2(SO_4)_3 \rightleftharpoons 2\ \underline{AlPO_4} + 3\ (SO_4)^{2-}$$
$$PO_4^{3-} + FeCl_3 \rightleftharpoons \underline{FePO_4} + CO_2$$

- Sulfur dioxide from stack gas:

$$SO_2 + CaCO_3 + {}^1\!/_2O_2 \rightleftharpoons \underline{CaSO_4} + CO_2$$

- Water softening by lime and soda ash precipitation:

$$Ca(HCO_3)_2 + Ca(OH)_2 \rightleftharpoons \underline{2CaCO_3} + 2H_2O$$

$$Mg(HCO_3)_2 + Ca(OH)_2 \rightleftharpoons \underline{Mg(OH)_2} + \underline{2CaCO_3} + 2H_2O$$

$$CaSO_4 + Na_2CO_2 \rightleftharpoons \underline{CaCO_3} + 2Na^+ + SO_4^-$$

$$MgSO_4 + Na_2CO_3 + Ca(OH)_2 \rightleftharpoons \underline{Mg(OH)_2} + \underline{CaCO_3} + 2\ Na^+ + SO_4^-$$

- Alum flocculation for clarifying water:

$$Al_2(SO_4)_3 \cdot 14H_2O + 3CA(HCO_3)_2 \rightleftharpoons \underline{Al(OH)_3} + 3CaSO_4 + 6CO_2 + 14H_2O$$

Typical Heavy Metals Found in Industrial Wastewaters

Heavy metals are common contaminants of industrial wastewaters. Because of their toxicity, they must be typically removed prior to wastewater discharge. The most common heavy metal contaminants are

- Arsenic
- Mercury
- Barium
- Nickel
- Cadmium
- Selenium
- Chromium
- Silver
- Copper
- Zinc
- Lead

Precipitation of Heavy Metals

- The heavy metal contents of wastewaters can be effectively removed to acceptable levels by precipitating the metal in an insoluble form.
- Heavy metals are typically precipitated from wastewater as
  - hydroxides
  - carbonates
  - sulfides or sometime sulfates
- Metal coprecipitation during flocculation with iron or aluminum salts is also possible for some metals (e.g., arsenic).

Precipitation of Heavy Metals as Hydroxides

- Precipitation by hydroxide formation is the most common heavy metal precipitation method.
- Lime or other caustic materials are typically used for this purpose.
- Some heavy metals are amphoteric. Therefore their solubility reaches a minimum at a specific pH (different for each metal).

Precipitation of Heavy Metals as Carbonates or Sulfides

- Heavy metal precipitation as carbonates are subject to the same constraints as hydroxides.
- Sulfide precipitation has the advantage of producing lower residual concentration in the wastewater because of the very small solubility of most sulfides.
- Sulfide sludges are harder to dewater than hydroxide sludges.
- Sulfide fumes are toxic.

Arsenic in Industrial Wastewaters

Arsenic is present in the wastewaters of a number of industries producing

- Metallurgical products
- Glassware and ceramic
- Tannery products
- Dye stuff
- Pesticides
- Synthetic chemicals
- Petroleum refinery products

Precipitation of Arsenic from Industrial Wastewaters

- Arsenic can be removed by precipitation as sulfide through the addition of sodium sulfide or hydrogen sulfide to the wastewater. The effluent concentration is 0.05 ppm.
- Arsenic can also be removed by coprecipitation with $FeCl_3$ when a $Fe(OH)_3$ floc is formed. The effluent concentration is 0.005 ppm.

Barium in Industrial Wastewaters

Barium is present in the wastewaters of a number of industries producing

- Metallurgical products
- Glassware and ceramic
- Dye stuff
- Explosives
- Rubber products

Precipitation of Barium from Industrial Wastewaters

- Barium can be removed by precipitation as sulfate adding any sulfate ion source.
- The solubility of barium sulfate is 1.4 ppm.
- Even lower residual barium concentrations (0.5 ppm) can be obtained using an excess of sulfate ions.

Cadmium in Industrial Wastewaters

Cadmium is present in the wastewaters of a number of industries producing

- Metallurgical products
- Ceramics
- Electroplated products
- Photographic products
- Pigments
- Textiles
- Synthetic chemicals

Precipitation of Cadmium from Industrial Wastewaters

- Cadmium can also be removed by precipitation as carbonate. The pH required in this case is between 7.5 and 8.5. The effluent concentration is comparable to that obtained through hydroxide precipitation at high pH.
- Cyanides interfere with any of these processes and must be removed prior to cadmium precipitation.

Precipitation of Cadmium from Industrial Wastewaters

- Cadmium can be removed by precipitation as hydroxide at pH ranging from 8 (solubility: 1 ppm) to 11 (solubility: 0.05 ppm).
- Cadmium can be removed by precipitation as sulfide. The effluent concentration is 0.05 ppm.
- Cadmium can also be removed by coprecipitation at pH 6.5 with $FeCl_3$ when a $Fe(OH)_3$ floc is formed. The effluent concentration is 0.008 ppm.

Chromium in Industrial Wastewaters

Chromium is present in the wastewaters of a number of industries producing

- Steel manufacturing
- Chrome-plated products
- Tannery products
- Dye stuff
- Paints

Precipitation of Chromium from Industrial Wastewaters

- Hexavalent chromium (CR^{6+}) is reduced to trivalent chromium (Cr^{3+}). Compounds such as $FeSO_4$, $Na_2S_2O_5$ (sodium bisulfite), or sulfur dioxide are used as reducing agents. The reaction is conducted at low pH (<3).
- Trivalent chromium is precipitated as $Cr(OH)_3$. Lime is typically used for the precipitation reaction reaction. The effluent concentration is 0.2 ppm at pH 7.5.

Precipitation of Chromium from Industrial Wastewaters

The reaction involved in chromium precipitation are

Reduction reaction (at pH<3):

$$H_2Cr_2O_7 + 3\,SO_2 \rightarrow Cr_2(SO_4)_3 + H_2O$$

(i.e., 1.85 ppm SO_2/ppm Cr) or, alternatively,

$$H_2Cr_2O_7 + 6FeSO_4 + H_2SO_4 \rightarrow Cr_2(SO_4)_3 + 3FE_2(SO_4)_3 + 7H_2O$$

Precipitation reaction (at pH of 8–9):

$$Cr_2(SO_4)_3 + 3Ca(OH)_2 \rightarrow 2Cr(OH)_3 \downarrow + 3CaSO_4$$

(i.e., 2.13 ppm $Ca(OH)_2$/ppm Cr).

Copper in Industrial Wastewaters

Copper is present in the wastewaters of a number of industries producing

- Chemicals using copper salts
- Chemicals using copper catalyst
- Metal-processing products
- Metal-plated products

Precipitation of Copper from Industrial Wastewaters

- Copper can be removed by precipitation as hydroxide at pH ranging from 9 to 10.3 (solubility: 0.01 ppm as cupric oxide).

- Copper can be removed by precipitation as sulfide at pH 8.5. The resulting effluent concentration is 0.01–0.02 ppm.
- The presence of cyanide or ammonia may interfere with copper precipitation. In such a case, activated carbon can be used to remove copper cyanide.

Lead in Industrial Wastewaters

Lead is present in the wastewaters of a number of industries producing

- Batteries
- Pigments
- Printing Products

Precipitation of Lead from Industrial Wastewaters

- Lead can be removed by precipitation as hydroxide (lime) at pH 11.5. The effluent concentration is 0.02 to 0.2 ppm.
- Lead can be removed by precipitation as sulfide at pH 7.5 to 8.5.
- Lead can also be removed by precipitation as carbonate. The pH required in this case is between 7.5 and 8.5. The effluent concentration is comparable to that obtained through hydroxide precipitation at high pH.

Mercury in Industrial Wastewaters

Mercury is present in the wastewaters of a number of industries producing

- Chlor-alkali
- Explosive
- Electronic products
- Pesticides
- Petrochemical products

Precipitation of Mercury from Industrial Wastewaters

- Mercury can be removed by precipitation as sulfide through the addition of sodium sulfide or hydrogen sulfide to the wastewater. The effluent concentration is 0.01 ppm.
- Mercury can be removed by coprecipitation with alum. The effluent concentration is 0.001–0.01 ppm.
- Mercury can be removed by coprecipitation with $FeCl_3$ when a $Fe(OH)_3$ floc is formed. The effluent concentration is 0.00005–0.005 ppm.

Nickel in Industrial Wastewaters

Nickel is present in the wastewaters of a number of industries producing

- Metal products (e.g., aircrafts)
- Steel
- Chemicals

Precipitation of Nickel from Industrial Wastewaters

- Nickel can be removed by precipitation as hydroxide at pH ranging from 10 to 11 (solubility: 0.12 ppm).
- Nickel can also be removed by precipitation as sulfate or carbonate.
- The presence of cyanide may interfere with nickel precipitation.

Hydroxide Precipitation

Hydroxide precipitation is used in treating aqueous metal wastestreams. This technology offers an effective alternative for removing heavy metals from spent process wastewaters. The most common precipitating agents used in this process include quick lime (CaO), hydrated lime ($Ca[OH]_2$), and caustic soda (NaOH). Variations of hydroxide systems which have shown potential for use as precipitants are copper hydroxide ($Cu[OH]_2$), magnesium hydroxide ($Mg[OH]_2$), and iron oxyhydroxide (or ferrihydrite).

Hydroxide precipitation with lime is the most economical and common reagent used for the precipitation of aqueous metal wastewaters. The process converts soluble metal ions into less soluble hydroxide compounds where metals can be separated from the supernatant via sedimentation and/or filtration. Lime in the form of dry CaO or slurry $Ca(OH)_2$ is added to the waste as a source of hydroxide ion, having the effect of raising the pH to a suitable level for optimum precipitation of the metal hydroxides. Since the optimum solubility curves are different for each metal ion, treatment of mixed-metal aqueous wastes may require some degree of adjustment. Typically, hydroxide precipitation is optimum at a pH between 9.5 and 12.

Some of the hazardous wastes that have been treated by lime precipitation include metal finishing wastewater and sludges following cyanide destruction, spent pickle liquor from steel finishing operations, and leachates from secondary lead smelting.

Disadvantages encountered when employing lime precipitation include voluminous sludge production which tends to increase costs of sludge disposal; limitation of metals removal due to solubility constraints attributed to matrix effects; and interferences associated with complexing agents (e.g., cyanide, ethylenediaminetetra-acetic acid [EDTA], citrate) when stabilizing metal hydroxide sludges.

Sodium hydroxide (NaOH) is used to precipitate heavy metals from aqueous wastestreams, as well as to neutralize strong acids by forming sodium salts. Some advantages associated with the use of caustic soda precipitation include homogeneity of feed solution; production of soluble by-products under low pH conditions; induces a rapid reaction rate providing available hydroxyl ions; and provides excellent neutralization efficiency. Like most precipitants, sodium hydroxide has some limitations. For example, it will not effectively precipitate sulfate wastestreams due to the solubility of sodium sulfate; floc is generated which requires larger settlers and dewatering apparatuses; it does not impart buffering capacity to industrial wastestreams; and as a monohydroxide, two parts hydroxide are required to precipitate divalent metals, such as nickel, making caustic soda more expensive on an equivalent basis.

Although process configurations for caustic soda vary as a function of waste type, volume, and raw waste pH, a typical system employs two reactors. The first stage is used for reagent addition and mixing, whereas the second stage is used for pH adjustment compensating for undershooting/overshooting circumstances. This tendency can be attributed to the inability of caustic soda in providing solution buffering capacity. For adequate process control, the interval between the addition of sodium hydroxide and the initial pH change should be less than 5% of the reactor residence time.

Magnesium oxide, MgO, or magnesium hydroxide, $Mg(OH)_2$, precipitation systems have shown to be effective in treating metal-bearing wastestreams which contain lower concentrations of dissolved metals ($\leq$50 mg/L). Such precipitants offer some distinct advantages over lime and caustic soda for precipitating heavy metals: MgO tends to be insoluble and lower sludge volumes are produced as the hydroxide sludge becomes more compact. This type of hydroxide sludge will compact together on standing, hindering resuspension of the metal ions. This phenomenon can be attributed to the fact that the positive surface charge of MgO complements the negative surface charge associated with most heavy metal hydroxides. As a result, water is expelled from the spaces between the particles which yields a more condensed solid. The resulting sludge is then easier to dewater.

Sulfide Precipitation

Similar to hydroxide precipitation, sulfide precipitation is a process that converts soluble metal ions into soluble sulfide compounds (MS) through contact between a metal ion (M^{2+}) and a sulfide ion (S^{2-}):

$$M^{2+} + S^{2-} = MS$$

Some of the advantages associated with this process include the ability significantly to remove metal constituents at a single pH; reduce chromates and dichromates without prior reduction of the hexavalent chromium to the trivalent state; and its insensitivity to the presence of most chelating agents (e.g., NTA, EDTA, EBTA, CDTA and citrate). Sulfide precipitation may be an effective alternative to hydroxide precipitation due to the low solubility of metal sulfides yielding high metal-removal efficiencies.

Limitations are associated with sulfide precipitation which reduce its application. Sulfide reagents can produce hydrogen sulfide (H_2S) when in contact with acidic wastewaters. Also, the sulfide ion itself is toxic and must be rigidly controlled to minimize an excess of sulfide in the treated waste to avoid post treatment.

Carbonate Precipitation

Carbonates such as calcium carbonate ($CaCO_3$), sodium carbonate (Na_2CO_3), or carbon dioxide (CO_2) have been used to precipitate heavy metals from aqueous metal-bearing wastestreams. Solubilities of most metal carbonates fall between hydroxide and sulfide solubility limits. Owing to the crystalline nature of the precipitate formed, this process offers improved settleability, reduced sludge volumes, and easier to filter precipitates. Other advantages associated include buffering capability, superior handling characteristics, and its availability. Some of the disadvantages encountered when employing carbonate precipitation systems include longer retention times ($\geq$45 min), lower solubilities (20% by weight), and difficulty of mixing.

Calcium carbonate has widespread availability and high reaction rate. Owing to the formation of soluble magnesium sulfate, less sludge is produced during precipitation. However, limestone is limited in its ability to treat wastewaters outside the operational pH range of 5–7, especially, for CrIII and FeII or FeIII. Efficiency falls off when treating wastestreams with acid concentrations greater than 5,000 mg/L.

Sodium carbonate (soda ash) has been used for treating acidic aqueous metal wastestreams which lack buffering capacity, such as acid bath rinse waters. This reagent has the advantage of providing buffering capacity to the wastewater while generating less sludge because of the characteristic of sodium-based endproducts being more soluble than calcium-based products. However, sludges generated from this process do not filter well as the less soluble calcium-based sludges.

Carbon dioxide (CO_2) has also been used to treat metal-bearing wastestreams. The CO_2 generates carbonic acid after being introduced into the reactor chamber, where it reacts with the available hydroxide ions to form less soluble carbonates. Initially, the CO_2 is vaporized within a heat exchanger or flash valve and sparged into the bottom of the reactor. Treatment requires a slow moving influent being fed into a reactor tank of sufficient depth to ensure adequate absorption of the CO_2 into solution. Since carbonic acid is formed, one of the disadvantages with the application involves neutralization required prior to final discharge.

In some cases, carbonate precipitation is preferred over other precipitation systems (i.e., hydroxide) for its ability to yield lower metal residuals in the final effluent and potential for recovery of metals. Cadmium will preferentially precipitate as a carbonate salt provided that high background carbonate levels exist or that supplemental carbonate has been added. Cyanide plating rinse waters, resulting from high plating bath pH values, will acquire background carbonate from the uptake of atmospheric CO_2. Optimum treatment efficiency for cadmium can be obtained at neutral pH in the presence of sufficient carbon. However, interferences such as cyanide, high pH, and low carbonate levels will diminish the effectiveness of carbonate precipitation. Carbonate systems will also produce lower soluble lead concentrations than those that can be achieved in a hydroxide system, especially at pH ≤ 8. Lead carbonate tends to be more crystalline than lead hydroxide which results in better filterability. In the pH range of 5.0–8.5, it will precipitate predominantly as carbonate in a lead–carbon dioxide water system as follows:

$$Pb^{2+} + CO^{2-} \rightarrow PbCO_3$$

$$3Pb^{2+} + 2CO^23 + 2H_2O \rightarrow Pb_3(CO_3)_2(OH)_2 + 2H$$

With carbonate alkalinity maintained at 200 mg/L (as $CaCO_3$), lead solubility has been observed to be ≤ 0.1 mg/L up to pH 8.6. At higher pH values, lead solubilities abruptly increased. Carbonate precipitation systems have also shown to be successful in treating copper, nickel, and zinc within prescribed pH limits and contact times. Figure 1 shows a flowsheet configuration using precipitation flocculation and sedimentation for sludge removal.

Reduction

Chemical reduction has been used to reduce complexed metals such as hexavalent chromium, mercury, copper, soluble lead, and silver. The overall reaction can be defined as a change in oxidation states where the reducing agent donates an electron in a oxidation-reduction (redox) reaction. Redox reactions typically occur in aqueous metal wastes that are relatively devoid of organic compounds. An application of this technique is used to reduce hexavalent chromium, as a chromate, CrO_4^{2-}, or dichromate, $Cr_2O_7^{2-}$, to the easier to precipitate trivalent form, Cr^{3+}. Common reducing agents for chromate reduction are sodium sulfite salts such as metabisulfite and bisulfite, sulfur dioxide, and ferrous sulfate. Other reducing agents which have recently shown potential for reducing aqueous metal wastes are dithiocarbonate, hydrazine, formaldehyde, and sodium borohydride ($NaBH_4$). The latter has shown promise for reducing mercury, soluble lead, and organo lead salts generated from tetra-alkyl lead manufacturing wastes. Table 1 shows waste treatment applications of reduction. Table 2 lists some conventional waste treatment reduction reactions. Figures 2 and 3 are flowsheets for chemical reduction applications.

Chromium Reduction

Chromium reduction, as a waste-treatment process, is a widely practiced and a developed technology. Sulfur-based reduction techniques are the methods widely used for treating chromate wastes. Hexavalent chromium, Cr^{6+}, is reduced from

Table 1 Waste Treatment Applications of Reduction

Waste	Reductant
Chromium (VI)	Sulfur dioxide (often flue gas)
	Sulfite salts
	sodium bisulfite
	sodium metabisulfite
	sodium hydrosulfite
	Ferrous sulfate
	Waste pickle liquor
	Powdered waste iron
	Powdered waste aluminum
	Powdered metallic zinc
Mercury	Sodium borohydride ($NaBH_4$)
Tetra-alkyl lead	Sodium borohydride
Silver	Sodium borohydride

a valence state of six plus to three plus, followed by the precipitation of the trivalent chromic ion, Cr^{3+}. In order to meet more stringent effluent discharge standards, other techniques such as ion exchange, evaporative recovery, and activated carbon adsorption have received closer attention.

Hydrosulfite reduction is one of several aqueous based processes which use soluble sulfite salts such as sodium bisulfite, metabisulfite (the anhydride of bisulfite), or hydrosulfite after lowering the wastestream's pH to a range of 2–3 with sulfuric acid. When using sodium bisulfite the reaction is as follows:

Table 2 Conventional Waste Treatment Reduction Reactions

Cr^{6+} to Cr^{3+} – using sulfur dioxide:

$$4SO_2 + 4H_2O \rightarrow H_2SO_3 + 2CrO_3 + 3H_2SO_3 \rightarrow Cr_2(SO_4)_3 + 3H_2O$$

sulfur dioxide, water, sulfurous acid, chromic, chromic sulfate

Cr^{6+} to Cr^{3+} – using bisulfites:

$$4CrO_3 + 6NaHSO_3 + 3H_2SO_4 \rightarrow 2Cr_2(SO_4)_3 + 3Na_2SO_4 + 6H_20$$

chromic acid, sodium bisulfite, sulfuric acid, chromic sulfate, sodium sulfate

Cr^{6+} to Cr^{3+} – using ferrous sulfate:

$$2CrO_3 + 6FeSO_4 + 6H_2SO_4 \rightarrow 3Fe_2(SO_4)_3 + Cr_2(SO_4)_3 + 6H_2O$$

chromic acid, ferrous sulfate, sulfuric acid, ferric sulfate, chromic sulfate

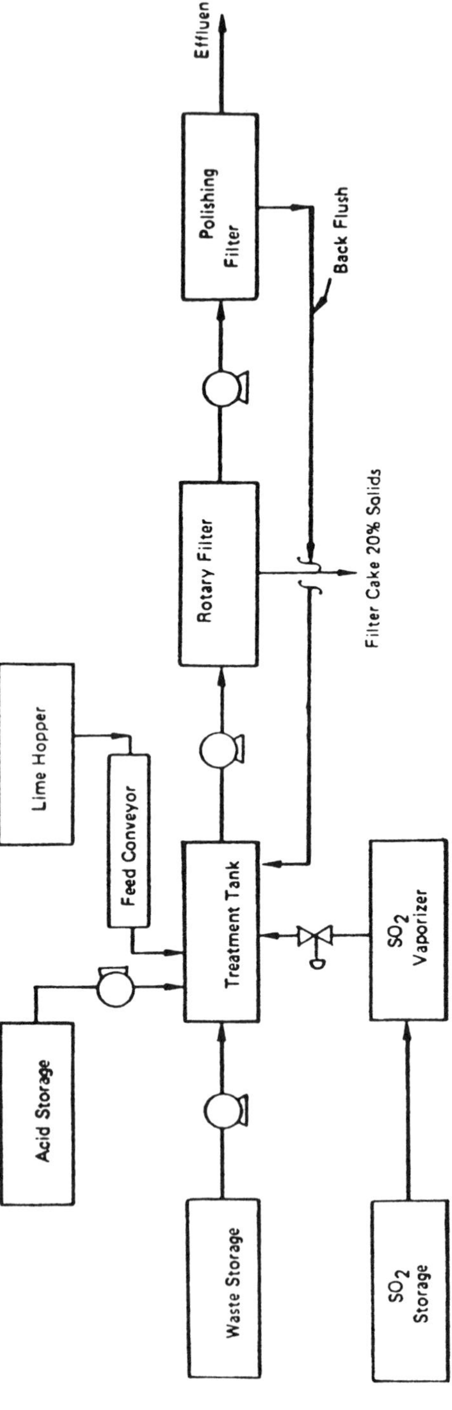

Figure 2 Typical process flowchart for chemical reduction.

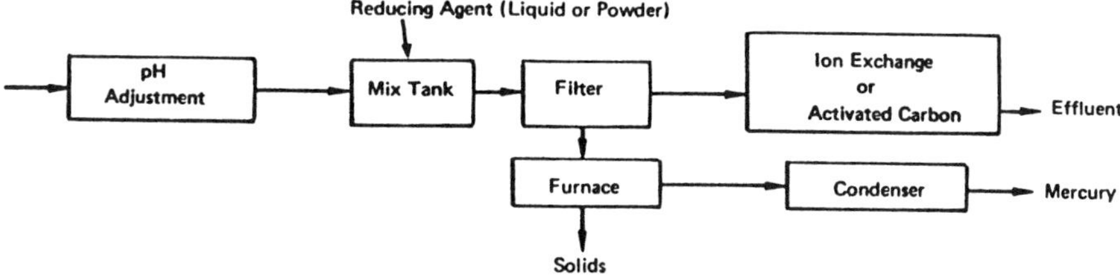

Figure 3 Mercury removal using chemical reduction.

$$Cr_2CrO_4 + 6NaHSO_3 + 3H_2SO_4 \rightarrow 2\,Cr_2(SO_4)_4 + 3Na_2SO_4 + 10H_2O$$

Trivalent chromium can be precipitated with sodium hydroxide after the solution pH is adjusted to the alkaline side:

$$Cr_2(SO_4)_3 + 6\,NaOH \rightarrow 2\,Cr(OH)_3 + 3\,Na_2SO_4$$

Excess consumption of reducing agents due to the presence of dissolved oxygen in chromium wastestreams can limit the application of this treatment process. The amount of residual nonreduced Cr^{+6} depends on reaction time, pH of the reaction mixture, concentration, and type of reducing agent. As a result, complete chromate reduction may not result.

Sulfur dioxide (SO_2) is another alternative which is the most widely used reducing agent for treating chromium waste streams. Gaseous sulfur dioxide is added under pressure to the stream. Both chromic acid and sulfurous acid are involved in a reaction which proceeds as follows:

$$3H_2SO_3 + 2H_2CrO_4 \rightarrow Cr_2(SO_4)_3 + 5H_2O$$

Although sulfur dioxide reacts with water to produce sulfurous acid, sulfuric acid is added to maintain the pH between 2 and 3. Theoretical chemical requirements to reduce 1 lb of chromium are 2 lb of SO_2 plus an excess for each liter of wastewater to be treated.

Ferrous sulfate systems are another alternative for treating acidic chromate wastestreams (pH 2–3). The ferrous ion (Fe^{3+}) sulfate. The reaction proceeds as follows:

$$Cr_2(SO_4)_3 + 3Fe_2(SO_4)_3 + 24\,NaOH \rightarrow 2Cr(OH)_3 + 6Fe(OH)_3 + 12Na_2SO_4$$

Advantages for using this system are the abundant inexpensive supply of ferrous sulfate, derived from the pickling of steel surfaces in the steel industry. A major disadvantage associated with this process is the increase of generated sludge resulting from the precipitation of the ferric ion. A variation of this process is ferrite coprecipitation, where the ferrous ion will coexist with heavy metal ions in solution. Alkali is added to neutralize the acidic solution to form a dark green hydroxide. In the presence of air, dissolution and complex formation occurs yielding a black ferrite.

Although the above treatment processes have shown to be successful in reducing chromate wastestreams, the addition of sulfides and/or sulfites to the effluent will require further treatment. Also, these reduction processes are limited to treating wastestreams since surface contact is diminished when slurries, tars, or sludges.

Mercury Reduction

Ionic mercury can be converted to the metallic form through reduction and separation techniques. Common reducing agents include zinc, hydrazine, stannous chloride, and sodium borohydride. Further treatment might be required to reduce mercury levels below 100 μg/L.

Sodium borohydride ($NaBH_4$) $+ 4M_2 + 2H_2O \rightarrow 4M^o + NaBO_2 + 8H^+$

Adsorption

Adsorption (or sorption) is a process which involves the contact of a free aqueous phase with a rigid particulate phase which has the propensity selectively to remove or store one or more solute(s) (e.g., metal species) present in the solution (wastestream). Usually, the sorbent has a fixed total uptake whereby one solute is exchanged for another (as in ion exchange processes). Once this capacity has been fulfilled, backwashing or regeneration becomes necessary, since desorption will occur leaching potential pollutants back into solution. A wide range of adsorbents and ion exchange resins have been used commercially in treating aqueous metal wastestreams. Some have been used for recovering valuable metals for the purpose of reuse; others have been employed to polish final effluents for discharge.

CYANIDE TREATMENT

Cyanides can pose great environmental concern and treatment of cyanide becomes necessary. Cyanide (CN) can be reactive in the presence of other compounds and wastes; especially those exhibiting acidic or low pH properties. Not all processes will treat cyanide wastes to levels suitable for discharge to a publicly owned treatment work (POTW) or land disposal.

Cyanide wastes are dominantly generated by electroplaters and metal finishers, steel manufacturers, paint producers, photographic operations, and mining processes. The choice of treatment or destruction technology depends primarily on the form of cyanide, its concentration, the presence of cocontaminants and specific industry-related practice.

Cyanides contain the CN functional group that can take many different forms. Some of the categories cyanides are classified into include

- Free cyanides, CN
- Simple cyanides, consisting of an alkali or metal bound to the CN ion, which can either be soluble (e.g., NaCN, KCN) or insoluble (e.g., $Zn[CN]_2$, $Cd[CN]_2$, CuCN, $Ni[CN]_2$, AgCN)
- Weak complexes (e.g., $Zn[CN]^{2-}$, $(CD[CN]^{2-}O)$
- Moderately strong complexes (e.g., $Cu[CN]^{2-}$, $Cu[CN]^{2-}Ni[CN]^{2-}$, $Ag[CN]^{2-}$)
- Strong complexes (e.g., $Fe[CN]^{2-}$, $Fe[CN]^{4-}$, $Co[CN]^{4-}$)

Other cyanide species come in the form of inorganic compounds such as thiocyanates, SCN^-, and cyanates, OCN^-. The stronger the complex, the more difficult they are to treat, since these complexes are very stable and resist oxidation.

As already mentioned, one of the major sources of cyanide waste is the electroplating industry. Cyanide baths are used to hold metallic ions in solution. Spent process baths, cleaning solutions, and dragout into rinse baths from plating operations containing cyanide ions produce wastewaters requiring treatment. Figures 4 and 5 are typical cyanide waste-treatment flowsheets. Table 3 lists cyanide compounds and some uses.

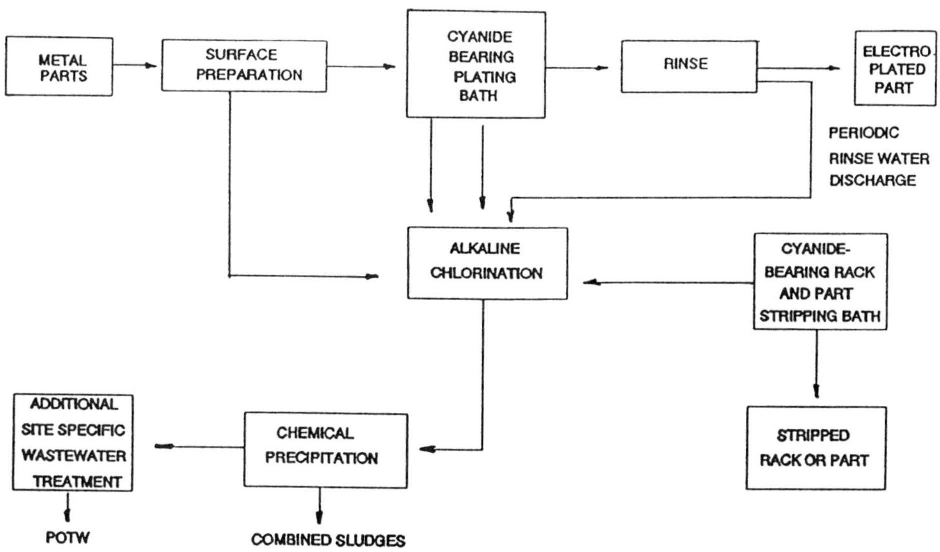

Figure 4 Typical cyanide-bearing electroplating process.

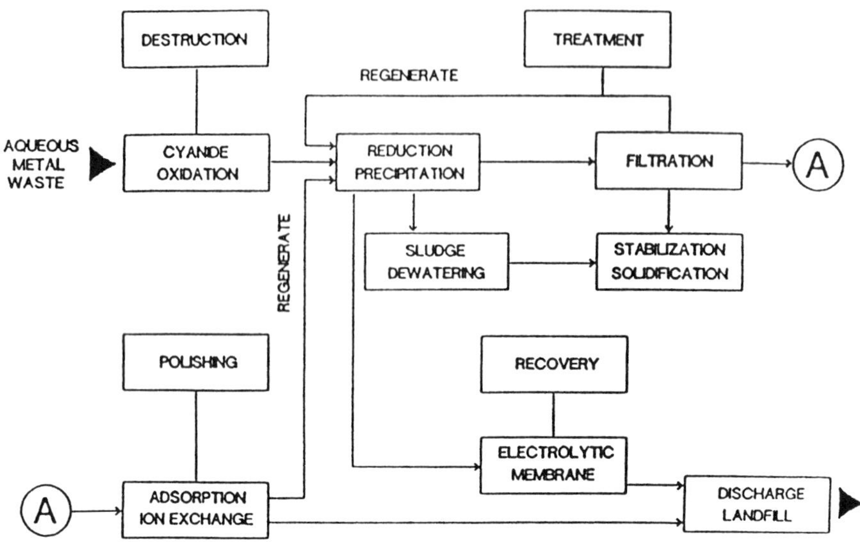

Figure 5 General treatment approach for aqueous metal/cyanide-bearing wastes.

Table 3 Industrial Uses of Cyanide Compounds

Compound	Use
Cadmium cyanide, $Cd(CN)_2$	Electroplating
Calcium cyanide, $Ca(CN)_2$	Ore cyanidation, froth flotation, fumigation, HCN production, ferrocyanide production, cement stabilizer
Cuprous cyanide, CuCN	Electroplating, medicine, insecticide, underwater paint
Cyanogen bromide, CNBr	Gold extraction, pesticides
Hydrogen cyanide, HCN	Electroplating, chelating agents, pharmaceauticals, rodenticide insecticide
Lead cyanide, $Pb(CN)_2$	Insecticide, electroplating
Nickel cyanide, $Ni(CH)_2 \cdot 4H_2O$	Electroplating
Potassium cyanide, KCN	Electroplating, steel hardening, extraction of metals from ores, silverpolish, photography
Potassium ferricyanide, $K_3Fe(CN)_6$	Photography, blueprints, metal tempering, electroplating pigments
Potassium ferricyanide, $K_4Fe(CN)_6$	Tempering of steel, process engraving, pigments
Silver cyanide, AgCN	Electroplating
Sodium cyanide, AgCN	Metal treatment, electroplating, synthesis, ore extracting, photography

Various treatment technologies have been employed for removal of cyanide compounds from spent process effluents. Although natural degradation including volatilization, biodegradation, and photodecomposition will reduce cyanide in wastestreams given enough time, alternative treatment methods must be used to more expeditiously treat cyanide. These can be grouped into several categories: chemical oxidation, electrolytic and thermal processes, precipitation, and ion exchange.

Chemical Oxidation

A majority of techniques used to treat cyanide in wastestreams involve the chemical oxidation of cyanide to a less toxic form. Catalysts such as Cu^{2+} or ultraviolet (UV) light are used in combination with chemical reagents for cyanide conversion to cyanate or to break metal cyanide complexes. Examples of processes include alkaline chlorination, ozonation (with UV light), titanium oxide/UV, sulfur dioxide/air, and hydrogen peroxide.

Alkaline chlorination is perhaps the most widely used chemical oxidation method employed in the metal-finishing industry for the treatment of cyanide wastestreams. The process has shown to be successful in destroying free (amenable) cyanide solutions generated from electroplating processes. The mechanism responsible for the destruction of cyanide is shown as follows:

$$CN + H_2O + OCl^- = Cl^-CN + 2OH$$

where CN is converted to dissolved cyanogen chloride in the presence of hypochlorite. Then, through hydrolysis of cyanogen chloride:

$$Cl^-CN + 2OH = OCN + Cl^- + H_2O$$

The final reaction results in carbon dioxide, nitrogen, and chlorine:

$$2OCN + 3OCl^- + H_2O = 2CO_2 + N_2 + 3Cl^- + 2OH^-$$

Figure 6 shows an oxidation process flowsheet. Table 4 lists some conventional waste treatment oxidation reactions.

The destruction process is a redox reaction where one or more electrons are transferred from the cyanide complex to the oxidizing agent. The process is usually performed in two steps. During the first step, the pH is raised to a value of 10 or higher. Hydrolysis of the cyanogen chloride complex is rapid with 80–90% conversion to cyanate (within 2 min depending on the wastestream composition). Sodium hypochlorite (or chlorine gas) is used to oxidize the cyanide to cyanate (OCN^-). During the second stage, the pH of the solutions is adjusted lower, to a value of 8.5, completing the oxidation of cyanate to carbon dioxide, nitrogen, and chlorine. This last step in the reaction sequence is completed within 1 hr to ensure complete destruction of the amenable cyanide.

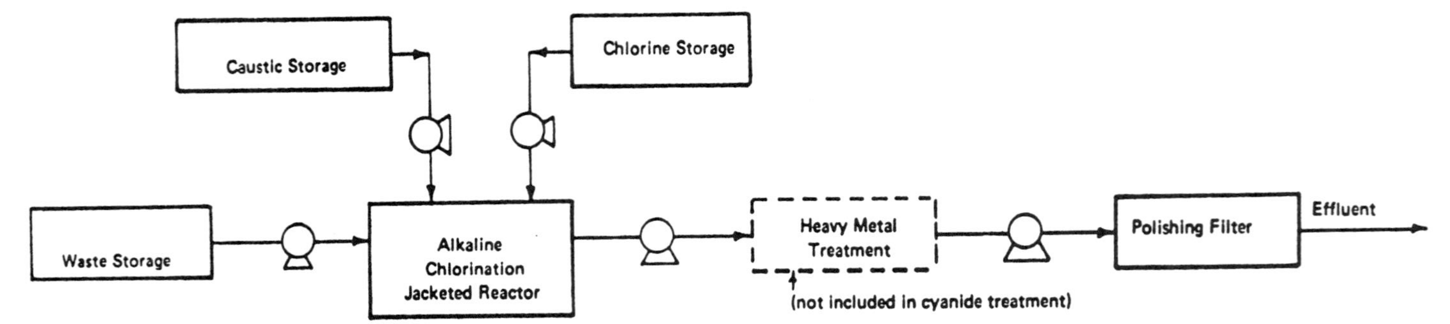

Figure 6 Oxidation process flowchart.

Table 4 Conventional Waste Treatment Oxidation Reactions

Cyanide destruction using chlorine gas:						
$2NaCN$ +	$5Cl_2$ +	$12NaOH$ →	N_2 +	$2Na_2CO_3$ +	$10NaCl$ +	6H2O
sodium cyanide	chlorine	sodium hydroxide	nitrogen	sodium carbonate	sodium chloride	water
Cyanide destruction using hypochlorites:						
$2NaCN$ +	$5NaOCl$ +	H_2O →	N_2 +	$2NaHCO_3$ +	$5NaCl$	
	sodium hypochlorite			sodium bicarbonate		
$4NaCN$ +	$5Ca(OCl)_2$ +	$2H_2O$ →	$2N_2$ +	$2Ca(HCO_3)_2$ +	3CaCl2 +	$4NaCl$
	calcium hypochlorite			calcium bicarbonate		
Conversion of cyanide to cyanate using permanganate:						
$2NaCN$ +	$2KMnO_4$ +	KOH →	$2K_2MnO_4$ +	$NaCNO$ +	H_2O	
	potassium permanganate					
Conversion of cyanide to cyanate using chlorine gas:						
$NaCN$ +	Cl_2 +	$2NaOH$ →	$NaCNO$ +	$2NaCl$ +	H_2O	
			sodium cyanate			
Conversion of cyanide to cyanate using hypochlorites:						
$NaCN$ +	$NaOCl$ →	$NaCNO$ +	$NaCl$			
	sodium hypochlorite	sodium cyanate				
Conversion of cyanide to cyanate using hydrogen peroxide:						
$NaCN$ +	H_2O_2 →	$NaCNO$ +	H_2O			
	hydrogen peroxide	sodium cyanate				

Disadvantages are associated with the use of this process. Cost of alkaline chlorination depends on the type of wastestream to be treated. For example, if thiocyanates are present, costs can be multiplied, since the demand for chlorine is significantly increased. Another disadvantage encountered involves the incomplete conversion of cyanate to its endproducts. Although cyanate is reported to be only one thousandth as toxic as cyanide, under certain conditions it can hydrolyze to form the ammonion ion:

$$2CNO^- + 2H^+ + 4H_2O = 2NH_4^+ + 2HCO_3$$

Ammonia is a complexing agent and can combine with cadmium, copper, and nickel to produce complexes not easily removed by conventional precipitation processes. Alkaline chlorination will not effectively oxidize strongly complexed cyanides such as ferrocyanide. These complexes can remain untreated and either pass into precipitation processes which can generate a metal/cyanide hydroxide sludge or are discharged with the final effluent. It is preferable to destroy cyanide completely rather than to have it mixed with metal hydroxide sludges, which might be disposed in landfills.

Alkaline chlorination can be performed either in batch or continuous-flow systems. Provided that amenable cyanide feed concentrations are below 1,000 mg/L, effluent cyanide levels of <1 mg/L can be achieved. Figure 7 shows a flow sheet of cyanide treatment using alkaline chlorination.

Some typical cyanide wastes include

- Cyanide-bearing plating solutions and sludges (some treated sludges may be reprocessed to recovery metals; sludges of no value may require subsequent treatment by the encapsulation process)
- Sodium cyanide
- Potassium cyanide
- Calcium cyanide
- Barium cyanide (the treated waste requires subsequent treatment by the sulfate precipitation process)
- Hydrogen cyanide (in aqueous solution)
- Nickel cyanide
- Potassium silver cyanide (sludges generated may be reprocessed to recover silver)
- Silver cyanide (sludges generated may be reprocessed for silver recovery)
- Miscellaneous cyanides

An alternative chemical oxidation process is ozonation. O_3 has been used for disinfection of potable water. More recently, this technology has been applied to the treatment of electroplating and photoprocessing wastestream containing cyanide. Ozonation is a chemical oxidation process which uses on-site generation of ozone as the oxidizing agent. Some of the advantages associated with the use of ozone include:

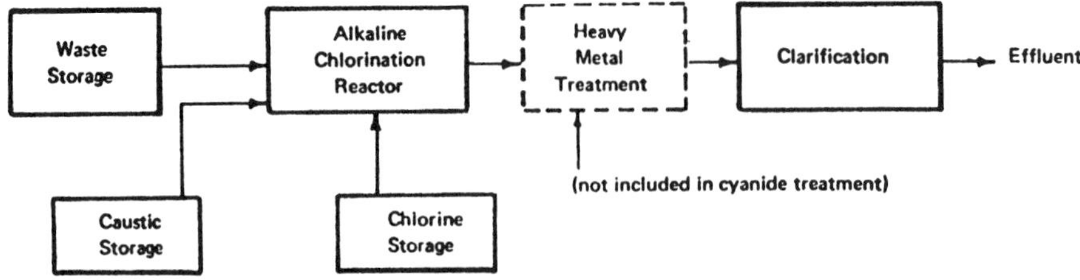

Figure 7 Cyanide waste treatment using alkaline chlorination.

- Rapid and complete destruction of dilute (< 1%) free cyanide in wastewaters.
- Ozone does not produce toxic byproducts.
- It can be used in combination with ultraviolet (UV) light (photolysis) and other reagents to break down difficult-to-treat cyanide metal complexes.

Symbolic representations for both ozonation and photolysis reactions are as follows:

- Free cyanide reactions with ozone are as follows:
  Rapid conversion to cyanide:

  $$2CN^- + 2O_3 = 2CNO^- + 2O_2$$

  Slower conversion to bicarbonate and nitrogen:

  $$2CNO^- + H_2O + 3O_3 = 2HCO_3 + N_2 + 3O_2$$

- Photolysis/ozone destruction of metal 1/CN complexes are as follows:

  $$MCN + \text{UV light} = M^+ + CN^-$$

  $$MCN + 2O_3 = M^+ \; CN^- + 3O_2$$

In the presence of thiocyanate (SCN), more ozone is required to convert the cyanide, since thiocyanate reacts very rapidly with ozone. Typically, cyanide conversion requires 1.85 g O_3/g CN, whereas thiocyanate requires 3.3 g O_3/g SCN.

The primary disadvantage associated with the use of ozonation (and/or photolysis) is the high energy consumption and capital cost for generating equipment and process monitoring controls. Another limiting factor is the mass transfer of ozone gas to the liquid phase.

Not all metal/cyanide complexes (such as ferric cyanide, iron blue, and paint/ink formulation sludges) are easily treated with chemical oxidation processes.

Hydrogen peroxide, H_2O_2, has been used in applications to oxidize free and weakly complexed cyanides according to the following reaction:

$$CN^- + H_2O = CNO^- + H_2O$$

Although costly, some of the benefits associated with this technology are that H_2O_2 produces few toxic residuals and does not add salt to the final effluents. To use hydrogen peroxide successfully, accurate measurement of chemical doses are required.

Sulfur-based technologies have gained attention owing to the advantages these applications have over other chemical oxidation processes. For example, alkaline chlorination has the potential to generate undesirable toxic residuals such as chlorinated aliphatic hydrocarbons while ozonation and wet air oxidation tend to be costly. One of the more popular SO_2/air applications is a process based on the oxidation of cyanide with sulfur dioxide (SO_2) and air for the destruction of cyanide in gold mill wastestreams.

A disadvantage associated with this process is that inefficient conversion of cyanide can produce toxic levels of cyanide complexes in precipitated sludges. Consequently, sludge dissolution under low pH conditions and sunlight has the potential to cause equilibrium reversal reactions freeing toxic cyanide and metals to the environment.

Electrolytic/Thermal Processes

Various electrolytic and elevated-temperature, pressurized processes have been employed to treat high-strength cyanide wastestreams (>1%). These include electrolytic hydrolysis, cyanide hydrolysis, and wet air oxidation.

Aqueous metal-bearing wastes containing high concentrations of free and/or complexed cyanide (>1%) can be successfully treated by use of electrolytic hydrolysis. Best results for cyanide removal were found when sodium chloride was added as a catalyst. A typical process will subject a concentrated cyanide wastestream to electrolysis at temperatures in excess of 200°C for several days. Cyanide is converted to cyanate with eventual decomposition to CO_2 and NH_4. A limiting factor associated with this technique is short circuiting. As the solution becomes more dilute, the waste becomes less capable of conducting electricity, where further treatment may be required. The presence of sulfates will also hinder the complete conversion of cyanide. As a result, this process has not been used very extensively by industry.

Cyanide hydrolysis is an aqueous-based process which has been used to destroy high-strength cyanide wastes under conditions of elevated temperature and pressure. A hydrolysis reactor is operated at temperatures ranging from 450 to 500°C under pressures ranging from 620 to 825 psig from 1 to 2 hr. By-products of the reaction are ammonia, ammonium hydroxide, and metal formate. Effluent concentrations of less than 1 mg/L total CN have been achieved from wastestreams containing as much as 40,000 mg/L CN (>99.99% destruction). However, posttreatment for metals removal may be required to meet discharge standards.

Precipitation

Soluble and insoluble cyanide complexes (e.g., ferrocyanide) can be precipitated to form stable, relatively nontoxic sludges which can be further treated by incineration or reused. When ferrous salts ($FeSO_4$, $FeCl_3$) are added to solutions containing cyanide, a precipitate is formed which is commercially referred to as Prussian blue, $Fe[Fe(CN)_6Fe]2$. Solutions containing free cyanides and heavy metal cyanides (e.g., $Cu[CN]^{2-}$, $Zn[CN]^{2-}$, and $Cd[CN]^{2-}$) have been in batch studies. The first step involves breaking the complexed metal cyanides for subsequent conversion to Prussian blue.

Ion exchange technologies have shown to be effective in selectively removing/polishing complexed cyanides from aqueous metal/CN–bearing wastestreams.

Neutralizing Toxic Electroplating Waters

High concentrations of cyanides and chromates can be removed from electroplating rinse waters by either physical, chemical, or—in the case of cyanide concentrations below 30 ppm—by biological methods.

Basically, electroplating processes require an electrolytic solution, soluble or insoluble anodes, a metallic bar used as a cathode, and a source of direct current. The plating solutions used for the electronic industry are alkaline gold, silver, and copper solutions (containing high concentrations of cyanides) and acidic chromate solutions.

A necessary part of electroplating processes are the water rinses. The abatement of pollution from such wastes includes recovery of raw material or total destruction in a waste-treatment plant. Recovery of precious metals is mandatory because of their high costs. Processes such as ion exchange, evaporation, dialysis, and electrodialysis and reverse osmosis are used to accomplish high reclamation of precious and heavy metals.

In response to the toxicity of cyanides and chromates and the discharge standards set by the US Environmental Protection Agency (EPA), total destruction of cyanides and chromates by chemical processes has been achieved using batch treatment, thus eliminating the toxic potential of the waste effluent.

In the electronics industry, electrolytic deposition of metals (electroplating) is essential. The process is used to reduce corrosion, to correct dimensions for finishing and soldering, and to improve wearing qualities. After electroplating is completed, the dragout from the plating solutions that cling to the parts must be eliminated through water rinses that are highly concentrated in cyanides and chromates. These rinses represent a pollution potential in terms of toxicity. Wastes from small plating shops also contain low concentrations of metals such as copper, iron, lead, nickel, silver, and cadmium.

Cyanides in waters cause toxic effects on the biological activity, altering the development of the normal aquatic life. The threshold limit of toxicity at infinite time for fish is 0.1 mg/L as cyanide. For microorganisms, toxicity has been shown at cyanide concentrations of 0.3 ppm and above. Cyanides are much less toxic for humans than fish. In doses of 10 mg or less, they are converted to cyanates in the human body. Cyanate is 100% less toxic—fatal effects occurring only when the detoxifying mechanism is overwhelmed.

Water rinses containing chromates are highly toxic, causing irritations by ingestion or inhalation. A limit concentration of 0.05 mg/L is considered safe.

Contributing to the enhancement of the industrial water-pollution control, the federal government through the EPA has promulgated regulations for industrial waste disposal. Industry is required to minimize the dumping of pollutant rinses via best practical technology. Plating shops with less than 33 m^2/hr of production or less than 2,000 A of installed DC are exempt from meeting the 1983 zero discharge requirements. However, they must adjust the pH of the discharge, reduce and precipitate all chromium salts, and destroy all cyanide in their effluents.

Cyanide Waste Treatment

The common method of cyanide destruction in wastes from electroplating processes is alkaline chlorination oxidation by sodium hydroxide addition or chlorine gas plus sodium hydroxide addition to the waste. Cyanide oxidation by chlorination occurs by two separate chemical reactions. In the first stage, cyanide is oxidized to cyanate. In the second stage, cyanate is oxidized to carbon dioxide and nitrogen.

First stage

$$Cl_2 + NaCN \rightarrow CNCl + NaCl$$
$$CNCl + 2NaOH \rightarrow NaCNO + NaCl + H_2O$$

The first reaction is instantaneous and occurs at all pH levels. The second reaction is a hydrolysis that converts cyanogen chloride to cyanate and is very much pH dependent: Hydrolysis is not complete at pH levels below 7.5; hydrolysis rate increases as the pH increases. Oxidation of cyanide is accomplished most completely and rapidly at pH 10.5. An oxidation period of 30 min to 2 hr is usually allowed. It is important that the hydrolysis process be completed as quickly as possible, since cyanogen chloride is volatile and toxic.

Free cyanide reacts fast. As its concentration falls, the metal cyanide may become insoluble or metal hydroxide floc may envelope the cyanide solution. To avoid this interference, rapid mixing should be provided during treatment. In this process, temperature is held above 20°F.

Second stage: The cyanate is further oxidized by adding chlorine to carbon dioxide and nitrogen:

$$2NaCNO + 4NaOH + 3Cl_2 \rightarrow 2CO_2 + 6NaCl + N_2 + 2H_2O$$

The reaction requires an excess of chlorine and caustic in order to go to completion. The reaction rate increases as pH decreases.

Because the first stage is performed at a pH of 10.5 and the second stage at a pH of 8.5, it is convenient to separate the two oxidation processes.

When a single-step chlorination process is used, an intermediate pH value must be selected so both stages of oxidation occur simultaneously. If the solution pH is too low, carbon dioxide and nitrogen gases are formed. Then, the hydrolysis of cyanide to cyanate will be incomplete and the intermediate product, cyanogen chloride, will be carried out by the nitrogen gas. This will create a hazardous condition in closed working areas. If the pH is too high, cyanate oxidation will require long detention periods and may never go to completion.

Sulfuric acid is used to lower the pH to 8.5. An oxidation-reduction-potential (ORP) system is used to control both chlorine addition and the pH. pH level control is by caustic or sulfuric acid addition. Complete oxidation is ensured if the pH and retention time are maintained at the proper levels. The reaction is completed with 7.35 parts Cl_2/part CN and 6.5 parts of NaOH/part CN.

In small plating shops, the rinse bath cyanide concentrations may be as high as 600 mg/L. For cyanides and total wastes up to 3,000 gpd (based on one shift per 8-hr day), a complete destruction of cyanide to CO_2 and N_2 can be achieved in a single-stage oxidation. The oxidation requires less than 1 hr. A constant pH of 8.5 during chlorination should be maintained. An excess of chlorine must be added to speed up the hydrolysis of cyanide to cyanate and to avoid liberation of the highly toxic cyanogen. When cyanide is oxidized to cyanate, toxicity is reduced to a thousand times.

Batch treatment (Fig. 8) has been used for alkaline chlorination when average flows do not exceed 15 gpm. This treatment has also been used for waste flows up to 30 gpm. Larger volumes of cyanide wastes require specially designed

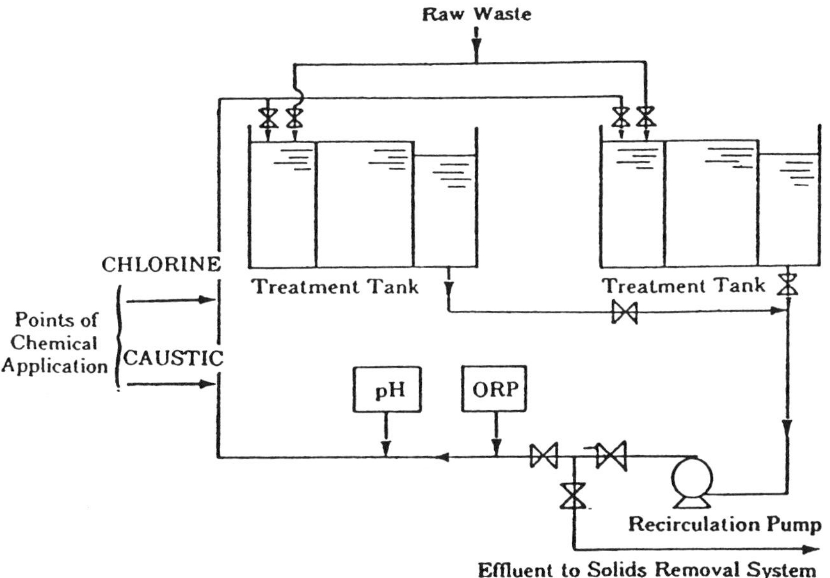

Figure 8 Instrumental batch cyanide treatment, chemical chlorination.

and constructed treatment plants. The exact design depends on the type of industrial wastes and the concentration of contaminants. Provisions must be made for handling concentrated solutions discharged at intervals. To maintain optimum pH, separate tanks are used for each stage of the reaction.

Electrolytic Decomposition

Electrolytic oxidation has been used successfully for treatment of wastes containing high concentrations of cyanide.

Concentrated wastes are subjected to anodic electrolysis at high temperatures (125–200°F) using carbon steel for the anode and cathode. Cyanide is completely oxidized to nitrogen and carbon dioxide. When the waste-electrolyte becomes weak and its conductivity decreases, the reaction of oxidation of cyanates may not be completed and final chlorination will be required.

Wastes containing high concentration of cyanide ranging from 45,000 to 100,000 mg/L require treatment periods from 7 to 18 days. Final concentrations are less than 0.5 mg/L. Sulfate causes interference. It forms heavy scaling at the anode, prevents further electrolysis, and allows cyanide reduction to 695–750 mg/L. To improve the electrolytic decomposition of cyanides at lower concentrations (i.e., below 200 mg/L) and to increase solution conductivity, sodium chloride can be added to the cyanide solution. At the anode, chloride is oxidized to chlorine gas (available as an additional oxidant).

Electrolytic decomposition is also an effective method to destroy cyanide metal complexes (i.e., nickel, copper, iron) plated out at the cathode. These complexes are difficult to destroy by means of alkaline chlorination.

Electrolytic oxidation treats concentrated cyanide wastes from spent plating baths, cyanide strip solutions, and preplating cyanide dips.

This procedure is impractical in dilute solutions because the low conductivity results in poor current efficiency.

Ozonation

In cyanide treatments ozonation has been used as follows:

$$CN + O_3 \rightarrow CNO + O_2$$
$$2CNO + H_2O + 3O_3 \rightarrow 2HCO_3 + N_2 + 3O_2$$

This is an expensive method for destroying cyanide. Excluding capital costs, the amount of ozone per pound of cyanide is less than that required for chlorination (less than 5 lb/lb CN, for completion oxidation). Zinc, nickel, and copper cyanide complexes are easily destroyed but the cobalt cyanide complex has been found resistant.

Ozone diffusion into the wastewater is usually accomplished by a bubble plate column or diffusion chamber.

The oxidation of cyanide to cyanate at pH 9–12 takes place in 10–15 min and is instantaneous when traces of copper are present. Oxidation of cyanate to final products is slow, and catalysts such as copper, iron, or manganese may be required. Because of the toxic effects of ozone, the system must be very well ventilated.

Hexavalent Chromium Waste Treatment

Industrial wastes containing hexavalent chromium, in the form of chromate, and dichromate are generated by the metal plating industry. The major source of these wastes is the chromic acid bath and rinse water after plating operations. For wastes from small plating shops (i.e., volume of waste <10,000 gpd), the chromate concentration ranges from 10 to 500 mg/L.

Reduction and precipitation processes are used to treat such wastes. Chromium is reduced through an oxidation-reduction reaction from a valence state of +6 to +3, followed by precipitation as an insoluble trivalent chromium hydroxide. The reducing agents are ferrous sulfate, sodium bisulfite, and sulfur dioxide. These reactions are very pH dependable and occur usually at a pH of <3. The reduced waste is then neutralized with lime slurry or caustic, causing the precipitation of trivalent chromium.

Ferrous Sulfate

In an oxidation-reduction reaction, ferrous ion reacts with hexavalent chromium, reducing the chromium to a trivalent state and oxidizing the ferrous ion to the ferric state. At a pH of <3, the reaction occurs rapidly. Sulfuric acid is added to lower the pH. The following reactions indicate the reducing process:

$$2H_2CrO_4 + 6\ FeSO_4 + 6HSO_4 \rightleftharpoons Cr_2(SO_4)_3 + 3Fe_2(SO_4)_3 + 8H_2O$$

Precipitation:

$$Cr_2(SO_4)_3 + 3Ca(OH)_2 \rightarrow 2Cr(OH)_3 \uparrow + 3CaSO_4$$

$$Fe_2(SO_4)_3 + 3Ca(OH)_2 \rightarrow 2Fe(OH)_3 \uparrow 3CaSO_4$$

The theoretical quantities of chemical compounds required to reduce and precipitate 1 mg/L of hexavalent chromium are

16.03 mg/L of FeSO
6.01 mg/L of H_2SO_4
9.48 mg/L of 90% lime

Sludge production is as follows:

0.38 mg/L $Fe(OH)_3$ from 1 mg/s of FeSO
1.84 mg/L $CaSO_4$ from 1 mg/L of lime
1.98 mg/L $Cr(OH)_3$ from 1 mg/L of Cr

Disadvantages of this method are:

- Sludge contamination by ferric hydroxide when the alkali is added. Recovery of chromic oxides from chromite ores is not economically feasible because of basic problems in the separation and removal of iron.
- Poor sulfate, reducing power at a pH 3 or higher, requiring the presence of strong acid (sulfuric) to lower the pH. This increases the amount of alkali needed for precipitation of the trivalent chromium.
- Need for an excess of two and a half times the theoretical dosage of ferrous sulfate to complete the reaction.

The apparent low-price advantage of this method vanishes when the quantities required for treatment and the problems associated with sludge recovery are considered.

Sodium Bisulfite

Sodium metabisulfite is used to attain sodium bisulfite. The following reactions occur:

$$Na_2S_2O_5 + H_2O \rightleftharpoons 2NaHSO_3$$

$$NaHSO_3 + H_2O \rightleftharpoons H_2SO_3 + NaOH$$

Acid is required to neutralize the NaOH. To reduce hexavalent chromium to the trivalent state, the pH of the reaction must be less than 3. A pH of 2 is recommended.

The overall reaction is as follows:

Reduction

$$2H_2CrO_4 + 3NaHSO_3 + 3H_2SO_4 \rightleftharpoons Cr_2(SO_4)_3 + 3NaHSO_4 + 5H_2O$$

Precipitation

$$Cr_2(SO_4)_3 + Ca(OH)_2 \rightarrow 2Cr(OH)_3 \uparrow + 3CaSO_4$$

The theoretical quantities of chemical compounds needed to reduce and precipitate 1 mg/L of hexavalent chromium are

2.81 mg/L of $Na_2S_2O_5$ (97.5%)
1.52 mg/L of H_2SO_4
2.38 mg/L of lime (90%)

To complete the reaction, 75% of bisulfite must be added in excess dosage of the theoretical demand.

Sulfur Dioxide

Sulfur dioxide is very effective in reducing hexavalent chromium due to strong acidic conditions of the liquid sulfur dioxide. Sulfur dioxide has good reducing power within the pH range of 3.0–5.5. The following reactions take place in the reduction and precipitation:

Reduction

$$SO_2 + H_2O \rightleftharpoons H_2SO_3$$

$$H_2CrO_4 + 3H_2SO_3 \rightarrow Cr_2(SO_4)_3 + 5H_2O$$

Precipitation

$$Cr_2(SO_4)_3 + 3Ca(OH)_2 \rightarrow 2Cr(OH)_3 + 3CaSO_4$$

The theoretical quantities of chemical compounds added to reduce and precipitate 1 mg/L of hexavalent chromium are

1.85 mg/L of SO_2
2.38 mg/L of lime (90%)

To account for the oxidation of SO_3 to SO_2 resulting from the presence of dissolved oxygen in rinse waters, an excess dosage of 35 mg/L SO_2 should be added. Sulfur dioxide is a more economical process than ferrous sulfate and sodium bisulfite. Even when the pH of the chromate waste is higher than 5.5, sulfur dioxide is used to bring the pH below 5.5 where theoretical proportions can be used. Sulfuric acid should be considered for lowering a pH above 8.5 to <5.5, thus cutting the cost of sulfur dioxide.

Hexavalent Chromium Waste Control Using Oxidation Reduction

When a platinum electrode and a reference electrode such as calomel are immersed in a solution containing both a reducing substance and its oxidation product (or vice versa), a potential is developed, which is a measure of the concentrations ratio of the reductant and the oxidant.

The measured potential varies 60 mV when the number of transferred electronic is 1 for every 10:1 change in the concentration ratio of the oxidant to the reductant.

Ferrous sulfate: When ferrous sulfate is used as a reducing agent in the treatment of hexavalent chromium, the redox potential is maintained at some voltage in order to secure complete chromium reduction. Also, the pH of the waste must be kept between 2.0 and 2.5 to satisfy the chemical reaction.

Sodium bisulfite: For complete reduction of hexavalent chromium with sodium bisulfite, a redox potential of 380 mV must be achieved at a pH of 2.5.

Sulfur dioxide: When SO_2 is used as a reducing agent, the redox equilibrium potention has been reported as 0.045 V. With a 10% excess of SO_2, a potential of 105 mV is used. With 1% excess of SO_2, a potential of 165 mV is used.

The chemical requirements and the redox control system usually are determined experimentally because plating rinse waters may contain other contaminants (i.e., metals and chemical complexes) that could interfere with the process. An advantage of redox control in waste treatment is that it is related to the ratios of oxidant and reductant concentrations and not to the actual concentrations. Thus, when a set endpoint is chosen, the redox control automatically adjusts the feed of the reducing agent to complete the reduction of the chromium regardless of its concentration.

Batch Treatment

Usually small plating shops have a total daily flow of less than 30,000 gpd, necessitating batch treatment. Two tanks are required, each with the capacity of 1 day's flow in order to use one for treatment and the other for filling or vice versa. The resultant sludge can be dewatered on sand-drying beds or hauled to disposal. Two pH recorder-controllers and ORP recorder-controllers are needed for automatic control (see Fig. 9).

Continuous Treatment

For volumes that exceed 30,000 gpd, continuous treatment is recommended and requires three tanks: acidification and reduction, a mixing tank for lime addition, and finally a settling tank. The time in the reduction tank should be four times the theoretical in order to assure complete reduction. Twenty minutes are required for flocculation and the maximum overflow rate in the settling tank is 500 gpd/ft^2. Automatic controls must be installed: pH and redox control for

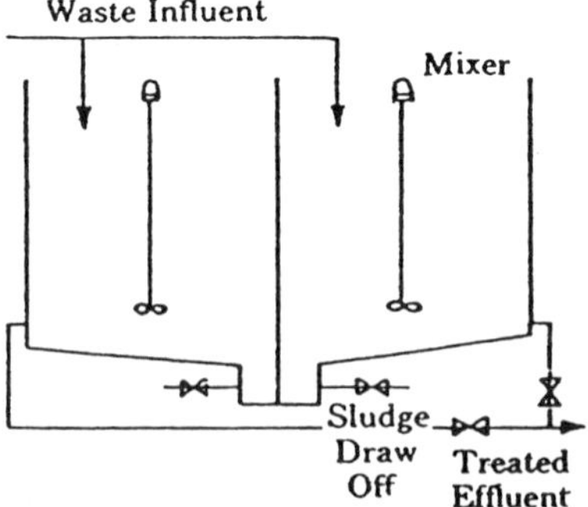

Figure 9 Manual batch treatment of chromium wastes.

the ıeduction tank and a pH recorder to control the lime addition in the second tank.

CEMENTATION PROCESS

Cementation is a spontaneous process which involves the reduction of more electropositive species by more electronegative metals such as Fe, Zn, and Al. A major advantage of such a reduction is that noble metals are obtained in an easily recoverable metallic form in a simple one step process. This makes the process favorable as compared to competing precipitation in forms of metallic hydroxides or sulfides which result in massive production of hazardous sludge.

Cementation is a very complex process in which rate and stoichiometry depend on an intricate way on a multitude of variables. These include concentrations of noble species, P admixture, pH, temperature, mixing rates, and others. Based on experiments conducted by Gould and co-workers, as well as other investigators, a new mechanism and core kinetic model have been used to interpret key features of the cementation process.

Mechanisms

Cementation is a heterogeneous physicochemical process which involves a number of interdependent transport and chemical and/or electrochemical steps. In order to evaluate the process mechanism, the principal active species, including intermediate, should be identified and a system structure reflecting the species interaction should be developed. Additionally, relevant relationships such as species distribution and system potential and mechanism of deposit formation, if any, should be developed.

Metals are arranged in the order of their activity in a series commonly called the electromotive series. The order of activity metals follows:

- Potassium
- Strontium
- Calcium
- Sodium
- Magnesium
- Aluminum
- Zinc
- Iron
- Cadmium
- Nickel
- Tin
- Lead
- Hydrogen

- Copper
- Silver
- Mercury
- Platinum
- Gold

The activity refers to various types of reactions involving the metals, among which is the relative activity of the metals in displacing hydrogen from acids and water. The farther an element is above hydrogen in this series, the more energetic is the displacement. The metals below hydrogen in the series do not displace it from acids. The relative activity of the elements in the series applies not only to displacement of hydrogen but also to the displacement of any other element in the series. Iron is above copper in the series. Thus if metallic iron is placed in a solution of copper sulfate, the copper is displaced by the iron, metallic copper is set free, and iron dissolves. Silver, which is below copper in the series, does not displace it from copper compounds. In general, a metal will displace any other element from its compound if the former is above the latter in the series.

The oxides of the metals vary widely in stability. Some, such as silver oxide, are easily decomposed by heat, whereas others, such as magnesium or calcium oxide, are not decomposed at the temperature of the electric furnace. This relative stability of the oxides of the metals is related to the relative activity of the metals. Thus, the active metals combine readily to form stable oxides, that is, oxides which are difficult to decompose, whereas the less active metals form oxides which are readily decomposed. The electromotive series also explains why only those metals below hydrogen are found in nature in the uncombined state. Natural waters contain small amount of acids, which would react with the free metal in nature, leaving only those metals which do not displace hydrogen, the metals below hydrogen in the series. Only in exceptional cases are metals which precede hydrogen found in nature—for example, iron in meteorites.

Owing to the nonuniform property of metal surfaces, multiple electrolyte and galvanic cells are spontaneously induced. Electrolytic cells include anodic sites, a boundary layer at anodic sites, the bulk solution and cathodes located opposite of the cathodic sites, and Cg on the sacrificial metal surfaces.

At the initiation of cementation, an electrical double layer is formed at cathodic sites. The outer Helmholtz layer comprises hydrogen ion, noble cation, and indifferent ions. As a result of interaction between hydrogen ions and electrons, hydride ions are formed at and emitted at the cathodic site, whereas normal ions diffuse from the bulk solution from the cathodic site. Thus, two opposite fluxes are formed: a hydride flux and a counterflux of normal ions. The species in this flux react with each other.

At the onset of the process, either the hydride flux or the noble cation flux is greater. In the former case, the reactor between the ions in these fluxes results in the disappearance in the noble ion from the layer adjacent to the cathodic site. And the reaction front moves to a distance from the cathodic site.

CONCLUSIONS

The behavior of the cementation system in the course of a simultaneous reduction of several noble species is similar to that in processes with external source of electric current. However, in the migrational cementation regimens, the potentials in the galvanic cell driving the entire process change when concentrations of noble species are changed. The change in these potentials explains the deviations of the cementation data from similar data on other electrochemical processes.

Anionic noble ions, for example, pentavalent arsenic and antimony, can be reduced only at the outer border of the boundary layer. This increases the migrational current within the boundary layer and, therefore, accelerates the main reduction process.

Neutral noble species, for example, oxygen, can be reduced either at the outer border of the boundary layer or at the metal surface. At both places, oxygen would consume hydrogen and hydride ions, thus reducing the process rate. However, the portion of oxygen which diffuses across the boundary layer contributes to the total electric current in the circuit and causes the excitation of the current due to main noble species, thus accelerating their reduction.

12
Disinfection

INTRODUCTION

Water treatment for a public water supply was almost nonexistent until after the London cholera epidemic in 1854. It was discovered that a water well was the source of the contamination and as a consequence slow sand filtration was incorporated at the site. It was not until the "germ theory of disease" postulated by Pasteur and Koch, however, that disinfection began to be recognized. In 1881, Koch demonstrated that chlorine could kill bacteria, but it took the typhoid fever outbreak in London in 1905 before a continuous chlorination system for the public water supply began. In the United States, regular disinfection practices were used in Chicago and New Jersey by 1908.

Disinfection is a process whose objective is the destruction or inactivation of pathogenic microorganisms. A good disinfectant should meet the following criteria:

- Should be a good germicide on whatever number of pathogens are found in the water.
- Should destroy all pathogens in the retention time required.
- Should not be rendered inefficient even during water fluctuation.
- The temperature and pH range in which the disinfectant is required to function must be adequate.
- Must not make the water toxic or unpalatable.

- Should be safe and easy to handle.
- Should be able to be monitored easily.
- Should provide residual protection against recurring contamination.
- Should be readily available at reasonable cost.

The mechanisms that control the effectiveness of the disinfection are

- Oxidation or rupture of the cell wall of microorganisms
- Diffusion into the cell wall with consequent cellular disintegration

Table 1 shows the standard potentials of some of the more commonly used chemical disinfectants. The higher the oxidation potential, the easier the compound is able to oxidize organic materials. If oxidation was the only governing factor, ozone would be the first choice, chlorine dioxide or chorine as the most effective disinfectant. The selection is more difficult, since germicidal properties and cell permeability are dependent on molecular weight, charge on chemical species, and other factors. In the halogen series, the diffusion order is iodine>bromine>chlorine, just the opposite of the oxidation potentials. It becomes evident that other considerations must be a guide in selecting the proper disinfectant for a particular application. Water quality characteristics other than the microbiological components which influence the efficiency of disinfectants are turbidity, organics, pH, and temperature.

Turbidity or cloudiness is caused by the suspended solids present in water, as has been discussed earlier. It has been demonstrated to interfere with disinfection because the suspended particles which are responsible for the turbidity surround and shield microorganisms. Normally, a maximum of 1 turbidity unit is recommended for a water supply unit that is going to be disinfected.

Organic compounds by adhering to the cell surfaces can hinder disinfectant efficiency, forming weaker germicidal properties or eliminate germicidal properties completely. Some inorganic compounds such as iron, manganese, hydrogen sulfide, cyanides, and nitrates can also decrease disinfectant efficiency because they become oxidized by some disinfectants.

Some disinfectants work best if the pH is on the acidic side, whereas others

Table 1 Standard Potentials of Disinfectants

Compound	Formula	Potential	Reactions
Chlorine	Cl_2	1.36	$Cl_2 + 2e^- \rightleftharpoons 2Cl^-$
Bromine	Br_2	1.09	$Br_2 + 2e^- \rightleftharpoons 2Br^-$
Iodine	I_2	.54	$I_2 + 2e^- \rightleftharpoons 2I^-$
Ozone	O_3	2.07	$O_3 + 2e^- \rightleftharpoons O_2 + H_2O$
Chlorine dioxide	ClO_2	.95	$ClO_2 + e^- \rightleftharpoons ClO_2^-$

are not affected at all by pH. Temperature affects the reaction of certain steps in the disinfection process, such as diffusion of the disinfectant through cell walls or the reaction rate with enzymes.

MICROORGANISMS AND DISEASE

Wastewater disinfection is the destruction or inactivation of pathogenic organisms and is carried out to minimize public health concerns. Destruction is the physical disruption or disintegration of a pathogenic organism, whereas inactivation is the removal of a pathogen's ability to infect.

A pathogen is any biological species that can cause disease in the host organism. Such organisms or agents fall into four broad categories: viruses, bacteria, parasites, and fungi. Within the parasite category, there are protozoa, nematodes, and helminths. Viruses, bacteria, and parasites are primary pathogens that are present at some level in sludge as a result of human activity upstream from the wastewater-treatment plant. Fungi are secondary pathogens and are only numerous in sludge when given the opportunity to grow during some treatment or storage process.

Pathogens enter wastewater-treatment systems from a number of sources:

- Human wastes, including feces, urine, and oral and nasal discharges
- Food wastes from homes and commercial establishments
- Industrial wastes from food processing, particularly meat-packing plants
- Domestic pet feces and urine
- Biological laboratory wastes such as those from hospitals

Additionally, where combined sewer systems are used, ground surface and street runoff materials, especially animal wastes, may enter the sewers as storm flow. Vectors such as rats that inhabit some sewer systems may also add to the pathogen population. Viruses, bacteria, parasites, and fungi differ in size, physical composition, reproductive requirements, occurrence, and prevalence in wastewater.

Viruses

Viruses are obligate parasites and can only reproduce by dominating the internal processes of host cells and using the host's resources to produce more viruses. Viruses are very small particles whose protein surface charge changes in magnitude and sign with pH. In the natural pH range of wastewater and sludges, most viruses have a negative surface charge. They will adsorb to a variety of material under appropriate chemical conditions. Different viruses show varying resistance to environmental factors such as heat and moisture. Enteric viruses are acid resistant and many show tolerance to temperatures as high as 140°F (60°C).

Many of the viruses that cause disease in humans enter the sewers with feces or other discharges and have been identified, or are suspected of being, in sludge. Major virus subtypes transmitted in feces are listed in Table 2 together with the disease they cause. Viruses are excreted by humans in numbers several orders of magnitude lower than bacteria. Typical total virus concentrations in untreated wastewaters are 1,000–10,000 plaque-forming units (PFU) per 100 ml; effluent concentrations are 10–300 PFU per 100 ml. Wastewater treatment, particularly chemical coagulation or biological processes followed by sedimentation, concentrates viruses in sludge. Raw primary and waste-activated sludges contain 10,000 to 100,000 PFU per 100 ml.

Bacteria

Bacteria are single-celled organisms that range in size from slightly less than 1 μm in diameter to 5 μm wide by 15 μm long. Among the primary pathogens, only bacteria are able to reproduce outside the host organism. They can grow and reproduce under a variety of environmental conditions. Low temperatures cause dormancy, often for long periods. High temperatures are more effective for inactivation, although some species form heat-resistant spores. Pathogenic bacterial species are heterotrophic and generally grow best at a pH between 6.5 and 7.5. The ability of bacteria to reproduce outside a host is an important factor. Although sludge may be disinfected, it can be reinoculated and recontaminated.

Bacteria are numerous in the human digestive tract; humans excrete up to 10^{13} coliform and 10^{16} other bacteria in feces every day. The most important of the pathogenic bacteria are listed in Table 3 together with the diseases they cause.

Parasites

Parasites include protozoa, nematodes, and helminths. Pathogenic protozoa are single-celled animals that range in size from 8 to 25 μm. Protozoa are transmitted

Table 2 Pathogenic Human Viruses

Adenoviruses	Adenovirus infection
Coxsackie virus, group A	Coxsackie infection, viral meningitis; AFRI,[a] hand, foot, and mouth disease
Coxsackie virus, group B	Coxsackie infection, meningitis; viral carditis, endemic pleurodynia, AFRI
echovirus, (30 types)	echovirus infection; meningitis; AFRI
Poliovirus (3 types)	Poliomyelitis
Reoviruses	Reovirus infection
Hepatitis virus A	Viral hepatitis
Norwalk agent	Sporadic viral gastroenteritis
Rotavirus	Winter vomiting disease

[a]AFRI = acute febrile respiratory illness.

Table 3 Pathogenic Human Bacteria

Species	Disease
Arizona hinshawii	*Arizona* infection
Bacillus cereus	*B. cereus* gastroenteritis; food poisoning
Vibrio cholerae	*Cholera*
Clostridium perfringens	*C. perfringens* gastroenteritis; food poisoning
Clostridium tetani	Tetanus
Escherichia coli	Enteropathogenic *E. coli* infection; acute diarrhea
Leptospira spp	Leptospirosis; Swineherd's disease
Mycobacterium tuberculosis	Tuberculosis
Salmonella paratyphi, A, B, C	Paratyphoid fever
Salmonella sendai	Paratyphoid fever
Salmonella spp (over 1,500 serotypes)	Salmonellosis; acute diarrhea
Salmonella typhi	Typhoid fever
Shigella spp	Shigellosis; bacillary dysentery; acute diarrhea
Yersinia enterocolitica	*Yersinia* gastroenteritis
Yersinia pseudotuberculosis	Mesenteric lymphadenopathy

by cysts, the nonactive and environmentally insensitive form of the organism. Their life cycles require that a cyst be ingested by a human or another host. The cyst is transformed into an active organism in the intestines, where it matures and reproduces, releasing cysts in the feces. Pathogenic protozoa are listed in Table 4 together with the diseases they cause.

Nematodes are roundworms and hookworms that may reach sizes up to 14 in. (36 cm) in the human intestines. The more common roundworms found in humans and the diseases they cause are listed in Table 4. They may invade tissues other than the intestine. This situation is especially common when a human ingests the ova of a roundworm common to another species such as the dog. The nematode does not stay in the intestine but migrates to other body tissue such as the eye and encysts. The cyst, similar to that formed by protozoa, causes inflammation and fibrosis in the host tissue. Pathogenic nematodes cannot spread directly from one human to another. The ova discharged in feces must first embryonate at ambient temperature, usually in the soil, for at least 2 weeks.

Helminths are flatworms, such as tapeworms, that may be more than 12 in. (30 cm) in length. The most common types in the United States (listed in Table 4) are associated with beef, pork, and rats. Transmission occurs when a human ingests raw or inadequately cooked meat or the eggs of the tapeworm. In the less serious form, the tapeworm develops in the intestine, matures, and releases eggs. In the more serious form, it localizes in the ear, eye, heart, or central nervous system.

Table 4 Pathogenic Human and Animal Parasites

Species	Disease
Protozoa	
Acanthamoeba spp	Amebic meningoencephalitis
Balantidium coli	Balantidiasis, balantidial dysentery
Dientamoeba fragilis	Dientamoeba infection
Entamoeba histolytica	Amebiasis, amebic dysentery
Giardia lamblia	Giardiasis
Isosora bella	Coccidiosis
Naegleria fowleri	Amebic meningoencephalitis
Toxoplasmosis gordi	Toxoplasmosis
Nematodes	
Ancyclostoma dirodenale	Ancylostomiasis, hookworm disease
Ancylostoma spp	Cutaneous larva migrans
Ascaris lumbricoides	Ascariasis, roundworm disease, *Ascaris* pneumonia
Enterobius vermicularis	Oxyuriasis, pinworm disease
Necator americanus	Necatoriasis, hookworm disease
Strongyloides stercoralis	Strongyloidiasis, hookworm disease
Toxocara canis	Dog roundworm disease, visceral larva migrans
Toxocara cati	Cat roundworm disease, visceral larva migrans
Trichuris trichiura	Trichuriasis, whipworm disease
Helminths	
Diphyllobothrium latum	Fish tapeworm disease
Echinococcus granulosis	Hydrated disease
Echinococcus multilocularis	Aleveolar hydatid disease
Hymenolepis diminuta	Rat tapeworm disease
Hymenolepis nana	Dwarf tapeworm disease
Taenia saginata	Taeniasis, beef tapeworm disease
Taenia solium	Cysticercosis, pork tapeworm disease

Fungi

Fungi are single-celled nonphotosynthesizing organisms that reproduce by developing spores, which form new colonies when released. Spores range in size from 10 to 100 μm. They are secondary pathogens in wastewater sludge, and large numbers have been found growing in compost. The pathogenic fungi, listed in Table 5, are most dangerous when the spores are inhaled by people whose systems are already stressed by a disease, such as diabetes, or by immunosuppressive drugs. Fungi spores, especially those of *Aspergillus fumigatus*, are ubiquitous in the environment and have been found in pasture lands, hay stacks, manure piles, and the basements of most homes.

Most microorganisms found in water are harmless, in fact, many are even beneficial. Some of the helpful bacteria consume organic detritus and are

Table 5 Pathogenic Fungi

Species	
Actinomyces spp	Actinomycosis
Aspergillus spp	Aspergillosis, Aspergillus pneumonia otomycosis
Candida albicans	Moniliasis, candidiasis oral thrush

incorporated into some wastewater-treatment plants for that reason. Others decrease foul tastes and odors of some influent waters, whereas still others kill or destroy the growth of some pathogens by the products they produce. Some bacteria, although not harmful, are a nuisance. The actinomycetes impart musky odors and taste to water. Others feed on sulfur-containing materials and produce hydrogen sulfide, the familiar rotten egg smell, whereas some produce acids that can corrode piping.

Generally, pathogenic bacteria are best suited to their environment inside a warm-blooded host animal and usually do not multiply and are readily killed by disinfecting chemicals. There are some exceptions, however, and total pathogen die-off should never be taken for granted. Spore-producing pathogens can resist chemical disinfection, and although they are not direct human-waterborne microorganisms, they should be noted.

Bacillus anthracis: The anthrax pathogen is picked up by animals and transmitted to humans.

Clostridium tetani: These spores introduced to bathers via deep wounds could conceivably cause tetanus (lockjaw).

Clostridium botulinum: Produces botulism toxin (the most powerful poison known). These spores will not affect human beings directly but uncooked foods sealed in spore-laden water can be deadly.

Some of the waterborne bacterial diseases are

Cholera: The most serious waterborne disease with potentially fatal results can be spread via polluted water. The organism *Vibrio cholerae* can persist for weeks in very turbid waters. Turbidity also protects it from some disinfectants.

Salmonellosis: There are several hundred species of the genus *Salmonella* known to attack humans. Their effects range in severity from typhoid fever to the common acute intestinal upsets (food, ptomaine, poisoning). The source is direct or indirect fecal contamination from practically any warm-blooded animal.

Shigellosis: This is the most common waterborne cause of acute diarrhea in the United States. There are many genus types, with *Shigella dysenteriae* being the most serious cause of dysentery. The malady known as "Montezuma's revenge" or "turista" is caused by regional variants of *Escherichia coli*. It is harmless to natives but often affects visitors.

Tuberculosis: This is a lung disease commonly thought to be spread through the air, but it also can be transmitted via swimming in or drinking contaminated water.

In addition to viruses, fungi, and bacteria, other organisms found in water are molds, algae, and single-celled animals (protozoa).

Algae are small often single-celled organisms which manufacture their own food. Although no waterborne diseases are attributed to algae, they can cause problems. Several species impart musty or fishy tastes to water. Others interfere with water-treatment plants by producing slime or by periods of rapid growth (blooms) which can clog filters.

Like fungi, molds can also impart tastes and colors to water as well as cause turbidity and clog filters.

Although viruses are very small and are present in water in far fewer numbers than bacteria, they are more virulent and more resistant to disinfectants. Both polio myelitis and hepatitis viruses have been specifically traced to a water source in causing disease.

Most of the protozoa found in water are beneficial. They consume large quantities of organic waste materials and they keep both algae and bacteria in the bounds of a balanced ecology. There are two major pathogenic protozoa, however, which are worth noting. *Entomaela histolytica* causes amebic dysentery via contaminated drinking water. It is highly resistant to disinfection in the cyst stage but can be removed through fine filtration. *Giardia lamblia*, another chlorine-resistant protozoan, causes severe diarrhea but can also be removed by filtration.

CHLORINATION

Chlorine was first discovered by the reaction of manganese dioxide and hydrochloric acid over 200 years ago by the Swedish chemist Scheel. In 1810, Sir Humphrey Davy recognized that chlorine was an element and named it from the Greek word *chloros* meaning green.

In North America, the first continuous municipal application of chlorine to water was used to disinfect the Boonton reservoir supply of the Jersey City, NJ, water utility in 1908. The results of this disinfection were quite dramatic. Not only did the bacterial count significantly decline but the taste and color of water improved. As the practice of chlorination became more widespread, the incidence of waterborne diseases declined rapidly. By 1930, typhoid had dropped from among the tenth leading cause of death to the twenty-sixth. The number of waterborne diseases declined steadily as chlorination of potable water rose.

Chlorine is the disinfecting agent used extensively to treat water for municipal and individual supplies. Its popularity is owed to the fact that it meets most of the criteria outlined for an ideal disinfectant earlier. In municipal supply systems, chlorine is used basically in two forms; a gaseous element or as a solid or liquid chlorine containing hypochlorite compound.

In most modern water-treatment plants, chlorine is supplied as a liquified gas in cylinders or drums and is injected into the water supply through a chlorinator. The chlorinator reduces the pressure of the gas leaving the cylinders, controls the rate of flow, mixes the gas with the water, and delivers it to the pump or injector which forces it into the filtered water.

In its gaseous state, however, chlorine is not safe to handle and the equipment needed to deal with it are too expensive for treating individual water supply systems. For small water supply systems of less than 5,000 persons or in large systems where safety concerns related to handling the gaseous form outweigh economic concerns, hypochlorite forms have been used.

Calcium hypochlorite, a dry bleach contains at least 70% available chlorine and has from 3 to 5% lime. It can be obtained as a powder or in tablet form which will not deteriorate if properly handled. Although used widely for sterilizing small water supplies, its high cost and corrosive properties make it unpopular in larger systems.

Sodium hypochlorite contains 12–15% available chlorine. It is unstable to some extent and will deteriorate more rapidly than calcium hypochlorite. Although the relative safety of sodium hypochlorite is well known, the large quantities that must be used present a problem. A water-treatment plant using 6 tons of chlorine per day requires a storage of 83 tons of hypochlorite per day because of the maximum concentrations of 15%. This requires transportation and storage costs of 30,000 tons per year. Such costs can become substantial given the fact that liquid sodium hypochlorite has a limited shelf life.

The stability of hypochlorite solutions is affected by heat, light, and pH and in the presence of heavy metals. Solutions will deteriorate at various rates depending on

- The higher the concentration, the faster the deterioration.
- The higher the temperature, the faster the deterioration.
- Presence of iron, copper, nickel and cobalt catalyzes the rate of deterioration of hypochlorite.

Regardless of such drawbacks, systems which can produce hypochlorite electrolytically from sodium chloride have been used. Raw materials required are salt (brine solution or seawater), power, and water and the process is achieved by an electrolytic cell. The cell converts the chloride ion to hypochlorite ion as follows:

$$NaCl + H_2O^e \rightarrow HOCl + H_2\uparrow$$

CHLORINE REACTIONS

When chlorine is dissolved in water between 49 and 212°F, it reacts to form hypochlorous and hydrochloric acids.

$$Cl_2 + H_2O \rightleftharpoons HCl + HOCl$$

The hypochlorous acid ionizes or dissociates almost instantaneously into hydrogen and hypochlorite ions:

$$HOCl \rightleftharpoons H^+ + OCl^-$$

The hypochlorite forms also ionize in water and yield hypochlorite ion in equilibrium with hydrogen ions:

$$CaOCl_2 + 2H_2O \rightarrow 2HOCl + Ca(OH)_2$$
$$NaOCl + H_2O \rightarrow HOCl + NaOH$$

HOCl is a most effective germicide of all the chlorine residual fractions. It is known in the industry as the free available chlorine residual. Germicidal efficiency of HOCl is due to the relative ease with which it can penetrate cell walls. This penetration is comparable to that of water and can be attributed to its low molecular weight and its electrical neutrality.

All things being equal, the germicidal efficiency of free available chlorine residual is dependent on the pH of the solution which establishes the amount of dissociation of HOCl to H^+ and OCl^- ions. Lower pH solutions as with lower temperatures suppresses the dissociation of HOCl. The disinfecting efficiency of free available chlorine residual decreases significantly as the pH rises. At a pH above 9, there is little disinfecting power. At this pH level and at 20°C, 96% of the free available chlorine will consist of the OCl^- ion, which is an indication of the low germicidal efficiency of this ion.

CHLORINE AND NITROGEN COMPOUNDS

If water to be treated did not contain nitrogenous compounds, chlorination would be a much simpler process. But since inorganic and organic nitrogen appears in most natural waters and reacts quickly with chlorine, it complicates the procedure. The reaction of chlorine with any compound containing nitrogen atom with one or more hydrogen atoms attached will form a relatively weak germicide known as chloramine.

The reactions are as follows:

$$NH_4 + HClO = HN_2Cl + H + H_2O \text{ monochloramine}$$
$$NH_2Cl + HClO = NHCl_2 + H_2O \text{ dichloramine}$$
$$NHCl + HClO = NCl_3 + H_2O \text{ trichloramine}$$

The relationship between the amounts of the three types of chloramine depends on pH and the amount of NH_4 in the water. As trichloramine can only be formed at very low pH values, the other two prevail in water treatment, dichloramine being much the more powerful bactericide. The total quantity of dichloramine and monochloramine is referred to as combined available chlorine.

BREAKPOINT CHLORINATION

Breakpoint chlorination applies to water already chlorinated through a water treatment and simply means that the chlorine dose is added initially to oxidize any reducing compounds present, then increased to form chloramines with any ammonia present, increased still further to destroy the chloramines, and increased finally to build up the highly bactericidal free chlorine residual.

As it passes through the third stage (destruction of the chloramines), there is a noticeable dip in the chlorine residual present, which is known as the breakpoint. When this point is passed, the free residual rises again, more or less with the chlorine being added and a low free residual chlorine will rapidly disinfect any clean water. Chlorine demand of any given water is the amount of chlorine required to take the reaction through the breakpoint. All waters need a minimum of about 0.2 mg/L of free residual chlorine to effect sterilization (about twice as much at high pH values), and to obtain this it is unlikely that less than 0.5 mg/L of chlorine gas by weight would have to be added.

The practice of adding ammonia to chlorinated water to achieve a NH_4 to Cl ratio is used. This ammonia-chlorine process has the merit of being persistent and maintaining a toxic dose far into the distribution system without creating taste problems to the same extent as does chlorine alone. However, it is less powerful than chlorine and acts more slowly. At doses of 0.3 mg/L of chlorine and 0.1 mg/L of ammonia, contact time of about 20 min is essential.

The ammonia-chlorine process is accepted by the EPA as a primary disinfectant provided contact time is adequate and that there is proof of disinfection. This is largely due to the fact that the addition of ammonia before chlorine prevents the formation of trihalomethanes (THMs), which are cancer-causing compounds. There are some disadvantages to this process, with one being the addition of ammonia to water does provide nutrients sufficient to produce algal blooms in reservoirs and an increase in distribution system for the bacteria population.

Breakpoint chlorination is a chemical treatment for ammonium ion removal. In the process, chlorine is added to a wastewater containing ammonium ion in mixing tank, where practically all the ammonium ions are oxidized to nitrogen gas. The amount of chlorine addition is precisely adjusted to a level (the breakpoint) which is sufficient for the oxidation and results in minimal residual chlorine and by-product formation. Hydrochloric acid is coproduced during the oxidation and must be neutralized by adding lime or caustic soda.

Equipment needs are relatively simple, but control requirements for chlorine dosage and pH adjustment are sophisticated and important.

A downstream dechlorination step for the removal of residual chlorine is usually adopted. This can be a SO_2 addition. Sodium hypochlorite (NaOCl) may be used for the oxidation instead of chlorine with no HCl coproduction. In this case, no lime or caustic soda addition is needed. It is economically attractive for wastewater with low ammonium ion concentrations (<5 mg/L) and can be employed as a polishing step following other ammonium ion removal processes. It is especially attractive in a cold weather location. The process is rated low in capital costs but high in operating costs, especially at ammonium ion concentrations above 16 mg/L. The potential for formation of chlorinated hydrocarbons in the effluent is present.

The process can reduce ammonium ion concentration to 0.1 mg/L or less and convert to nitrogen gas and to insignificant amounts of by-products (nitrate at 0.2–0.45 mg/L and NC13 0–0.25 mg/L) under normal operation. Performance is not affected by temperature fluctuation or toxic compounds. However, pH and chlorine dosage have significant effects on by-product formations.

DECHLORINATION

Dechlorination is the practice of removing all or a specified fraction of the total combined chlorine residual. In potable water practice, dechlorination is used to reduce the residual to a level at a point where the water enters the distribution system. In some cases where taste and odor control is a problem, control is achieved by complete dechlorination followed by rechlorination. This removes the taste-producing residuals and prevents the formation of NCl_3 in the distribution system. Dechlorination of wastewater and power plant cooling water is required for elimination of chlorine residual toxicity which is harmful to the aquatic life in the receiving waters.

The most practical method of dechlorination is by sulfur dioxide and/or aqueous solutions of sulfite compounds. Other methods used are granular activated carbon and hydrogen peroxide.

Both free chlorine and chloramine residuals are toxic to fish and other aquatic organisms. Dechlorination involves the addition of sulfur dioxide to the wastewater, whereby the following reactions occur:

$$SO_2 + HOCl + H_2O = SO_4^{2-} + Cl^- + 3H^+ \text{ (for free chlorine)}$$

$$SO_2 + NH_2Cl + 2H_2O = SO_4^{2+} + Cl^- + 2H^+ + ND_4^+$$

(for combined chlorine)

As can be seen, small amounts of sulfuric and hydrochloric acids are formed; however, they are generally neutralized by the buffering capacity of the wastewater. Dechlorination can also be used in conjunction with superchlorination. Since superchlorination involves the addition of excess chlorine, dechlorination is required to eliminate this residual. Sulfur dioxide is the most common chemical used. It is fed as a gas using the same equipment as chlorine systems. Because the reaction of sulfur dioxide with free or combined chlorine is practically instantaneous, the design of contact systems are less critical than that of chlorine contact systems. Detention of less than 5 min is quite adequate, and in-line feed arrangements may also be acceptable under certain conditions.

Metabisulfite, bisulfite, or sulfite salts can be used. Automatic or manually fed systems can also be used. If chlorine is used at the site, sulfur dioxide is preferred, since identical equipment can be used for the addition of both chemicals. Alternative dechlorination systems include activated carbon, hydrogen peroxide (H_2O_2), and ponds (sunlight and aeration). The technology of dechlorination with sulfur dioxide is established but is not in widespread use. It can be used whenever a chlorine residual is undesirable. This usually occurs when the receiving water contains aquatic life sensitive to free chlorine. Dechlorination is generally required when superchlorination is practiced or stringent effluent chlorine residuals are dictated. It will not destroy chlorinated hydrocarbons already formed in the wastewater. It has been reported that about 1% of the chlorine ends up in a variety of stable organic compounds when municipal wastes are chlorinated. Available chlorine residuals can be reduced to essentially zero by sulfur dioxide dechlorination.

Process requirements include

Contact time: 1–5 min
Sulfur dioxide feed rate: 1.1 lb per pound of residual chlorine
Sodium sulfite feed rate: 0.57 lb per pound of chlorine
Sodium bisulfite feed rate: 0.68 lb per pound of chlorine
Sodium thiosulfate feed rate: 1.43 lb per pound of chlorine.

Sulfur dioxide addition for dechlorination purposes is reasonably reliable from a mechanical standpoint. The greatest problems are experienced with analytical control, which may lower the process reliability. This process requires very little use of land, and no residuals are generated. It is used to eliminate the environmental impact of chlorine residuals. Overdosing can result in low pH and low DO effluents.

ADVANTAGES AND DISADVANTAGES OF CHLORINATION

Chlorine is the most popular disinfectant for wastewater and potable water treatment in the United States today. The chlorine residual that remains after water has left the disinfection unit provides continuing antibacterial protection. The amount of this residual can be readily determined with simple inexpensive tests. Chlorine is a highly effective disinfecting agent that is readily available at a reasonable price.

There are, however, some drawbacks to chlorination. THMS and possibly other carcinogenic chlorinated organics may be produced when certain organic materials combine with chlorine. Turbidity in the water can also reduce the effectiveness of chlorine. Additionally, variations in water quality may affect the degree of bacteriological protection of the treated water. Because of these variations in effectiveness, doses often exceed necessary levels, producing high residuals which may impart a noticeable taste and odor to the water. Chlorine can be a potentially dangerous gas if not handled properly at the treatment site.

OTHER DISINFECTANTS

The objective of disinfection is to destroy microorganisms found in water. Some microorganisms only impart bad tastes and odors to water, whereas others can cause diseases, many of which can be fatal. The factors which influence the efficiency of a disinfectant are turbidity, organics, pH, and temperature. A disinfectant should therefore be used with these water characteristics in mind.

Chlorine whether in gaseous form or as a hypochlorite is the most popular choice for a disinfectant in the United States. This is primarily due to the fact that it meets the criteria outlined for an ideal disinfectant. Chlorine residual that remains after water has left the disinfectant unit provides continuing antibacterial protection and this residual can be easily determined. Chlorine is not only an effective germicide but it is inexpensive and readily obtained. One of the drawbacks of chlorine is the formation of THMs. As a result the practice of adding ammonia prior to chlorination is becoming popular.

Other disinfectants that have been used with varying success are

- Iodine: This agent is used in rural areas or undeveloped countries or for emergency treatment for small supplies.
- Ultraviolet light: This is used in special-purpose supplies to sterilize potable water where the level of bacterial concentration is highly sensitive.
- Ozone: Although gradually becoming more popular in the United States, it is basically a European disinfectant method. Some of its advantages are
  - Eliminates tastes and colors in water
  - Strongest germicide available for use
  - Purifies naturally with no chemical residuals
  - Effective over a wide range of pH

Iodine

Compared with free chlorine residuals iodine is inferior. Free chlorine is 5 times more cysticidal than HIO, 200 times more viricidal, and 2 times more cysticidal than I_2. Owing to side reactions, 45 mg/L iodine is needed to achieve the same level of disinfection as 8 mg/L chlorine. Iodination, however, can be used where chlorination or more complete treatment of water is either impossible or impractical. It is thus limited to rural areas or undeveloped countries where little or no adequate supervision and expertise are available in handling chlorine. The most popular application is in tablet form or the tincture of iodine solution for emergency treatment of small supplies, such as water supplies for troops on bivouac or travelers in foreign countries where "traveler's diarrhea" is an occurrence. This type of emergency treatment has proven to be both reliable and effective, probably because of the characteristics of this agent, which include

- Bactericidal capacity is not greatly influenced by pH except at very low temperatures.
- Ammonia and organic nitrogenous impurities have little effect on disinfection efficiency.
- Action depends less on contact time and temperature than does chlorine.
- It is highly effective against pathogenic organisms, including spores and viruses in short contact times.

One of the major disadvantages of iodine is its cost (20 times more costly than chlorine) and its lack of availability. This with the uncertainty of the long-term effects of iodine on humans will keep iodine from ever becoming a popular water disinfection technique.

Bromine

All bromine species used in water and wastewater treatment revert to bromides after being consumed in the oxidation process. Although this in itself is not an issue, when a potable water treatment plant chlorinates water containing bromides, they are oxidized to hypobromous acid and bromamines if ammonia is present. These compounds react with natural precursors in the water to form yet another series of THMs which are considered to be carcinogenic. The use of bromine as an alternative to chlorine is not considered practical.

Ozone

Ozone is a form of oxygen having three atoms per molecule rather than the two atoms typical of atmospheric oxygen. Ozone has greater germicidal effectiveness against bacteria and viruses than chlorine. It also reduces iron, manganese, lead, and sulfur concentrations in water and eliminates most tastes and odors. Furthermore, potency is not affected by pH, temperature, or ammonia content.

Although there is some worldwide use of ozone for water disinfection, it is not common in the United States even though use has been increasing. This is principally due to the US desire to maintain residual germicidal power in the distribution system and to avoid the higher equipment and operating costs of ozonation systems.

Because ozone (O_3) is an unstable molecule, it quickly reverts to normal oxygen (O_2). Therefore, ozonation must occur at the point of use. The most practical way of generating ozone is to pass oxygen through a corona discharge, which is produced by applying high voltage across two electrodes with a dielectric and an air gap in between.

Ozone (O_3) may be used for the final disinfection step in a wastewater-treatment process. As a disinfectant, dosages of 3–10 mg/L are common. Ozone is an effective agent for deactivating common forms of bacteria, bacterial spores, and vegetative microorganisms found in wastewater, as well as eliminating harmful viruses. Additionally, ozone acts to chemically oxidize materials found in the wastewater and can reduce the BOD_5 and COD, forming oxygenated organic intermediates and endproducts. Further ozone treatment reduces wastewater color and odor.

Ozone breaks down to elemental oxygen in a relatively short period of time (half-life about 20 min). Consequently, it is generated on site using either air or oxygen as the raw material, The ozone-generation process utilizes a silent electric arc or corona through which air or oxygen passes yielding a certain percentage of ozone. Automatic devices are commonly applied to control voltage treatment, frequency, gas flow, and moisture, all of which influence the ozone generation rate. Ozone injection into the wastewater flow may be accomplished by mechanical mixing devices, countercurrent or cocurrent flow columns, porous diffusers, or jet injectors. Ozone acts quickly and consequently requires a relatively short contact time. Ozonation has been used as a tertiary treatment process to reduce BOD_5 and COD.

Ozone use is fully demonstrated but not widely used in the United States because of relatively high cost of ozone. Recent developments in ozone generation have lowered the cost and thus make it more competitive with other disinfection methods. Applicable in cases where chlorine disinfection may produce potentially harmful chlorinated organic compounds. If oxygen-activated sludge is employed in the system, ozone disinfection is economically attractive, since a source of pure oxygen is available facilitating ozone production. Ozone disinfection does not form residuals that will persist and can be easily measured to assure adequate dosage. Ozonation may not be economically competitive with chlorination under nonrestrictive local conditions. Effluents containing high levels of suspended solids may require filtration to make ozone disinfection more cost effective.

Easily oxidizable wastewater organic materials consume ozone at a faster rate than disinfection; therefore, effectiveness of disinfection is inversely corrected with effluent quality but directly proportional to ozone dosage. When sufficient ozone is introduced, ozone is a more complete disinfectant than chlorine. Air or pure oxygen may be used as the raw material for the ozone generation.

Mechanically ozone generation is highly reliable and highly reliable in deactivating microorganisms. Ozone has been found to be a good oxidant for removal of cyanide, phenol, and other dissolved toxic organic materials. Combination of ozonation and activated carbon treatment can achieve 95% chloroform and other trihalomethanes removals. Ozone is an air pollutant which can discolor or kill vegetation coming in contact with it. Residual ozone is off-gas streams must be processed for ozone decomposition prior to release. Ozone is toxic when inhaled in sufficient concentrations.

Ultraviolet Light

When UV treatment is used, the water is usually passed under the lights in a very shallow layer so that all of it is exposed to the UV rays. This requires an elaborate installation to treat reasonably large volumes of water. It is not in general use in public water supplies and is limited mostly to small, special-purpose supplies.

It has been found that for most bacteria, fungi, and some viruses, the optimum killing wavelength of UV light is at 2,650 Å. The germicidal action of UV light results from its effects on nucleic acids. Figure 1 shows the bactericidal curve for UV light at various wavelengths. The relative 100 bactericidal effectiveness of 2,650 Å and the 85 relative effectiveness of 2,537 Å (the energy produced by germicidal lamps) shown in Figure 6 should not be confused with the possibility that a 100% absolute kill with UV lamps is attainable.

A rapid drop in bactericidal action from wavelengths less than 2,500 Å is also characteristic of the absorption of UV light by nucleic protein and is representative of practical germicidal effects in which all the liquid and gaseous elements in the environment of an organism absorb the UV of shorter wavelengths and protect the organism. Studies suggest that for practical purposes, the killing action of UV light ends above 3,600 Å.

Mercury vapor sources of ultraviolet light for practical and experimental uses consist of

- Commercially available low-pressure (0.004–0.02 mm Hg) mercury lamps, (germicidal lamps)
- High-pressure (400–60,000 mm Hg) photochemical, therapeutic, and filtered sunlamps
- Special experimental lamps of limited availability

Low-pressure mercury arcs are 5–10 times more efficient at inactivating pathogens than high-pressure arcs in envelopes of the same transmission. Consequently, most UV sources for germicidal action are low-pressure mercury vapor lamps. High-pressure quartz-mercury arcs, however, may be of practical use in situations where high UV intensity is required and the exposure time is

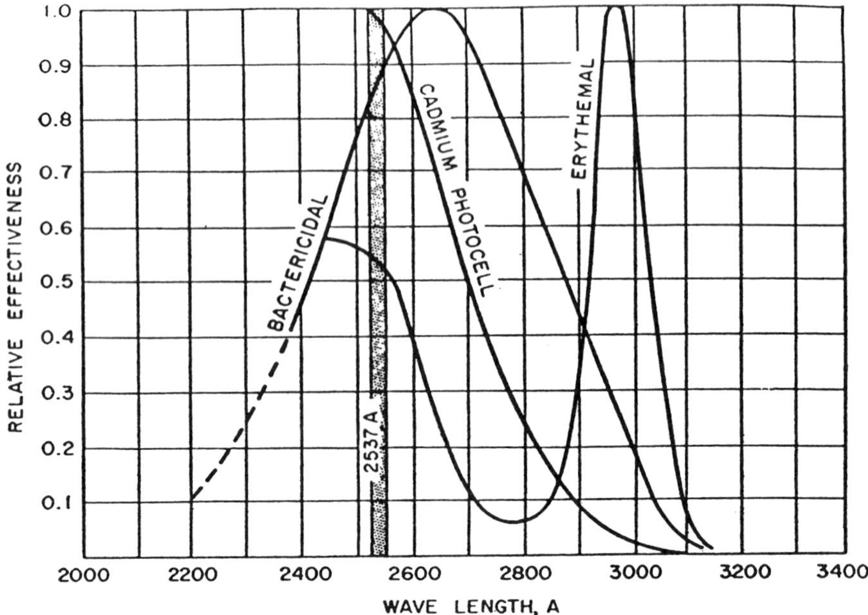

Figure 1 Bactericidal and erythematous action curves.

limited. Approximately 95% of the UV energy emitted by germicidal lamps is at 2,537 Å.

The bacterial kill from UV exposure can be experienced as a product of intensity, I, and time, T, in the following relationship:

$$K = IT$$

where K is the death rate of the microorganisms. This relationship stipulates that a given kill rate can be achieved by increasing the intensity and reducing the time or vice versa. However, under practical wastewater-treatment conditions, time is generally dictated by the flow rates of the wastewater stream.

Important factors in the effectiveness of UV disinfection include

- The dose, which is affected by the intensity of the lamps, the residence time (in seconds), and the clarity of the wastewater
- The presence of interfering constituents in wastewater, especially iron, manganese, and hardness (calcium and magnesium)
- Short circuiting in the contact chamber, which must be minimized
- The water in the contact chamber, which needs to be maintained at constant levels to ensure sufficient exposure to radiation
- Various monitoring devices to ensure that effective disinfection occurs

In wastewater-treatment facilities where chlorination-dechlorination is required, UV systems are much cheaper to install and their operating costs are comparable.

Although UV is an energy-intensive system, some steps can be taken to control the operating costs. Since the effectiveness of the UV system and the dosage required for disinfection are controlled by the amount of suspended solids in the effluent, a well-run treatment process and filtration prior to UV disinfection will ensure optimum energy efficiency.

As disinfection and water quality requirements become more stringent along with the need to protect receiving waters and their inhabitants, research into alternative forms of disinfection becomes increasingly important. UV disinfection is an effective method for killing bacteria found in wastewater. Because it does not provide residuals, its application is limited to wastewater treatment and some drinking water treatment derived from ground water. Also, care must be taken to ensure that cysts are not present. UV systems are relatively inexpensive to install, safe and easy to operate, and are comparable in operating costs to chlorination-dechlorination systems.

THERMAL TREATMENT METHODS

Thermal treatment methods are used successfully in a wide variety of operations where volume reduction, sterilization, detoxification, or complete destruction of wastes are required (mainly sludge). The decision to use thermal treatment in the case of waste treatment is based on cost, environmental regulations, exposure to liability, and energy recovery opportunities. Thermal treatment costs can be reduced by using mechanical methods such as filtration for separating liquids from solids prior to thermal treatment.

Thermal treatment methods can be generally separated into two main categories—low-temperature systems and high-temperature systems. Many low-temperature treatments such as spray dryers, rotary dryers, vacuum dryers, belt dryers, and sterilizers are modified processes used for the production of commercial products. Low-temperature thermal systems operate at less than 1,000°F. Systems operating at greater than 1000°F such as incinerators are considered high-temperature treatment. High-temperature systems include rotary kilns, liquid injection incinerators, multiple hearth furnaces, fluidized bed combustion, and pyrolysis systems.

Environmental regulations, public policy issues, and community activism often determine which, if any, thermal treatment system will be used. Reasons for opposition include concern for spills, releases, odors, health impacts, site selection, and distrust of owners and government.

Several thermal treatment processes for disinfection, wastewater sludge conditioning, destruction of hazardous wastes, and sterilization are available.

Thermal treatment can be as simple as sterilizing water by boiling or a complex arrangement of equipment and processes required to operate a modern incineration facility. As advancing technology allows for more efficient and environmentally acceptable processes, thermal treatment methods will become more prevalent in application.

The most important goals of thermal treatment are detoxification, volume reduction, and permanent disposal. Issues involved with thermal treatment are high capital and operating costs, long-term liability, waste characterization, and increased public environmental awareness.

LOW-TEMPERATURE PROCESSES

Low-temperature processes are routinely used for sterilization and drying operations. Sterilization and drying operations as applied to water, wastewater, and sludge treatment and focus on the destruction of pathogenic and other disease-causing microorganisms and volume reduction. A brief description of low-temperature processes follows. A summary of low-temperature thermal processes is presented in Table 6.

Sterilization

Simple sterilization of water can be accomplished by boiling water for 10–20 min. Batch sterilization of water is accomplished by holding the heating solution at 115 and 121°C for 30 and 15 min, respectively. The purpose of sterilization is to kill all microorganisms. Autoclaves combine steam and pressure for sterilization when temperatures above 110°C are required.

Pasteurization

Pasteurization is a thermal process that kills all pathogenic bacteria. Pasteurization is accomplished when the material being disinfected (whether milk or sludge) is held at 157–161°F for 30 min. Sludge can be pasteurized prior to land application to kill pathogens.

Evaporators

Thermal evaporators can be used to concentrate salts and other solids from solutions. The volume of solutions containing high salt concentrations can be reduced in tube evaporators or submerged combustion evaporators. In the tube evaporator, steam is used to produce heat. In the submerged combustion type evaporator, the combustion gases from a natural gas or fuel oil burner are discharged into the solution. Salt concentrations build up in the evaporator and have to be drawn off as sludge. Solutions with high dissolved organics are not suitable for evaporation processes. Reject streams from reverse osmosis processes

Table 6 Low-Temperature Thermal Processes

Process	Purpose/use	Operating conditions	Characteristics
Sterilization	Canning/medical	115°C at 15 min 120°C at 30 min	Kills *all* microorganisms
Pasteurization	Sludge conditioning; food processing	161°F at 30 min	Kills pathogens; aids dewatering; produces odor and storage problems
Autoclave	Sterilization for medical/food industry	110°C at high pressure	Higher temperatures acheivable; moist heat penatration of cells
Evaporators	Volume reduction of inorganic solutions	110°C at atmospheric pressure	Concentrates disolved salts; produces sludge; no organics
Spray dryers	Liquids and slurries that cannot be dewatered mechanically	300–500°F; 100–1,000 psig at nozzle	Continuous feed; feed must be pumpable; product is powdery, sphereical; used with heat-sensitive materials; low detension
Drum dryers	Liquids, slurries, sludges; continuous operation	50–100 psi steam	Capacity depends on drum speed and surface area; odors; used with hood; high maintenance
Rotary dryers	Liquids, slurries, sludges; continuous operation	200–300°F direct heat application	Feed into rotating cylinder through which hot gas flows in opposite direction; high cost
Belt dryer	Pastes, sludges, large solids, special shapes	200–300°F direct heat application	Includes conveyor/tunnel dryer; drying rate controlled by belt speed
Vacuum dryer (rotary)	Slurries and pastes, free-flowing material; batch operation	200–300°F low heat under vacuum	Suitable for nonsticking, heat-sensitive materials; solvent recovery
Tray dryer	Free-flowing, nonsticking materials; batch operation	Less than 500°F	Material supported in shallow, stacked trays; can be used with vacuum

which are high in salt concentration are good candidates for volume reduction by evaporation.

Dryers

Wet solids and slurries are frequently dried in a variety of low-temperature dryer processes. These processes can be used as product recovery systems or waste treatments. In the spray drier, a slurry is sprayed under pressure through a nozzle into a chamber where it is contacted with hot air. Dry granules form as the water is evaporated from the slurry.

There are several types of drum driers where slurries are pumped into a shallow tray through which a drum with a hot surface temperature rotates. The liquid part of the slurry evaporates on the surface of the hot drum, leaving the solids on the drum. The solids are then scraped off for disposal.

A rotary drier operates similarly to a kiln except the temperature is lower. Slurries are fed into the drier countercurrent to the direction of hot gases that enter the opposite end. The solids are dried as they tumble through the rotary drier and are removed into a hopper at the opposite end.

Slurries such as sludge can also be dried on belt driers. Sludge is applied to an endless belt that pass through a heated chamber or tunnel.

Composting

Composting is the biodegradation of organics in solid wastes and wastewater sludges. The three principal methods of composting are the window method, static pile, and the in-vessel method. This can be used for sludge and other waste solids.

Windrows are 12-ft long by 6-ft high piles of compostable material which are turned over regularly to aerate the pile.

Aerated static piles combine compostable material and bulk materials such as wood chips. The static piles are placed over perforated pipes attached to blowers that pull fresh air into the pile to sustain the biological decomposition of the organic material.

In-vessel composting accelerates decomposition by controlling the flow of air and moisture. The systems resemble drums which can be rotated by computer control.

Temperatures can reach 158°F in the compost pile, effectively destroying most pathogens. As the organic material is consumed by the biological reaction, the pile will begin to cool. The entire process takes from 14 to 21 days.

HIGH-TEMPERATURE PROCESSES

High-temperature treatment processes are generally combustion processes that involve municipal wastewater solids, hazardous wastes, or trash. In order to make these processes efficient and environmentally acceptable, more upstream preparation such as solids dewatering and segregation of noncombustibles is necessary. The criteria for selecting high-temperature processes are

- Limited land available for other disposal methods
- Landfill expansion prohibited or restricted
- Destruction of toxic materials is required
- Energy recovery is possible

High-temperature processes provide the maximum volume and weight reductions and detoxification of pathogens and toxic chemicals.

Table 7 lists a summary of commonly used high-temperature thermal treatments. The five predominant thermal treatments are combustion processes which usually require supplemental fuel to sustain the process. The following systems will be discussed:

- Rotary kiln
- Fluidized bed
- Liquid injection
- Multiple hearth
- Fixed hearth

Lesser used and emerging technologies are listed in Table 8. Many of the so-called emerging technologies have actually been under development for several years, including the molten salt and high-temperature fluid wall processes shown in Figure 3.

Many of these newer technologies, so far, have only limited practical application. Their development is generally toward destroying the most toxic and most difficult to treat wastes. High capital costs and high operational costs restrict their general application.

Thermal treatment processes can handle a wide variety of wastes and contaminants. The thermal treatments shown in Tables 7 and 8 are for the most part proven technologies. However, each category, either high or low temperature, has advantages and disadvantages. Issues to be considered when selecting a thermal treatment method are

- Environmental acceptability
- Costs—capital and operational
- Liabilities
- Energy and resource recovery

Issues have been raised by the public concerning the environmental acceptability of thermal treatments. Questions about risks and potential hazards

Table 7 General High-Temperature Processes

Type	Process principle	Application	Combustion temperature, °F (°C)	Residence time	Excess air, % stoichiometric
Rotary kiln	Waste is burned in a rotating refractory cylinder.	Any combustible solid, liquid, or gas	1200–2500 (650–1370)	Seconds for gases; hours for liquids and solids	50–250
Single chamber/liquid injection	Wastes are atomized with high-pressure air or steam and burned in suspension.	Liquids and slurries that can be pumped	1300–3000 (700–1650)	0.1–1.0 s	120–250
Multiple hearth	Wastes descend through several grates to be burned in increasingly hotter combustion zones.	Sludges and granulated solid wastes	1400–1800 (760–980)	Up to several hours	200–400
Fluidized bed	Waste is injected into an agitated bed of heated inert particles. Heat is efficiently transferred to the wastes during combustion.	Organic liquids, gases and granular or well-processed solids	1400–2000 (760–1100)	Seconds for gases and liquids; minutes for solids	100–150
Fixed hearth	Waste is ram fed into primary starved air chamber. (pyrolysis)	Smaller onsite hazardous waste incineration	Primary 1200–1800	0.5–1.0 hr	50–80
	Vaporized organics are burned in secondary chamber.		Secondary 1400–2000	0.25–2.5 s	100–200

Table 8 Lesser Used and Developing Thermal Processes

Process	Characteristics
Cement kilns	Used for hard to burn wastes that require good mixing, very high temperatures, and long residence times.
Pyrolysis	Pyrolysis is air starved combustion that avoids vaporizing inorganic compounds such as heavy metals.
Molten salt	Molten salts such as sodium carbonate at 1500–2000°F act as catalysts to promote heat transfrer. Air Pollution control equipment is minimum. High ash wastes and corrosives can be problems. The technology is not fully developed.
Wet air oxidation	Used for treatment of aqueous wastes with less than 5% organics which are too dilute for incineration but too toxic for biological treatment. The process operates at 400–650°F and 1200–1800 psig. No air emissions is one advantage. The process cannot treat chlorinated organics. Treated effluent may be dilute enough for biological treatment.
Super critical water	Suitable for chlorinated aqueous wastes too dilute for incineration. The temperature and pressure of water is elevated above its critical point (347°F and 3206 psi) where the water has exceptional solvent properties. Oxygen can be injected to enhance organics decomposition to carbon dioxide and water. there are no air emissions.
Infrared incinerator	Infrared radiation volatilizes organic material which is then thermally destroyed in a secondary combustion chamber. The process has relatively low fuel requirements and is designed to treat solids. Process has not been fully demonstrated.
High-temperature fluid wall	Waste is exposed to 4000°F as it passes through an electrically heated cylinder. Heating is by radiation. The cylinder wall is protected by an envelope of inert gas such as nitrogen.
Plasma-arc vitrification	Electric conductance by a plasma torch is used to heat solid and liquid organic wastes to high temperatures. Liquids are vaporized while metals and solids remain in vitrified residue.

associated with air emissions from thermal treatments are routinely asked. Odors and potentially harmful emissions from thermal treatments are possible and need to be addressed.

High capital and operating costs are a distinct disadvantage of thermal treatment processes. Construction of these processes usually invites considerable public and political attention. Strict environmental emissions and treatment standards add significantly to the costs. Thermal waste-disposal costs average from 5¢ to $3 per pound, depending on the BTU, ash, and hazardous materials content of the waste.

One advantage of thermal treatments, particularly high-temperature processes, is that long-term liabilities are reduced by destroying wastes that otherwise may have been stored indefinitely or landfilled or disposed of improperly. Improperly designed and poorly operated thermal processes can create unwanted liability for facility owners if risks to the public health and safety are perceived.

Energy and waste heat recovery is the politically correct and preferred design option for any thermal treatment system. The sale of steam and electricity to off-site users must be able to offset the additional operations and capital equipment costs associated with energy recovery. The cost versus benefits analysis must be part of the initial facility design consideration.

Major issues frequently raised by communities include

- Hazardous materials spills
- Environmental and health impacts
- Poor selection of site
- Distrust of incinerator owners and operators
- Inability of government agencies to enforce compliance

INCINERATION TECHNOLOGY

The five most common incinerator designs (Table 7) are liquid injection, rotary kilns, fixed hearth, multiple hearth, and fluidized bed. Selecting incineration systems is a complex process which must consider many factors. In addition, proper incinerator designs must incorporate four major subsystems. These systems are

- Waste preparation and feed
- Combustion chambers
- Air pollution control
- Residue and ash handling

Rotary kilns (Fig. 2) can heat treat solid wastes, slurries, liquid waste, and containerized waste. A rotary kiln is a horizontal rotating cylinder. The kiln rotation promotes mixing and high residence times. Volatilization of the wastes is enhanced. Kilns are equipped with afterburners to combust the volatilized wastes.

Fluidized bed incinerators (Fig. 3) are used for liquids, sludges, and shredded solids. All waste feeds must be reduced to a size less than 2 in. in diameter. Combustion air is forced through a distribution manifold in the bottom

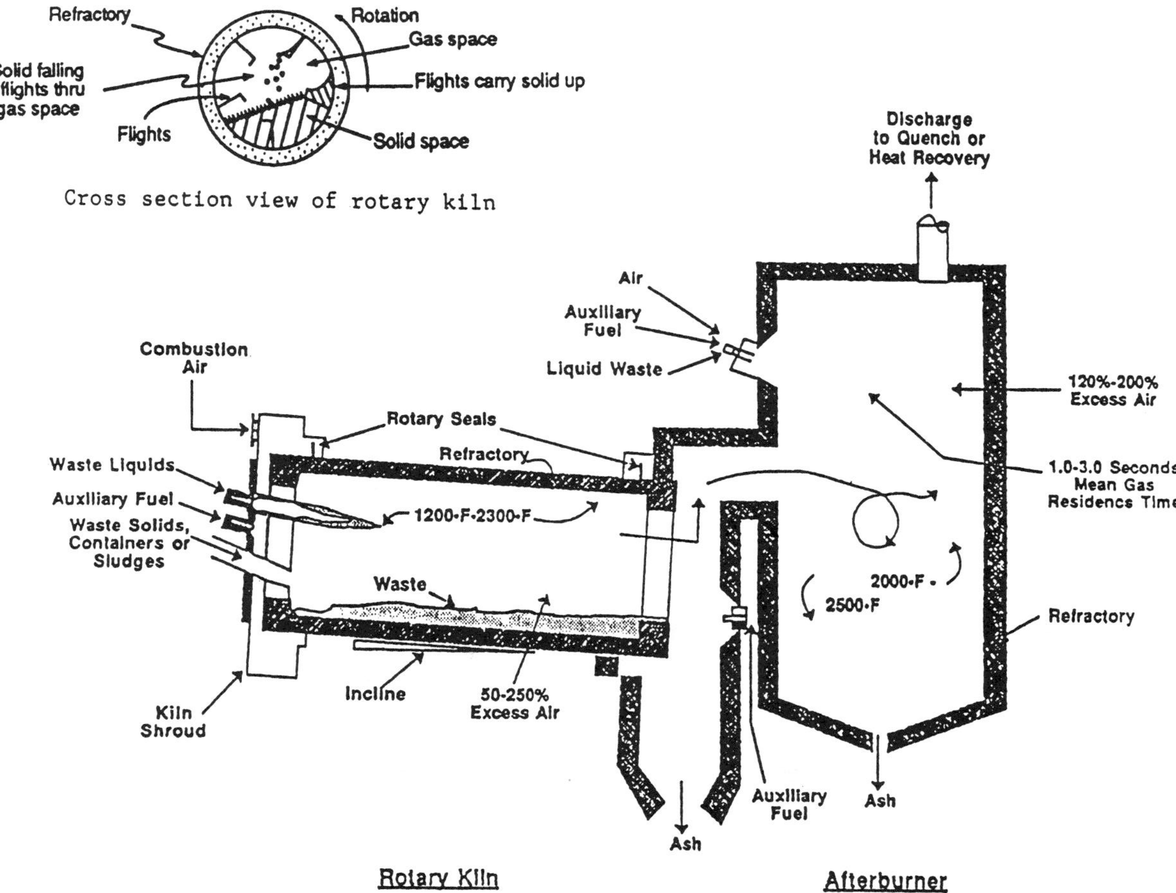

Figure 2 Rotary kiln.

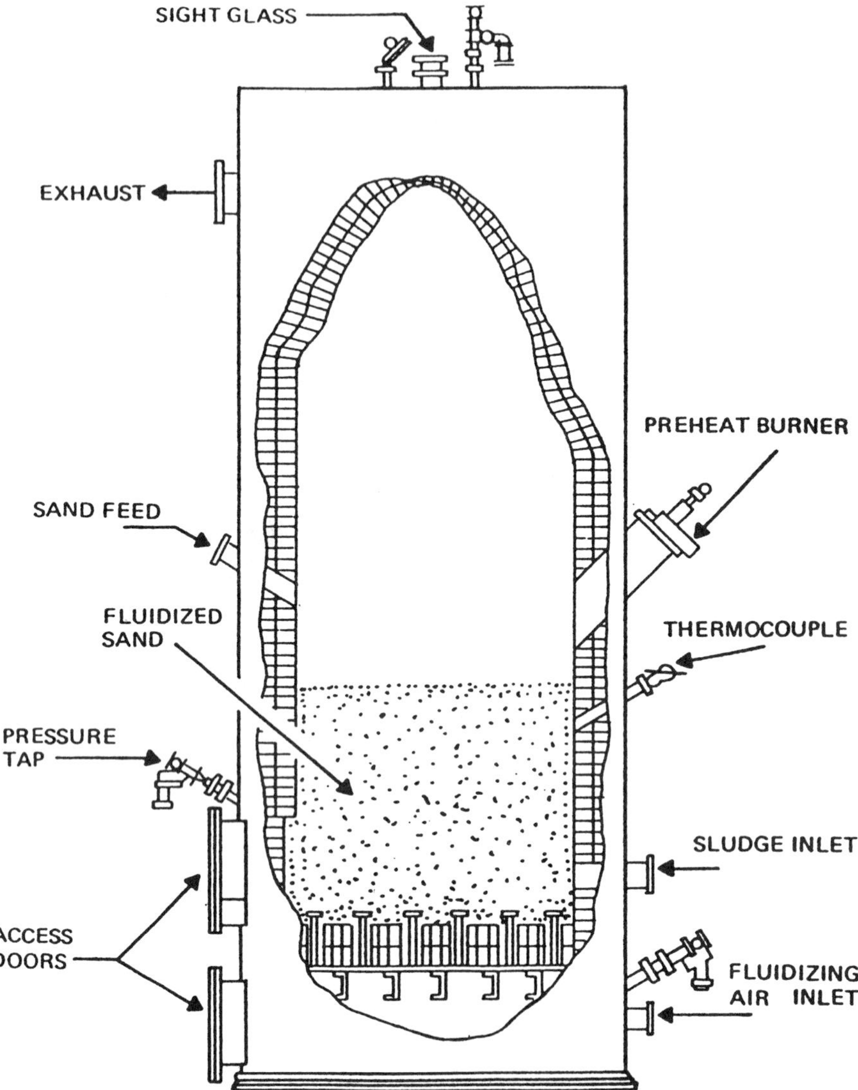

Figure 3 Fluidized bed.

of the incinerator into a bed of sized wastes and heat-transfer media such as sand. Movement of the gas through the bed appears to move like a fluid. Fluidized beds are subject to agglomerate formation which upsets waste movement, heat transfer, and the fluidization process.

Liquid injection incinerators are applicable almost exclusively for "pumpable" liquid wastes. Liquid wastes are pumped into a combuston chamber through a nozzle. Good atomization is critical to achieving high waste destruction efficiency. Generally, the smaller droplet size achieved by the atomizing nozzle, the more efficient the operation. Liquids with high salts and high ash contents are injected downward, whereas wastes with low ash content are usually injected horizontally. Figure 4 shows a typical liquid injection incinerator configuration.

Multiple hearth furnaces (Fig. 5) are used mostly for incineration of wastewater-treatment plant and industrial sludges. Sludge enters the top of the multiple hearth and is moved downward through three heat zones, or hearths, by rotating rabble arms. Combustion air and gases move upward. Multiple hearths are easy to operate, are of simple design, and can accommodate a wide variety of sludges. Mechanical problems because of the internal moving parts at high temperatures are a disadvantage.

Fixed hearth incinerators (Fig. 6) are also called controlled air or pyrolytic incinerators. This is a two-stage thermal treatment where wastes are pumped or rammed into an air-starved primary combustion chamber. Volatile fractions of the waste are vaporized and subsequently burned in the secondary chamber. Fixed hearths are a specialized technology suitable primarily for smaller waste streams for on-site incineration.

Thermal treatment methods are commonly used in industrial production, waste volume reduction, municipal waste destruction, and medical waste destruction. A significant need for growth in thermal processes is forecasted despite negative public opinion, high costs, complex permit issues, and waste minimization initiatives in industry and communities. It is expected that most expansion of incineration and other emerging technologies will be directed toward hazardous waste treatment. Thermal treatment methods which include combinations of liquid injection, rotary kilns, pyrolysis, multiple hearths, and fluidized bed technologies will continue to be the mainstay in thermal treatment until the emerging, more specialized processes listed in Table 8 are more fully developed.

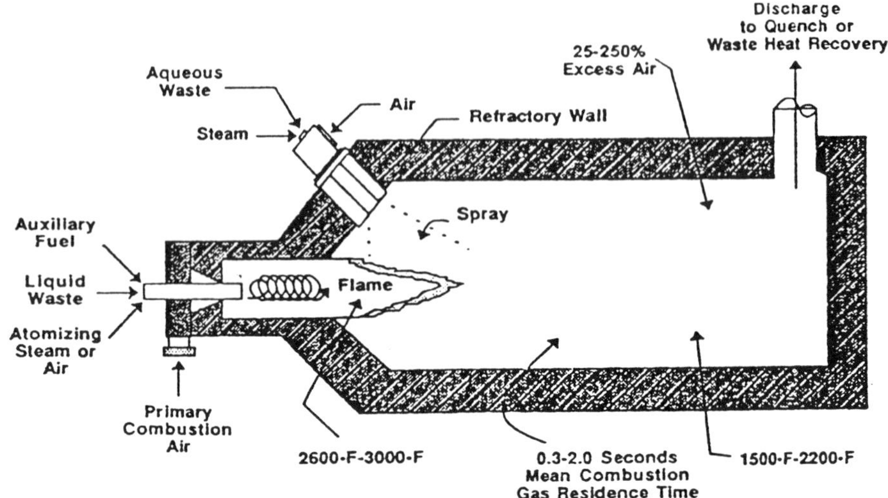

Figure 4 Liquid injection incinerator.

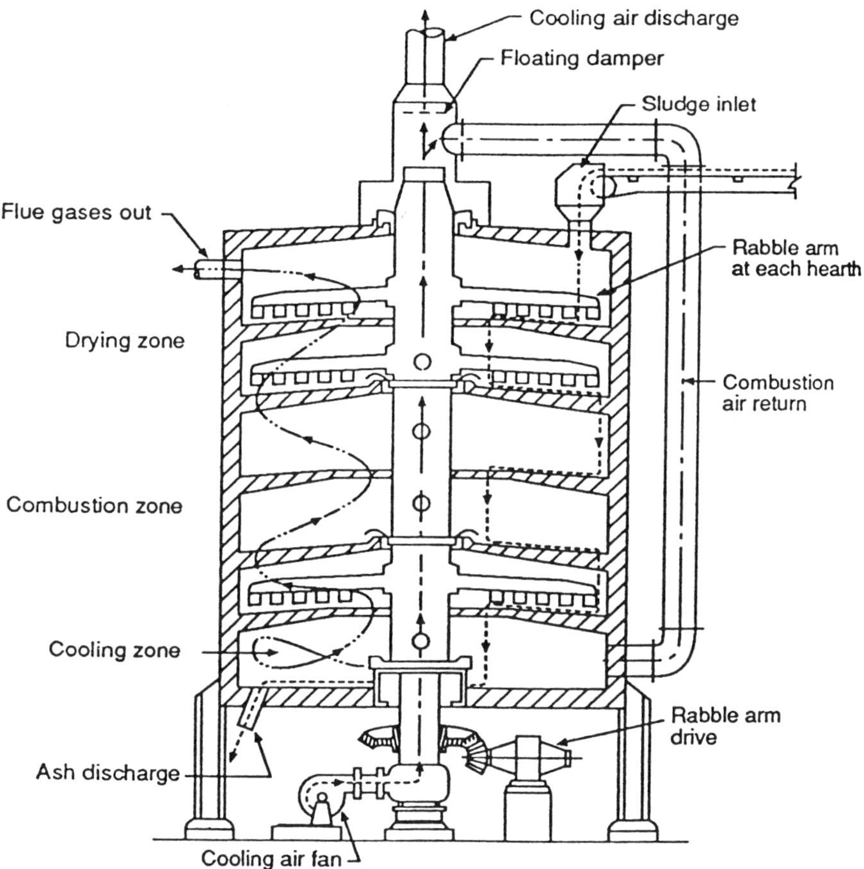

Figure 5 Multiple hearth incinerator for sludge conditioning and drying.

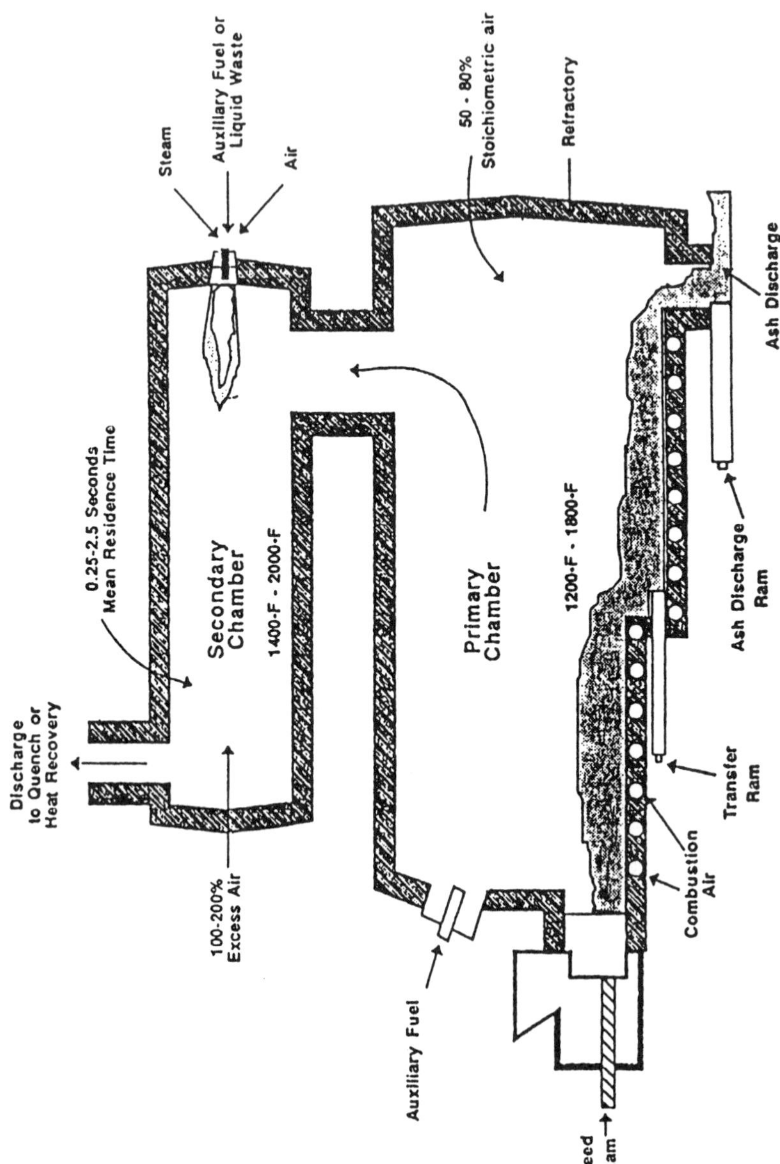

Figure 6 Fixed hearth incinerator.

13
Membrane Filtration

INTRODUCTION

Processes known as reverse osmosis (RO), ultrafiltration (UF), and electrodialysis may, in general, be characterized as material separation processes. Through these processes, dissolved substances and/or finely dispersed particles can be separated from liquids. These processes rely on membrane transport; that is, the passage of solutes or solvents through thin, porous polymeric membranes. A comparison of the three membrane processes is given in Table 1. The most widely used of these processes is reverse osmosis.

Membrane filters are classified as surface or screen filters. Particles are retained on the surface of the filter or within a depth of 10–15 μm. It is this characteristic that distinguishes membrane filters from depth filters or filter aids, which trap particles within the filter matrix. Cellulosic membrane filters are approximately 150 μm thick with a myriad of conically shaped pores arranged in a tortuous pattern. A typical membrane filter has 400–500 million pores per square centimeter of surface area. The pores in cellulosic filters may be envisioned as a spongelike construction and comprise about 80 of the filter's volume. The extreme thinness of membrane filters compared to depth filters gives them a distinct advantage. Because microorganisms are retained on or near the membrane surface, they can absorb nutrients through the filter and produce visible colonies.

Table 1 Classes of Membrane Processes

Name	Feed	Separating agent	Products	Separation principle	Application example
Electrodialysis (ED)	Liquid	Anionic and cationic membranes; electric field	Liquids	Tendency of anionic membranes to pass only anions, etc.	Desalination of brackish waters
Reverse osmosis (RO)	Liquid solution	Pressure gradient (pumpimg power) + membrane	Two liquid solutions	Different combined solubilities and diffusivities of species in membrane	Seawater desalination
Ultrafiltration (UF)	Liquid solution containing large molecules or colloids	Pressure gradient (pumping power) + membrane	Two liquid phases	Different permeabilities through membrane (molecular size)	Wastewater treatment; protein concentration; artificial kidney

Membrane filters are capable of retaining microorganisms and particles by mechanisms other than a simple sieving action. Microbes and particles may be adsorbed by the filter, react with the membrane itself, or may be retained on the membrane by coagulation. The fundamental mechanisms of filtration may be considered as sieving modified by adsorption and blocking arising from the large ratio of pore length to pore diameter. The ratio of pore length to width in a 0.2 m pore size-rated filter is approximately 750. This ratio, in combination with the geometrical configuration of a tortuous pore structure, heavily influences the retention of particles by membranes. Variation in cellulosic esters produces membrane filters that selectively adsorb viruses when the pore diameter greatly exceeds the size of the virus, whereas a different combination of cellulosic esters allows viruses to pass through. Basic guidelines are: nitrocellulose adsorbs viruses; cellulose triacetate repels viruses.

When used to retain viruses, membrane filters, with a thickness of about 5000 virus diameters, function as depth filters. Viral particles trapped in the matrix of a nitrocellulosic filter can be eluted from the filter.

Introduction of the membrane filter into microbiology revolutionized the analysis of water and other liquids for microorganisms. Using membrane filters, it became possible to filter large volumes of liquids containing low concentration of microorganisms and to cultivate them in situ.

REVERSE OSMOSIS

Reverse osmosis (RO) is an advanced unit operation in water treatment. RO membranes are capable of removing at least 90% of the dissolved solids in water as well as organics, bacteria, and other impurities.

RO has proven its commercial feasibility in the treatment of medium to high total dissolved solids (TDS) waters, including brackish waters, and many installations are successfully operating. The availability of RO has opened up the use of heretofore unavailable water supplies. This is especially true in states like Florida and California where water supplies have been unusable due to saltwater intrusions.

RO has also been used by industry as a pretreatment for ion exchange demineralization. RO acts as an economical roughing demineralizer, bringing down the overall cost and improving the life of ion exchange resins and operation of equipment.

RO has found application in waste treatment, chemical separation, and food and drug processes.

Osmosis

Osmosis can be defined as the spontaneous passage of a liquid from a dilute to a more concentrated solution across an ideal semipermeable membrane which allows the passage of the solvent (water) but not the dissolved solids (solutes) (Fig. 1a). The transport of the water from one side of the membrane to the other continues until the head or pressure (P) is large enough to prevent any net transfer of the solvent (water) to the more concentrated solution. At equilibrium, the quantity of water passing in either direction is equal, and the pressure (P) is then

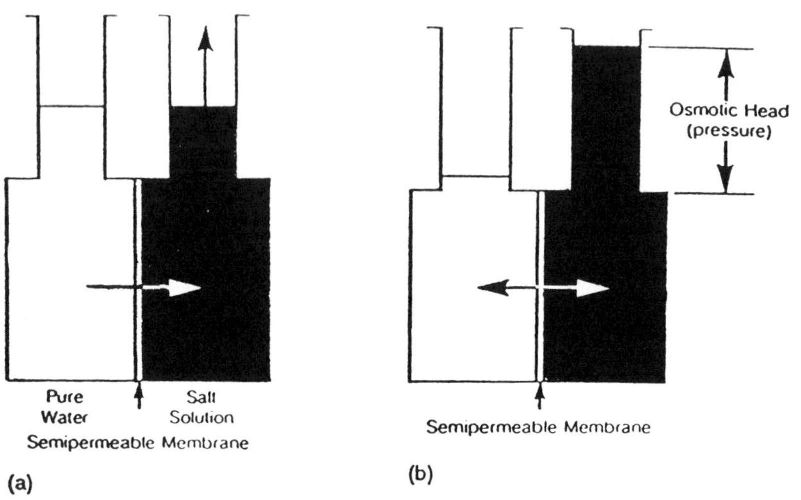

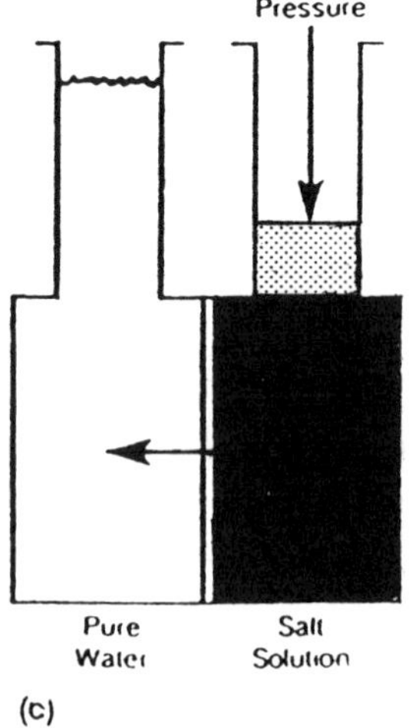

Figure 1 (a) Osmotic flow. (b) Osmotic equilibrium. (c) Reverse osmosis.

defined as the osmotic pressure of the solution having that particular concentration and makeup of dissolved solids (Fig. 1b).

If a piston is placed on the more concentrated solution side of a semipermeable membrane (Fig. 1c) and a pressure, P, is applied to the solution, the following conditions can be realized: (1) P is less than the osmotic pressure of the solution and the solvent still flows spontaneously toward the more concentrated solution; (2) P equals the osmotic pressure of the solution and solvent flows at the same rate in both directions; that is, no net change in water levels; (3) P is greater than the osmotic pressure of the solution and solvent flows from the more concentrated solution to the "pure" solvent side of the membrane leaving a higher concentration behind. Condition (3), shown in Figure 1c, represents the phenomenon of reverse osmosis.

A simplified flow diagram (Fig. 2) of a typical RO system shows how the process works. Feedwater (usually pretreated) is supplied to the concentrate side of the membrane by a high-pressure pump (normally 300–1000 psi). Product water permeates the membrane, since the operating pressure is greater than the feedwater osmotic pressure, leaving most of the dissolved solids behind in the concentrate or reject. The product is continuously withdrawn, usually at 20–50 psi. The concentrate is also continuously withdrawn at 10–30 psi below the operating pressure of the feedwater. It is in the concentrate stream that the dissolved solids rejected by the membrane are flushed from the unit.

The following equations apply (see Fig. 2 for explanation of terms).

Flow Balance

$$Q_F = Q_p + Q_C$$

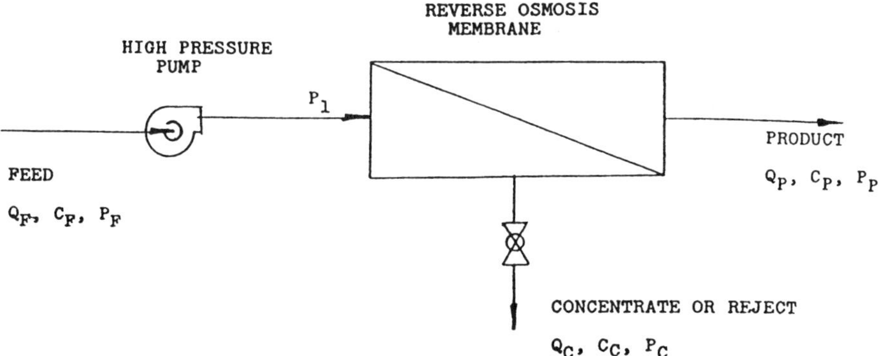

Figure 2 Simplified flow diagram of typical reverse osmosis system. Q = flow rate, C = concentration of dissolved solids and other impurities, and P = pressure. The subscripts F, C, and P indicate feed water, concentrate or reject, and product stream, respectively.

Conversion or Recovery (%)

$$Y = \frac{Q_P}{Q_F} \times 100QP \times 100$$

Material Balance

$$C_F Q_F = C_R Q_r + C_P Q_P$$

Salt Rejection (%) and Salt Passage (%)

$$SR = \frac{C_R}{C_F} \times 100 \times 100 \qquad SP = 9\frac{C_P}{C_F} \times 100 = 1 - SR$$

The feed concentration of the various ions present is usually known from the feedwater analysis.

The product concentration is usually calculated from published data of rejection (%) of various ions by the membrane manufacturer.

The concentrate concentration is calculated. The concentrate concentration can be estimated conservatively by assuming the product concentration to be zero. In such a case:

$$C_F \times Q_F = C_R \times Q_R$$

This latter equation gives a good estimate of the concentrate concentration and may be used to determine if any disposal or precipitation problems will be encountered.

Figure 2 can be considered to show one membrane module or a series of membrane modules in parallel with a single pump supply and a single product and concentrate header. Best performance is achieved when the flow through the feed concentrate channel is sufficient to maintain good distribution. For this reason, minimum concentrate flow rates have been established for various modules. As a result, there is a maximum conversion or recovery which can be obtained by a single stage arrangement as shown in Figure 2.

In order to obtain higher conversions or recoveries, the modules are staged (either two or three stages) (Fig. 3). In staging, the concentrate from the first stage becomes the feed for the second stage. Products from each stage are combined into a single product header. The concentrate to waste is only the concentrate from the last stage. As a general rule, two stage systems can obtain 70–75% recovery and three stage systems can obtain 80–85% recovery.

OSMOTIC PRESSURE

The osmotic pressure, which is a property of a solution, increases with the solution's concentration. A rule of thumb, based on sodium chloride solution,

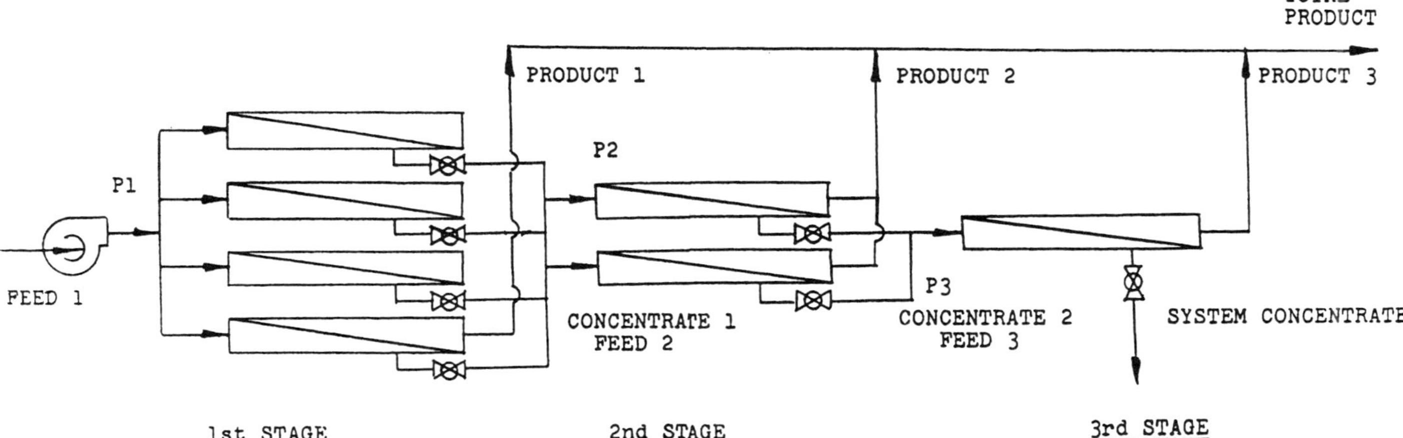

Figure 3 Simplified flow diagram for typical staging of reverse osmosis system. P2 and P3 are the interstage pressures and are lower than the initial feed pressure by the pressure drop (feed to concentrate). System Conversion = (Total Product Flow)/(Feed 1 flow). Feed 1 = Product 1 + Concentrate 1. Note the number of membranes in each state and the flow rate depend on the particular application.

is that the osmotic pressure increases by approximately 0.01 psi for each milligram/liter. This approximation works well for most natural waters. However, high molecular weight organics produce a much lower osmotic pressure. (For example, sucrose gives approximately 0.001 psi for each milligram/liter.)

Several methods are available for measuring the osmotic pressure. It can be calculated from the depression of the vapor pressure of the solution, by depression of the freezing point, and by the equivalent of the ideal gas law equation. Some calculated values for common components are listed in Table 2. Several devices are commercially available for direct measurement of the osmotic pressure which measure the pressure necessary to stop the flow of water through a membrane.

Attempting to measure the osmotic pressure of a solution directly by operating at a pressure just sufficient to obtain zero flow is not practical, since the membranes are not perfectly semipermeable. This technique would measure the difference in osmotic pressure between the feed and the product water. At low pressures, the salt rejection is relatively poor, so that a false osmotic pressure somewhat lower than the real value would be determined.

A procedure used to measure the osmotic pressure of a solution measures the water flux through a module under operating conditions at several pressures and a plot of water flux versus pressure is extrapolated to a zero water flux; the intercept is the osmotic pressure. This gives the effective osmotic pressure, including any concentration polarization. Care must be taken either to maintain constant recovery or correct for the variation in concentration.

The driving force for operation of a reverse osmosis unit is the difference between the operating pressure and the osmotic pressure.

Table 2 Typical Osmotic Pressures at 25°C (77°F)

Compound	Concentration (mg/L)	Concentration (mols/L)	Osmotic pressure (psi)
NaCl	35,000	0.6	398
NaCl	1,000	0.0171	11.4
$NaHCO_3$	1,000	0.0119	12.8
Na_2SO_4	1,000	0.00705	6
$MgSO_4$	1,000	0.00831	3.6
$MgCl_2$	1,000	0.0105	9.7
$CaCl_2$	1,000	0.009	8.3
Sucrose	1,000	0.00292	1.05
Dextrose	1,000	0.00555	2.0

Note: Based on the data from table for common ionic species, a useful rule of thumb for estimating the osmotic pressure of a natural water supply requiring demineralization is 10 psi per 1,000 mg/L (ppm).

Osmotic pressure is also used to calculate the equivalent NaCl concentration which in turn is used to establish the salt passage correction factor (SPCF). Published data on salt passage is usually taken at a specific NaCl concentration, so that the SPCF is used to calculate actual salt passage at the solutions osmotic pressure or equivalent NaCl concentration. It is important to calculate the osmotic pressure of the feed for a given application.

Also note in Table 1 that the osmotic pressure for salt water (35,000 mg/L) is about 350 psi higher than that of brackish water. Operating pressure for brackish water treatment by reverse osmosis is about 400 psi in order to maintain good flux. For sea water treatment, the operating pressure requirement for equivalent (economical) operation is about 800 psi.

Osmotic Pressure Calculation

The osmotic pressure of a feed may be calculated from the following equation:

$$\pi = 1.12\ (T + 273) \sum \overline{Mi}$$

where

π = osmotic pressure, psi
T = temperature, °C
$\Sigma \overline{Mi}$ = sum of molalities of ions and nonionic compounds

The equation assumes ideal solution behavior, but should give values which deviate by no more than ±10% from actual values.

MEMBRANE CONFIGURATIONS

Problems encountered in making up an RO unit include

- Membrane structures themselves are relatively weak and fragile. They require support designs for operation at 300–1000 psig differential pressure, feed to product side.
- A reliable seal is required to prevent intrusion of the high-pressure feed or concentrate into the low-pressure product.
- Product flow through a membrane is directly proportional to the surface area of the membrane and inversely proportional to the thickness. A module design containing the greatest possible area of the thinnest possible membrane offers the advantage of low pressure vessel cost and smallest required RO plant size.
- A controlled feed and concentrate flow path is required to minimize concentration polarization and fouling.

There are four different designs which are in current use. These are plate and frame, tubular, spiral-wound, and hollow fine fibers.

Plate and Frame Design

This design is similar to a conventional filter press (Fig. 4). Circular membrane sheets are placed or cast directly on each side of porous supporting plates. Several such discs, each with two membranes, are stacked with a spacer between each pair of discs. The feed and concentrate flow takes place in the space between the discs, whereas the product permeates through the membrane into the porous supporting plates and flows radially outward to a product collecting system.

The plate and frame device is simple and rugged. It has poor concentrate flow patterns and is very inefficient due to the small membrane area per unit volume of module (approximately 150 ft^2/ft^3). For these reasons, this device has little application in water-treating plants. However, it is widely used for laboratory and developmental purposes.

Tubular Design

A semipermeable membrane is either inserted into or coated onto the inside surface of a porous tube, which is designed to withstand the operating pressure. Feed is introduced into the end of the tube at the required operating pressure.

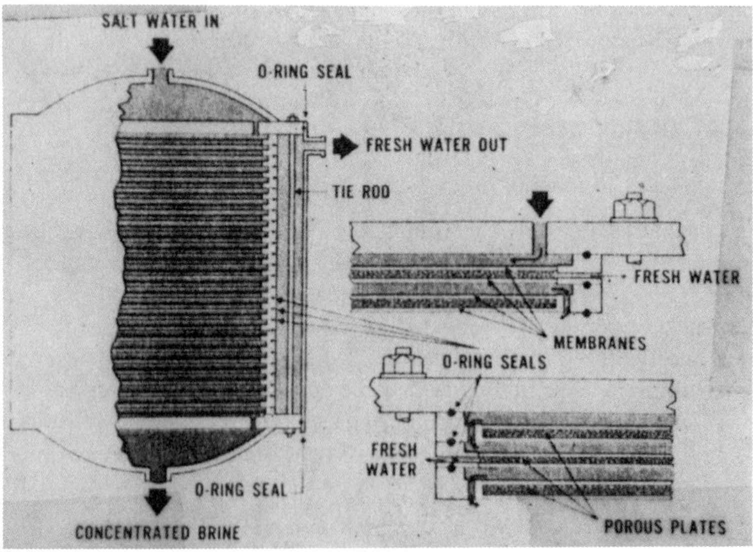

Figure 4 Schematic of plate and frame reverse osmosis.

Product permeates through the membrane and porous tube and is collected on the outside. The concentrate is discharged from the far end of the tube (Fig. 5).

This arrangement was used successfully during the late 1960s, especially in non-water applications, such as chemical separation, concentration, and food and drug processing. However, owing to its small membrane area per unit volume (approximately 100 ft^2/ft^3), it is not economical for water-treatment applications.

Spiral-Wound Design

Sheet-type cellulose acetate membranes are cast directly on each side of a porous backing material. The backing material serves as a support for the membranes against the operating pressure as well as a flow path for collection of product water (Fig. 6).

The above flat sheet is sealed around three edges with a water-resistant adhesive to prevent the feed or concentrate from contacting the product. The fourth edge is sealed to a hollow plastic tube which is perforated inside the edge seal area so that product water can be collected from the porous backing material through the plastic tube. The flat sheet is then rolled up around the central plastic tube in the form of a spiral along with a mesh spacer which separates the facing membrane surfaces and promotes turbulence of the feed as it passes through that section (Fig. 7). Several modules or elements are connected in series (Fig. 8),

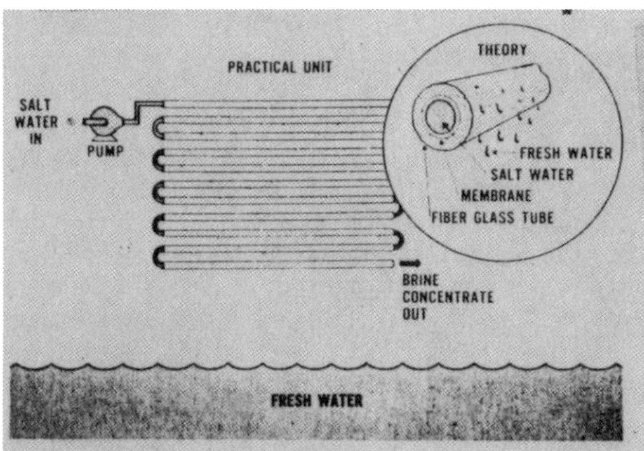

Figure 5 Tubular reverse osmosis schematic.

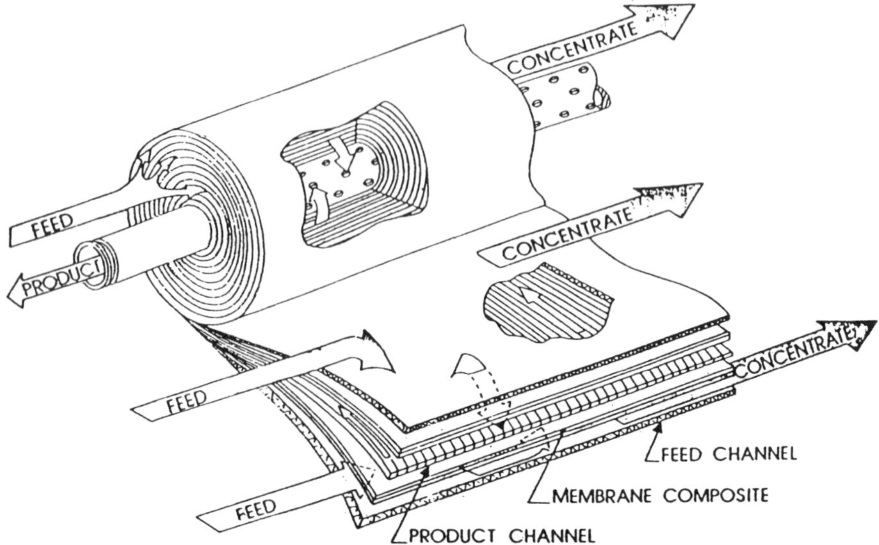

Figure 6 Spiral-wound membrane module partially unrolled.

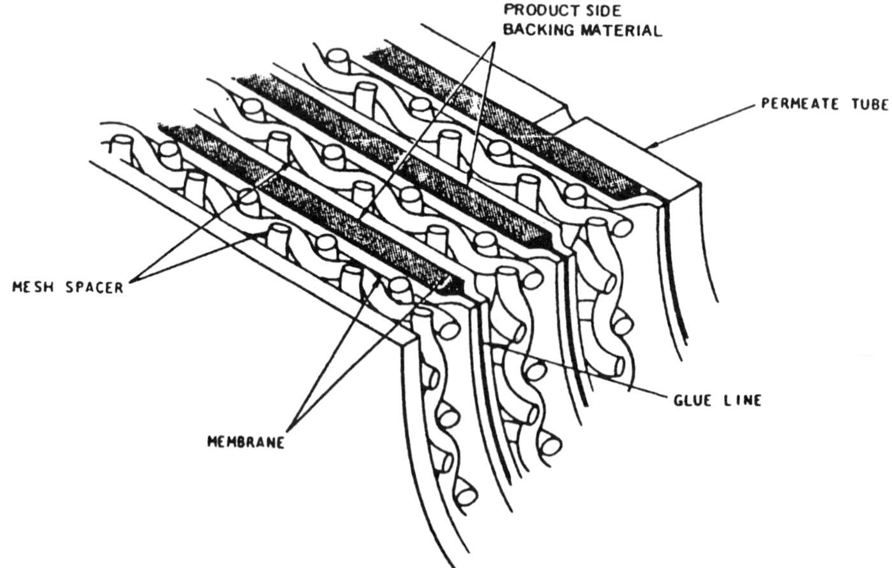

Figure 7 Detail of spiral-wound membrane module cross section.

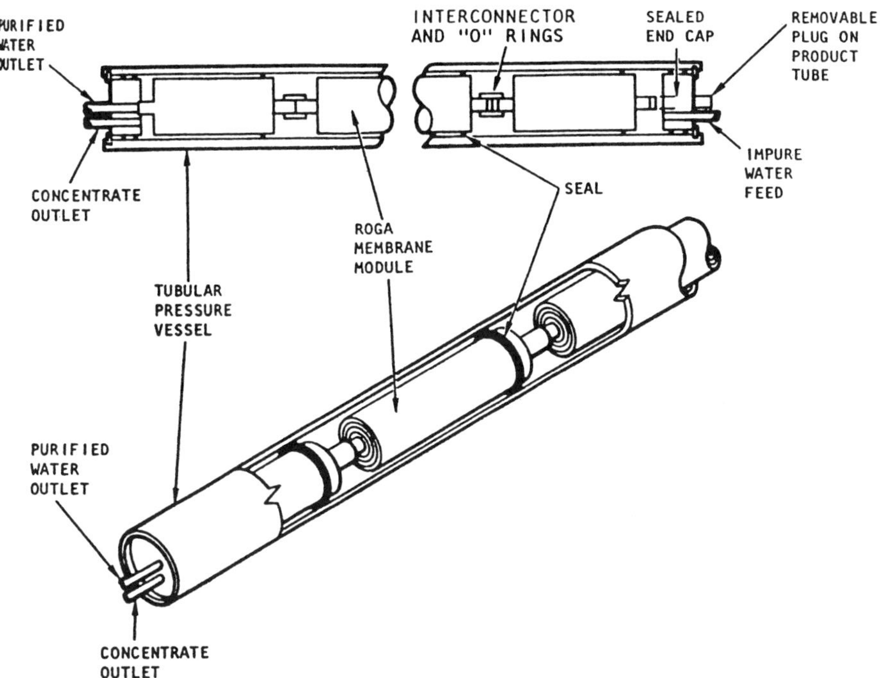

Figure 8 Spiral-wound membrane module pack assembly.

equipped with a peripheral seal, and slipped into standard steel pipe which acts as the pressure vessel. For corrosion resistance, elements in fiberglass tubes are also available.

The feed enters the upstream end of the first element through the mesh spacer channel and flows axially through the element. Product water permeates through the membrane surface and down the porous backing material to the central tube from which it is collected. The concentrate (including a higher concentration of TDS than the feed due to removal of TDS from product water by the first element) exits from the first element and enters as feed to the second element in series. The peripheral seal eliminates the possibility of short circuiting of feed around the element. The feed passes through each of the elements connected in series and the final concentrate exits from the unit through a special fitting in the end closure. The product from each element is combined and collected in the central tube and exits through a second special fitting in the pressure vessel end closure. This arrangement has been used in water treatment applications with economical and reliable results.

Hollow Fine Fiber Design

A hollow fine fiber reverse osmosis device is a compact bundle of thousands of longitudinally aligned hollow fine fiber membranes surrounding a feedwater distribution core. The distribution core runs the entire length of the permeator. Feedwater at operating pressure (usually 400 psi) enters one end of the distribution core and moves radially from the porous or slotted core around the outside of the fibers and toward the outer shell, still at the operating pressure. The pressure (above feed osmotic pressure) forces essentially pure water (product) through the fiber walls into the bore of the fiber. This product flows along each fiber bore to the tube sheet end where the fibers have been cut to allow the product to exit. The concentrate flows toward the outer perimeter and is removed through the concentrate port.

The fibers have a very thin, dense skin at the outer surface which is the semipermeable membrane (0.1–1.0 μm thick). The membrane is backed up by a thick, porous layer which provides support and owing to its high porosity allows product to flow freely to the center bore after permeating the thin membrane skin. The hollow fibers have a ratio of outside to inside diameter of 2:1 and as such act as thick wall cylinders with sufficient strength to withstand the high operating pressures. This arrangement has been used successfully in water-treatment applications.

The major advantage of the hollow fine fiber device is the extremely high packing density and membrane area per unit volume of module (approximately 5000 ft^2/ft^3). This allows for very compact and simple RO plant design and low-pressure vessel cost per unit flow.

FACTORS AFFECTING OPERATION

The performance of an RO system, regardless of configuration, is determined by the following three factors:

- Capacity (flux) in terms of product flow
- Performance in terms of salt passage (salt rejection)
- Stability in terms of operating life

The above factors are affected by

- Characteristics of feed water such as pH, temperature, and dissolved solids (quantity and makeup)
- Characteristics of the membrane material and structure
- Operating conditions such as pressure and conversion

Commercially available membranes are rated for initial product flow and initial salt passage under a set of standard conditions.

Although standard test conditions involve operation at 75% conversion, it is not normal practice to operate individual modules at that conversion because of the minimum concentrate flow requirements necessary for good flow distribution.

In general (for all configurations), passage of monovalent ions will be greater than passage of divalent ions which will be greater than passage of trivalent ions. Gases such as CO_2 completely pass through the membrane.

If the actual operating conditions vary from the standard rating conditions, the product flow and salt passage must be adjusted to the actual conditions. This is done by using curves, tables, and equations published by the various membrane manufacturers.

The product flow through an RO membrane is defined by the following equation:

$$Q_{Pi} = K_{wi} (\Delta P - \Delta \pi)$$

where

Q_{Pi} = initial product water flow through membrane
K_{wi} = initial membrane water permeability constant
ΔP = applied pressure differential ($P_{f-c} - P_p$; average feed-concentrate pressure in module minus product pressure)
$\Delta \pi$ = osmotic pressure differential
($\Delta P - \Delta \pi$ is the driving force for the product water flow)

The salt flux through an RO membrane is defined by the following equation:

$$Q_{si} = K_{si} (\Delta C)$$

where

Q_{si} = initial salt flow through membrane
K_{si} = initial membrane salt permeability constant
ΔC = salt concentration differential across membrane
(ΔC is the driving force for the salt flux)

The water and salt permeability constants (K_{wi} and K_{si}) are characteristic of the membrane used and can be varied by changes in material and manufacturing procedures.

The above equations show that the product flow increases with increasing operating pressure while the salt flow remains constant.

The feed water temperature has a significant affect on an RO membrane's performance. K_{wi} and K_{si}, previously defined, are constants only at a given temperature. Product flow will increase with increased temperature. The salt rejection does not vary significantly with temperature because the salt flux varies by about the same amount as the water flux. The increased product flow with increased temperature is not as advantageous as one might think, because increased operating temperatures give rise to a decrease in membrane life as well as other negative affects. Cellulose acetate membranes will hydrolyze to cellulose and acetic acid. The rate at which this hydrolysis occurs is a function of feed temperature and pH. Hydrolysis results in a reduced membrane salt rejection capability. The rate increases with increased temperature.

Hydrolysis should be taken into account when predicting salt passage and membrane life. Aromatic polyacid membranes are very stable and do not hydrolyze. Salt passages remain relatively constant throughout the life of the membrane.

Salt passage will vary depending on the ion mix. Membranes will reject trivalent ions better than divalent ions and divalent ions better than monovalent ions.

All reverse osmosis membranes are subject to declines in product flow with time in operation, independent of feedwater characteristics on pretreatment precautions to prevent fouling. This decline is due to compaction of the membrane structure under the stresses of operating pressure. The higher the operating pressure and temperature, the greater the effects of compaction on product flow. The term *flux decline* is used in preference to the term *compaction* because, in actuality, the decline in flux with time is usually an effect of compaction plus some slight fouling.

The number of membrane modules and operating conditions are usually selected to give the required performance after 3 years of operation. Initial operation is usually at reduced pressure (350 psig) and this pressure is increased as required to meet performance objectives. Theoretically, at the end of 3 years, the operating pressure should be 400 psig. Operating at the reduced pressure initially will increase membrane life and reduce the affects of compaction.

PRETREATMENT, POSTTREATMENT, AND CLEANING

Pretreatment

Reverse osmosis equipment must be considered as only a building block in a complete water-treatment system. Without proper pretreatment, posttreatment, and membrane cleaning procedures, successful economical operation of the system will not be achieved. In fact, membranes can be permanently damaged with improper pretreatment and controls.

Influent waters are usually separated into three general classes depending on their sources:

- Deep well
- Municipally treated well water
- Surface water (including shallow wells)

Deep well waters are usually the easiest to treat because their composition is relatively constant and usually do not contain suspended or colloidal solids. Problems can be encountered if the water is corrosive.

With municipally treated waters, some variations in characteristics may occur from day to day. With manual type controls, a relatively constant feed is of advantage.

The most difficult class of water to treat is surface water because of seasonal variations in water quality and the water is usually high in suspended and colloidal solids. Organics may also be present which can cause problems in RO membranes.

Some pretreatment requirements based on membrane considerations are

- pH adjustment: An important limiting factor in the life of cellulose acetate membranes in reverse osmosis is the rate of membrane hydrolysis. Cellulose acetate will hydrolyze to cellulose and acetic acid. The rate at hydrolysis occurs is a function of solution or feed pH and temperature. As the membrane hydrolyzes, both the amount of water and the amount of solute which permeate the membrane increase and the quality of the product deteriorates. The rate of hydrolysis is at a minimum of about a pH of 4.7 and it increases with both increasing and decreasing pH. As in any chemical reaction, the hydrolysis is strongly influenced by temperature and increases with increasing temperature. Good practice dictates that adjustment of pH and moderation of temperature be done where appropriate to achieve an economically satisfactory membrane and module life.

  The aromatic polyamide membranes do not hydrolyze.

  pH adjustment is achieved by the addition of a chemical feed system which pumps acid or caustic as required into the feed stream. As will be discussed below, pH control may also be essential in controlling precipitation of scale-forming or membrane-fouling minerals.
- Temperature adjustment: In some cases where the feed temperature is appreciably lower than design temperature, it may be economical to increase the temperature to increase the water flux. It may also be necessary to lower the feed temperature if it is above the maximum operating temperature in order to obtain the rated membrane life.
- Chlorine residual control: For cellulose acetate membranes (cellulose acetate [CA] and cellulose triacetate [CTA]) the maximum value tolerable for the chlorine residuals is 1 ppm. For the aromatic polyamide membrane, the manufacturer recommends complete removal of chlorine residual but lists 0.1 ppm as the maximum tolerable level. Permanent damage to the membranes will occur if these limits are exceeded.

Dechlorination or lowering of total chlorine residual can be achieved by passing feed through a bed of activated carbon or by the addition of a chemical feed system which pumps sodium bisulfite into the feed stream to oxidize the chlorine. The later technique is more popular today.

The primary purpose of a membrane is to act as a barrier to the passage of dissolved solids. It also acts as a barrier to the passage of suspended solids. Owing to their very nature, membranes are susceptible to plugging or fouling by various materials. The major pretreatment requirement for any RO unit is the prevention of membrane fouling. Because of the nature of RO membrane operation, fouling always occurs to some extent. Pretreatment is required when the fouling interferes with economical operation of the RO system.

Fouling involves the trapping of some type of material within the RO module or on the surface of the RO membrane. The following are five types of fouling:

- Membrane scaling
- Fouling by metal oxides
- Plugging
- Colloidal fouling
- Biological fouling

Membrane fouling is accompanied by a reduction in flux and an increase in pressure drop and salt passage. Membrane cleaning procedures can remove foulant and restore membrane to its original state. If cleaning is required more than once a month in order to maintain performance, the pretreatment should be considered inadequate.

Membrane Scaling

Membrane scaling is caused by the precipitation of some of the salts dissolved in the feed water. The salts in the feed water are concentrated in the RO process; two times at 50% conversion and four times at 75% conversion. This alone can cause their solubility limits to be exceeded and precipitation to occur. Concentration polarization and unequal distribution within the module can cause greater concentrations to occur in some places. Concentrate flowrates must be maintained above minimum values to minimize those effects and conversion rates kept at reasonable levels.

The most common scales encountered are calcium carbonate ($CaCO_3$) and calcium sulfate ($CaSO_4$) but other compounds such as silica, strontium, and barium sulfates can also cause scaling.

There are three basic pretreatment techniques used to control scale:

- Conversion control to avoid exceeding solubility limits. Scaling can be avoided by operating at conversions wherein the solubility limits of $CaCO_3$ and $CaSO_4$ (in the concentrate) are not exceeded.
- Removal of the ions responsible for a scale-forming compound. The removal of calcium ions from the feed by sodium cycle ion exchange softening can be used to prevent calcium carbonate or sulfate scaling. The sodium compounds

formed are highly soluble; however, this method is not economical in any but small installations. It does allow operation at increased conversions.

Calcium carbonate scaling can be avoided by adjusting the pH of the feed with the addition of acid (usually to around pH of 6). The acid reduces the carbonate ion concentration by converting it to bicarbonate and/or carbon dioxide. The CO_2 is not removed by the membrane and must be removed from the product by degasification to reduce the loading on demineralizers or to increase the pH if the product water is for a potable supply.

- Inhibiting the crystal growth of a scale-forming compound. A chemical feed system is added to pump sodium hexametaphosphate (HMP) to the feed stream. The HMP inhibits precipitation of $CaSO_4$.

The solubility limits for $CaSO_4$ are not usually exceeded at normal conversions (i.e., 75) unless the feed TDS exceeds 1500 ppm. $CaCO_3$ scaling on the other hand is a problem with most natural waters.

Based on the ion analysis, pH and temperature of the feedwater and the required operating conversion, it can be determined if $CaCO_3$ or $CaSO_4$ precipitation problems will be encountered and what pretreatment is necessary. For $CaCO_3$, Langelier index calculations are used to determine if precipitation will be a problem. For $CaSO_4$, solubility product (Ksp) calculations are used to determine if precipitation will be a problem.

Scaling should not occur if the feedwater analysis and the projected operating conditions (conversion) are evaluated and the proper pretreatment is selected and is adequately monitored.

Metal Oxide Fouling

Soluble manganese and iron in the feed can be oxidized in the RO system ahead of the modules or in the modules themselves to form insoluble species which can then deposit in the modules.

Iron oxide fouling is the most common type. The ferrous ion being oxidized to the ferric and ferric hydroxide precipitate occurs in the module. There are two pretreatment methods used to eliminate this problem.

Removal of the iron (or Mn) from the feed: Two methods for this removal include

Oxidized ferric: clarification and/or filtration processes

Soluble ferrous: aerator followed by sand filtration, manganese greensand filtration; sodium cycle softening

Prevention of oxidation from the ferrous to the ferric state: If the system can be controlled to prevent this oxidation, it will present no problems to the RO system because it will be rejected with the other cations.

From the analysis of total iron, total dissolved oxygen, and pH, the potential for fouling can be determined and the appropriate pretreatment selected.

Plugging

Plugging is caused by mechanical filtration in which particles too large to pass through the feed-concentrate passage are trapped by the membrane. This problem is most critical for the hollow fibers membrane configuration.

The minimum pretreatment used by any RO system is a 5- or 10-μm cartridge filter. This offers adequate protection from plugging for any type of RO configuration. Where cartridge replacement is too frequent, additional pretreatment may be required. This would involve clarification and/or sand filtration.

Colloidal Fouling

Colloidal fouling is caused by the entrapment of colloids on the membrane surfaces. duPont research has shown that this type of fouling is caused by the coagulation of the colloids during the RO process.

The colloids are usually in the 0.3- to 1.0-μm size range and are in suspension because of their electrical charges.

The rate of coagulation, which controls the rate of colloidal fouling, depends on two main parameters:

1. *Concentration of the colloids*: To determine the concentration of colloids, a measurement called the silt density index (SDI) is used. The SDI is calculated from the rate of pluggage of a 0.45-μm filter paper at 30 psig applied pressure. Although SDI is not an absolute measurement of colloid concentration, it yields an excellent correlation with the rate of RO membrane fouling. Well waters usually have a SDI of less than 3 and present no problem of colloidal fouling. Surface waters have SDIs ranging from 10 to 175 and can cause severe fouling problems unless pretreatment lowers the SDI to more reasonable levels (i.e., <12).
2. *Stability of the colloids*: The measurement of zeta potential is an effective guide to colloid stability and tests have shown that if the zeta potential can be increased from –10 to –30 MV or more, colloidal fouling will be reduced.

Pretreatment techniques used to eliminate colloidal fouling are aimed at reducing the concentration of colloids (SDI) and reduction of colloid stability is not attempted. The colloidal concentration can be lowered to acceptable limits by

Filtration, using sand, carbon, or other media (reduction by a factor of 2)
For SDI >50, coagulation in clarifiers with the use of alum or iron salts

aided by polyelectrolytes and followed by sedimentation and gravity sand filtration

For SDI <50, in-line coagulation with same chemicals followed by pressure filtration

The SDI should be monitored on a regular basis to determine if the pretreatment procedure is doing its job or if changes in feed requires changes in chemical feed dosages. At this time, there is no automatic SDI monitoring device available and samples must be taken at regular intervals and the SDI test run.

Biological Fouling

Biological fouling is caused by the growth of microorganisms in the RO module. Three factors must be considered:

- Biological attack of the membrane: The aromatic polyacid membranes are impervious to bacterial attack. Although rarely encountered in practice, CA membranes may be subject to attack by the enzyme systems of some organisms. The most commonly used bactericide in the United States is chlorine. Chlorine reacts with the aromatic rings of the polyacid membrane and if the reaction proceeds far enough, the membrane will be destroyed. duPont recommends that the feed supplies containing chlorine be dechlorinated prior to entering the RO membrane. The CA and CTA membranes can tolerate 1 ppm maximal residual chlorine, so feed can be chlorinated if necessary. However, careful monitoring of chlorine residual is required.
- Biological fouling: This has never been a problem with aromatic polyacid membranes, however, precautions should be taken to minimize the potential problem; that is, chlorinate the feed storage tanks and treat modules with formaldehyde solution if plant is shut down for any extended periods of time. For CA and CTA membranes, feed chlorination and prefiltration should eliminate any potential biological fouling.
- Bacterial content of product water: The membranes will reject nearly 100% of all bacteria; however, seals can leak slightly, fibers can be broken, etc. If a sterile product water is required, posttreatment is necessary. Ultraviolet radiation has been used successfully by the electronics industry. Potable water supplies should be postchlorinated.

Posttreatment

Treatment of the product or concente may be necessary before the product water or concentrate can be used or disposed of. Each problem must be evaluated in terms of the proposed use. However, as far as the product water is concerned, the methods of treatment are mostly conventional water-treatment procedures.

Degasification and pH Adjustment

Since the feed is usually adjusted below pH 7, depending on its alkalinity, an excessive amount of carbon dioxide is formed but most remains in solution. The carbon dioxide passes through the membrane into the product. It may affect the taste of the water, cause corrosion problems, or otherwise interfere with the use of the water. Most of the carbon dioxide can be removed down to the 2–10 ppm/L concentration by degasification using a packed tower.

Air diffusion, spray degasification, and vacuum and steam degasifiers are also available, depending on the application. Residual carbon dioxide may be removed and a final pH adjustment may be made by the use of lime or sodium hydroxide. It should be recognized that in final neutralization, a small amount of dissolved solids or hardness may be added to the water.

Degasification or pH adjustment of the concentrate can be carried out by the same procedures. If the concentrate is a waste to be disposed of, increasing the pH may result in the precipitation of a number of constituents and it may be necessary to separate the solids by sedimentation and/or filtration before final disposal. The most frequently used sources of alkali are lime and caustic soda.

In cases where ground waters containing hydrogen sulfide are processed, a portion of the hydrogen sulfide will pass through the membrane and appear in the product water. Degasification may remove all or part of the hydrogen sulfide, but chlorination may also be required.

Chlorination and Disinfection

Product water intended for use as potable water or other purposes requiring the highest degree of sanitation will have to be chlorinated or disinfected by an approved method before its distribution and use. When the feedwater to the reverse osmosis loop is chlorinated (CA and CTA membranes only), the product water may contain the required chlorine residual. Provision should be made for product water disinfection where required.

Interferences to chlorination or disinfection of a water are completely or partly removed during reverse osmosis and, in general, disinfection is rendered easier and more effective on the product water. One exception is phenol and some of its analogues which permeate the membrane and on chlorination will create taste and odor problems.

Cleaning

A separate cleaning system, including a chemical tank and pump, is usually supplied with each RO system. For small systems, the RO booster pump instead of a separate pump is piped up to handle the cleaning chemicals. Instrumentation to monitor pH, temperature, and flow of cleaning chemicals is usually provided. Cooling coils may be required to keep the recirculating cleaning chemicals from exceeding the temperature limits of the membrane.

The required cleaning chemicals are mixed in the chemical tank. The cleaning pump discharge is hooked up to a valved connection on the feed line to one bank or train of RO modules. The hook-up may be rigid pipe or may be removable flexible connections. Cleaning is done at lower than operating pressure (60–120 psig). Most of the cleaning chemical passes through the module and exits the concentrate valved cleaning connection. Some chemical permeates the membrane and exits the product valved connection. Both lines are returned to the chemical tank to close the recirculation loop. Cleaning may continue for 1–4 hr and then the next bank is cleaned (flushed). If cleaning is required more often than once a month, pretreatment is inadequate.

Table 3 lists conditions that indicate cleaning is required for a RO membrane and the probable cause of the problem.

Table 3 Conditions that Indicate Cleaning Is Required for RO Membranes

Symptom	Probable cause
High permeator or bank pressure drop	1. Concentrate flow through the module is high 2. Module fouled
Product flow decreases (1st or 2nd stage)	1. Feed temperature low 2. Low feed pressure 3. Salt concentration too high (high osmotic pressure) 4. Fouling
High product conductivity plus high concentrate conductivity	1. Plugging of concentrate tubing or control valve 2. Internal restriction of concentrate flow (fouling)
High product conductivity plus high concentrate conductivity plus high pressure drop across modules	1. Fouling of module causing a restriction in concentrate flow
High product conductivity plus high pressure drop with concentrate flow normal, low product flow	1. Fouling of modules
High pressure drop plus low product flow plus small increase in product conductivity	1. Plugging of module feed path (feed tube in hollow fiber module)

BASIC REVERSE OSMOSIS SYSTEM

The design of an RO system, including selection of membrane configuration, pumps, pretreatment, posttreatment, materials of construction, degree of automation, and instrumentation, depends on the particular applications of the raw water, end use of product water, size of plant, and operator skill as well as many other factors.

At a particular installation, the system may be simpler or more complex. The manual system described is typically simple, trouble-free operation. Daily monitoring and control is required by a reasonably skilled operator. Figure 9 is a system consisting of a two-stage arrangement schematic diagram of a basic RO system. The feed water, Q_F, enters the suction side of the high-pressure pump after passing through a final micron filter. Pretreatment and posttreatment are not shown. The feed is then raised to the required operating pressure and delivered to the membrane array. The two-stage system shown is very common. The feed, at the inlet pressure of approximately 400–500 psi, enters the first stage and passes through the membrane assembly into a common concentrate header. First-stage product water which permeates through the membrane is also withdrawn through a common header.

The first stage concentrate stream is fed into the second stage (feed), passes through the membrane assembly and out the second stage concentrate header (Q_c) and is transported to waste. The second stage product water permeates through the membrane and is combined with first stage product water and withdrawn for use.

The pump throttle valve (PTV) is used to control the water volume to be fed from the high-pressure pump by controlling the pump discharge pressure.

The concentrate control valve (CCV) is used to control concentrate flowrate. The balancing of the two valves (PTV and CCV) will establish the operating parameters of the system (i.e., Q_F, Q_c, P, Y). Once the system is adjusted, only minor changes will be required over periods of many months. The RO systems are basically very stable because they are large masses of water seeing very limited and very slow load changes (i.e., flowrate, temperature, water analysis).

The basic system may be broken down into four sections for further examination:

1. Pumps
2. Membrane packaging
3. Controls
4. Miscellaneous components

Pumps

The high-pressure booster pump is the second most important component in a membrane system. Multistage centrifugal pumps are most frequently used

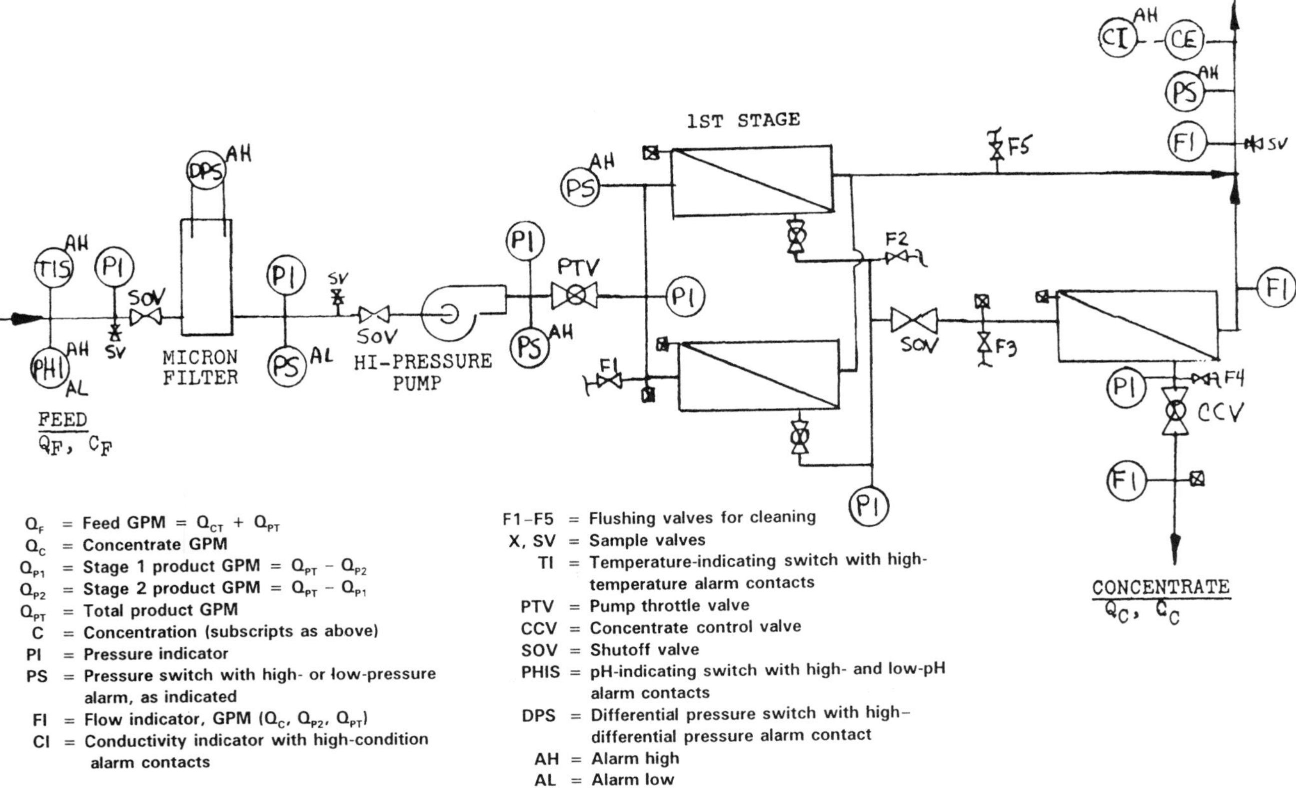

Figure 9 Schematic diagram of basic reverse osmosis system.

because of their operating curve characteristics, smooth flow characteristics, economy, and throttling ability. The pump is selected to provide the operating flow (feed gallons per minute) at the operating pressure (400–500 psi).

Materials of construction are selected to be compatible with the corrosion characteristics of the feed to be handled. Niresist, all bronze, and stainless steel construction are most frequently selected.

Pump efficiency is important, since the cost of electricity is a significant part of the overall operating costs. Several different types of pumps may be selected depending, for example, on flow, pressure, and initial cost.

MICROFILTRATION

Microfiltration involves pressure-drive processes requiring the retention of particulates, organisms, colloids, and viruses generally in the 0.02- to 10.0-μm size range (>300,000 molecular weight). Microfiltration membranes were developed before the advent of ultrafiltration and reverse osmosis membranes which also operate in pressure-drive processes. Microfiltration membranes are discussed in this section with respect to types and performance, advantages and disadvantages, and applications.

Types of Microfiltration Membranes

There are two basic types of microfiltration membranes that are fundamentally different in their manufacture and structure—the tortuous-pore membrane and the capillary-pore membrane. Their differences are reflected in the different applications in which these membranes are used.

Tortuous-Pore Membranes

Tortuous-pore membranes consist of a polymer matrix surrounding numerous interconnected vacuoles that results in convoluted pathways through the membrane matrix material. The most common technique for manufacture of tortuous-pore microfiltration membranes involves a phase-inversion casting process in which the relative amounts of polymer, solvent, and water as well as drying rates are all controlled to determine the number and size of the vacuoles in the polymer matrix.

An important consequence of this type of membrane casting is the characteristics of the channels arising from interconnected vacuoles. Any given channel is not uniform in its width, and there are constrictions which may occur anywhere along the channel path. It is the diameter of the channels at these constrictions which limits the size of the particle or material which can pass through the membrane. In addition, these channels are not linear but rather follow convoluted and interconnected pathways, hence the name "tortuous pore."

Capillary-Pore Membranes

A novel technique that produces straight-channel cylindrical pores with uniform diameter in a dielectric film was developed at General Electric in the mid 1960s. These membranes are produced by the Nuclepore Corporation and are manufactured in a two-stage "track-etch" process. In the first stage, a film of polycarbonate, polyester, or polypropylene is subjected to bombardment by massive energetic nuclei which leave a thin trail of radiation-damaged material through the film. In the second stage, the film is subjected to an etching process that selectively dissolves that portion of the film that has been ionized. The etching process leaves cylindrical, straight pores through the membrane material.

This two-stage process provides a high degree of control over the nature of the membrane. The density of pores, or the number of pores per unit area, is a function of the number of high-energy nuclei which bombard the film. The etching process is independent of the radiation exposure and, by varying the temperature, concentration, and residence time in the etch bath, the pore size can be controlled.

Uses

For many potential applications, both of these microfiltration membrane types have similar desirable characteristics. Both types of membranes exhibit absolute retention of all particles larger than the specified pore size and are thus amenable to use for sterilization or other applications that would require complete particle or organism removal. In addition, both types of membranes can be autoclaved at least once. The capillary-pore membranes made of polycarbonate can be autoclaved repeatedly.

The differences between capillary-pore membranes and tortuous-pore membranes may be important in specific applications:

- Tortuous-pore membranes have a porosity of about 80%, whereas capillary-pore membranes generally have a porosity of only 10%; the thin nature of the base film of capillary pore membranes necessitates a low porosity to maintain strength.
- Capillary-pore membranes are only about one-twelfth the thickness of the tortuous pore membranes.
- Capillary-pore membranes are highly flexible and thus can be pleated into cartridges.

The high porosity of tortuous pore membranes is of particular utility in

- Uses where organism must be cultured and collected on the membrane surface particularly when nutrients are to be transmitted through the pores from below the membrane

- Membrane electrophoresis in which molecules must migrate across the membrane diameter in an electric field

The low porosity and uniformly flat surface of the capillary pore membranes are particularly advantageous in applications such as

- Microscopic analysis of material collected on the surface of the membrane (which would tend to be imbedded below the surface in tortuous-pore networks)
- Where staining is required to count organisms such as bacteria (since the polycarbonate film does not stain)

The thin and straight channels in capillary-pore membranes are important in several types of applications:

- Concentration and or purification of biological agents, such as viruses, is more cost effective with capillary-pore membranes because far less material is lost due to adsorption on the walls of the capillary tube than would adsorb to the tortuous channels.
- Capillary-pore membranes are used in diffusion and migration studies because of their shorter path length.
- Capillary-pore membranes are particularly useful in cases where one wants to fractionate a sample containing particles of different sizes. The small particles have less surface area for adsorption in the capillary pores than they would in the tortuous pores where a significant portion of the sample could be lost.

Operational Considerations

There are a number of operational considerations that must be made in choosing a microfiltration membrane. These considerations include pore size determination, effect of particle deformation, effect of pore structure and configuration, effect of adsorption, and various particle capture mechanisms.

Determination of pore size as well as testing of membrane integrity can be accomplished by means of the "bubble-point" test. This test is based on the gas pressure required to displace liquid from the pores of a wetted membrane. As the pressure of a gas applied to one surface of a wetted membrane is increased, the first gas bubbles to pass through the membrane represent water displacement from the largest pore(s) in the membrane. This passage corresponds to the maximum pore size of the membrane. Passage of bubbles through a membrane at pressures far below that indicated by the maximum rated pore size is indicative of potential defects in the membrane itself.

The ability of some particles, especially biological materials, to change their shape is an important consideration in relating pore size to size of particle retained by, or transmitted through, a membrane. In the case of red blood cells, normal cells will pass through pores with diameters that are smaller than the cell diameter, but hardened cells cannot pass through a membrane unless the pore diameters are larger than the cell diameter.

Given a membrane with a particular pore size for filtration of a sample containing variable-sized particles, different capture mechanisms apply to the different particle sizes. Very small particles, having relatively high diffusion coefficients, are captured on the membrane surface or on the internal pore surface by adsorption following diffusion to the membrane surface. For relatively large particles, impaction on the pore lip is the dominant mechanism. Impaction on the pore lip or inner surface of the pore near the lip results when the mass of the particle is sufficiently large to carry it across the fluid viscous-flow stream lines of material through the pore.

Advantages and Disadvantages

To a large extent, the relative advantages and disadvantages of microfiltration membranes of both the capillary pore and the tortuous pore type have been referred to in the previous discussion. Advantages and disadvantages depend on the specific application and type of membrane. However, the most important considerations include the following:

- A relatively high initial cost of microfiltration systems which is offset by lower operating costs owing to a reduced need to replace filter elements compared to ultrafiltration and reverse osmosis.
- Microfiltration membranes tend to show a sharper molecular weight cutoff than other membrane types and can effectively retain all of a given particle size. These features permit applications requiring sample fractionation and/or sterilization.
- Capillary-pore membranes, because of their smooth surface and straight cylindrical pores, have a much lower capacity to retain particles than do tortuous-pore membranes. However, since capillary-pore membranes can be pleated and inserted into cartridges which provide a large surface area in a small space, this lack of capacity can be offset.

Applications

The most common applications of microfiltration membranes are dictated by their ability to retain all of the particles above a specified size and by their retention of particles <300,000 molecular weight and in the 0.02- to 10-μm size range. Microfiltration membranes are used in applications where organisms and particulate material removal is of primary concern, such as sterilization, harvesting of cellular material in the pharmaceutical industry, and removing waterborne particulates such as asbestos and radioactive waste. Most of these specific applications are small-scale uses and laboratory applications, with the exceptions of brewery, wine, and soft drink sterilization and in filtration of particulates from water prior to washing microelectronic components.

Initially, microfiltration got its start in microbiology with the development of microfiltration membranes that would retain bacterial organisms which were cultured and counted on the surface of the membrane. This application has a widespread use in laboratories and in water-quality analyses.

Industrial applications of microfiltration are largely extensions of its laboratory application. Specific examples include

- Removal of particulates and other material causing turbidity in serum
- Removal of stroma and particulate material during the collection of hemoglobin
- Concentration of yeast and bacteria used in the biological production of enzymes and pharmaceuticals
- Fractionation of enzymes
- Removal of radioactive material during effluent cleanups and/or decontamination of nuclear reactors or other facilities handling radioactive material

Large-scale applications of microfiltration membranes to remove either particulates or microorganisms are usually associated with other separation processes, such as

- Removal of particulates from deionized water used for rinse bath in the manufacture of electronic parts
- Removal of particulate and microorganisms from water before or after purification by reverse osmosis or electrodialysis
- Yeast removal from liquid sugar

A promising use for microfiltration systems is in the cold sterilization of beer and wine. Yeasts and bacteria are removed from the product without the use of heat which could alter the taste of the beer or wine.

The large-scale industrial applications of microfiltration processes are probably limited by the potential for membrane fouling, particularly in tortuous-pore membranes. As a result of the resistance of anisotropic ultrafiltration membranes to fouling, they are often the membrane of choice in those applications where complete removal of microorganism or particulates is not required.

ULTRAFILTRATION

Like microfiltration, ultrafiltration is a pressure-driven process that employs a porous membrane. Ultrafiltration membranes, however, generally retain particles in the 0.001- to 0.02-μm range with a molecular weight of 300–300,000. The process has generally been used for the retention of colloids, viruses, and macromolecules in solution. This section briefly discusses the nature of ultrafiltration membranes, some of their advantages and disadvantages, and specific applications.

Types of Ultrafiltration Membranes

The original ultrafiltration membranes were amorphous structures much like the tortuous pore microfiltration membranes but with smaller diameter pores. These membranes were not very effective owing to extremely low flux, clogging, and relative nonselectivity. (As a result of operational difficulties, there were relatively few technological advances in ultrafiltration until the development of the thin-skinned anisotropic ultrafiltration membrane.)

Anisotropic ultrafiltration membranes are cast using a three-phase system of polymer, such as cellulose acetate; solvent, such as a mixture of formamide and acetone; and a water-precipitating phase. As the polymer precipitates in the water phase, a very thin porous skin is first formed with pores in the 0.001–0.02 μm diameter range. As the polymer precipitates, its concentration in the solvent decreases, the precipitation rate decreases, and the pore spaces became larger trapping more water. The end result is a membrane with a thin skin (approximately 0.1 μm thick) and very small pores and with a lower basal layer approximately 120 μm thick with relatively large pores. This thick skin provides strength and support to the ultrafiltration membrane without impairing flux of permeate through the membrane material. In addition, the relatively large pores in the lower portion of the membrane are not subject to internal clogging. In some cases, these membranes are even further strengthened by an additional spongy backing.

Anisotropic membranes are formed either in flat sheets or in 200- to 500-μm inside diameter microtubules with the thin skin on the inside of the tubule. The flat sheets are generally used in large-scale industrial applications in stacked-plate, thin-channeled, or spiral-wound modules. The microtubules are generally packed linearly in cylindrical cartridges which are potted at the ends with epoxy. Proprietary techniques are used in spinning the hollow fibers.

Operational Characteristics

There are a number of characteristics of ultrafiltration membranes that govern their use in industrial applications. Some of these characteristics include membrane pore size; configuration of the molecules to be retained or passed; chemical characteristics of the membrane; effects of environmental variables of pressure, temperature, and pH; and concentration polarization at the membrane surface.

The size of ultrafiltration membrane pores generally ranges from 0.001–0.02 μm and allows for the passage of molecules in the 300–300,000 molecular weight range. Unlike microfiltration membranes, ultrafiltration membranes tend to have a diffuse molecular weight cutoff as indicated in Figure 10. As a result,

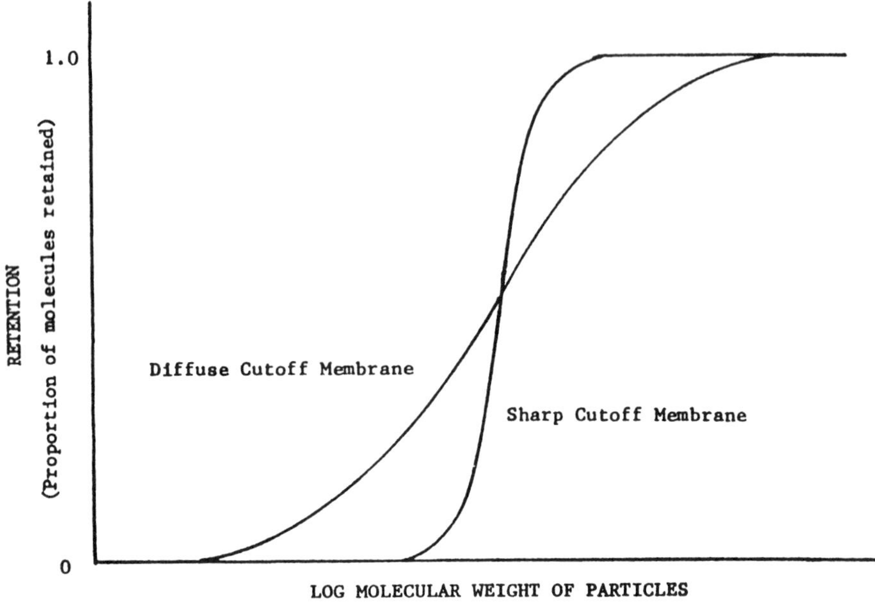

Figure 10 Comparison of sharp and diffuse cutoff membranes.

ultrafiltration membrane retentions are specified in product literature by spherical molecules retained at 90% for a given molecular weight. This has significance for those applications where absolute retention is required, as in sterilization, since bacteria can and do pass through ultrafiltration membranes. The ability of a membrane to pass or retain a molecule depends on the shape of that molecule in addition to its molecular weight relative to pore size.

The polymer of which the membrane itself is cast is an important determinant for various uses and handling characteristics. Although in most cases the actual polymer or its formula is proprietary, membrane manufacturers' technical literature specify such information as

- Whether or not the membrane pores will collapse if the membrane is allowed to dry
- Permissible and/or optimal temperature and pH range
- Membrane resistance to chemicals which may be a normal constituent or contaminant of a process stream, or a potential cleaning agent
- Retentivity of the membrane to specific chemical substances

Several environmental variables can significantly affect membrane performance. Of particular importance are temperature, pressure, and pH. Generally, flux for an aqueous filtrate in ultrafiltration systems increases linearly with increasing temperature at a rate of 3% increase in flux with each degree Celsius increase in temperature. Operating pressure can have two effects on membrane performance. If a feedstream is passed through an ultrafiltration module at a constant pressure, the flux of ultrafiltrate through the membrane will decrease with time owing to the formation of a boundary layer by means of concentration polarization, as described in the next paragraph. To maintain a constant rate of flux, one can increase the operating pressure. Increased pressure, however, results in an increase in the compaction of the membrane and can ultimately result in membrane collapse. In addition, under higher pressure, more of the retentate molecules trapped at the boundary layer are forced through the membrane, resulting in a lower reaction rate for a given molecular species as pressure increases.

The formation of a boundary layer by concentration polarization at the membrane is a major potential problem in most membrane applications (microfiltration, ultrafiltration, reverse osmosis, and electrodialysis). Fortunately, most of the potentially deleterious effects of concentration polarization can be avoided by proper system design and process control. Much of the design innovation in filtration systems has been aimed toward eliminating this problem. The boundary layer is formed when fluid permeates through the membrane leaving behind the larger retentate molecules. These molecules tend to increase in concentration at the membrane boundary as long as back-diffusion of molecules is less than transport of molecules toward the membrane. Thus the concentration gradient of the permeate across the membrane is lowered. If the concentration of retentate molecules gets too high, it can exceed the solubility concentration and the retentate will precipitate out as a cake or slime on the membrane surface, further reducing membrane flux. This problem can be exacerbated in some cases where two molecular species are involved, such as albumin and gamma-globulin. Although albumin is normally passed through a 100,000 molecular weight cutoff filter when alone in solution, if gamma-globulin is present as well, it tends to form a dynamic barrier which in turn prevents passage of albumin through the membrane. The primary mechanisms to prevent or reduce the effects of concentration polarization involve techniques to increase the rate of back-diffusion of molecules from the membrane boundary. In laboratory applications, magnetic stirrers suspended just above the membrane suffice to prevent the formation of boundary layers. For industrial membrane applications, mechanisms have been devised to provide cross-flow over the membrane surface by use of thin-channel, spiral-wound, and hollow-fiber modules. Two potential secondary problems result, however; an increased need for energy to drive the pumps results in an increase in operating costs; and some organic molecules such as enzymes are adversely affected by the shear-force of the resulting turbulent flow within these membrane modules.

Advantages

Ultrafiltration processes have a number of advantages that they share to some degree with microfiltration processes:

- Ultrafiltration is a thermal process and, if necessary, can operate at relatively low temperature to preserve heat-sensitive materials.
- No phase change, such as freezing or evaporation, is required for separation or concentration of materials by ultrafiltration. As a result, this process is not only gentle on the materials to be separated or concentrated, but in most cases it also saves the energy that would otherwise be required to accomplish the phase change.
- Relatively low hydrostatic pressures are used in ultrafiltration processes, particularly compared to reverse osmosis, since high pressure can lead to membrane compaction or even rupture. These lower pressures are relatively gentle and use less energy.
- No chemical reagents or catalysts are required to operate ultrafiltration processes. Although in many instances detergents, chlorine, sodium hydroxide, or other chemical agents are used to clean fouled membranes, it is possible to backflush the membrane with permeate.
- Ultrafiltration processes permit the simultaneous concentration and purification of process and/or product streams, thus reducing the complexity of process design and equipment and, in many cases, reducing energy requirements.
- Ultrafiltration processes maintain a constant pH in the concentrate stream and (unlike reverse osmosis) also maintain a constant ionic strength.
- Economics are generally favorable for ultrafiltration systems. Microfiltration systems usually have more surface area and longer membrane life than ultrafiltration systems, but this longer life is a result of lower flux of permeate through the membrane. The two prime economic considerations are membrane replacement costs and utility costs. Although reverse osmosis modules are among the cheapest available, the rising world energy costs are such that the cost of utilities is rapidly becoming a key factor in process consideration.

Applications

Unlike microfiltration, ultrafiltration membrane systems have been applied to a wide variety of industrial uses. Two of the largest applications, in terms of installed capacity and sales, are in the electrocoat paint industry and in oil-water separation applications. A significant application for ultrafiltration is in food processing. Other emerging applications include recovery of textile sizing, metallic colloid recovery, pharmaceutical production, pollution abatement, and particulate removal from electronic parts rinse water.

The electrocoat industry has had applications for ultrafiltration technology and was the first viable commercial application of ultrafiltration.

In electrocoat paint processes, an electrically charged metallic workpiece is reversed in a bath of ironically charged paint, and the paint particles are uniformly deposited on the metal. As the paint is removed from the bath by subsequent workpieces, the concentration of amine solubilizer in the bath increases. The use of a closed-loop ultrafiltration system (Fig. 11) allows both the maintenance of ionic balance of the paint bath and selective removal of water and solubilizer. The ultrafiltrate is used to rinse the excess paint dragged out of the paint bath by the workpiece. This closed-loop rinse cycle decreases the cost of providing deionized rinse water and eliminates the need for potentially costly waste-treatment systems for discharged effluent.

Oil–Water Separation

Ultrafiltration systems have been used in oil-water separation applications such as

- Separation of water-soluble coolants, cutting and grinding oils, and lubricants in metal-machining operations
- Separation of oil from aqueous discharges from parts wash, rinse water, and floor wash in metal-working facilities
- Separation of lubricating and cooling oils used in rolling and drawing of metals in foundry operations

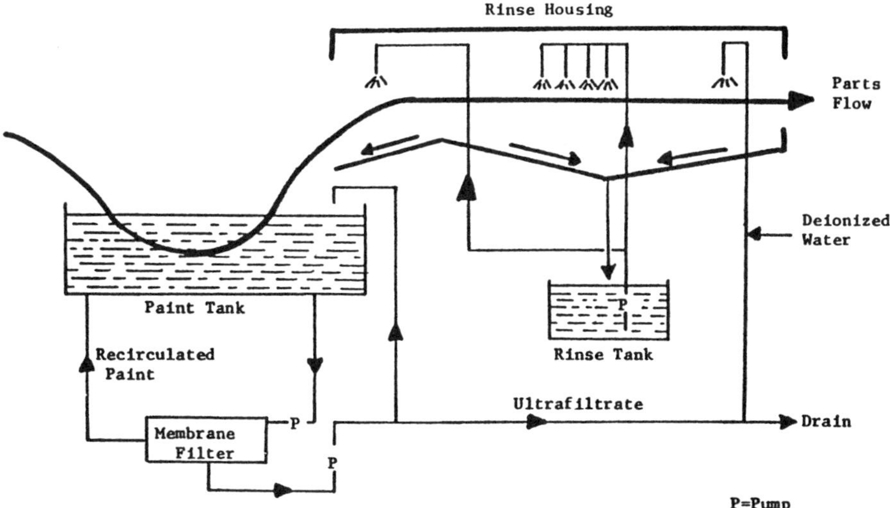

Figure 11 Process schematic for electrocoat paint regeneration.

- Separation of emulsified and water-soluble oily material collected from a variety of industries
- Separation of natural fats and oils from animal and plant processing wastes
- Separation of oil from aqueous discharges resulting from cleaning railroad tank cars
- Separation of oils from wool-scouring and fabric-finishing processes

Recovery of oils from these processes serves valuable functions in most industrial applications: (1) the oil recovered often has economic value either for reuse in the process or as a source of process heat, (2) removal of oils from process effluent is a pollution abatement function.

Dewatering of Metallic Colloids

Metallic colloids can be removed from process and wastewater streams by means of ultrafiltration in a variety of specific industrial applications:

- In the metal-plating industry, metal ions that would normally be disposed of with the waste rinse water can be reacted to form a colloid which is concentrated by means of ultrafiltration and recycled to the plating bath.
- Gelatinous ferric hydroxide can be removed from boilers and from nuclear power plant cooling water by ultrafiltration.
- With the increasing value of silver, the recovery of colloidal silver halides from photographic process wastewater not only reduces pollution but also recovers a valuable resource.

Pollution Abatement

The use of ultrafiltration systems in pollution abatement applications has particular utility in the pulp and paper industry and in sewage treatment. In the pulp and paper industry, the initial focus of interest was reverse osmosis membrane applications. However, the most valuable molecules in the process waste stream are the high molecular weight lignins and lignosulfonates which also impart the color often associated with pulp and paper mill wastes. Ultrafiltration can be used to remove these compounds, effecting 99% removal of color, 64% removal of chemical oxygen demand, and 43% removal of the total dissolved solids.

GLOSSARY OF TERMS

Terms commonly used in membrane filtration water treatment are:

Aromatic polyamide: Membrane material used for reverse osmosis membranes.

Cellulose triacetate (CTA): A modified cellulose wherein all available reactive sites have been acetylated. Material used for hollow fiber reverse osmosis membranes.

Compaction: Tightening of a membrane structure due to compressive stress of operating pressure.

Concentrate (*reject, brine, waste*): The waste stream from the reverse osmosis module which contains most of the dissolved solids of the feed in a concentrated form.

Conductivity: A measure of salt concentration due to the ability of dissociated ions to conduct electricity, usually expressed as microhms per centimeter.

Feed (*feedwater*): The water to be treated by the reverse osmosis module pumped into the module under high pressure.

Flux (*productivity*): The measurement of rated output (product) expressed in gallons per 24-hr day under a given feedwater temperature, salinity, and operating pressure.

Hydrolysis: Chemical reversion of cellulose acetate back to cellulose by removal of acetate groups occurring most rapidly at very high or low pH.

Ion: An electrically charged atom, radical, or molecule.

Cation: Positively charged ion, such as Ca^{2+}, Mg^{2+}, Fe^{2+}, or Na^{2+}.

Anion: Negatively charged ion, such as CO_3^-, SO_4^-, HCO_3^-, Cl^-, or NO_3^-.

Modified cellulose acetate: Membrane material used in reverse osmosis elements.

Osmosis: The diffusion of a solvent (such as pure water) from a dilute saline solution into a more concentrated saline solution through a semipermeable membrane separating the two solutions.

Osmotic equilibrium: Osmotic equilibrium is reached when no further water transport takes place owing to equalization of the solute concentration on both sides of the membrane.

Osmotic pressure (*head*): Osmotic pressure (head) is the change in static head resulting from water transport through the semipermeable membrane separating two saline solutions of different concentrations.

Polyelectrolyte: Water-soluble polymers containing ionizable groups as part of their structure; often useful in causing agglomeration of fine suspended solids. Used often in pretreatment of feed.

Product (*permeate*): The purified portion of the feed passing through the membrane wall and exiting via the product collector tube or channel and collected at a port provided on the module end plate.

Recovery (*conversion*): The percent of feed converted into product:

$$\frac{\text{Feed Rate} - \text{Concentration Rate}}{\text{Feed Rate}} \times 100$$

Reverse osmosis: Application of pressure in excess of the inherent osmotic pressure which causes the solvent from the more concentrated solution to flow through the membrane into the more dilute solution.

Salt rejection: The percentage of the dissolved solids held back by the membrane. A measure of membrane effectiveness:

$$\frac{\text{Feed Concentration} - \text{Product Concentration}}{\text{Feed Concentration}} \times 100$$

Semipermeable membrane: Impedes passage of solute but allows solvent flow.

Solute: Material (salts) dissolved by solvent (water).

Total dissolved solids: Total dissolved inorganic and organic materials present in the RO streams. Usually expressed as parts per million (ppm) or milligrams per liter.

14
Steam Production and Cooling Tower Water Treatment

INTRODUCTION

The objectives of modern boiler water treatment are to maintain the integrity and performance of steam boiler components. Although proper water treatment practices alone can not assure meeting these goals, the lack of proper water treatment will assure that they will not be obtained. There has been an increasing worldwide awareness that industrial growth and energy production from fossil fuels are accompanied by the release of potentially harmful pollutants into the environment. In addition to the obvious air pollution associated with steam generation, water pollution can take the form as a secondary pollutant by-product of, for example, an air-cleaning operation, a thermal pollution caused by condensing steam, blowdown, or fuel-handling runoff.

STEAM-GENERATING SYSTEMS

Before considering the specifics of boiler water treatment, it is helpful to focus briefly on the functional components of a typical steam-generating system and explain the importance of a water-treatment system on performance and maintenance requirements.

Boiler feedwater, regardless of the type and extent of external treatment, may contain contaminants that can cause deposits, corrosion, and carryover.

Deposits directly reduce heat transfer, causing higher fuel consumption, high metal temperatures and eventual material failures. Deposits, although most serious in the boiler, may also cause problems in the preboiler or after-boiler systems.

Corrosion not only results in component failure but also produces serious levels of metal oxide contamination which may deposit elsewhere. As in most water-handling systems, the problems of deposit formation and corrosion are so closely related that both must be jointly corrected or prevented in order to achieve satisfactory results.

Carryover results in superheater, turbine, and condensate system deposits and corrosion and/or erosion problems. Very serious losses in efficiency, especially when superheated steam is used for power generation, result from loss of superheat, turbine blade deposits, erosion, or corrosion. Even extremely low levels of carryover may cause failures and system outages.

The goals of modern water technology are to maintain the integrity and performance of steam boiler components. Although proper water treatment practices alone can not assure meeting these goals, the lack of proper water treatment will assure that they will not be obtained. Proper water treatment can result in a number of important benefits:

- Minimal water side deposits of impurities and corrosion products. If permitted to grow uncontrollably, such deposits can lead to appreciable acceleration of corrosion and unacceptable pressure drops.
- Limit corrosion rates so that boiler components meet their expected lives without premature damage or failure.
- Prevention of excessive carryover of impurities with the steam. Such carryover may result in damage to superheaters, steam turbines, or downstream process equipment.

All parts of the steam-water system are interdependent. In deciding on the treatment to be used, the entire system must be considered as a whole. Mechanical designs which minimize cracks, crevices, and high stress zones reduce the likelihood of accelerated corrosion attack. Proper pretreatment during startup can minimize initial boiler surface deposits.

System Components

Steam is the primary medium used to produce electricity. Even though the fuel used to produce heat may vary, the basic process for generating steam is virtually the same. Figure 1 illustrates a typical high-pressure steam system.

External Treatment Equipment

The extent to which pretreatment is undertaken is determined by raw water characteristics, the systems operating pressure, required steam quality, average

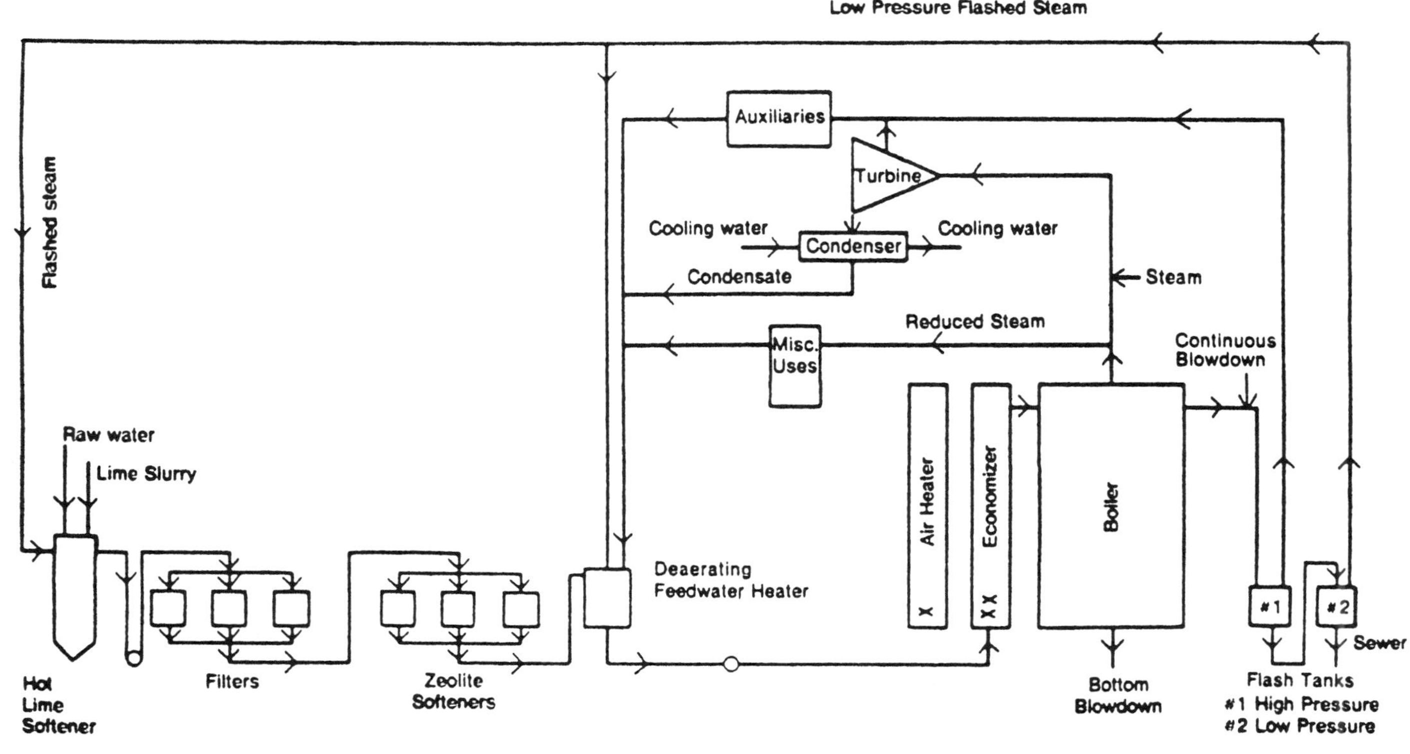

Figure 1 Typical high-pressure steam system.

steam load, amount of returned condensate, type of boiler, fuel cost, the degree of heat treatment, and overall capital cost.

External treatment is normally of minimal significance in low makeup, low-pressure boilers (>100 psig). Medium pressure boilers (100–600 psig) generally use sodium zeolite softening and occasionally also use hot or cold process softening. High-pressure boilers (>600 psig) usually employ demineralization. Condensate polishing for iron and copper removal are frequently used in systems where no return-line corrosion control is possible or desirable such as the case in certain pharmaceutical plants and breweries. Refineries, vegetable oil plants, and foundries often remove oil from condensate as part of their pretreatment. The primary objective of most pretreatment systems is the reduction of deposits in boilers and superheaters.

Deaerating Feedwater Heater

Deaerating heaters serve two purposes in a steam-generating system: to increase feedwater temperature by direct contact with low-pressure exhaust steam and to remove undesirable dissolved gasses. Early designs allowed for a countercurrent flow by simply bubbling steam in an open vessel. These primitive deaerators achieved modest reduction in oxygen. Current designs can remove dissolved oxygen to at least 0.005 ml/L. The improved performance is due to more efficient hardware such as tray- and spray-type deaerators, or a combination of the two.

Tray-type deaerators promote efficient scrubbing by reducing water to a fine spray as it flows freely over several trays or baffles. Spray-type deaerators work on the same general principles, however. Nozzles located in the top section of the unit spray water into a plenum for direct contact heating and primary deaeration. Secondary deaeration takes place in a second-stage spray valve or tray through which incoming steam passes. Most deaerating feedwater heaters provide a storage tank which can be advantageous in introducing chemicals.

Economizers

Simply stated, an economizer is a heat exchanger placed in the gas passage between the stack and boiler which recovers heat of combustion by preheating feedwater. It should be noted that increased temperature aggravates and accelerates corrosion and deposition on the waterside of the economizer. As a result of this situation, feedwater treatment is extremely important. A typical economizer is illustrated in Figure 2.

Blowdown Systems

Blowdown is the purging from a system of a small portion of the concentrated boiler water in order to maintain and balance the maximum level

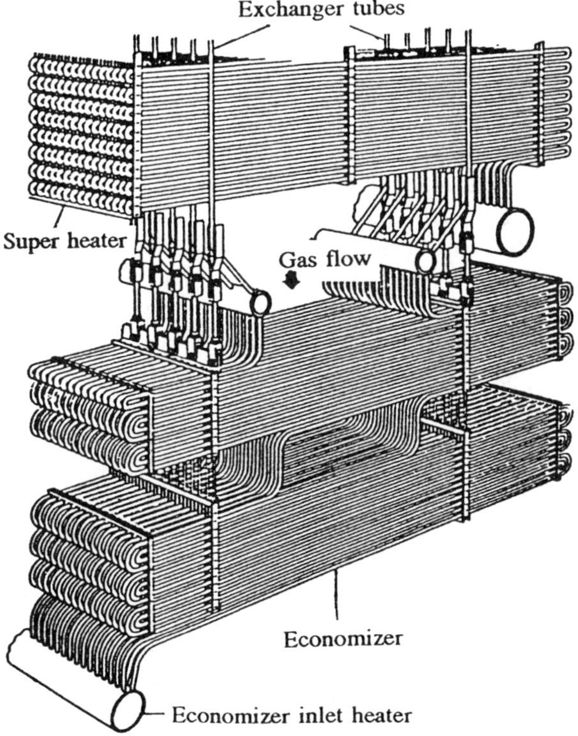

Figure 2 Economizer.

of dissolved and suspended solids in the system. Blowdown may be continuous or intermittent.

Continuous surface blowdown offers the most economical and consistent control of total dissolved solids or any other specific dissolved solids. Such blowdown exits the boiler through a perforated line positioned about 6 in. below the normal operating water level in the steam drum. The flowrate through this line is controlled by the use of an adjustable orifice, a needle valve, or a V-notch valve. Automatic control of continuous blowdown can be accomplished using a conductivity monitor to actuate the process.

Manual blowdown, also known as bottom blowdown, usually is present even in a system that uses continuous blowdown to control cycles of concentration. Settled sludge is removed by manual blowdown. Frequent manual blowdown of short duration is typically more effective than occasional blowdown of longer duration. Special precautions must be taken to assure that flashed steam is vented safely. In situations where the blowdown flow is large enough, the water flow from the flash drum may contain enough heat to justify passing it through a heat-recovery exchanger.

Boilers

The most commonly used boiler is the water tube boiler. In this type of boiler water is converted to steam inside tubes while hot gases pass on the outside. The tubes are interconnected to common water channels, drums, and steam outlets. Tube banks are generally constructed with a series of baffles which channel the combustion gases across the heating surface to obtain maximum heat absorption. Water tube boilers are available in a wide range of designs that vary, for example, in operating pressure, capacity, quality of steam produced, type of fuel, and installation and start-up cost. A diagram of a typical water tube boiler is shown in Figure 3.

The need for increased boiler efficiency in electric utility boilers has led to extremely high operating temperatures. At 3,206.2 psia (the critical pressure)

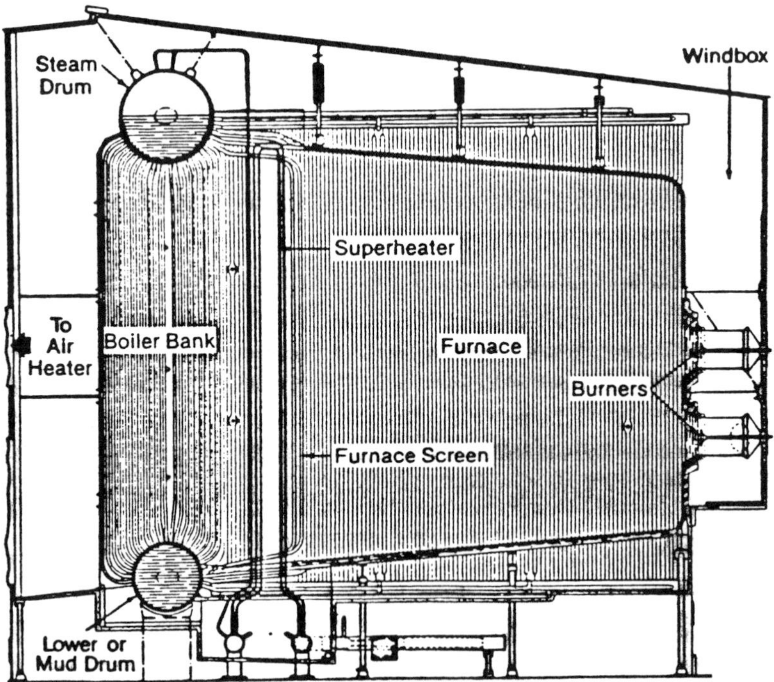

Figure 3 Water tube boiler.

and above, the water and steam in the boiler become a single phase. Boilers designed for this pressure are built on a once through monotube design. Monotube boilers are actually composed of many tubes, connected by headers, to provide water supply and steam collection. To ensure equal distribution of fluid in the tubes, the header is equipped with orifices. The water entering the tubes leaves as a single phase, "supercritical" fluid which then flows through a superheater before going to the turbine. As a result boiler tubes are suspectable to failure as a result of corrosion. Boiler tube failure ranks as the number one equipment problem in fossil power plants. To prevent deposition of impurities in these tubes, the feedwater must be of extremely high quality. Stringent control of water chemistry is required at all times.

Steam Drum Internals

The main purpose of the steam drum in a water tube boiler is to provide sufficient volume and low enough velocity to allow separation of steam from water. This can be assisted significantly by the addition of mechanical devices within the drum, causing the steam to travel greater distances on its path to the outlet header. Such devices can greatly reduce the mechanical entrainment of water droplets in steam (mechanical carryover) but do not affect vaporous materials dissolved in steam (volatile carryover).

The problem of designing efficient drum internals is intensified by increasing the pressure at which systems operate. This primarily derives from changes in the physical properties of water and steam when temperature is increased. Both surface tension and the difference in density between steam and water fall as pressure and temperature rise.

Although separation equipment has a marked effect on the boiler, other factors are also significant. Carefully selected and controlled internal water treatment plus pretreatment of feedwater reduce the tendency of boiler water to carry over. The steam drum in most designs is not only the withdrawal point for continuous blowdown but the entry point for boiler feedwater and internal chemical treatment. Special designs and multiple drum boilers may have these lines installed in different drums. Improper placement and distribution can easily cause problems. In general the following arrangement is desirable:

- Continuous blowdown line in the area of maximum concentration of boiler water (near the risers).
- Feedwater lines arranged to supply downcomers equally, without short-circuiting to the continuous blowdown.
- Chemical feed line, if on a phosphate program, located to distribute treatment evenly to the downcomer side of the feedwater, but only after the feedwater has mixed with the alkaline boiler water. It does not short-circuit to the continuous blowdown.

Superheaters

Saturated steam leaving the drums of large industrial and utility boilers is commonly directed through superheaters before leaving the boiler. In a superheater the steam is heated above the saturation temperature for the particular boiler pressure. Superheated steam contains more BTUs per pound than saturated steam. Superheated steam also has the additional benefit of containing less moisture thus reducing corrosion problems. Superheater tubes have steam on one side and hot combustion gases on the other. Surface temperatures are consequently higher than in a boiler tube.

Whether carbon steel or alloy steel is used, it is critically important that proper steam flow by maintained in each superheater tube to avoid overheating. Operation at steam temperatures above design for the tube metal employed, even though not sufficiently high to result in tube failure, can result in excessive iron oxide formation inside the tube. This oxide tends to scale from changes in temperature during startup and shutdown and may cause severe abrasion of a steam turbine's nozzle block and first stage blades. It is equally important that tube surfaces be cleaned internally and externally and that carryover be minimized. Otherwise, internal deposits of water solids become a distinct danger resulting in overheating and possible failures. Where reheaters are employed, the same precautions apply as previously discussed.

Turbines

Turbines are rotary devices, powered by boiler steam, usually superheated. Their function is to drive some other rotating piece of equipment. Most often they are used to drive generators which produce electric power. In most units, the maximum energy in the steam is utilized to produce power before the exhaust steam is condensed and returned to the boiler as feedwater. In some steam electric utility plants and in many industrial power plants, steam energy may be partially used to generate electricity. The remaining energy, as lower pressure steam, is exhausted by a topping turbine. This exhausted steam can be used for various industrial applications. This simultaneous production of electricity and process steam is referred to as cogeneration.

Requirements for steam purity, or absence of carryover, vary widely according to how the steam is used. As a general rule, highest purity is required when the steam is used to drive turbines. This usually coincides with boiler operation at relatively high pressures.

The economy and performance of a turbine depend upon the design and construction of its blading. A schematic of a steam turbines blading arrangement is presented in Figure 4. Since turbines are built with very close tolerance, stationary and moving parts have little clearance. Any vibration, therefore, has a damaging effect. Of special concern is the erosion of turbine blades by

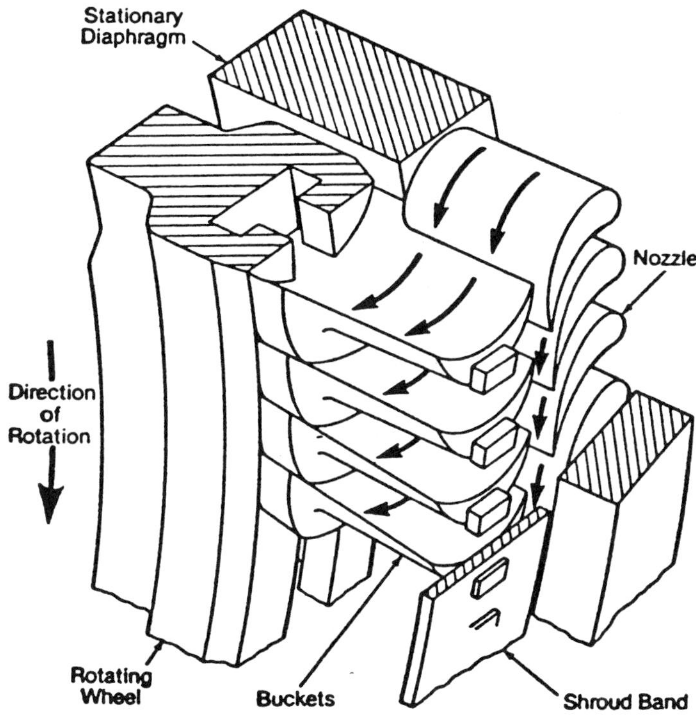

Figure 4 Steam turbine blading view.

condensation. Deposits distort turbine blades and nozzle shapes. This produces a rough surface and increases steam flow resistance. Secondly, heavy blade deposits can cause the turbine to become unbalanced, producing intolerable vibration.

Generally, turbine deposits are caused by either carryover or corrosion. Boiler water carryover can be caused by a high level of total solids in the boiler water, high boiler water alkalinity, feedwater quality, and mechanical difficulties. Generally, the deposit that forms on the turbine blades as a result of such carryover can be rinsed from the blades with the condensate formed during low load or unloaded operation.

Corrosion caused by insufficient feedwater treatment when the system is out of service or by improper boiler and turbine layup procedure can also be a serious problem in turbines. Partial or complete reblading is a frequent consequence of failure to observe good out-of-service procedures.

Condensers

Exhaust steam leaving a turbine must be used in the plant or be condensed before it reenters the boiler as feedwater. Condensers operate with a vacuum in order to condense steam. The vacuum reduces back pressure on the turbine and thus greatly increases the unit's efficiency. The surface condenser is most commonly used. It consists of a closed vessel filled with a tube bundle. Cooling water flows through the tubes while steam flows around the outside. The coolant flowing through the tubes can either be used on a once-through basis or a recirculated through a cooling tower. Condensate flows out at the base of the condenser to a hot well where it is ready to reenter the water steam cycle. Noncondensible gases and air which have entered through inleakage are removed by ejectors or vacuum pumps from a special air-removal section of the condenser. Air and oxygen removal not only reduce back pressure on the turbine but also minimize iron and copper pickup corrosion in the preboiler system.

In addition to corrosion, condensers are subject to erosion and vibration on the steam side. Condensate droplets in the exhaust steam, at sufficiently high velocity, can erode portions of the top two rows of tubing to the point of failure. Shields or grids of a more resistant alloy are placed on or ahead of the first row of tubing to correct the problem. Tube vibration can also result in rapid tube failure by fretting or fatigue. Factors responsible are inlet steam velocity and space between the tube support plates in excess of the critical values for tube material, diameter, and wall thickness. Elimination of vibration in existing condensers often requires reducing turbine capacity.

CORROSION AND DEPOSIT CONTROL

Corrosion and deposits in boiler systems arise from a number of sources: preboiler component corrosion products, condenser cooling water contaminant in leakage, makeup water impurities, preexisting oxides found in new components, and corrosion products from the boiler materials themselves. Water treatment addresses these issues through the following activities:

- Raw makeup water treatment
- Preoperational cleaning and procedures
- Internal water treatment for boilers
- Condensate treatment
- Chemical cleaning

Raw Makeup Water

Pure water rarely exists and all natural waters contain varying amounts of dissolved and suspended matter. The type and amount of impurities vary with the source, such as lake, river, well, or rain, and with the location of the source. Rain brings atmospheric gases such as oxygen, nitrogen, and carbon dioxide into the water. Water dissolves and picks up minerals as it percolates through soil. Surface water frequently contains organic material which must be removed. Solids appear in both a suspended and dissolved form in water.

Operating power plants in areas where limited water sources exist require a "zero discharge" facility design. This design allows for minimal water waste by maximizing water treatment incorporating a multitude of preliminary treatment technologies.

Preliminary Treatment

Suspended solids such as mud, silt, and clay can be removed through filtration. This process can be accelerated by coagulation. Coagulation is a process by which finely divided materials are combined by the use of chemicals to produce large particles capable of rapid settling. Typical coagulation chemicals are alum and iron sulfate. Preliminary treatment also involves chlorination of water for the destruction of living organisms. A schematic of a preliminary pretreatment assembly is illustrated in Figure 5.

Following coagulation and settling, the water is passed through filters. Activated filters may be necessary to remove traces of organics and excess chlorine.

Sodium Cycle Softening

After removing the suspended materials, scale-forming constituents still remain and require further treatment. Sodium cycle softening uses resin materials (denoted by R) that can exchange sodium (Na) for the hardness constituents, calcium (Ca), and magnesium (Mg). The process continues until the sodium ions in the resin become depleted or the resin can no longer absorb the calcium and magnesium. A typical reaction for the exhaustion of sodium zeolite is as follows:

$$2R^- + Na + Ca^{++} + CO_3^= \rightleftharpoons R_2Ca + 2Na^+ + CO_3^=$$

where $CO_3^=$ is the carbonate ion.

The forward reaction is chemically favored. Regeneration is accomplished by passing a high concentration of salt solution (NaCl) through the resin. This increases the Na concentration on the right side of the equation. The reaction is then overwhelmed with sodium, forcing it to the left. This is known as the mass action effect.

A commonly used process in boiler water treatment is the hot lime zeolite softening. In this process, hydrated lime reacts with the bicarbonate alkalinity of the raw water. The precipitate, calcium carbonate, is filtered from the solution. To reduce silica, the natural magnesium of the raw water can be precipitated or magnesium hydroxide ($Mg[OH]_2$) can be added as a natural silica absorbent. Reactions are carried out in a tank that is just ahead of the zeolite softener tank. The effluent from this tank is filtered and then introduced into the zeolite softener. There is always some residual hardness leakage from the hot process softener that must be removed in the final zeolite process. This system is shown in Figure 6.

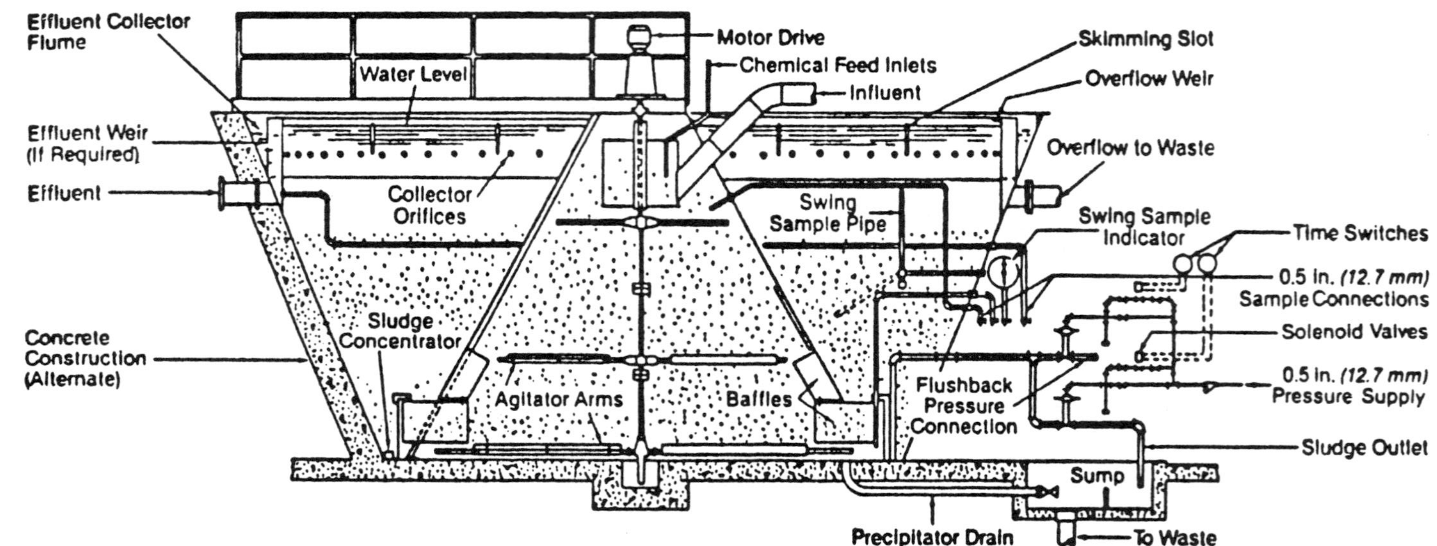

Figure 5 Preliminary treatment assembly.

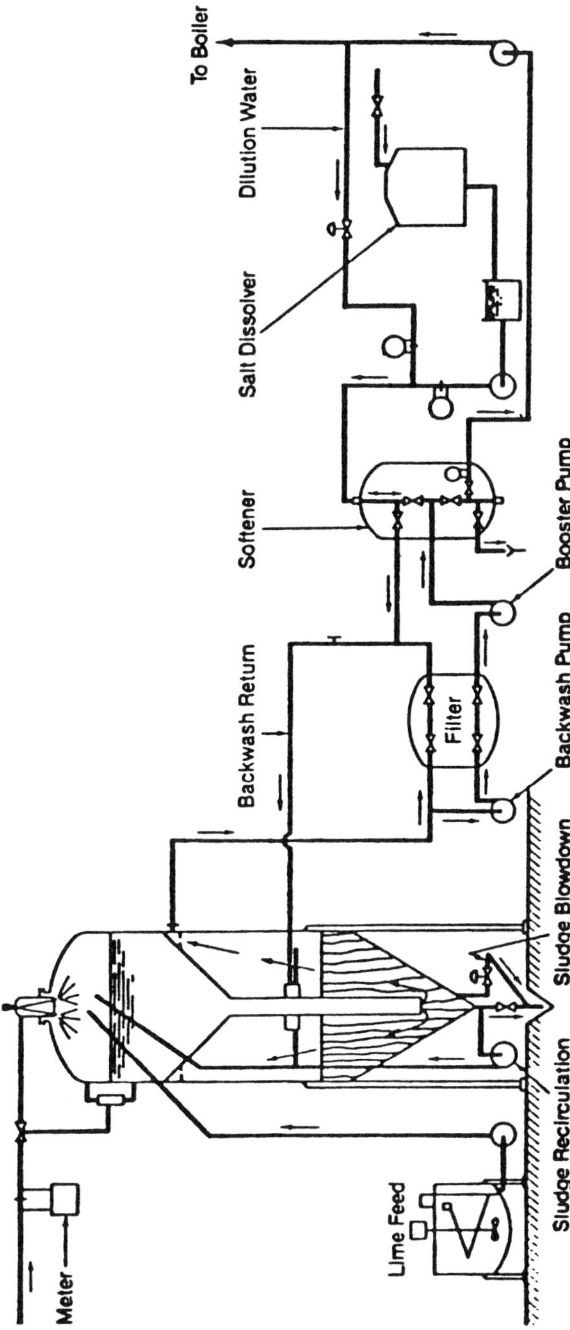

Figure 6 Hot lime zeolite softening process.

Demineralization

At boiler drum pressures above 1,000 psi, demineralization or evaporate distillation of makeup water is desirable. Treated makeup water that closely approaches theoretical chemical purity can be obtained by either evaporation or ion exchange and are discussed elsewhere in this book.

Evaporation is a distillation process and the typical evaporator consist of a steel shell and tube heat exchanger and an external condenser. The tubes are submerged in makeup water. Steam supplied to the inside of the tubes give up heat to the surrounding water. The heat causes the makeup water to evaporate, forming a vapor which leaves the evaporator. The vapor is recondensed forming purified water.

Ion exchange is similar to zeolite softening. Cations in the solution are replaced with hydrogen ions by the process describe for the zeolite system. In addition, the anions in the solution are replaced by hydroxide ions through the contact with an anion resin regenerated in the hydroxide form. For makeup water treatment, two tanks are normally used in series in a cation-anion sequence. Effluent from demineralization is approximately neutral (pH = 7) with nearly all salts removed.

Other Techniques

Several other processes are also available to purify water. Although not extensively used, applications for these methods can be desirable.

Osmosis is the process in which solvents flow through a semipermeable membrane from a less concentrated solution to one with higher salt concentrations. This normal osmotic flow can be reversed by applying hydraulic pressure to the more concentrated solution to produce a high-purity water. Reverse osmosis in conjunction with demineralization has proved to be extremely effective in the removal of dissolved solids. A schematic of a reverse osmosis unit is provided in Figure 7.

Electrodialysis is a membrane process where an applied electrical charge draws impurity ions through permeable membranes to create high purity feedwater streams. A schematic of an electrodialysis stack for water purification is shown in Figure 8.

Ultrafiltration is another process that forces water through a filtering membrane by means of a pressure gradient, leaving large particles behind in a concentrated reject stream.

Preoperational Cleaning

In general, all boiler systems receive an alkaline boilout—hot circulation of an alkaline mixture with frequent blowdown and final draining of the unit. Many systems receive preoperational chemical cleaning. Chemical cleaning of the

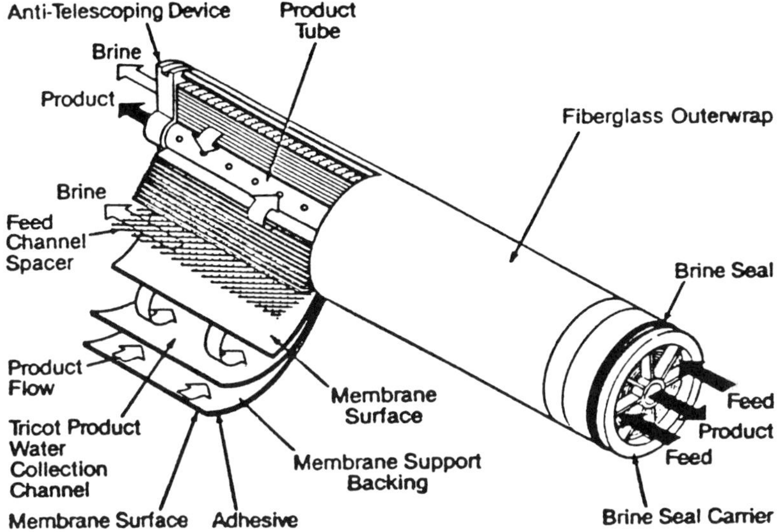

Figure 7 Reverse osmosis unit.

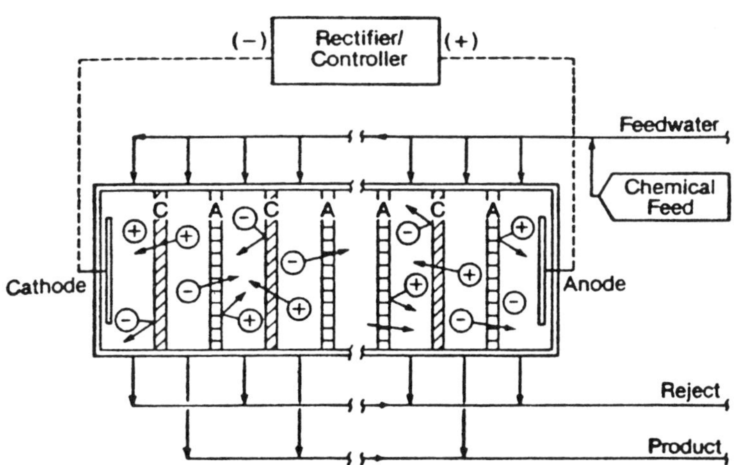

Figure 8 Electrodialysis stack.

superheater and the reheater is not necessary for initial operation. However, these surfaces should be given a conventional steam blow—a period of high-velocity steam flow which carries debris from the system.

Alkaline Boilout

All new boilers should be flushed and given an alkaline boilout to remove debris, oil, grease, and paint. The boilout can be accomplished with a combination of trisodium phosphate and disodium phosphate, with a small amount of surfactant added as a wetting agent.

Chemical Cleaning

After boilout and flushing are completed, corrosion products may remain in the feedwater system and boiler in the form of iron oxide and mill scale. Chemical cleaning removes some of these corrosion products. Different solvents and cleaning processes are frequently used and include

- Inhibited 5% hydrochloric acid with 0.25% ammonium biflouride and a corrosion inhibitor
- 2% Hydroxyacetic/1% formic acids with 0.25% ammonium biflouride and a corrosion inhibitor
- Inhibited ammonium salts of ethylenediaminetetra-acetic acid

Internal Water Treatment for Boilers

Direct boiler water treatment, usually referred to as internal treatment, is used to prevent deposit formations caused by hardness constituents and to provide pH control for corrosion prevention.

The permissible amount of contaminated and treatment chemicals entering the boiler decrease with rising boiler pressures. As boiler operating pressures increase, the boiler water chemistry may have a significant impact on the transport of contaminated solids to the superheater and turbine. Such impurities in the boiler water have a significantly detrimental effect if too high. Solids are carried into the steam in one of two ways:

- Mechanical carryover of water droplets in steam
- Vaporous carryover of constituents such as silica in the steam

Figure 9 shows a relationship between dissolved solids and average solids in steam at various operating pressures. The relationship depends on the on the boiler water chemistry.

There are various methods of internal treatment used. These include hydroxide, phosphate (conventional, coordinated, congruent, and equilibrium), volatile, chelant, and polymer. The treatment method is generally dictated by the pressure range of the unit and the nature of the feedwater.

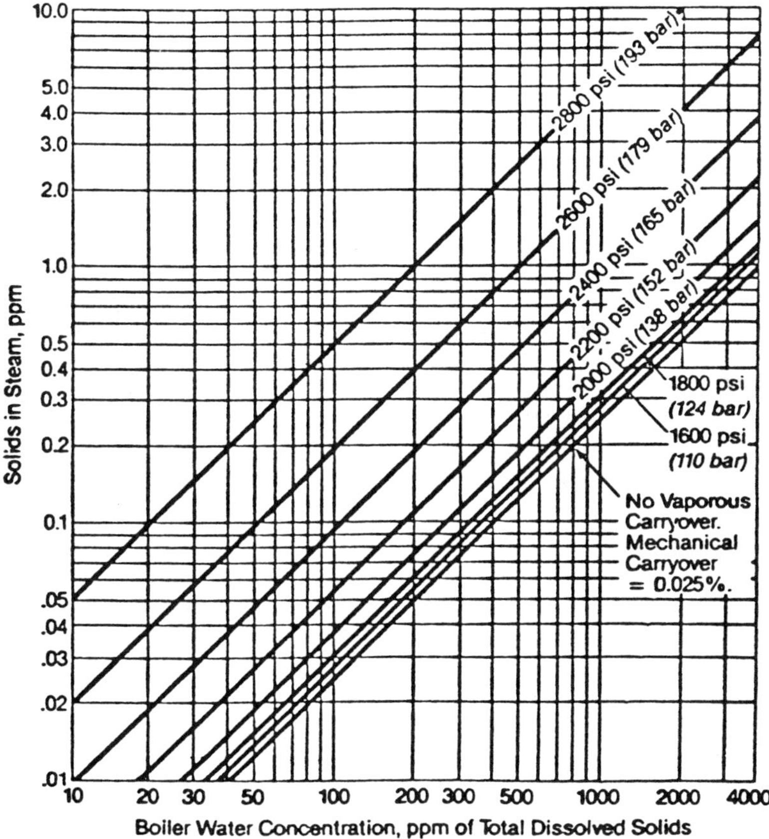

Figure 9 Solids in steam versus dissolved solids.

Hydroxide Treatment

This treatment is used in very low-pressure boilers. In this method, a few parts per million of NaOH are maintained in the boiler water to control boiler water pH in the range of about 10.0–11.5. Because NaOH is very soluble at operating temperatures, accumulation during normal operation is not a problem, and operators find it easy to control within established limits. The alkalinity present provides a limited amount of neutralizing capacity for mineral acids which are generally formed from salts, such as magnesium and calcium chlorides, and sulfates which enter during periods of condensate leakage.

Conventional Phosphate Treatment

This method, in which phosphates and caustic are added to the boiler water, is common for industrial boilers operating below 1,000 psi. The primary purpose of phosphate addition is to precipitate the hardness constituents. Calcium reacts with phosphates under the proper pH conditions to precipitate calcium phosphate as calcium hydroxyapatite; this adheres less to boiler surfaces than simple tricalcium phosphate, which is precipitated below 10.2 pH. Caustic reacts with magnesium to form magnesium hydroxide or brucite, which is formed in preference to magnesium phosphate above 10.5 pH.

Coordinated and Congruent Phosphate Treatment

Coordinated phosphate pH treatment was designed to provide boiler water alkalinity with mixtures of disodium and trisodium phosphate to minimize free sodium hydroxide and its corrosion potential when concentrated under deposit or in a crevice location. The control parameters involved maintaining a Na to PO_4 ratio at or slightly below 3:1. The precipitate from a concentrated solution of trisodium phosphate ($NaPO_4$) at elevated temperatures has a sodium to phosphate ratio of less than 3, and the supernatant liquid is correspondingly rich in sodium hydroxide (NaOH). The sodium hydroxide can destroy the magnetite protective film and boiler surfaces.

Equilibrium Phosphate

This method involves determining the level of phosphate which is stable at maximum operating pressures and not exceeding this phosphate level. Chemical injection practices as well as the loss of phosphate from the boiler water should be monitored by material balance calculations. The concentration of additives should be controlled by the feed rate.

Volatile Treatment

In this method, no solid chemicals are added to the boiler or preboiler cycle. As a result, the volatile carryover of solids is eliminated and turbine deposits are avoided. Cycle pH is controlled between 9.0 and 9.5 with a volatile amine such as ammonia. Hydrazine, or a suitable substitute, is added as an oxygen scavenger. With volatile treatment the feedwater must not contain hardness or condenser leak constituents.

Chelants

These organic agents react with the residual divalent ions such as calcium, magnesium, and iron in the feedwater to form soluble complexes. The resultant complexes are removed through continuous blowdown. The chelant feed should be based on known quantities of hardness in the feedwater; the objective is to maintain a residual approximating 1 ppm of chelant. If sludge deposits accumulate, the chelant feed should be increased by 1 or 2 ppm. Thermal decomposition of chelanting agents must also be considered in order to maintain proper chelant levels.

Polymer Treatment

Synthetic polymers have been used in all polymer treatment programs to control both hardness and iron oxide deposition. Their limitation is thermal stability and are normally used above 1,200 psi.

Condensate Treatment

In many cases, condensate does not require treatment prior to reuse. Makeup water is added directly to the condensate to form boiler feedwater. In some cases, however, especially where steam condensate is contaminated by corrosion products or by inleakage of cooling water or substances used in process.

The presence of acidic gases in steam makes the condensate acidic, with consequent corrosion of metal surfaces. In such cases, the corrosion rate can be reduced by adding chemicals (neutralizing or filming amines) that produce alkaline gases or form a protective barrier in the condensate system.

Where significant quantities of magnetic oxide or other magnetic species are present in the condensate feedwater system, an electromagnetic filter (EMF) can effectively remove these suspended solids. Advantages of the EMF are a low-pressure drop through the filter, minimal required quantity of flush water, a short backwash process period (minutes), and no chemical requirements.

Chemical Cleaning

The internal surfaces of boiler water side components accumulate deposits even though standard water-treatment practices may be followed. These deposits are generally classified as hardness-type scales or soft porous-type deposits. A regular chemical cleaning program should be developed to alleviate this problem. Carefully planned chemical cleaning of superheater and reheat surfaces has been effective in reducing the number of steam blows and limiting system corrosion. In general, four steps are required in a complete chemical cleaning process:

- Internal heating surfaces are washed with a solvent containing an inhibitor to dissolve or disintegrate the deposits.
- Clean water is used to flush out loose deposits, solvent adhering to the surfaces, and soluble iron salts. Corrosive or explosive gases that may have formed in the unit are displaced.
- Unit is treated to neutralize the heating surfaces. This treatment produces a passive surface.
- Unit is flushed with clean water to remove any remaining loose deposits.

The two generally accepted chemical cleaning methods: (1) continuous circulation of the solvent and (2) filling the unit with solvent, allowing it to soak and then flushing the unit.

WASTEWATER TREATMENT

Steam-generating facilities produce significant waste streams which can be potentially hazardous to the environment. One method of combating this problem is effective water treatment. Several waste streams, associated with steam generation, have been identified and are discussed below.

Once-through cooling water: Water from rivers, lakes, or oceans is used to absorb heat from a steam condenser. This water is treated with chlorine to prevent fouling before entering the condenser and the cooling tower. Concern over the total residual chlorine to living organisms exists.

Cooling tower blowdown: The removal of suspended solids through blowdown may contain chemicals for control of biofouling, corrosion inhibitors, chemicals for scale control, and products of corrosion.

Ash handling water waste: Ash is removed from a hopper with sluice water. This water will thus contain pollutants associated with the fossil fuel they remove.

Coal pile runoff: Water and oxygen from air react with the minerals in coal to produce a leachate contaminated with ferrous sulfate and sulfuric acid.

FGD blowdown: An aqueous blowdown discharge saturated with calcium sulfate, calcium sulfite, and sodium chloride results from this process.

Metal-cleaning waste: These include discharge from ion exchange water treatment, evaporator blowdown, boiler blowdown, and floor drains.

Wastewater generated by an industrial facility can differ greatly from that of a fossil-fired electric-generating facility. Small industrial plants can either send the entire wastewater stream to sanitary sewer or segregate the waste stream. When segregated, the municipal waste goes to the sanitary sewer and the industrial waste is discharged into a local waterway after appropriate treatment. Treatment can be a simple clarification process or an integrated treatment process including neutralization, filtration, oil removal, and/or biological degradation.

All power plants produce waste streams containing chemicals in various concentration that must be treated. The common treatment in coal-fired plants is the use of ash settling ponds. Ion exchange regeneration wastewater streams are normally mixed in retaining sumps and are neutralized prior to discharge into an ash pond. Ash pond overflow is normally sufficiently low in suspended solids and can be discharged into a local waterway. Settled solids in the pond are periodically removed and shipped to solid waste–disposal facilities.

The effective treatment of water significantly reduces corrosion, deposits, and carryover and increases plant performance and equipment life. Treatment of raw makeup and condensate water combined with internal boiler treatment and regular cleaning result in an overall effective treatment process. Enhanced plant performance and equipment life results in considerable cost savings. The treatment of waste streams is important. Although cost savings are not realized through wastewater treatment, the operators of steam-generating facilities are aware of the waste produced and can take effective measures to eliminate or reduce them.

COOLING TOWER WATER TREATMENT

Water has experienced extensive use in cooling operations both because it is an excellent cooling medium and because it is abundant and inexpensive. All natural waters do, however, contain dissolved solids, gases, and a variety of suspended matter in different amounts. These contaminants can be the source of varying operating problems. Bicarbonates and sulfates of calcium, sodium, and iron are the most common dissolved solids. The amount of each will depend on their abundance in the earth at the source of the water. Carbon dioxide is the most common of the dissolved gases found in water and the highest concentrations exist in waters from shallow wells and lakes due to decay processes. Suspended solids may consist of silt and a variety of organic constituents. All water systems are capable of developing algae and slimes in varying degrees if environmental conditions are proper.

The presence of suspended and dissolved matter can lead to precipitation under the proper conditions causing severe scaling or fouling problems in process equipment and distribution systems. High concentrations of suspended matter can result in erosion. Both problems can be translated into lost dollars because of costly maintenance and downtimes for equipment replacement. In the case of organic constituents, the ideal environment for microorganisms to grow can exist, posing serious health hazards in certain applications.

Because of the inherent problems of scale formation and potential health hazards in some applications, cooling tower water treatment must be carefully considered in the overall cooling system design.

Problems

In addition to the common minerals absorbed from the soil, natural waters can also be affected by industrial drainage, often resulting in acidic conditions. Faulty processing equipment may introduce a variety of contaminants such as oils, fats, acids, alkalies, and hydrocarbons directly into the cooling system. Undesirable airborne contaminants, such as hydrogen sulfide and acid vapors released by processing equipment and fly ash from coal-burning furnaces, may be drawn into the tower and dissolved in the circulating water. Without proper control, the presence of any of these materials may cause corrosion of metal parts, wood deterioration, or loss in thermal performance throughout the entire cooling system.

There are five types of cooling water problems encountered in cooling tower systems. These are scale formation, corrosion, organic growths (algae and bacteria slime), suspended matter (e.g., sand, mud, silt), and oil leakage. With the exception of oil leakage, these problems can be controlled to a certain extent by standard water-treatment techniques. Different types of treatment have been employed with various degrees of success. Treatments employed include using a circulating system with a small quantity of treated makeup water, with or without the addition of chemical inhibitors to the circulating water. Another method of controlling contaminations and scale formation is the use of alloy tubes. Unfortunately, this is an expensive solution and often one that is difficult to justify economically.

In addition to the first four problems, which are of water origin, oil leakage into the water also causes problems. Oil will interfere with other treatments employed. It is therefore desirable to eliminate oil leakage as much as possible by repairing leaks as soon as they develop.

Owing to evaporation, salts contained in the water tend to concentrate and could precipitate, causing scale in the system. The scaling tendency of the circulating water can be controlled by an appropriate blowdown to lower the salt content and by the addition of treating chemicals. These chemicals are inhibitors that prevent precipitation from occurring. The corrosion problems encountered in evaporative cooling water systems concern the cooling circuit (e.g., exchangers, piping) and the cooling tower itself.

Oxygen, carbon dioxide, and various chemicals used to reduce scaling can cause corrosion. Corrosion control is provided largely by the use of inhibitors such as chromates, polyphosphates, silicates, and alkalies.

Corrosion within the tower itself is mainly due to the particular conditions existing therein (air, humidity, and temperature) and also to the chemical treatment of the water. All construction materials exposed to these conditions must be selected carefully. Hardware and piping for distribution headers have been successfully made with hot-dipped galvanized steel, coated steel, stainless steel, and silicon bronze.

Scaling and corrosion are related phenomena. The properties of water influencing both are the calcium hardness, alkalinity, total dissolved solids, pH, and temperature. Theoretically, the above conditions can be controlled, so that the water is in equilibrium and neither corrosion nor scaling results. In practice, however, this equilibrium is difficult to achieve, since it is a border condition, and a delicate balance must be maintained.

Water corrosion of iron and steel is simply oxidation of the metal forming iron oxide by galvanic action. The rate of oxidation is faster at higher oxygen concentrations. This is why corrosion is more of a problem in recirculating cooling water systems than in once-through systems. Likewise, the rate of attack is higher for waters of higher acidity because a low-pH water is a better electrolyte. Therefore, increasing the pH up to the equilibrium point decreases corrosion. However, increasing the pH further causes scale formation. The principal scale-forming material in cooling systems is calcium carbonate, which has a solubility of about 15 ppm and is formed by the decomposition of calcium bicarbonate.

Scaling results when the solubility limit of calcium carbonate is reached, at which point precipitation onto tube surfaces occurs. The extent of calcium carbonate precipitation is a function of the composition of the water and the temperature. The alkalinity, dissolved solids, and pH determine the scaling characteristics. Decreasing the pH by the direct addition of acid or by carbonization will decrease the scaling tendencies of the water within limits. If a water is on the scaling side of equilibrium, increasing the temperature will increase the scale deposition.

Calcium carbonate scale is objectionable because of the resistance offered to heat transfer in heat exchanger equipment. It is of interest to point out that increasing the temperature will increase the rates of scaling and corrosion. Scaling and corrosion are not likely to occur simultaneously, although it is possible owing to temperature differences in the system. The water could be on the corrosive side at the inlet and on the scaling side at the outlet. If water is close to this equilibrium condition, the rates of corrosion and scaling would exist, but are likely to be very small.

Corrosion is less of a problem in once-through cooling water systems, where the oxygen content is relatively low. Likewise, scaling is less of a problem in once-through systems than in recirculating systems because the water has not been concentrated, as in the recirculating case. In recirculating systems, the water is reaerated in cooling towers, which makes it more aggressive from the standpoint of corrosion. The purpose in using circulating cooling systems is to conserve makeup water; however, this is difficult to achieve because of the dissolved solids concentrate buildup from evaporation losses.

Summarizing the above discussion, the dissolved mineral matter in most natural waters consists mainly of calcium in the form of bicarbonate (temporary hardness) and chlorides and sulfates (permanent hardness). The tendency of the water to deposit scale when made alkaline by heating or to attack metals corrosively depends on the balance of these various constituents.

The scale formed under moderate temperatures is usually due to temporary (bicarbonate) hardness being converted into calcium carbonate, which occurs on heating or increase in alkalinity sufficient to result in calcium carbonate saturation. The solubility of calcium carbonate also affects corrosion, since the alkalinity of dissolved carbon dioxide in the water is greatly reduced as the saturation equilibrium is approached. Ideally, at equilibrium the various forms of carbon dioxide (free CO_2, bicarbonate, and carbonate) are so balanced that they cause neither scale nor corrosion.

Pretreatment

The prevention of scale formation and corrosion is common to all heat-transfer equipment, not just cooling towers. The need for protecting metal surfaces against corrosion in cooling water systems is essential to achieving maximum system efficiency and equipment life. Corrosion that is inadequately controlled can lead to irreversible equipment damage and costly unscheduled unit outages for cleaning operations or equipment replacement. Unscheduled shutdowns can seriously undermine plant efficiency and productivity.

Effective corrosion control programs are essential in reducing unit downtimes. To be effective, the program must address not only specific corrosion problems but anticipate and prevent them as well. Consequently, effective pretreatment in addition to other corrosion control measures is important.

Corrosion control of metal surfaces depends on the formation and maintenance of a protective corrosion inhibitor film on the exposed metal surface. This protective film may be established during normal application of a corrosion-inhibitor program; however, there will be some lag time before the film is completely built up. Metal surfaces that are exposed to the cooling water before the film is completed may become candidates for accelerated corrosion during the initial system operation. Normally, localized corrosion or pitting is common during these early stages of operation.

Allowing a unit to undergo no treatment for periods of time before being placed into operation can result in severe damage to exposed metal surfaces. In addition to the loss of the metal and shortened equipment life, voluminous and porous corrosion by-products may form and actually act as a barrier to the formation of the protective inhibitor film.

In dealing with metal water-cooling systems, today's trend is toward the use of nonchromate-based treatment chemicals. Nonchromate applications rely on less tenacious films for corrosion protection rather than conventional chromate systems. As such, it is extremely important that the corrosion protection film be established very early in the operation.

We can define pretreatment as the initial conditioning period whereby a corrosion inhibitor is applied to the metal surfaces of the cooling system. Pretreatment conditions must be conducive to the rapid formation of the protective barrier. The conditioning procedure should involve (1) the cleaning and preparation of metal surfaces and (2) the actual application of higher than normal inhibitor concentrations.

The cleaning and passivation can be done separately or in a combined step.

There are several procedures that can be employed to clean metal surfaces. Common techniques include hydroblasting, treatment with a mild inhibited acid cleaner and/or alkaline cleaner, and the use of special surfactants during cleansing. The system must be flushed thoroughly after the cleaning stage to minimize undue metal attack by residual concentrations of cleaning chemicals.

Chemical passivation should be started as soon as possible after the cleaning of metal surfaces. Accumulation of new corrosion products can occur if it is not initiated soon after cleaning. It may be achieved by treating equipment either on- or off-line.

On-line passivation involves elevating the corrosion inhibitor concentration as high as three times normal maintenance levels. At higher concentrations, the rate at which the protective film forms is accelerated. This, in turn, reduces the degree of initial corrosion on clean but unprotected metal surfaces. The rate at which corrosion protection takes place depends on the temperature, pH, and inhibitor used.

Off-line passivation involves treatment of equipment currently out of service. Treatment levels are typically higher; consequently, passivation is completed more quickly. Passivation of nonchromate treatment generally uses either a polyphosphate, zinc, molybdate, or other nonchromate-based inhibitor in combination with various surface-active cleaning agents. The passivation solution should be disposed of after the pretreatment stage rather than dumped back into the cooling system where the potential for fouling can exist owing to the precipitation of pretreatment compounds such as zinc or phosphate. Table 1 outlines both on-line and off-line pretreatment procedures.

The first methods of cooling tower corrosion control involved adding several hundred parts per million of sodium chromate, as chromate is capable of excellent anodic corrosion control at these dosages. However, these early programs were both inefficient and expensive. The advent of synergized zinc chromate–polyphosphate treatments not only made corrosion control more effective but also lowered its cost. Excellent corrosion control requires only 30–60 ppm of inhibitor instead of a concentration one to two orders of magnitude higher.

Polyphosphates are also used in cooling systems to attain sufficient corrosion control. Cooling towers are operated in a pH range of 6.0–7.5 to provide optimum stability for the polyphosphate. The feasibility of cooling tower operation at higher pH levels, in which the potential for corrosion is decreased, has increased the popularity of low-chromate programs.

CORROSION DETECTION

Corrosion detection plays an important role in any corrosion control program. Most of the methods employ nondestructive test methods and include hydrogen

Table 1 Pretreatment Procedures

On-line pretreatment procedures

1. Increase inhibitor concentration to two to three times its normal level.
2. Circulate the high inhibitor concentration slurry for 4–12 hr. Maintain pH between 6 and 7 and temperatures between 49 and 60°C. If ambient temperature must be used, increase the pretreatment period to 24–48 hr.
3. After passivation, the system should be deconcentrated. Reduce the inhibitor concentration to normal maintenance levels.
4 Initiate normal treatment program.

Off-line passivation procedures

1. Thoroughly clean system of all dirt, oil, scale, organics and corrosion by-products.
2. After system cleaning, refill with fresh water. Add the pretreatment formulation to the required concentration level.
3. Circulate the solution throughout the unit, maintaining pH levels between 6 and 7. Circulation should continue for 2–12 hrs at temperatures between 49 and 60°C.
4. After passivation, remove the pretreatment solution and replace it with normally treated cooling water.
5. Place unit back on-line and resume normal service.

evaluation, radiography, dynamic pressure, corrosion probes, strain gauges, and eddy current measurements. Of these, the methods employed in cooling tower practice are *hydrogen evaluation* and *corrosion probes*.

Hydrogen evaluation is used to detect corrosion in closed systems at low or slightly elevated temperatures in aqueous environments. Sensitive detectors are available to detect the presence of hydrogen, which is a by-product of most aqueous corrosion processes. This method cannot locate the corrosion but can predict the approximate total corrosion rate.

Corrosion probes detect and measure the amount of corrosion occurring at a given point in a system and can be used to estimate the total amount of corrosion and the type of corrosion anticipated. Probes are available for use in a wide variety of temperature and pressure conditions.

COOLING WATER INHIBITORS

There are three methods available for evaluating cooling water inhibitors. These are laboratory methods, service exposure, and sample exposure.

Laboratory Methods

In general, these methods are unreliable and often give misleading results.

Actual Service Exposure

The most reliable evaluation can be obtained by this method; however, it can be very expensive and usually only few materials can be evaluated.

Exposing Sample Materials

This method involves the assembly of several specimens in the form of a corrosion test spool. Test specimens are weighed before and after exposure in the actual service where data are urgently needed.

Unfortunately, there are no commonly accepted standard procedures for securing reliable corrosion data. Any data collected from any test methods will have value only if they can be interpreted properly. Many corrosion environments can vary widely from day to day and even from hour to hour. Even slight variations in operating procedures can drastically affect the corrosion characteristics of the cooling tower. Therefore, it is important to establish methods on how the corrosion data are to be accumulated, evaluated, and put to use.

LANGELIER AND RYZNAR EQUATIONS: SATURATION AND STABILITY INDEX

A convenient method of interpreting water analysis for the purpose of determining the calcium carbonate solubility equilibrium conditions is embodied in the Langelier equation. The Langelier equation can be used to determine the carbonate stability or corrosive properties of a cooling water for a specific temperature when the contents of dissolved solids, total calcium, total alkalinity, and pH values are known.

The saturation index is the difference between the actual measured pH and the calculated pH_s at saturation with calcium carbonate, determined by

$$pH_s = (9.3 + A + B) - (C + D) \tag{1}$$

where

A = total solids, ppm
B = temperature, °F
C = calcium hardness, expressed as ppm $CaCO_3$
D = alkalinity, expressed as ppm $CaCO_3$

The Langelier equation for the saturation index is then

$$I \text{ (Saturation Index)} = pH \text{ (actual)} - pH_s \tag{2}$$

If the saturation index is 0, water is said to be in chemical balance. If the saturation index is positive, scale-forming tendencies are indicated. Finally, if the saturation index is negative, corrosive tendencies are indicated.

The Ryznar equation was developed to provide a closer correlation between the calculated prediction and the quantitative results obtained in the field.

$$\text{Stability Index} = 2\ \text{pH}_s - \text{pH (actual)} \tag{3}$$

For Eq. (3):

1. If the stability index is 6.5, the water is scale forming.
2. If the stability index ranges from 6.5 to 7.0, the water is in a good range.
3. If the stability index is 7.0, the water is corrosive.

The optimum value for the stability index is 6.6.

However, these convenient indexes must serve strictly as guides rather than as absolute control methods; the reason being that uneven temperatures exist throughout a cooling system. Because of this, some exchangers will scale, some will be protected, and still others will corrode.

Note that the pH (actual) is the log of the hydrogen ion concentration.

Organic Growths

Organic matter also aggravates scaling and fouling conditions in cooling systems by combining with silt and/or calcium carbonate to plug up or scale up equipment, thus reducing the effectiveness of the heat transfer surface. Microbiological growths on heat exchangers retard cooling, cut plant efficiency, and increase maintenance cost. Iron- and sulfur-reducing bacteria are often a direct cause of corrosion. Algal growths can occur in all types of heat exchangers. Chlorine and chlorinated organic compounds are the most commonly used chemicals to prevent attack from bacteria and algae. Fungi and other forms of microorganisms can biologically attack the wood inside the cooling tower. This problem can be minimized by the use of impregnated cooling tower lumber.

Undissolved solids or suspended matter plug up cooling and condensing equipment, as well as fill up the cooling tower with silt and mud, which can lead to pumping problems. In addition, suspended matter aggravates scaling conditions in cooling water because silt and mud combine physically with the calcium carbonate to produce a thicker and softer scale than would be formed by calcium carbonate alone. This interferes with heat transfer and water flow. Normally, these are eliminated by continuous filtration.

Legionnaires' Disease

Following the American Legion Convention at a Philadelphia, Pennsylvania, hotel in July 1976, the public first became aware of a new type of disease (legionnaires' disease). Of the 221 cases of legionnaires' disease, 34 resulted in death. Since then, the Centers for Disease Control in Atlanta, Georgia, confirmed that a bacterial microorganism had produced the illness in Philadelphia and at least 11 other locations.

The legionnaires' disease organism was discovered breeding in cooling water at several locations where the disease broke out. The conditions for this bacterium to turn lethal were created when the energy crisis of 1973 imposed conservation measures on water usage.

More and more water is now being recycled. Reducing cooling tower blowdown increases the volume of suspended solids, minerals, and salts, and as the cycles of concentration become higher, the pH and nutrient levels for biological growth increase. The temperature of hot water return to the tower, normally around 120–130°F, creates an ideal condition for the breeding and rapid reproduction of different organisms.

The legionnaires' disease organism may exist in the ground water and even in the air. The bacterium is in a dormant state until ideal life conditions appear. Once the bacterium enters the environment of the cooling tower, it can reproduce very rapidly by binary fission, creating a potential disease outbreak. It can be carried away by blowdown or windage droplets, and unless the water-treatment expert develops a comprehensive program to minimize the possibility of the bacterium breeding in the cooling system, another outbreak is inevitable. Biocides are required and must be monitored continually to ensure that the proper rate is maintained in the cooling water.

Water Analysis and Treatment

For control of scaling, corrosion, and algal and bacterial growth, the cooling tower water supply must be analyzed and properly treated. Water analysis covers three areas: water hardness, alkalinity, and detection of inerts. Hardness can be distinguished in the following:

- Carbonate hardness: The presence of calcium (Ca) and magnesium (Mg) carbonate and bicarbonate.
- Noncarbonated hardness: The presence of other salts of Ca and Mg.
- Total hardness: The sum of the carbonate and noncarbonated hardness.
- Temporary hardness: The presence of Ca and Mg bicarbonate and can be eliminated by boiling the water to transform bicarbonate into insoluble carbonate. Temporary hardness is slightly different from carbonate hardness because it does not take into account the presence of carbonate, which is only slightly soluble.
- Permanent hardness: Ca and Mg residue in the water after boiling and differs from noncarbonated hardness because it also measures the carbonate remaining in solution.

The normal units used to measure water hardness are

- French degrees = g $CaCO_3$/100 L water
- German degrees = g CaO/100 L water
- English degrees = g $CaCO_3$/L imperial gal water

The value of the different degrees are

- 1° French = 10 ppm $CaCO_3$
- German = 1.78° French = 17.8 ppm $CaCO_3$
- English = 1.43 French = 14.3 ppm $CaCO_3$

With respect to total hardness, makeup water can be classified as follows:

	Total Hardness (ppm $CaCO_3$)
Very soft	15
Soft	15–50
Medium hard	50–100
Hard	100–200
Very hard	200

Alkalinity is a measure of the concentration of all electrolytes that give basic reaction when hydrolyzed in water; that is, salts of strong bases and weak acids (e.g., hydrates, carbonates, bicarbonates, phosphates, silicates, borates, sulfites). Chlorides and sulfates do not contribute to alkalinity. The evaluation of alkalinity is made by titration, and the results are reported in ppm of $CaCO_3$.

Sometimes the alkalinity is reported as centimeters of the acid (HCl or H_2SO_4) used for the titration of 100 cc of water. To convert to parts per million of $CaCO_3$, use the following relationship:

$$\text{cc of 0.1N acid} \times 50 = \text{ppm of } CaCO_3 \tag{4}$$

$$\text{cc of 0.02N acid} \times 10 = \text{ppm of } CaCO_3 \tag{5}$$

The number of centimeters of 0.1N acid used for titration of 100 cc of water is frequently referred to as millivalents.

Other analysis data needed for the makeup water are pH, suspended solids (ppm), chlorides (ppm Cl), sulfates (ppm SO_4), and silica (ppm SiO_2).

It is interesting to note that when the total alkalinity is less than the total hardness, then calcium and magnesium are present in compounds other than carbonates, bicarbonates, and hydrates. In this case, the amount of hardness equivalent to total alkalinity is the carbonate hardness; the remainder is the noncarbonated hardness.

The solubilities of the more common salts at approximately 120°F are as follows:

	Chloride (ppm)	Carbonate (ppm)	Sulfate (ppm)
Sodium (Na)	270,000	290,000	310,000
Magnesium (Mg)	270,000	125	330,000
Calcium (Ca)	520,000	17	2,200

When the number of concentrations of the circulating water is in the order of 3–7, some of the salts dissolved can exceed their solubility limits and precipitate, causing scale formation in pipes and coolers. The purpose of the treatment of the cooling water is to avoid scale formation. This is achieved by the injection of sulfuric acid to convert Ca and Mg carbonates (carbonate hardness) into more soluble sulfates. The amount of acid used must be limited to maintain some residual alkalinity in the system. If the system pH is reduced to far below 7.0, it would result in an accelerated corrosion within the system. As stated earlier, scale formation and/or corrosion tendency is defined by the saturation index (Langelier index) and stability index (Ryznar equation).

If the saturation index is positive (which implies that the stability index is <6.5), then the water has scale tendency and the addition of sulfuric acid in appropriate quantities would be required to prevent scaling formation. The following example illustrates the estimation of the required amount of acid.

Example 1

A cooling tower is operating with the following makeup water composition:

Ca hardness, ppm $CaCO_3$	85
Mg hardness, ppm $CaCO_3$	33
Total hardness, ppm $CaCO_3$	118
Total alkalinity, ppm $CaCO_3$	90
Sulfates, ppm SO_4	20
Chlorides, ppm Cl	19
Silica, ppm SiO_2	2

It is clear that the total hardness is greater than the total alkalinity. Assume, for instance, that the number of concentration in the circulating water to reduce blowdown is maintained at 5. Also, assume that the temperature of the hot water entering the tower is 120°F. Sulfates in circulating water are $5 \times 20 = 100$ ppm SO_4 or

$$100 \times \frac{136}{96} = 142 \text{ ppm } CaSO_4$$

where

136 = molecular weight of $CaSO_4$
96 = molecular weight of SO_4

The $CaSO_4$ solubility limit is 2200 ppm $CaSO_4$, so the additional sulfate formation permissible is 2200 – 142 = 2058 ppm $CaSO_4$ or

$$2058 \times \frac{96}{136} = 1452 \text{ ppm as } SO_4$$

The alkalinity in the circulation water, if not converted into sulfates, is 5 × 90 = 450 ppm $CaCO_3$.

Assume that 10% of the alkalinity is left unconverted to avoid corrosion, then 450 × 0.9 = 405 ppm $CaCO_3 \rightarrow CaSO_4$.

$$\text{Sulfate formed} = 405 \times \frac{96}{100} = 338 \text{ ppm as } SO_4$$

where

96 = molecular weight of SO_4
100 = molecular weight of H_2SO_4

Thus, the sulfuric acid concentration required is 405 × 96/98 = 397 ppm.

The composition of blowdown water in this case will be as follows:

Hardness, ppm $CaCO_3$	5 × 118 = 590
Alkalinity, ppm $CaCO_3$	0.1 × 5 × 90 = 45
Sulfates, ppm SO_4	5 × 20 + 388 = 488
Chlorides, ppm Cl	5 × 19 = 95
Silica, ppm SiO_2	5 × 2 = 10

In this example, the cycles could have been carried much higher because we have 488 ppm of SO_4 versus 1450 allowed.

For sulfuric acid injection, a storage drum and a proportioning pump must be provided. Carbon steel is a suitable material for the concentrated sulfuric acid drum, providing that moisture does not enter the drum. For safety purposes, it is suggested to avoid glass level gauges. It is best to install a floating-type level gauge.

The injection point of the sulfuric acid is in the pump bay or as near as possible to water intake. The sulfuric acid pump is normally a motor-driven proportioning pump, and an electric motor is connected to a pH analyzer installed on the cooling water supply header so that the pump starts and stops, depending on the pH in the circulating water. Table 2 summarizes various chemical treating agents for cooling water towers.

An example for estimating the required amount of different chemicals follows:

Table 2 Chemical Treating Agents for Cooling Water Towers

Chemical and common name	Water treatment use	Quantity (ppm in circulating water)
Inorganic chromate salts	Corrosion control	300 + 500 ppm of CrO_4
Inorganic and organic phosphates and polyphosphates	Scale and corrosion control	2 + 10 ppm of PO_4
Chromate and phosphate combination treatment	Corrosion control	CrO_4 10 + 40 PO_4 20 + 50
Lignin and tannin organic	Scale and corrosion control	20 + 50 ppm
Organic chromates		5 + 20 ppm
Chlorine and chlorinated phenols	Algal and bacterial slime	1 ppm 4 hr/day
Quaternary ammonium copper complexes	Algal and bacterial control	220 ppm intermittent
Sulfuric acid	Solubility control	As necessary to maintain same residual alkalinity

Example 2

Calculate the chlorine and phosphates requirements for a tower operation:

1. *Chlorine*. As stated in Table 2 for algal and bacterial control, the normal quantity of chlorine needed is 1 ppm every 4 hr daily, which represents

$$1 \text{ ppm} \times \frac{4 \text{ hr/day}}{24 \text{ hr/day}} = 0.2 \text{ ppm}$$

continuously.
Suppose a tower operates with 100,000 gpm of circulating flow:

$$\text{Chlorine (lb/day)} = \frac{100{,}000 \text{ gal/min} \times 1440 \text{ min/day}}{7.48 \text{ gal/ft}^3/62.4 \text{ lb}}$$

$$\times \frac{0.2 \text{ lb } Cl_2}{1{,}000{,}000 \text{ lb } H_2O} = 240 \text{ lb/day}$$

2. *Phosphate*. From the table, the requirements of phosphate are 2–10 ppm of PO_4. The loss of phosphates will be due only to slowdown and windage. To calculate the phosphate requirements:

$$\text{Phosphate (lb/day)} = \frac{10 \text{ lb } PO_4 \times (W+B) \text{ gal/min} \times 1440 \text{ min/day}}{1{,}000{,}000 \text{ lb } H_2O \times 7.48 \text{ gal/ft}^3 \times \text{ft}^3\text{/62.4 lb}}$$

where

W = windage losses
B = blowdown losses

All other chemicals can be calculated in the same way.

Plastic Cooling Towers

Corrosion problems and costly water treatment can be minimized in many applications through the use of plastics. Since about 1970 the use of industrial-grade plastics has become widely accepted in prepackaged, factory-assembled cooling tower units. There are numerous advantages to component construction, including polyethylene shell, ABS wet decking and drift eliminator system, and PVC distribution assembly, which have proven superior to steel and wood construction in many applications. There are several advantages of plastics construction over wood or steel. Plastics have a seamless, leakproof one-piece shell, are noncorrosive, nonbrittle, nonporous when wet, and lighter. Further, they require less maintenance and give longer service.

Operating Conditions

Water quality and environmental conditions in the vast majority of HVAC and light-to-medium industrial cooling tower applications permit acceptable service life from standard cooling tower construction.

Standard tower design under normal conditions assumes a maximum of 120°F hot water to the tower, including system upset conditions. Temperatures over 120°F, even for short durations, may impose damaging effects on PVC fill, many thermoplastic components, galvanizing, and plywood. Those rare applications demanding hot water in excess of 120°F usually should be reviewed with the tower manufacturer to assure that appropriate materials changes from the standard configuration are included in the initial purchase specification.

Normal circulating water chemistry falls within the following limits:

- pH between 6.5 and 8.0, although pH down to 5.0 is acceptable if no galvanized steel is present. Low pH attacks galvanized steel, concrete and cement products, fiberglass, and aluminum. High pH attacks wood, fiberglass, and aluminum.
- Chlorides (expressed as NaCl) below 750 ppm.
- Calcium (as $CaCO_3$) below 1200 ppm—except in arid climates where the critical level for scale formation may be much lower.

- Sulfates below 5,000 ppm—if calcium exceeds 1200 ppm, sulfates should be limited to 800 ppm (less in arid climates) to prevent scale formation.
- Sulfides below 1 ppm.
- Silica (as SiO_2) below 150 ppm.
- Iron below 3 ppm.
- Manganese below 0.1 ppm.
- Langelier saturation index between –0.5 and +0.5—negative LSI indicates corrosion likely; positive indicates $CaCO_3$ scaling likely.
- Suspended solids below 150 ppm if solids are abrasive—avoid film-type fills if solids are fibrous, greasy, fatty, or tarry.
- Oil and grease below 10 ppm or loss of thermal performance will occur.
- No organic solvents.
- No organic nutrients which could promote growth of algae or slime.
- Chlorine (from water treatment) below 1 ppm free residual for intermittent treatment; below 0.4 ppm free residual for continuous chlorination.

These conditions usually define normal *circulating* water, including the chemical concentrating effects caused by recirculating the water to some predetermined number of concentrations.

Most of the heat transfer in a cooling tower occurs through evaporation of a portion of the circulating water. The evaporated water leaves the cooling tower as purified vapor, leaving behind dissolved solids which concentrate in the circulating water. Most operators control the number of concentrations by dumping a calculated fraction of the circulating water into a holding tank or sewer, a process commonly called blowdown.

Makeup water, usually city water supply, will be concentrated on the tower. The net effect is that the tower will be exposed to concentrations of corrosive agents which are multiples of their concentration in the makeup supply. The result of this concentrating effect may lead to conditions which indicate that some materials should be more corrosion-resistant than those used on standard towers.

Deviations from normal water are usually regional, and necessary adjustments reflect area conditions. Most tower users operate their towers within the range of three to five cycles of concentrations. However, regional concerns such as water availability or sewer usage restrictions may dictate unusually high circulating water concentrations. Anyone unfamiliar with the impact of these conditions on cooling towers will benefit from discussing the specific application with one or more knowledgeable cooling tower suppliers.

Common environmental conditions which may dictate use of nonstandard materials of construction include proximity to bodies of salt water, the presence of corrosive vapors, and the presence of unusually dense air pollution in the form of SO_x, hydrogen sulfide (H_2S), or potentially corrosive particulates.

Towers which operate beyond normal limits require greater care at the specification stage. Solutions for regional concerns such as unusual local water chemistry or proximity to saltwater are usually well known. More unusual conditions may demand even greater attention to details in order to assure that the cooling tower will operate successfully over its service life.

Corrosion Protection

Corrosives are those elements or compounds whose tendency is to react chemically or electrolytically with a metal, given the opportunity and proper circumstances. Principal among these is oxygen. Oxidation is the interaction with other elements or compounds and rust is its particular reaction with iron, which is the primary ingredient of the carbon steel typically utilized for various cooling tower components.

Corrosion has always been a primary concern when using carbon steel. Where atmospheres tend to be dry and cool, the concern can be minor. However, where atmospheres are warm and humid, the potential for corrosion increases dramatically and must be addressed.

Constituents of the atmosphere have effects on corrosion that can contribute to its acceleration. The increase in industrialization, and the gaseous by-products thereby generated, has created a changed, typically corrosive atmosphere.

Add to this an oxygen-containing vehicle such as water in which atmospheric gases can be absorbed and concentrated, cause it to come into intimate contact with the steel, and provide continuous aeration—along with heat—and you have increased the opportunity for corrosion.

The aspects of flow and evaporation must be considered. In stagnant water, steel usually has the opportunity to form a self-protecting surface film of oxidation which tends to reduce the rate of corrosion. Under flow conditions, however, this protective film can erode away, continuously exposing new material to deterioration. With evaporation, as occurs in a cooling tower, pure water vapor leaves the system, concentrating the remainder into a highly aggressive bath which tends to accelerate the corrosion within an already susceptible system.

As applied to cooling towers, protective coatings can be categorized into two basic types: barrier type and sacrificial type. Both are used extensively throughout the cooling tower industry, occasionally both at the same time.

Barrier-type paint coatings are intended to form a protective barrier between the steel and the agent of corrosion. Most are applied in liquid form by brush, roller, or spray. Some are applied in powder form, by electrostatic deposition, followed by application of heat to promote bonding.

Materials used for barrier-type coatings in the cooling tower industry are usually unaffected by the environment typically encountered. All protective coatings are, however, permeable (porous) to a greater or lesser degree. Some have greater porosity than others. Permeability can be decreased by increasing the applied thickness. Nevertheless, in no commercial formulation or applied thickness can barrier-type coatings be classified as impervious to the intrusion of moisture and/or atmospheric gases. It is, however, a matter of time before the barrier layer is penetrated, exposing substrate metal to elemental corrosion.

Powdered epoxy coatings, electrostatically applied, for example, achieve a proper bond only under stringent quality assurance procedures. Not only is the metal preparation critical in regard to the temperature, concentration, and application time of the cleaner, but repelling electrical charges, which naturally form in corners and angles, virtually preclude the application of a uniform coating thickness. Hangers, normally reused in successive coating applications, provide progressively reduced grounding capability, and ultimately bonding of the applied coating depends on the curing which takes place within a period of time in an oven at a *specific* temperature. Obviously, multistep processes such as this, in which each step is subject to precise control, introduce considerable margin for error.

Corrosion, fed by oxygen entering through the coating's natural porosity, gradually undermines the coating and gains increasing access to the metal. Such corrosion often goes undetected until too late to make proper repairs. Telltale blisters usually tend to give this condition away. The apparent integrity of the coating sometimes disguises it, permitting concentrated corrosion to proceed unchecked. Barrier-type coatings alone are considered inadequate for proper corrosion protection in cooling towers.

Sacrificial or Galvanic Protection

Cooling tower manufacturers make use of a *sacrificial*-type coating, such as galvanizing. In the galvanizing process, steel is submerged in a bath of molten zinc at approximately 850°F. It emerges from this bath with several layers of iron/zinc alloy topped by a coating of pure zinc; the effective thickness of the coating is governed by the time in the bath.

Although oxygen will combine with virtually all known elements, it has a distinct order of preference. Given an equal opportunity to react with either iron (carbon steel) or zinc, for example, it will avoid the iron in favor of the zinc. Therefore, as to corrosion, zinc is considered sacrificial with respect to carbon steel. It does not have to form an impenetrable barrier—it merely has to be there. Barrier-type coatings may permit steel corrosion to begin shortly after contact with water; galvanizing, by nature, will not. As long as zinc exists in proximity to steel, and is allowed freely to contact the water, the steel is protected against progressive corrosion.

Galvanizing of sheet metal is now performed by the steel-producing companies themselves. It is referred to as "mill galvanizing" and is a hot-dip process. Continuous sheet steel (previously rolled to gauge) is annealed in the galvanizing line and conducted through a bath of molten zinc. It then proceeds through wiping dies, steam jets, or jets which establish the required zinc thickness and uniformity.

Galvanizing protects steel sacrificially and its protection longevity is directly related to the thickness of zinc applied in the galvanizing process. The greater the amount of zinc applied, the more years will be required for it totally to react with the elements of corrosion. Galvanizing also offers the unique property of protecting substrate metal by sacrificial reaction which radiates in all directions from the point of initial corrosive attack. Any agent that impedes this effect, such as the imposition of a barrier coating on top of the galvanizing, may actually reduce the time necessary for corrosive activity fully to penetrate through the zinc to the substrate metal.

15

Controlling Sewage Odors

INTRODUCTION

Annoying odors in the vicinity of sewage-treatment plants has long been a recognized problem. Land adjoining these sites may become subdivided and subsequently developed for industrial and residential purposes. This is especially true in metropolitan areas where land is markedly scarce.

Many states have codes for minimum distances between housing developments and sewage- and refuse-treatment areas. However, even though a new sewage-treatment facility may meet the minimum distance requirement, along with some additional landscaping to avoid objections about the visual effects from the surrounding residents, it is entirely possible that the odors produced can become a substantial nuisance and source of complaints.

DOMESTIC SEWAGE ODORS

Odors from domestic wastewater usually result from some sort of biological activity in the sewer collection system and wastewater-treatment plants. Gases that generally arise from domestic sewage are inorganic. Sewage containing industrial wastes may have odor problems compounded by organic gases from any waste chemicals added to the sewer system.

Of the inorganic gases, hydrogen sulfide (H_2S) and ammonia (NH_3) are odorous, whereas dissolved oxygen (DO), carbon dioxide (CO_2), methane (CH_4), nitrogen (N_2), and hydrogen (H_2) are odorless. Gases originating from organic sources by anaerobic decomposition usually contain nitrogen and sulfur. Many different combinations of gases can occur at one time and can compound or change the description of the odors that would exist individually. Tables 1 and 2 summarize the substances that are the odor-causing agents.

ODOR MECHANISMS

Fresh domestic sewage containing dissolved oxygen has a slight odor, which is usually described as musty. Natural aerobic breakdown of the sewage, whether it occurs in a sewer system or wastewater-treatment plant, depletes the DO. Of course, in a wastewater-treatment plant the DO is being replaced by mechanical means. Only in the digestion tanks should the odor process of anaerobic decay be allowed. Once sewage is depleted of dissolved oxygen, the odors of H_2S and other malodorous compounds are given off and a general septic condition exists. Any chemicals added from industrial processes will cause even more problems. Since most of the odorous gases that arise from sewage are caused by bacteria, this process allows these odors to escape into the atmosphere.

Bacteria attack the organic matter to gain energy for their normal life

Table 1 Odor Characteristics and Threshold Concentrations

Substance	Formula	Threshold odor (mg/L)	Remarks
Allyl mercaptan	$CH_2{:}CH{\cdot}CH_2{\cdot}SH$	0.00005	Very disagreeable, garliclike odor
Ammonia	NH_3	0.037	Sharp, pungent odor
Benzyl mercaptan	$C_6H_5CH_2{\cdot}SH$	0.00019	Unpleasant odor
Chlorine	Cl_2	0.010	Pungent, irritating odor
Chlorophenol	$Cl{\cdot}C_6H_4{\cdot}OH$	0.00018	Medicinal odor
Crotyl mercaptan	$CH_3{\cdot}CH{:}CH{\cdot}CH_2SH$	0.000029	Skunk odor
Diphenyl sulphide	$(C_6H_5)_2S$	0.000048	Unpleasant odor
Ethyl mercaptan	$CH_3CH_2{\cdot}SH$	0.00019	Odor of decayed cabbage
Ethyl sulfide	$(C_2H_5)_2S$	0.000025	Nauseating odor
Hydrogen sulfide	H_2S	0.0011	Rotten egg odor
Methyl mercaptan	CH_3SH	0.0011	Odor of decayed cabbage
Methyl sulfide	$(CH_3)_2S$	0.0011	Odor of decayed vegetables
Pyridine	C_6H_5N	0.0037	Disagreeable, irritating odor
Skatole	C_9H_9N	0.0012	Fecal odor, nauseating
Sulfur dioxide	SO_2	0.009	Pungent, irritating odor
Thiocresol	$CH_3C_6H_4{\cdot}SH$	0.0001	Rancid, skunklike odor
Thiophenol	C_6H_5SH	0.000062	Putrid, nauseating odor

Table 2 Selected Odor Threshold Concentrations

Chemical	Symbol	Concentration for threshold odor (ppm by volume)
Carbon disulfide	CS_2	0.21
Acetaldehyde	C_2H_4O	0.21
Hydrogen sulfide	H_2S	0.00047
Nitrogen compounds		0.00021–100.0
Skatole	C_6H_6N	0.019
Mercaptans		
ethyl	C_2H_6S	0.001–0.00026
methyl	CH_4S	0.041–0.0021
Chlorine	Cl_2	0.314
Ammonia	NH_3	46.8
Perchloroethylene	C_2Cl_6	4.68
Phenol	C_6H_6O	0.6

sources and new cell production. The first step taken by the bacteria is to remove hydrogen atoms from the organic matter and gain energy (dehydrogenation). An inorganic or organic substance then becomes a hydrogen acceptor, causing the following reactions to occur in the order shown:

H acceptor	*Reduced product*	
$O_2^+ + 4H^+$	$= 2H_2O$	(1)
$2NO_3^- + 12H^-$	$= N_2 + 6H_2O$	(2)
$SO_4 + 10H^-$	$= H_2S + 4H_2O$	(3)
Oxidized organics + xH^+	= reduced organics	(4)
$CO_2 + 8H$	$= CH_4 + 2H_2O$	(5)

The sources of odors are created in Equations (3) and (4). The odor coming from Equation (4) depends on what product produced it (e.g., organic acids, aldehydes, ketones, amines, sulfides, mercaptans, indoles, skatoles). Also, there are many different types of bacteria involved in each reaction. Equation (1) is the only aerobic reaction. The result of the reactions can be caused by either anaerobic or facultative organisms.

Hydrogen sulfide gas is the most prevalent odor associated with domestic sewage. Besides odor production, crown corrosion of concrete sewers is another effect. Sulfate ions normally exist in domestic wastewater in concentrations ranging from 30 to 60 mg/L. Higher quantities may exist depending on how many times the water has been reused.

The anaerobic process of creating H_2S starts out with the sulfur-containing proteins being broken down to form SO_4 ions. The reaction continues as shown in Equations (6) and (7) and Figure 1.

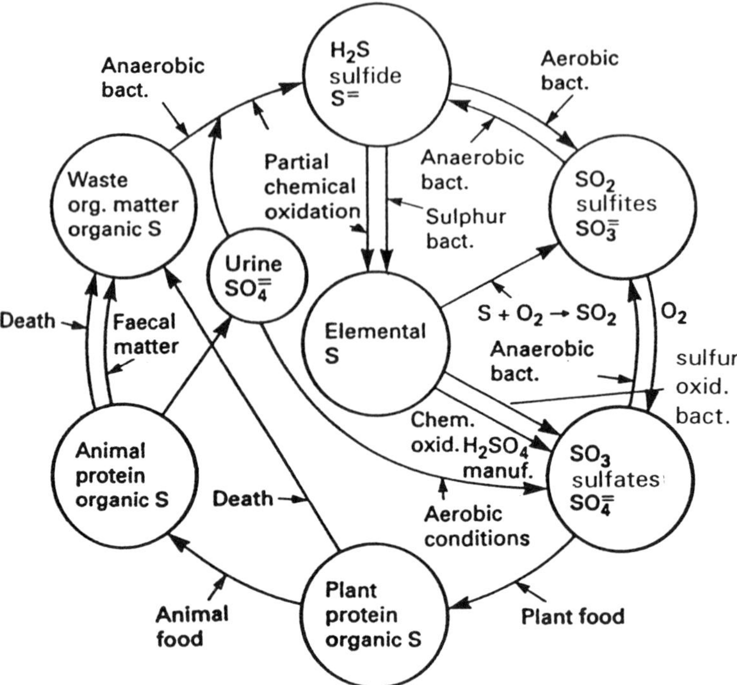

Figure 1 The sulfur cycle.

$$SO_4^{2-} + \text{organic matter} \xrightarrow[\text{bacteria}]{\text{anaerobic}} S^{2-} + H_2O + CO_2 \quad (6)$$

$$S^{2-} + 2H^+ \rightarrow H_2S \quad (7)$$

The amount of bacteria, the pH, and the temperature of the wastes are crucial factors in such reactions. The number of sulfur-reducing bacteria ranges from 60 to 600/ml of sewage and up to 25,000/ml for sludges. The effects of pH should be apparent from Equation (7). Since the gaseous state causes the odor, an increase in H^+ ions will cause more H_2S production. This effect is summarized in Figure 2, where the optimum pH range is 7.5–8.0. The optimum temperature for H_2S generation is around 30°C. Typical sulfate-reducing bacteria are *Vibro desulphurications*, *Spirillum desulphurican*, and *Microspira aestauri*.

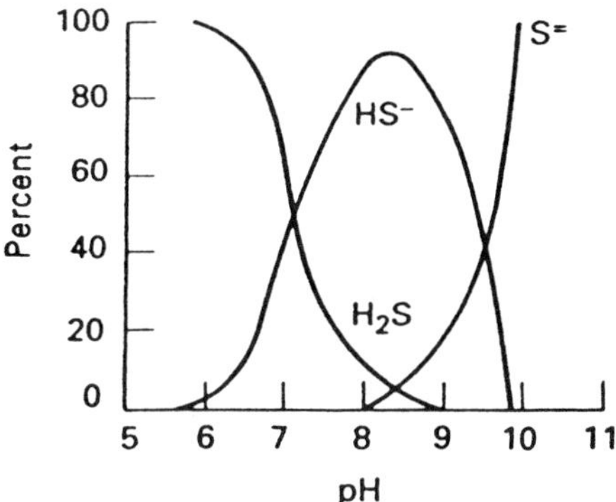

Figure 2 Effect of pH on hydrogen sulfide–sulfide equilibrium (10^{-3} molar solution, 32 mg H_2S/L).

THE SEWER SYSTEM

Raw sewage can become very odorous under anaerobic conditions, which can occur only when the dissolved oxygen of the sewage is used up, thus allowing the anaerobic bacteria to predominate.

Anaerobic conditions must exist for sufficient time in a sewer system before producing H_2S. The natural consequence is the observation that most odor problems arising from sewers occur on those lines that have relatively flat grades.

The odor effects in sewers are usually most prevalent during warm weather conditions, 20°C and higher, which favor a more rapid multiplication of the anaerobic bacteria. A result of a flat sewer grade is the need for a pumping station, which stores the sewage for a longer period of time and thus serves as a concentrated source of odors. Sulfide production reinforces the odor danger of flat-graded sewer systems and their resulting pump stations by showing that sulfides are produced only by slimes on the submerged surface of the sewer and by the deposited sludge, since a sewer constructed on a flat slope has a greater flow depth versus a steeper sloped sewer of the same size.

A pump station will then receive septic sewage and provide it with additional surface area in the wet well for further decomposition of the slimes. Increased use of garbage grinders provides even more solids that settle to the bottom of the sewer. A prime example of the influence of garbage grinders on odor conditions can be found on the many military bases built on flat terrain where garbage grinders are the rule rather than the exception, and severe odor problems are the result.

The control of sewer odors can be accomplished by physical and/or chemical means. If at all possible, avoid flat grades where the velocity of a sewer flowing full approaches 0.6 m/s. Most sewers are designed with this value as the minimum acceptable. There is one serious error that occurs by constantly discussing the sewer design parameters at full or even half-full conditions. The error is that most sewers, especially laterals, never approach the full condition, since the design flowrate capacity of the sewer at 0.6 m/s far exceeds the average discharge that can be expected. Therefore, it is important for the sewer designer to check the velocity of the sewage at the average flowrate (usually two to three times more than the design flowrate) through the use of a partial flow diagram that is available in most references dealing with sewer design. Even the average flowrate may be meaningless on a time basis for many sewers. It is quite possible that some approximation can be made concerning the velocity at the minimum flowrate. On the other side of odor control through good sewer design should be the avoidance of excessive turbulent conditions such as deep manholes. At these locations the previously tranquil sewer flow is energized by a sudden change in elevation, thus allowing the sewage to release any stored-up gases. Vortex manholes avoid this condition by allowing for a nonturbulent change of elevation.

Sewer cleaning may need to be increased to reduce odor problems by removing the sludge and slime layers where the H_2S and other gases originate. Pump stations especially need to clean the slime that builds up on wall surfaces. Small, packaged pump stations are rarely cleaned, but they should be monitored for odors more frequently as they receive a low sewage flowrate.

Trapping of gases has been successful where U sections are provided for house connections and turned-up bends for mains and laterals.

Venting of sewage by natural and forced draught is practical. The only problem is the need to obtain a disposal point for the ejected gases that will not itself become a point source of odor.

Addition of air to sewage force mains will prevent large columns of odor-laden gas at the free discharge end. This problem can be especially acute if the force main is long followed by a discharge into a flat-graded gravity sewer. Air injection at the wet well of a pumping station tends to keep the slime formation from producing gases.

Chemical control of odors can be accomplished by the following methods. Economic considerations and individual problems are the controlling elements of choice.

As in most sewage-treatment applications, chlorine has been found to be the most popular chemical for inhibiting or disinfecting the anaerobic bacteria causing H_2S generation in sewers. Since chlorine is a strong oxidizing agent, there are many possible reactions. The main ones occur with H_2S, NH_3, and C_6H_5OH (phenol).

$$H_2S + 4Cl + 4H_2O \rightarrow H_2SO_4 + 8HCl \quad (8)$$
$$NH_3 + Cl_2 \rightarrow NH_2CL + HCl \quad (9)$$
$$HN_2Cl + Cl_2 \rightarrow HNCl_2 + HCl \quad (10)$$
$$HNCl_2 + Cl_2 \rightarrow NCl_3 + HCl \quad (11)$$

Equations (9), (10), and (11) are the typical textbook equations associated with chlorination in the presence of free ammonia. Ammonia is an odor-causing agent, but it must be noted that the reaction should not be allowed to proceed to the trichloramine form (NCl_3), which is malodorous. It is quite common for chlorine to react with other organic compounds contributed from industrial sources, such as phenol, and create an intolerable product such as chlorophenol, with its very low threshold odor value. The effectiveness of chlorine and its economic desirability lies in the fact that inhibition of the odor-causing bacteria is all that is needed and not complete disinfection. The usual range of chlorine dosage for inhibition purposes is 10–50 mg/L, with 8.87 mg of chlorine required to oxidize completely 1 mg of sulfide. In many cases, a much higher dose of chlorine may be needed effectively to treat denser slimes and sludges on the sewer invert. Mechanical sewer cleaning, combined with chlorine or one of the other oxidizing agents discussed below, is the usual method of attack. Care must be taken to decide where on the sewer system is the best starting point. By applying the chlorine at points farthest away from the treatment plant, the most effective use of the chemical is achieved.

Sodium hydroxide can be applied to control H_2S odor by causing a shift in the equilibrium of equation (3). This method is suggested for use at a temporary storage area such as a pump station wet well. A lime slurry can also be just as effective when applied to sewer lines.

ODOR CONTROL

Treatment plant operations involve many sources within the plant where foul odors can be generated. Many times the odor problem can be solved by simple cleaning operations. Other plants will require more elaborate physical and/or chemical methods. Whatever methods are employed, the design and operating engineer must be aware of the increasing problem of odors. The problem should increase substantially in the future as many plants become hydraulically overloaded, leading to odorous conditions. Other modern treatment devices, such as ammonia stripping, will be an obvious source of odors. One must be careful to avoid an air pollution problem while treating a water-related one.

Basic sewage-treatment mechanisms beyond primary treatment or simple settling revolve around allowing a volume of sewage to achieve maximum surface area contact with air. This is accomplished in the activated-sludge process by diffusing air through a bubbling mechanism or by forcing air through a mechanical mixer into a volume of sewage that has settled. The other basic, biological alternative to this is to pass the sewage over a large surface area such as gravel (trickling filter). Both of these methods are basic sewage-treatment principles and will not be discussed further.

By forcing air into the sewage, small bubbles in the order of 1,000–1,500

μm are produced. These bubbles burst and dispense small droplets into the air. Most of these droplets have their water part evaporated, leaving a small (10–15 μm) solid particle suspended in air. Depending on weather conditions, these solid particles can be spread over the surrounding area, causing respiratory irritation and possibly infection.

Gases and vapors are the odor-causing agents in the sewage-treatment plant. For many years, design engineers never considered the sewage-treatment process as giving off odorous gases and vapors. The principal odor-causing compounds are methyl sulfides, amines, indoles, skatoles, and H_2S. Certain hydrocarbons, such as the methane produced by sludge digestion, can be a large point source of air pollution if they are not collected and disposed of properly. Any addition of industrial wastes increases the production of gases. Figures 3 and 4 show sources of odors in typical treatment plants.

Emissions

Human fecal waste contains etiological agents that cause intestinal disease. What happens to these bacteria? Referring to Figures 3 and 4, the raw sewage (1) entering a plant may have already developed odors, as discussed. Screening and comminuting devices (2) can create odor if an accumulation of grease is allowed build up and if the bar screens used do not have a self-cleaning attachment. The accumulated debris may be coated with or be made of organic material that may quickly give off an odor. Usually these objects are landfilled once a day or even less frequently, with a good chance definite odor problem developing. Grit chambers (3) are another possible point source of odors because of the organic coating on most of the sand, metal pieces, and so forth, that are collected. Again, it is a matter of timely housekeeping to avoid an odor problem.

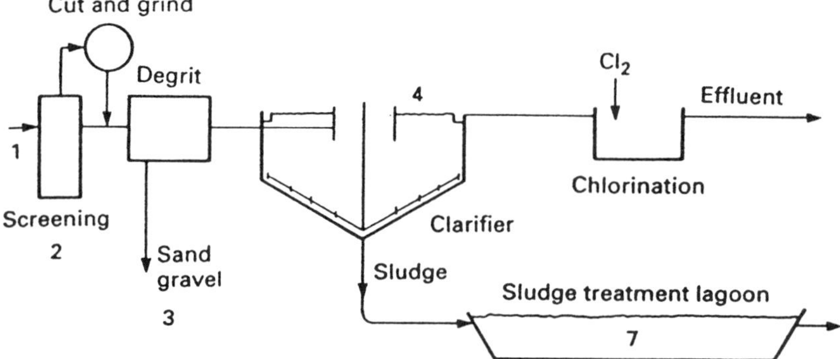

Figure 3 Simple wastewater-treatment plant.

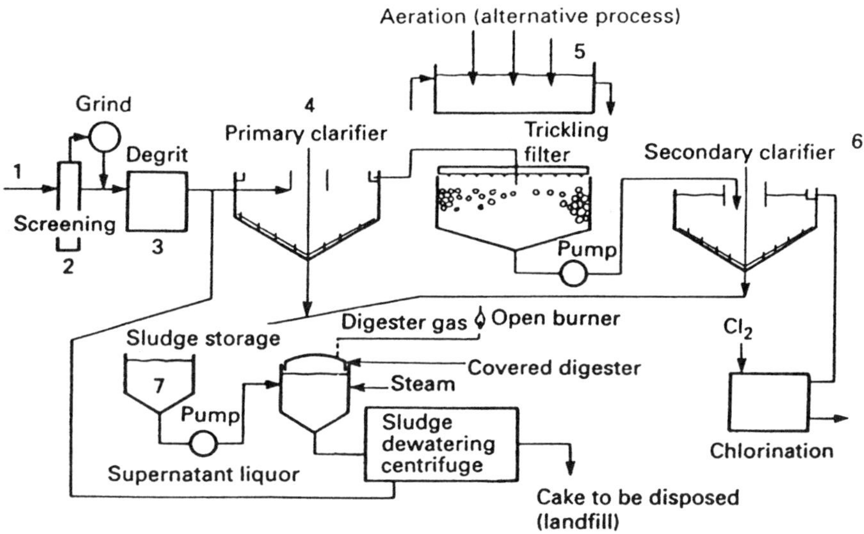

Figure 4 Primary and secondary treatment with sludge digestion.

When the bulk of the sewage reaches the primary settling tank (4), there are usually influent and effluent weirs. The effluent weir especially can cause the emission of odorous gases from septic sewage. The weir allows the sewage to be dropped 0.3–0.6 m to a collecting trough and then possibly an additional 0.3–0.6 m to a main collecting bin between two parallel tanks. Also, any scum-collection device concentrates highly odorous materials on the air-water face of the tank. The sewage may then be pumped or gravity fed to the next unit process with little chance for any more odor generation, since the sewage is usually confined in an enclosed vessel.

If secondary biological treatment (5) follows, the problem of odor-causing aerosols arises, as has already been discussed. It must be added, however, that in the activated process, water spraying of dry foam formation is a common practice. By quickly bursting the bubbles that form, a more rapid formation of gases is achieved and the overall quantity of odorous gases may be increased.

Secondary settling tanks (6) may cause odor problems if the sludge is not removed on schedule. There should be little odor production at this point in the process, since aerobic treatment preceded the settling process.

The sludge-treatment process, which can include thickening, digestion, and sludge dewatering (7), is a viable and many times vigorous source of odors. The sludge is constantly in an anaerobic state, except in certain sludge dewatering processes such as vacuum filtration. Sludge thickeners usually receive both septic primary and secondary sludges.

The digestion process gives off gases that contain methane, volatile organics, and other odorous gases such as H_2S. In most plants these gases are confined and flared to destruction (8). Since the sludge must be heated to about 32–35°C for efficient anaerobic digestion, a transfer of the odor-laden sludge by plunger-type pumps gives additional opportunities for point sources of odorous compounds.

Digested sludge that is either air dried, mechanically dewatered, or burned (9) contains ammonia as its predominant odor, with small amounts of volatile organic acids adding to the problem. Odor problems increase in sludges if they are improperly treated and then exposed to the atmosphere. Lime treatment of raw sludge can push the equilibrium equation for ammonia toward the gaseous state, with an odorous condition resulting.

Sludge processing can also include incineration, with multiple-hearth furnaces being the most popular. This operation consists of first having the moisture removed in the upper furnace and the next several hearths operating at higher temperatures to oxidize the remaining organic materials. If the combustion furnace is operated at temperatures ranging from 650 to 760°C, no odors will be given off. Any particulate matter is normally removed by a wet scrubber.

Oxidation ponds, serving as independent and complete treatment units or as tertiary treatment devices, can be a source of a multitude of odors. The organisms employed—algae, bacteria, and actinomycetes—oxidize the organic matter discharged into the pond for food. The aerobic and anaerobic processes going on in the pond can transfer to the atmosphere algae, bacteria, oxides of sulfur and nitrogen, ammonia, methane, hydrogen sulfide, and hydrocarbons. Hydrocarbons can reach the atmosphere by simple evaporation before the biological breakdown process takes place. Methane may be prevalent, and it may be produced at a rate of 0.6 m^3/kg of BOD destroyed.

Odors from nidole, sulfides, skatole, mercaptans, cadaverine, amines, metabolic products of algae, and actinomycetes have been associated with oxidation ponds. Algae are probably the main organisms associated with odor production. The algae involved—diatoms, blue-green algae, and pigmented flagellates—can in themselves cause the odor problem or provide the food necessary for the odor-causing actinomycetes. The actinomycetes produce odor-causing by-products such as aromatic amines, aldehydes, saturated fatty acids, ketones, and unsaturated aromatics.

Odor Prevention or Removal

Treatment plant prevention or removal of odors can involve many methods. There are many proprietary methods. In most cases, it is difficult to identify the cause and nature of an odor, and trial and error methods have to be employed. Many factors in selecting the best method involve consideration of local weather conditions, treatment method, previous experience with odors, ability of the plant operators, duration and frequency of each type of odor, and the available funds for correcting the problem. Three general areas of odor control in wastewater-treatment plants are

- Changes in present plant operation, with consideration being given to new treatment techniques
- Chemical treatment
- Collection and treatment of odorous gases

Changes in Plant Operation

This is probably the cheapest and, in many cases, the most effective means of odor control. Timely maintenance preventing offensive odors from reaching the surrounding areas may be all that is needed for most small plants.

Grit Chamber

Most of the odors in this unit process are from H_2S. When odors become frequent, the operator should check flowthrough velocities in the system, determine the total volatile solids of the grit, analyze the raw sewage for total and dissolved sulfides, lead acetate/H_2S/tiles in the chamber, and check for debris accumulation, including grease. Corrective methods can include a daily washing of the grit chamber to remove any overabundance of organic solids, occasional dosing of the chamber with hyperchloride or other chemical preparation, and sealing the chamber so that the odors can be removed by a gas scrubbing apparatus.

Primary Settling Tanks

The presence of H_2S requires tests for sulfides, pH, and dissolved oxygen. The most important factor is the amount of total solids in the tank. Overloaded primary tanks are common, as they often receive solids from other plant processes.

Prechlorination and pH Control

These measures are suggested for minor odor problems. More intense odors are probably due to excess solids. These odors can be controlled by making sure that excess total solids from other plant processes do not exceed 5% of the daily inflow. Regular removal of sludge and scum layers may be all that is needed for most settling tanks for odor control.

Aeration Tanks

Any odor problems in this process are due to excessive turbulence in the bulk of the water. Under windy conditions, an odorous spray can be spread throughout the area. Also, water spraying of foam build-up makes more gas pockets available, but any odor problem should be in the immediate vicinity of the tank.

Trickling Filters

Odor problems caused by H_2S can be significant, because as the slime layer builds up around the filter rocks, it is natural that the inner layer of organic material becomes anaerobic. This anaerobic condition eventually causes a sloughing off of the built-up layer with a release of gases. By increasing the recirculation rate in a high-rate filter, the dissolved oxygen will be increased and quicker sloughing will occur. The vents and underdrain system should be checked to avoid concentrated odors after one of these items becomes clogged. More expensive measures include covering the entire bed, with the odors fed to a wet scrubber.

Sludge

There are very few changes that can be performed on most modern anaerobic sludge digesters and associated unit processes. Most well-designed digesters destroy the odors by burning off the noxious gases. Most odors come from sludge thickening tanks and from discharging incompletely digested sludge to sand-drying beds or dewatering units. With incineration of raw sludge, odors from sludge handling is increased.

Oxidation Ponds

Odors from oxidation ponds are caused by hydrogen sulfide from an accumulation of solids on the bottom of the pond. If the pond is used as a tertiary unit, odors should be negligible, unless there is a substantial accumulation of blue-green algae. Ponds used as secondary treatment devices contain substantially more solids and can give off odors, especially during warm weather conditions. In either case, when odor problems arise, the pond should be checked for scum accumulation, pH of influent and effluent, dissolved oxygen of the pond at several points and depths, and total and dissolved sulfides in the pond. Control measures can involve the use of mechanical aerators, prechlorination, addition of sodium nitrate, and odor-masking chemicals.

Chemical Treatment

Chemical treatment can provide a convenient method of odor control, especially when the odors originate from organic sources. This odor-removal mechanism usually involves oxidation coupled with a pH adjustment. Chemicals used for odor control are chlorine and its related compounds, potassium permanganate, nitrates, metallic ions, spray counteractants, and proprietary chemicals under various trade names. The adjustment of pH to prevent H_2S formation is usually done by lime and in some cases by caustics.

Chlorination is the most popular chemical odor-treatment method, because it is readily available at most plants for treating sewage. As in sewers, the chlorine retards biological activity and oxidizes odorous sulfur compounds to the free colloidal sulfur. A dose of 10 mg/L at the maximum flow is recommended for prechlorination.

Chlorine dioxide has been used on an emergency basis at various plants when the aeration tank does not supply sufficient oxygen because of overloading or maintenance.

Nitrates have been applied to tidal basins and lagoons for suppression of odors. The mechanism of odor control chemically combined oxygen which prevents the reduction of sulfates. As long as some oxygen and nitrates are present, other more odorous decomposition products are not allowed to form. The cost-effectiveness of nitrate addition is low.

Metallic ions such as copper, mercury, zinc, and iron combine with sulfide to form insoluble precipitates. The use of these metallic ions has been researched since the 1930s. Ferric iron (Fe^{3+}) is more effective for sulfide precipitation than ferrous (Fe^{2+}). When some oxygen is present in the wastewater, a combination of Fe^{3+} and Fe^{2+} is most effective. It was also found that Fe^{2+} works best when the water is devoid of oxygen. However, unless there is a cheap source of iron, this method is not economical.

Counteraction of odors involves overwhelming a foul odor with one that is supposedly more pleasant. Complete counteraction of an odor is unlikely because of the many variables involved. Attempts at counteraction often lead to masking, which may present new problems, depending on the chemical used. Odor counteractants are vaporized and usually introduced at ground level, on the roof of the plant structure, and in the vicinity of the odor source to provide complete vertical coverage of the odor. The counteractant is vaporized by drawing the chemical from a drum and spraying it through a compressor.

Collection and Treatment of Gases

Confinement of odors through the use of, for example, domes or enclosures has become a popular practice for odor control at large treatment plants. Physical confinement involves covering most unit processes, since odor creation varies with location and time. The interior atmospheric pressure must be kept slightly below that of the exterior to ensure positive collection of the odors. Ventilation ducts, needed in areas where workmen perform constant maintenance, have to be connected to the central collection system. The major cost of this type of odor treatment is the confinement structure. Once the gases are confined, there are three general methods of treatment:

- Simple or catalytic combustion
- Ozonation
- Chemical oxidation or absorption

Combustion

One of the more desirable techniques of odor control is combustion. Complete combustion produces carbon dioxide, oxides of nitrogen, water, and sulfur dioxide. If complete combustion is not achieved, the endproducts can be more odorous than the original gas. Simple combustion usually involves a separate system when large volumes of gas need to be destroyed. The key to complete combustion is maintaining the proper temperature and contact time in the combustion chamber. Temperatures between 730 and 820°C have been used at many plants, with contact times ranging from 0.3 to 3.0 s.

Catalytic combustion allows the use of a substantially lower incineration temperature (320°C) through the use of a platinum metallic screen which accelerates the combustion process. Operating costs can be high, since natural gas may be required for complete combustion if there is not sufficient digester gas available.

Ozonation

Ozone has been used with success in domestic wastewater. Owing to its high generation costs and its high reactivity, the addition of ozone to a large flow is not practical. However, ozone can be applied economically to odors that are collected by covered facilities. Basically, ozone will oxide sulfides and amines to nonodorous sulfoxides and amino-oxides. A dose of 1 mg/L of ozone by volume is needed to handle up to 10 mg/L of sulfide. Detention time in the blower assembly should be at least 15 s and preferably longer.

Ozone has been applied to grit and screening chambers, aeration tanks, vacuum filters, and sludge storage tanks. Odor problems if these unit processes are controlled by ozone usually demands that the plant be under one roof with exhausts going through a single stack. Single units are easily sized by knowing the frequency of air change and the dose of ozone needed.

Absorption

The treatment of odorous gases by absorption can be applied to various covered unit processes. Wet scrubbing of gases with spray devices or scrubbing equipment, packed columns, adding ozone to the gas, or passing the gas through activated carbon have all been used. The liquid used in the wet scrubbing process depends on the gas involved, with alkaline solutions being preferred for hydrogen sulfide removal. Hypochlorite, lime, permanganate, and chelated iron have been used in scrubbing, but their use tends to raise costs considerably.

ODOR CONTROL OPTIONS

The general principles of odor control, including the physical and chemical mechanisms of their operation and examples of their specific applications, deserve discussion. Certain operating procedures have been designed to eliminate the conditions which tend to enhance the production of odorous materials.

1. Good housekeeping in collection systems includes the prevention of conditions that would allow the system to become anaerobic as well as to minimize bottom deposits and slime layer formation. Venting the system by natural or forced draft is helpful in preventing anaerobic conditions from developing. The injection of compressed air into pumping stations of force lines is also helpful. Care must be taken that the ejected air does not become a point source of odors. It is not necessary for the production of H_2S that the entire system be anaerobic, only that anaerobic conditions be present in the slime and accumulated bottom deposits. Therefore, periodic cleaning of lines and pumping station walls is essential.

2. Good housekeeping procedures within the treatment plant include prompt (at least daily) disposal of solids and scum from the grit chamber and settling tanks. Odor problems from trickling filters can occur when the built up slime layer on the filter media is sloughed off, since the inner layer of the slime coating becomes anaerobic. Increasing the recirculation ratio will promote quicker sloughing and diminish odor. When existing biofiltration facilities are overloaded, increasing recirculation is not always possible. Indeed, most serious odor problems reported from trickling filters are in connection with overloaded facilities. Vents and underdrains must be kept free from obstructions. Overloading any type of treatment facility will tend to increase odor problems by increasing solids and air requirements and decreasing hydraulic retention time and thus efficiency in each of the unit operations.

3. Good housekeeping of oxidation ponds and lagoons consists of periodic cleaning and disposal of bottom sediments as well as the skimming of any floating algae mats or sludge mats.

Design Modification

Some design modifications have been helpful in odor abatement.

Collection Systems

Most sewer lines are designed for a minimum velocity of two feet per second at maximum flow. However, the average flow may be two or three times less than the design flowrate, permitting the accumulation of sludge deposits.

At the same time, excessive turbulence, which would tend to release trapped gases, should be avoided. Trapping gases in U sections has been reported to be successful.

Treatment Facilities

Preaeration of raw sewage has been used when sewage arrives at the plant in an anaerobic condition. Preaeration times of 10–45 min are recommended and frequently consist of increased detention time in aerated grit chambers. This modification is particularly applicable to overloaded systems, since the preaeration step effects a certain amount of BOD removal.

Substituting oxygen for compressed air in aeration basins is reported as having helped to abate odors in an overloaded facility.

Odor Modification

Odor modification is a phenomenon in which odor intensity is reduced by adding a nonchemically reactive controlling agent to a malodor. Odor modification has frequently been divided into subgroups called odor cancellation, odor counteraction, and odor masking, but the mechanisms are somewhat overlapping and the distinctions unclear. They all rely on one or more of the following principles:

- Temporary inactivation of the olfactory nerves, lessening the response, the mechanism of which is probably similar to the "key that fits the lock but won't turn" blocking action of some antibiotics.
- In some cases, it has been shown that when two substances are mixed in a given ratio, the resulting mixture may have an odor less intense than that of the separate components. Examples of these pairs are ethyl mercaptan and eucalyptol, skatole and coumarin, and butyric acid and oil of juniper. Best results can only be obtained with specific ratios of the two components, which is generally difficult to effect in the open.
- In odor masking, a malodor is incorporated into an overall odor mixture that is pleasant. For example, oil of jasmine contains skatole and indole, and a synthetic jasmine, prepared without indole, is said to be effective in masking these compounds. Odor masking may also take the form of simply attempting to overcome an unpleasant odor with a pleasant one.

The odor-modifying agent may be sprayed into the air at the odor source or along the property line, added to the liquid, or incorporated into a floating chemical screen through which gaseous emissions from the liquid surface must pass.

The effects of an odor-modifying agent will be variable depending on temperature, pH, and wind direction and velocity. In many instances, additives are blends of aromatic oil that will decompose quickly on hot days or in the presence of strong acids. Even in traces, there is also the chance that after a period of time the odor-masking agent will become as objectionable as the original odor.

Odor modification is, in general, less expensive and less effective than other means of odor control. Odor modification should not be used to mask malodors that are toxic (e.g., H_2S).

Chemical Additives

The addition of chemicals can be an effective and convenient method of odor control in collection systems and treatment facilities, as already discussed. The most commonly used chemicals are chlorine (Cl_2) and related compounds, hydrogen peroxide (H_2O_2), nitrates (NO_3), metallic ions, calcium carbonate ($CaCO_3$), and sodium hydroxide (NaOH).

Some of the basic reaction mechanisms and applications of chemical additives should be discussed.

Chlorination is probably the most popular method for odor treatment. Chlorine reacts with water to form hypochlorous and hydrochloric acids:

$$Cl_2 + H_2O \rightleftharpoons HOCl + H^+ Cl^- \quad (12)$$

The HOCl is a weak acid which dissociates:

$$HOCl \rightleftharpoons H^+ OCl^- \quad (K_{eq} = 2.7 \cdot 10^8) \quad (13)$$

Chlorine may also be added in the form of sodium of calcium hypochlorite which ionizes in water:

$$Ca(OCl)_2 \rightarrow Ca^+ + 2OCl^- \quad (14)$$

and

$$NaOCl \rightarrow Na^+ + OCl^- \quad (15)$$

The hypochlorous ions formed in Equations (14) and (15) then enter into the equilibrium shown in Equation (13).

Chlorine and hypochlorous acid react with a wide variety of substances including H_2S, unsaturated hydrocarbons, and ammonia in the following manner:

$$H_2S + 4Cl_2 + 4H_2O \rightarrow H_2SO_4 + 8HCl \quad (16)$$

$$\text{—}\underset{H}{C}\text{=}\underset{H}{C}\text{—} + HOCl \rightarrow \text{—}\overset{Cl}{\underset{H}{C}}\text{—}\overset{OH}{\underset{H}{C}}\text{—}$$

$$NH_3 + HOCl \rightarrow H_2O + NH)2Cl \text{ (Monochloramine)} \quad (18)$$

$$NH_2Cl + HOCl \rightarrow H_2O + NHCl_2 \text{ (Dichloramine)} \quad (19)$$

$$NHCl_2 + HOCl \rightarrow H_2O + NCl_3 \text{ (Trichloramine)} \quad (20)$$

The mono- and dichloramines have significant disinfecting power, but the reaction should not be allowed to proceed to the trichloramine [Eq. (20)], which is malodorous. With mole ratios of chlorine to ammonia 1:1, both monochloramine and dichloramine are formed.

Chlorine can also react with phenols to form mono-, di-, or trichlorophenols, which can impart tastes and odors to water.

For odor control in raw wastewater it is only necessary to add enough chlorine to react with the H_2S and for bacterial inhibition, not disinfection. The usual range of chlorine dosage for inhibition in collection systems is 10–50 mg/L and 10 to 20 mg/L for prechlorination at the plant before primary settling. Prechlorination at that level has no detrimental effect on biological treatment.

Chlorine compounds are often used in collection systems such as emulsified orthodichlorobenzene (chloroben). The recommended chemical cleaning method consists of shock loading combined with retention for trouble spots, followed by a rotating schedule of shock dosing for upstream lines. Care should be taken that o-dichlorobenzene concentration not exceed 20 ppm in biological treatment operations.

The application of hydrogen peroxide (H_2O_2) as an oxidant for sulfide in the water has been known for many years. Recently, this oxidizing ability has been applied to control odor and corrosion problems caused by domestic wastewater. The H_2O_2 is believed to act in three different ways to control sulfide production and resulting odors:

1. Oxidant action:

$$H_2S + H_2O_2 \xrightarrow[\text{neutral}]{\text{acidic or}} H_2O + \text{elemental sulfur} \tag{21}$$

$$\text{2. } H_2S + H_2O_2 \xrightarrow{\text{alkaline}} H_2O + \text{Sulfates} \tag{22}$$

Oxygen producing:

$$\text{3. } 2H_2O_2 \xrightarrow{\text{catalase}} O_2 + 2H_2O \tag{23}$$

H_2O_2 is bactericidal to the sulfate-reducing bacteria. Mechanisms two and three act to prevent formation of the odiferous sulfide compounds, whereas mechanism one acts to remove the already existing sulfide compounds.

H_2O_2 doses between 1 and 2 mg are required for each milligram of sulfide in collection systems, and doses from 15 to 40 mg/L have been found necessary for odor control at an overloaded trickling filter plant.

H_2O_2 is generally used in the form of 50% solutions. Although it is available at 70%, the higher concentration is unstable and hazardous to store. The use of H_2O_2 for odor and corrosion control has the advantages of increasing the dissolved oxygen between 1 and 2 mg/L and causing no problem of reaction by-products.

With some success, sodium nitrate has been added to collection facilities and oxidation ponds. Dosage to sewers is 10 lb of $NaNO_3$ for each pound of sulfide. Evidently the nitrate acts as a stimulant to agar growth in ponds as well as an oxygen source. As long as nitrates are present, they will be reduced preferentially to nitrites.

Commercial products containing a proprietary mixture of air, oxides of nitrogen, and other gases have also been applied to pump stations and mildly sloped sewers.

Metallic ions, zinc and ferric in particular, have been used to reduce the sulfide concentration by precipitation, according to the general reaction

$$S^{=} + MSO_4 \rightarrow MS\downarrow + SO_4^{=} \quad (24)$$

The hydrogen sulfide–sulfide equilibrium is extremely sensitive to pH. There is very little H_2S present at pHs above 8. Therefore, it is not uncommon to add chemicals which will increase the pH in order to control the odor of H_2S_2. Sodium hydroxide (NaOH) and lime ($CaCO_3$) have been used for this purpose. When lime is added to sludge, it shifts the equilibrium

$$NH_4^{+} = NH_3 + H^{+} \quad (25)$$

to the right, and ammonia then becomes the predominant odor of the sludge.

Ammonia gas has been proposed as an odor-controlling agent at a sludge transfer station and on transport vessels. The theory is to elevate the pH in the sludge with ammonia gas to a level that will preclude the evolution of H_2S, mercaptans, and skatol. Ammonia itself, however, can present substantial odor problems.

Odor Collection

The classic physical layout of sewage-treatment facilities has favored open channels and uncovered tanks. Nevertheless, many of these facilities have elected to cover the tanks and to collect and treat the odors evolved.

The four most common methods of treating collected odors are incineration, ozonation, wet scrubbing, and adsorption.

Incineration is an effective method of odor control. However, with rising fuel costs it is becoming more and more expensive.

It is important that the combustion be complete, since the intermediate steps in the oxidation of organic materials are aldehydes, ketones, and organic acids, which are often stronger odorants than the original chemicals. In order to assure complete combustion, proper temperatures (1350–1500°F) and contact time (0.3–3 s) must be maintained. Catalytic combustion allows the use of lower incineration temperatures (600°F).

Only sludge gas (from a digester) does not require supplementary fuel, since it contains 60–75% methane by volume. This digester gas may be used as a portion of the supplemental fuel required at other plant sites.

Ozone (O_3) will oxidize sulfides and amines to nonodorous sulfoxides and amino oxides. One part per million by volume of ozone is needed to react with 10 ppm of sulfide. Detention times should be at least 15 s, and the ozonated air must be in motion and mix well with the pollutant. Electricity requirements are approximately 10 kwh/lb of ozone. The generation of ozone and its mixing with collected gases from grit and screening chambers, aeration tanks, vacuum filters, and sludge storage tanks has proven to be an effective odor-control device.

Chemical absorption in the form of wet scrubbing is considered by some to be the most effective and economical odor abatement procedure.

The most common type of absorber is the countercurrent packed tower. Gases enter at the bottom, pass through the packed bed, which is irrigated with a scrubbing liquid, through a mist eliminator, and is exhausted from the unit by a fan located at the clean air side. The scrubbing liquid is usually a solution of an oxidizing agent such as potassium permanganate ($KMnO_4$) or sodium hypochlorite (NaOCI) or an alkaline material such as sodium hydroxide (NaOH) or calcium hydroxide (Ca[OH]).

A scrubbing solution of 1–4% $KMnO_4$ oxidizes mercaptans, aldehydes, unsaturated ketones, hydrocarbons, phenols, amines, H_2S, and SO_2. The manganese is reduced to manganese dioxide (MnO_2) under basic conditions and to the manganous ion (Mn^{2+}) under acidic conditions.

Reactions of hypochlorites and caustics with odorous gases have been discussed earlier.

Activated carbon is the most commonly used adsorbent in odor control. Owing to its nonpolar surface, activated carbon has the ability to absorb organic and some inorganic materials in preference to water vapor. In general, organics having molecular weights over 45 and boiling points over 0°C will be readily adsorbed. Adsorption takes place when the attraction between the gas molecules and the solid (surface) molecules is greater than the intermolecular attraction in the gaseous phase. This will then condense even at partial pressures 0.01 that of saturation pressures. Carbon adsorption is therefore particularly useful in removing odors caused by low concentrations of organic gases.

The carbon may be regenerated thermally in a furnace or by passing a stream of hot as or steam through the bed.

Activated carbon was found to be effective on persistent exhaust odors from hypochlorite scrubber units treating odorous air from covered biofiltration units of a 70-MGD plant in Sacramento, California. Carbon canisters have also been installed on enclosed primary and secondary clarifiers, grit chambers, and pumping stations.

Physical Barrier Method of Odor Control

These are methods designed to minimize the effects of the odor-causing substances rather than to prevent their formation or to remove them by, for example, oxidation, neutralization, or adsorption, as discussed in previous sections.

Concrete collection pipes have been sprayed with various vinyl and resinous compounds in order to prevent the sulfuric acid from attacking and weakening the concrete (crown corrosion). These coatings have had varying degrees of success. Results using coated concrete were reported as unsatisfactory at Corpus Christi, Texas. They have since installed polyethylene slip lining in the system. It is interesting to note that their engineers had experimented with H_2O_2 addition to the collection system with good results, but they consider the H_2O_2 process only a temporary measure, since federal funding is available for slip lining but not for chemical additives.

Other "physical barrier" methods include pouring melted paraffin and cetyl alcohol on the surface of ponds and floating small balls (diameters of 0.75 to 6.0 in.) on the surface of ponds or lagoons. The odor control effected is presumably due to the decrease in exposed surface area.

AIR POLLUTION PROBLEMS RELATED TO SEWAGE TREATMENT

Two related air pollution problems often caused by sewage-treatment facilities are bacterial aerosols and insects.

Bacterial Aerosols

The formation of aerosols in the treatment process by spraying, splashing, and particularly by air injection has been discussed. Since human pathogenic microorganisms are known to be present in wastewater in large numbers at any stage of handling, airborne microorganisms resulting from aerosolization may represent a potential health hazard.

Bacterial samples near the source of emission have generally revealed that relatively high counts were present, whereas much lower counts were found as the downwind distance was increased. Evidently a very rapid initial decay rate is followed by a much slower one. The decay rate is attributed to organism die-off from the stress of droplet evaporation. Nevertheless, bacterial aerosols significantly above background were encountered 200 m from the spray nozzle. Buffer zones of 500–1000 m have been proposed.

There is a definite need for further studies to ascertain the health significance of wastewater aerosols and the risks involved for plant workers and those involved in land application operations as well as nearby residents.

Insects

Air pollution often results from the emergence of large numbers of flies from trickling filters and sludge-drying beds. Psychoda commonly called "filter fly," is found throughout North America. The female fly lays her eggs on the film covering the meter media. At 70°F, the eggs hatch in less than 2 days. Psychodae may cause a nuisance within a radius of 1 mile from the treatment plant; however, inhabited places within distances of 500–1000 ft have experienced serious air pollution problems caused by the filter fly.

Certain midges (*Chironomidae*) breed prolifically in ponds, with adults becoming a nuisance around residential areas because of their intolerable numbers. The midge larvae are most abundant in the bottom spongelike algal mats in the shallow areas (to 2.5 ft deep) of the ponds. The more shallow area in the pond, the greater the midge productivity.

Several factors have combined to produce a need for reduced air pollution in the form of odors from sewers and treatment plants. These factors include expansion of collection systems, frequent overloading of treatment facilities, urban encroachment on sewage-treatment plant sites, and public demand for improved environmental quality in general.

There is a need for further studies to evaluate the potential health hazards resulting from bacterial aerosols and wastewater-bred insects. Studies to date are inconclusive and their results somewhat contradictory.

As discussed, there are many possible sources and causes of odor problems. For each particular problem there are a multitude of corrective measures to be considered, ranging from revisions in operating procedure or "better housekeeping" practices to enclosing the operation and collecting and treating the odorous gases. One of the most effective methods of odor control is collection and wet scrubbing. Chemical treatment with H_2O_2 is also becoming popular and has been used with good results.

It is possible that federal funding regulations encourage the installation of structural remedies when chemical treatment would be adequate, since federal funds are available for necessary structural changes but not for chemicals.

The degree of treatment required will also be a factor in the selection of a method. Solutions to existing problems must be handled on a case by case basis. Sources and cause of odor must be identified, and the best practical method of abatement will have to be selected in accordance with the particular conditions and requirements.

Future treatment facilities should be designed for maximum odor control at the lowest cost. The general plant layout as well as each unit operation should be engineered with odor-control considerations and abatement equipment either built in or easily added on.

AMMONIA STRIPPING

Ammonia stripping is a simple desorption process used to lower the ammonia content of a wastewater stream. In the process, wastewater at elevated pH is pumped to the top of a packed tower with a countercurrent flow of air drawn through the bottom openings. Free ammonia (NH_3) is stripped from the falling water droplets into the air stream which is then discharged to the atmosphere.

Lime or caustic soda is added prior to the stripping to raise the pH of the wastewater to the range of 10.8–11.5, converting essentially all ammonium ions to ammonia gas which can be stripped by air. Process controls required for the operation are the proper pH adjustment of the influent wastewater and maintenance of proper air and water flows.

Ammonia-removal efficiency is highly dependent on air temperature and air/water ratios. As the air temperature decreases, the efficiency drops significantly. The most common operating problem of this process is the occasional formation of calcium carbonate scale. The influent should always be clarified before stripping.

The process entails operation of the stripping gas in a closed system with an ammonia adsorption unit for removal of the CO_2 from the stripping gas stream to reduce scaling problems; reclamation of ammonia from the closed-cycle absorption unit; the use of high-pH holding ponds, followed by a cross flow spray tower and final removal of the residual ammonia by breakpoint chlorination. Applications are for wastewater with high ammonia content (more than 10 mg/L). For higher ammonia content (more than 100 mg/L), it may be economical to use alternate ammonia-removal techniques.

Poor efficiency in cold weather locations (0–10°C) is a problem. It cannot be operated in freezing conditions (unless sufficient heated air is available). Ammonia is discharged to atmosphere usually at low levels (6 mg/m^3). This may be objectionable in certain locations. Nitrite, nitrate, and organic nitrogen are not removed. There is poor efficiency when ammonia concentration is low (less than 10 mg/L). Scale formation can be removed hydraulically in most cases but not in all, resulting in a need to pilot test at most locations.

The stripping tower closely resembles a conventional cooling tower. The operation is unaffected by toxic compounds which can disrupt the performance of a biological system. However, volatile toxics will be stripped during the process. Operating efficiency is highly dependent on air temperature:

Air temperature (°C)	NH_3 Removal efficiency (%)
10	75
20	90–95

Efficiency may be reduced by severe scaling in the tower; under normal operating conditions, residual ammonia concentrations are in the 1–3 mg/L range. Lime or caustic soda is needed to raise the pH of the wastewater to the range of 10.8–11.5. For wastewater with a high calcium content, an inhibiting polymer may be added to ease the scaling problem. Effluent from the stripping may need pH readjustment to neutral condition with an acid (H_2SO_4 at 1.75 parts for one part of lime added) or recarbonation followed by clarification.

Typical operating conditions include

- Wastewater loading: 1–2 gal/min/ft^2
- Packing material: Plastic or wood
- Stripping air flow rate: 300–500 ft^3/gal
- Packing spacing: Approximately 2 in horizontal and vertical
- Packing depth: 20–25 ft
- Providing: Uniform water distribution
- pH of wastewater: 10.8–11.5
- Providing: Scale removal and clean-up
- Air pressure drop: 0.015–0.019 in of water/ft
- Land requirement: Small

Figure 5 shows a countercurrent air-stripping process.

Ammonia Removal and Recovery

Ammonia removal and recovery consists of two packed towers for stripping and absorption. In the stripping tower, wastewater flows downward against an upflow gas stream. Ammonia in the wastewater is stripped into the gas stream. The gas stream is then directed to the absorption tower, in which an absorption solution is sprayed downward. With good countercurrent contact, most of the ammonia transferred to the gas stream is absorbed by the solution. The gas stream is then recycled back to the stripping tower for reuse.

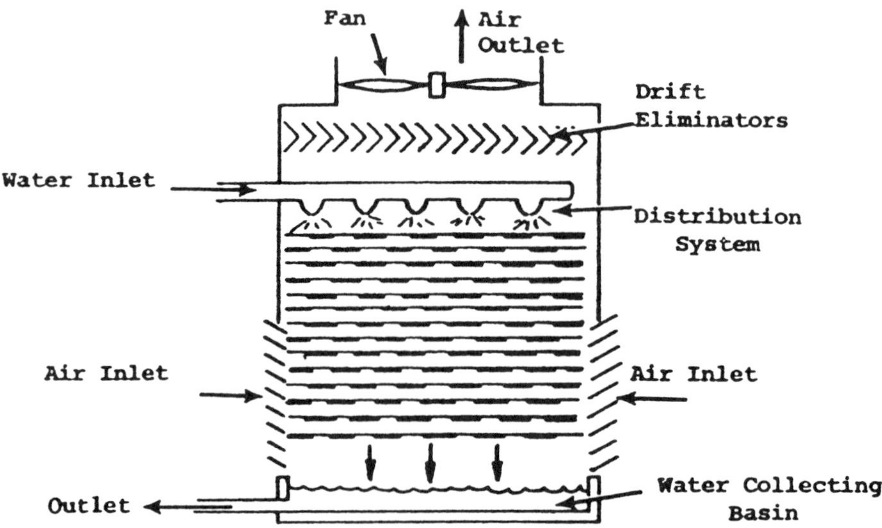

Figure 5 Schematic of a countercurrent air stripper.

Lime or caustic soda is usually added to the wastewater prior to ARRP to convert the ammonium ion in the wastewater to free ammonia. Air is used as the stripping gas. Water or a dilute acid (sulfuric acid) is frequently selected as the absorption solution, so that the process produces an aqueous ammonia solution or an ammonium sulfate solution.

For wastewaters with high ammonium ion concentrations (>300 mg/L), steam may be economically used as the stripping gas. Steam is injected at the bottom of the stripping tower and is condensed as it exits. A wastewater feed-effluent heat exchanger is often used to minimize energy consumption. Steam stripping and absorption operations are commonly used in chemical and fertilizer industries. In wastewater treatment, air stripping is considered fully demonstrated but is not widely used. The process is economically attractive for treatment of wastewater with a high ammonium ion concentration (>100 mg/L). This approach is being used for stripping ammonia from selective ion exchange regenerant. It may produce a waste ammonia stream with some value. The process is less competitive as ammonium ion concentration decreases. It highly susceptible to the ammonia market to become cost effective.

Ammonia-removal efficiency can be expected to be higher than with ambient air stripping towers in the colder climates, since the stripping gas temperature approximates the wastewater temperature. Removal efficiencies are projected to range from 90 to 95% with water temperature of 20°C to 75% at water temperatures of 10°C. Scaling problems are reduced when compared to NH_3 air stripping towers.

Chemicals required include sulfuric acid (H_2SO_4) at 2.72 parts per 1.0 part of ammonium ion recovered, if an $(NH_4)_2SO_4$ solution is the desired product.

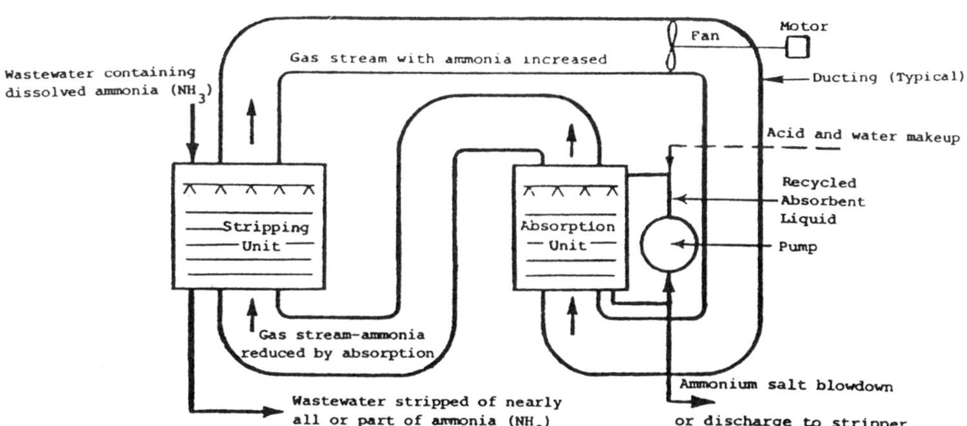

Figure 6 Ammonia removal and recovery process.

No sulfuric acid is needed if water is used as the absorption solution. Sufficient lime (CaO) to raise pH to 10.8 to 11.5. Acid for pH readjustment may be needed for neutralization of the residual alkali.

Typical operating conditions include

Stripping operation	Absorption operation
Wastewater loading: 1–2 gal/min/ft^2 (air stripping) 7 gal/min/ft^2 (steam stripping)	Product solution: 1–30% (aqueous ammonia) to 50% $(NH_4)_2SO_4$ solution
Gas flow rate: 300 to 500 ft^3 (steam stripping)	Tower diameter: 50–75% flooding velocity Degree of recovery: about 90%
Packing depth: 20 to 25 ft	Packing depth: 15–20 ft
Wastewater pH: 10.8 to 11.5	Gas pressure drop: 2–3 in. of water

Figure 6 shows schematic for ammonia removal and recovery.

16
Ground Water Remediation

INTRODUCTION

Ground water contamination is a serious environmental problem. This contamination results from agriculture as well as industry. Contaminants, being either organic or inorganic, exhibit particular characteristics. Clean up is specific to the contaminant and site. Conventional physical methods of remediation are proven, yet have some limitation. In contrast, there are a number of emerging technologies which show potential. Emerging technologies include biological and chemical in situ techniques. The potential benefits include a more complete treatment and lower capital costs than conventional treatment. Of the emerging technologies, biological treatment has received greater research focus and as such there is more data available for biological treatment than for others. Research suggests that combining chemical and biological in situ techniques with physical pump and treat methods optimize the remediation process.

The single most important aspect of ground water is its value as a potable water supply. Some sources estimate that more than 50% of the US population uses ground water as its source of potable water. Many areas depend on ground water as the sole source of water.

Contamination of ground water supplies has long been recognized as a serious and widespread problem. Once contaminated, ground water is difficult to clean up. Owing to the mechanism of ground water flow, the rate of fluid movement is slow. It can take a seemingly endless amount of time to flush contaminated water through an aquifer. The contaminants can migrate through the system via various pathways. Contaminants can sorb onto the soil and slowly leach into the ground water or they can move quickly into the ground water. The contaminants can float on the top, sink to the bottom, and/or mix with the ground water.

Increasing numbers of individual wells or well fields have become contaminated by toxic compounds. The contamination can generally be attributed to nearby industrial discharges, agricultural operation, or landfill leachate. The organic contaminants of concern include organic solvents, volatile organic compounds (VOCs), petroleum products, pesticides, and nitrosamines. The inorganics include heavy metals, arsenic, nitrates, and total dissolved solids (TDS). These inorganic compounds are found as contaminants in ground water less frequently and in lower concentrations. Still, these compounds render the ground water unusable.

The US Environmental Protection Agency (EPA) has identified more than 100 priority pollutants. It is agreed that low levels of these compounds could have long-term health effects. Many of these are carcinogenic, mutagenic, or teratogenic. Many of the organic priority pollutants are VOCs/solvents which are subject to rigorous regulation by the EPA and tolerated only in amounts approaching the limits of detectability.

The majority of contaminants that have been identified in ground water have already been encountered in wastewater. For decades, physical, chemical, and biological treatment has been used on wastewater. Ground water remediation has developed largely within the recent past. Although there are some similarities between treatment of ground water and wastewater, application of such conventional treatment methods to the problem of contaminated ground water is not so straightforward. The physical methods include air stripping, carbon adsorption, ion exchange, and membrane separation. Chemical methods include precipitation and oxidation/reduction. All methods have their limitations and have been covered in earlier chapters.

Regulations by government and increased public awareness of environmental problems has led to a search for new treatment technologies. The application of biological principles to ground water remediation shows much promise. Bioremediation uses microorganisms to detoxify, degrade, or destroy toxic compounds. Bioremediation is particularly attractive, because it has the potential to remove contaminants at lower capital and operating costs as compared to conventional technologies.

The physical methods are exclusively above-ground systems, whereas the chemical and biological methods are applicable above ground or in situ. No single unit operation or process is capable of treating every contaminant encountered in ground water. Frequently, two or more unit operations are combined into an overall treatment train.

AQUIFERS

Surface water percolates slowly down through the soil and rock until it reaches a layer which it cannot penetrate. Water accumulates above this impermeable layer and the soil matrix becomes saturated with water. This saturated zone marks the ground water table. The zone of soil layers which contain both liquid and vapor is the vadose (unsaturated) zone.

An aquifer is a geological formation (soil, sand, gravel, rock) which contains ground water and can deliver it in sufficient, usable quantities. An aquifer can be shallow to the surface, situated in an area of highly permeable layers. This constitutes an unconfined aquifer. In such a case, seepage from the surface into the ground water is largely unimpeded. This contrasts the confined aquifer which is bounded above by a relatively impermeable confining layer. The ground water flows through the void spaces (pores) in the soil matrix of the aquifer. A measure of the resistance to this fluid flow within the aquifer is the permeability. A concept related to permeability is that of hydraulic conductivity, which defines the volume of water that passes through a unit cross-sectional area in a unit time.

Ground water has a significant influence on the engineering properties of soils and rocks. The interrelation between ground water flow and geological conditions is the basis of the study of hydrogeology.

Ground water flows under the influence of the gradient between discharge and recharge. Manipulation of the flow of ground water can be achieved by creating a hydraulic gradient through draw-down and recharge pumping. There is a zone of influence created around a well within which the natural water level and flow are modified. As water is withdrawn, a cone of depression is formed. As water is injected, a recharge mound is formed.

The flow of ground water is not solely a function of the aquifer orientation. The surface topography strongly influences the ground water flow. The regional hydrological and hydrogeological systems are interrelated components of the overall ground water flow system.

When considering the problem of ground water contamination, a thorough understanding of the hydrogeology of the surrounding area is essential if we hope to predict the fate of the contaminants in the environment. A detailed understanding of the concept of the ground water flow system is essential in determining the direction, rate, and areal extent of contaminant migration in the ground water. Equally important is an understanding of the position of contaminant sources relative to the ground water flow system.

Before contaminated ground water can be cleaned up, the overall problem must be evaluated in order to determine the best approach to a solution. The treatability study addresses this concern. The ultimate goals of the treatability study are to:

- Determine what combination of physical, chemical, and biological treatment most efficiently removes the contaminants
- Collect sufficient data for full scale process design

This feasibility study is an essential aspect of the remediation project. Characteristics specific to the site affect the removal of the contaminants. In order for any ground water treatment process to be successful, it must be designed specific to these site characteristics. The parameters defining a contamination site are related to the distribution of contaminants and the factors controlling their transport. Therefore, characterizing a site involves analyzing past behavior in order to predict future conditions.

The extent of the contamination must be assessed accurately. Ground water quality, specific contaminants, and subsurface soil characteristics are determined by taking subsurface samples. Monitoring wells are used to characterize the rate and extent of contaminant plume migration in the subsurface environment (see Table 1).

Once the specific contaminants are identified, treatment alternatives can be developed and evaluated. A laboratory scale pilot plant frequently follows. If physical or chemical treatment appears to be most appropriate, then parameters of the particular unit operation are further defined. For example, air-stripping packed tower loading tests might be run or adsorption isotherm tests might be conducted. If biological techniques are indicated, such data are collected.

A biotreatability study provides data concerning the degree of biological treatment achievable, including organic removal rates, oxygen requirements, oxygen transfer characteristics, and biokinetic constants. Of primary concern is the biodegradation rate of target compounds.

Analysis includes evaluation of the pH condition, dissolved oxygen level, subsurface microorganisms, and nutrient availability. A geochemical evaluation of the site will determine which chemicals, if any, are deficient. Determination is made as to whether the indigenous microbes can sufficiently degrade the contaminants. In addition to assessing the contaminants' toxicity to microbial growth, techniques for enhancing the natural biodegradation are investigated. These techniques include addition of supplemental microbes, addition of nutrients, addition of cosubstrate, pH adjustment, and oxygen addition.

CONVENTIONAL TREATMENT ALTERNATIVES

Stripping

Stripping is a unit operation of mass transfer which enhances the separation of volatile compounds from solution. The process exploits the difference between the actual concentration and the equilibrium concentration of the dissolved compounds (gases, vapors) in water. The basic concept in a stripping process is to provide a liquid-vapor interface, the point at which phase change occurs. The

Table 1 Data Needed as Part of the Site-Specific Evaluation

Contaminant characteristics
solubility
volatility
degradability
toxicity/health risks
number of competing compounds
Contaminant source characteristics
point/nonpoint
spill/long-term release
areal extent
Soils properties
type (clay, sand, loam)
heterogeneity
formation porosity and permeability
hydraulic conductivity
ion chemistry
Plume delineation
physical/chemical character of the contaminants
fate of contaminant
contaminant transport mechanism and pathways
areal extent, depth, and amount of contaminants
Hydrogeological characterization
soil profile (physical/chemical properties)
regional hydrological flow system
aquifer: confined/unconfined
ground water level and fluctuation
ground water flow pattern
gradient/recharge areas

volatile compounds leave the liquid phase at the liquid surface. The rate of mass transfer depends on transfer area (interface surface area) and driving force (the concentration differential between liquid and vapor). Maximizing the overall separation can be achieved by increasing the interface surface area, increasing the mass flow rate through the system, and/or increasing the driving force.

In the stripping process, the temperature is an important variable. As the temperature increases, so does the volatility resulting in a greater rate of mass transfer of the dissolved target compounds (typically VOCs). High-temperature or steam stripping is used in some cases to capture this benefit; however, operating costs (of heating) are greatly increased.

For air stripping, the equipment could be as simple as a spray nozzle which releases a stream of water into the air. The more commonly used equipment is the packed tower.

A significant problem with the process of air stripping is the potential release of the volatile compounds into the atmosphere. This is merely relocating the pollutants from the water into the air.

Adsorption

Adsorption is the process of capturing molecules of dissolved solids, liquids, or gases on the surface of certain active solids. This can be a physical or chemical mechanism occurring at the molecular level. For physical adsorption, the attraction is related to the surface tension of the active solid (the adsorbent). For chemical adsorption, chemical bonding occurs at the surface of the adsorbent.

Adsorption on activated carbon follows the physical mechanism. The great surface area that activated carbon has available for the adsorption process is attributed to its porous internal structure. Activated carbon has a matrix of micropores which yield a surface area of up to 1400 m^2/gr of material.

There is much tabulated data available specifying the adsorption capacity of activated carbon for a great number of compounds. However, when designing a system for a particular situation, a column pilot test is needed to optimize the operation. Often, there are multiple compounds in solution which compete for adsorption sites. Activated carbon can more easily collect and hold compounds of lower volatility.

Adsorptive capacity increases as solubility (of the target compound) decreases and also increases as temperature decreases.

Carbon adsorption has been around for decades. It has been used traditionally to remove taste and odor from drinking water. Activated carbon treatment is used to remove organic and inorganic compounds from water, although the process is more common to organics removal. Carbon adsorption is often used in combination with other treatments as a polishing step.

Ion Exchange

Ion exchange is the process of removing unwanted ionic species from solution by replacing them with a different species of ion. All ions can be displaced/removed by this process. This includes, but is not limited to, nitrites, nitrates, heavy metals, alkali metals, sulfites, sulfates, chlorides, fluorides, and bicarbonates.

The charge of the ion removed is specific to the exchange resin. There are applications for positive charge cations and negative charge anions. For complete removal, a mixed bed unit or a multistep system is used. In the mixed bed exchanger, both cation and anion resins are contained in a single unit. A multistep system alternately arranges cation and anion exchangers is series flow.

There are some disadvantages with ion exchange. For waters with high concentrations of dissolved ionic solids, the exchange capacity is rapidly exceeded. This requires frequent regeneration or replacement of the resin.

Pretreatment may be required to avoid plugging or fouling of the resin bed by suspended solids. Certain residual organic compounds can deteriorate the resins.

Reverse Osmosis

This is a separation process which removes dissolved compounds from solution by filtration through a semipermeable membrane at high pressure. The product stream is clear permeate and the waste stream is a concentrate of mostly dissolved metal salts. Depending on the specific membrane composition, the dissolved solids rejection rate can range from 50 to more than 99%. Although some organic compounds are removed, the rate is low. It is generally not acceptable for organics removal.

Chemical Treatment

As with biological treatment, chemical treatment is applicable above ground or in situ. Above-ground chemical-treatment systems are the same as conventional wastewater-treatment systems. The ground water is extracted and treated and then discharged on the surface or recharged to the aquifer. Treatments include oxidation and precipitation.

Chemical oxidation/reduction is the loss or gain of electrons by atoms. As applied to organics, oxidation is the chemical process of degrading the compounds ultimately to carbon dioxide and water. The two chemical-oxidizing agents commonly used in ground water treatment are hydrogen peroxide and ozone.

Chemical precipitation has often been used to remove metals and other inorganics from wastewater. Metals can generally be removed by precipitation as the metal hydroxides, carbonates, or sulfides. Following the chemical reaction which forms the insoluble solids, these precipitates must be removed by a physical separation. This is typically filtration or sedimentation.

EMERGING TECHNOLOGIES

Biological Techniques

Over the past decade, interest in harnessing natural biological processes for the destruction of hazardous wastes has resulted in much experimental data on a variety of organic compounds. Compounds which were previously believed to be resistant to biodegradation (refractory) prove to be amenable to biological treatment.

Bioremediation began decades ago with the application of microbes which were originally developed for the production of enzymes used in detergents. Since then, modified strains of bacteria were developed which could resist the toxic effects of a significant number of the organics on the EPA list of priority pollutants. Basically, any indigenous microorganism could adapt to and degrade any synthetic organic compound. Selective development consists of identifying a microorganism known to have some activity in the presence of the desired toxic compound, adapting it to progressively higher concentrations, selecting the most active colonies, and preserving these for later application.

Most organics which have toxic properties are still biodegradable and can serve as carbon and energy sources for microbial growth under select environmental conditions. Priority pollutants can generally be biodegraded under laboratory conditions using specially acclimated microbes. This acclimation must be carefully evaluated for any biological process treating specific inhibitory compounds. In order to maintain a contaminant-degrading microorganism population, a minimum level of the target compound must be maintained in the influent to the bioreactor. Very low concentrations (<50 ppm) of inhibitory organic compounds may be insufficient to stimulate production of the specific enzymes necessary to degrade these compounds.

Microbes metabolize the organic compounds producing energy and/or new cellular material. The three mechanisms which microbes use to produce energy are

- Fermentation
- Anaerobic respiration
- Aerobic respiration

Fermentation is the process of metabolism via a series of enzyme reactions which do not involve an electron transport chain. Anaerobic respiration is the metabolic process involving an electron transport chain with the terminal electron acceptor being nitrate, sulfate, or carbon dioxide. Aerobic respiration is the metabolic process involving an electron transport chain with the terminal electron acceptor being oxygen. Respiration is the more efficient mechanism, and aerobic respiration is the most often used system.

Biodegradation is a kinetically limited biochemical process. The biokinetic characteristics of inhibitory compounds must be investigated when designing a system. Substrate inhibition occurs when the growth-supporting substrate becomes inhibitory at high concentrations. That is, while the cell growth rate increases with increasing substrate concentration for low concentrations, the growth rate decreases with increasing substrate concentration for higher concentrations. In contrast to noninhibitory substrates, when degrading inhibitory compounds, cell growth rates typically attain a peak value and then decrease.

Most instances of contaminated ground water reveal multiple organics. There is a great range in metabolic responses of microbes to mixtures of organic compounds where more than one substrate can serve as carbon and energy sources but which are still inhibitory. These responses range from diauxic growth to concurrent utilization of several substrates.

The organics should be the limiting factor to the specific growth rate; however, often in ground water environments, the dissolved oxygen level is identified as the growth rate limiting factor. Oxygen dissolves in water only to a limited degree. When dissolved compounds other than oxygen are present in the water, the rate of oxygen transfer is typically decreased along with the overall saturation value. The use of mechanical aerating devices for oxygen transfer limits the dissolved oxygen level to approximately 10 mg/L. Such limitations of oxygen transfer can be avoided by the use of chemical oxidizing agents such as hydrogen peroxide. The concentration of the oxidizing agent needs to be carefully determined. The concentration has to be high enough to maximize the microbial growth rate, yet not so high as to chemically oxidize the biomass.

Environmental parameters that influence microbial growth and metabolism are temperature, pressure, pH, salinity, inorganic nutrients, oxygen, and the absence of high concentrations of toxic/inhibitory compounds. Biotreatment systems need to provide for acceptable levels of these constituents as required by the microbial population for a given substrate. Inorganic nutrients which include nitrogen, phosphorus, sulfur, potassium, magnesium, calcium, iron, and manganese substantially reduce the length of time for acclimation of the microorganisms whose natural transition period would otherwise be several months.

Bioaugmentation is the process of adding nonindigenous microbial supplements to a bioreactor for the purpose of artificially increasing the microbial population diversity and activity. This, in turn, should enhance the treatment rate and/or reduce acclimation time. Often it is found that such supplements fail to accomplish these desired results owing to a difficulty in metabolizing the target compounds.

Some organics are so highly resistant that microbes cannot directly degrade them. Cometabolism is the process of degrading compounds without metabolizing them. Microorganisms produce enzymes for the utilization of specific substrates, which can also initiate the oxidation of a wide range of nonspecific organic compounds; however, the microbes do not consume these nonspecific compounds as a source of energy and/or carbon.

One example of this process is the cometabolism of halogenated aliphatics, such as trichloroethane (TCA), using methane as the primary substrate. Methanotrophic bacteria cometabolically oxidize the halogenated compounds. Once initiated, heterotrophic bacteria continue the degradation to stable end products.

Anaerobic biodegradation is slower and used less frequently, but it is sometimes the most suitable approach to remediation. There are some organic compounds which require the strong reducing conditions of anaerobic degradation. Anaerobic reductive dehalogenation is recognized as the mechanism in the direct metabolism of halogenated aliphatics such as TCA, TCE, and PCE.

Process Design

Prior to actually treating the contaminated ground water, it is essential to remove the source of contamination, control the contamination plume, and control the movement of the contaminated ground water. Control is commonly exercised through pumping. The zone of influence of the draw-down wells need to encompass the contaminant plume. A contaminant plume in ground water is usually an elongated body of fluid moving with the prevailing ground water flow. The desired cone of depression around the well may require very high water-removal rates.

In a remediation project, there are numerous options from which to select while formulating a final design. For above-ground treatment systems, the processes are analogous to that of treatment of wastewater. In situ methods effectively provide a reaction vessel in the ground.

These "pump and treat" above-ground systems pump contaminated ground water to the surface and feed it into the treatment equipment such as an air-stripping tower, a carbon adsorption column, and chemical or biological reactors. The treated effluent is either recharged to the aquifer or discharged at the surface.

There are several designs for a biological treatment system. It can be aerobic or anaerobic; however, aerobic systems seem to be more common. These systems can be designed as suspended growth (as the activated sludge process) or fixed film (as trickling filter or RBC).

In suspended growth systems, biomass is suspended in the liquid phase in the reactor. The most common example is an activated sludge process. Although such a process would theoretically produce the highest quality effluent, control of the process is relatively complex.

In the fixed film systems, the biomass is attached to an immobile carrier medium within the reactor. The fixed film systems are particularly suited to degradation of inhibitory contaminants. These systems have a high biomass volume and a high solids retention time which allows the microbes to become acclimated to the inhibitory compounds in the ground water while minimizing the hydraulic detention time. This approach is straightforward to control but has some disadvantages, such as the possibility of clogging the media with biomass, greater sensitivity to temperature, greater cost for capital equipment, and less operating flexibility than an activated sludge process.

An emerging technology for the treatment of contaminated ground water is that of in situ treatment. This "in place" treatment method relies on the introduction of external agents into the subsurface environment of an aquifer. The supplements are introduced via wells and circulated through the treatment zone by pumping one or more draw-down wells.

The subsurface soil must have sufficient porosity to allow for mixing and transport of the external agents. These agents interact with the contaminants and detoxify them without transfer of the contaminated ground water to the surface. Complete mixing of treatment reagents with the ground water is a necessity of in situ treatment. The reagents must be delivered to the target compounds at sufficient rate to assure against decomposition or reaction with nontarget compounds. The treatment zone must be completely contained so that no contaminants spread beyond its limits.

Treatment reagents in solution can be applied on the surface by means of flooding ponds or trenches partially filled with gravel. An alternate to this is a shallow subsurface drain. These are gravity flow delivery systems and may be suitable for treatment of shallow ground water contamination.

Generally, gravity flow methods are better suited to disposal of extracted and treated ground water. By design these methods can be used to create a fluid flow loop which flushes contaminants from the soil in the vadose zone into the ground water for subsequent treatment. This arrangement can be used to quickly modify the flow pattern of shallow water table aquifers.

The more commonly used delivery scheme is the forced injection method. This is a pressurized well through which the chemical reagents are pumped into the subsurface. For deep or confined ground water formations, forced injection is the only method applicable.

Two specific requirements of the in situ methods are

- It must be operated within the natural environmental conditions (pH, temperature, precipitation conditions).
- It must allow for monitoring of the treatment progress. Generally, in situ treatment is most effective when treating localized aquifer contamination.

Advantages of in situ remediation include systems with less equipment and lower cost than a corresponding pump and treat system. In situ methods as such are not time tested as are the more conventional technologies. However, there are a number of in situ techniques which have been used successfully and have the potential for continued development. There are, however, disadvantages. The injection of the external agents may be considered as the unlawful application of hazardous waste. There is the potential for clogging of the injection well with biomass or suspended sediments. Subsurface conditions can cause chemical precipitation of external agents.

In Situ Biological Treatment

Of the three types of microbial metabolism, aerobic respiration is the most feasible for application to in situ biological treatment. As such, the most research has been done in this area. These aerobes degrade organics faster and more completely than anaerobes.

In situ biological treatment manipulates subsurface environmental conditions in order to nurture the naturally occurring microbes which degrade or destroy the organic contaminants. An understanding of the subsurface environment and how it interacts with the distribution and growth of the microbes is prerequisite to development of effective biological treatment techniques.

This method allows natural selection to control the population of microorganisms, thus limiting external control to that of manipulating such parameters as loading rates, nutrient addition, and supplemental microbial cultures.

Microbial activity relies on intimate contact of microbes, nutrients, and contaminants. The aquifer permeability must be sufficient to provide for a relatively short residence time of the chemical supplements and thereby ensure complete mixing throughout the entire biologically active treatment zone. Complete mixing is not automatic but must be considered in the design of the chemical supplement delivery system.

The flowrate of ground water through the treatment zone should be high enough to provide for multiple flushing cycles over the designed treatment period. This may require pumping to manipulate the natural flow. In situ treatment requires approximately two to six recirculations of water through the treatment zone of the aquifer. This necessitates that the ground water flow be manipulated to flow through the aquifer as fast as possible.

A passive method of in situ treatment involves the application of a solution of active microorganisms and nutrients via surface trenches which then percolates down from the surface into the subsurface environment.

For in situ treatment, there is a minimum amount of above-ground equipment that is needed for supplying nutrients, oxygen, and microorganisms. Often in situ treatment is combined with a few pump and treat operations. A central draw-down well is located at the center of the contaminant plume with recharge wells at the periphery of the central well zone of influence. The contaminated ground water is withdrawn, aerated, and mixed with supplements. The pH and temperature are adjusted for optimal growth conditions. As the microbes consume the organics, the biomass increases. The nutrient and dissolved oxygen level is monitored and maintained such that the recharge water has sufficient levels of each to continue the residual clean up in the subsurface biologically active treatment zone.

Generally, biological treatment only partially degrades refractory compounds. Pretreatment by chemical oxidation enhances the ability of microbes to biodegrade refractory organics.

Oxygen has a limited solubility in water of approximately 8–10 mg/L depending on the temperature under critical conditions. The water in the aquifer can supply a limited amount of oxygen. Therefore, water rich in oxygen must be recirculated quickly through the treatment zone. Oxygen can be supplied by:

- Bubbling/sparging compressed air into wells
- Diffusing pure oxygen into wells
- Introducing more reactive oxidizing agents

At concentrations in ground water of 100 ppm or less, the cytotoxic properties of hydrogen peroxide are not sufficient to destroy the active microbial cultures.

The simplest active method of supplying oxygen and chemical supplements to the treatment zone consists of pumping a low continuous dose. However, supply of nutrients and oxygen can create a zone of high activity around the injection well. A single large dose could cause the biomass in the immediate vicinity to deplete the dissolved oxygen and upset the overall treatment process. Alternatives include multiple point injection and pulsed injection. The pulsed injection process alternately pumps nutrients and oxygen for specific time durations. The reagents mix gradually in the subsurface and as a result the growth of biomass is less dense around the well. Both of these methods reduce biofouling and provide for more uniform distribution of biomass growth throughout the treatment zone.

In Situ Chemical Treatment

In situ chemical treatment involves the introduction of specific chemical agents into the subsurface in order to degrade, immobilize, or increase the mobility of the contaminants. The scheme of fluid flow is analogous to biological in situ treatment. External agents are introduced via injection wells and circulated through the contamination treatment zone via draw-down pumping.

Many of the chemical in situ methods are based on studies and laboratory data. These methods are in development as field data is scarce. Practical limitations are imposed by incomplete mixing of contaminated ground water with treatment reagents. The chemical treatments include precipitation, oxidation-reduction, polymerization, and neutralization.

Precipitation is aimed at immobilizing inorganic heavy metals. Many of the heavy metals can be treated by precipitation. All divalent metal cations are subject to precipitation by sulfide, phosphate, hydroxide, or carbonate. Addition of the reagent (e.g., caustic, lime) raises the pH to above 7, at which point precipitation of metals starts. There are, however, some difficulties to the generality. Not all metals precipitate at the same pH and some have an optimum point above which their solubilities increase. Some metals must be in a particular valence state to precipitate, whereas others require the presence of a tertiary component. The agents are sodium sulfide, calcium carbonate, sodium hydroxide, calcium oxide/hydroxide, or sodium phosphates which result in precipitating the corresponding species of the heavy metal compounds.

The metal sulfides would be the most stable over a wide pH range. Metal sulfides generally have lower solubilities (than hydroxides or carbonates) and thus sulfide treatment has greater removal capability. However, difficulties in chemical and sludge handling negate any benefits.

Precipitation as the metal phosphate may be subject to interference from naturally occurring calcium in the subsurface.

The metal hydroxides and carbonates have an optimum pH range for precipitation which is narrow and specific to the particular metal. Outside of this range, solubility increases. The most commonly used process is hydroxide.

Removal of heavy metals using chelating agents is primarily applied to contaminated soils. However, the chelating agent increases the mobility of the heavy metal in solution resulting in a stable metal complex which resists degradation. This treatment may be used in combination with above ground processes for removal of heavy metals.

Oxidation of organic compounds include a wide range of organics (aromatics, aldehydes, ketones, halogenated aromatics) which are acted on by ozone or hydrogen peroxide. The contaminants are either detoxified or their toxicity is reduced. Chemical oxidation can make certain organics more amenable to biodegradation.

Oxidizers can be used with catalysts, such as ferrous ions (or ultraviolet light for above ground), to increase the reaction rate. Without catalysts, partial oxidation is a common result. There are three oxidizing agents which can be used for in situ treatment of organics: ozone, hydrogen peroxide, or hypochlorites.

The hypochlorites have limited application, as there is the possibility of forming chlorinated hydrocarbon compounds. These chlorinated organics can be more hazardous and or difficult to treat than the original compounds.

Ozone has greater, but still limited, application. Ozone is highly reactive and relatively unstable. It must be generated as needed on site, and it rapidly decomposes on application to ground water.

Hydrogen peroxide seems to be the best suited for straightforward application to a wide range of treatment systems. It is more stable and simpler to handle than ozone. Hydrogen peroxide does not react to yield chlorinated organic products. It is commercially available in stabilized aqueous solutions of varying concentrations. Hydrogen peroxide as the oxidizing agent has the benefit of providing dissolved oxygen for the subsequent biodegradation. One drawback of hydrogen peroxide is that it tends to increase the sorptive capacity of soils. This could adversely affect contaminant removal.

Oxidation of inorganics focuses on arsenic, iron, and manganese. The oxidizer of choice is potassium permanganate, which results in precipitating the metal oxides.

The reduction of organics is applicable to nitro aromatics and chlorinated aromatics and aliphatics. The agents are catalyzed metal powders which result in removal of the nitro or halo side chain or saturation of the aromatic structure. This detoxifies the compound for subsequent biological treatment.

Reduction of inorganics is limited to hexavalent chromium and selenium. The reducing agent is ferrous sulfate, which acts to reduce these metals to the trivalent species.

Neutralization is a straightforward chemical reaction balancing the pH to 7. Often this step is integrated into other chemical treatment operations.

Polymerization focuses on a few very specific organic compounds. Contamination by monomers of vinyl chloride, isoprene, or acrylonitrile can be converted to the corresponding polymers by applying a suitable catalyst agent. This immobilizes the contaminants into a gel-like mass.

Widespread application of polymerization is questionable. Numerous injection wells are required to introduce the catalyst. There is difficulty in assuring sufficient mixing between the monomer and the catalyst. The catalyst can be considered a hazardous substance thereby restricting application to the ground water.

Hydrolysis focuses on a few specific types of organics: esters, amides, carbamates, organophosphorus esters, and certain pesticides. The reaction is base catalyzed and the reagent is either caustic soda or lime. So far, this is purely theoretical as no field data yet exists.

Although data on demonstrated chemical in situ systems are limited, there is one process which is proven. The Vyredox method is designed to remove iron and manganese. The system injects highly aerated water in a zone encircling the supply well. Iron is precipitated first, and the solids are retained in the strata furthest from the supply well. Closer to the well, the conditions are maintained as favorable for bacteria which preferentially oxidize (precipitate) manganese. Thus, the water entering the well is free of iron and manganese.

REMEDIATION SUMMARY

Contaminated ground water remediation begins with a preliminary assessment of the overall problem. Next, the site characteristics are investigated; field samples are collected if necessary. Then specific treatment alternatives are developed. Suitable technologies must be site and waste specific. Contaminated soil frequently accompanies the ground water contamination. An integrated strategy must be developed for concurrent remediation. Pilot plant studies often follow. Final selection of treatment process design, in addition, takes into consideration the project cost and time limitations. For example, a pump and treat air stripper with carbon adsorption may be the quickest treatment but the cost may be excessive. In contrast, biological in situ treatment may be an effective technology yet the time required may be excessive.

Pump and treat systems using physical treatment methods still account for the majority of remediation projects. Conventional pump and treat methods are used for the purpose of contaminant containment at a majority of the Superfund sites where hopes for complete remediation have been abandoned.

Inorganics are generally treated above ground by conventional physical and chemical methods. Heavy metals, in particular, are effectively removed by chemical precipitation.

Volatile and semivolatile organics are typically treated by air stripping and/or carbon adsorption. Biological methods are becoming increasingly acceptable for treatment of organics.

Although biological in situ treatment has been successfully demonstrated, there is limited field data available. Chemical in situ treatment has received much less research and, as such, is still largely experimental. Analogous to combining different unit operations/processes in series, the use of in situ techniques is frequently combined with above ground systems as part of an overall treatment train.

For remediation projects where biological methods are used, aerobic systems account for a majority of the installations. In one case, ground water was pumped to the surface and fed along with nutrients into an aerated completely submerged attached growth bioreactor. The clarified effluent was discharged to percolation trenches.

There are, however, successful pilot studies using anaerobic systems. In the subsurface environment associated with ground water contaminated by toxic compounds, an indigenous microorganism population develops which is unique. It is the result of biochemical adaptation to the characteristics of the inhibitory compounds. In one case of contamination by TCA, the contaminated ground water was pumped to the surface, chemical supplements added, and fed into an anaerobic bioreactor. The effluent, which was carrying an enhanced population of anaerobes, was recharged into the subsurface. A portion of the effluent was applied at the surface as a barrier to oxygen, thus protecting the subsurface anaerobic zone.

These two cases used biological treatment to reduce the organic contaminants to below the desired limit. A third case of contamination with organics required carbon adsorption after the biological treatment in order to meet the required effluent levels.

Table 2 is a summary of treatment technologies.

A variety of remediation technologies are available for treatment of contaminated ground water. Bioremediation has emerged as an effective alternative to treat contaminated ground water while maintaining relatively low capital costs as compared to conventional methods. In addition, it minimizes site disruption and reduces or eliminates the costs associated with transportation, handling, and disposal of the recovered contaminants. Among the biological methods, a trade-off exists. The in situ approach is generally less costly but does not offer stringent biological process control. When subsurface conditions do not provide a conducive environment for the indigenous microbes, treatment in place may not be efficient or even possible. The pump and treat systems, although more capital intensive, are able to be directly controlled.

Many physical and chemical methods do not destroy the hazardous

Table 2 Summary of Treatment Technologies

Technology	Application	Process Description
Physical		
Stripping	VOCs, ammonia	Pump and treat system removes compounds from water into the air.
Adsorption	High MW organics	Pump and treat system captures compounds on activated carbon.
Ion exchange	Ionic cmpounds (typically inorganics)	Pump and treat system captures charged ions on specific resins.
Reverse osmosis	Metals, heavy metals	Pump and treat system filters water to produce clean water and a concentrated waste stream.
Chemical		
Precipitation	Metals, heavy metals	Reagents react with contaminants forming insoluble compounds which fall out of solution. Applicable in situ but has been used largely above ground.
Oxidation/ reduction	Inorganics	Reagents react with compounds changing their oxidation state. Resulting compounds are removed by subsequent means. Selectively applicable in situ. Generally used above ground.
	Organics	Reagents convert the organics to less toxic compounds. Typically used as a pretreatment to biological treatment. Applicable in situ and above ground.
Biological	Organics	Naturally occurring microbes are stimulated to biochemically degrade the organic compounds. Applicable in situ and above ground.

compounds; they simply concentrate and relocate them. In most cases, the biological methods can convert these hazardous compounds into safe and acceptable products. Since the concentrations of toxics which can be degraded are relatively low, bioremediation is slow and requires sufficient time.

TYPES OF GROUND WATER CONTAMINANTS

All ground water contains some impurities, even if it is completely unaffected by human activities. The types and concentrations of natural impurities depend on the nature of the geological material through which the ground water moves and the quality of the recharge wager. Water undergoes chemical changes as it passes through the soil and the underlying aquifers. Ground water moving through igneous rocks dissolves relatively small quantities of mineral matter, because the elements that compose such rocks are relatively insoluble. Sedimentary rocks and soils, however, typically contain a wide range of compounds such as magnesium, calcium, and chlorides that are more soluble. Some aquifers also have high natural concentrations of other dissolved constituents such as arsenic, boron, and selenium. Because of their slow rate of movement, ground waters generally contain higher concentrations of such dissolved constituents than surface waters. And the higher the acidity of the recharge water, the more capacity it has to dissolve many minerals. In addition to these natural variations, ground water can pick up contaminants from many human sources. Major types of biological and chemical contaminants have been found in ground water.

Biological Contamination

Biological contaminants of ground water include bacteria, viruses, parasites, and other biological agents. There is very little information about which specific pathogens are actually present in ground water. The most common bacteria tested for is *Escherichia coli*, a type of bacteria that, although not necessarily harmful in itself, is commonly found in human and animal wastes and may indicate the presence of other types of more harmful bacteria. In spite of the very limited testing available, reports of several different types of bacteria and viruses in ground water, including those that cause typhoid, tuberculosis, cholera, and hepatitis, have been made.

Inorganic Substances

Inorganic substances contaminating ground water include metals, salts, and other compounds that do not contain carbon. The Office of Technology Assessment has identified 37 inorganic substances that are known to occur in ground water, including some 27 metals (Table 3). The EPA has established primary drinking water standards for 10 inorganic compounds; between 1975 and 1985, these were exceeded in an estimated 1500–3000 ground water supplies. The most commonly exceeded standards were for fluoride (in 1000–2000 supplies) and for nitrate (in 500–600 supplies). The EPA also found contamination by a variety of inorganic substances in wells serving rural households.

Although natural causes are responsible for much of this inorganic contamination, human sources can create significant amounts as well.

Selenium and arsenic were the toxic elements most often reported to exceed drinking water standards in ground water. Excessive levels of mercury, lead, cadmium, silver, and selenium in 5–25% of the wells serving rural households

Table 3 Inorganic Substances Found in Ground Water

Substance	Concentration (mg/L)
Aluminum	0.1–1,200
Ammonia	1.0–900
Antimony	—
Arsenic	0.01–2,100
Barium	2.8–3.8
Beryllium	<0.01
Boron	—
Cadmium	0.01–180
Calcium	0.5–225
Chlorides	1.0–49,500
Chromium	0.06–2,740
Cobalt	0.01–0.18
Copper	0.01–2.8
Cyanides	1.05–14
Fluorides	0.1–250
Iron	0.04–6,200
Lead	0.01–5.6
Lithium	—
Magnesium	0.2–70
Manganese	0.1–110
Mercury	0.003–0.01
Molybdenum	0.4–40
Nickel	0.05–0.5
Nitrates	1.4–433
Nitrites	—
Palladium	—
Potassium	0.5–2.4
Phosphates	0.4–33
Selenium	0.6–20
Silver	9.0–330
Sodium	3.1–211
Sulfates	0.2–32,318
Sulfites	—
Thallium	—
Titanium	—
Vanadium	243.0
Zinc	0.1–240

Source: From the Office of Technology Assessment, U.S. Congress, *Protecting the Nation's Groundwater From Contamination*. Washington, D.C., U.S. Government Printing Office, 1984, pp. 29–30.

were found, although there notable regional variations. Iron and manganese, which are not known to have any toxic effects, are also relatively common, particularly in the north-central region of the United States. Table 4 summarizes some findings for inorganic elements of an EPA rural water survey. Chromium, barium, magnesium, and arsenic were found to exceed standards in less than 1% of rural wells. These national statistics, however, may understate conditions in certain local settings.

Organic Substances

Many organic substances occur in nature. Tens of thousands of synthetic organic compounds have been developed in laboratories in the past. These are used in such common products as dyes, food additives, detergents, plastics, solvents, and pesticides. For years, there was little ground water monitoring for contamination from organic chemicals. However, in recent years, the discovery of synthetic organic chemicals in ground water supplies has caused major concern. Table 5 lists concentrations of organic compounds found in drinking water wells and surface water.

Table 4 Summary of Inorganic Elements EPA Rural Water Survey

Element	Level exceeded (mg/L)	In % of rural households				
		nationwide	west	north-central	northeast	south
Mercury	0.002	24.1	10.4	31.8	22.0	25.0
Iron	0.3	18.7	7.0	28.2	16.0	17.0
Cadmium	0.01	16.8	27.1	20.7	1.6	17.3
Lead	0.05	16.6	16.9[a]	10.8[a]	9.6[a]	23.1[a]
Manganese	0.05	14.2	4.7	19.9	16.9	12.3
Sodium	100	14.2	15.0	19.2	6.0	14.1
Selenium	0.01	13.7	41.3	25.7	0.0	2.1
Silver	0.05	4.7	2.1	3.7	4.8	4.8
Sulfates	250.0	4.0	11.7	7.4	0.5	0.7
Nitrate-N	10.0	2.7	4.0	5.8	0.3	1.3
Fluoride	1.4	2.5	6.2	1.8	0.0	2.7
Arsenic	0.05	0.8	2.1	1.8	0.0	0.0
Barium	1.0	0.3	0.0	0.0	0.0	0.7
Magnesium	125.0	0.1	0.5	0.1	0.0	0.0
Chromium	0.05	b	0.0	0.0	0.0	0.0
Boron[c]						

[a]May be distorted upwards.
[b]Not detected.
[c]Not tested.
Source: From the U.S. Environmental Protection Agency,Office of Drinking Water, *National Statistical Assessment of Rural Water Conditions, Executive Summary*. Washington, D.C., U.S. Environmental Protection Agency, June 1984.

Table 5 Concentrations of Toxic Organic Compounds Found in Drinking Water Wells and Surface Water

Chemical	Ground water concentration (ppb)[a]	Highest surface water concentration reported (ppb)[a]
Trichloroethylene (TCE)	900–27,300	160
Toluene	55–6,400	6.1
1,1,1-Trichloroethane	965–5,440	5.1
Acetone	3,000	NI
Methylene chloride	47–3,000	13
Dioxane	2,100	NI
Ethyl benzene	2,000	NI
Tetrachloroethylene	717–1,500	21
Cyclohexane	540	NI
Chloroform	67–490	700
Di-*n*-butyl-phthalate	470	NI
Carbon tetrachloride	135–400	30
Benzene	30–330	4.4
1,2-Dichloroethylene	91–323	9.8
Ethylene dibromide (EDB)	35–300	NI
Xylene	69–300	24
Isopropyl benzene	290	NI
1,1-Dichloroethylene	70–280	0.5
1,2-Dichloroethane	250	4.8
bis (2-Ethylhexyl) phthalate	170	NI
DBCP (1,2-dibromo-3-chloropropane)	68–137	NI
Trifluorotrichloroethane	35–135	NI
Dibromochloromethane	20–55	317
Vinyl chloride	50	9.8
Chloromethane	44	12
Butyl benzyl-phthalate	38	NI
Gamma-BHC (Lindane)	22	NI
1,12-Trichloroethane	20	NI
Bromoform	20	280
1,1-Dichloroethane	7	0.2
Alpha-BHC	6	NI
Parathion	4.6	0.4
Delta-BHC	3.8	NI

[a]ppb = parts per billion; 1 ppb = 1/1000 ppm; 1 ppm = 1 mg/L.
NI, not investigated.
Source: From the Council on Environmental Quality, *Contamination of Ground Water by Toxic Organic Chemicals* (Washington, D.C.: U.S. Government Printing Office, 1981), Table 8, pp. 36–39.

Radionuclides

Almost all ground water contains some amount of naturally occurring radioactive substances. The types and levels of these substances vary geographically depending on local geology. Approximately 20 radionuclides are known to occur in ground water, originating from both natural radiation and from radioactive substances resulting from human activity.

Activities associated with nuclear power generation can release both natural and human-produced radiation into the environment. Atmospheric testing of nuclear weapons and the use of radionuclides in medical and other scientific research provide additional conduits for the release of radioactivity. Radionuclides are also a natural by-product of other activities such as phosphate mining and the production of phosphoric acid.

Radioactive materials are mixed and distributed in water by the same forces as other soluble or suspended materials, but each radioisotope tends to have its own route and rate of movement. If it has a relatively slow decay rate, a radioactive contaminant can retain its toxicity for long periods of time.

SOURCES OF GROUND WATER CONTAMINATION

A wide variety of human activities and natural processes can lead to ground water contamination. Because available data on the specific causes and extent of ground water pollution are sparse, only a partial picture of the contamination problem can be seen. Some sources create high levels of toxic contaminants in geographically limited areas and others can cause lower levels of contamination over a more widespread area. The threat posed by a given source will vary from site to site depending on the particular set of circumstances that exists. Some of society's essential activities are also among the most common sources of ground water contamination: waste disposal, materials handling and storage, mining, and agriculture.

Waste Disposal

Probably the greatest threat to ground water quality and human health are posed by contamination resulting from the disposal of wastes. Substantial amounts of solid, liquid, and gaseous waste are generated in the United States, perhaps as much as 29 metric tons per person per year. Among the waste-disposal technologies posing the greatest threat to ground water are on-site sewage disposal, underground injection wells, surface impoundments, land application, and landfills. Hazardous and radioactive wastes—two waste types requiring special handling and disposal—are also of increasing concern.

Modern economic activity requires the transportation and storage of large volumes of materials used in manufacturing, processing, construction, agriculture, and other activities. Along the way, some of these materials can be lost through spillage, leakage, or improper handling. Where the spills are large or the leakage remains undetected, serious ground water contamination can result. Recently, substantial attention has been focused on the loss of potential contaminants from leaking underground storage tanks. Much less attention has been given to other types of storage and the transport of materials.

The mining of fuel and nonfuel minerals can create many opportunities for ground water contamination. Problems most commonly stem from the mining process itself, the disposal of mining wastes, and the processing of the ore and the wastes it creates. Oil and gas drilling and production pose the threat of contamination as well. Approximately 73,000 mining and drilling establishments were in operation in 1982, 60,500 of which were related to oil and gas production. Of the remaining 12,500, approximately 1500 were metal mines, 5000 were coal, and 6000 were nonmetallic minerals.

Modern agricultural practices have the potential to cause serious environmental problems, including ground water contamination. The primary ground water contamination threats from cropland appear to result from the application of pesticides and fertilizers, seepage of irrigation water into the ground, and wastes from livestock operation. The tendency for agricultural chemicals to leach into ground water depends on soil properties, the chemical characteristics, local weather conditions, and farming practices.

Sometimes the simple conversion of natural grass or forest lands to agriculture can cause a problem. The conversion of grasslands in some northern Great Plains areas to a winter wheat/summer fallow crop rotation is an example. During the fallow season, rainfall that under natural conditions would be taken up by the grasses now seeps into the ground and dissolves salts in the soil. This seepage increases the salinity of shallow ground water up to the salt concentration of seawater. It often seeps out of the ground at lower elevations killing any vegetation growing there. Proposed efforts to control nonpoint-source pollution of surface waters may also increase the threat of ground water contamination by agricultural chemicals. Many of the cropping methods and structures used to reduce storm water runoff from agricultural lands result in increased infiltration of water into the ground. This increased seepage may carry with it both the residue of pesticides and fertilizers that were formerly carried off into surface waters and the chemical residue that formerly remained on the surface.

Although waste disposal, materials handling, mining, and agriculture are usually identified as the major causes of ground water contamination, a number of other sources pose the potential for problems in certain areas or under certain conditions. Some of the more common of these sources are polluted surface waters, seawater, urban runoff, de-icing salts, atmospheric pollutants, and a number of activities associated with urban and rural living.

GROUND WATER INVESTIGATION

Steps to establish a ground water monitoring network include

- Review all available background data. Look at any regional ground water studies which have been made. Review at the EPA any ground water studies of surrounding facilities. Look for past well driller's logs.
- Establish the local aquifer characteristics and ground water flow directions using the hydrological and hydrogeological data previously obtained, geophysical surveys, borings, soil sampling, and so forth.
- Identify all potential off-site upgradient sources that could have an impact on background quality. Determine the chemical composition of any waste constituents suspected of being in the ground water. Look for their potential for movement. Are the constituents floaters or sinkers? Will they get tied up in the soil matrix? Will they break down into other constituents?
- Site a sufficient number of upgradient wells to detect any facilities that might contribute to background quality. It may be necessary to put wells off property boundaries.
- Site downgradient wells to completely define the plume both its vertical and horizontal extent. Again, it may be necessary to put wells off property boundaries.
- Select the proper size and composition of well casings and screens. For example, PVC casings may be affected by chlorinated solvents and create a false impression of contamination.
- Use approved well designs.
- Use a appropriate sampling and analysis protocols.

There are a wide variety of data on site characteristic which can be obtained without drilling wells. Table 6 provides some of these parameters and Table 7 provides a list of the methods that can be used to determine these parameters.

Table 6 Hydrological and Hydrogeological Parameters Determined Without Using Wells

Storage capacity
overall
perched ground water table
Transmissivity of liquid wastes
flux
infiltration rate
velocity
Pollutant mobility
unsaturated chemical and microbial properties of liquid wastes
saturated chemical and microbial properties of liquid wastes
Salinity

Table 7 Methods of Obtaining Information

Storage capacity	Transmissivity
depth to water table	infiltration rate
water level maps	water budgets
sounding in wells	infiltrometers
driller's logs	seepage meters
test wells	test plots
available porosity	percolation rates
grain size data	instantaneous rate
test well cuttings	lab column
neutron logging	permeameter
stratigraphy	grain size
driller's logs	flux
test wells	gravimeter
natural gamma radiation	neutron moisture logging
perched ground water	gamma ray attenuation
driller's logs	tensiometers
test wells	hygrometer/psychrometer
neutron logs	heat dissipation sensor
Pollutant Mobility	resistor/capacitor
chemical properties	sensor
batch tests	remote sensing
columns	heat pulse tests
field plots	install flowmeters
auger samples	velocity
	suction cup lysimeter
	vadose zone wells
	dye tracers
	Salinity
	salinity sensors
	four-probe method
	EC probe

Surface Geophysical Techniques

Developed in the oil and gas and geothermal energy fields, geophysical techniques are finding widespread use in the ground water contamination studies. Such techniques generally offer an economical advantage versus traditional well installation and sampling techniques. A complement to the physical sampling method, they can be used to cover wide areas of suspected contamination. As one part of an integrated approach, they should be combined with soil and water sampling and downhole geophysical techniques. Table 8 provides a listing of the more common surface geophysical techniques.

Surface techniques can determine the likelihood of the horizontal extent of

Table 8 Surface Geophysical Techniques Commonly Used in Ground Water Studies

Electrical
resistivity
self-potential
vertical sounding
Electromagnetic
induction
metal detection
magnetometer
ground-penetrating radar
Seismic
reflection
refraction
Remote sensing
aerial photography
soil gas analysis

any contamination plume and locate the areas of greatest potential for contamination transport. They can be used to determine where monitoring wells should be placed and where downhole techniques can be used to determine the vertical extent of any contamination plume. They can locate where fractures and geological changes may contribute to plume migration. Surface techniques provide neither concrete evidence of contamination plume concentrations nor the types of waste constituents in the plume.

Electrical resistivity surveys involve placing two sets of electrodes in the ground and then measuring the response to an imposed electrical field from the subsurface: Differing geologic materials show differing resistances to current flow. Varying the spacing of the electrodes produces a vertical profile; moving the electrodes at a constant spacing across the area under study. Resistivity can measure soil, rock, and the ground water parameters. Most soil and rock materials are highly resistive, whereas water is highly conductive.

These methods can be used to delineate contaminant plumes, to detect changes in natural hydrogeological conditions, and to measure the depth and thickness of differing geological layers. Buried metal pipes, buildings, fences, and other metal objects affect the results. Electrical resistivity cannot identify specific geolologic materials.

Electromagnetic methods measure the electrical properties and conductivity of subsurface soil, rock, and ground water. These methods can operate rapidly and do not require ground contact. They induce a current in the ground through an alternating current coil and then measure any changes in magnitude and phase of the individual currents. These methods are useful in assessing natural hydrogeological conditions and determining the extent of contaminant plume/ trench boundaries, buried wastes, drums, and metallic utility lines.

Variations in conductivity both laterally and with depth can identify potential problem areas rapidly. Time series measurements are useful in determining gradients within and at the edges of the plume and migration rates; sounding methods can be used to determine vertical variations; and profiling methods can be used to determine lateral variations. The use of computers to help analyze the data can make the process simpler and less time consuming. Of the electromagnetic methods, ground-penetrating radar provides the best resolution of all the surface methods. An antenna moving across the ground's surface radiates high-frequency electromagnetic waves downward. These waves reflect back to the receiving antenna, and return signal variations are continuously recorded. This process profiles a continuous cross section of the shallow subsurface conditions and detects differences in soil or rock layers, and buried waste materials, drums, pipes, and cables. This method's effectiveness varies with site conditions. Silts and clays cannot be penetrated as easily as dry, sandy, or rocky areas. Typical penetration is approximately 30 ft. Ground water in finer grained materials or at greater depths than 30 ft may not be detected. Resolution depends on frequency of the waves (the higher the frequency, the better the resolution, but the shallower depth) and the speed of moving the antenna (the slower the movement, the better the resolution).

Magnetometry measures the strength of the earth's magnetic field. Because it can locate ferrous metals, it is used primarily to locate buried drums or abandoned wells. Magnetic minerals in the soil, passing vehicles, and time-variable changes in the earth's magnetic field can all adversely affect the results. Discrete and continuous results can be obtained depending on the type of magnetometer used. Drums buried to a depth of 30 ft or more can be detected by this method, whereas metal detectors are limited to 20 ft for massive piles of drums.

Seismic techniques, reflective and refractive, are other surface methodologies. Both methods measure the travel time of seismic waves transmitted by an acoustic source through the subsurface. These waves can be reflected or refracted by the subsurface material. Both methods require a seismic source, such as a sledge hammer or explosive charge, geophones to translate the seismic vibrations into electrical signals, and a seismograph to record the electrical signals. These methods are susceptible to error from outside vibratory sources.

The refraction method is used for detecting waste pits, trenches, and the depth of burial sites or landfills. It is effective to a few hundred feet. The spacing of the geophones and size of seismic source determines detection resolution and the depth of measurement. This method cannot detect underlying low-velocity layers when higher-velocity layers are above them and is time consuming and costly. The reflection method requires less energy to reach greater depths and can provide relatively detailed geological sections. It is useful in mapping the top of bedrock, any bedrock channels, alluvial and fluvial deposits, and fractures and cavities. It can provide a cross section as detailed as ground-penetrating radar.

Soil gas analysis is a common remote sensing method, and it is used principally in locating areas contaminated by volatile organics. In this method, a shallow hole is drilled or a sample tube is injected into the soil. Volumes of soil gas are evacuated to the surface for collection and analysis at a remote laboratory or measured on-site by a laboratory-quality vapor analyzer. This method can be used to analyze cuttings from well drilling operations as well. Soil gas analysis is dependent on the pore spacing within the soil and is less reliable in tightly packed soils such as clay. It also cannot be used to detect nonvolatile organic compounds and inorganic compounds. A complement to surface techniques are downhole geophysical techniques.

Downhole geophysical techniques involve obtaining subsurface information by physically entering the subsurface either through existing wells or newly drilled wells and measuring or logging specific physical parameters. Table 9 lists some of the basic downhole techniques.

Most of these methods involve lowering an instrument into a bore hole or well. The instrument usually imposes some type of stress on the subsurface materials and measures the corresponding response. A log of the vertical stress responses is then obtained. Downhole techniques can be combined with surface techniques. The obtained data must then be interpreted by a trained expert.

Induction logging can be used to identify soil and rock types, geological correlations, soil and rock porosity, and pore fluid conductivity. Resistivity logging is effective in identifying soil and rock types, geological correlations, soil and rock porosity, pore fluid resistivity, and secondary permeability such as the locations of fractures and solution openings. Natural gamma logging can assist in the positioning of wells and casings, provide clay or shale content, grain

Table 9 Typical Downhole Geophysical Techniques Used in Ground Water Studies

Electric log	Tracer dye studies
Temperature log	Flowmeter
Neutron log	Television
Gamma-gamma log	Pressure log
Fluid conductivity log	pH
Fiberoptics	Induction log
Natural gamma log	Acoustic log
Resistance log	Radar
Water level	Sonar
	Packers

size, pore fluid resistivity, and soil and rock identification. Gamma-gamma logging will help to position cementing of the well casings and total porosity or hulk density. Neutron logs will provide moisture content above the water table, total porosity below the water table, specific yield of confined aquifers, the location of the water table outside of the casing, chemical and physical properties of the water, and the rate of moisture infiltration.

Temperature logs help provide the chemical and physical characteristics of the water, source and movement of the water in the well and dilution, and dispersion and movement of the waste. Television will help to locate fractures and cavities and provide a visual guide to the well or bore hole.

Tracer studies can be used to determine the source and movement of water in a well, rate of moisture infiltration and direction, and velocity and path of ground water flow.

Lithology of aquifers and associated rocks and the stratigraphic correlation of aquifers and associated rocks can be determined by using electrical, acoustic-sonic, radiation, or caliper logs. Electrical logs can help to determine the water level location or levels of perched water. Pump tests can be used to determine hydrological properties.

Surface and downhole techniques are used to characterize the geological and hydrogeological conditions underlying the area of study and to determine the extent of contamination. The most accurate way of studying the ground water conditions is through the installation of wells.

Monitoring Wells

Before construction of a monitoring well, it is necessary to determine what permits are necessary, if any. It is not uncommon for states, counties, or cities to require permits for monitoring wells. Different, or additional, permits may be required if wells are constructed for water production (as in a pump and treat remediation system). Many states require that water well drillers be licensed and may require filing of completion reports for each well constructed. It should be verified that the drilling contractor is properly licensed, that all required permits are obtained, and that all necessary reports are filed with the appropriate agencies. Application for permits will require time and labor effort and may require payment of a fee; these factors must be taken into account in project planning and scheduling.

Field work may include drilling of exploratory borings and installation of monitoring wells. Experience has shown that certain items tend to be overlooked or taken for granted in planning monitoring well installation. Included are

- Obtaining permits (if necessary)
- Researching which supplies are available for purchase near the job site
- Ensuring that drillers know what equipment and supplies they are expected to provide

A list of equipment and supplies typically used in monitoring well installation is shown in Table 10.

Geophysical techniques described provide a valuable insight into the subsurface environment. However, these techniques do not provide actual samples of ground water being studied nor direct access to this water. The only method for acquiring this information is to penetrate the subsurface into the aquifer being studied. The geophysical techniques are key in siting the locations of monitoring wells to be installed. Knowing the horizontal extent of any contamination plume is critical to deciding the number and location of wells that will provide the needed information on the ground water environment such as waste constituent concentrations, and plume migration rate.

Most water well drilling techniques are adaptable to drilling monitoring wells. Table 11 provides a list of such techniques; however, only three find prevalent use in the installation of monitoring wells: auger drilling, rotary drilling, and cable tool drilling. Rotary drilling methods include hydraulic rotary, reverse circulation, air rotary, air-percussion rotary, and diamond bit coring. Cable tool methods include cable tool, hollow-rod drilling, and jet drilling. Auger methods include solid stem, hollow stem, and bucket auger.

In rotary drilling, a bore hole is cut by means of a rotating bit and the cuttings are removed by a fluid such as water, a water/bentonite mix, or forced air. In cable tool methods, a string of pipes is lifted and dropped on a cable. A bit at the bottom of the tool string crushes and breaks the formation materials. Cuttings are removed by the use of a fluid or baled. Augered wells are drilled by pushing a string of pipes through rotation. A bit at the end of the string cuts the material. The cuttings are removed by a spiral flight attached to the drill string.

Hydraulic rotary methods can be used in any formation, can be used to any depth, do not require a casing in which to drill, and are relatively inexpensive. However, the fluids used can interact with the ground water and result in problems when sampling. If bentonite is used in the fluid, it can absorb metals. This method can also circulate contaminants. In some rock formations, diamond bit drilling may be used.

When the fluid interaction is critical, air rotary methods may be used. They are best for hard rock formations. It is easy to determine where the water table is because when it is hit, water is expelled along with the air. This method does require a casing, especially when drilling through soft, caving formations.

Cable tool methods are excellent for obtaining samples and can be used in all formations but are best for coring, large gravel-type formations. Cable tool methods are slower than rotary methods, require a casing with a minimum of 4 in. in diameter, and are more expensive than rotary drills. In many places, rotary drilling has replace these methods.

Auger methods are used primarily in soils and soft rock. They are fast,

Table 10 Typical Equipment and Supplies for Well Installation[a]

Equipment	Supplies	
Air lift pump	Aluminum foil	Fuel
Bailer	Bentonite	Gloves
Camera	Blank casing	Hard hats
Cement mixer	"Blue ice"	Hose/cable for pumps
Drill rig and support equipment	Buckets	Ice chest
Generator	Casing centralizers	Padlocks
Grout tank/grout pump	Cement/bentonite grout	Paper towels
Photoionization detector (and calibration gas) or flame ionization detector	Chain-of-custody forms	Plastic bags
Safety equipment	Decontamination detergents/solvents	Plastic sheeting
Shovels	Distilled water	Ready-mix cement
Soil sampler	Drinking water containers	Rope
Steam cleaner	Drums of waste	Rubber boots
Submersible pump	Duct, strapping, clear, and electrical tape	Safety glasses
Surge block	Electrical extension cords	Sample containers
Tanks or drums for storage of development water	End caps	Sample labels
Tape measure	Field notebook	Scrub brushes
Tool kit	Film	Slotted casing (screen)
Trowel	Filter sand	Surface casing or traffic-rated box
Truck	Fire extinguishers	Tremie pipe
	First aid kit	Tyvek suits
	Flashlights	Waterproof marking pens
		Well log forms

[a]Not all equipment and supplies would be needed for all projects or for all phases of one project.

Table 11 Common Water Well Drilling Techniques

Cable tool
Rotary
hydraulic rotary
reverse circulation
air rotary
air-percussion rotary
diamond bit
Augering techniques
hollow stem
solid stem
bucket

mobile, and inexpensive. They are limited in depth with solid-stem augers limited to 150 ft and hollow-stem augers limited to 15 ft. No casing is required, but when the auger is removed, the bore hole may collapse.

Many regulatory agencies require that samples of the material being drilled be obtained and categorized. Common sampling methods include split-spoon samplers used with hollow-stem augers and Shelby or push-tube samplers for soils. Rock samples may be obtained by using the drilling pipe and bringing up the samples. This method is generally used with a rotary core drilling system. Another method is to examine the cuttings, but this method may not tell the depth of formation changes.

Any drilling should be under the supervision of a geologist experienced in drilling. Regulatory agencies require that subsurface conditions be logged as the drilling progresses.

Completing the well involves placing a casing within the bore hole, setting a screen at the ground water table, sealing the well, and developing the well. The type of casing selected depends on the expected contaminants. Casing materials may be PVC, Teflon, polyethylene, polypropylene, stainless steel, carbon steel, galvanized steel, Viton, Tygen, and ABS plastics. Teflon is the preferred material, but it is relatively expensive. PVC is the most commonly used material, but it is susceptible to bleeding when in contact with certain organic compounds. The deciding factors include what impact the casing will have on the ground water to be sampled, the length of the casing, costs, and longevity. The casing may be placed inside a bore hole or within the drill core when the drill removed. A screen is placed usually at the area of ground water contact. The screen can be of the same casing material or different. Generally, Teflon screens are preferred. They come in 10-ft lengths. Teflon screens can comprise a major cost of any well. It may be possible to place several screens at different depths to investigate different aquifers. However, problems arise in sampling different layers, and regulatory agencies do not look favorably on multiple screens with one well. A cluster of wells is preferred, but this is considerably more expensive when checking different depths at one particular location. Completing the well involves placing materials around the casing with in the wells. A layer of packed sand is placed around the screen to allow for ground water infiltration. A seal of bentonite is usually placed above the sand to prevent infiltration of water from sources above the aquifer to be studied. This seal is usually 2 ft thick. The remaining length of the casing up to the surface is grouted with cement or a cement/bentonite mixture. A cap is then placed over the well. The cap may be flush mounted where traffic may encounter the well or above ground at some height to make it visible in remote areas. The well elevations should then be measured and recorded.

The final step in completing a well is to develop it. Developing a well involves removing all foreign material that permeated it during well construction and completion. Development usually involves pumping water out of the well until all evidence of mud, foreign material, and so forth stops. This process can take several hours to several days. With the well system in place, it is possible for ground water samples to be taken and analyzed.

Ground Water Sampling

Prior to any sampling, a sampling and analysis plan should be developed that outlines sampling procedures, discusses preservation of the samples, and establishes chain-of-custody procedures and laboratory and analysis procedures for the constituents who are being monitored. Before samples are drawn, the depth from the top of the well casing to the water table should be measured and recorded. The well should be evacuated by removing from three to five times the calculated volume of water in the well and then allowing the well to recover.

A wide variety of sampling devices are available for drawing and collecting samples. Table 12 is a list of the most commonly used well sampling equipment.

Bailers are the oldest and simplest method. A bailer is open at the top and has a ball-type seal at the bottom. The bailer is lowered into the well on a line until a specific depth is reached. The weight of the water opens and closes the ball seal as the bailer is lowered and lifted. Bailers are most commonly made

Table 12 Common Well-Sampling Equipment

Bailers
Syringes
Suction pumps
Submersible electric pumps
Positive displacement pumps
Gas-driven devices

of PVC, Teflon, and stainless steel. The bailer may be cleaned between wells or dedicated to a specific well. They are inexpensive, simple to operate, easily cleaned, and portable. Care must be taken when sampling volatile organics, and the sampling method is time consuming. Bailers can be used in very small wells and to a variety of depths.

Mechanical and pneumatic devices include syringes. They generally require little or no power. They can be made of inert materials, can be easily cleaned, are relatively inexpensive, and are fairly portable. However, they are prone to breakdown and are relatively inefficient for collecting large volumes of samples. These devices can be used in relatively small wells and to a variety of depths. Pneumatic devices are suitable for both organics and inorganics. Mechanical devices are suitable for inorganic and some organics.

Vacuum pump (suction-lift) sampling devices are used in many instances. Centrifugal and peristaltic pumps are primarily used. These devices are simple to operate, portable, easily cleaned, relatively inexpensive, provide a continuous and variable flow, and can be used in wells to 0.5 in. in diameter. However, they may alter the chemical composition of the water, are capable of sampling only to 20–25 ft, and may require priming. Suction lift pumps are not recommended for gasoline and some inorganic compounds. If carefully used, these devices can sample organic parameters. Positive displacement methods include submersible centrifugal pumps, submersible piston pumps, and gas-squeeze pumps. Used where degassing and volatilization of contaminants are of prime consideration, they can be used to flush wells. Pumps are relatively costly can be difficult to clean, and they generally require larger well diameters. The minimum well diameter is 2 in., but submersible centrifugal pumps may require 12 in. Centrifugal pumps can pump large volumes of water. Gas-driven pumps have low power requirements and can be used to depths of 200–400 ft. The gas-driven piston and gas-squeeze pumps can be used for organics, gasoline compounds, and inorganics. The other positive displacement pumps are suitable for sampling for inorganics.

Gas-driven devices operate by pressurizing the sample container before lowering it into the well. The pressure is released and water is allowed to flow into the sample container. When the container is filled, pressure is again applied and the unit is removed from the well. These types of devices can be used for wells of 1.5 in. and larger, are highly portable, can provide multiple depth sampling, can be used to depths of 300 ft, and are relatively inexpensive and easy to clean. If air or oxygen is used, they might impact the collected sample. They require an air compressor or pressurized air tank, which makes transport to remote locations difficult. These devices are suitable for inorganics but not organics.

After the sample is drawn from the well, it must be transferred to a sample bottle. Transfer must be done with caution, particularly when volatile compounds are involved. The samples should be preserved according to accepted standards.

GROUND WATER RECOVERY AND TREATMENT

Control and Recovery

A variety of ground water control and recovery techniques are available. These include

- Volatilization
- Vitrification
- Solidification
- Biodegradation
- Capping
- Excavation/treatment/disposal
- Leaching
- Land treatment
- Thermal treatment

One basic strategy for minimizing the increased effects of continued contamination of the ground water is to reduce the flow through the vadose zone. In many cases, the ground water is threatened by contamination in the soil above the water table. As rainfall infiltrates the soil, it permeates the contamination before entering the ground water. Several techniques can control or remove the vadose zone contamination and include

- Stabilization/solidification
- Capping
- Sheet piling walls
- Well systems
- Physical barrier
- Grout walls
- Slurry walls
- Interceptor trenches

A vadose zone technique called stabilization/solidification isolates the contamination from the ground water by chemically fixing the soil to prevent further leaching of the contamination. Chemical fixing methods can include pH adjustment, cement addition, lime or pozzalonic material addition, polymer addition, encapsulation in an inert coating material, and vitrification.

Chemical stabilization can be conducted in situ or by excavating the material, mixing it in an above-ground plant, and backfilling the hole with the stabilized material. Of these two approaches, in-plant mixing provides a more stabilized material. Stabilization does nothing to treat the ground water but can help prevent it from getting worse. One of the most common methods of controlling ground water movement is with recovery wells. Wells are used to manipulate the subsurface hydraulic gradient. Water can either be withdrawn or injected to accomplish this. To be effective, the well system must be adequately designed. Figure 1 is a schematic diagram of a deep well.

Design begins by placing monitoring wells and performing certain hydraulic studies through these wells. These studies should be used to determine the hydrogeological characteristics of the aquifer. Given this information, the radius of influence and pumping rates can be determined so the ground water contaminant plume can be captured and movement arrested. There are many computer models available to conduct these calculations

Types of well systems that can be used for ground water control include well-point systems, deep well recovery systems, and injection well systems. With a well-point system, a number of closely spaced, shallow wells are placed and connected to a main header pipe and then a lift pump. These systems are used for areas with a relatively high water table.

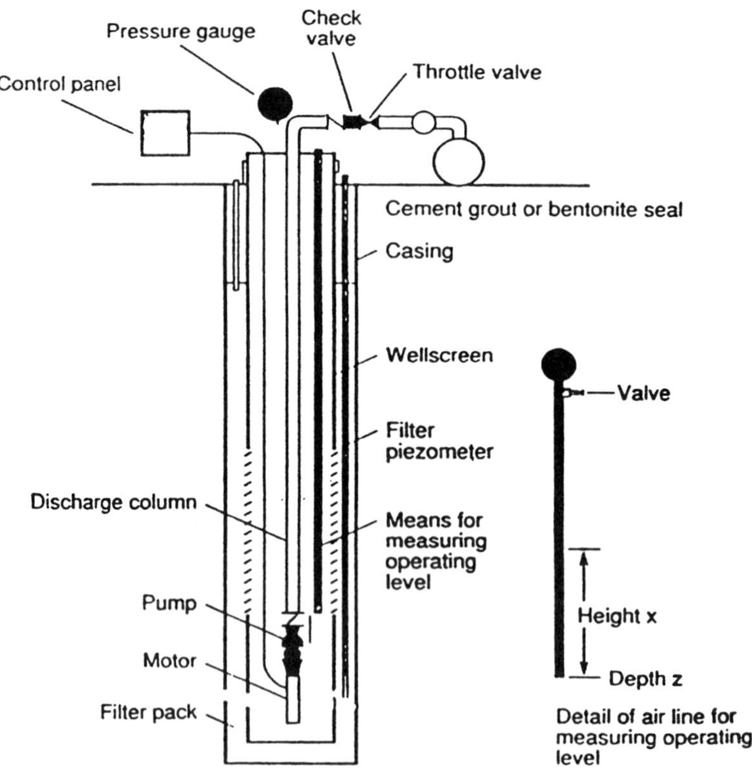

Figure 1 Components of typical deep well.

Deep well systems consist of wells placed to deeper depths, spaced at greater distances, and designed to recover ground water from specific cones of influence. Each well will usually have a dedicated pump, which may or may not discharge to a central system.

Another system consists of one or more injection wells. The water is injected through them to form a mound of water around the contaminant plume. This system is placed on the perimeter of the ground water plume.

Construction of a recovery well system is similar to installation of a monitoring well system. The basic operation involves drilling the well hole, installing the casing and liner, grouting and sealing the annular space, installing well screens, fittings and gravel pack, installing any recovery pump system, and developing the well.

Another method of recovering ground water is with an interceptor trench. A trench is excavated 3–4 ft below the water table. A perforated pipe is usually inserted in the trench with sand or gravel packed around it to promote water flow. The more permeable materials surrounding the pipe cause the water to flow into it.

This system can be used where the hydraulic gradient is relatively flat or on the down gradient side of a contaminant plume. The water is collected in the pipe and then routed away through a pipe header network to a sump where it is pumped to a treatment system. This type of system is often used to collect leachate from beneath a landfill. In some instances, the trench is on the land surface. Design of these systems is similar to many open channel or pipe storm systems. Figure 2 interceptor trench applications.

Another method used to control the movement of contaminants from the vadose zone into the aquifer is to cap the surface. The cap is intended to be nearly impermeable. It is designed to prevent rainfall from penetrating the vadose zone and carrying contaminants into the aquifer. A typical cap is shown in Figure 3. A cap consists of a plastic liner above the contamination, a 2-ft layer of low-permeability soil, a second liner, a sand layer to divert rainfall percolation, a fabric filter to keep the top soil out of the sand layer, a 1- to 2-ft layer of top soil, and a vegetation cover to prevent erosion. A drainage system is constructed around the perimeter to control run-on and run-off. This method is generally used when a facility, such as a landfill or surface impoundment, is closed.

Design considerations are usually of the hydraulic conductivities of the liner layers, liner material, type of clay, sand and top soil to be used, and design of the rainfall collection system. Installation costs can range from $20,000 to $50,000 per acre. Maintenance costs are minimal, usually cutting the grass. This type of system does nothing to affect any contaminated ground water. A supplemental system must be used.

Another physical control method to be discussed involves a physical barrier

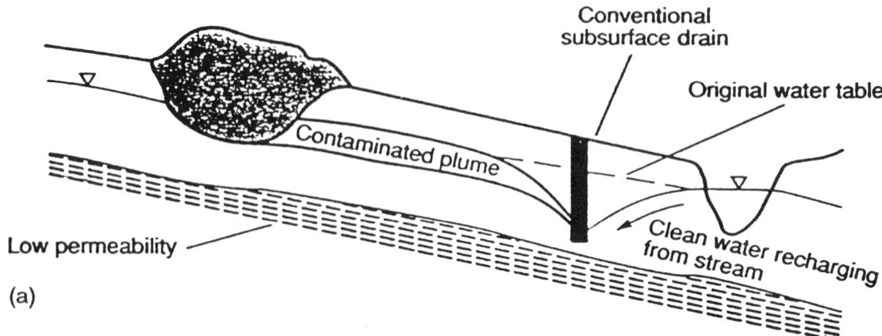

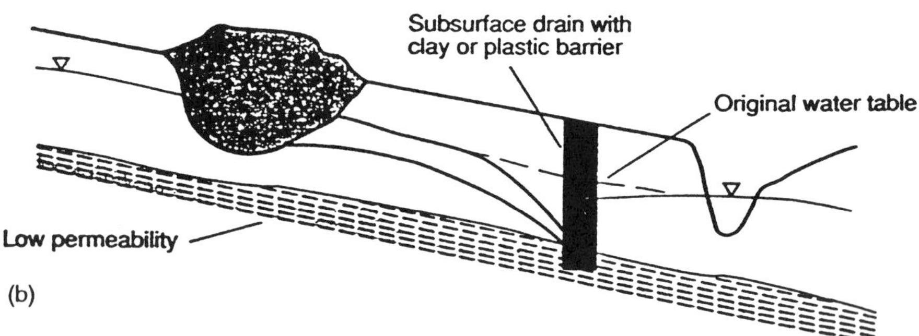

Figure 2 Typical interceptor trench applications: (a) the conventional subsurface drain receives recharges from the stream as well as the leachate plume, resulting in larger collection and treatment, (b) One-sided drainage reduces flow to drain.

down gradient of the contaminant plume. Several types of barriers can be constructed. One of these involves driving sheet piling into the ground. This system forms a thin impermeable barrier to flow. It should be driven below the water table to a sufficient depth to arrest the plume flow. This type of system is more effective against contaminants that float on the water or dissolve in the upper level of the water table.

Another type of barrier is a cut-off wall. This system is similar to the first, except a grouting agent is used to from the barrier. In grouting, a liquid, slurry, or emulsion is injected into the soil under pressure. The fluid occupies the pore spaces and over time solidifies. The type of grout used depends on the contaminants present.

Grouts are generally Portland cements or polymers. Construction of these

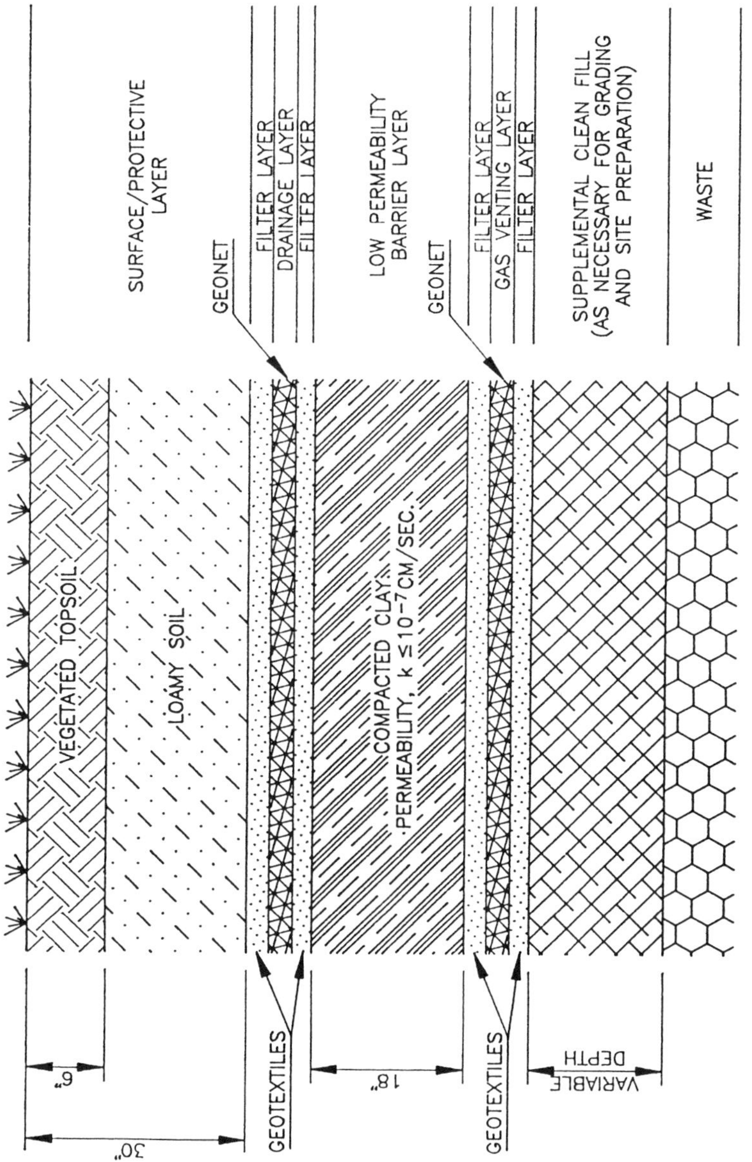

Figure 3 A cap prevents rainfall from penetrating the vadose zone and carrying contaminants into the aquifer.

walls requires the injection of the grout via wells drilled into the geological seam closest to the surface. After sufficient grout has been injected, the holes are drilled further to the next lowest seam and so forth until the desired depth is reached. Another method of inserting the grout is to drill the holes to the desired depth and use a packer through which the grout is injected at a lower depth.

A major design involves the pressure used to inject the grout. Excessive pressures can cause weakening of the rock strata. Pressure should be sufficient to ensure the grout penetrates to the desired distance away from the well. Another method of building the cut-off wall is with slurry wall. A trench is dug either up gradient to prevent water movement into a contaminant plume or down gradient to prevent the migration of contaminated water. A slurry mixture is backfilled into the trench, where it hardens and creates an impermeable barrier. Design considerations include selection of the most appropriate slurry, given the contaminants. This system can be cost effective only when natural barriers exist at lower depths. Figure 4 provides examples of slurry wall options.

Treatment Technologies

The ground water that is recovered via the recovery well system may require some type of treatment or disposal. The easiest, and sometimes least expensive type of disposal, is to transport it off-site to an existing treatment facility. This can be done if the quantities are small by trucking from an on-site holding tank. The other possibility is to install a pipeline to an existing treatment facility. Pretreatment still may be required and could be as simple as separation or air stripping.

Another treatment option is to construct an on-site facility. The treatment of inorganics in water has been an ongoing process that is well documented and tested. The treatment of organic compounds in water is relatively new. The most effective systems for removal of organics is air stripping and adsorption using activated carbon. Both of these systems have been used in some form by municipal and industrial water-treatment plants for many years. Treatment systems must be designed after thorough pump testing, bench testing, and water analyses have been performed. Treatment technologies, particularly bioremediation, have been reviewed earlier in this chapter.

Small packaged wastewater- and water-treatment facilities can be used for treatment of ground water collected by recovery systems. Small package plants are an excellent method of expediting the construction and reducing capital and operating cost. The last and most inexpensive option is no treatment, but a blending of the contaminated water with the effluent from an existing wastewater plant if permitted.

All treatment processes are best evaluated by considering all of the data obtained in the investigation stage and factoring in space limitations, operating and capital cost, and regulatory compliance.

Keyed-in slurry wall

Hanging slurry wall

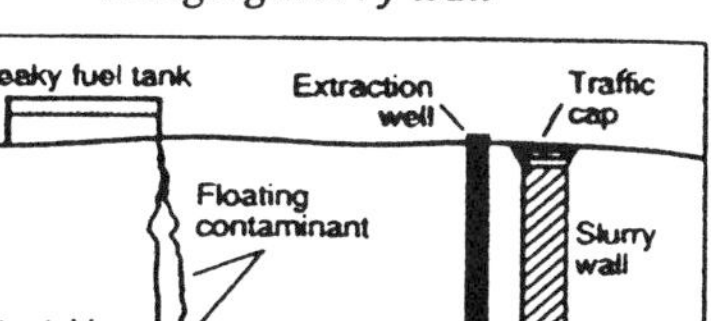

Cut-away cross section of circumferential wall placement

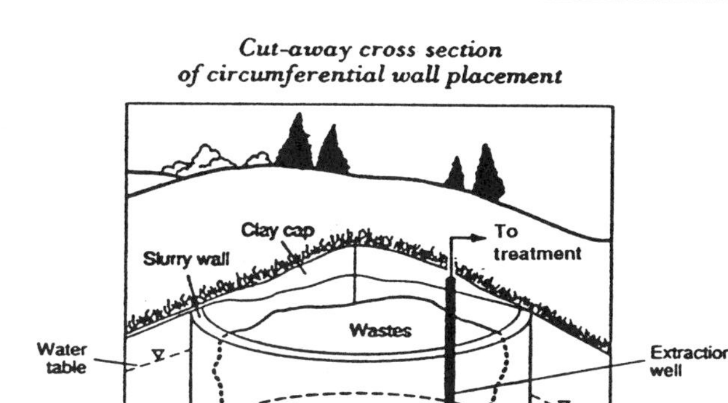

Cut-away cross section of downgradient placement

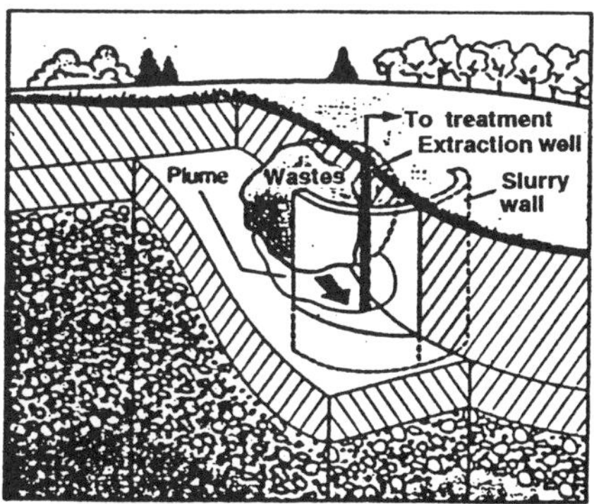

Cut-away cross section of upgradient placement with drain

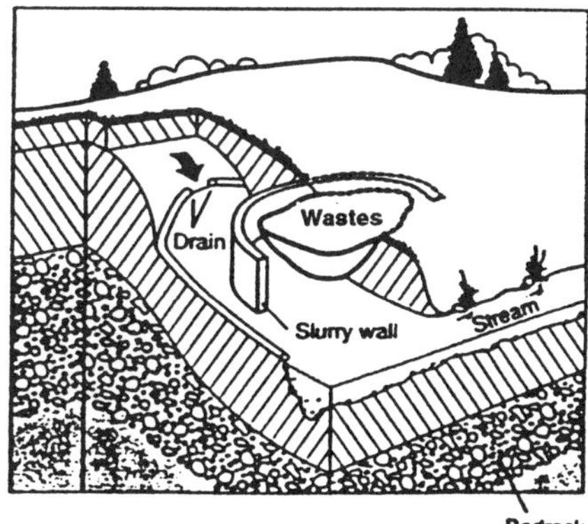

Figure 4 Examples of slurry wall placement options. Slurry wall systems only arrest the movement of ground water temporarily and require a well system to reduce the hydraulic gradient near the wall.

Ground water–treatment options include

- Air stripping
- Chemical precipitation
- Ultrafiltration
- Carbon adsorption
- Ion exchange
- In situ biological treatment
- Biological treatment
- Reverse osmosis

Treatment of Organics

Organic chemicals are manufactured or created by varying combinations of oxygen, carbon, hydrogen, and nitrogen. Many of these compounds are volatile; that is, have a high enough vapor pressure to be easily moved from the liquid state into the gaseous state.

Air Stripping

In an air-stripping operation, substances in solution in water are transferred to a solution in gas. A limited number of compounds are capable of being treated by this method. However, many of the most common solvents, such as benzene, toluene, and xylene, are treatable.

There are four basic types of air strippers: countercurrent packed columns, diffused aeration systems, cross slow towers, and coke tray aerators.

17

Instrumentation and Test Methods

INSTRUMENTATION

Instrumentation comprises devices used directly or indirectly to measure or control a variable. Instrumentation includes primary elements (sensors), final control elements (control valves and pumps), and switches, pushbuttons, controllers, annunciators, and related devices used to manipulate the variable.

Water and wastewater facilities cannot be operated properly without proper and adequate instruments. Control of processes and environmental parameters utilizes analytical factors that are involved in measurement and applying such measurements as standards of comparison and regulation. Making such measurements is important for a number of reasons which include

- Analytical data may be required for designing control equipment.
- Instrumentation may be required for process control and determining efficiency.
- Data may be required for submission to governmental enforcement agencies for compliance with source measurement requirements.
- As legal evidence of compliance.
- Protection of health and property for both workers and the public at large.

The instruments used in water- and wastewater-treatment plants are many and varied. According to what each instrument measures or controls, instrumentation can be categorized, for example, into the following subgroups: analytical measurement, flow reasurement, level measurement, pressure measurement, velocity measurement, pumps and different control valves, variable-speed drives, control valves, and control valve actuators.

ANALYTICAL MEASUREMENT

Chlorine Residual Analyzer

The most common method of disinfecting water and wastewater is chlorination. Free chlorine gas, or hypochlorite, acts as an agent to destroy microscopic organisms that are disease producing or otherwise objectionable. Chlorine residual analyzers measure the residual chlorine indirectly. Calorimetric, amperometric, and polarographic are some of several measurement methods used for chlorine residual, of which amperometric is the more common used method.

The amperometric measurement method uses two dissimilar metals held in a solution or electrolyte. A voltage is applied to the two metals which act as electrodes. Electrons flow from the negative electrode to the positive electrode generating a current. Figure 1 shows the amperometric cell. The amount of current flowing between the electrodes is proportional to the amount of chlorine present in the solution.

Figure 2 shows the basic amperometric chlorine residual analyzer which consists of an inlet sample tank and flow regulator, reagent solutions with metering pumps, measurement cell, and electronic signal converter. The metered sample stream acts as the electrolyte as it flows through the measurement cell. Since chlorine in the sample can exist in different chemical forms, the sample

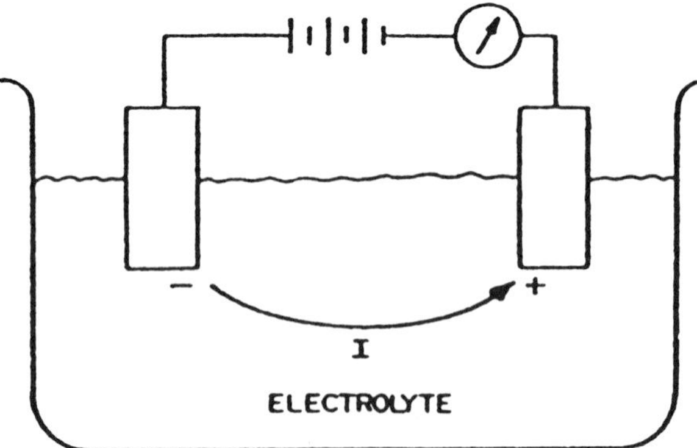

Figure 1 Amperometric measurement, where I is proportional to Cl concentration.

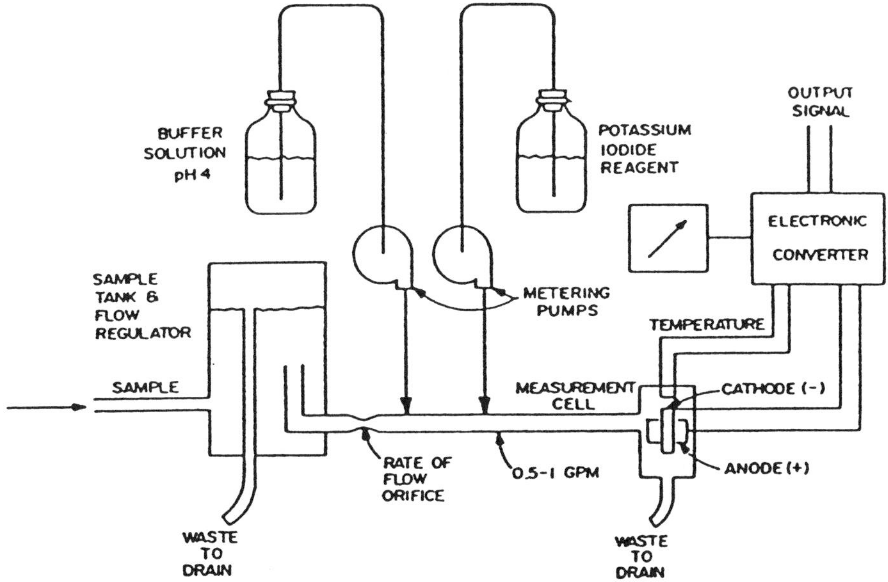

Figure 2 Amperometric total chlorine residual analyzer.

is conditioned with other chemicals in order for the cell to measure all chlorine present in the stream.

Chlorine residual analyzers are normally housed in free-standing enclosures. The sample is piped from the chlorine contact basin to the analyzer. The sample system is a cortical element for a successful analyzer application. There should be sufficient contact time between the chlorine and effluent stream for disinfection to occur. A commonly accepted disinfection period is 30 mins.

The sample is delivered to the analyzer 30 min after adding chlorine. Time in the contact tank plus the time to deliver a sample to the analyzer is the total contact time, as shown in Figure 3. A sample line and pump are required to deliver the sample to the analyzer. Figure 4 shows features of a sample transport assembly.

The pump should be capable of delivering 20–40 L/min (5–10 gallons per min [gpm]). Pipe should be sized for a sample velocity of 1.5–3.0 m/s (5–10 ft/s). Length of sample line should be determined such that it will provide the desired transport time. A valve should be installed next to the analyzer so samples can be taken for calibration checks on analyzer. A source of clean water and valves should be provided so the sample line can be backflushed to prevent plugging. A filter should be installed if solids are present.

Disadvantages of this system are:

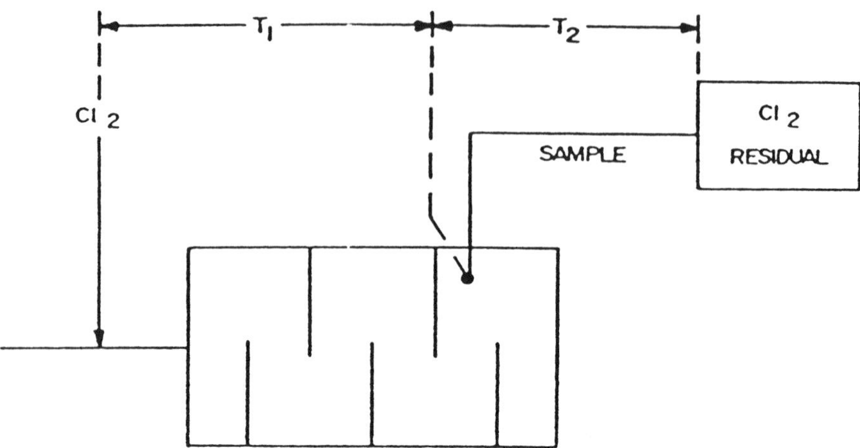

Figure 3 Sample point location. Total contact time = contact tank time (T_1) + sample transport time (T_2).

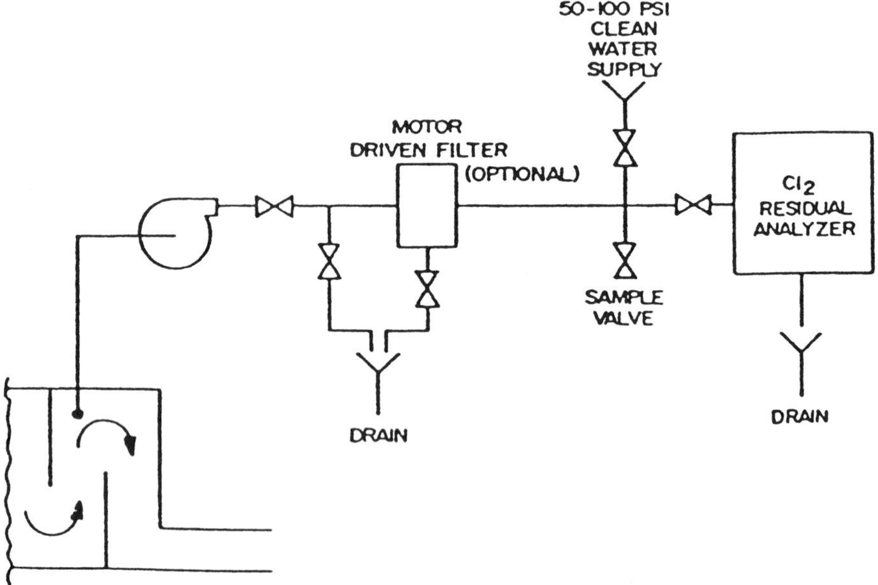

Figure 4 Sample transport.

- The sample point does not provide a representative mixed sample. This may be due to poor mixing, sample point location, or contact tank design.
- Contact time from the point of chlorine addition to the analyzer is too long.
- Sample lines plug and cannot be backflushed.
- No provision for taking sample at the analyzer for calibration checks.
- Using reagents of wrong concentration.

DISSOLVED OXYGEN

Dissolved oxygen (DO) is essential for the biological respiration of microorganisms. Generally, DO meters in wastewater plants provide an approximate measurement of the oxygen available to support biological activity. In receiving waters, the DO meters monitor one parameter of water quality. DO meters are generally not used in water plants. Recommended DO meter applications include

Aeration tank
Oxygenation basins
Mixed liquor streams
Secondary effluent
Plant effluent
Sample systems

Nonrecommended DO meter applications are

Chlorine contact tank
H_2S bearing streams

DO meters consist of an electrochemical cell, the probe, and a signal conditioner or transmitter. The two principal types of electrochemical cells used in DO probes are the galvanic cell and the polarographic cell.

Galvanic and polarographic cells have very similar operating principles. Both cells consists of an electrolyte and two electrodes, as shown in Figure 5.

In most commercially available probes, the electrolyte is contained by a gas-permeable membrane, and oxygen is brought into contact with the electrodes by the action of diffusion. Whether the probe is of the membrane or nonmembrane type, oxygen is reduced at the cathode, where the half-cell reduction reaction is

$$O_2 + 2H_2O + 4\ \text{Electrons} \rightarrow 4\ OH^-$$

and at the anode, the anode metal is oxidized. The result of this oxidation/reduction process is a flow of electrons from the cathode to the anode proportional to the oxygen dissolved in the process stream.

The rate of this oxidation/reduction process is strongly affected by temperature. Accurate temperature measurement and compensation is essential to accurate DO measurement. Temperature is usually monitored by a thermistor

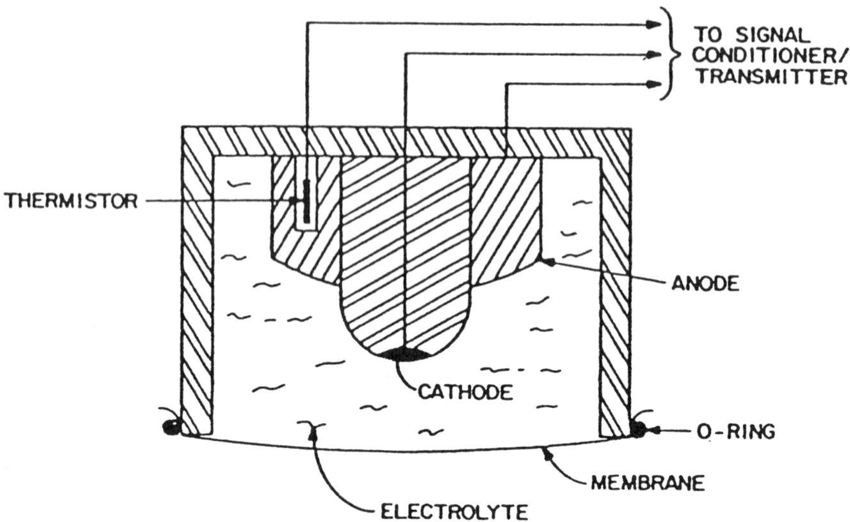

Figure 5 Membrane-type DO probe.

located in the probe, and compensation is made in the signal conditioner/transmitter electronics.

In most cases, open tanks have a convenient guardrail on which to mount the probe and transmitter is appropriately monitored. The probe holder must be rigidly supported, but it must also be readily removable for probe maintenance. Figure 6 shows a typical DO meter setup.

In closed tanks, probe holders can be inserted in the process stream through a flanged opening in the tank cover. The holder must be rigidly fixed to the tank cover by a flange or quick connect sample port cover that is removable for probe maintenance. A stilling well can be installed to provide gas seal and a lateral support for an extended probe length. Probe placement in the process fluid is the same as in open tanks.

In pipelines probes are usually mounted in a tee in the pipeline with either a seal on the probe or a bypass line to allow removal of the probe for maintenance. If the probe is part of a sample system, care must be taken to have a short transport time from the process to the probe. Direct measurement at the point of intersect is preferable to transporting a sample for DO measurement. Disadvantages include

- Agitator or cleaner becomes fouled with hair or fibers.
- Probe becomes fouled within a few hours because of process stream characteristics such as grease or slime growth.

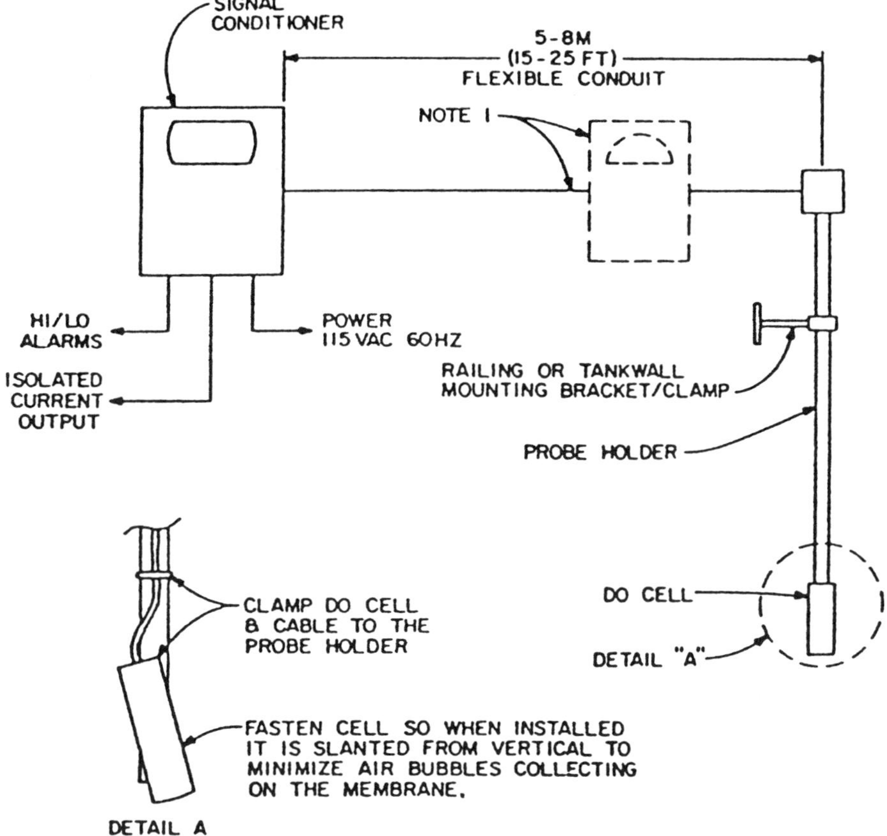

Figure 6 Dissolved oxygen meter configuration

- Probe cannot be withdrawn from the process stream because of mounting.
- Probe is placed in a dead area of the tank. Poor mixing with the rest of the tank results in a false signal that does not show the true state of the process.

pH SENSORS

pH measurement is one of the most important and frequently used tests in water chemistry. Practically every phase of water supply and wastewater treatment, such as acid-base neutralization, water-softening precipitation, coagulation, disinfection, and corrosion control, is pH dependent. Additionally, pH is used in alkalinity and carbon dioxide measurements and many other acid-base equilibria. Recommended applications for continuous monitoring with pH sensors include

Raw water
Plant influent
Primary effluent
MLSS (if applicable)
Plant effluent

Nonrecommended application:

Digester sludge

pH sensors employ a glass membrane electrode that develops an electrical potential varying with pH in the process fluid. A reference electrode is used to measure the potential generated across the glass electrode. Figure 7 illustrates a typical pH sensor arrangement.

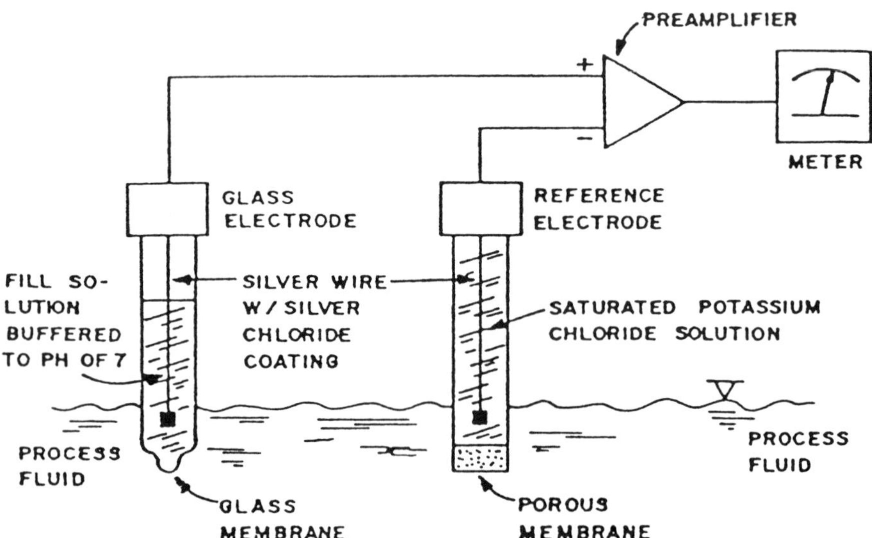

Figure 7 Typical pH sensor.

Figure 8 shows an equivalent electric circuit of the pH sensor in Figure 7. Voltage at the input of the amplifier is

$$E_i = E_g - E_r$$

where

E_i = amplifier input, mV
E_g = glass electrode potential, mV
E_r = reference electrode potential, mV

Installation of a pH meter, along with other on-line analytical instruments, is part of the sample system shown in Figure 9. This locates the pH meter with other high-maintenance instruments for ease of service. Buffer solutions needed for standardization can be stored with other analytical instrument reagents.

For in-tank open-channel installations, use a submersion-type electrode assembly with an integral preamplifier. Figure 10 shows a typical installation.

Disadvantages of pH systems include

- Electrodes become coated with grease or sludge.
- Plugging of reference electrodes.
- Calibration is difficult.

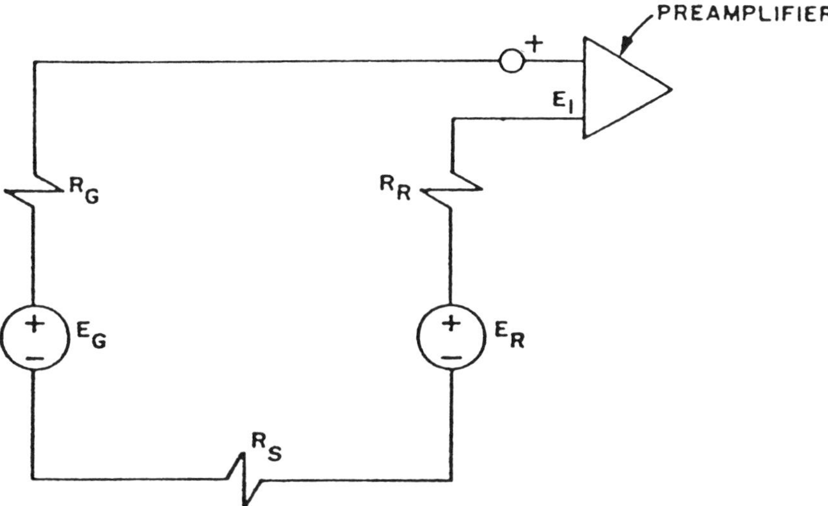

Figure 8 Equivalent circuit, where R_G = resistance of glass electrode, R_S = resistance of process fluid solution, R_R = resistance of reference electrode, and E_G, E_R, and E_I are constants.

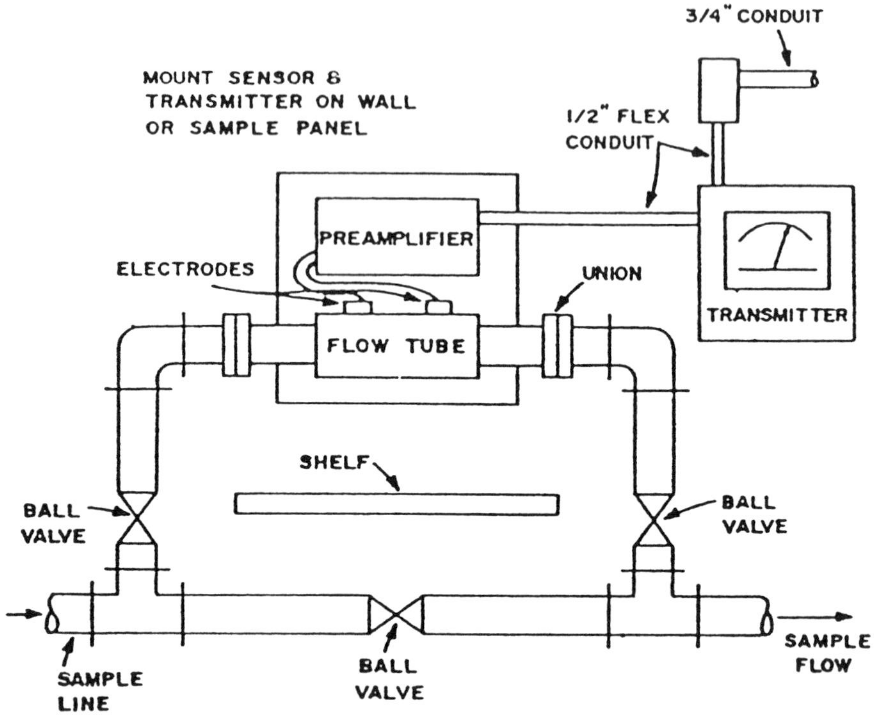

Figure 9 Flowthrough pH sensor installation.

SUSPENDED SOLIDS ANALYZERS

Suspended solids analyzers can be used in wastewater-treatment plants to measure continuously the concentration of solids in various process streams. Concentrations of interest range from effluent quality to thickened sludge of several percent solids. Application guidelines for suspended solids are as follows:

Recommended—optical analyzers
- Solids concentrations from 20 mg/L—8%
- Raw water
- Return activated sludge
- Waste activated sludge
- Mixed liquor
- Plant effluent/finished water

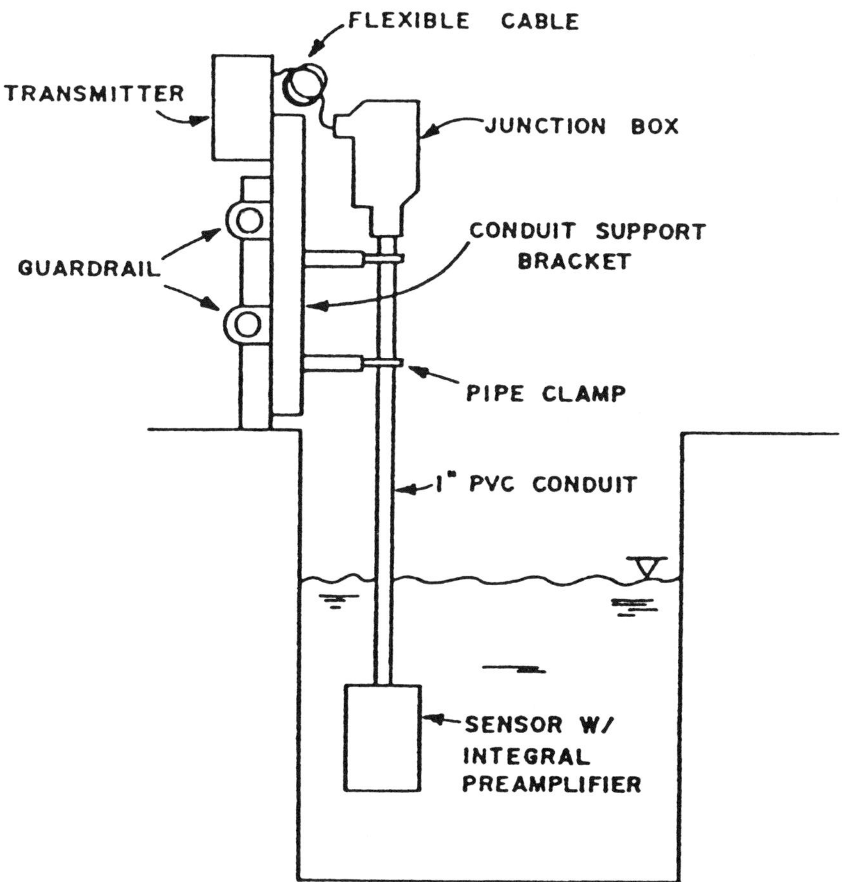

Figure 10 Submerged pH sensor installation.

Gravity thickened sludge
Filter effluent
Centrifuge supernatant

Recommended—nuclear analyzers
Thickened sludge with concentrations >8%
Centrifuged sludge

Recommended—ultrasonic analyzers
Solids concentration from 1–8% solids

Primary sludge
Waste-activated sludge
Return-activated sludge
Gravity-thickened sludge

Not recommended—optical analyzers
Primary solids
Flotation-thickened sludge
Solids concentration >8%

Not recommended—nuclear analyzers
Streams with solids concentrations <15%
Streams with entrained air bubbles
Line sizes >35 cm (14 in.)
Line size <15 cm (5 in.)

Not recommended—ultrasonic analyzers
Mixed liquor
Secondary effluent
Plant effluent
Pipe sizes >30 cm (12 in.)
Pipe sizes <10 cm (4 in.)

Suspended solids instruments are based on the attenuation or scattering of a beam of radiation. The type of beam used can be light, ultrasound, or nuclear.

Optical techniques for measuring suspended solids are based on scattering of a beam of light by the suspended particles as shown in Figure 11.

The portion of light scattered is a function of the number and size of the suspended particles. Light transmitted through the stream is reduced in proportion to the light that is scattered. An instrument which can measure the scattered light, the transmitted light, or both provides a measure of the suspended solids present.

The optical type of suspended solids analyzer consists of a lamp which acts as a source of light and a photocell which measures the transmitted or scattered light, as shown in Figure 12.

NUCLEAR RADIATION

Nuclear density gauge is a noncontact measurement of solids density. It does not measure percent solids directly but rather measures the specific gravity of the material. If the specific gravity of the liquid and solids is constant, then a correlation can be made between measured specific gravity and percent solids concentration. Figure 13 shows a nuclear solids analyzer.

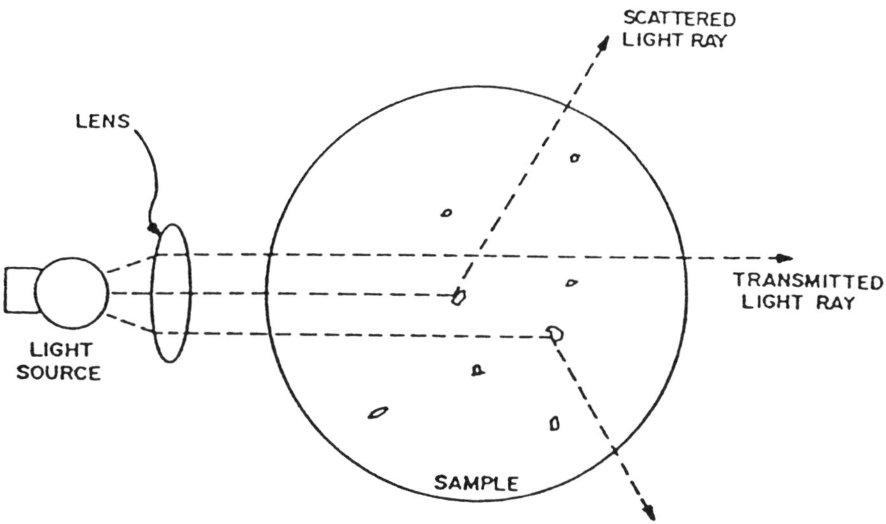

Figure 11 Light scattering.

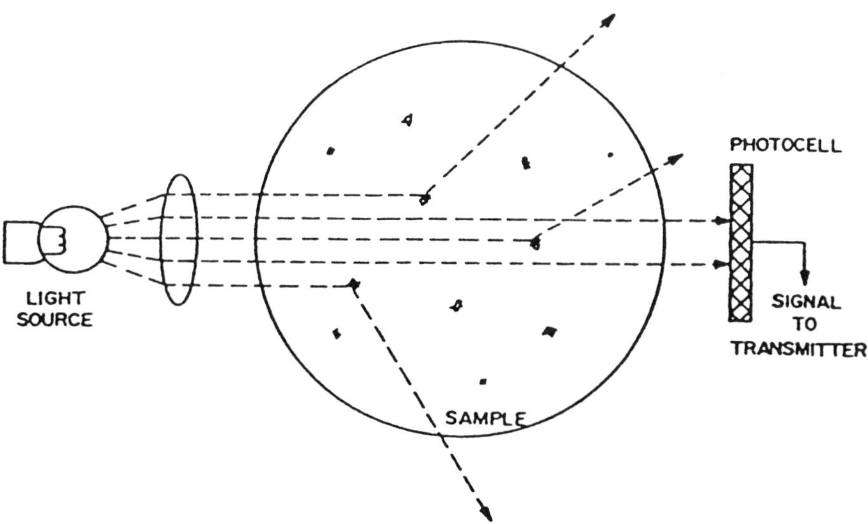

Figure 12 Transmissive-type optical suspended solids analyzer.

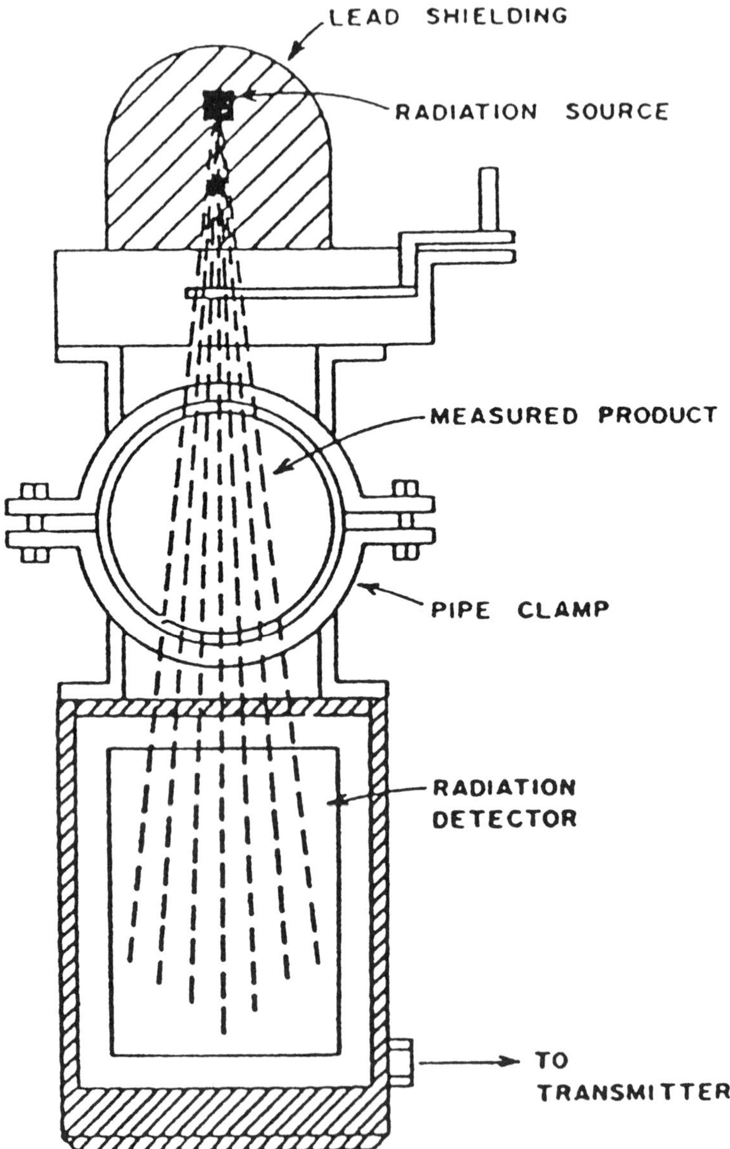

Figure 13 Nuclear solids analyzer.

TURBIDITY

Turbidity can be classified as forward scatter or side scatter measurement types. In forward scatter turbidimeters, the measurement is in Jackson turbidity units (JTU). The JTU was derived from the Jackson candle turbidimeter shown in Figure 14. In this instrument, the sample is poured into the glass tube until the candle flame is seen to disappear leaving a uniform field of light. At this transition point, the height of the column is read and converted into JTUs from a standard table.

Figure 12 shows a forward scattering–type turbidimeter. This instrument measures the amount of light scattered by particles in the forward direction from the light beam. By establishing and maintaining a ratio of scattered light to the transmitted light, the effects of color changes can be eliminated and a direct measurement made of the particulates.

In side scatter turbidimeters, the turbidity is determined by measuring the amount of light scattered at some angle (usually 90 degrees) from the light path by particles suspended in the sample. Figure 15 illustrates two styles of turbidimeters which use the side scatter method of measurement. The units for

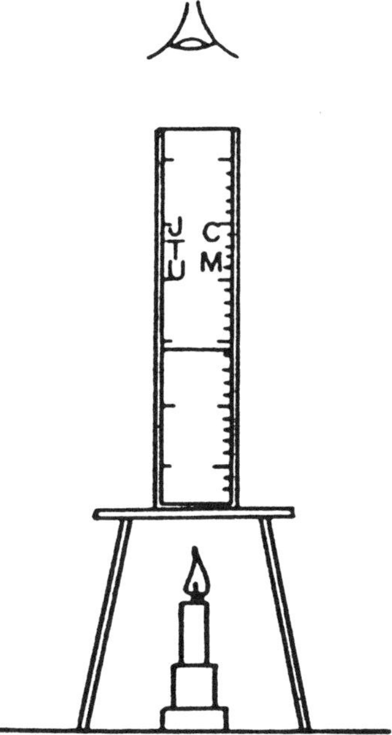

Figure 14 Jackson candle tubidimeter. (JTU = Jackson turbidity units.)

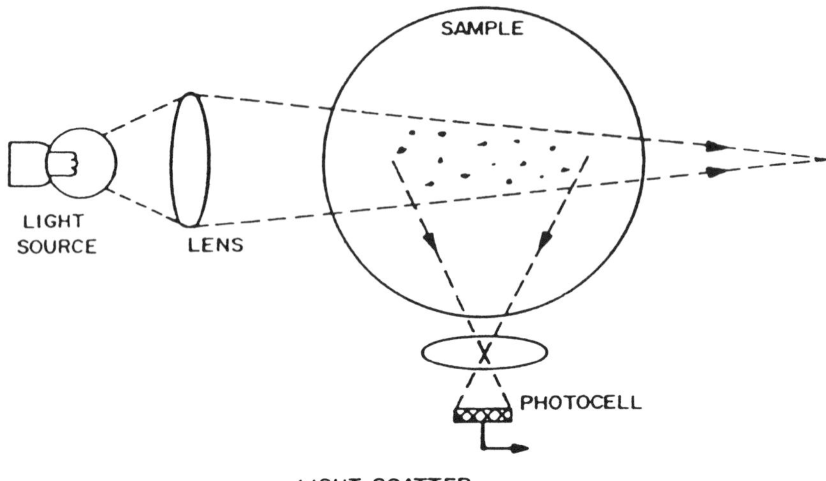

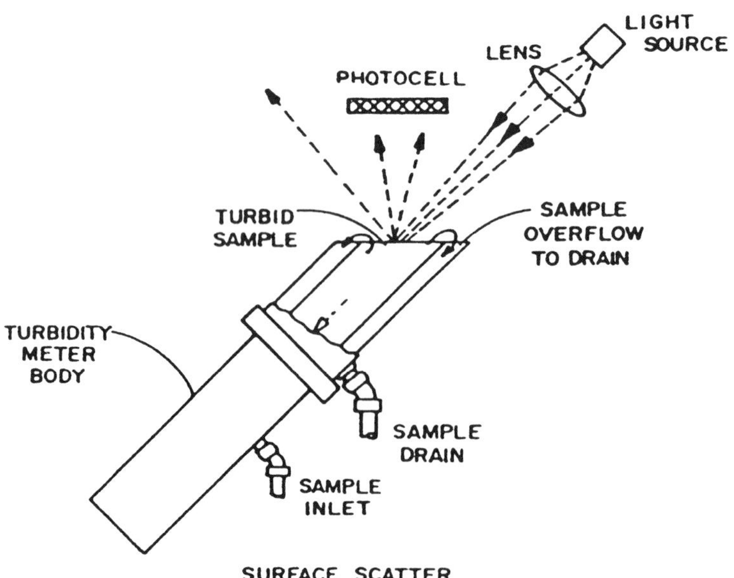

Figure 15 Side scatter turbidimeters.

side scatter turbidimeters are nephelometric turbidity units (NTUs). The word *nephelometric* describes the optical technique of measuring scattered light at an angle to the light path.

Formalin polymer has gained acceptance as the turbidity reference suspension standard. It is easy to prepare and is reproducible in its light-scattering properties. Although a sample of formalin suspension measured by forward scatter (JTUs) and side scatter (NTUs) turbidimeters will read approximately the same, they are not identical. A poor correlation exists when measuring a sample because of the variation in the absorption and optical scattering properties of the suspended particles. Because of this, turbidity units are not interchangeable between different types of turbidity meters. JTUs or NTUs can be correlated to suspended solids for a specific application.

Installation details for solids analyzers are unique to each manufacturer. The variations of installation are too numerous to list here. Manufacturers' installation manuals should be obtained and used when designing for a solids analyzer installation. Considerations in installing solids analyzers include

- Solids analyzers require frequent attention and calibration checks. Provide space for servicing and locate the sensor so it can be easily reached.
- If sample lines are required, make sure they are large enough and that flow velocity is high enough to minimize line plugging.
- Provide flushing water for the instrument and sample valve.
- Provide a sample valve next to the sensor so samples can be taken to check analyzer calibration.
- Mount the transmitter within sight of the sensor.
- Locate sensors or sample line taps where air bubbles are least likely to be present. Preferably a vertical line with an upflow.

FLOW MEASUREMENT

Open-Channel Measurement

The most commonly used flowmeters to measure open-channel flow are weirs and flumes.

Weirs

A weir is a dam or bulkhead placed across an open channel with an opening on the top through which the measured liquid flows. The opening is called the weir notch. Its bottom edge is called the crest. Normally the notch is cut from a metal plate and attached to the upstream side of the bulkhead. This is done to prevent the water form contacting the bulkhead and is known as a sharp-crested weir. A weir with a plate where the water contacts the bulkhead is known as a sharp-crested weir and is shown in Figure 16.

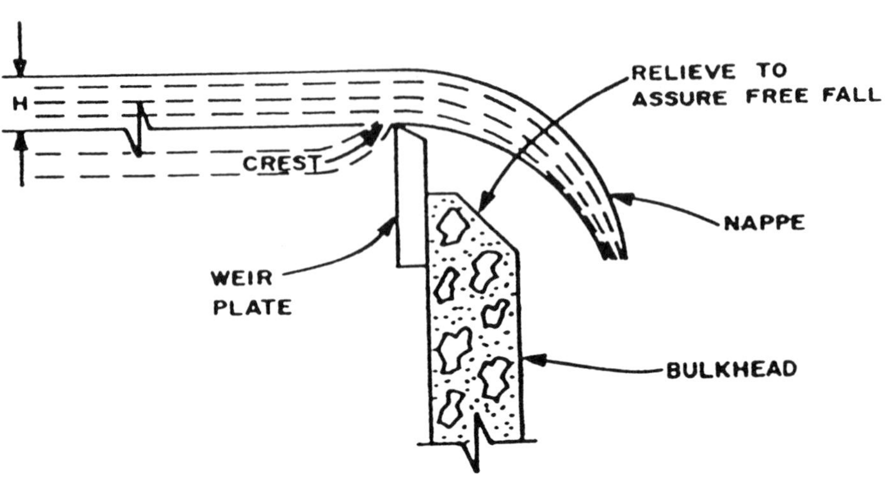

Figure 16 Sharp-crested weir.

The water depth measured at a prescribed distance upstream can be used to determine the discharge through the weir. Characteristic head versus flow relationships are governed by the weir geometry. All level measurements are made relative to the crest elevation.

Weir openings are normally fabricated in rectangular, trapezoidal, or V-notch shapes (Fig. 17). A trapezoidal weir with a side slope of 4:1 is known as a Cipolletti weir. V-notch weirs are suitable for flows up to 17 kl/min.

A weir is used to measure flow in open channels where the water is relatively free of suspended solids. Weirs are suitable for metering flow under the following general conditions:

- Flow stream should have less than 50 mg/L suspended solids.
- Sufficiently hydraulic head exists so a weir can be used. Typically, the head loss of a rectangular weir is four times that of a Parshall flume of equal size at the same flow.
- Flowrates vary over a large range. A range of flows of 20:1 can be tolerated by most weirs. For weirs larger than 2.5 m, ranges of 75:1 are reported. These wide ranges are not recommended.
- The approach conditions ensure that at all flowrates the flow is tranquil, free of eddies or surface disturbance. Under maximum flow, approach velocities in the upstream channel should not exceed 10 cm/s.

Recommended applications for weirs are with raw water, finished water, secondary effluent, and primary effluent. Applications not recommended include raw sewage, mixed liquor, and sludge.

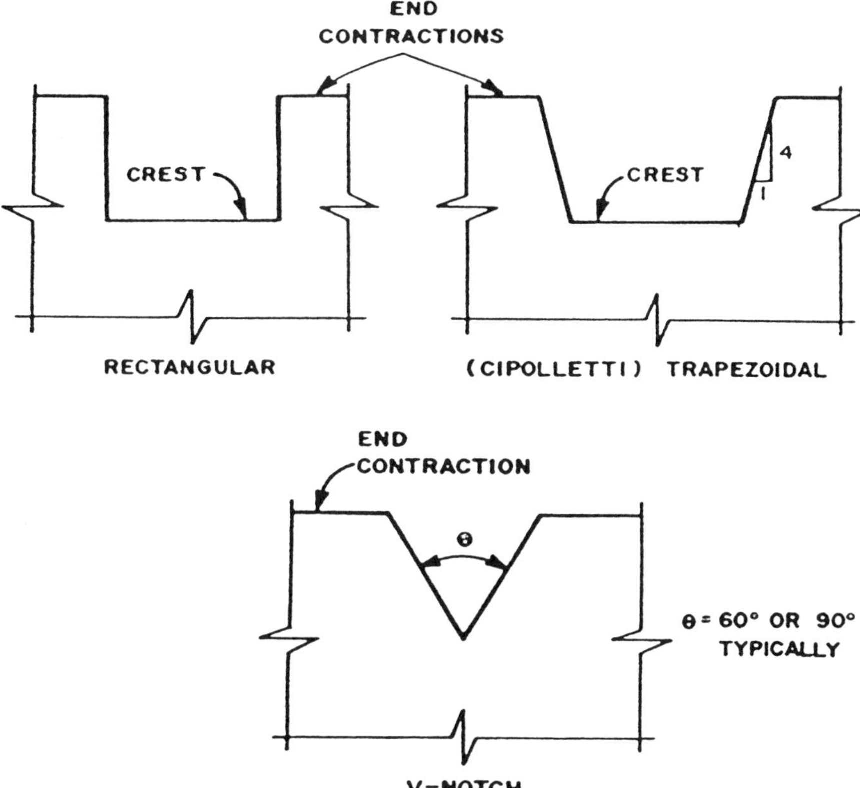

Figure 17 Weir shapes.

The weir is a primary element. The accuracy of flow indicated or recorded flow is also dependent on the secondary elements, the level sensor, and flow converter. For the best accuracy when designing and installing a weir, the following should be used as guides:

- Minimum head should be 6 cm or greater.
- Maximum head should be less than one-half the height of the weir.
- In using rectangular or Cipolletti weirs, the maximum head is less than one-half the crest length of the weir.
- Rectangular and Cipolletti weir crests should be level.
- A V-notch weir for low flow measurement should be used.

For proper installation of a weir, it is important for measurements to be accurate and reliable. The following installation procedure should be used:

- Make the upstream face of the bulkhead and weir plate smooth, and install it in a vertical plane perpendicular to the axis of the channel.
- Ensure the crest is level for rectangular and Cipolletti weirs. For a V-notch weir, ensure the bisecting line of the V is vertical.
- Cut the V-notch weir angle precisely and mount the plate so the angle is bisected by a vertical line.
- Machine or file the weir edges to be straight and free of burrs. Chamfer the trailing edge to obtain a crest thickness of 1–2 mm.
- Install the weir so the distance from the weir crest to the bottom of the approach channel is the >30 cm or two times the maximum head.
- Design the weir so the end contractions on each side (except the suppressed weirs) will be a minimum of 30 cm or two times maximum head.
- Provide air vents under the nape on both sides of a suppressed rectangular weir.
- Position a rectangular weir so that sides are straight up and down.
- Slope the side of a Cipolletti weir outward one horizontal to four vertical.
- Make the crest length of rectangular and Cipolletti weirs at least three times the maximum upstream head.
- Construct the bulkhead opening approximately 8 cm larger on all sides than the weir notch.
- Slope the top of the bulkhead down to assure that the nape falls free without hitting the bulkhead.
- Locate the level sensor next to the sidewall so it can be reached easily.
- Position the level sensor upstream of the weir at least four times the maximum head to avoid the effect of the draw down.
- Install the depth gauge and level sensor so the zero reference elevation is the same as the weir crest elevation.

To maintain and calibrate a weir properly, a depth gauge should be mounted adjacent to the level sensor. Periodic level readings and determination of the flow by calculation or from a table should also be performed. A comparison of remote flow readings with the value determined from the visual inspection as a conformance check on the calibration of the level sensor and flow converter is also recommended.

Problems that have been encountered in weir installations include

- Insufficient lead during low-flow conditions so there is no free air space under the nape.
- Suppressed rectangular weirs without air vents under the nape.
- Insufficient relief on the bulkhead so the nape strikes the bulkhead interfering with the free fall.
- Pool level downstream is too high so insufficient free fall exists.

- Weir notches cut from metal plate stock and installed without the edges finished to proper thickness, shape, or straightness.
- Rectangular or Cipolletti weirs installed without leveling the crest. Incorrectly cut angles and V-notch weirs.
- Level sensors located too close to the weir.
- Level sensors or gauges not zero referenced to the bottom of the weir notch.

Parshall Flume

The Parshall flume is a device for measuring liquid flow in open channels. The Parshall flume is a constriction of the channel that develops a hydraulic head which is proportional to flow. Figure 18 shows the shape and sections of a Parshall flume. Parshall flume sizes refer to the width of the throat section. Flumes are available in sizes from 0.025 up to 15 m. Large flumes are constructed on site, but smaller flumes can be purchased as prefabricated structures or as lightweight shells which are set in concrete.

If a Parshall flume has been constructed to standard dimensions and

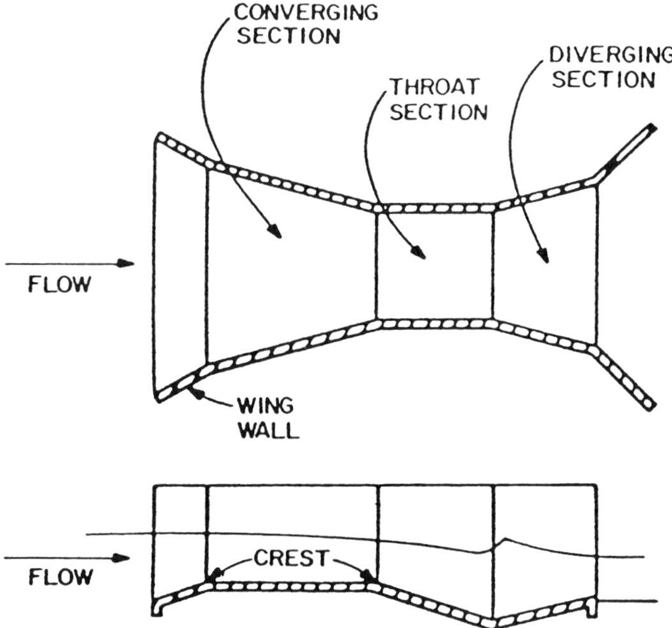

Figure 18 Parshall flume flow element.

properly set, it is possible to calculate flow through the flume by measuring level at a single point. The location for the level measurement shown as H_a in Figure 19 is approximately proportional to the three halves power of the hydraulic head.

Two flow conditions can exist is the Parshall flume: free flow and submerged flow. Free flow exists when the only restriction is the throat width and the water is not slowed by downstream conditions. If the flow through the flume increases sufficiently, the downstream channel level may rise and impede the discharge from the flume, thereby slowing the fluid velocity. This is known as submerged flow.

The flume discharge is not necessarily reduced as soon as the tail water level exceeds the elevation of the crest. Free-flow conditions can still exist even with some degree of submergence. Two elements are involved in obtaining a flow measurement with a Parshall flume: the primary element is the Parshall flume structure and the secondary element is the level measuring device. Accuracy of a measurement is derived using a Parshall flume depending on the combination of the accuracies of the primary (flume) and secondary (level measurement) elements.

Additional sources of error that can decrease the accuracy of the flow measurement include:

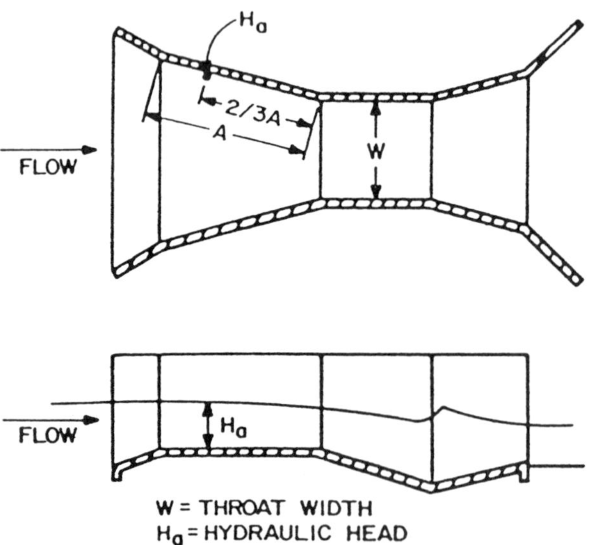

Figure 19 Head and width parameters for a Parshall flume.

- Deviations of the throat width from standard dimensions.
- Longitudinal slope of the floor in the converging sections. Tests on a 0.75-m flume demonstrated that a downward sloping floor produced added errors of 3–10% from low- to high-flow conditions.
- Transverse slope of the flume floor.
- Approach conditions which do not produce a smooth flow with uniform velocity distribution parallel to the center line of the flume.
- Incorrect zero reference of the level measurement device to the center line elevation of the crest.
- If a stilling well is used, the connector hole is improperly sized.
- Incorrect zero reference of the level measurement device.

Parshall flumes are suitable for metering flow under the following general conditions:

- Flow is in an open channel.
- Hydraulic head loss must be minimized. Typically, head loss for a Parshall flume is approximately 25% of a weir of equal capacity.
- Sediment or solids in the measured stream (velocities in the flume tend to scour and flush away deposits).
- Anticipated flowrates will vary widely. Depending on flume size and the accuracy of level measurement, a maximum to minimum flow range of 20:1 is reported for Parshall flumes.
- Approach conditions upstream of the flume will ensure that the entering flow is tranquil and uniformly distributed.

The recommended applications are with raw water, finished water, filtered water, raw sewage, primary and secondary effluent, plant final effluent, and mixed liquor. Applications not recommended include sludges, chemicals, and high-viscosity fluids.

Proper installation of a Parshall flume is important for measurements to be accurate and reliable. The following installation procedure is recommended:

- Construct or install the flume so that the floor section is level longitudinally and transversely.
- Establish the flume floor's elevation to prevent submergence conditions at maximum flow.
- Plan the flume installation to allow access for inspection of the flume to ensure correct elevation and leveling of the floor.
- Provide an approach channel long enough to create a symmetrical, uniform velocity distribution and a tranquil water surface at the flume entrance. A general rule is that 10 channel widths of straight run should exist upstream of the flume inlet.

In order to maintain a Parshall flume properly, a depth gauge should be mounted on the converging section so manual readings and flow calculations can be made to check remote flow indicators or recorders.

The depth gauge should be checked, preferably on a weekly basis, with other level or flow indicators for the flume to determine if calibration of the secondary system is required. Also, flume walls should be wiped down to remove slime or other build-up monthly or as necessary. The zero of the reference depth gauge and the flume surfaces should be checked for signs of deterioration and wear periodically.

Typical problems encountered in Parshall flume installations include

- Insufficient straight channel upstream resulting in nonuniform velocity distribution through the throat.
- Sloping floor in the converging section.
- Low flume floor elevation (relative to downstream charmel level), resulting in submerged flow condition.
- Measuring the depth of the hydraulic head at the wrong location.
- Using an incorrect equation for calculating flow from level.
- No provisions for purging the stilling well when measuring a solids-bearing stream build-up of deposits in the stilling well renders the level sensor inoperable).
- The foundation is not water tight allowing leakage under or around the flume.
- Flume sized too large for the operating flow range.

CLOSED CONDUIT LIQUID FLOW MEASUREMENTS

There are various types of flowmeters. Three flowmeters that are discussed in this section are magnetic, sonic, and turbine. These meters can be divided into three different categories, insert, full-bore, and clamp-on, depending on how the meter is mounted.

Insertion meters, such as magnetic, some sonic and ultrasonic, target, small turbines, vortex shedding, and pitot tubes are located in the pipe through a tap in the pipe wall.

Full-bore meters, such as vortex shedding, target, positive displacement, turbines, orifice plates, Venturies, and magnetic meters, are placed in the process piping. Removal of these meters can be a difficult process.

Clamp-on meters, such as sonic and ultrasonic, are affixed to the outside of the pipe.

Magnetic Meters

Magnetic flowmeters (mag meters) operate using Faraday's principle of electromagnetic induction in which the induced voltage generated by an electrical conductor moving through a magnetic field is proportional to the conductor's velocity. Figure 20 shows a magnetic flow meter.

Commercial power is applied to the meter, and the coil driver energizes the magnetic coils which encase the spool pipe creating a magnetic field. If the process liquid has enough conductivity, it will act as an electrical conductor and will induce an electrical voltage. This voltage is a sum of all the incremental voltages developed within each liquid particle occupying the magnetic field and is proportional to the field strength, pipe diameter, and conductor velocity. The more rapid the rate of liquid flow, the greater the instantaneous value of electrode voltage.

The induced voltage is received by the two electrodes mounted 180 degrees apart in the meter. This signal is sent to the converter/transmitter where it is summed, referenced, and converted from a magnetically induced voltage to the appropriate scaled output. The mag meter output is proportional to flow.

Two basic types of magnetic meters are available: the AC mag meter and the DC mag meter. With the AC meter, line voltage is applied to the coils and a continuous flux is created producing a continuous low-level AC electrode voltage. With the DC meter, the magnetic coils are periodically energized (pulsed) producing two induced-electrode voltages, one energized, then the other

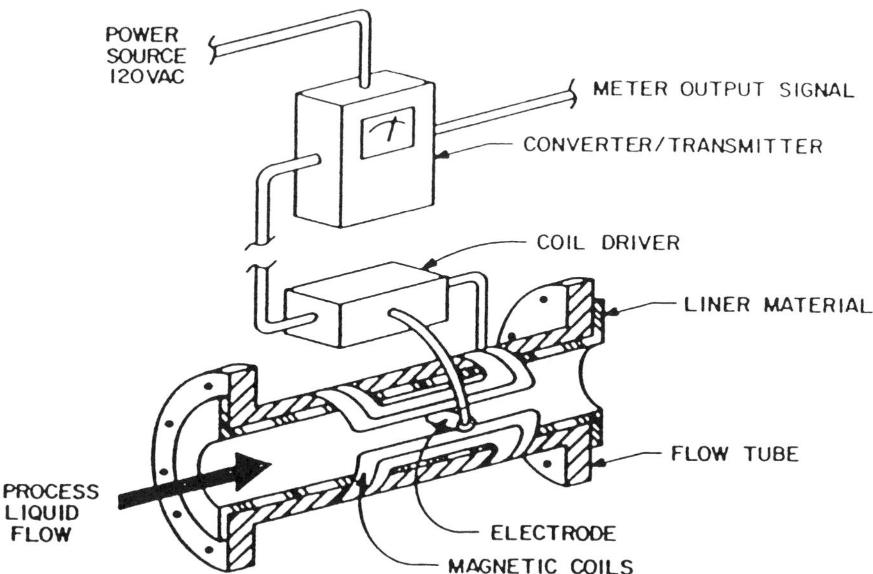

Figure 20 Magnetic flowmeter.

energized. The energized electrode voltage represents only noise. The difference between the two voltages is measured yielding a "clean" signal. Because of this operating scheme, the pulsed DC magnetic meters are zeroed every cycle, whereas AC meters require stopping the flow for periodic rezeroing.

DC magnetic meters are generally less bulky than AC magnetic meters and cost about 35% less. This is particularly true for the water style DC mag meters. The DC mag meter consumes less power than the AC mag meter and is somewhat easier to install and wire.

Where process flow is expected to change rapidly, the AC magnetic meter will follow the changes much better, since it operates at 60 Hz, whereas a DC magnetic meter operates from about 3–30 Hz. AC magnetic meters handle slurries better, because the hard particles in the slurry can create low frequency noise which affects the DC magnetic meters more than the AC magnetic meters.

Magnetic flowmeters are suitable for the following applications and conditions:

- Minimum head loss is desired.
- The process fluid has a conductivity greater than 5 μohms/cm.
- The liquid is corrosive or abrasive.
- The liquid has solids concentrations less than 10% by weight.
- The pipe always flows full.

Magnetic flowmeters are not recommended for

- Nonconducting liquid process streams
- Gas streams
- Streams with powdered or granular dry chemicals
- Liquid streams with a solids concentration greater than 10% by weight

Table 1 identifies applications guidelines for using magnetic meters.

The accuracy of mag meter is ±1% of full scale, and ±3% of indicated flow when operating in the lower one-third of the meter range. The repeatability of the meter is ±0.5% of full scale.

There are various conditions that will degrade these levels of operation. Flow-disturbing piping obstructions located too near the meter inlet and outlet may add an additional 1–10% of uncertainty to the measured flow. The location of valves, gates, tees, elbows, pumps, and severe reducers and expanders (>30-degree inclined angle) should not be placed less than five pipe diameters to the meters inlet or outlet.

Proper installation of a magnetic meter is vital for the resultant measurements to be accurate and reliable. The following installation guidelines should be observed:

Table 1 Applications Guidelines for Magnetic Meters

Service	Liner material	Gasket material
Raw sewage	polyurethane	rubber, neoprene
Raw water	polyurethane, rubber	rubber, neoprene
Settle sewage	polyurethane	rubber, neoprene
Primary sludge	polyurethane or Teflon	Teflon
Mixed liquor	polyurethane	rubber, neoprene
Return activated sludge	polyurethane	rubber, neoprene
Waste activated sludge	polyurethane	rubber, neoprene
Thickened sludge	polyurethane or Teflon	Teflon
Digester sludge	polyurethane or Teflon	Teflon
Digester supernatant	polyurethane or Teflon	Teflon
Polymer solutions	Teflon, rubber	Teflon
Clean (process) water	polyurethane, rubber	rubber, neoprene
Strongly corrosive	Teflon or Kynar	Teflon

- Locate the meter on the discharge side of pumps and on the upstream side of throttling valves.
- Locate the meter in a straight run of pipe free of valves or fittings with a minimum of five diameters upstream and downstream length.
- The process conduit must flow full of liquid.
- Meter sizing is critical.
- The meter must have self-cleaning electrodes, ultrasonic, or heated, for all applications except where process water is equivalent to, or better than, secondary effluent quality.
- Install the mag meter so it can be taken out of service (Fig. 21) for calibration and/or maintenance without disrupting the associated process.
- Properly ground all mag meters using stainless steel grounding rings and grounding straps supplied by the meter manufacturer.
- Avoid locating mag meters near heavy induction equipment because it causes meter operational problems (100-hp motors and larger, no closer than 20 ft).
- Provide sufficient space to facilitate calibration, in-line maintenance, or meter removal.
- Orient the meter so the electrodes lie in a plane parallel to the floor.
- Wall-mount the transmitter/converter within sight of the meter in a NEMA 4 enclosure (NEMA 6 if possible submergence), or flush panel mounted so the cable length from the meter does not exceed 60 m.
- Use driven-shield signal leads and route them between the transmitter and meter through dedicated 2-cm conduit, and route power wiring in separate conduit.

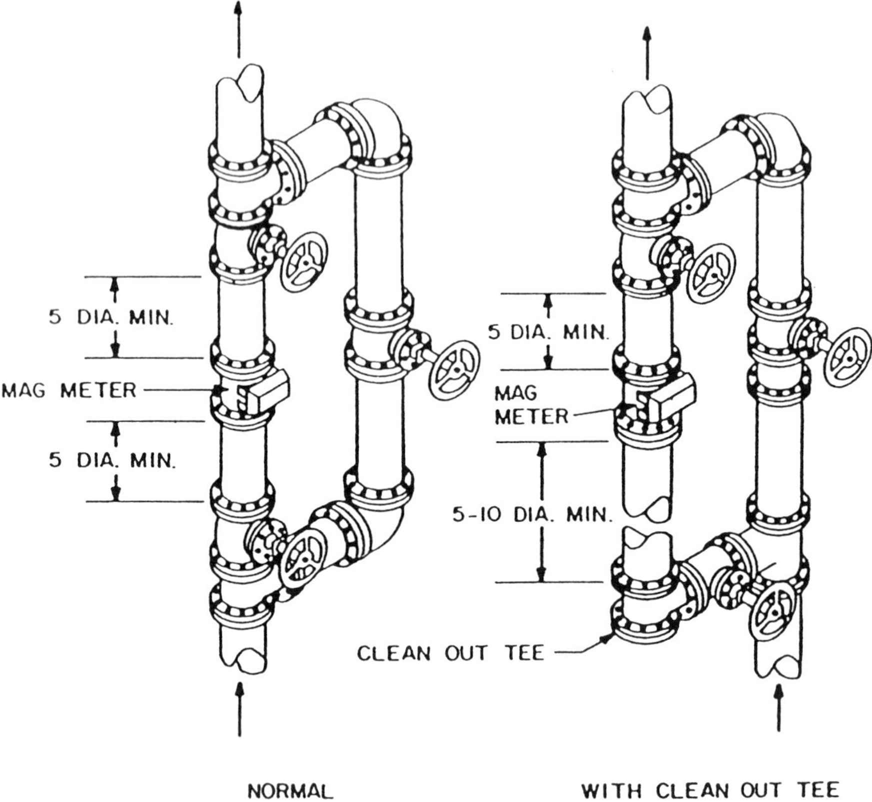

Figure 21 Bypass pipe installation for magnetic meter.

- Torque flange connections to the manufacturer's installation specifications.
- Wire power to the transmitter/converter and the coil driver through the same dedicated circuit.
- Mount the meter in a vertical pipe run with the flow direction upward and install air bleed valves for meters mounted horizontally.
- Metered lines should not self-drain when shut down.
- Provide for flushing and filling with clean water in sludge applications where intermittent operation is expected.

The proper maintenance and calibration for the meters includes calibrating the transmitter on regular basis and as well as the flow.

Sonic Meters

Sonic meters are available in two basic types; the transmissive (through-beam) type and the reflective (frequency-shift), or Doppler, type.

The transmissive sonic flow meter, Figure 22, also called through-beam or time-of-travel meter measures fluid velocity by measuring the difference in the time required for a sonic pulse to travel a specific distance through the fluid in the same general direction as fluid flow and the time required for a sonic pulse to travel the same distance in the opposite direction. This meter is available in two types: pipe section with integral well-mounted transducers and a direct-mounted version with the transducers mounted externally to an existing pipe. Both types use the same operating principle.

Some transducers are energized alternately by electrical pulses and emit sonic pulses across the flow. The pulse whose directional component is downstream traverses the pipe in a shorter time than the pulse traveling against the flow upstream. This time difference is proportional to the flow velocity, and an output signal linearly proportional to the flowrate is computed in the meter transmitter.

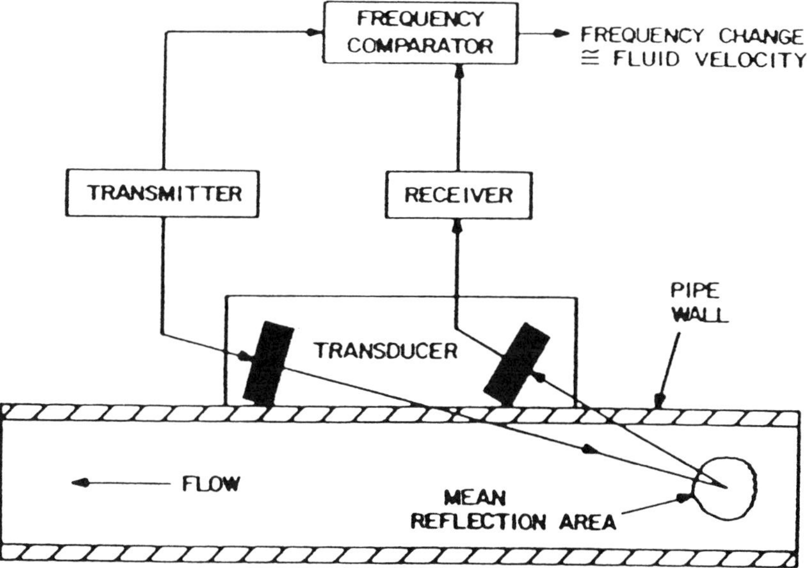

Figure 22 Transmissive sonic flowmeter.

Transmissive-type sonic flowmeters are suitable for the application as follows:

- Minimum head loss is desired.
- The pipe always flows full.
- The amount of suspended solids and entrained air bubbles in the process liquid together are "equivalent" to not >0.3% suspended solids by weight.
- Process liquid temperatures range between 0 and 80°C.
- Line size is small enough so the sonic signal attenuation does not cause a problem. Consult meter manufacturer about applications in lines larger than 100 cm.

Transmissive sonic flowmeters have various applications. Recommended uses are for raw water, primary effluent, mixed liquor, secondary clarifier effluent, plant final effluent, finished water, and process wash waters. Uses not recommended include raw sewage, primary sludge, thickened sludge, return-activated sludge, and waste-activated sludge.

The operation of the reflective, or Doppler, sonic flowmeter is based on a principle different from the transmissive type. The single transducer used is mounted on the external wall of the pipe. A signal of known frequency is sent into the fluid where it is reflected back to the transducer by suspended particulates or gas bubbles. Because the reflective matter is moving with the process stream, the frequency of the sonic energy waves is shifted as it is reflected. The magnitude of the frequency shift is proportional to the particle (flow) velocity and is converted electronically to the meter output signal linear to flow.

Reflective-type sonic flowmeters (Figure 23) are suitable for various applications under the following conditions:

- Minimum head loss is desired.
- The pipe always flows full.
- The amount of solids and the entrained air bubbles in the process liquid must be equivalent to a suspended solids concentration >25 mg/L but less than 4% by weight.
- Flow velocities at the transducer must be maintained between 1 and 9 m/s.
- Pipe wall thickness must be <5 cm.
- The pipe is not constructed of or lined with an aggregate material.
- The thickness of the pipe wall is exactly known.
- A portable measurement device is desired.
- Penetrating the process piping is to be avoided.
- Low cost is a prime consideration.

Reflective sonic flowmeters have a variety of applications. Recommended uses are for raw water, primary sludge, thickened sludge, return-activated sludge, and waste-activated sludge. Uses not recommended include raw sewage,

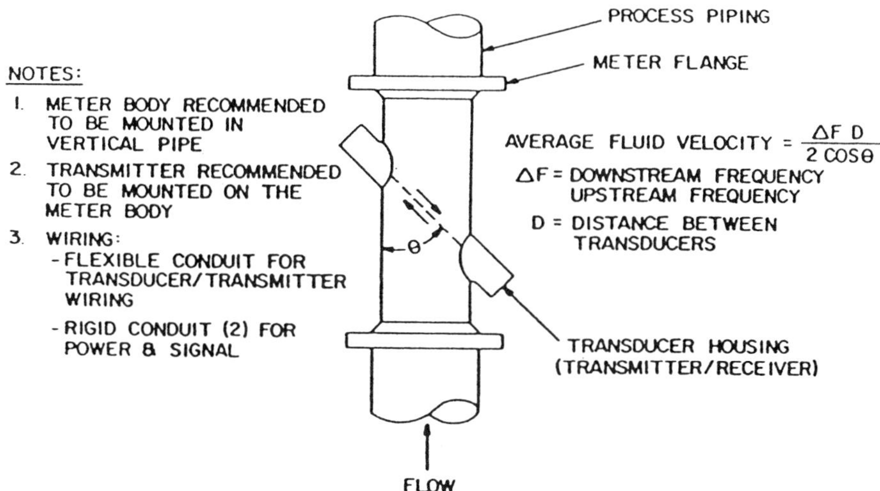

Figure 23 Reflective sonic flowmeter.

secondary clarifier effluent, plant final effluent, process (wash) water, and finished water.

Several factors which can degrade the accuracy of the meter are valves (modulating and isolating), gates, elbows, tees, pumps, and severe reducers and expanders. These parts should not be located nearer than 7–10 pipe diameters from the meter inlet, or 5 pipe diameters from the outlet (on flow-tube, transmissive meters) or within these distances from the external transducer of the Doppler-type meter.

Skewing of the velocity profile will result if the recommended straight lengths of pipe are not provided upstream and downstream of the meter. Skewing will cause errors in the flow measurement. Meter orientation leading to a nonfull pipe or resulting in material build-up or deposition will severely degrade meter accuracy.

Accuracy of Doppler meters depends on the velocity profile; amount, size, variation, and distribution of sound reflectors (solids or bubbles); line size; and flowmeter design characteristics. If the meter can be wet calibrated under actual operating conditions, $\pm 1\%$ accuracy can be obtained.

Proper installation of a magnetic meter is important for the resultant measurements to be accurate and reliable. The following installation guidelines are recommended:

- Install transmissive sonic flowmeters having wetted transducers so the meters can be taken out of service for calibration or maintenance without disrupting the associated process.

- Locate meters on the discharge side of pumps and on the upstream side of throttling valves if these devices are near the required meter location.
- Flow velocities through the meter should be maintained between 1 and 9 m/s.
- Provide straight runs of pipe upstream and downstream of the meter.
- Orient spool-piece–type meters in or locate clamp-on type meters on vertical process piping where possible, with flow direction upward only.
- Metered lines should not self-drain when shut down.
- Locate the meter in accessible location with sufficient space for calibration, in-line maintenance, and meter removal.
- Install clamp-on transducers according to the manufacturer's suggested procedures.
- Use separate conduit to wire line power and signal wiring.
- Follow precisely the manufacturer's guidelines for aligning sonic transducers to the pipe.
- Install the transmissive meter as a spool piece for pipes ranging in size from 7.6 to 91 cm.
- If the process piping is such that a clamp-on type meter is not practical, consider an insert-type Doppler.

Sonic flowmeters should be calibrated every two months to maintain proper maintenance and calibration.

Problems that have been encountered in existing meter installations include

- Piping obstructions located too near to meter cause accuracy problems.
- Meter sizing is such that adequate flow velocities are not maintained.
- Installation resulting in nonfull pipe during low flows.
- Solids coating due to low flow velocity or intermittent flow.
- Meter or transmitter located so that calibration and maintenance accessibility is difficult.
- Infrequent calibration.
- No provisions for flow rate testing/calibration.
- Solids concentration or entrained air greater than acceptable for transmissive type resulting in poor accuracy or an unacceptable signal.
- Grease and scum build-up on pipe walls and wetted transducers.

Turbine Meters

Turbine flowmeters consist of a pipe section with a multibladed impeller suspended in the fluid stream on a free-running bearing as in Figure 24. The direction of rotation of the impeller is perpendicular to the flow direction, and the impeller blades sweep out nearly the full bore of the pipe. The impeller is driven by the process liquid impinging on the blades. Within the linear flow

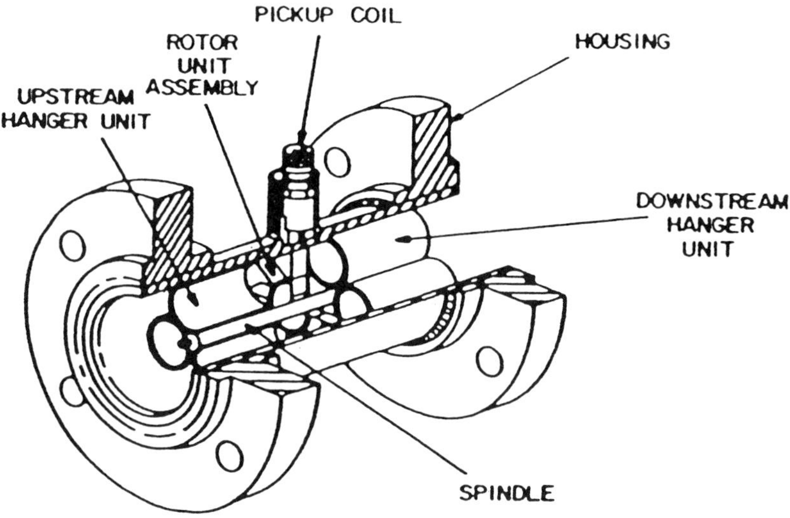

Figure 24 Turbine flowmeter.

range of the meter, the impeller's angular velocity is directly proportional to the liquid velocity which is, in turn, proportional to the volumetric flowrate. The speed of rotation is monitored by an electromagnetic pickup coil which operates either on a reluctance or inductance principle to produce a pulse. The output signal is a continuous-voltage pulse train, with each pulse representing a discrete volume of liquid. Associated electronics units then convert and display volumetric flow (flowrate) and/or total accumulated flow. Owing to the nature of their linear-to-flow relationship, turbine meters must be properly sized by volumetric flow rate. A meter size to a specified range or linear flowrate measurement should not be used for flowrates out of that range.

Turbine flowmeters are suitable for the following applications:

- The typical head loss through a turbine meter of 21–35 kPa can be tolerated.
- The process pipe flows full under all conditions
- The process liquid is relatively "clear"; for example, a solids concentration $<0.1\%$ by weight and is free of fibrous materials and/or debris.
- A maximum meter range of 10:1 is acceptable.
- An intermittent flow may be expected.
- The process fluid's viscosity range is 2–15 $\mu m^2/s$.

The accuracy and repeatability characteristics of turbine flowmeters, when properly applied and installed, should be:

Accuracy ±0.25% of actual flow, within the linear range of the meter
Repeatability ±0.05% of actual flow, within the linear range of the meter

Each turbine flow meter has a unique "K" factor (the number of pulses per unit volume) which is determined during factory calibration. This factor is adversely affected by two conditions: the liquid viscosity is significantly greater than that of clean water, and the moving components become impaired by build up of solids and/or fibrous materials.

Proper installation considerations of an accurate and reliable magnetic meter include the following:

- Flow-disturbing piping obstructions severely affect turbine meter accuracy. Figure 25 shows the recommended installation piping and details, including a flow-straightening element.
- When a flow-straightening element is used, the flow-disturbing effects of valves, gates, tees, elbows, and severe reducers and expanders will be adequately damped in a minimum upstream distance of 10 pipe diameters (including the straightener).
- Locate piping obstructions no nearer than 5 pipe diameters downstream from the meter.
- Install the meter in a horizontal pipe run.

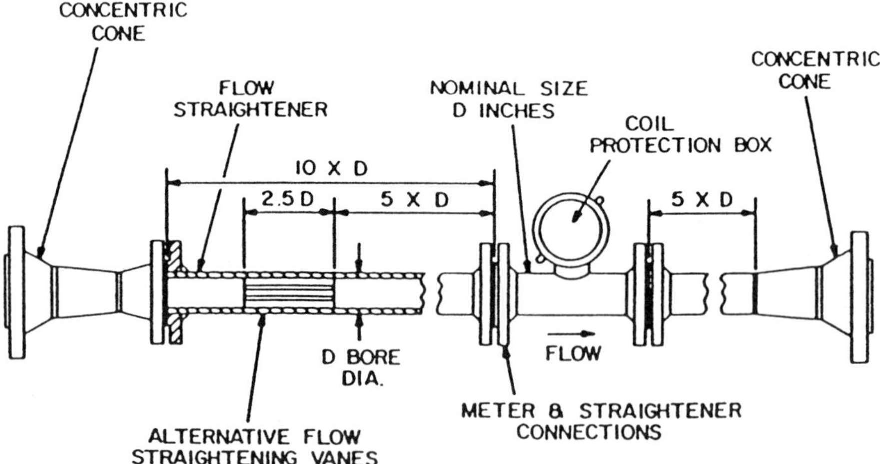

Figure 25 Turbine meter mounting.

- Install the meter on the discharge side of pumps and on the upstream side of throttling valves.
- Shield the cable between the turbine meter and electronics. Minimize cable length and do not route through areas of high electrical noise.

Turbine meters normally do not require periodic calibration. When accuracy becomes questionable, it should be examined to determine maintenance requirements or, if none are required, a new "K" factor by hydraulic testing should be determined.

The meter constant, K, in clean water is determined by the manufacturer prior to meter shipment. If the intended application process liquid has physical characteristics that significantly differ from clean water, the manufacturer should be consulted for additional testing data.

Problems that have been encountered with magnetic meter installations include

- Inadequate upstream and downstream straight run piping resulting in poor meter accuracy
- No flow-straightening vans resulting in poor accuracy
- Meter sized too large resulting in nonfull pipe and/or poor accuracy at low flows
- Meter sized correctly, but reducers located too near inlet and outlet
- Meter applied to a process liquid with an excessive solids concentration.

VENTURI TUBES AND FLOW TUBES

Recommended applications for Venturi and proprietary flow tubes as primary elements include nearly all water- and wastewater-treatment process streams. The most critical aspect of proper application is the type of pressure-sensing system used to measure the differential pressure produced by the primary tube.

Therefore, the recommended applications for these instruments are listed by type of pressure-sensing system. The major types of sensing systems are

1. Open connection, without flushing (including piezometric rings)
2. Open connection, with flushing (excluding piezometric rings)
3. Diaphragm sealed connections

Additional requirements for venturi and flow tube applications are

1. The metering tube must flow full.
2. A maximum to minimum measuring range of 4:1 is acceptable.
3. The Reynolds number of the process flow at the meter should be >150,000.
4. Do not use Venturi or flow tube meters in line with a positive displacement pump. The resultant flow pulsations will produced excessive signal noise and measurement inaccuracy.

Venturi Tube

A Venturi tube operates on the principle that a fluid flowing through a pipe section that contains a constriction of known geometry will cause a pressure drop at the constriction area. The difference in pressure between the inlet and the constriction area (throat) is proportional to the square of the flow rate. Figure 26 shows a cutaway of a typical Venturi tube.

General equation:

$$W = \frac{353Yd^2}{hr/(1-\beta^2)^{1/2}}$$

where

d = throat diameter of pipe
h = differential produced (in inches of water)
Y = net expansion factor
β = d/D (Beta ratio)
r = specific weight
W = flow rate (pounds/hr)

Several manufacturers provide differential-causing flow tubes which are modified versions of the classic Venturi tube. These devices operate on the same principle as the classic Venturi. They provide features which make them more attractive for some applications; for example, less space is required for installation, less overall head loss, and lower installed cost. Figure 27 shows three commonly used flow tubes.

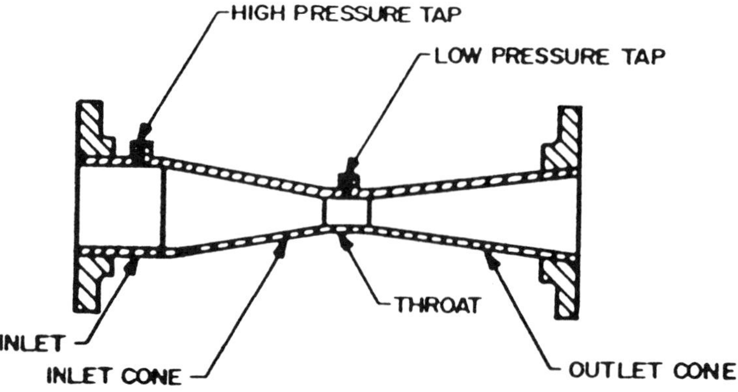

Figure 26 Classic Herschel Venturi tube.

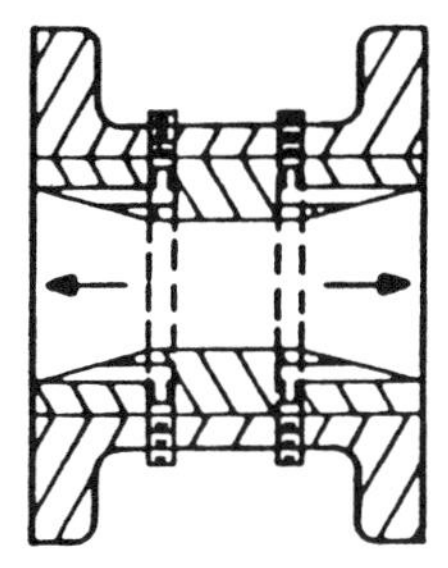

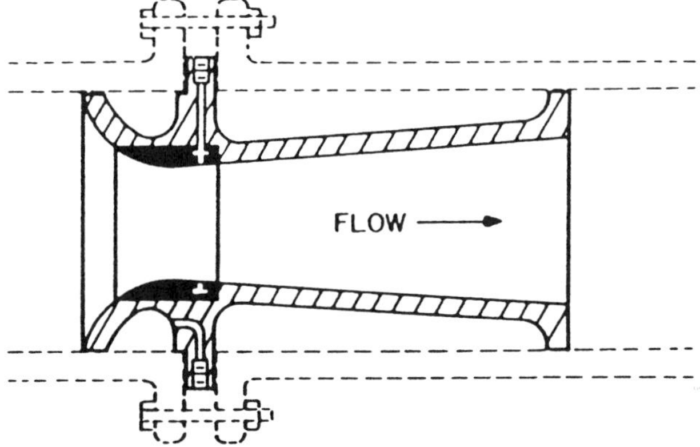

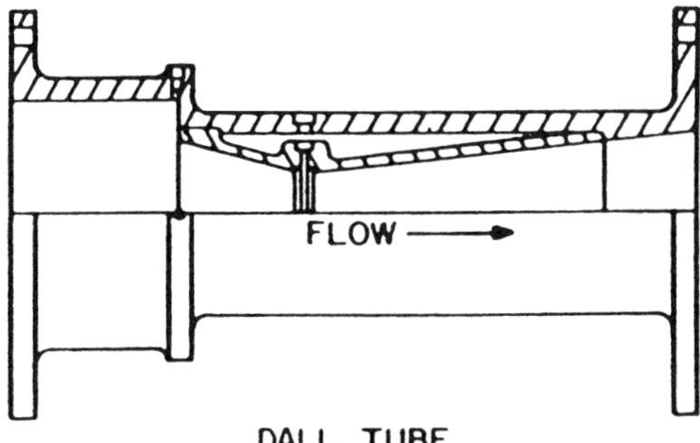

Figure 27 Flow tubes.

At Reynolds numbers below 40,000, most differential head elements deviated significantly from the square root relationship between flow and differential pressure. The V-element meter creates a differential pressure that conforms to the square root relationship down to Reynolds numbers as low as 500. The meter can be used for viscous liquids and low flows. Figure 28 shows a V-element meter. The V-shaped restriction of this element creates the differential pressure. The V-element has no critical surface dimensions or sharp edges that must be kept within strict tolerances. Its upstream face is slanted to minimize damage from the impingement of solids.

The Venturi tube, the proprietary flow tubes, and the V-element are primary sensing elements and require a secondary element to measure pressure differential.

The differential pressure created by a Venturi or flow tube is generally measured by connecting a differential pressure (DP) transmitter to the sensing taps with pipe or tubing (tap lines). The discussion here focuses on three common DP transmitter connection methods.

Open Connection

For this method, the Venturi tube pressure taps are connected directly to the DP transmitter and the tap lines are allowed to fill with process liquid. To avoid tap line clogging, do not use this method for liquids with >30 mg/L solids.

Providing a flushing water system for this sensing method allows the measurement of process liquids containing solids which would normally clog the sensing lines. There are two methods of operating flushing water systems. In one, flushing water is applied intermittently to purge solids from the tap line (measurement is interrupted during the purge cycle). In the second, a continuous

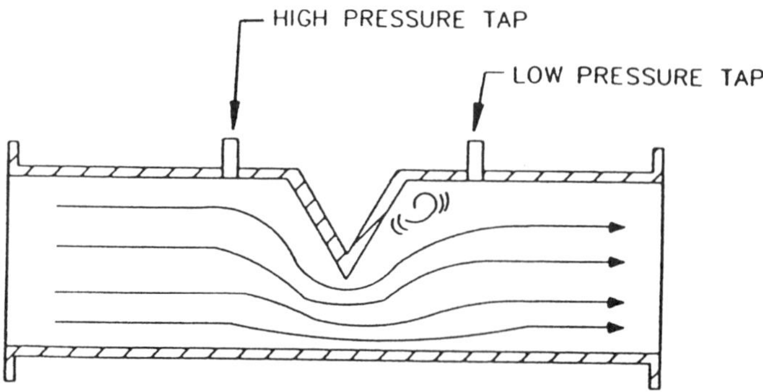

Figure 28 V-element meter.

equal flow of purge water is applied to both taps to act as a barrier to solids. As the purge water back pressure is measured in the latter method, it is critical that purge flows are equal. Direct and flushing water connections are illustrated in Figure 29.

Piezometric Rings

Piezometric rings may be used to sense inlet and throat pressures. These are normally used in very large diameter tubes where an average pressure is required to compensate for velocity profile variations. The rings consist of several holes for each tap (in a plane perpendicular to flow) connected to an annular ring. They should be used only on clean liquids. Flushing water systems cannot be used because the purge water short circuits within the annular ring to the nearest tap hole.

The diaphragm sealed sensor method allows solids-bearing liquids to be measured without a tap flushing system. The process liquid is separated from the tap lines and transmitter by a diaphragm.

Installation for primary systems:

- Venturi and flow tubes may be installed in any position to suit the requirements of the application and piping as long as the meter flows full.
- For best accuracy, flow-disturbing obstructions (fittings) should not be located too near the meter inlet. The following guidelines indicate the minimum upstream distance of straight pipe recommended between the fitting and meter inlet:

Reducers	8 diameters of reduced pipe size
Expanders	4 diameters
Fully open valve	5 diameters
Check valve	12 diameters
Throttled gate or ball valve	20 diameters
90-degree bend(s)	4 diameters

- If a flow control valve is required in the line, it should be placed a minimum of 5 pipe diameters downstream of the meter tube. To place the valve upstream requires a much greater straight run of pipe.
- Locate all downstream pipe fittings a minimum distance of 4 pipe diameters downstream of the throat tap(s).
- Place the meter tube a minimum distance of 10 pipe diameters downstream from the pump discharge. The meter tube can be located on the suction side of a centrifugal pump only if subatmospheric pressure can be avoided.
- Install the primary and secondary flow elements in an accessible location with suitable space for maintenance and calibration.
- Orient single-tap meters (at inlet and throat) in the process piping so that the taps lie in the upper half of the meridian plane.

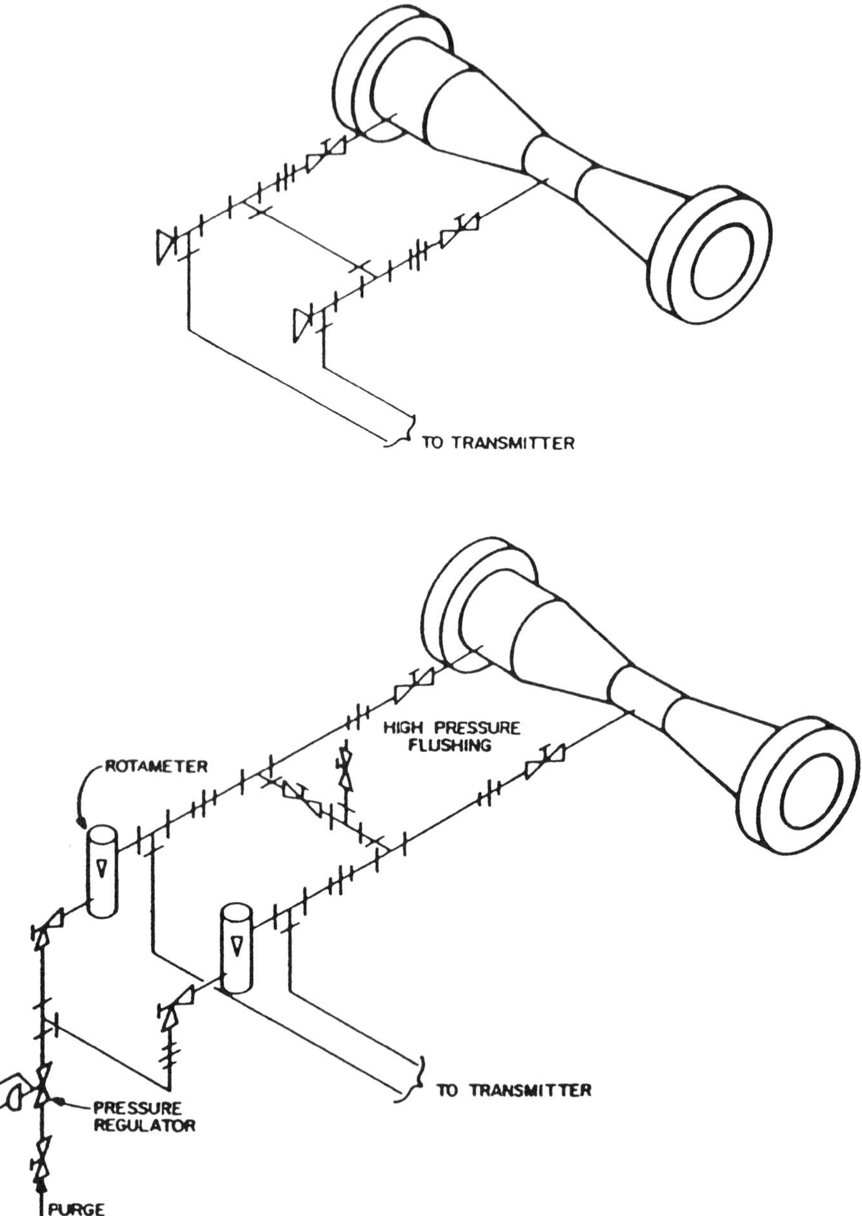

Figure 29 Typical differential piping.

For secondary systems:

- Place the differential pressure transmitter below the hydraulic grade line to facilitate positive gas bleeding.
- Place an indicator gauge (DP) near the primary element for convenience in calibration and performance checking.

The following refer to all installations except those having diaphragm-sealed sensors (see Figure 29).

- If there is a possibility of the tap lines freezing, use insulation and heat tape to wrap them.
- Install tap line (connecting) tubing so that it has a minimum downward slope from the meter of 1:12.
- Install a bleed valve or gas collector at the highest point in the tap line run.
- Provide valves to isolate the transmitter for calibration.
- Provide a flushing water system if the process liquid contains >30 mg/L of solids.
- Use connecting tubing no smaller than 1 cm (0.375 in.) in diameter.

The following refer to installations where continuous flushing is required (see Figure 29):

- The head loss in the tubing between the flushing water connection and the sensor tap should be the same in both lines so the pressure differential is unaffected.
- The flushing water supply pressure should be at least 70 kPa (10 psi) higher than process pressure.
- Equip the flushing water supply line for each tap with a rotameter for visual inspection and adjustment of purge flow.

VORTEX SHEDDING

Vortex shedding flowmeters are suitable for use under the following general conditions:

1. Medium head loss is acceptable.
2. The process fluid is a relatively clean liquid or gas.
3. The pipe always flows full.
4. The Reynolds number is >10,000.
5. The process fluid does not contain air bubbles.
6. To replace an orifice meter.

The unit can also be used for clean gases and steam. Vortex shedding flowmeter application guidelines can be found in Table 2.

The wetted parts of a vortex shedding flowmeter consist of a bluff body

Table 2 Vortex Shredding Flowmeter Applications Guidelines

Recommended	Not recommended
Raw water	High viscous fluids
Secondary effluent	Mixed liquor
Finished water	Dirty liquids which can coat or wear the bluff body
	Raw sewage
	Sludges

and a vortex sensing element. The bluff body is an upstream obstacle producing vortices which are detected by the downstream sensing element.

During flow, a slow-moving boundary layer of fluid is formed on the outer surface of the bluff body. The sides of the bluff body shed this boundary layer alternately forming low pressure eddies or vortices behind it, as shown in Figure 30.

The frequency at which the vortices are shed is directly proportional to the fluid velocity. A flagpole is an example of a bluff body. Vortices shed by the pole cause the flag behind it to flap in the wind.

Ways to sense the vortices shed from the bluff body include the following:

- A piezoelectric sensor inside the bluff body but outside the process fluid. It can sense the change in stress on the bluff body caused by the changing low- and high-pressure zones.
- A dual bluff body arrangement with diaphragms imbedded on either side of

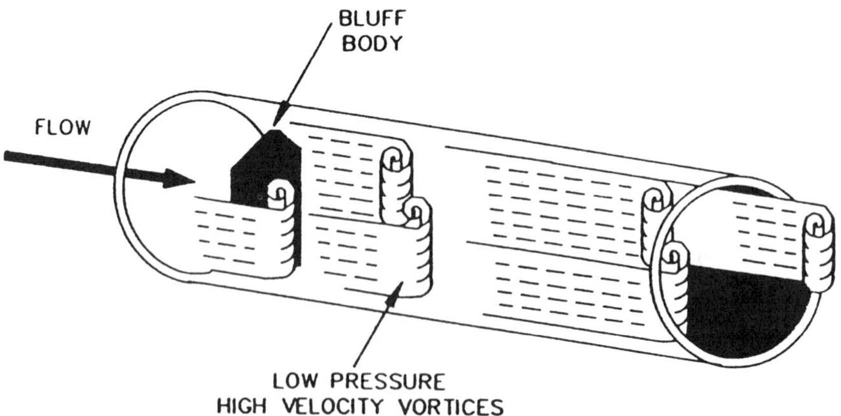

Figure 30 Vortices formed by bluff body.

the second bluff body that senses the pressure fluctuations (vortices). The diaphragm movement can be converted to either variable strain or variable capacitance readings.

- An ultrasonic beam downstream from the bluff body is modulated by the vortices.

Installation should be as follows:

- Locate the meter on the discharge side of pumps and on the upstream side of throttling valves.
- Locate the meter in a straight run of pipe free of valves or fittings with a minimum of 15 pipe diameters upstream to elbows and fitting, 50 diameters upstream to control valves, and 5 diameters downstream to obstructions. If a flow straightener is used, mount the meter 8 diameters downstream. See Figure 30 for additional details.
- The process conduit must flow full of liquid. Vertical pipes with liquid flowing up is ideal. In horizontal runs, ensure full pipe flow. See Figure 31 for meter orientation.
- Pipe sizing is critical. Size the pipe to provide a fluid velocity of 1–9 m/s (3–30 ft/s).

LEVEL MEASUREMENT

Techniques used for level measurement include pressure, capacitance, and sonic sensors. This chapter includes bubbler capacitance probes, floats, and sonic and ultrasonic level sensors.

BUBBLERS

Bubbler level measurement instruments are used in water- and wastewater-treatment processes for measuring both liquid level and differential liquid level. Bubblers are frequently used to sense the hydraulic head created by flumes and weirs in open-channel flow measurement. A special signal converter indicates flow based on the level sensed by a bubbler. This section addresses the use of bubblers in open tanks which is applicable to open-channel flow measurement. Bubbler level measurement applications are listed in Table 3.

An open-ended pipe, called the bubbler tube or dip tube, is connected to an air supply and positioned in the process so that the open end is set at a reference level. A constant air rate-of-flow regulator is used to maintain air in the tube with enough excess to bubble out the open end continually. Thus, the air pressure in the pipe is equal to the head of the process liquid above the reference level.

A pressure transmitter connected to the bubbler tube measures the pressure

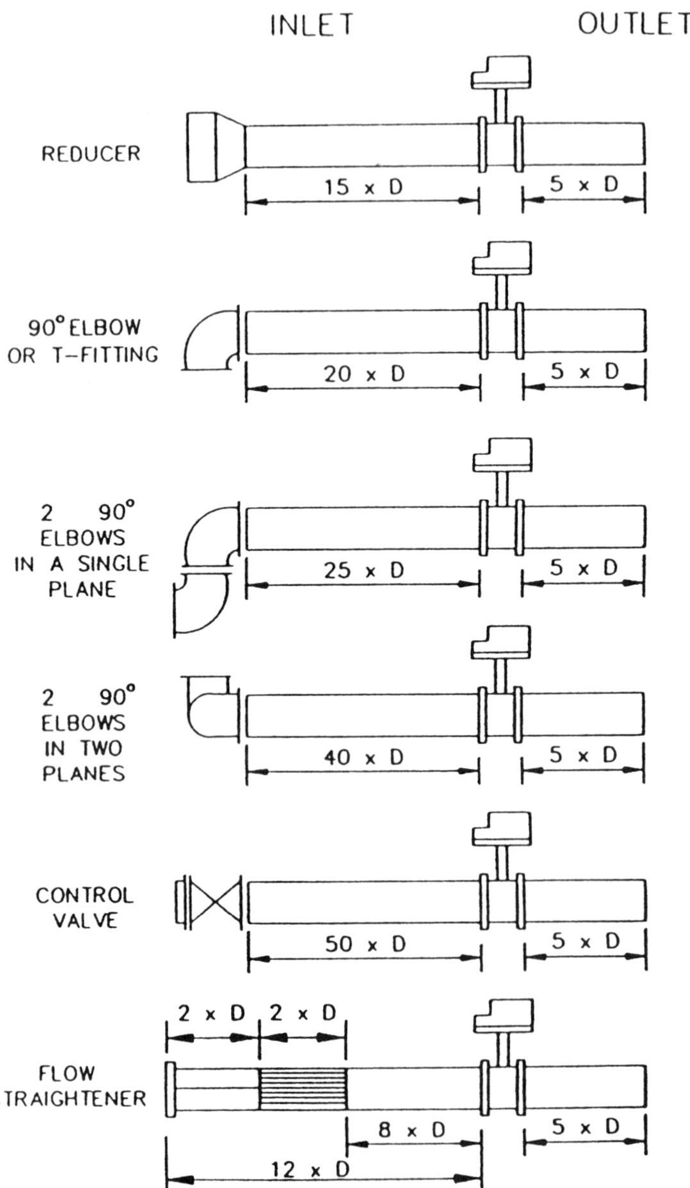

Figure 31 Recommended inlet and outlet distances.

Table 3 Bubbler Level Measurement Applications

Recommended	Not recommended
Liquid treatment processes	Digesters Volatile chemical storage tanks

of the dip tube. For water, the level is equal to the pressure sensed by the transmitter. For measuring other liquids, the transmitter must be calibrated for that liquid's specific gravity. For closed tanks, a differential pressure transmitter is used, with the high-pressure port connected to the bubbler tube and the low-pressure port connected to the gas space in the top of the tank. Typical bubbler applications are shown in Figure 32.

Because of head loss caused by airflow in the tube and connecting pipe, pressure at the transmitter will not be exactly in the same as at the open end of the bubbler tube. Difference in pressure necessitates minimizing pipe and fittings between the rate-of-flow air regulator and the dip tube.

Airflow head is affected by bubble formation. To minimize errors, the bottom of the tube usually has a notch or an angular cut to produce a continuous stream of small equally sized bubbles. Since build-up of process solids on the end of the tube will alter bubble formation, the tube end must be kept clean.

The air supply rate is controlled by a pressure regulator and a flow control valves. Typical airflow rates are 8–30 cc/s (1–4 cfh). Frequently a purge/rotameter is used to adjust the airflow. The air supply can be from instrument air, plant air, compressed gas tanks, or dedicated bubbler compressor. For applications requiring infrequent level readings, a hand-operated pump can be used.

The bubbler tube should be rigidly supported at a convenient location in the tank. The opening of the tube is the lowest level that can be detected, so set the tube depth at or below the lowest level at which a measurement is needed. Notch the tube opening to produce a continuous flow of small bubbles.

Fabricate the bubbler tube from 1.25-cm (0.5-in.) diameter stainless steel tubing or galvanized pipe. Properly supported, this makes a rigid installation which can withstand turbulence and wave action. A tee with one branch plugged, when installed on top of the bubbler tube, provides an opening for a cleaning rod when the high-pressure air purge cannot remove bubbler tip restrictions.

The bottom of the tube should be at least 8 cm (3 in.) from the tank bottom to avoid solids build-up on the tank floor. This offset must be included in the zero reference level for the liquid in the tank. An exception to this is a bubbler installed in a flume or ahead of a weir. In flume applications, the bubbler tip must be at the same elevation as the flume floor, or if it is elevated, the degree of elevation must be compensated for in the flow calculation. In weir applications, the bubbler tip must be at the same elevation as the bottom of the notch, or if

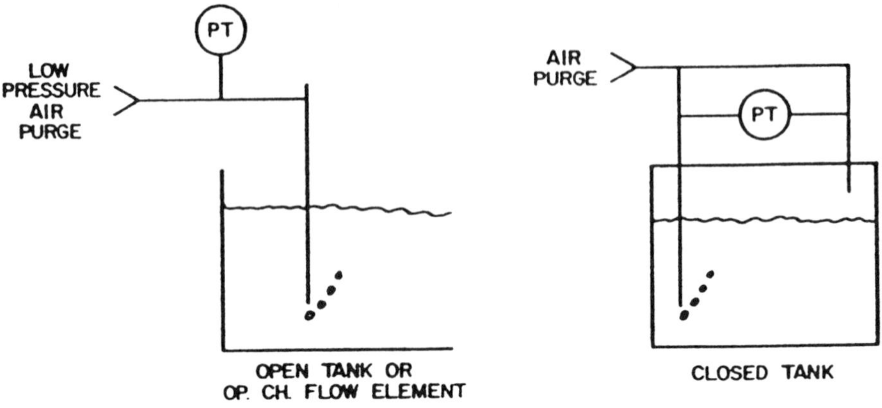

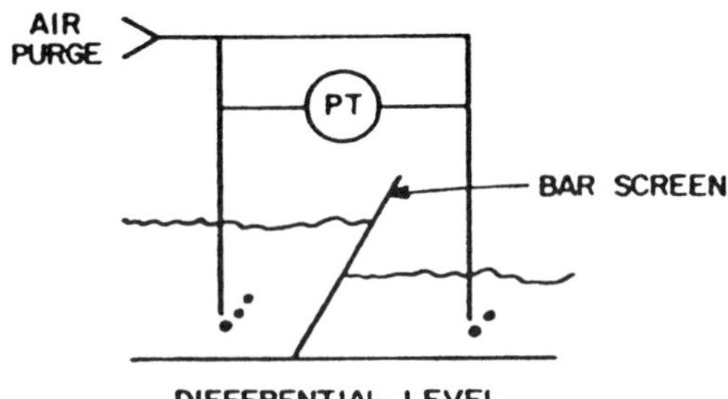

Figure 32 Typical bubbler applications.

it is below the notch, the degree of offset must be compensated for in the flow calculation.

To minimize level measurement errors caused by airflow head loss, the airflow controller must be mounted as close to the dip tube as possible and connected with a minimum of fittings and tubing. For 1-cm (0.25-in.) tubing, the distance from the air purge regulator to the bubbler tube should not exceed 15.38 m (50 ft).

To make sure that the air-purge tubing is free from traps where moisture condensate can collect, install the tubing with a continuous downward slope from the pressure transmitter and the airflow controller to the bubbler tube.

In open tanks and flumes, for periodic reference checks and to facilitate recalibration if the tube is removed for cleaning or replacement, a depth (staff) gauge should be installed in the tank at a location visible from the dip tube. Zero on this gauge must correspond to the bubbler tube's zero reference.

A typical installation schematic is shown in Figure 33. Maintenance access is needed for the clean-out tee and for the bubbler system enclosure.

FLOATS

Float-type level indicators are often used in treatment plants if remote readout is not needed. Floats can be connected to transmitters for monitoring; however, this arrangement is seldom used. Some other type of level meter such as bubbler, capacitance, or sonic is used instead. Float-type level switches are generally used for alarms and equipment on/off control. Applications for float level indicators are listed in Table 4.

Float level indicators consist of a float, an attached rod with pointer, float guide, and indicator scale. These components are shown in Figure 34. As the float rides up or down on the liquid surface, the pointer indicates the level. Another type of float level indicator is shown in Figure 35. In this case, float movement is indicated by the counterweight position.

Float switches depend on the liquid's buoyant force to activate the switch. In one switch position, the float is buoyed up by the liquid; in the other position, the float hangs down in the absence of liquid. A wide variety of float devices exist which translate the float position into electrical signals. Of these devices, the majority of water and wastewater applications use the tilt switch type shown in Figure 36. Each switch is a buoyant bag with a mercury switch inside. When the bag is tilted from one position to the other, the mercury switch opens or closes an electrical circuit. The circuit activates a relay to provide the contacts necessary for local controls and remove monitoring.

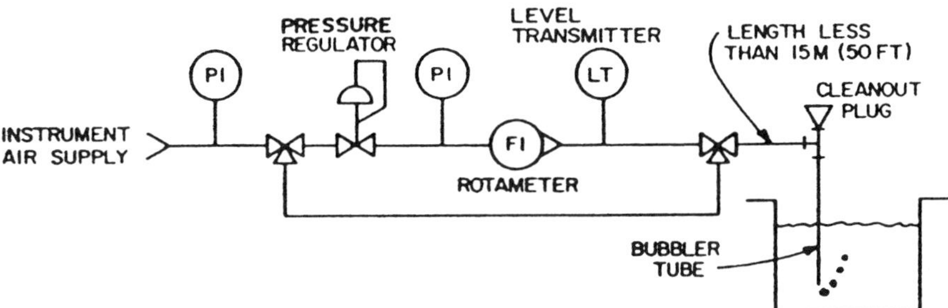

Figure 33 Open-tank bubbler installation.

Table 4 Float Level Indicators Applications

Recommended	Not recommended
Finished water	Liquids where grease or solids buildup could occur
Tertiary effluent	Polymer solutions
Raw water	Turbulent processes
Potable water tanks	

Installation

A recommended installation method for float switches is shown in Figure 37. If turbulence is expected, a stilling well should be installed. The installation figure shows the floats permanently fixed to a specific level which is true of most float switch installations. The switch setting is not easily changed, so careful thought must go into selecting the proper location and design.

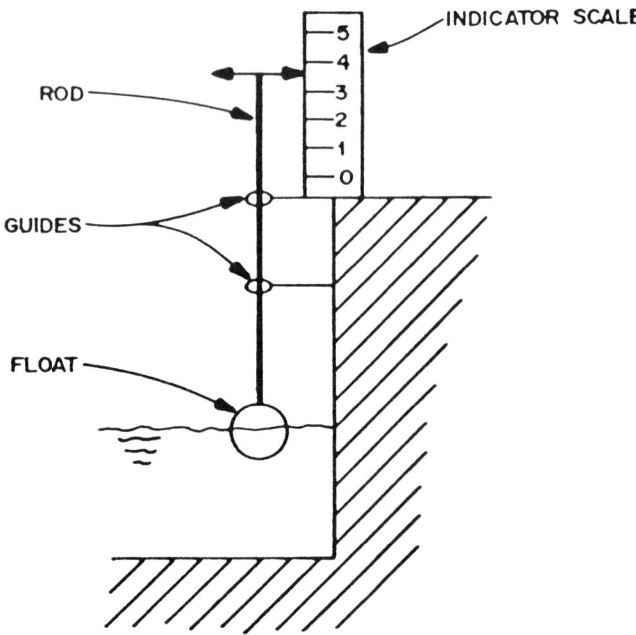

Figure 34 Simple float level indicator.

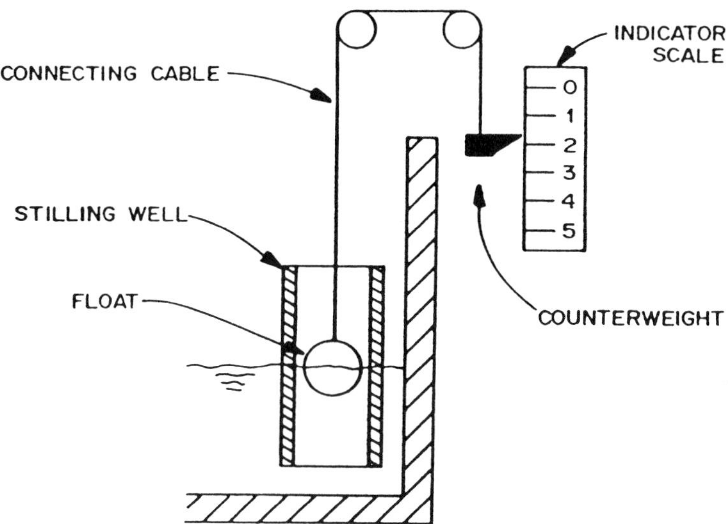

Figure 35 Counterweight float level indicator.

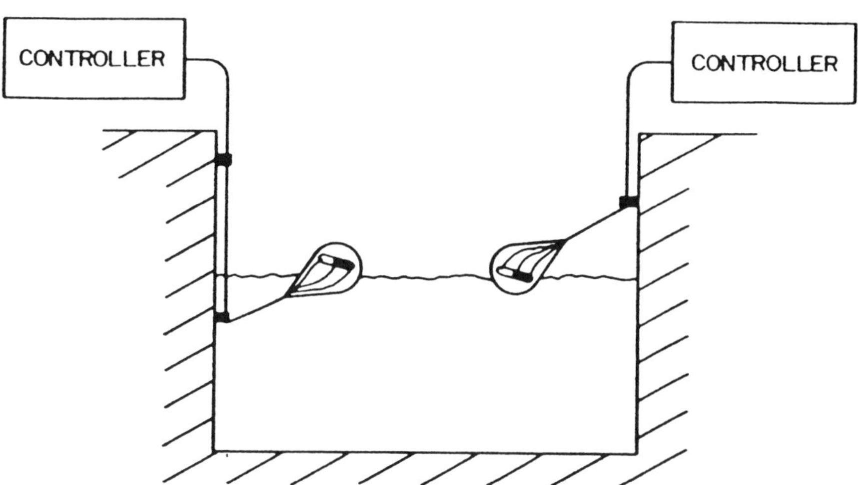

Figure 36 Float switches.

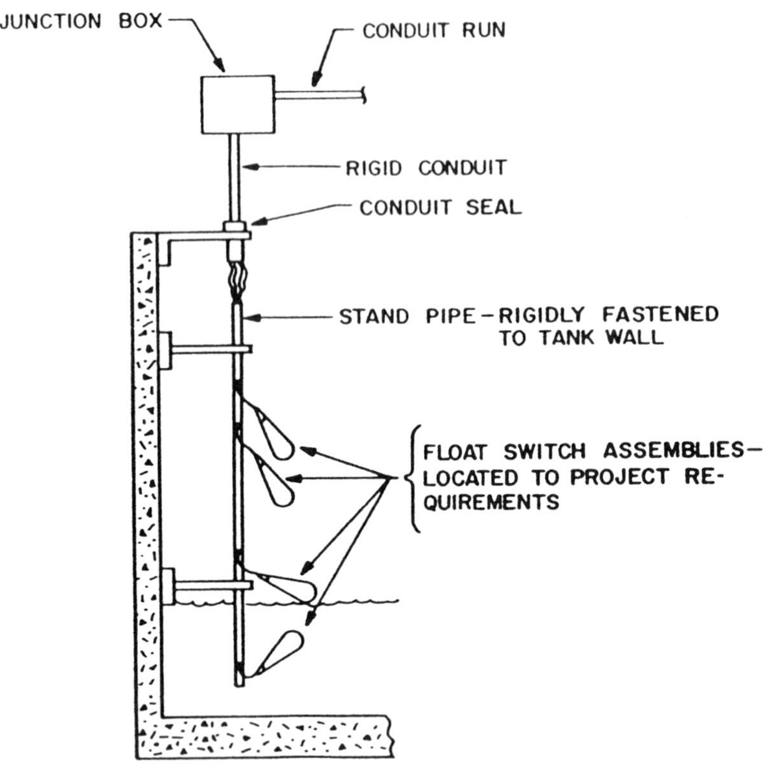

Figure 37 Float switch mounting installation.

PRESSURE MEASUREMENT

Differential Pressure

Differential pressure transmitters, DP cells, are used with primary elements to measure flows, gauge pressure, and liquid level. With the aid of isolation diaphragms or purge systems, DP cells can be applied successfully to any water- or wastewater-treatment process. Applications for differential pressure transmitters are in Table 5.

The three most common elements used to indicate pressure are Bourdon tubes, bellows, and diaphragms. In each case, the element moves in proportion to differential pressure. This motion is amplified by mechanical linkage to a pointer and dial. Diagrams of each type are shown in Figure 38.

Table 5 Pressure Measurement Applications

Recommended	Recommended with isolation diaphragm
Air	Chlorine
Oxygen and ozone	Wastewater with solids
Digester gas	Sludge
Water (raw and filtered)	
Secondary effluent	

Bourdon Tube

A Bourdon tube is a curved tube sealed at the tip. As process pressure increases inside the tube, the tube straightens causing the tip to deflect.

The deflection is indicated on a dial by mechanical linkage. Besides a C-shaped tube, Bourdon tubes are available in spiral, twisted, and helical forms, round, oval, or rectangular in cross section.

Bellows

Bellows elements are deeply corrugated metal cylinders closed at one end. Process pressure applied to the high side of the bellows causes it to expand. Bellows expansion is converted to pointer and dial indication. Bellows are also configured to contract on increasing pressure. In some cases, restoring springs are added to increase operating range or reduce element wear.

Diaphragms

Diaphragms are metal disks either flat or concentrically corrugated. High and low process pressures are applied to opposite sides of the diaphragm. This causes the diaphragm to deflect. A mechanical linkage connects the diaphragm to a pointer for dial indication. Corrugated diaphragms allow larger deflection and better linearity than flat diaphragms.

Electromechanical Elements

Electrical signals proportional to differential pressure are obtained by mechanically connecting an electrical component such as a capacitor, strain gauge, or inductor to a diaphragm. Deflection of the diaphragm will change the associated electrical property; for example, distance between plates in a capacitor, piezoelectric response, and loop reluctance.

In some elements a restoring force is applied to the diaphragm to keep it undeflected. This eliminates nonlinearity due to diaphragm deflection. The restoring force is measured electrically and converted to differential pressure. These elements are called force-balance transducers.

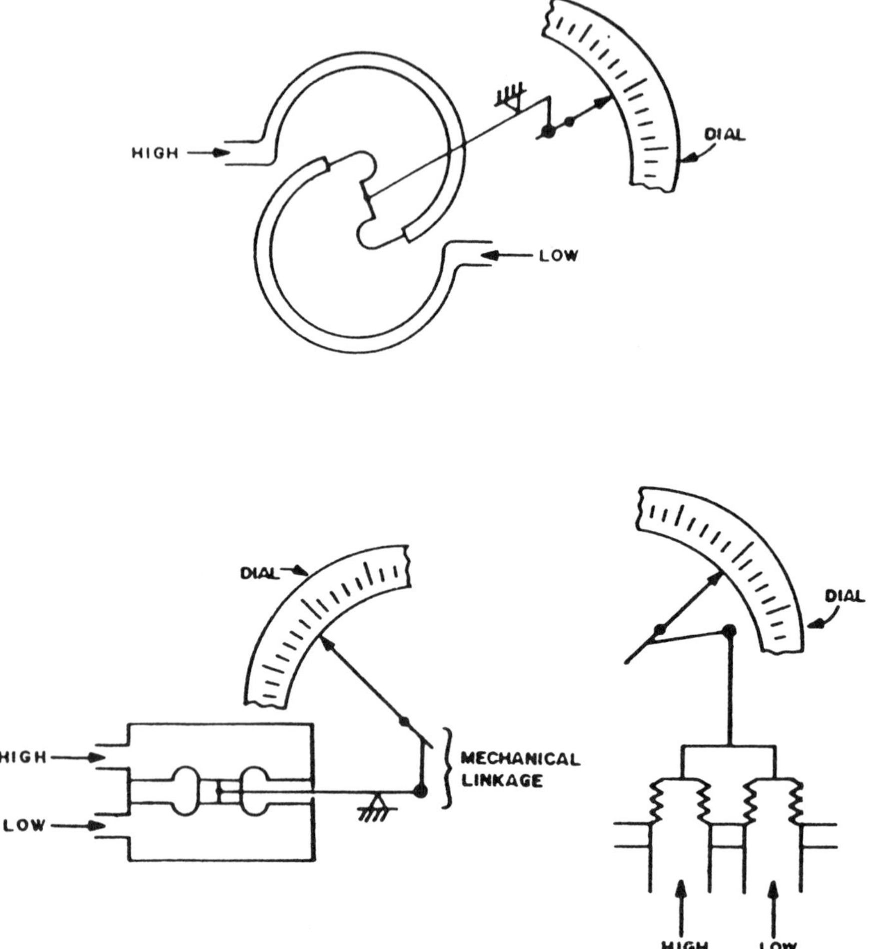

Figure 38 Mechanical differential pressure elements.

Absolute Pressure Elements

Absolute pressure elements are differential pressure elements with the low-pressure side evacuated to –101.3 kPa and sealed from the atmosphere.

Installation should follow the following requirements:

- Install the transmitter in an environment recommended by the manufacturer. This is usually –20–65°C (0–150°F) and 0–95% relative humidity. Zero and span will shift with changes in temperature, so avoid temperature extremes.

- Install the transmitter as close as possible to the process measurement site to reduce response time, which can be important in flow control or level control applications. The installation must allow easy access for maintenance. In some cases, it will not be practical to install the transmitter to meet both nearness and maintenance criteria. In these situations, control requirements must be given first priority.
- For solids-bearing liquid process lines, connect meter runs horizontally. Do not connect meter runs to the upper quadrant of the pipe. This will minimize the amount of solids and gas entering the connection. Entrapped gas will decrease response time, and solids may plug the meter connection. Slope meter runs 8 cm/m (1 in./ft) of run, so that gas bubbles bleed back into the process line.
- Connect meter runs to gaseous process lines at the top of pipes or tanks to minimize the amount of solids and moisture entering the connection. Slope meter runs at least 8 cm/m (1 in./ft) of run, so that condensation will drain into the process line. Low spots in meter runs should be avoided; they cannot add Entrapped liquids will affect meter accuracy and may cause accelerated corrosion.
- Special precautions must be taken in steam applications to prevent overheating of the manifold and transmitter. Side mount pressure taps to allow steam into the tap while still allowing drainage of excess condensate back into the process pipe. A condensate pot should be installed on each meter lead. To avoid overheating, blowdown valves should not be incorporated in the manifold. In steam applications, a water seal is required between the condensate pot and the manifold. This prevents steam from reaching the manifold and transmitter and prevents uncontrolled buildup of condensate in the meter leads. The condensate pots must be identical in size, the same height above the transmitter, and self-draining to the process pipe.
- For applications involving solids-bearing liquids, flushing provisions or diaphragm isolation may be needed. Diaphragm connections to the process should be a minimum of 2.5 cm (1 in.) for sludge lines.
- Install an isolation valve on all meter runs at the process measurement site (pressure tap) and, except for very short tap lines, at the transmitter.
- Materials recommended for harsh environments are:
  Chlorine—Hastelloy
  Digester gas—316 stainless steel
- Differential pressure transmitters are usually installed with valves that enhance some combination of instrument calibration, blowdown of accumulated material in the meter piping, and isolation of the transmitter. An economical way to provide the desired functions is to use factory-made valve manifolds. Some common manifolds are shown in Figure 39. Available manifold options are:

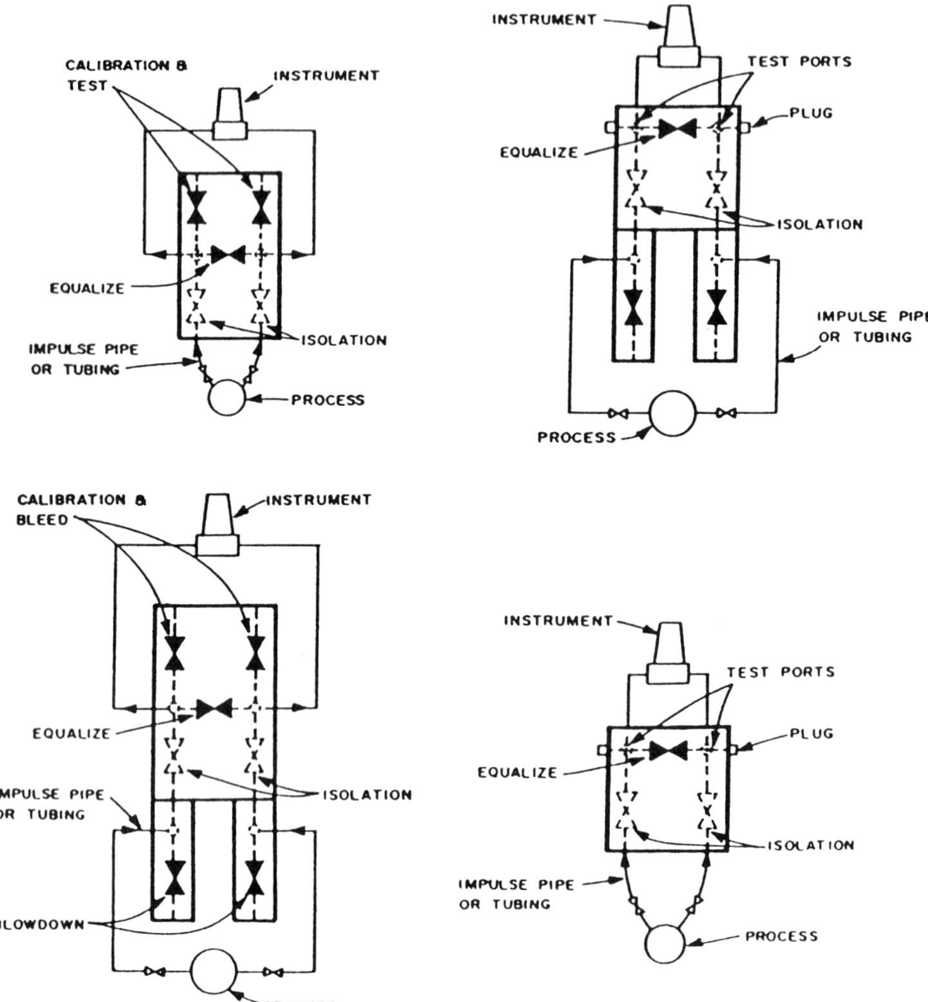

Figure 39 Manifolds.

1. Number and configuration of valves:
   a. Three-valve: isolation and equalization
   b. Five-valve: isolation, equalization, and calibration
   c. Five-valve: isolation, equalization, and blow-down
   d. Seven-valve: isolation, equalization, calibration, and blow-down
2. Process connections:
   a. Pipe: 1/2-in. NPT female and 3/8-in. NPT female

   b. Tube: 0.375 and 0.5 in.
3. Transmitter connections:
   a. Pipe: 0.5-NPT female
   b. Direct-flanged
   c. Tube: 0.375 and 0.5 in.
4. Materials of construction: Same selection as for transmitter
5. Remove zeroing: Motor-operated three-valve manifold

A check of meter zero is sufficient and a check of the span is not required at a frequent enough interval to justify the expense of a more complex manifold. Span calibration test frequencies of 3–6 months for most applications confirms this practice. Where DP cells are part of a flow measurement system used as a standard or for billing purposes, the frequency of calibration and testing may be much more often. In such cases, it is recommended that five-valve manifolds be installed to reduce calibration setup time.

- A common use of DP cells is with primary flow elements to measure pressure drop for flow calculation. Special installation practices for flow applications are presented with the primary device (see orifice meters, Venturi meters).
- Another common usage of DP cells is to measure liquid levels. Two general methods are used: hydrostatic head and bubbler (dip tube). Figure 40 shows typical hydrostatic level installation for an enclosed tank. The tank is covered, so the DP cell must use the pressure of the vapor phase as a reference. The DP cell is at the same level as the bottom pressure tap, and the connection to the top tap is vapor filled or "dry." Thus, the difference in pressure is proportional to liquid level. If the tank were open, a plumbing connection to the reference side of the DP cell would not be necessary. The DP cell would just need to be at the same ambient pressure; that is, in the same room with the tank or both the DP cell and the tank outdoors.

A "wet leg" configuration for level measurement is shown in Figure 41. The reference side of the DP cell is filled with a liquid. It does not have to be the same as the tank liquid. If the liquid is not the same, the difference in specific gravity of the two liquids must be used to correct the meter calibration. The liquid in the wet leg prevents unwanted accumulation of condensate at the reference side. In this configuration, the reference (wet leg) is connected to the low-pressure side of the DP cell, just as in a dry leg setup. As shown, the meter would read 100 at "0" liquid level and 0°/0 at 100% liquid level. This is corrected during calibration by suppressing zero to provide a correct tank level indication.

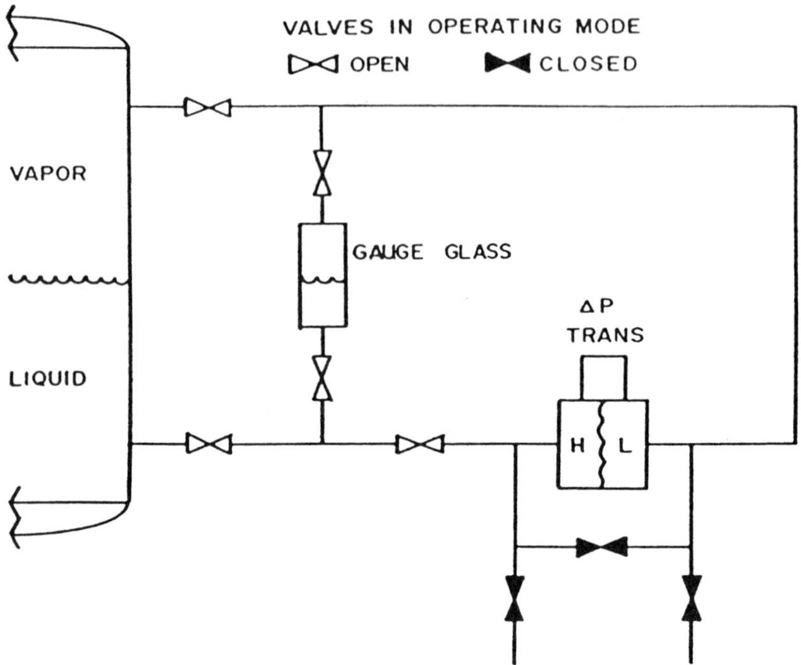

Figure 40 Dry leg.

TEST METHODS

To supply the data needed for chemical decision-making, test results are needed in the areas of

- Physical/chemical properties
- Degradability
- Accumulation potential
- Acute and chronic toxicity
- Carcinogenic/mutagenic/teratogenic potential

Data for specific applications may not be needed in all areas, but more complete test results are needed as a chemical's use and hence the potential for exposure increases.

The test methods to be reviewed here were selected based upon considerations such as

- Applicability to analysis of toxic chemicals including limits of application
- Accuracy and reproducibility of results

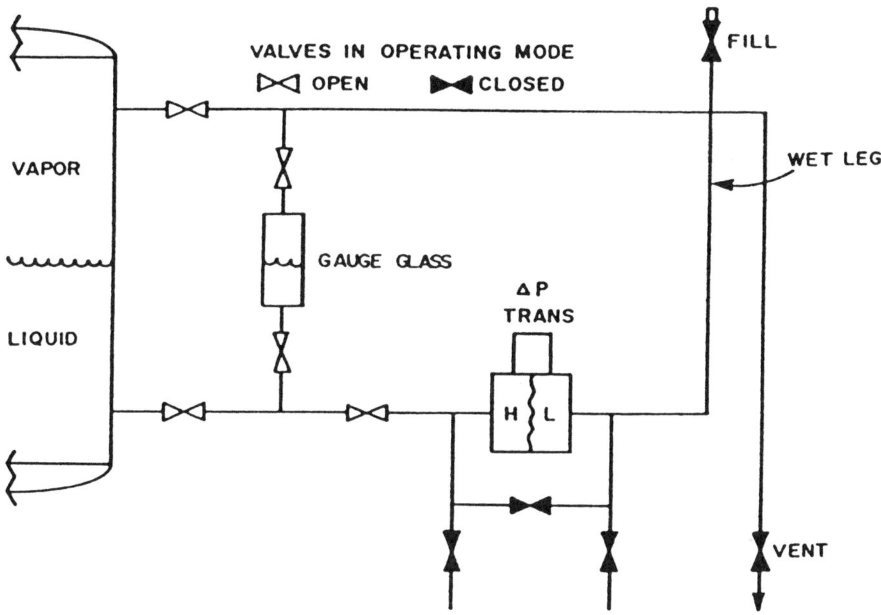

Figure 41 Wet leg.

- Sensitivity of the test method
- Validation across a range of chemicals
- Acceptance
- Potential for standardization

Short- and mid-term methods are given preference, but a number of long-term methods have been considered. Individual test methods are described.

Since most of the commonly used test methods for determining a compound's potential adversely to affect the environment or human health are covered here which should provide an initial reference point of methods for general use. This may include alternates of the presented methodologies which may make them more useful for an even wider range of chemicals.

Test Methods to Determine Selected Physical and Chemical Properties

In general, physical/chemical properties are relatively easy to measure and the results of these measurements can be used to predict a compound's behavior in the environment and/or in biological systems. Test methods to measure four of these properties are:

- Water solubility
- Partition coefficients
- Dissociation constants
- Volatility

Water Solubility

Water solubility is an important characteristic of chemical substances, since the degree of water solubility often determines the fate of a substance in the environment. If a compound is soluble in water, it tends to be widely distributed and will come into contact with more ecosystems. Similarly, it will occur at a lower specific concentration. Soluble compounds are more likely to be biodegraded, since they will come into contact with more varieties of microorganisms. Water solubility is a function of temperature and hydrogen ion content (pH), as well as concentration, and these should be taken into account when designing test protocols for specific compounds. Test methods have been included to measure water solubility of both hydrophilic and hydrophobic compounds.

Partition Coefficients

Partition coefficients are defined as the ratio of a compound's solubility in a solvent divided by its solubility in water at equilibrium. Certain solvents such as alcohols and amines can act as both hydrogen donors and acceptors which increases their versatility. The n-octanol/water-partitioning system has been the most widely used and is a good predictor of uptake of a chemical by biological systems. Water solubility shows a high inverse correlation with bioconcentration; this reflects the contrasting lipophilic and hydrophilic molecular properties. Earlier workers used olive oil, cottonseed oil, fish oil, and other oils to represent the lipophilic phases, but n-octanol is easily obtained and gives more reproducible results than the natural oils listed. Partition coefficients are usually expressed as $\log_{10}P$.

Once a chemical enters a living system, its distribution within that system is governed by its partition characteristics; that is, water solubility, molecular polarity, and affinity for lipids. It has been demonstrated and is generally accepted that the n-octanol/water partition coefficient is related to both the toxicity and electronic parameters of a compound. Chemicals with lipid/water partition values greater than 1000-fold may be expected to cause some sort of environmental problems. The partition coefficient, therefore, may be used as a first approximation in predicting bioaccumulation of a chemical substance. Extensive accumulation testing should not be required for chemicals with low partition coefficients.

Partition coefficients can be measured using both direct and indirect methods. The standard method for determining a substance's octanol/water partition coefficient as well as variations using high pressure and a method for gases are included. Results using such methods correlate with a calculation method which relies on a knowledge of the compound's chemical structure and a library of partition coefficient values for other similar compounds and substitution groups. This nonlaboratory method might be useful to provide a first approximation of the partition coefficient in a short time frame at a minimal cost. If a potential for bioaccumulation was suspected, more exact measurements by a direct method could be made.

Dissociation Constants

The dissociation constant is a way of expressing the equilibrium between a chemical substance and its dissociation into fragments in aqueous solution. The value obtained is obviously related to the compound's solubility in water. This property is required in order to evaluate the environmental fate of chemicals and treatment processes for their removal from water. Knowledge of the distribution of the undissociated chemical and ionic species with pH (predominantly under acidic conditions) can be useful in ascertaining the availability of the substance to enter into biologically mediated oxidation reduction reactions, dissolution, ion exchange, coagulation, or adsorption.

Volatility

Volatilization may result in the distribution of a chemical into several media. Under natural conditions, this process is dependent on particle size or solids, climate, interactions with environmental substrates, and interactions with other formulation components. These factors modify the rate of transfer of a pure chemical from a solid or a liquid surface to diffuse in still air layers.

Vapor pressure provides a good indication of a pure substance's relative tendency to volatilize under laboratory conditions. For a large number of substances, the vapor pressure (p) is related to evaporation by the relationship:

$$p = K\,(\text{Evaporation Rate})^n$$

where k and n depend on such environmental factors as temperature and wind speed.

Volatilization from aqueous systems depends not only on a compound's vapor pressure but also on its solubility. A saturated aqueous solution is considered to have the same vapor pressure as the pure chemical.

The problem is soils is that adsorption is a competing reaction. The actual volatility of a substance from soil will depend on the moisture conditions of the soil surface, the depth to which the chemical has penetrated the soil, and a compound's tendency to adsorb onto the soil particles.

Alternative methods for measuring volatility have been included which rely on measuring vapor pressure to predict the volatility: using a hollow fiber mass spectrometry method to measure volatility directly, measuring volatility from soil; and a method relying on a knowledge of several chemical properties to predict volatility according to a predetermined mathematical relationship.

Biodegradation Determination

All living systems must obtain sources of carbon, other essential elements, and energy from their environments. For nonphotosynthetic organisms, this involves adsorption of organic compounds and the metabolism of these compounds to produce energy. In different organisms, the metabolic pathways may be significantly different. For an organism to be able to degrade, the following conditions must be met:

- The chemical substance must be able to reach the organism or the enzyme site.
- The compound must not be lethal.
- Enzymes necessary to transform the chemical must be present or be able to be induced.
- Environmental conditions must permit operation of the enzymes.

There is usually a lag between exposure to a chemical and the initiation of degradation or transformation. This lag is referred to as acclimation, in which the biological system adjusts to the presence of the compound, and the enzymes to degrade it are induced. The most significant biological systems involved in biodegradation to CO_2, H_2O, and other inorganic compounds are bacteria and fungi. Because of the ubiquitous nature of these microorganisms in the environment, they are by far the most likely living form to come into contact with a chemical substance. They also have a broad spectrum of metabolic activity, with a single organism being able to metabolize a wide variety of carbon sources.

Since bacteria and fungi are most likely to serve as the major factors in the ultimate degradation of synthetic organic chemicals in the environment, test methods to demonstrate biodegradation should measure the ability of these microorganisms to metabolize or transform the substance.

Biodegradation in the Soil

Synthetic chemicals find their way into the soil both through direct and indirect application. Chemicals may be applied to soil as pesticides or fertilizers or be present in sludge or municipal and industrial wastes. Indirect application results from industrial discharge into the air and subsequent settling, from translocation by animals and plants, and from the flooding of polluted water over the soil.

Soil is a complex system of surfaces, catalysts, and reactors. Synergistic reactions can occur in soil and are highly dependent on its composition, pH, and moisture content. Microorganisms seem to be the major contributors to the degradation of these chemicals in soil, far outstripping photodegradation or chemical degradation for most substances.

Existing methods for determining biodegradation in soils can be classified into three categories:

- Laboratory tests
- Greenhouse studies
- Field studies

Distinct methods for measuring the ability of soil microorganisms to degrade a synthetic chemical substance are a manifold assembly soil method, a biometric flask method, a spectrophotometric method, a soil percolation method, and an open incubation method.

Biodegradation in Water

Chemical substances enter aquatic environments in a number of ways, including runoff from land, accidental dumping, direct application to control pests, discharge from industrial waste, and sewers. The microorganisms present in the natural waters will acclimate to the presence of a toxic substance, and as long as its concentration is sufficiently low to avoid toxic effects and the physical conditions such as temperature and pH are adequate, they will generally begin to degrade the substance.

Two basic types of tests are used to measure microbial degradation in aquatic systems. Static tests use low bacterial concentrations and simulate natural water conditions. They usually are more time consuming because of the low concentrations involved but give more reliable results than the dynamic tests which use high bacterial concentrations. In both types of tests, biodegradation is usually measured by analyzing the incubation mixture of microorganisms and test compound periodically for disappearance of the test compound, formation of degradation products, oxygen uptake, or CO_2 evolution.

Photodegradation

Photodegradation of a chemical results from its absorption of energy (photons) from solar radiation leading to degradation of the light-absorbing molecule and to radical chain reactions. Photodegradation results in oxidation, hydrolysis, or a combination of the two depending on the reactants and prevalent conditions in the area over which a chemical is distributed. Important types of photoreactions are

- Photo-oxidation
- Photoisomerization
- Photoionization
- Photodimerization (polymer formation)
- Free radical formation
- Photoreduction

Near the surface of the earth, the radiation is predominantly over 290 nm. Only those molecules absorbing light in this range can react to natural sunlight. Some classes which are reactive include polycyclic aromatic hydrocarbons, heteroaromatics, monocyclic heteroaromatics, aromatic and aliphatic ketones and aldehydes, peroxides and hydroperoxides, azo compounds, chelates with transition metals, aromatic amines, nitroderivatives, aliphatic nitrates and nitrites, phenols, quinones, disulfides, conjugated multiple bonds, nitrous compounds, and iodides. The rates of photodegradation of the above compounds vary over many orders of magnitude, but all are sensitive to some degree.

Most of the photodegradation reactions occur in the atmosphere, but others occur near the surfaces of soil and water and on various surfaces such as buildings and leaves. Because each of these media impose special parameters on the photochemical reaction, each will be discussed separately.

Atmospheric Photodegradation

The rate and extent of photochemical reactions in the atmosphere are influenced by a number of factors, including concentration of water vapor, humidity, concentration of substrates, oxidant concentration, the presence of competing photochemicals, and the temperature, intensity, and wavelength of the irradiation.

Experiments for measuring photodegradation involve a light source, a reaction vessel, and an analytical procedure. The light source may be natural sunlight or it may involve a fluorescent or mercury arc lamp to imitate the natural sunlight. This is especially important if irradiation is to continue for more than a few days, since the intensity of natural sunlight varies appreciably.

These experiments are generally performed in the presence of oxygen, and many of the reactions which occur are in fact photo-oxidations. Within the reaction vessel, temperature and humidity can be varied to approximate natural conditions. Chemicals are often adsorbed onto fine dust particles suspended in the atmosphere, and may undergo photodegradation in this adsorbed state.

Following irradiation of the sample substance under one of these conditions, the resulting photoproducts as well as the disappearance of the parent compound are monitored. These two methods are long-path infrared spectrometry and measurement of photo-oxidation using a gas chromatograph.

Photodegradation in Liquid Media

Although water is transparent to ultraviolet (UV) light, the intensity decreases with increasing depths and photochemical reactions proceed at extremely slow rates a few meters below the surface. For this reason, the density of any liquid substance relative to water must be considered. Other factors which affect the rate of photodegradation in liquid media are

- Light source intensity
- Concentration of reactants
- Temperature
- Hydrogen ion concentration
- Solvent
- Need for and presence of sensitizers

Generally, increases in temperature within natural climatic ranges result in increases in the reaction rate of photodegradation. The acidity or basicity of the solvent may affect specific reactions depending on the solvent choice and the substance being tested. The choice of solvent is important for interpreting results. Water is obviously the preferred solvent, but methanol, n-hexane, or cyclohexane (H-donor solvents) are often used when the substance being tested is insoluble in water. The products of the photoreactions may vary, however, since the H-donor solvents result only in the substitution of hydrogen for a halogen, whereas in water both hydrogen and hydroxyl substitutions will occur. Therefore, extrapolation of photochemical effects from other solvents to water should be done with some caution.

Most reactions in liquid media are done in the presence of oxygen and do not attempt to distinguish photo-oxidation from other categories of photo-reactions. In nonoxidation reactions, it is possible to use a nitrogen atmosphere.

Certain compounds can act as sensitizers for a photochemical reaction of a second substance. This means they absorb the photochemical energy, become excited, and transfer this energy to a receptor chemical which then degrades. Riboflavin and chlorophyll are two natural sensitizers. The effectiveness of the sensitizer depends, for example, on the wavelength at which it absorbs light, the efficiency of the conversion of its singlet to triplet state, the efficiency with which it transfers energy to the acceptor molecule, and its concentration. Other substances such as benzene may be used as a sensitizer in photochemical reactions in liquid media. These factors should all be taken into account in designing test methods to measure photodegradation in liquid media.

Photodegradation in Soil

Many of the chemical substances that are released into the environment, especially less volatile ones, may come to reside in water and soil. In soil, only those substances at or within a few centimeters of the surface are susceptible to photoreactions.

In many cases, the chemical substances coming in contact with soil particles may be adsorbed to their surfaces. Methods exist which attempt to mimic this adsorption of compounds onto a glass plate covered with a thin soil layer.

Reactions may be followed spectrophotometrically or chromatography may be used to separate the species. In addition, fluorescence or radioautographic techniques work well with soil samples.

Photodegradation on Solid Surfaces

Chemical substances also become deposited on various surfaces through such processes as condensation, rainfall, and atmospheric dispersion. These surfaces provide an inert medium for photodegradation. Again such factors as light intensity, temperature, and the presence of other substances will be important in determining whether photodegradation will occur.

The surfaces available in the environment vary from natural materials such as leaves and bark to man-made roofs of buildings, roads, and windows.

Thermal Degradation

Thermal degradation is the breakdown of a molecule into smaller molecules responding to a change in temperature. Thermal degradation under natural conditions takes place from approximately –20 to +70°C. Thermal degradation in the natural temperature ranges is a far less important cause of chemical breakdown than biodegradation, photodegradation, or chemical reactions. However, it should not be ignored.

Available methods to measure thermal degradation involve rapid heating, usually to relatively high temperatures. Some of these methods, such as those which use a gas chromatograph or a gravimetric balance, could be adapted for use at the lower environmental temperatures.

Chemical Degradation Test Methods

Environmental transformation of organic compounds is known to result from at least three important chemical reactions: hydrolysis, oxidation, and reduction. Such reactions occur independently of any biological or photochemical reactions. It is important to consider degradation of a chemical substance in all possible media by all likely mechanisms. It is necessary to consider chemical reactions even though they may not be the major source of degradation under most circumstances. Products which are yielded by chemical reactions also differ in many instances from those resulting from photochemical or biological degradation.

Hydrolysis

Hydrolysis is an important chemical process that occurs in aquatic and non-arid soil environments. Hydrolysis is used to describe reactions in which a bond of the molecule is broken and a new bond is formed with the oxygen atom of a water molecule. Hydrolytic reactions are generally mediated by acid and/or base. The extent to which these catalytic effects occur is dependent on the type of reaction and the chemical structure of the compound. The pH of the solution may therefore have a profound effect on the hydrolysis rate. It is also important that natural pH ranges be considered in establishing test conditions.

Temperature effects vary for different reaction pathways and the magnitude of these effects varies with chemical structure. In fact, temperature may affect not only the rate of the reaction but the products as well.

The presence of metal contaminants such as Cu^{2+} and particulates found in natural waters have also been shown to influence the rate of hydrolysis as have the concentrations of the chemical substance.

Compounds likely to undergo hydrolysis include esters, amides, carbamates, and compounds with good leaving groups located at positions that would stabilize a carbonium ion (e.g., allylic, benzylic). The OH radical is by far the major species depleting hydrocarbons in the atmosphere. The high rates of hydrolysis observed for aromatic hydrocarbons have significant implications for the control of photochemical air pollution.

Oxidation/Reduction Reactions

Oxidation is defined as electron loss and can be brought about by HO, O_3, or RO· radicals where R is an organic group. In the atmosphere, the major oxidizing species are the HO radical and O_3, whereas only $RO_2\cdot$ is likely to be an important oxidizer in an aquatic system.

The half-life in the environment for various reactions can be estimated if the concentration of the oxidizing species and the rate constant for a reaction are known. Values for the half-lives of the various oxidizing species have been calculated. The values for HO_2O_3 are probably accurate to a factor of 3 for the gas phase reactions while the RO_2^- values in aqueous phase are accurate only to an order of magnitude. Although the rates of attack of the oxidizing species can be estimated, the exact products of the reactions cannot be predicted. Certain generalizations can, however, be made:

- Aromatic compounds will oxidize in air to give mixtures of phenols, quinones and side chain ketones, alcohols, and aldehydes. In the atmosphere, ring oxidation will probably dominate (>60%).
- Aliphatic compounds will oxidize in air to give alcohols, ketones, alkyl nitrates, and mostly cleavage products resulting from intermediate alkoxy radicals.
- Oxidation in water of both aromatic and aliphatic compounds will probably proceed through RO and RO_2 radicals to give alcohols, ketones, and hydroperoxides.

Another factor is the presence of naturally occurring antioxidants such as phenols and aromatic amines with readily extractable hydrogen atoms. These compounds react with free radical oxidation intermediates to form nonreactive free radicals and halt the chain propagation of the overall reaction sequence.

The effect of chemical oxidation independent of the photo and biological reactions has received little attention. There are several test methods used to measure oxidation in both aquatic and atmospheric environments.

Reduction, the reverse of oxidation, has been defined, for example, as the gain of electrons, loss of oxygen, and addition of hydrogen. To reduce organic compounds, an environment of low redox potential is required. Reducing reactions under natural conditions, therefore, only occur to any measurable extent in water and soil at lower depths where oxygen is limited.

Some of the reducing reactions which occur with organics have been described by such terms as *hydrogenation*, *halogenation*, and *coupled oxidation-reduction* and generally require catalysts. The rates of these reactions, like those of oxidation reactions, are influenced by temperature as well as pH.

Adequate test methodologies do not currently exist to separate the chemical reduction reactions from those mediated by biological organisms. More research will be required to distinguish between these two phenomena in an in situ environmental situation.

Mobility and Transport

Chemicals are discharged from a point source or a disposal site into one or a combination of media: air, water, and/or soil. Rarely do materials reside permanently in the medium into which they were originally discharged. Transport to other media is related to such factors as meterological conditions, flow of surface waters, location and flow of ground waters, tidal action, topographical characteristics of land surfaces, and soil characteristics.

Air, water, and soil can in turn be more finely divided. The altitude above the earth and the relative humidity greatly affect the properties of the atmosphere as chemicals enter or react with it. Fresh, saline, and estuarine waters have very different properties as do many surface and ground water bodies. There are thousands of soil types which differ widely both with regard to structure and composition. Each type has unique binding and surface properties when it interacts with a chemical substance.

Differences in the media to which a chemical substance is exposed will greatly affect its availability for biological and chemical interactions and conversely the extent to which it will persist or accumulate. Differences also mean that no simple test scheme can predict absolutely the behavior of a chemical in the environment or even exactly how long it will reside in a particular medium. The tests can, however, estimate where a particular pollutant is likely to reside and a rough estimate of its concentration and reactivity.

Abiotic accumulation is related to a compound's adsorption potential and its resistance to degradation. The very act of a compound becoming adsorbed to the surface of a soil particle may render it no longer available for degradative reactions.

Accumulation is also defined as persistence, but often carries with it the connotation of increased concentrations of the test compound in certain partitions of the environment. In biological systems, this may be an enzymatically mediated active uptake, whereas in abiotic systems, only chemical forces are involved. Accumulation may also occur in response to an increased pollutant concentration.

Increased levels of adsorption may not be reversed when the external pollution concentration is decreased and thus resulting in a net concentration of the pollutant in an adsorbed state.

Adsorption to Soil Particles

The transport, degradation, and bioavailability of chemicals is greatly affected by the adsorption of chemicals to soil surfaces. Adsorption reduces the concentration of chemicals in solution and thus reduces the quantity of chemicals free to undergo reactions.

The relative distribution between adsorbed and solution phases depends on a number of factors including

- Chemical/physical properties of the chemical substance such as dissociation constant, size, shape, water solubility, and vapor pressure
- Soil properties such as the clay and organic matter content, surface area, soil structure, and soil pH
- Temperature
- Moisture content of the soil
- Presence of inorganic salts

The organic matter content of the soil is probably the most important of the above factors, since a high positive correlation exists between the soil's organic matter content and its ability to adsorb.

A number of intermolecular interactions are involved in adsorption of chemical substances to the surface of soil particles.

The following types of interactions may occur which involve electrostatic forces:

- van der Waals–London interactions
- Charge transfer
- Hydrogen bonding
- Ligand exchange
- Ion exchange
- Direct and induced ion–dipole and dipole–dipole forces
- Chemisorption

Other interactions such as hydrophobic bonding and magnetic bonding may occur which do not involve electrostatic forces. The relative effects of these various forces vary depending on the composition of the soil and the environmental conditions.

Many factors therefore affect the adsorption process in natural soils. It is extremely difficult to determine the extent to which any one factor is influencing the results relative to the others for a given test compound and soil sample under a given set of conditions. Rather it is important to realize that many forces are at work, and that any test procedure designed to determine a substance's potential to accumulate abiotically in soil is a sum of all these effects. A number of test methods which measure chemical adsorption under laboratory conditions have been included in this chapter. They involve such techniques as equilibrium dialysis, slurry, and continuous-flow systems.

Adsorption to Particles in Air

Depending upon the meterological conditions, a large number of fine particles will be suspended in the air. Such particles may result from soil or from other sources such as pollen and industrial exhaust. When chemical substances are adsorbed onto soil which becomes airborne, they may remain adsorbed. Chemicals that vaporize may come into contact with the particles and become attached. The result of these two phenomena is the same—the chemical is transported with the particle. The particles may settle because of weight or with rainfall or may now be inhaled. The adsorption to dust particles therefore becomes significant for estimating exposures due to inhalation. Few test procedures are available to measure adsorption onto particles or aerosols in suspension, and those that do exist are not well developed. The topic has received little research, although its significance is acknowledged.

Adsorption to Particles in Water

Chemical substances also adsorb to particles suspended in water and to sediments. These particles with the adsorbed substance may then be deposited owing to flooding or silting out. The properties which govern adsorption to soil are much the same except for the effect of water. The chemical must have a stronger attraction for the particle to remain adsorbed in an aqueous media.

Leaching/Desorption

Abiotic accumulation must be assumed to represent a dynamic equilibrium. Not only are substances adsorbed to particles, but the reverse reaction occurs. The conditions which govern desorption are important for they determine a compound's availability for biological action. Very often the factors which govern adsorption also govern desorption. This is especially true for moisture, salt content, temperature, and pH.

Leaching, the removal of chemical substances which are adsorbed to soil, implies subsequent movement of these substances through the soil. Most biological reactions occur in the top 20 cm of soil; therefore movement below this depth would reduce the potential for biodegradation or bioaccumulation. Of environmental concern is the fact that the unbound chemical may contaminate ground water or become part of drainage effluent. Leaching tests provide a semiquantitative index of the inherent mobility of chemical substances within a soil matrix. Tests using thin-layer chromatography and soil columns to measure leaching potential are included in this report. These methods have been widely used to estimate the leachability of pesticides.

Desorption can also occur in aquatic environments. Chemicals adsorbed to sediments or absorbed by biota may be released at later times.

Synergistic Effects

Movement of chemicals from one medium to another within the environment occurs by a number of pathways. Some of the pathways include several transfer mechanisms. Movement is governed by meteorological conditions, the composition of the media, and the physical/chemical properties of the chemical substance. It is possible that certain conditions and properties will function synergistically to affect the transport of a substance.

Biological Accumulation Testing

Accumulation of toxic materials in biological tissues has received increasing attention in the past few years. Much of this interest originally was due to the implications of toxic heavy metal and pesticide contamination of commercially important fish and shellfish. Discovery that residues of DDT and its principal metabolites could be transferred through aquatic and terrestrial food webs in increasingly higher concentration further enhanced research into the process of biological accumulation.

Many factors have significant effects on the results and reproducibility of tests for biological accumulation. Factors which such tests include are

- Selection of test organisms which can be cultured under laboratory conditions
- Degree of acclimation of organisms to test conditions
- Appropriate concentrations and dosages of test substances
- Duration of exposure

Procedures which address these factors will probably rely substantially on established protocols for acute and chronic ecotoxicity studies.

Biological accumulation tests are described separately for terrestrial and aquatic biota. These procedures for terrestrial and aquatic organisms differ in design because of the gross physical and chemical differences in their respective environments. Uptake of environmental pollutants by terrestrial animals may occur by inhalation through the lungs, by dermal exposure, by food intake, and by drinking water. On the other hand, aquatic organisms are completely surrounded by a liquid media in which dissolved and suspended toxicants are in constant contact with the organisms. Absorption of pollutants directly from water is often the dominant route of entry of toxic substances into aquatic organisms. Furthermore, since water is the universal solvent, many other trace elements and compounds which may have synergistic or antagonistic effects are present in the aquatic environment. Dissolved oxygen, temperature, hardness, pH, and nutrient concentrations are other water parameters which may have significant effects on the rate of uptake of test materials in aquatic systems. It is not surprising that the problems, protocols, and terminology differ substantially between aquatic and terrestrial biological accumulation tests.

Owing to the nature of the aquatic environment, however, test procedures do not differ significantly from organism to organism regardless of the trophic level they occupy.

Terrestrial Environment

Mobility of substances in the terrestrial environment is usually more restricted than in aquatic systems. Because many compounds are bound to both organic and inorganic surfaces, their movement into biota is governed, in part, by the physical and chemical environment surrounding the organism. The mode of uptake varies in the four categories considered (microorganisms, invertebrates, vertebrates, and vascular plants).

In biological accumulation tests with soil microorganisms, cultures are isolated and exposed to the test chemical in solution. Mixed or pure cultures are utilized and test procedures can be applied to a wide variety of substances. Active uptake and metabolism are distinguished from adsorption by comparing results with live and dead cells. This approach is applicable to virtually all bioaccumulation tests.

At the higher trophic levels, accumulation occurs by exposure to sublethal concentrations in drinking water and ambient environment as well as food. Tests for uptake, metabolism, and tissue accumulation are described for earthworms, rats, and birds. Although the organisms are fed during the tests, exposure to the chemical is by direct stomach intubation. The doses are sublethal, and after a set time period, the animals are killed and individual tissues are analyzed. The results of these tests provide a basis for further tests used to define food web pathways and associated ecotoxicological effects.

The uptake of various heavy metals by terrestrial plants and their ability to accumulate other classes of compounds is not as well understood. In addition to accumulation from the soil, plants can absorb substances through foliar surfaces. Preliminary tests should avoid complicating soil processes such as ion exchange by the use of nutrient solution and vermiculite cultures. If results indicate a potential for accumulation, tests can be performed with standard soils.

Aquatic Environment

Bioconcentration is the entry of a test material directly from the water into a test organism (generally via gills or epithelial tissues). Bioaccumulation includes bioconcentration as well as accumulation of the test material via dietary uptake. Biomagnification refers to the process by which concentrations of the test material increase as they pass through two or more trophic levels of the aquatic food chain.

The relationship between concentration and exposure time is an important consideration in the design and interpretation of tests for biological uptake. In tests for aquatic organisms, the decision whether to use a static or continuous-flow system is partially dependent on the stability of the test compound over time. Whether the substance is introduced into the environment in a continuous fashion or as a pulse may also influence the decision between static or continuous systems which prevent the build-up of toxic or oxygen consuming metabolites and waste products in test chambers. Long-term tests often require flowthrough systems. A series of subexposures can be employed to account for decreases in concentration without utilizing continuous flow.

Tests for accumulation in microorganisms and microinvertebrates are relatively simple to perform due to the ready availability of test organisms and their suitability for culture in the laboratory. Multiple concentrations of biomass, compound, and various incubation periods should be used. Tests using aquatic macroinvertebrates are generally similar in design to tests for fishes, except special handling and feeding techniques vary considerably from one species to another. Organisms selected as test species should be somewhat resistant to the test material so that toxic responses do not interfere with the test results and so that long-term survival of prey is assured in the case of food chain studies.

Tests for uptake by aquatic plants are very straightforward in design. As with terrestrial plants, it is recommended that sediment be eliminated from initial tests. Tests should be run long enough to monitor any loss of test compound to the water after the initial uptake.

Toxicity

The field of toxicology has coincided with a growing concern for environmental effects of pollutants on ecosystems and health.

Toxicity tests involve a stimulus, the test substance applied to a subject on the test organism or tissue culture over a set time period. Tests are initially classified according to the duration of exposure:

- Acute—a few hours to 2 weeks
- Subchronic—a few weeks to a year
- Chronic—a full life cycle

The time required in each of these categories is dependent on the life history of the organisms tested. The toxicity of a substance is determined by the relationship between the dose received and the response of the test organisms. Response is measured as death or immobility in acute tests and changes in various measurable characteristics in subchronic and chronic tests such as weight, histopathology, or behavioral symptoms of stress.

In evaluating a broad range of compounds, all toxicity test procedures must be standardized and documented. This includes culturing practices, exposure conditions, measurements, observations, and statistical analyses. The experimental design of a toxicity test is directly related to the calculations and statistical procedures used to define toxicity. Important considerations are

- Concentrations tested (range and specific values)
- Number of organisms at each concentration and in the control groups
- Frequency of measurements and observations
- Number of replicates

These will vary with the statistical model chosen as well as the type of test such as acute or chronic.

The dose-response relationship can typically be best described by a logarithmic effect pattern. It is, therefore, most efficient to use a logarithmic series of concentrations ranging from that causing zero mortality or effect to that causing complete mortality. The resulting scatter of points in a plot of log concentration versus mortality is relatively symmetrical about the 50% mortality line. Any calculation of median toxicity should include a measure of this dispersion; confidence limits (95%) are the preferred measure because they are in the same units as the toxicity value.

Although the experimental design complements the calculation method, data which do not fit a particular model should not necessarily be considered unacceptable. There are a number of generally accepted calculation procedures based on various assumptions and constraints; the theoretical and practical considerations in choosing one are covered in the references listed at the end of this volume. Some rules which apply to most toxicity tests (especially acute and subacute tests) are

- At least 10 organisms per concentration
- No more than 10 percent mortality in controls
- Low variability between replicate tests

One of the objectives in any toxicity protocol is to reduce variation which is not inherent to the biological response. The confidence limits of the median toxicity value should reflect the natural variability of the test organisms rather than fluctuations in test conditions or toxicant concentration. Replicate tests must be conducted to define the precision of a particular procedure. Increasing the sample size (up to 30 organisms per concentration) significantly decreases the standard error in median toxicity level estimations. Any curve fitting should be done by a statistical method rather than "eye" estimation.

Of the numerous indices of toxicity available, the median lethal or sublethal toxicity is the usual method of reporting results. This computed figure reflects the concentration or dose that effects the average or typical organism. The symbols LC_{50} and LD_{50} are used to reflect the median lethal concentration and the median lethal dose (amount actually received inside the body), respectively. When effects of a substance are measured at a level below that which directly causes death, an effective concentration or dose (EC_{50}, ED_{50}) is reported. The duration of the test is usually included in the median toxicity value (e.g., 96 hr LC_{50}). Results of chronic dosage studies can be reported as a 90-day median toxicity or a chronicity factor.

Toxicologists have devised a number of indices which identify acceptable toxicant concentrations in the environment. The application factor is the most common approach to defining biologically safe levels. The object is to integrate the effects of variable species sensitivity, length of exposure, and environmental conditions on toxicity. Data required from a few representative species are

- 96-hr median toxicity level (LC_{50}, LD_{50})
- Maximum acceptable toxicant concentration (TC), i.e., concentration that caused no significant effects on reproduction, growth, and survival of test organisms during a full life cycle

Once the basic dose-effect relationship has been established in acute and chronic tests with a compound, more involved toxicity measurements can be made. Among these are

- Toxicity of mixtures of two or more compounds
- Effects of various time-concentration curves
- Direct measurement of physiological or reproductive functions in response to a toxicant

Since exposure conditions in the first two measurements and response measurements in the latter differ from standard toxicity procedures, the analysis and presentation of results differ also.

Toxicity tests can be broken down into two major categories: ecotoxicity and human health toxicity. The ecotoxicity tests center on key microbial, plant, and animal species from aquatic and terrestrial environments. Methods of test substance administration, conditions, and duration of exposure vary from rapid and simple screening tests to lengthy and complex feeding and microcosm studies. The ecotoxicity literature is weighted toward aquatic animal tests; the number of method summaries presented in this category reflects this.

Toxicity to Aquatic Organisms

Owing to the nature of the aquatic environment, exposure to a potentially toxic substance is often more direct and predictable than in terrestrial systems. Aquatic organisms are in constant contact with toxicants dissolved or suspended in the water column. Since exposure conditions are relatively uniform, tests procedures are standardized and readily modified to include species of interest in a particular region. The laboratory equipment and culturing systems required to conduct aquatic toxicity tests are well documented in the references listed with each method summary and in the bibliography.

Standard tests of acute, subchronic, and chronic toxicity to fish are screening tests with *Daphnia* and various microorganisms. Such tests serve both as screening procedures for toxicity to higher organisms and as direct measures of acute toxicity to important food chain and wastewater-treatment organisms.

Since chronic tests with fish are time consuming, difficult to perform, and very expensive, a series of simple, short-term tests have been developed which predict chronic effects. These involve monitoring of respiratory stress and swimming ability in response to sublethal exposure to a toxicant.

Toxicity Terrestrial Organisms

Exposure to toxicants in the terrestrial environment is more difficult to model under laboratory conditions. The actual dosage levels received are a function of the physical and chemical environment surrounding the organism as well as the mode of uptake (oral, dermal, or pulmonary). Development of standard procedures in terrestrial ecotoxicity has lagged behind aquatic toxicology. This is due to problems of simulating highly variable exposure conditions and the belief that toxicants are less of a hazard in the terrestrial environment owing to restricted mobility resulting in reduced exposure levels. There are a number of standard procedures for testing small mammals under laboratory conditions. Although designed to determine human toxicity hazards, the results of these tests should be included in an assessment of terrestrial ecotoxicity.

Owing to their high mobility and varied feeding habits, birds are an important part of an ecotoxicity testing program. Waterfowl are often key indicator organisms in biological residue monitoring, since they feed on aquatic organisms which accumulate high levels of certain toxicants. Tests are presented which measure direct oral, subchronic dietary, and dermal toxicity. Reproductive effects are measured in a sublethal chronic test and field exposure conditions simulated in small and large pen studies. The latter two are appropriate for testing of pesticides or other substances which are applied to large areas of natural habitat.

18
Pumps

INTRODUCTION

Transporting liquids to and from process equipment is an integral part of water treatment and distribution technology. Energy requirements depend on the height through which the fluid is mixed, the length and diameter of the transporting conduits, the rate of flow, and the fluid's physical properties (in particular, viscosity and density). In some applications, external energy for transferring fluids is not required. For example, when liquid flows to a lower elevation under the influence of gravity, a partial transformation of the fluid's potential energy into kinetic energy occurs. When transporting fluids through horizontal conduits, especially to higher elevations within a system, mechanical devices such as pumps are employed.

Methods for transporting fluids between process equipment include

- Centrifugal force inducing fluid motion
- Volumetric displacement of fluids, either mechanically or with other fluids
- Mechanical impulse
- Transfer of momentum from another fluid
- Electromagnetic forces
- Gravity induced

The first four methods are described in this chapter and emphasis is placed on mechanical devices for transporting incompressible fluids, namely, pumps.

CLASSIFICATIONS AND CHARACTERISTICS

Major types of pumps used in process plant applications are centrifugal, axial, regenerative turbine, reciprocating, metering, and rotary. These classes are grouped under one of two categories: dynamic pumps or positive displacement pumps.

Dynamic pumps include centrifugal and axial types and are operated by developing a high liquid velocity, which is converted to pressure in a diffusing flow passage. These pumps generally are lower in efficiency than the positive displacement types. However, they do operate at relatively high speeds, thus providing high flowrates in relation to the physical size of the pump. Furthermore, they usually have significantly lower maintenance requirements than positive displacement pumps.

Positive displacement pumps operate by forcing a fixed volume of fluid from the inlet pressure section of the pump into the pump's discharge zone. This is performed intermittently with reciprocating pumps. In the case of rotary screw and gear pumps, the action is continuous. Pumps in this category operate at lower rotating speeds than do dynamic pumps. A positive displacement pump also tends to be physically larger than an equal-capacity dynamic pump.

Four characteristics describe all pumps:

- *Capacity*, Q (m^3/s): The quantity of liquid discharged per unit time.
- *Head*, H (m): The energy supplied to the liquid per unit weight, obtained by dividing the increase in pressure by the liquid per unit weight, obtained by dividing the increase in pressure by the liquid specific energy. This specific energy is determined by the Bernoulli equation. Head may be defined as the height to which 1 kg of discharged liquid can be lifted by the energy supplied by a pump. Therefore, it does not depend on the specific weight, Y (kg/m^3), or density, ρ (kg/m^3), of liquid to be pumped.
- *Power*, N (kgf-m/s): The energy consumed by a pump per unit time for supplying liquid energy in the form of pressure. Power is equal to the product of specific energy, H, and the mass flowrate, YQ:

  $$N + \gamma QH = \rho gQH \tag{1}$$

  Effective power, N_e, is larger than N because of energy losses in a pump. Its relative value is evaluated by the pump efficiency, η_p,

  $$N_e = \frac{N}{\eta_p} = \frac{\rho gQH}{\eta_p} \tag{2}$$

- *Overall efficiency*, η: The ratio of useful hydraulic work performed to the actual work input. This parameter characterizes the perfection of design and performance of a pump. The value of η reflects the relative power losses in the pump and is expressed by the following equation:

$$\eta = \eta_v \times \eta_h \times \eta_m \tag{3}$$

where η_v is the volumetric efficiency defined as the ratio of liquid actually pumped to the liquid that theoretically should be discharged. In other words, it indicates the percentage of losses (or slip). In practice, slip should not exceed 5%. The hydraulic efficiency η_h is defined as the ratio of the actual head pumped to the theoretical head:

$$\eta_h = \frac{H}{H + \text{Hydraulic Losses}} \tag{4}$$

Hydraulic losses are those head losses in the suction and discharge sections of a pump. In the suction end, these losses comprise

- Velocity head
- Entrance head
- Friction head in the suction line
- Losses in bends and losses in suction valves

Discharge line losses include

- Losses in the discharge valves
- Velocity head
- Friction in the discharge piping

The mechanical efficiency, η_m, is the relation between the indicated pump horsepower and the actual power input from the drive. It characterizes mechanical losses in the pump; for example, in bearings and stuffing boxes.

Overall efficiency, η, depends on the pump design and varies from 50% for small pumps to about 90% for large sizes. The power consumed by a motor (defined as the nominal power of a motor), N_m, exceeds brake power by mechanical losses incurred in transmission and in the motor itself. These losses are accounted for in Equation (5) by including the efficiencies of a transmission, η_{tr}, and a motor, η_m:

$$N_m = \frac{N_e}{\eta_m \times \eta_{tr}} = \frac{N}{\eta_e \times \eta_m \times \eta_{tr}} \tag{5}$$

The product of η_e, η_m, and η_{tr} is the total efficiency of a pump and may be defined as the ratio of hydraulic power to the motor's nominal power:

$$\eta = \frac{N}{N-m} = \eta_p \, \eta_{tr} \, \eta_{mot} \tag{6}$$

From Equations (5) and (6), therefore, the total efficiency of a pump may be expressed by the product of five values:

$$\eta = n_v \times \eta_h \times \eta_m \times \eta_{tr} \times \eta_{mot} \quad (7)$$

The actual power of a pump motor, N_A, should be based on the energy required to overcome the fluid's inertia at startup so as to avoid overloading the unit.

$$N_A = \beta N_m \quad (8)$$

Coefficient β is determined from the size of the motor. Typical values are given in Table 1.

HEAD AND SUCTION HEAD

Head, H, characterizes the excessive energy (l = gH) added to 1 kg of liquid in a pump. Figure 1 illustrates a simple pumping scheme for which we will write the Bernoulli equation.

At suction (sections 1-1 and 1′1′)

$$\frac{p_1}{\rho g} + \frac{w_1^2}{2g} = H_s + \frac{w_s^2}{2g} + \frac{p_s}{\rho g} + h_{ls} \quad (9)$$

At discharge (sections 1′-1′ and 2-2)

$$\frac{p_d}{\rho g} + \frac{w_s^2}{2g} = H_d + \frac{w_2^2}{2g} + \frac{p_2}{\rho g} + h_{ld} \quad (10)$$

where

w_1 and w_2 = liquid velocities in tanks 1 and 2, respectively
w_s and w_d = liquid velocities in the suction and discharge pump nozzles
h_{ls} and h_{ld} = head losses in the suction and discharge pipings

If velocities $w_1 = 0$ and $w_2 = 0$ are assumed, then the total dynamic head, H, of a pump is the difference between discharge head, H_d, and suction head in the nozzles:

$$H = \frac{p_d - p_s}{\rho g} \quad (11)$$

From Equations (9) and (10), we obtain

$$H = \frac{p_2 - p_1}{\rho g} + \frac{w_s^2 - w_d^2}{2g} + H_d + H_s + H_{ld} + H_{hls} \quad (12)$$

Table 1 Typical Values of Coefficient as a Function of Motor Power

N_A (kW)	<1	1–5	5–50	>50
β	2–1.5	1.5–1.2	1.2–1.15	1.1

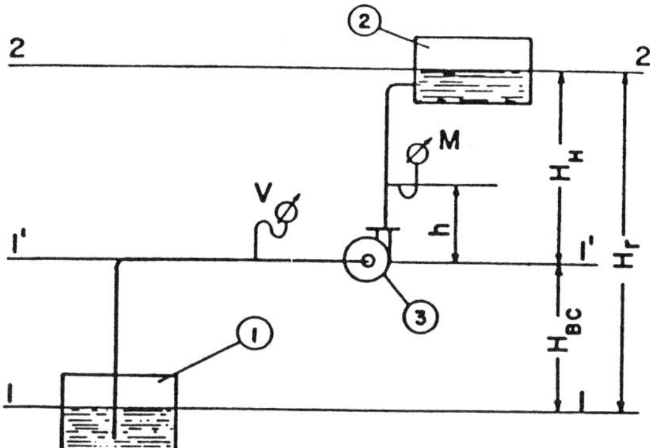

Figure 1 Scheme of pumping unit: (1,2) tanks; (3) pump; (M) manometer; (V) vacuum gauge.

If the suction and discharge nozzles of a pump have the same diameter, then

$$w_d = w_s$$

and

$$H_t = H_d + H_s \qquad \text{and} \qquad h_{ld} + h_{ls} = h_{lt}$$

Equation (12) then simplifies to

$$H = H_t + \frac{p_2 - p_1}{\rho g} + H_{lt} \tag{13}$$

Equation (13) states that the total head is expended in (1) lifting the liquid to height H_t; (2) overcoming the pressure difference in tanks 1 and 2; and (3) overcoming the hydraulic resistances in the suction and discharge pipings. If $p_1 = p_2$, Equation (13) becomes

$$H = H_t + h_{lt} \tag{14}$$

For a horizontal section of piping, $H_T = 0$,

$$H = \frac{p_2 - p_1}{\rho g} + h_{lt} \tag{15}$$

In the case in which $p_1 = p_2$ and $H_T = 0$,

$$H = h_{lt}$$

The total head of the operating pump may be evaluated from measurements obtained from the pressure p_m and vacuum gauges p_v.

Total suction head can be obtained from the measurement H_{sg} on the gauge on the pump's suction nozzle (corrected to the pump centerline and converted to feet of liquid) plus the barometer reading in feet of liquid and the velocity head, H_{vs}, (ft) at the point of gauge attachment:

$$H_s = H_{sg} + atm + H_{vs} \tag{17}$$

When the static pressure at the suction flange is less than atmospheric, the measurement obtained from a vacuum gauge replaces H_{sg} in Equation (17) and is assigned a negative sign.

Total discharge head, H_d, is obtained from a gauge, H_{dg}, at the pump's discharge flange (corrected to the pump centerline and converted to feet of liquid) plus the barometer reading and the velocity head, H_{vg}, at the point of gauge attachment:

$$H_d = H_{dg} + atm + H_{vg} \tag{18}$$

Again, if the discharge gauge pressure is below atmospheric, the vacuum gauge measurement replaces H_{dg} with a negative sign. Before installation, it is possible to estimate the total discharge head from the static discharge head, H_{sd}, and discharge friction, H_{fd}, as follows:

$$H_d = H_{sd} + H_{fd} \tag{19}$$

Static suction head, H_{ss}, is defined as the vertical distance (ft) between the free level of the source of supply and the pump centerline plus the absolute pressure at this level (converted to feet of liquid).

Total static head, H_{ts}, is the difference between discharge and suction static heads.

The suction generated by a pump is derived from a pressure difference between the suction source, p_1, and the pump, p_s, or is due to the action of the head difference

$$\frac{p_1}{\rho g} - \frac{p_s}{\rho g}$$

Suction height can be determined as follows:

$$H_s = \frac{p_1}{\rho g} - \left(\frac{p_s}{\rho g} + \frac{w_s^2 - w_1^2}{2g} + H_{ls}\right) \tag{20a}$$

or

$$H_s = \frac{p_1}{\rho g} - \left(\frac{p_s}{\rho g} + \frac{w_s^2}{2g} + H_{ls}\right) \tag{20b}$$

because $w_1 = 0$. These expressions demonstrate that the suction head increases with p_1 and decreases with increasing w_s and h_{ls}.

If liquid is pumped from an open tank, that is $p_1 = p_a$ (where p_a corresponds to atmospheric), the suction pressure, p_s must exceed pressure p_t (the pressure of the saturated vapor of the liquid at the pumping temperature [$p_s > p_t$]); otherwise the fluid begins to boil. When the pumped liquid vaporizes, the suction head goes to zero at the limit and flow stops. Consequently,

$$H_s \leq \frac{p_s}{\rho g} - \left(\frac{p_t}{\rho g} + \frac{w_s^2}{2g} + h_{ls}\right) \tag{21}$$

This expression shows that the suction head is a function of atmospheric pressure, fluid velocity and density, temperature (and correspondingly the liquid's vapor pressure) and the hydraulic resistance of the suction piping. When pumping from an open tank, the suction head cannot exceed the head of pumping liquid, which corresponds to atmospheric pressure (the value of which depends on the height of the pump installation above a specified datum, normally sea level). Thus, for example, if water is pumped at t = 20°C, the suction head cannot exceed 10 m at sea level. If the same pumping system is used at an elevation of 2000 m, the suction head cannot be greater than 8.1 m, which corresponds to the atmospheric pressure in meters of water column.

At temperatures approaching the boiling point of the liquid, the suction head becomes zero:

$$H_s = 0 \text{ at } \frac{p_a}{\rho g} = \frac{p_1}{\rho g} + \frac{w_s^2}{2g} + h_{ls}$$

In this situation, the pump must be installed below the suction line to provide a back liquid. This method is also used for pumping high-viscosity liquids. In addition to evaluating the friction head and local resistance losses, inertia losses (for piston pumps), H_i, and the effect of cavitation (for centrifugal pumps), h_k, must be accounted for in the overall suction head term.

Head losses due to overcoming inertia forces, H_i (in piston pumps), may be estimated by an expression that relates the pressure acting on the piston to the inertia force of a liquid column moving in the suction piping:

$$H_i = \frac{6 l f u^2 r}{20 g f_1} \tag{22}$$

where

l = height of liquid column in the piping (for pumps having a gas chamber—the distance between pump centerline and the liquid level in the chamber)
g = acceleration due to gravity
f and f_1 = cross-sectional areas of the piston and piping, respectively
u = circumferential crank velocity
r = crank radius

CAVITATION

Cavitation in centrifugal pumps arises from high velocities or when handling hot liquids under conditions of vaporization. It is a phenomenon caused by the formation and collapse of vapor cavities existing in a flowing liquid. Vapor cavities can form at any point in the fluid at which the local pressure approaches that of the liquid vapor pressure (at the operating temperature). At these positions, a portion of the liquid vaporizes to form bubbles or cavities of vapor. Low-pressure zones are generated in several ways: (1) by a local increase in velocity resulting in eddies or vortices near the boundary contours; (2) by rapid vibration of the boundary; (3) by separating or parting of the liquid due to water hammer; or (4) by an overall reduction in static pressure. Collapse of the bubbles initiates when they move into regions where the local pressure is higher than the vapor pressure. This often results in objectionable noise and vibration, as well as extensive erosion or pitting of the boundary materials in the immediate vicinity. Even more important, cavitation results in a decrease in pumping performance and efficiency. A dimensionless parameter called the cavitation number, σ_c, is used to correlate performance:

$$\sigma_c = \frac{p - p_v}{\rho w^2/2g_c} \tag{23}$$

where

p = static pressure (absolute) in the undisturbed flow (lb_f/ft^2)
p_v = liquid vapor pressure (absolute) (lb_f/ft^2)
ρ = liquid density (lb/ft^3)
w = free-stream velocity of the liquid (fps)
g_c = conversion factor (32.17 lb ft/lb_f s^2)

The physical significance of the cavitation number is the ratio of the net static pressure available to collapse a bubble to the dynamic pressure available to initiate bubble formation. The value of this dimensionless group at conditions of incipient cavitation, σ_{ci}, is dependent on the pump or equipment geometry.

A cavitation correction factor may be determined from the following correlation:

$$H_c = 0.019 \frac{(Qn^2)^{2/3}}{H} \tag{24}$$

where

Q = pump capacity (m^3/S)
n = number of revolutions (s^{-1})
H = fluid head on pump (m)

In practice, to avoid cavitation the suction head for pumping liquids with physical properties close to those of water should exceed the values in Table 2.

CENTRIFUGAL PUMPS

Types and Applications

Centrifugal pumps are used extensively because of their simplicity in design, low initial cost and maintenance, and flexibility of application. This type of pump is available in a wide range of sizes: capacities ranging from a few gallons per minute up to 100,000 gpm and discharge heads (pressures) ranging from a few feet up to several thousand pounds per square inch. Basically, a centrifugal pump consists of an impeller, which is a series of radial vanes of various shapes and curvatures rotating within a circular casing. Figure 2A illustrates the operation. The liquid from suction piping (1) enters at the axis of rotating impeller (2) into the pump chamber (3) and is thrown outwards by centrifugal action against the blades (4). The impeller's high speed of rotation causes the liquid to acquire kinetic energy. A pressure difference between the suction and discharge sides of the pump is produced by the conversion of kinetic energy of the liquid flow into pressure energy in the discharge piping. A reduction in pressure occurs at the entrance of the impeller, and the liquid is fed continuously into the pump from a supply tank. Without filling the pump chamber with liquid, the impeller cannot produce an adequate pressure difference, which is necessary for lifting liquid in the suction line.

Figure 2B shows a heavy-duty end-suction centrifugal pump used to pump large quantities of nonaggressive liquids. Pumps of this type are employed for general service applications in industry, irrigation, and municipal water supplies. This design is generally all iron or bronze-fitted construction with replaceable shaft sleeves and wear rings. A separate cover is usually provided, and the casing has integrally cast support feet. Ball bearings are oil lubricated.

Most centrifugal pumps are not self-priming and, hence, cannot evacuate vapor from the suction line, so that liquid can flow into the pump casing without external assistance. The impellers on centrifugal pumps are designed especially for efficient pumping and are not operated at high enough tip speeds to convert them into vapor compressors. The differential head that the pump impeller can deliver

Table 2 Typical Suction Heat Limits to Avoid Cavitation

Temperature (°C)	10	20	30	40	50	60	65
Suction head (m)	6	5	4	3	2	1	0

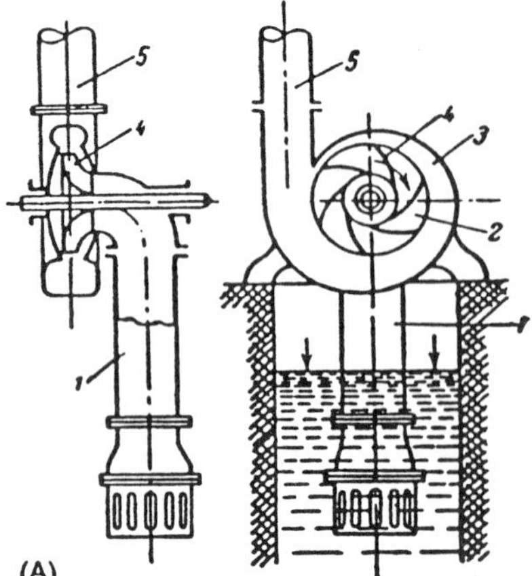

Figure 2 (A) Operating scheme of a centrifugal pump: (1) suction piping; (2) impeller; (3) casing; (4) vanes; (5) delivery piping. (B) Horizontal end-suction centrifugal pump.

is the same on the vapor as on the liquid. However, the equivalent differential pressure rise capability is typically much lower with vapor. To prime a centrifugal pump, both the suction line and pump casing must be filled with liquid. When the suction source is at positive pressure or is positioned above the pump, priming is accomplished by opening the suction valve and venting the trapped vapor from a valve connection on the pump casing or discharge line. Liquid then flows into the suction line and pump casing to displace the escaping vapor.

Centrifugal pumps are used in a multitude of applications throughout the chemical industry. Designs may consist of an open impeller system mounted on an externally adjustable shaft for handling clear liquids, slurries, or liquids with suspended solids. Also, they can be closed impellers for pumping clear liquids or light slurries. Figure 3 shows a cross section of a process pump.

Features of a process pump include

- Pumping casing: back pull-out, sell-venting top centerline discharge, vortex suppressing guide vane in suction nozzle, 1/8-in. corrosion allowance, rugged integral casted centerline supports to allow thermal or pressure expansions without causing misalignment and shank deflection.
- Positive alignment: positively and permanently achieved by full-circle registered fit on all mating parts. All such fits are away from liquid being pumped, preventing crevice corrosion.
- Fully confined gaskets: on wet and dry side of casing cover, as well as between impeller nut, impeller, and shaft sleeve, provide safety against leakage.

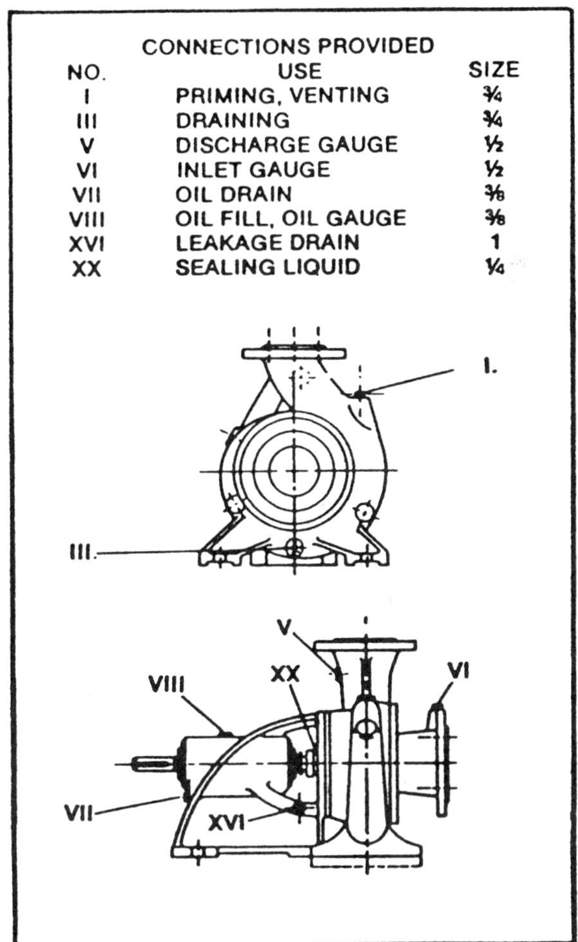

CONNECTIONS PROVIDED

NO.	USE	SIZE
I	PRIMING, VENTING	¾
III	DRAINING	¾
V	DISCHARGE GAUGE	½
VI	INLET GAUGE	½
VII	OIL DRAIN	⅜
VIII	OIL FILL, OIL GAUGE	⅜
XVI	LEAKAGE DRAIN	1
XX	SEALING LIQUID	¼

(B)

- Enclosed Impeller: for high efficiency and low NPSH, with back vanes for axial thrust balancing and low stuffing box pressure, keeping erosive impurities out of shaft seal area, keyed to shaft for positive fastening. Positioned by acorn-type impeller nut with Heli-Coil lock insert, it cannot come loose under reserve rotation.
- Bearing frame: five sizes or 27 models from 1.5- to 12.0-in. discharge. Designed to carry maximum load with under 0.02-in. shaft deflection and at least 2-year bearing life.
- Built-in casing heating or cooling jacket: increases the application scope

PART NO.	PART NAME	STD. MATERIAL	BRONZE FITTED
1	CASING	CAST IRON	CAST IRON
2	IMPELLER	CAST IRON	BRONZE
6	SHAFT	CARBON STEEL	CARBON STEEL
9	SUCTION COVER	CAST IRON	CAST IRON
13	PACKING	ASBESTOS	ASBESTOS
14	SHAFT SLEEVE	CAST IRON	BRONZE
16	BALL BEARING—INBOARD	STEEL	STEEL
17	GLAND	CAST IRON	CAST IRON
18	BALL BEARING—OUTBOARD	STEEL	STEEL
19	FRAME	CAST IRON	CAST IRON
24	IMPELLER NUT	STEEL	STEEL
24A	LOCKING PLATE	STEEL	STEEL
25	WEAR RING—SUC. COVER	CAST IRON	BRONZE
27	WEAR RING—STUFF. BOX	CAST IRON	BRONZE
29	LANTERN RING	CAST IRON	CAST IRON
32	IMPELLER KEY	STEEL	STEEL
35	BEARING COVER—INBOARD	CAST IRON	CAST IRON
37	BEARING COVER—OUTBOARD	CAST IRON	CAST IRON
40	DEFLECTOR	STEEL	STEEL
47	OIL SEAL—INBOARD	FELT	FELT
47A	CAP—INBOARD SEAL	STEEL	STEEL
49	OIL SEAL—OUTBOARD	FELT	FELT
49A	CAP—OUTBOARD SEAL	STEEL	STEEL
73A	GASKET—COVER	ASBESTOS	ASBESTOS
73B	GASKETS—BEAR. COVER	PAPER	PAPER
143	OIL GAUGE	STEEL	STEEL

CAST IRON PARTS ARE ASTM A-48-35.

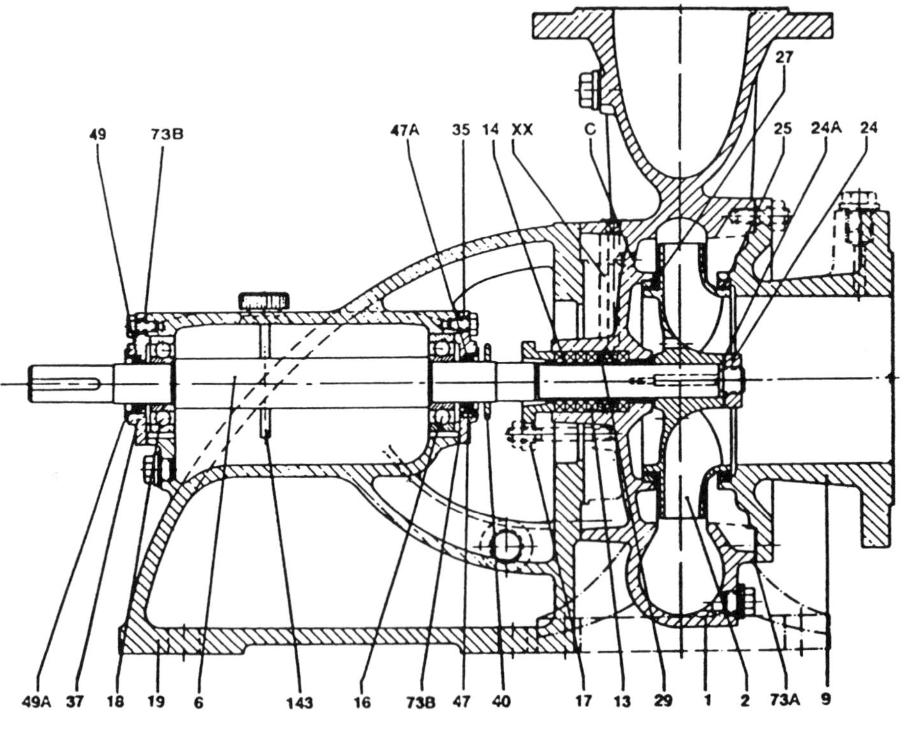

Figure 3 Cross-sectional drawing of the pump.

without need of added costs for optional or extra parts. Special intensive coding of stuffing box only is also available.
- Part interchangeability.
- Stuffing box: can be adjusted to particular application requirements. Replaceable shaft sleeve is provided with stuffing box packing or mechanical seal, providing maximum adaptability for various seal designs. Step design shaft sleeve simplifies accurate seal setting or long seal life.

Most pumps are built with casings cast in high-cost alloys. Casings frequently are foot supported or bearing bracket supported rather than centerline supported. Pumps are available in a wide range of operating conditions but most often are limited to low to moderate flows.

Many centrifugal pumps have single casings; that is, a single wall between the liquid under discharge pressure and the atmosphere. Double casings are used in horizontal, multistage, high-pressure pumps and in vertical pumps. In the former, a heavy barrel-shaped casing surrounds the stack of stage diaphragms. The stack of diaphragms constitutes the inner casing, whereas the barrel forms the outer casing. This type of arrangement is used most often in boiler feed pumps.

Casings may be joined on the same plane as the shaft axis (called axially split) or perpendicular to the shaft (called radially split). Axially split horizontal pumps most commonly are referred to as horizontally split.

Radial split horizontal pumps are commonly called vertically split. Radial joining is used on horizontal overhung pumps to allow ready removal of the rotor-and-bearing bracket assembly for maintenance. This design configuration also is employed in high-pressure multistage pumps because of structural problems associated with bolting together the halves of axially split casings exposed to high internal pressure.

The term single-stage overhung refers to the impeller mounting/support arrangement. The casings for these designs are supported at the centerline. Two shaft bearings are mounted close together in the same bearing bracket, with the impeller cantilevered or overhung beyond them. Normally, this type configuration utilizes top suction and discharge flanges, wearing rings both on the front and back of the impeller and casing, a single-suction closed impeller, and a single stuffing box fitted with a mechanical seal. These pumps are well suited to high-temperature operations and can be used for handling flammable liquids.

A two-stage overhung pump is a modification of the single-stage process pump and is capable of higher head. Usually the stuffing box pressure is approximately halfway between suction and discharge pressures.

Multistage centrifugal pumps generally are used for generating higher heads (pressures) than can be obtained by single-stage pumps. These pumps are available for pressures as high as 3,000 lb/in.^{2} at capacities greater than 3,000 gpm. The operation of this type pump is illustrated in Figure 4. As shown, the designs have impellers (A) in one aggregate casing (B), which are located in series on one shaft (C). Liquid discharged from the first impeller enters through the offtake (D) in the second impeller, where it acquires additional energy from the second impeller through the offtake in the third impeller, and so on. Thus, multistage pumps may be thought of as several single-stage pumps on one shaft, with the flow in series. Hence, the total head developed is the head of one impeller multiplied by the number of impellers (usually designs do not exceed five impellers).

Multistage pumps are employed in a multitude of processing applications. Examples include hydrocarbon processing and refining, boiler feed operations, descaling operations, mine dewatering, and hydraulic power recovery, in which excess plant energy is recovered to drive other equipment.

Vertical pumps are another orientation used widely in the process industries. In this type, a vertical cylinder buried in the ground houses the pumping element. Suction liquid enters the outer cylinder, flows to the bottom, and then up through the pumping element stages. The diaphragms of the stages in the pumping element constitute the inner casing. As *inline pumps*, the casings are designed to be bolted directly to the piping, much like a valve. Two basic configurations of inline pumps are coupled and close coupled. Service life and maintenance requirements for both are about the same. The total head is the sum of the discharge pressure measured at the discharge nozzle above the floor plate, the velocity head at the same location, and the vertical distance from the centerline of the pressure gauge to the liquid surface in the sump. These pumps generally are equipped with tail pipes, which allow pumping down below normal liquid level. The pump must be capable of operating under pumpdown conditions without allowing the throttle bushing to run dry.

Vertical multistage pumps can have 24 or more stages. High specific-speed

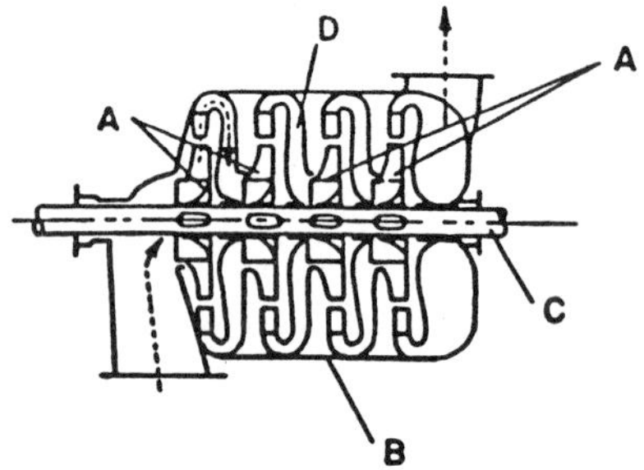

Figure 4 Operation of a multistage centrifugal pump: (A) impeller; (B) casing; (C) shaft; (D) offtake.

impellers often are used. The first stage is usually at the bottom of the assembly, below grade. These pumps require a large number of close-running clearances and, thus, are sensitive to damage by solids ingestion and by dry or two-phase flow conditions. This type of pump is employed in a broad range of applications. Examples include (1) fossil power plants, where they are used for condensate service in large power generating plants; (2) nuclear power plants, where they are used in condensate and feedwater heater drain service; and (3) desalination whose operating facilities require large-capacity pumping equipment that must perform with low net positive suction head (NPSH) available.

"Can" pumps are motor pump units with the rotating rotor and impeller housed entirely within a pressure casing. This type design eliminates the need for a stuffing box. The pumped fluid serves both as a lubricant for bearings and as a coolant for the motor. Designs typically are limited to low-flow, low-pressure, and low-temperature services.

Basic Equations for Centrifugal Machines

In centrifugal pumps, the liquid flows along the surface of the impeller vanes while the tip moves relative to the casing of the pump. To develop an expression of the virtual head developed by a centrifugal pump, we shall assume the path followed by a volume of liquid as it passes through the pump in relation to a stationary impeller, with the fluid having the same relative velocity as an actual rotating impeller. Figure 5 defines the system under consideration.

Defining C_1 and C_2 as the vector sums of the relative and tangential velocities of the fluid entering and exiting the impeller, respectively, then the relative velocity components along the vanes are w_1 and w_2, and the tangential components (tangent to the circumference of rotation) are u_1 and u_2. Further, a reference datum is defined to be the surface of the impeller in Figure 5. We then

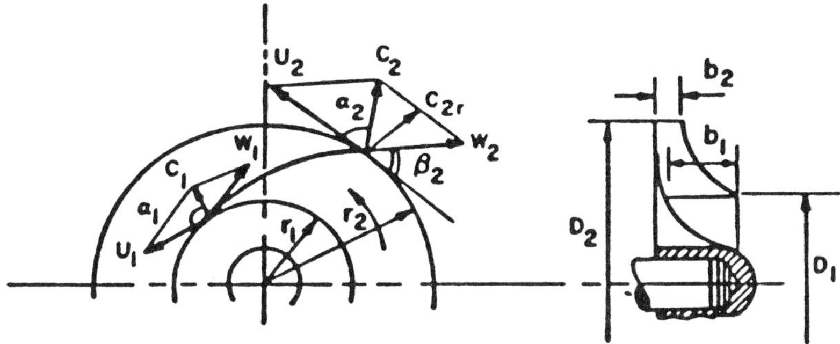

Figure 5 Derivation of equations for centrifugal machines.

may write an energy balance for the fluid passing through the impeller at $Z_1 = Z_2$:

$$\frac{p_1}{\rho g} + \frac{w_1^2}{2g} = \frac{p_2}{\rho g} + \frac{w_2^2}{2g} \tag{25}$$

When the impeller rotates, the liquid obtains an additional energy, A, which is derived from the work of centrifugal force along the path $r_2 - r_1$. Hence,

$$\frac{p_1}{\rho g} + \frac{w_1^2}{2g} = \frac{p_2}{\rho g} + \frac{w_2^2}{2g} - A \tag{26}$$

The centrifugal force, C, acting on the liquid particle of mass, m, is

$$C = m\omega^2 r = \frac{G}{g} 2r \tag{27}$$

where

G = weight of fluid particle
ω = angular velocity
r = moving radius of particle rotation

The work, A_g, derived from the centrifugal force by displacement of this fluid particle along the path $r_2 - r_1$ is determined as follows:

$$A_g = \int_{r_1}^{r_2} \frac{G}{g}\omega^2 r\, dr = \frac{G\omega^2}{2g}(r_2^2 - f_1^2) = \frac{G}{g}\left(\frac{u_2^2 - u_1^2}{2}\right) \tag{28}$$

The specific work per unit weight of liquid is equal to the specific energy obtained by the fluid in the pump:

$$A = \frac{u_2^2 - u_1^2}{2g} \tag{29}$$

Substituting Equation (29) into Equation (26) we obtain

$$\frac{p_2 - p_1}{\rho g} = \frac{w_1^2 - w_2^2}{2g} + \frac{u_2^2 - u_1^2}{2g} \tag{30}$$

The heads of liquid at the inlet and outlet from the pump are

$$H_1 = \frac{p_1}{\rho g} + \frac{c_1^2}{2g};\ H_2 = \frac{p_2}{\rho g} + \frac{c_2^2}{2g} \tag{31}$$

Hence, the head of the pump is equal to the difference of heads between the pump's inlet and outlet:

$$H_T = H_1 - H_2 = \frac{p_2 - p_1}{\rho g} + \frac{c_2^2 - c_1^2}{2g} \quad (32)$$

Substituting $(p_2 - p_1)/\rho g$ from Equation (30) into Equation (32) we obtain

$$H_T = \frac{w_1^2 - w_2^2}{2g} + \frac{u_2^2 - u_1^2}{2g} + \frac{c_2^2 - c_1^2}{2g} \quad (33)$$

From the geometry of Figure 5 we have

$$w_1^2 = u_1^2 + c_1^2 - 2u_1c_1 \cos \alpha_1; \; w_2^2 = u_2^2 + c_2^2 - 2u_2c_2 \cos \alpha_2 \quad (34)$$

Substituting the last set of expressions into Equation (33) results in an equation for the virtual head of a centrifugal pump:

$$H_t = \frac{u_2c_2 \cos\alpha_2 - u_1c_1 \cos \alpha_1}{g} \quad (35)$$

This equation represents the theoretical maximum head that could be developed for a specified set of operating conditions. Note that the liquid entering the pump usually moves along the impeller in the radial direction. In this case, the angle between the absolute velocity value of the liquid entering the impeller and the tangential velocity is $\alpha_1 = 90$ degrees, which corresponds to the liquid entering the impeller without any shock. Equation (35) then simplifies to

$$H_T = \frac{u_2c_2 \cos \alpha_2}{g} \quad (36)$$

From Figure 5, we may write

$$c_2 \cos \alpha_2 = u_2 - w_2 \cos \beta_2$$

Hence

$$H_T = \frac{u_2^2}{g}\left(1 - \frac{w_2}{u_2} \cos \beta_2\right) \quad (37)$$

From the width of the impeller, b, the length of the circumference, $2\pi r_2$, and the cross section of the flow leaving the impeller, $2\pi r_2 b$, the quantity of liquid being pumped is

$$V = 2\pi r^2 b w_2 \sin \beta_2 \quad (38)$$

Hence

$$w_2 \cos\beta_2 = \frac{V}{2\pi r_2 \tan \beta_2}$$

Substituting Equation (39) into Equation (37), we obtain the following relationship between the head, H, and the volumetric flowrate through the pump, V.

$$H = \frac{u_2^2}{g} - \frac{V}{g(2\pi r_2 b)\tan \beta_2} \tag{40}$$

For a given speed of rotation there is a linear relation between the head developed and the rate of flow. This relationship is illustrated for different vane configurations in Figure 6.

If the outlet vane angles are inclined backwards, β_2 is less than 90 degrees; hence, tan β is positive and, therefore, the head decreases as the throughput increases (curve A, Fig. 6). If β_2 is greater than 90 degrees, that is, the outlet vane is inclined forward, the head increases at higher throughputs (curve B, Fig. 6). Radial vanes provide a constant head (curve C, Fig. 6). When the flow is zero, $V = 0$, then regardless of the vane angle our head expression is

$$H_T = \frac{u_2^2}{g} \tag{41}$$

Figure 6 shows that the maximum theoretical head achievable in a centrifugal pump is with vanes curved in the direction of rotation of the impellers, whereas the minimum occurs with vanes curved in the opposite direction. However, pumps are fabricated with angles $\beta_2 < 90$ degrees because an increase in β increases hydraulic losses and decreases the hydraulic pump efficiency.

The actual head is always less than virtual head, for the following main reasons:

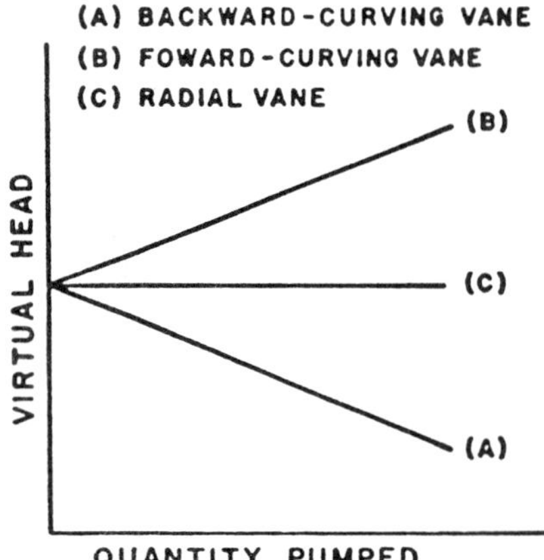

Figure 6 Plot of virtual head versus flow capacity for different outlet vane angles.

- The fluid circulating in the spaces between the vanes forms eddies.
- Frictional losses occurring in the suction port, the impeller, and the discharge nozzle increase with pump speed and liquid viscosity.
- Losses occur as the liquid is discharged from the impeller. The vane angles are correct only for the designed head and throughput. Deviation from these conditions causes an increase in the losses due to turbulence.
- Leakage reduces the head developed, especially at low discharge rates.

The actual head is equal to

$$H = H_T \eta_h \epsilon \tag{42}$$

where η_h is the hydraulic efficiency (with typical values between 0.8 and 0.95) and ϵ is a coefficient that accounts for the number of vanes, $\epsilon = 0.6–0.8$.

Referring back to Figure 5 the throughput, Q, corresponds to the liquid discharge through the channels between the impeller vanes having widths b_1 and b_2:

$$Q = b_1(\pi D_1 - \delta z')C_{1,r} = b_2(\pi D_2 - \delta z')C_{2,r} \tag{43}$$

where

δ = vane thickness
z' = number of vanes
b_1, b_2 = width of impeller at the internal and external circumferences, respectively
$C_{1,r}, C_{2,r}$ = radial components of absolute velocities at the inlet and outlet of the impeller ($C_{1,r} = C_1$)

The output and head of a centrifugal pump depend on the number of revolutions per unit time made by the impeller. As follows from Equation (43), throughput is directly proportional to the radial component of the absolute velocity at the exit from the impeller; that is, $Q \propto C_{2,r}$. If the number of revolutions is changed from n_1 to n_2 (thus changing from Q_1 to Q_2 correspondingly), the trajectories of the motion of liquid particles remain unaltered and the velocity parallelograms at any corresponding points will be geometrically similar. This geometry is illustrated in Figure 6. Consequently,

$$\frac{Q_1}{Q_2} = \frac{C'_{2\rho}}{C''_{2r}} = \frac{u'_2}{u''_2} = \frac{\pi D_2 n_1}{\pi D_2 n_2} = \frac{n_1}{n_2} \tag{44}$$

From Equation (37), the head is proportional to the square of circumferential velocity; that is,

$$\frac{H_1}{H_2} = \left(\frac{u'_2}{u''_2}\right)^2 = \left(\frac{n_1}{n_2}\right)^2 \tag{45}$$

The power developed by the pump is proportional to the product of volumetric flowrate, Q, and head, H. From Equations (2), (44), and (45) we then have

$$\frac{N_1}{N-2} = \left(\frac{n_1}{n_2}\right)^3 \tag{46}$$

Equations (44–46) are the *equations of proportionality* for centrifugal machines. These expressions state the following:

- A change in the number of impeller revolutions (from n_1 to n_2) causes the pump throughput to change in a manner that is directly proportional.
- The heads of the two systems are proportional to the number of revolutions raised to the squared power.
- The powers developed by pumps are proportional to the number of revolutions to the third power.

In a practical sense, these relationships do not hold rigorously. The proportionality between pump parameters is not strictly maintained when the number of revolutions is changed by more than a factor of two.

The above theoretical treatment assumes such factors as the absence of friction and eddies. In real pumps, all these complications exist and influence pump parameters. Now we shall examine commerical pump behavior, with attention given to backward-curved vanes. This type impeller configuration is used most extensively.

The main parameters that reduce the virtual head of a centrifugal pump (using backward-curved vanes) to its developed head are summarized in Figure 7. The dashed line represents the "virtual head," and thus defines the theoretical characteristics of this type of impeller, $\beta_2 < 90$ degrees. Physically, the virtual

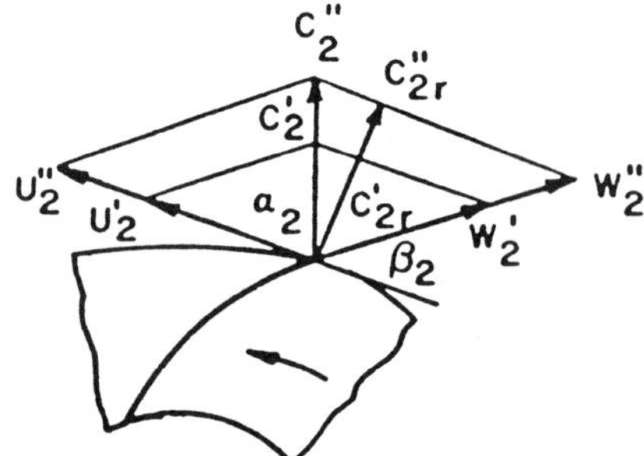

Figure 7 Virtual head curve showing that if the number of impeller revolutions may change, similarity of the velocity parallelograms is maintained.

head line represents an infinite number of vanes and only infinitesimal amounts of liquid exist between them. In an actual pump, finite volumes of liquid exist between the vanes and inertia effects arise. Consequently, the portions of liquid near the impeller periphery have a greater velocity than inside the impeller. This velocity creates circulation across the space between vanes. Hence, inertial effects cause the virtual head curve in Figure 7 to shift downward.

Now let us assume that the liquid velocity through the pump is increased but that the same number of impeller rotations is maintained. (This velocity increase can be accomplished by increasing a valve opening ahead of the pump.) The head in this case will decrease because of an increase in friction resistance. These losses are compensated for partially by a decrease in leakage, which a low throughputs is larger than at a high throughput. As shown in Figure 7, this counterbalancing effect is not that significant. The last factor noted in Figure 7 is responsible for decreasing head is turbulence or eddy formation on vanes. All these factors result in the "developed head curve," which approaches zero on attaining a maximum throughput. Such a curve must be constructed from experimental data.

Returning for a moment to the ideal pump, a maximum volumetric throughput corresponding to zero head, $H = 0$, exists for a constant n. Conversely, a maximum head exists for the case of no flow. In real pumps, however, there is always some head at the maximum flow and some small flow at the pump's maximum head. The behavior of real pumps between these extremes is best explained by performance curves. Note that the product of the developed head (in units of pressure) and volumetric flowrate represents the power absorbed by the pumped fluid. Because the head approaches zero at the maximum flowrate, power first increases from zero; that is, at $V = 0$, to a maximum and then decreases to zero at a maximum volumetric flowrate. This power profile is illustrated by the pump horsepower versus throughput curve in Figure 8.

Power needed to drive the pump is the same as that required to overcome all the losses in the system and to supply the energy added to the liquid. The losses include frictional losses at the impeller, as well as turbulent losses; the disk friction (or energy required to rotate the impeller in the fluid); leakage from the periphery back to the eye of the impeller; and mechanical friction losses in various pump components, such as bearings, stuffing boxes, and wearing rings. The sum of all these power consumption items produces the final brake horsepower curve shown in Figure 8. As shown, brake horsepower is required even when the volumetric flowrate is zero. With increasing flowrate, the brake horsepower increases even when the head is zero. In this case, the flowrate will be at a maximum. The brake horsepower will not drop to zero but will have a definite value.

An additional cross plot can be obtained from this last figure by dividing

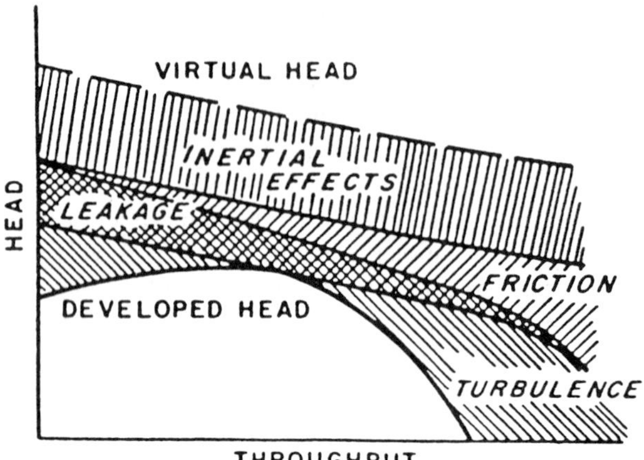

Figure 8 Plot of head versus throughput summarizing the major factors that reduce virtual head.

the fluid horsepower by the brake horsepower values (the definition of mechanical efficiency of a pump) and plotting efficiency versus throughput, as shown in Figure 10. All three performance curves—head (H–V), Figure 8; power (N–V), Figure 9; and efficiency (η–V), Figure 10—are combined conveniently into a single diagram (Figs. 11 and 12). Figures 11 and 12 are illustrative only. Manufacturers should be consulted for the specific performance data of a pump.

The plots given in Figure 13A contain several curves corresponding to different impeller diameters for a specific type of machine. Also shown are several lines of constant brake horsepower and constant efficiency, whose paths could be predicted from Figures 9 and 10. Such a diagram provides information on the characteristics of a pump for a definite heat at a specified liquid flowrate. By specifying coordinates H–V, we can interpolate among the curves of brake horsepower, impeller diameter, and efficiency to obtain all characteristic values of a pump under consideration.

It is important also to evaluate the influence of the number of impeller revolutions on pump performance. From Equation (45), we note that an increase in the number of rotations, n, is accompanied by an increase in head at a constant flowrate. This increase is illustrated graphically in Figure 12, showing the influence of number of impeller rotations on head (H–V), brake horsepower (N–$\dot{V}$), and efficiency (η–V).

Consider the characteristic curves of a typical centrifugal pump for liquids of different viscosities. Higher viscosity translates to higher resistance to flow and, consequently, greater frictional losses. Hence, we may expect a decrease

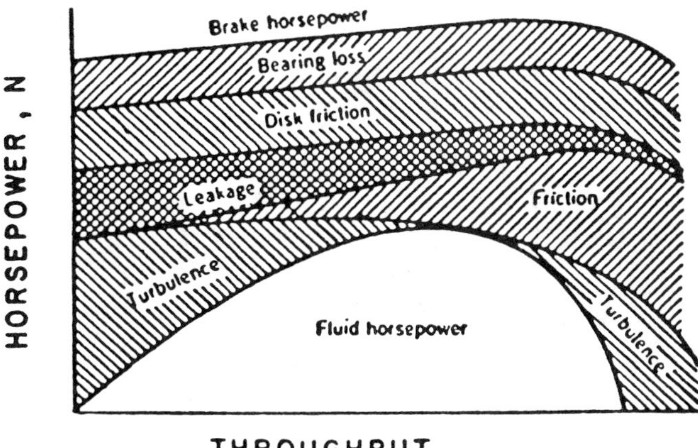

Figure 9 Plot of pump horsepower versus throughput summarizing factors affecting pump power.

in the head, an increase in brake horsepower, and a decrease in efficiency. Typical curves are shown in Figure 13B.

In selecting a pump, it is necessary to consider the entire pump system's characteristics; that is, the arrangement of piping, fittings, and equipment through which liquids flow. The characteristics of a pumping system express the relationship between flowrate, Q, and head, H, needed for liquid displacement through a given arrangement. Head, H, is the sum of the geometric height, H_g, and the head losses, h. Taking into account that V_{sec} = WS and $hl = (w_2/2g)$.

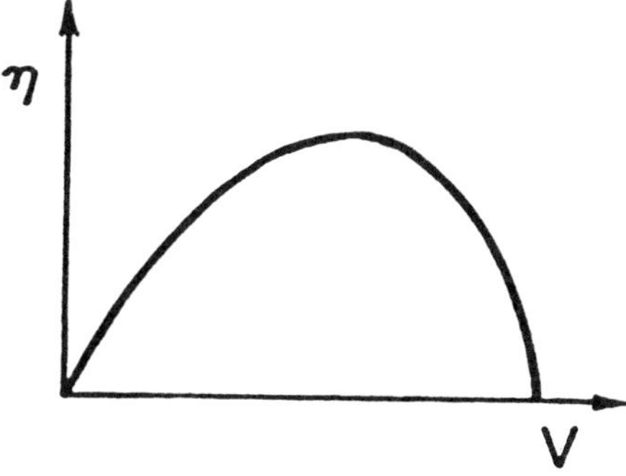

Figure 10 Typical curve of the mechanical efficiency of a pump (efficiency versus throughput).

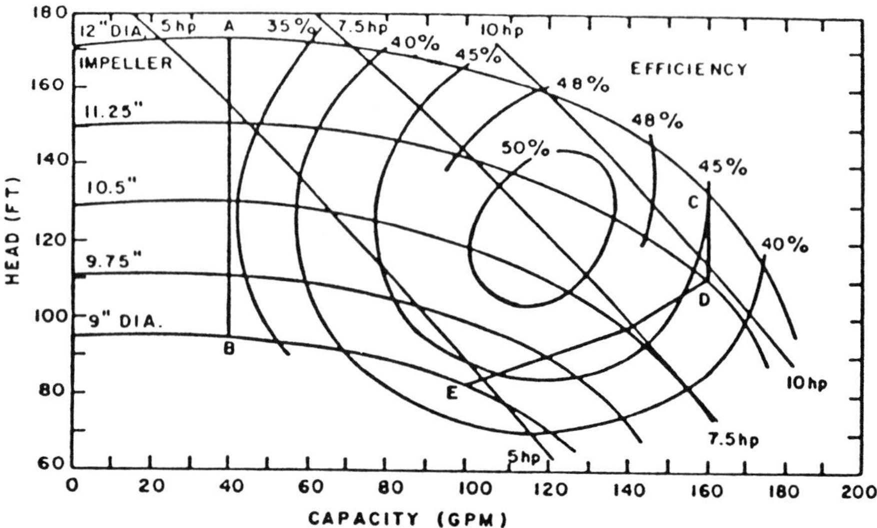

Figure 11 Total characteristics of a centrifugal pump.

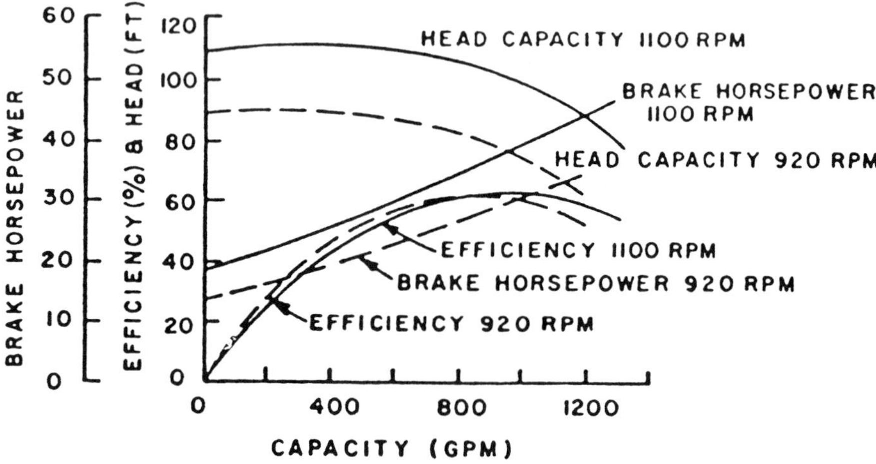

Figure 12 Head, efficiency, and brake horsepower plotted as a function of capacity for a centrifugal pump operating at two speeds with the same liquid.

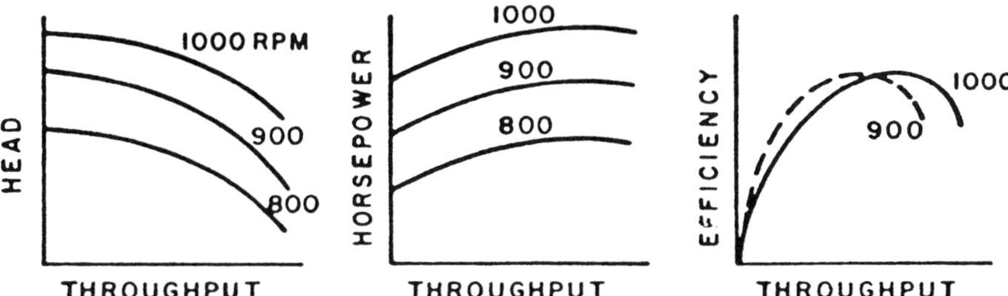

Figure 13A The influence of the number of impeller rotations on pump performance.

Denoting $V_{sec} = Q$, we determine that the head losses are proportional to the square of the flowrate:

$$hl = KQ^2 \tag{47}$$

where K is a coefficient of proportionality. Hence, the characteristics of a pump system may be expressed by the following parabolic expression:

$$H = H_g + KQ^2 \tag{48}$$

The characteristics of both the pump system and the pump can be represented on a common plot, as shown in Figure 14. Point A (the intersection of both characteristics curves) represents the operating point and corresponds to the maximum capacity of the pump, Q, while operating for a pump system. If a higher capacity is required, it is necessary either to increase the number of motor rotations or to change to a larger pump. Increased capacity can also be achieved by decreasing the hydraulic resistance of the pump system, h_l. In this case, the operating point is displaced along the pump characteristics toward the right. Hence, a pump should be selected so that the operating point corresponds to a desired head and capacity.

In *series pumping*, the pump head component of total available head is the sum of the heads developed by each pump at any given flow. Each pump must be selected to operate satisfactorily at the system design flow. Pumps operating in series are referred to as "pressure additive." This principle is illustrated in Figure 15. With one pump operating, the system flow will occur at point A. With both pumps operating, the system flow will occur at point B, which is the system design flow. In this operating mode, both pumps are developing equal head at the full system flow.

In *parallel pumping operations*, the pump component of total available head is identical for each pump and the system flow is divided among the number

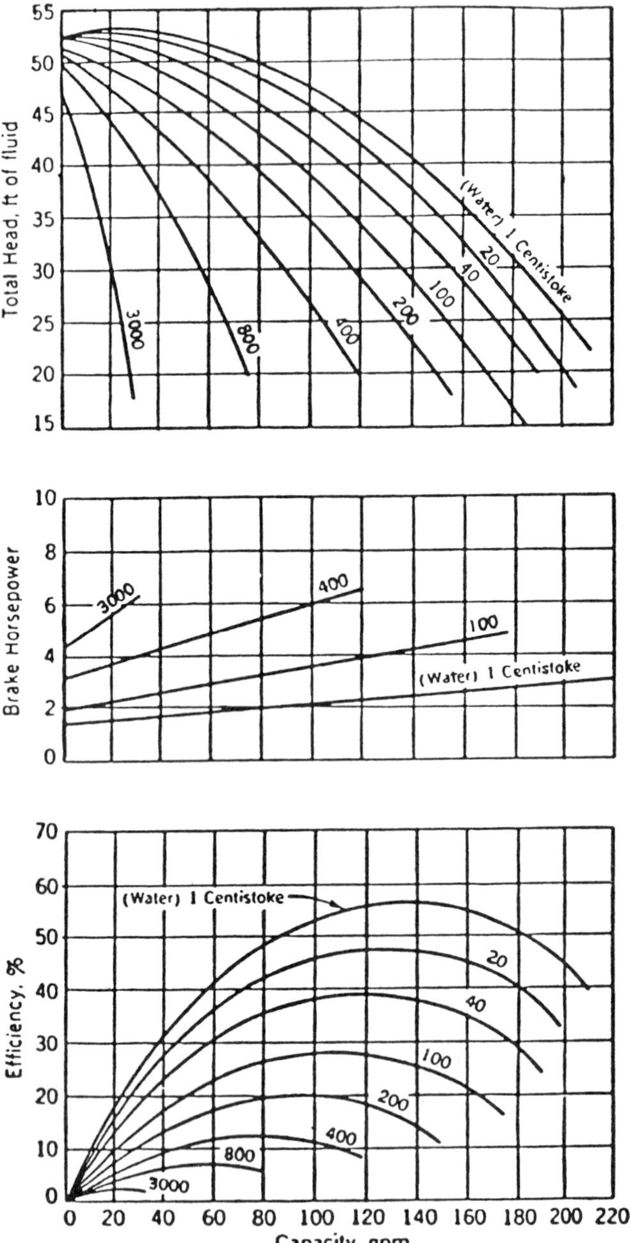

Figure 13B Characteristic curves of a typical centrifugal pump for liquids of different viscosities.

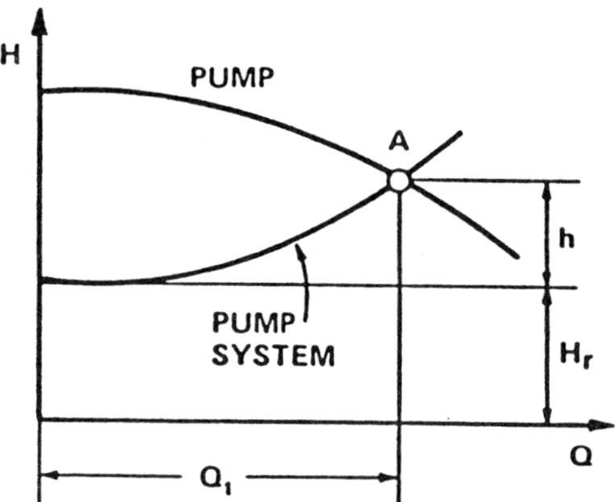

Figure 14 Pump and pump system characteristics represented on a single plot.

of pumps operating in the system. The flows produced by individual pumps can represent any percentage of the total system flow. Where series pumping is described as pressure additive, parallel pumping is described as "flow additive." When operating in parallel, pumps always develop an identical head value at whatever their equivalent flowrate for that developed head, and the sum of their capacities will equal the system flow. Parallel pumping characteristics are illustrated in Figure 16.

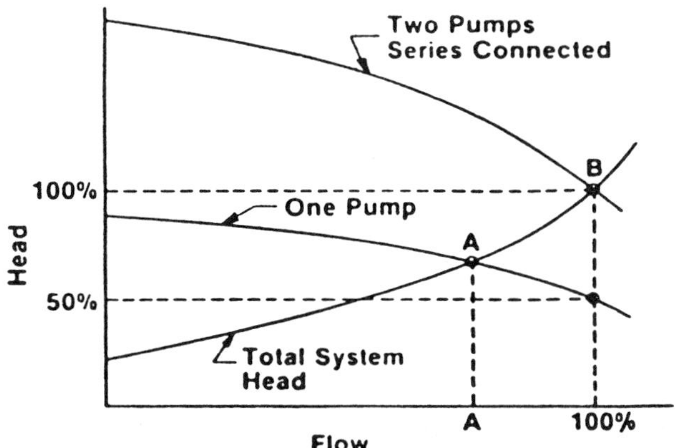

Figure 15 The effect of series pumping on the head curve.

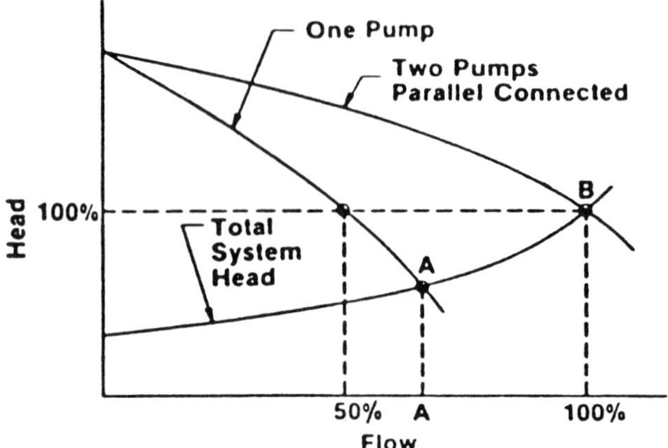

Figure 16 The effect of parallel pumping on the head curve.

In this example, each pump develops 50% of the total flow at 100% head. With one pump operating, the system flow will occur at point A; with both pumps in operation, the flow will occur at point B.

The *high-speed coefficient* (also referred to as the specific number of revolutions, n_s) is the number of revolutions of geometrically similar models of an impeller, which, at the same efficiency and capacity of 0.075 m^3/s, has a head of 1 m. The high-speed coefficient is a basic characteristic of a series of similar pumps having equal angles, α_2 and β_2, and coefficients, ϵ nd η_h. The high-speed coefficient is expressed by the following relationship:

$$n_s = \frac{3.65n \sqrt{Q}}{H^{3/4}} \tag{49}$$

where

n = the number of rotations (min^{-1})
Q = the maximum pump capacity (m^3/s)
H = the total pump head (m)

As follows from Equation (49), the high-speed coefficient increases with increasing capacity and decreasing head. Therefore, low-speed impellers generally are used for obtaining higher heads at low capacities, and high-speed impellers are used for creating high capacities at low heads. The impellers are divided into three groups depending on the value of the high-speed coefficient. Ranges of values are summarized in Table 3.

Table 3 Range of Values for the High-Speed Pump Coefficient

Type speed	n_s
Low	40–80
Normal	80–150
High speed	150–300

POSITIVE DISPLACEMENT PUMPS

Positive displacement pumps operate on the principle of forcing a fixed volume of liquid from the inlet pressure zone into the discharge zone of the pump. The two basic types of positive displacement pumps are *reciprocating* and *rotary*. Whereas the total dynamic head developed by a centrifugal pump is determined uniquely by the speed at which the impeller rotates, a positive displacement pump ideally will produce whatever head is imposed on it by the restrictions to flow on the discharge side.

Reciprocating Pumps

Reciprocating pumps produce pulsating flow, develop high shutoff or stalling pressure, display constant capacity when motor driven, and are subject to vapor binding at low NPSH conditions. There are three classes of reciprocating pumps: piston pumps, plunger pumps, and diaphragm pumps.

The *piston pump* consists of a cylinder with a reciprocating piston connected to a rod that passes through a gland at the end of the cylinder. The basic design is shown in Figure 17 and the operating principle is outlined in Figure 18. The liquid suction and delivery in the pump are derived from the reciprocating motion of the piston (1) in the pump cylinder (2) of Figure 18. When the piston moves toward the right, a vacuum develops in the closed space between head (3) and the piston. Owing to the pressure gradient between the suction tank and cylinder, the liquid is lifted through the suction piping and enters the cylinder via the suction valve (4). The delivery valve (5) is closed during the piston's motion to the right as it is subjected under liquid pressure from the suction piping. When the piston moves toward the left, the pressure generated in the cylinder closes one valve (4) and opens another (5). The liquid passing through the delivery valve enters into the discharge piping and then into the tank. Thus, both liquid suction and delivery by a single-acting piston pump vary with time because of the periodic motion of the piston. The volume delivery starts from zero at the instant the piston begins to move forward, reaches a maximum when it is fully accelerated at approximately the mid-point of its stroke, and then gradually falls off to zero. There will be a short time interval during the return stroke when liquid fills the cylinder and the delivery remains at zero (Figure 19).

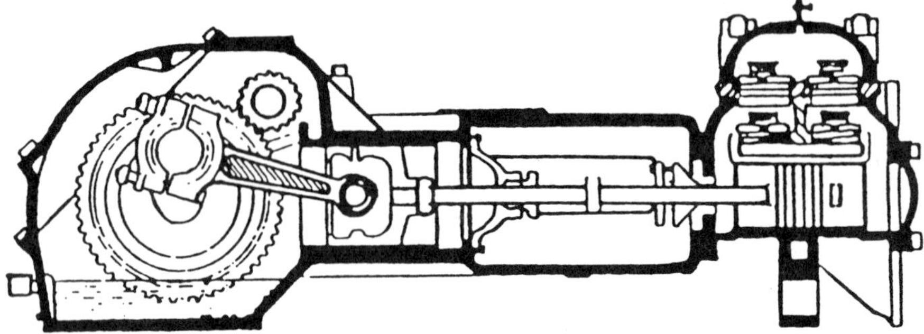

Figure 17 Basic design of the piston pump.

The piston shown in Figure 18 is driven by a crankshaft assembly (6), which converts the rotary motion of the drive wheel to the back-and-forth linear movement of the piston rod. Piston pumps are classified as single acting and double acting. In a single-acting piston pump, a single revolution of the crankshaft provides suction and delivery; in a double-acting pump, two strokes of a piston effect delivery/suction. As noted in Figure 18, the main working component of a piston pump is the piston itself, consisting of disks and seal rings (7) housed inside a polished cylinder.

A *plunger pump* differs from a piston pump in that it has a plunger reciprocating through packing glands causing the displacement of liquid from the cylinder. In this type pump, considerable radial clearance exists between the plunger and cylinder walls. Plunger pumps usually are thought of as single acting in the sense that only one end of the plunger is used for pumping liquid. Figure 20 provides details of the operating principle of a single-acting, horizontal plunger pump. As shown, a piston serves the role of the plunger (1), reciprocating in

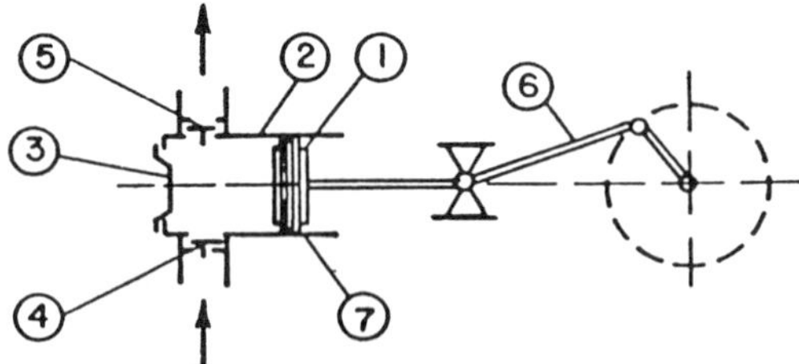

Figure 18 Operating principle for a horizontal single-acting pump: (1) piston; (2) cylinder; (3) head of cylinder; (4) suction valves; (5) delivery valve; (6) crank and connecting rod assembly; (7) seal rings.

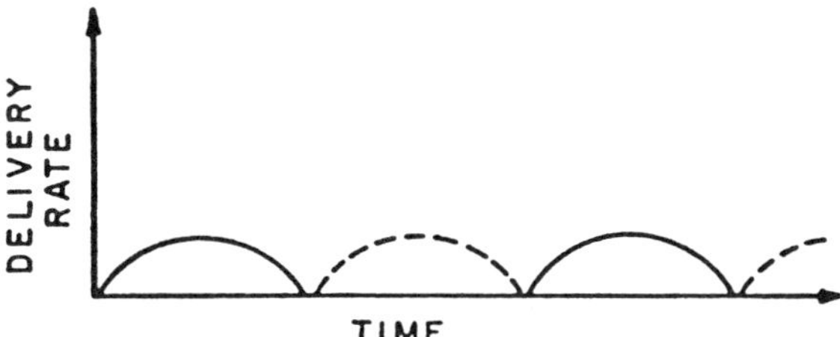

Figure 19 Liquid delivery rate for a simplex pump.

cylinder (2) and sealed with a packing gland (3). The internal cylinder surface of a plunger-type pump does not require as high a degree of finishing as does a piston pump. Also, leakage is readily minimized by tightening or replacing the packing gland. Plunger pumps are well suited to handling suspensions and viscous fluids because of their large radial clearances and ability to generate high pressures.

A more even delivery is achieved with piston and plunger pumps when operation is converted to double acting. The horizontal, double-acting plunger pump (Fig. 21) may be considered to be an aggregate of two single-acting pumps having four valves (two suction and two delivery). When the plunger (1) moves toward the right of Figure 21, liquid enters into the cylinder (2) through the suction valve (3) while liquid passes through the delivery valve (6) into the discharge piping. During the reverse stroke of the plunger, suction occurs in the right side of the cylinder through the suction valve (4) and delivery takes place through the valve (5). Hence, with a double-acting pump, suction and delivery take place during each piston stroke. Consequently, the capacity of these pumps is greater and delivery is smoother than with single-acting designs.

These designs are widely used because of the pulsation-free flowrate they

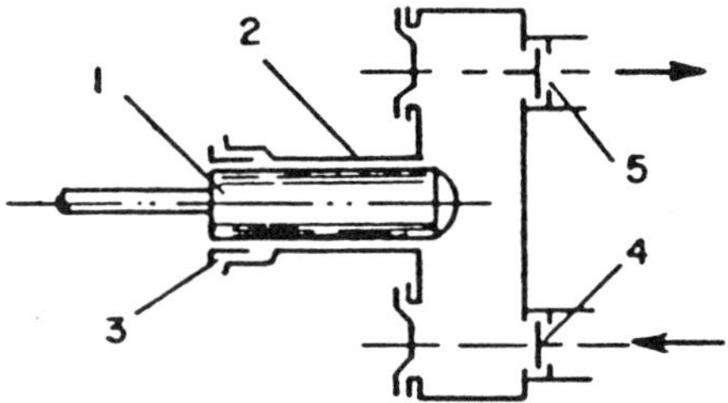

Figure 20 Operating principle behind a horizontal single-acting plunger pump: (1) plunger; (2) cylinder; (3) packing gland; (4) suction valve; (5) delivery valve.

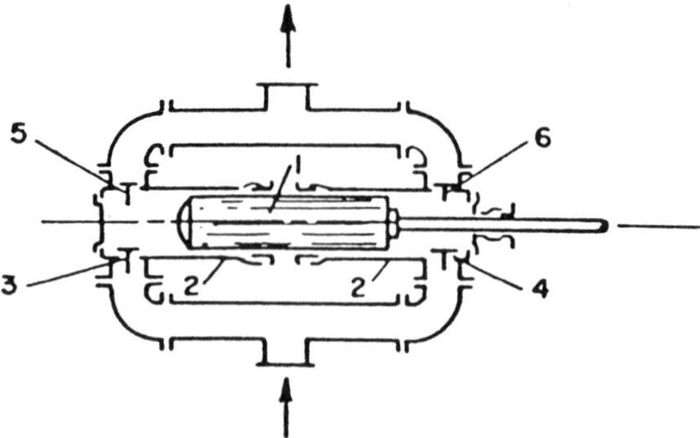

Figure 21 Operating principle behind a horizontal double-acting plunger pump: (1) plunger; (2) cylinder; (3,4) suction valves; (5,6) delivery valves.

provide, as well as their smooth torque at maximum stroke frequencies, which can exceed 1,500 strokes per minute (spm). The triplex pumps are single-acting tripled pumps having cranks located 120 degrees from each other. The total triplex-pump delivery is the sum of the deliveries of the individual single-acting pumps. For one rotation of the crankshaft, the liquid is suctioned and delivered three times.

Piston and plunger pumps may be actuated directly by steam-driven pistons or by rotating crankshafts through a cross head. The direct-acting steam pump consists of a steam cylinder end in line with a liquid cylinder with a straight rod connection between the steam and pump pistons or plunger.

Direct-acting steam pumps are available as simplex (one steam and liquid cylinder) and duplex (dual side by side) units. Duplex units are employed in large-capacity services and for reducing the flow pulsations below that of the simplex. Dual pumps are designed with an interconnecting steam valve linkage arrangement so that one side pumps when the other side reaches the end of its stroke. Steam pumps consist of a rod and piston design and are double acting, that is, each side pumps on every stroke. Consequently, a duplex pump will have four pumping strokes per cycle.

Power pumps convert rotary motion to low-speed reciprocating motion via speed-reduction gearing, a crankshaft, connecting rods, and cross heads. Plungers or pistons are driven by the cross-head drives. Rod and piston construction, similar to that in duplex double-acting steam pumps, is used by the liquid ends of the low-pressure, higher-capacity units. The higher-pressure units are normally single-acting plungers. This latter style generally employs three (triplex) plungers (Figure 22). Three or more plungers substantially reduce flow pulsations relative to simplex and even duplex pumps.

In general, the effective flowrate of reciprocating pumps decreases as

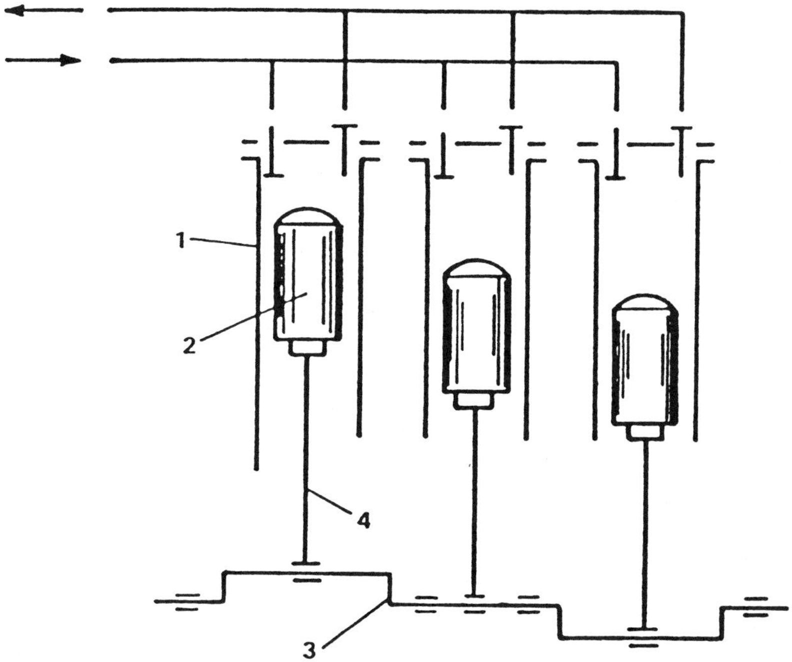

Figure 22 Operating principle behind a triplex pump: (1) cylinders; (2) plungers; (3) crankshaft; (4) connecting rods.

viscosity increases, because the speed must be reduced. High viscosity also leads to a reduction in pump efficiency. In contrast to centrifugal pumps, the differential pressure generated by reciprocating pumps is independent of fluid density. Rather, it is dependent entirely on the magnitude of force exerted on the piston.

Reciprocating pumps are used most often for sludge and slurry services, particularly where other pump types are inoperable or troublesome. Maintenance in such services tends to be high because of valve, cylinder, rod, and packing wear.

The theoretical delivery of a piston pump is equal to the total volume swept out by the piston in the cylinder times the number of strokes of the piston per unit time. Thus, the theoretical delivery in a single-acting pump is

$$Q_{th} = FSn \tag{50}$$

where

F = piston cross-sectional area
S = stroke length
n = number of crankshaft revolutions (or number of double strokes)

In the double-acting pump, there are two sections and two deliveries per crankshaft revolution. When the piston moves to the right of Figure 21, the liquid volume sucked from the left side is equal to FS and that delivered from the right side is (F – f)S, where f is the rod cross-sectional area. When the piston moves to the left, the volume FS is delivered from the left side into the delivery piping; from the right side, the liquid volume (F – f)S is sucked in from the suction line.

Consequently, for n rotations of the crankshaft, the theoretical delivery of a double-acting pump will be

$$Q_{th} = FSn + (F - f)Sn = (2F - f)Sn \quad (51)$$

The actual delivery may be less than the theoretical value because of leakage past the piston and valves or because of inertia in the valves. The actual delivery of the pump is

$$Q = Q_{th}\eta_v \quad (52)$$

where η_v is the volumetric efficiency, defined as the ratio of the actual discharge to the swept volume.

The volumetric efficiency of large pumps is typically 0.97–0.99; for medium flowrates ($Q = 20\text{–}300\ m^3/hr$), $\eta_v = 0.9\text{–}0.9S$; and for small flowrates, $\eta_v = 0.85\text{–}0.9$.

In some cases, however, the actual delivery may be greater than the theoretical value because of fluid momentum in the delivery line and sluggishness in the delivery valve operation resulting in continued flow.

Rotary Pumps

Rotary pumps combine the rotating movement of the working parts with the positive displacement of the fluid. There are a number of rotary pumps, but all operate on basically the same principle.

The rotating parts move in relation to the casing to create a space that first enlarges, drawing in the fluid through the suction line, sealing it, and then reducing the fluid volume and forcing it through the discharge port at a higher pressure. The rotating elements of the pump generate a reduced pressure in the suction line, thereby allowing the external pressure to force liquid into the pump. The capacity of a rotary pump is a function of its size and speed of rotation. This type of pumping equipment provides near constant deliveries in comparison to the fluctuating flows of reciprocating pumps. The main reason for selecting rotary pumps over centrifugals is to take advantage of their high-viscosity handling capability. In addition, rotaries are simple in design and efficient in handling flow conditions that generally are considered too low for economic application of centrifugals. Rotary pumps operate in moderate pressure ranges and have small to medium capacities.

Rotary pumps may be divided into five main types according to the character of the rotating parts: gears, screw, lobe, cam, and vane. There is, however, considerable overlap among these types.

The simplest rotary pump is the *external gear pump*, illustrated in Figure 23. Two gear wheels (2) operate inside the casing (1), which provides a snug fit to seal effectively the spaces between each pair of adjacent teeth. One of the gear wheels is driven by the driver and the other rotates in mesh, as indicated by the arrows. As the spaces between the teeth of the gear wheel pass the suction opening, A, liquid is impounded between them, carried around the casing to discharge opening, B, and then delivered outside through the opening. The straight teeth in gear wheel pumps produce pulsations in the delivery with a frequency equivalent to the product of the number of teeth on both gear wheels and the speed of rotation. Pulsation can be eliminated by the use of gear wheels having *helical* teeth with a particular angle.

Gear pumps handle liquids of a very high viscosity but cannot be used with suspensions. Generally, spacings between gear teeth are too close to handle solids without suffering significant erosion. The rotating parts in the casing create a space that draws in the liquid from the suction line. As the parts rotate, the liquid is trapped between them and the pump casing, forcing the liquid out through the delivery side of the pump at the necessary higher pressure. Typical performance characteristics of external gear pumps are shown in Figure 24.

Internal gear pumps are of two principal types. Both kinds employ a modified spur gear, which rotates inside a larger gear rotating around an axis that is parallel to that of the spur gear and displaced somewhat to mesh snugly at one point on the periphery. Figure 25 shows one internal gear pump design whose two gears maintain a continuous series of sliding seals with each tooth.

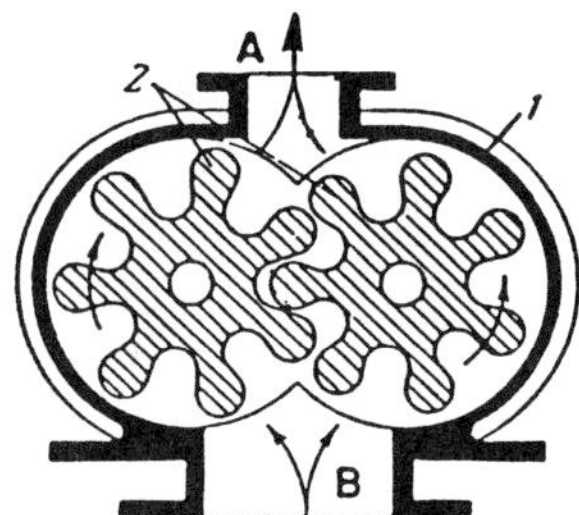

Figure 23 Operation of a gear-type rotary pump: (1) casing; (2) gear wheels.

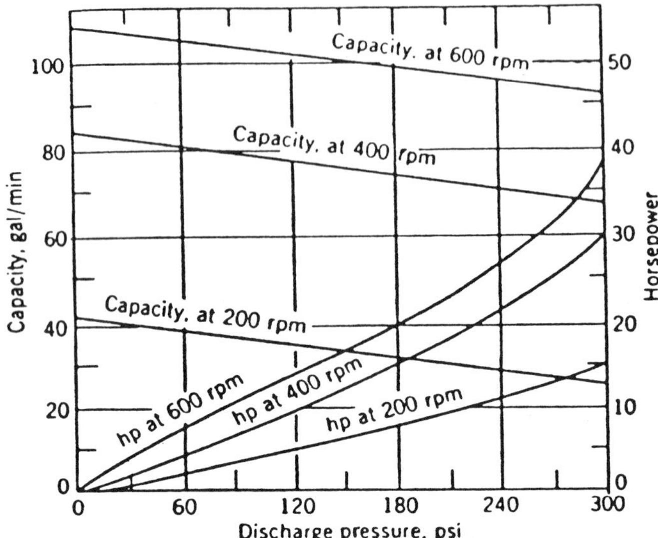

Figure 24 Typical performance characteristics of an external gear pump.

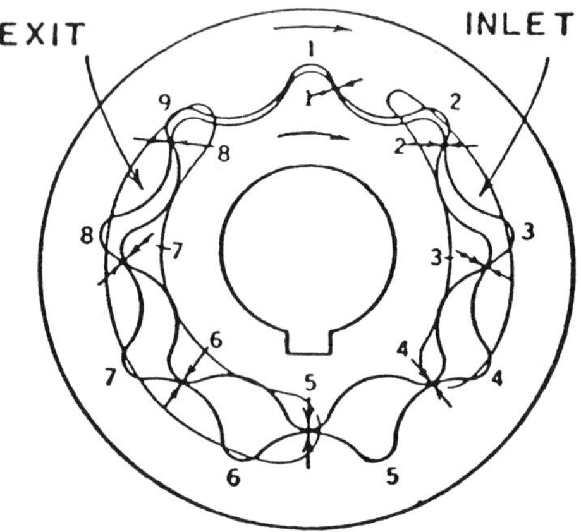

Figure 25 Internal gear pump with sliding seal.

This action forms pockets of fluid entirely between the gears. This type of pump can be equipped with any number of teeth in the spur gear (provided one more socket is included in the ring gear).

Screw pumps are modified helical gear pumps. They may have one, two, or as many as three screws turning along the pump axis, with liquid flowing between the screw threads and the casing. Both single-rotor and multiple-rotor screw pumps are commercially available. Pumping action is accomplished by progressing cavities, which advance the fluid along the rotating screw from inlet to outlet. This axial flow pattern minimizes vibration, producing a rather smooth flow.

A single-rotor screw pump (also called a progressive cavity pump) can handle slurries with relatively large particles. This type of pump comprises a rotor that revolves within a stator, executing a compound movement. The rotor revolves about its axis while the axis itself travels in a circular path. Hence, the rotor is a true helical screw, and the stator consists of a double internal helical thread. With each complete revolution of the rotor, the eccentric movement allows the rotor to contact the entire surface of the stator. Voids between the rotor and stator contain entrapped fluid, which is moved continuously toward the pump outlet. This pumping action provides continuous flow at low, smooth, and uniform rates. The action minimizes fracturing of particles as well as abrasion damage to the pump. Single-screw pumps are used extensively in the food processing and chemical industries for handling solid/liquid mixtures that are abrasive or require gentle handling of the solids.

Figure 26 shows a twin-screw pump consisting of two sets of screws that rotate and mesh in an accurately bored casing. Relatively tight operating clearances are maintained between the screws. This tight clearance usually is maintained by a pair of timing gears mounted on the shafts. The gears also transmit power from the drive shaft to the driven shaft. The body consists of a casing with two precision-machined bores, which house the rotating screws. Fluid passes from the inlet into the pumping chamber and then to the discharge zone.

Some screw pumps are operated without timing gears. In these units, a driver is used to turn the screws directly. The designs of twin-screw pump bodies and screw assembly arrangements are interrelated. Figure 27 illustrates typical body design.

The operation of a three-rotor screw pump is shown in Figure 27. The center rotor (1) is the driving member, whereas the other two (2) are driven. Liquid enters at the end of the rotors, is sealed between the rotors and casing, and is forced smoothly to the delivery end.

The mechanical displacement of liquid from inlet to outlet is generated by trapping a slug of fluid in the helical cavity (referred to as a “positive lock”) created by the meshing of the screws.

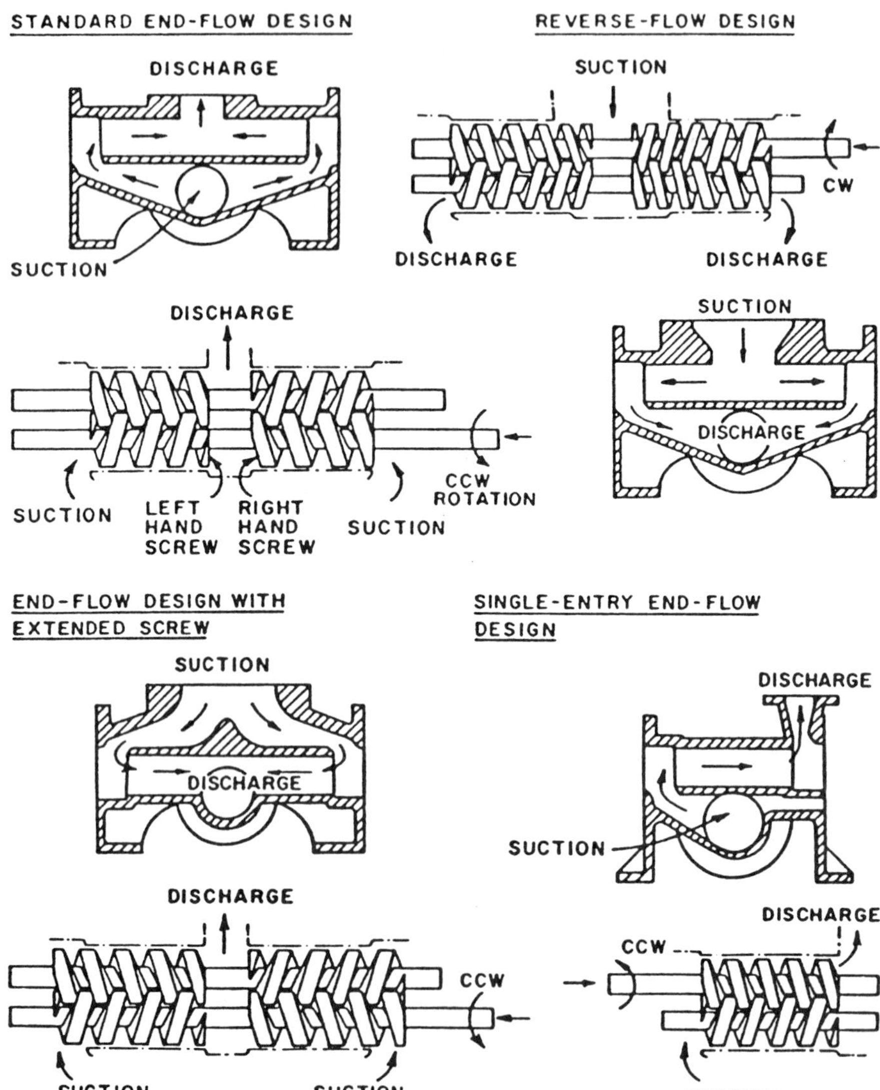

Figure 26 Typical twin-screw pump bodies and flow patterns.

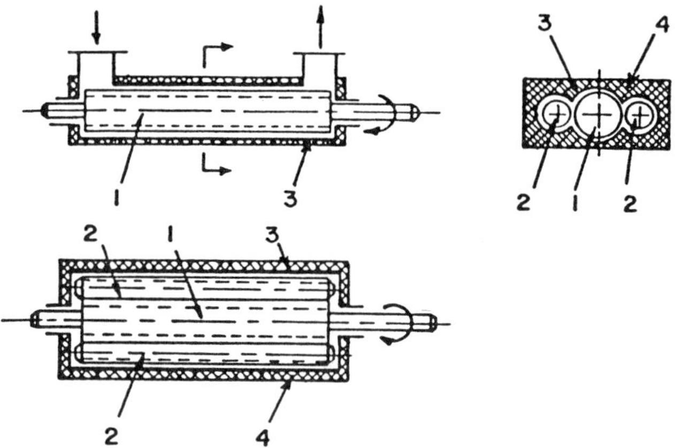

Figure 27 Operating scheme of a three-rotor screw pump: (1) driving screw; (2) driven screws; (3) sleeve, (4) casing.

Two- or *three-lobe pumps* are similar in design to gear pumps (Fig. 28), but gear wheels are replaced either by two or three lobes, which are driven separately through an external gearing mechanism. This arrangement makes it possible to avoid actual contact of the lobes with each other. Wear on the lobes and casing can be minimized by maintaining a small clearance between them. The clearances are a few thousandths of an inch, sufficient to reduce friction and wear but maintain minimum leakage from the delivery to the suction side. The characteristics of the lobe pumps are generally similar to those of the gear pumps.

Cam pumps consist of an eccentrically mounted circular rotor that sweeps a circle whose radius is the sum of the radius of the rotor and the eccentricity. The circular cam is fixed rigidly to the shaft and is housed inside the rotor ring, which is free to rotate about the cam. The plunger slides freely through a slide pin to act as a discharge valve. The plunger may be replaced by any form of vane, which seals the suction line from the discharge line. In general, contacts between surfaces are almost free of friction and wear except for the cam inside the rotor.

As the cam rotates, it expels liquid from the space ahead of it and sucks in liquid behind it. The characteristics again are similar to those of a gear pump.

Vane pumps (Fig. 29) include a massive cylinder (1) that carries rectangular vanes in a series of slots arranged at intervals around the curved surface of the rotor located eccentric to the casing (2). The tip of the vanes (3) is thrown outward by centrifugal force. As the cylinder rotates, the space behind a vane enlarges when it moves from the suction nozzle (S) to the vertical axis of the pump, resulting in the formation of a vacuum in space (4), thus drawing in liquid. At this point in the cycle, liquid is trapped between the vanes. When the vane

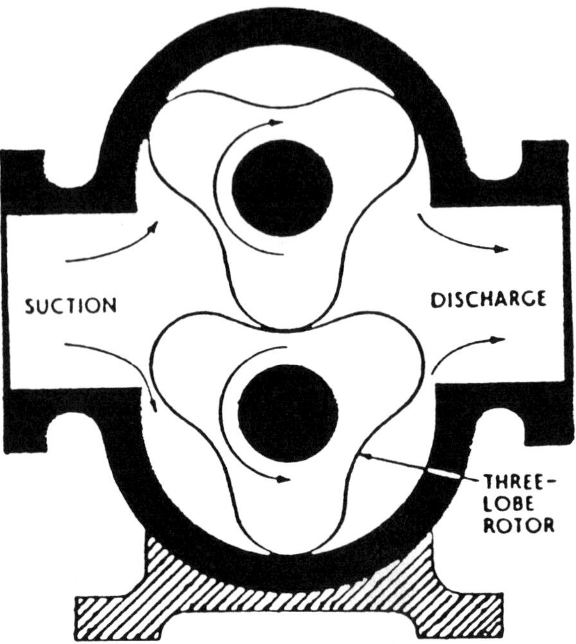

Figure 28 Sectional diagram of a three-lobe pump.

moves from the vertical axis in the direction of rotation, the volume of the chamber (space 4) decreases, and the liquid eventually is forced out through the discharge nozzle (6). Because of wear, the vane dimensions change; however, this wear is compensated for somewhat until the seal is broken. At that time, new vanes must be inserted.

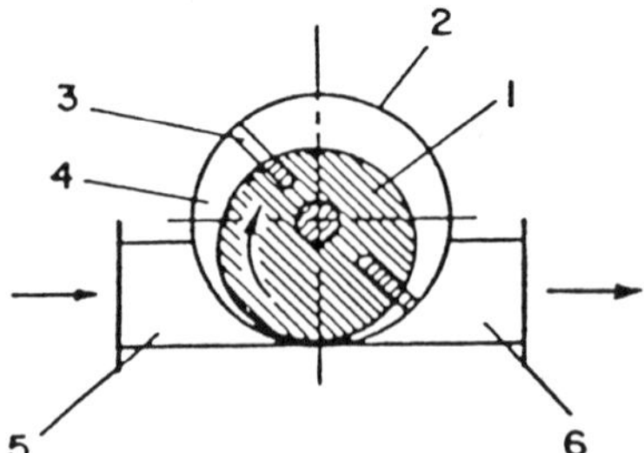

Figure 29 Sliding vane pump: (1) rotor; (2) casing; (3) vanes; (4) working space; (5) suction nozzle; (6) delivery nozzle.

Diaphragm Pumps

A *diaphragm pump* is a special type of positive displacement pump that operates by the periodic movement of a flexible diaphragm. It has the advantages of no stuffing boxes and high tolerance to abrasive slurries and chemically aggressive liquids. The operation is illustrated in Figure 30. The plunger (1) operates in a cylinder (2) in which a noncorrosive liquid is displaced. The movement of the fluid is transmitted by means of the flexible diaphragm (3) made from soft rubber or a special steel. When the plunger moves upward, the diaphragm is bent to the right and liquid is sucked into the pump through the globe valve (4). When the plunger moves downward, the diaphragm is bent to the left and liquid is discharged to the delivery piping through a delivery valve (5). The parts of the pump that are in contact with a liquid to be pumped are usually constructed of corrosion and erosion-resistant materials.

New and improved fully sealed diaphragm and bellows pumps have been developed to minimize leakage. Types range from micrometering pumps in the mililiter size to large diaphragm pumps having drives of several hundred kilowatts for high-pressure processes. These designs often achieve economic viability through reduction of maintenance and decontamination costs and improvement of process reliability.

Depending on the diaphragm geometry, the pumping characteristic is usually not quite linear. Metering accuracy is thus related to the tolerances of the operating conditions. Figure 31 shows the relationship between flow and pressure at various ratios of stroke length to delivery rate. The metering rate in relation to stroke frequency is linear.

Bellows-type metering pumps (Fig. 32) are built almost exclusively of glass and polytetrafluoroethylene (PTFE) components. Their pressure level is limited to less than 5 bar (70 psi) because of the glass parts. The output capacity is

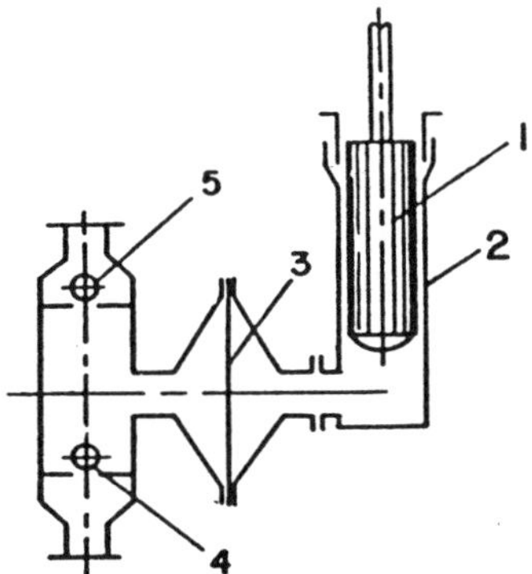

Figure 30 Operating scheme for a diaphragm pump: (1) plunger; (2) cylinder; (3) diaphragm; (4) suction valve; (5) delivery valve.

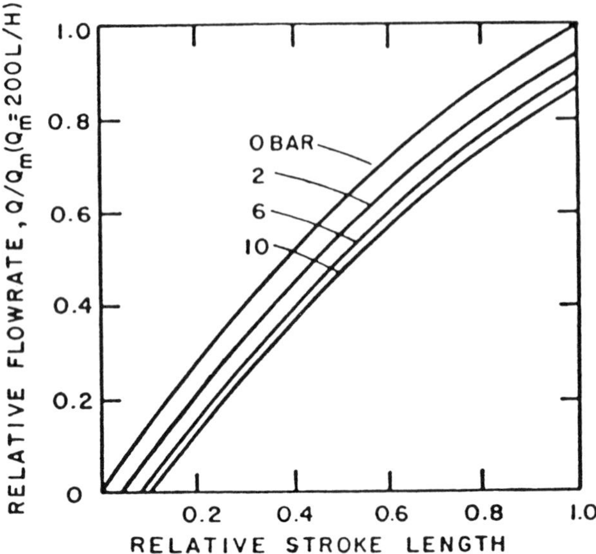

Figure 31 Diaphragm metering pump flow characteristics.

virtually unlimited, because there is no problem in producing fatigue-free bellows with diameters of several hundred millimeters.

Bellows require more care than do diaphragms during forming and manufacture but normally achieve lives of 5000–10,000 hr at maximum load. Because of radial stiffness of the bellows, the metering characteristic curve is purely linear and only slightly pressure dependent.

As in the case of the diaphragm pump, bellows failure is monitored via level or pressure sensors in the chamber below the bellows. This chamber is sealed from the drive by an auxiliary packing.

Plunger displacement by means of hydraulic fluid offers the benefit of uniform diaphragm support. The loading of the diaphragm then becomes solely the result of elastic deformation and not of resistance to pressure.

Operating limits and the advantages of various constructions are described briefly in Table 4. Critical, toxic, or abrasive media require the use of leak free metering or production pumps, with a realistic upper size limit of about 300 kW.

A widely used construction utilizes a mechanical diaphragm drive (usually limited to <85 psi). The metering rate usually is limited to less than 200–500 charges/hr, because the diaphragm loading becomes unfavorable with increasing diameter. The stroking rate also is restricted to less than 150 strokes/min because of acceleration shocks that occur at part-stroke settings with the widely used cam-and-spring-return or magnetic-drive units.

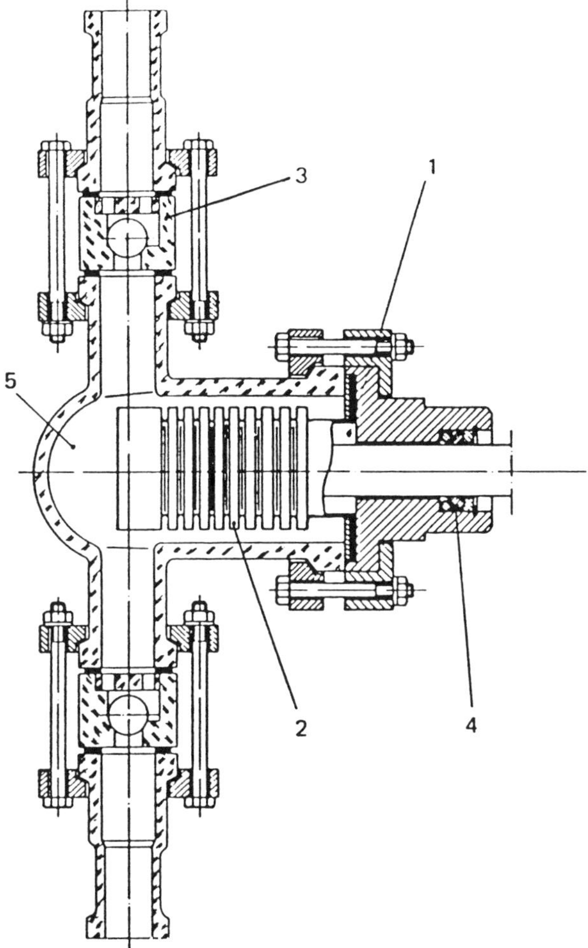

Figure 32 Bellow-type metering pump: (1) rupture chamber; (2) bellows; (3) check valve; (4) packed seal; (5) pump chamber.

For a given membrane geometry, diaphragm life depends on stroke length, pressure, temperature, and compatibility with the fluid being processed. The detailed design is empirical based on extensive fatigue trials. Diaphragm life easily can reach 3000 hr under maximum favorable conditions and will considerably exceed this value at, for example, lower pressures and shorter strokes.

Diaphragm failure can be signaled by a float switch that responds to the

Table 4 Operating Limits and Applications of Various Metering Pumps

Pump type	Limits	Application	Wetted parts made of . . .
Low-pressure diaphragm pump with direct mechanical drive	<6–10 bar (85–140 psi), <500/h	Metering, pumping	PVC or austenitic stainless steel; elastomer diaphragm
Low-pressure bellows pump with direct mechanical drive	<5 bar (70 psi)	Metering, pumping	Glass; PTFE bellows
High-pressure microdiaphragm metering pump with hydraulic drive	<700 bar (9.940 psi), <10/h	Metering	Acid-resistant steel
Diaphragm pump with hydraulic compression of tubular member	<50 bar (710 psi)	Metering, pumping	Acid-resistant steel; elastomer diaphragm and tube
Diaphragm pump with hydraulic drive	<350 bar (5,000 psi) for PTFE diaphragm <3,000 bar (42,600 psi) for metal diaphragm	Metering, pumping	PVC, PTFE, titanium, acid-resistant steel

presence of process fluid in the space behind the diaphragm. Another failure detection method involves two mechanically coupled diaphragms. A rise in pressure in the space between them indicates diaphragm failure and will cause simple sensors to trigger a warning signal.

Hydraulically driven pumps allow greater diaphragm displacements, but their operating conditions are restricted to about 5000 psi and 120°C maximum by the fatigue limit under compression for PTFE and by clamping effects.

The largest number of applications can be met by compact diaphragm metering pumps. Stroke-adjustable drives operate the piston via a connecting link; the piston slides in a liner tight enough to provide a seal. The lubrication and hydraulic systems are conventional. Plunger movement displaces the diaphragm, which can deflect between two perforated support plates. Topping up the space between plunger and diaphragm (the hydraulic space), when the diaphragm has reached the lower support plate, is controlled at negative pressure by vacuum-replenishing valve.

The hydraulic system also contains a relief valve for protection against excess pressure. This relief sometimes can take the place of an external safety valve that otherwise would be in contact with the pumped fluid. The upper support plate, together with this relief valve, protects the diaphragm against excessive deflection under certain process conditions. An automatic vent valve in the highest point of the hydraulic system provides a degassed hydraulic medium and, thus, a faultless delivery characteristic. Compact diaphragm pumps with common oil systems generally require the use of sandwich diaphragms.

The pump is ideally suited to slurries of all types and cleaning-in-place in the food industry. Topping up the hydraulic system is controlled via the gate that is diaphragm position–dependent and by a relief-and-snifter valve. The continuous vent can be seen at the highest point.

For services beyond the practical limits of PTFE, *metal diaphragms*, made of dead-parallel cold-rolled sheet are used. Pumps with metal diaphragms are especially suited to

- Pressures above 5000 psi and temperatures up to 200 °C (in special cases, 400°C maximum)
- Microdiaphragm dosing pumps because of the low compressibility of diaphragm and clamping
- Applications where all-metal construction is mandatory, usually to ensure radiation resistance

Because of the lower elasticity of metal diaphragms, the same stroke-displacement volumes require larger-diaphragm diameters and larger pump heads. For larger outputs, diaphragm pumps with metal diaphragms are more costly to manufacture than are pumps with nonmetallic diaphragms. In addition, because of the risk of notch formation and because of their lower deflection capability, thin metal diaphragms are not suitable for the unsupported type of diaphragm.

Important design considerations are an even flow distribution through the support perforations and the prevention of local adhesion. Because of diaphragm durability and adhesion avoidance, the dimensioning of the perforations and the profiling of the diaphragm support surface cannot be calculated mathematically. They are determined by experience gained through successful construction and installation of high-pressure diaphragm pumps.

In micrometering pumps for high pressures, the displacement volumes and the diaphragm pump head dimensions are very small. Therefore, the design parameters have a different rank of importance than for high-output pumps. The pump cavity is designed for minimum dead volume and maximum rigidity. For improved degassing, oil circulation in the hydraulic system has proved to be a satisfactory concept. This circulation is achieved by a plunger displacement aided by two nonreturn valves.

Hydraulic-linkage pipes (also referred to as "pendulum" or "remote-head" systems) serve to remove high temperatures and other troublesome and dangerous influences from the drive, making it feasible to extend the temperature range of application of diaphragm pumps (e.g., to 400°C). A familiar application is in the radioactively heated zone of atomic fuel recovery plants, where extreme demands are made for safety. Figure 33 shows the major modules of these metering pumps.

Loss in suction pressure (largely influenced by the suction valve) is much the same for diaphragm and plunger pumps. In particular, types having unrestricted forward chambers experience no additional internal pressure losses on the process fluid side. However, at high stroking frequencies, in excess of 300 strokes/min, it is essential that internal flow patterns be optimized.

Some types of diaphragm pumps require minimum suction pressures of 7–42 psi. These include pumps with metal diaphragms for larger output and faster stroking rates. Metal diaphragms with their lower stiffness are more sensitive to cavitation than are the heavier PTFE diaphragms.

Cavitation can be minimized by proper layout of the pipe system and pressure conditions. Upsets in metering accuracy and performance owing to cavitation can be tremendous. Delayed compression, with cavitation present, leads to sizable pressure and loading shocks.

Because of elastic influences, the instantaneous delivery flow is not in phase with plunger displacement. Hence, the delivery often starts jerkily at the end of the compression phase (final plunger speed) with a corresponding shock wave as the result. This shock problem, which occurs in all high-pressure pumps, can be overcome by shock dampers that have been sized in line with pulsation theories.

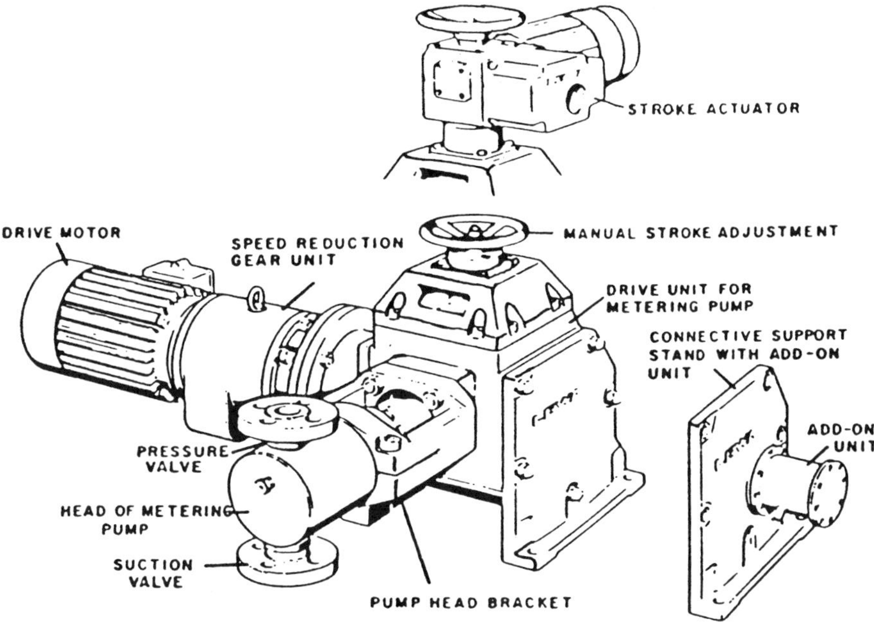

Figure 33 Modules of metering pumps.

PUMP SELECTION

Proper pump selection for an energy-efficient liquid transportation system demands an understanding of the principles of mechanics and physics affecting the pumping system and the fluid. Pump efficiency is strongly dependent on the behavior of the pumping fluid. Principles of pump operations defining head and flow relationships must be understood clearly before the system performance can be evaluated accurately.

General performance characteristics of commercially available pumps can be obtained from the manufacturers. Selection is made on the basis of desired capacity and head, which are calculated in accordance with overall piping system layout. The electric motor for a pump is chosen from the brake horsepower determined from the equations given earlier.

The overall procedure in choosing a pump for a particular application is as follows:

- Obtain information on the physical and chemical properties of the liquid at the intended operating conditions (i.e., specific gravity, viscosity, vapor pressure, corrosiveness, toxicity).

- Lay out the piping systems on paper, defining major flow resistances in the system. Calculate total heads for the system.
- Establish the capacity requirements in terms of a range. That is, define normal average capacity needs as well as system lows and peak flow required. If possible, estimate the time pumps will have to operate at peak loads.
- Based on the above information, select the class and type of pump. A more detailed specification can be made from examination of the manufacturer's literature.

Of the pumps described in this chapter, centrifugal's are the most versatile and widely used throughout the chemical process industries.

The following advantages of centrifugal pumps should be remembered when comparing different pump classes for an application:

- Simple in construction and, as a general rule, less expensive than many positive displacement types. They are available in a wide range of materials.
- Do not require valves for their operation.
- Operate at high speeds (4000 rpm or higher) and, therefore, can be coupled directly to an electric motor. In general, higher speeds typically mean smaller pumps and motors for a given duty.
- Give steady deliveries.
- Depending on the application, maintenance costs are lower than for any other type of pump.
- Typically smaller than other pumps of equal capacity. Therefore, they can be made into a sealed unit with the driving motor and directly immersed in the suction tank.
- Liquids having relatively high concentrations of suspended solids can be handled.

At the same time, centrifugal pumps have several disadvantages, the primary ones are

- Single-stage pumps cannot develop high pressures. Multistage pumps will develop greater heads but are much more expensive and cannot be readily constructed from corrosion-resistant materials without significantly higher costs due to their greater complexity. As a general rule, it is better to use very high speeds to reduce the number of stages required.
- High-efficiency operation is usually obtained over only a limited range of conditions. This limitation is especially true for turbine pumps.
- The vast majority of centrifugal pumps commercially available are not self-priming.
- A nonreturn valve must be installed in the delivery or suction line or the liquid will run back into the suction tank when the unit is not running.

- Centrifugal pumps have problems handling highly viscous materials. They typically operate at greatly reduced efficiencies.

For pumping problems requiring relatively small capacities and high heads (e.g., 50–1000 or more atm), piston pumps are recommended. These pumps are well suited in these ranges to pumping liquids of high viscosity that are flammable and of an explosive nature (steam pumps). In addition, they make excellent metering pumps.

Screw pumps are best suited for handling high-viscosity liquids, fuels, and petroleum products. These pumps are used for capacities up to 300 m^3/hr and pressures up to 175 atm at speeds of rotation up to 3000/min. The advantages of screw pumps are their high speed, compactness, and quiet operation. Pump capacity is practically independent of pressure, and efficiency is rather high (in the range of 0.75–0.80). The field of application of single-screw pumps is restricted by capacity up to 3.6–7.0 m^3/hr and pressures up to 10–25 atm. Their cost and maintenance are similar to those of centrifugal pumps of low capacity operating under pressures up to 3–5 atm. Screw pumps are considerably more economical when their delivery pressures exceed 10 atm. Single screw pumps are employed in handling dirty and aggressive liquids, solutions, and high-viscosity polymer solutions.

Sliding-vane pumps are used for the displacement of clean liquids (without solid particles) at moderate capacities and heads.

Gear pumps are best suited to pumping viscous liquids without solid particles at low delivery rates (not higher than 5–6 m^3/min) and high pressures up to 100–150 atm.

Jet pumps are typically employed in operations in which pumping requirements are intermittent, where an inexpensive standby unit is desirable or corrosion is important. They are used for the displacement of low-viscosity clean liquids at low delivery rates up to 40 m^3/hr and relatively high speeds (up to 250 m). Efficiencies are typically low (η = 20–50%).

Acid-egg pumps and airlifts are used in industries in which moving and friction parts are highly undesirable.

METERING PUMPS

In treatment plants, metering pumps are used, for example, to add polymer or other chemicals to clarifiers for aid in settling or for precipitation of pollutants, to add chemicals to boiler make-up water, to add chemicals for dewatering of sludges, and to add acid or base of pH control.

Metering pumps are typically positive displacement pumps of either the reciprocating or rotary type. Reciprocating metering pumps have two general methods to vary pump output: variable speed or variable stroke. Stroke on a variable-speed pump is adjusted by either varying the crank travel directly (amplitude modulation) or by varying the amount of fixed-crank travel transmitted to the piston.

Amplitude modulation can also be achieved by using a slider-crank in which the length of a pivot arm (or eccentric) is adjusted. Adjusting the pivot arm to zero length results in zero piston travel. Design of slider-crank mechanisms vary by manufacturer. An example is shown in Figure 34.

Another method to modulate amplitude is by the shift-ring drive in which the piston rotates in a ring that can be positioned. Fixed-crank stroke adjustments use a lost-motion drive that limits the piston or diaphragm travel. Motion can be lost in the transfer from crank to piston as shown in Figure 35.

The piston reciprocates as it follows the eccentric cam until the piston is stopped by the stroke adjustment pin. Piston travel is resumed when the eccentric rotates enough to pass the position of the piston stop. Another lost-motion drive is shown in Figure 36. Here motion is lost between the piston and the diaphragm by allowing some of the hydraulic fluid to escape to a reservoir.

Variable-speed drives on a metering pump can be eddy-current, silicon controlled rectifier (SCR), variable frequency, or belt drives using conical pulleys to vary the ratio of drive to pump speed.

Metering pumps can have variable speed, variable stroke, or both variable speed and variable stroke capabilities. The choice of variable drive depends on the pump's application. Another method of metering uses a recycle valve to vary pump output. Variable-rate pumps are used in automatic control to eliminate the cost and maintenance of a control valve as well as save energy.

VARIABLE-SPEED DRIVES

Variable-speed drives are used for pumping or other mechanical functions that cannot be properly accomplished with a reasonable number of constant-speed units. Variable-speed drives also offer increased flexibility in control. Improved system efficiency can also be obtained where periodic changes in the demand allow operating at reduced horsepower to save energy.

A variable-frequency drive (VFD) system consists of an induction motor where both the voltage and frequency supply are controlled by an electrical inverter to adjust the motor's speed.

Characteristics of a VFD system are as follows:

- Other than the addition of thermostats in two phases of the stator winding, a standard squirrel-cage alternating current (AC) induction motor can be used.
- All standard motor sizes are available from 4 to 400 kW (5–500 hp).
- A variable-frequency drive system is suitable for variable torque pumping or fan loads and is normally suitable for piston pumps and conveyors.
- Continuous operation at constant torque is available over a 3:1 speed range.

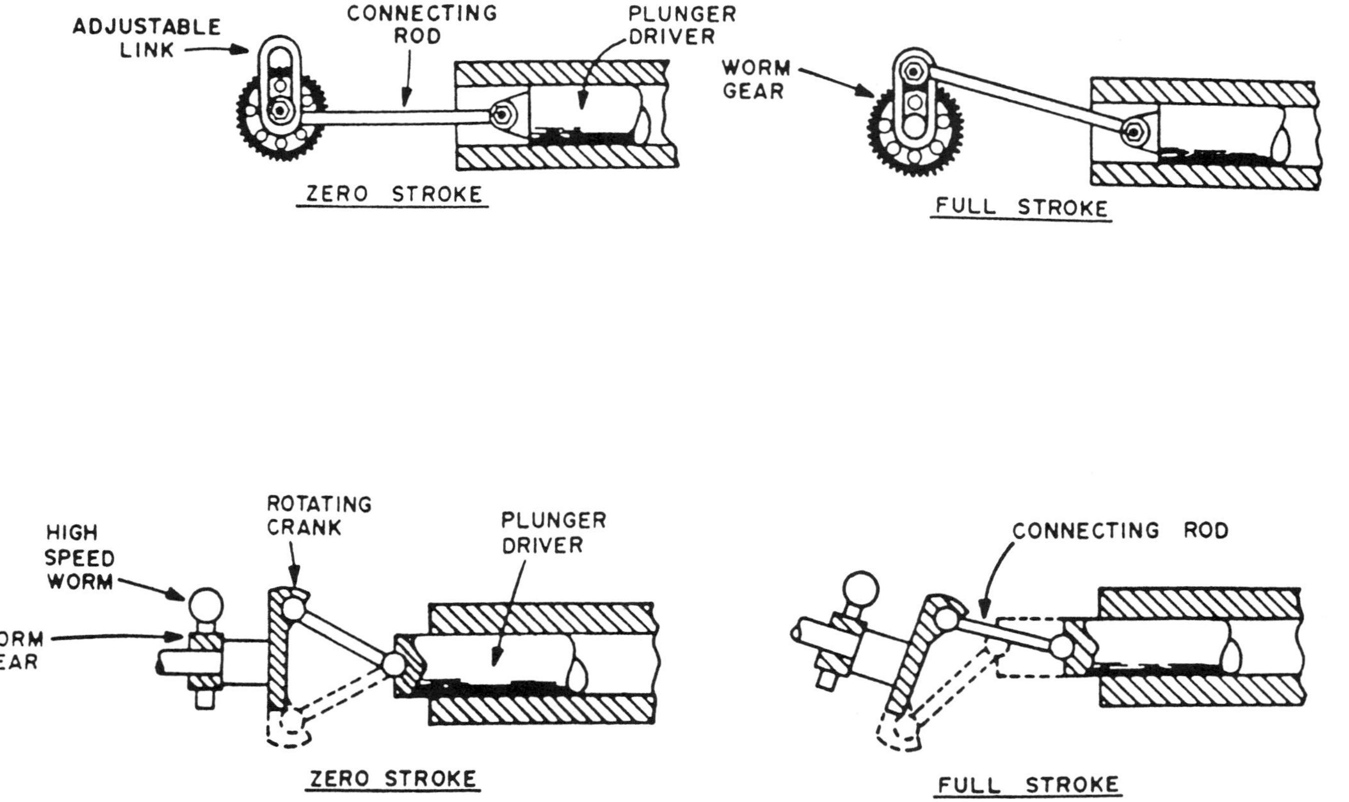

Figure 34 Slider-crank stroke adjustment.

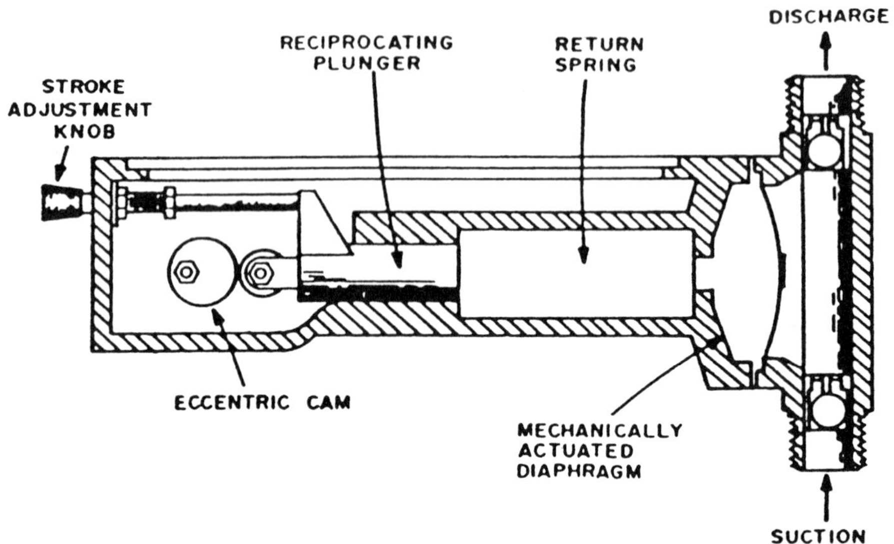

Figure 35 Eccentric cam lost-motion stroke adjustment.

- Overall efficiencies are about 83% at full speed down to approximately 75% at half speed.
- Multiple motors can operate off one common VFD simultaneously. Also, one VFD can control more than one motor where it is switched between motors to give a combination fixed-speed and variable-speed system.
- The VFD can convert existing constant speed motors to variable-speed operation when retrofitting existing installations.
- Starting current can be limited to less than full-load current.
- The VFD requires little maintenance. However, complex components and circuitry require an expert technician when problems do occur.

The variable-speed controller/power converter changes constant frequency, constant voltage line power to variable frequency, and variable voltage power to vary the drive motor's speed. Raising the frequency of the power applied to the drive motor is adjusted to control the motor's output power. The variable-speed controller compares the drive motor's speed to an adjustable setpoint value and outputs the required frequency and voltage to maintain the desired speed (Fig. 37).

VARIABLE PULLEY

Electromechanical variable-speed pulley belt drives use a variable sheave ratio principle to change the speed of the final element.

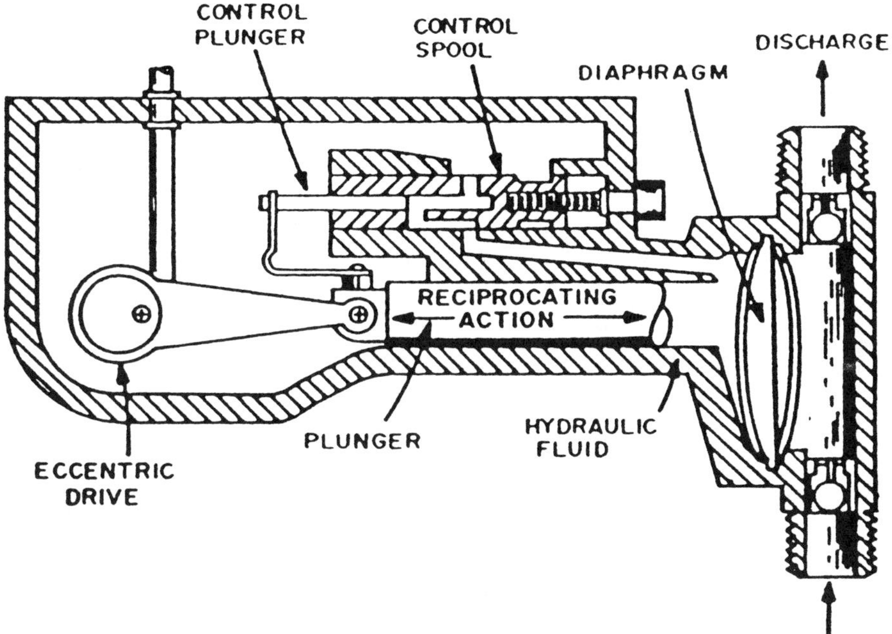

Figure 36 Hydraulic lost-motion stroke adjustment.

Characteristics of a variable pulley are as follows:

- It is available in all standard motor sizes from 0.2 to 75.0 kW (0.25–100.0 hp), but is normally used only in the lower ratings of 0.4–20.0 kW (0.25–25.0 hp).
- It is a constant torque drive.
- In general, a 10:1 speed range is available.
- Overall efficiencies are about 70% at maximum speed down to approximately 45% at half speed.
- Additional floor space or ceiling height is required as compared to most other drives.
- Belt life is typically 18 months.

Variable pulley or electromechanical drives use a constant speed motor coupled to a variable belt drive system for speed control. Output speed is varied by changing the diameter of the drive sheave or pulley. Drive sheave diameter is adjusted using a reversing gear motor connected mechanically to the drive system. When the gear motor operates in one direction, the sheave

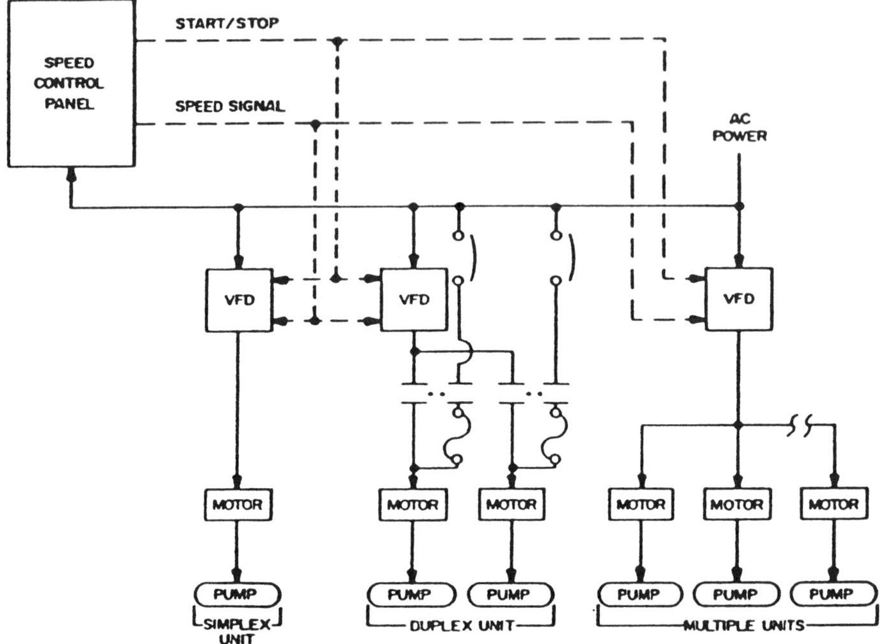

Figure 37 Variable frequency drive functional diagram.

diameter is increased and the drive speed increases. When the gear motor operates in the other direction, the sheave diameter decreases and the drive speed decreases.

DIRECT CURRENT

Variable-speed control of DC motors is achieved by varying the armature or field voltage to the motor or both.

Characteristics of direct current (DC) variable-speed drives are

- DC motors are considerably more expensive than comparable AC motors.
- Drives are available from 4 to 100 kW (5–150 hp).
- DC variable-speed drives can be used for variable torque operation and are usually suitable for piston pumps and conveyors.
- In general, a speed range of 60–100% is available.
- Reduced speed efficiency is very good.
- The DC motor has a commutator and brushes, both potential maintenance problems.

- Commutators on DC motors can cause problems if they are not well ventilated and properly maintained or if they are subject to a corrosive environment.
- Drive requires little maintenance, but complex components and circuitry require an expert technician when problems do occur.
- Starting current can be limited to less than full-load current.

A silicon-controlled rectifier (SCR) drive is a variable-speed drive that is based on the use of a DC motor. Such SCR drives include a speed controller that uses an SCR-controlled bridge circuit to rectify constant-voltage AC line power to variable DC voltage. Variable DC voltage is applied to the DC motor to regulate its speed. As the applied voltage increases, the motor speed increases. As the voltage decreases, the speed decreases. SCR speed controllers accept an input signal representative of the desired motor speed and provide a DC output sufficient to operate the motor at the required speed.

CONTROL VALVES FOR MODULATING SERVICE

Control valves for modulating service are used in all water- and wastewater-treatment processes. Control valves are made in a wide variety of designs.

Ball Valves

Valves are classified as either "rotary" or "linear." In a rotary valve, the ball, disk, or plug is rotated to open or close the flow stream. A linear valve lifts the gate, disk, or plug up or down to open or close the flow stream. Ball valves are rotary. Flow goes through a port in the ball. To shut off flow, the ball is rotated until the port is closed. The ball may be a complete sphere, full ball, as shown in Figure 38, or a partial sphere. Standard port diameters range from 80% of inside pipe diameter to 100%, or full port.

Ball valves have high friction resistance to rotation owing to valve body and trim contact with the ball surface. This gives a tight shutoff at the expense of needing larger actuators. In some designs, the ball is allowed to "float" in its seat, so that line pressure will assist keeping a tight shutoff by pushing the ball against the downstream seal ring. This freedom of movement introduces a deadband between the actuator and the ball. The deadband will make this design unusable for many control applications. For control valves, a rigid ball to actuator connection is preferred.

Butterfly Valves

Butterfly valves are rotary-type valves that use an axially pivoted disk to restrict or open the flow stream. The disk may be flat or contoured in shape and mounted to the steam (pivot) in several ways. An example is profiled in Figure 39.

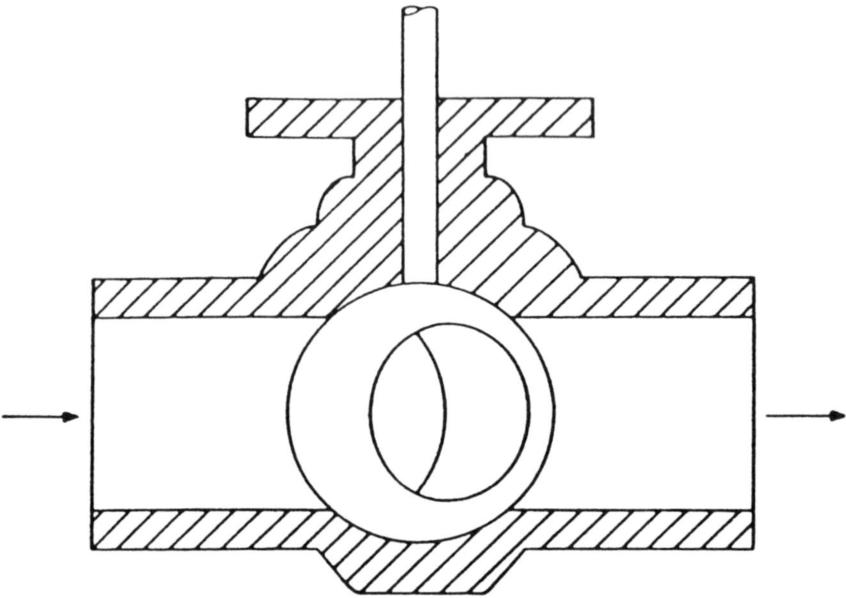

Figure 38 Ball valve.

One problem with butterfly valves is obtaining a tight shutoff. The seal must be made along the entire circumference of the valve body. Upper and lower half seats must be matched and the pivot point must be sealed. Liners and seals used for good shutoff increase the opening and closing torque requirements of the actuator.

Another problem encountered in butterfly operation is torque applied by the fluid. During rotation of the disk, torque reaches a maximum at about 70 degrees (from full open). This is shown in Figure 40. In some applications, this peak may exceed the opening or closing torques. To overcome this problem, disks are contoured for low torque in the near closed positions.

Gate Valves

Gate valves are linear valves in which the gate's disk is raised or lowered past a port by the actuator. The disk is a flat or wedged-shaped plate. The valve may have one or several ports for flow which can be sealed by the plate. A gate valve is shown in Figure 41. In knife-gate valves, plates are made with a sharp edges for service with solids-bearing streams including dry solids. Gate valves are not usually used for modulating service.

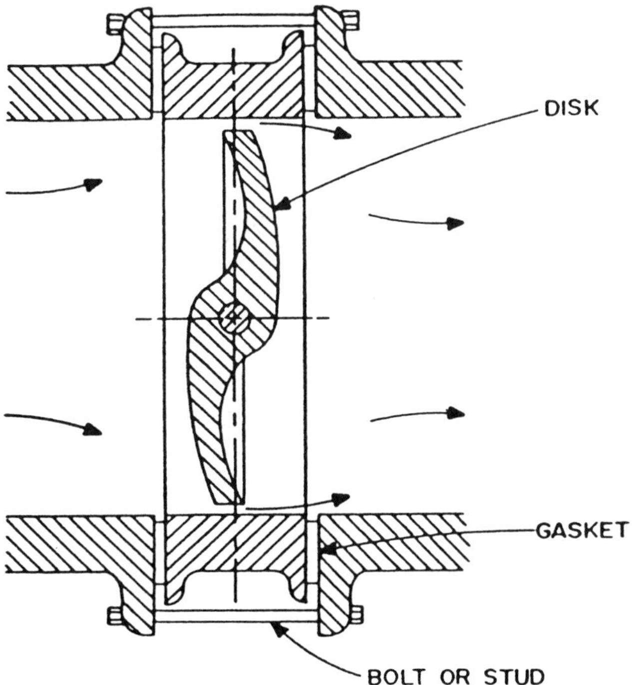

Figure 39 Butterfly valve, swing-through type with flangeless pipe connection.

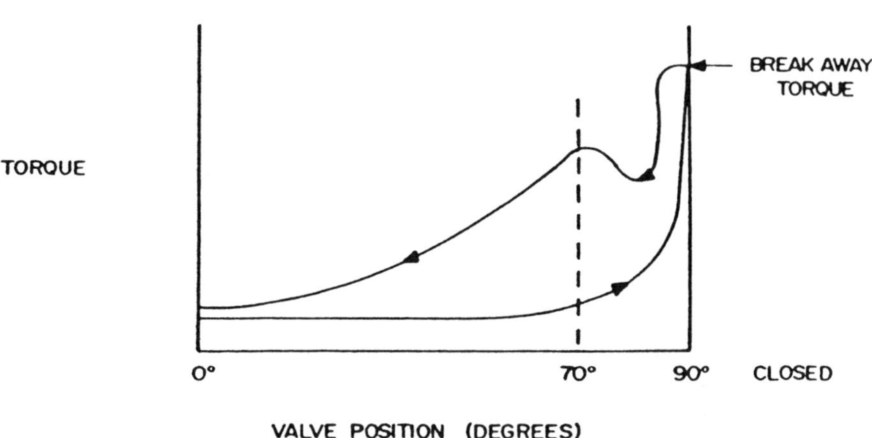

Figure 40 Butterfly torque.

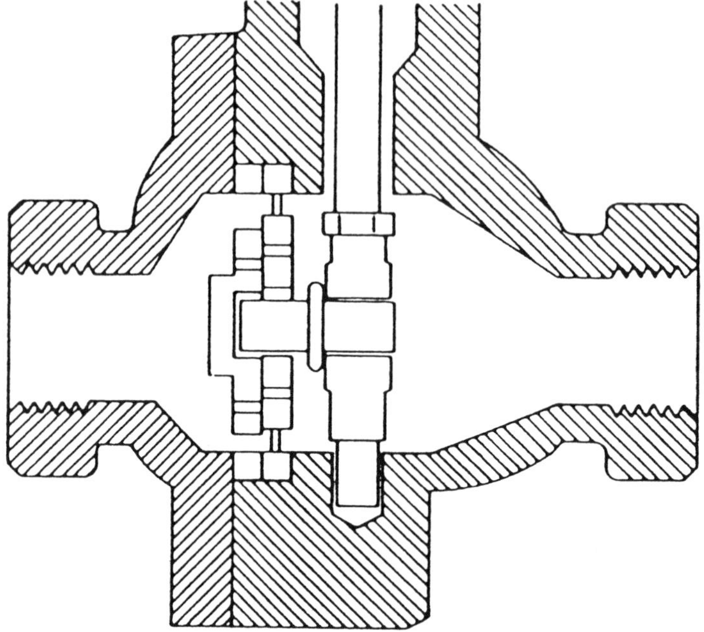

Figure 41 Gate valve, multiorifice.

Glove Valves

Glove valves are linear valves in which the plug moves up or down into a port. Some valves have two sets of plugs and ports; these are called double ported. In order to reduce stem size and to obtain better seating of the plug on the port, the stem or plug is mechanically guided. An example of a glove valve is shown in Figure 42.

Actuator size can also be reduced for double-ported valves with balanced flow. The ports and plugs are aligned such that the flowing stream tends to close one port and open the other, thus balancing the fluid forces.

Plug Valves

Plug valves are rotary-type valves with a conical or cylindrical shaped plug which has an orifice. Rotating the plug at a right angle to flow causes tight shutoff while full flow occurs when the orifice of the plug is parallel to the flow axis. A simple plug valve is shown in Figure 43. Also in Figure 43 is a popular version of the plug valve, the eccentric spherical plug or "camflex" valve. Just as a traditional plug valve is similar to a traditional ball valve, the eccentric spherical

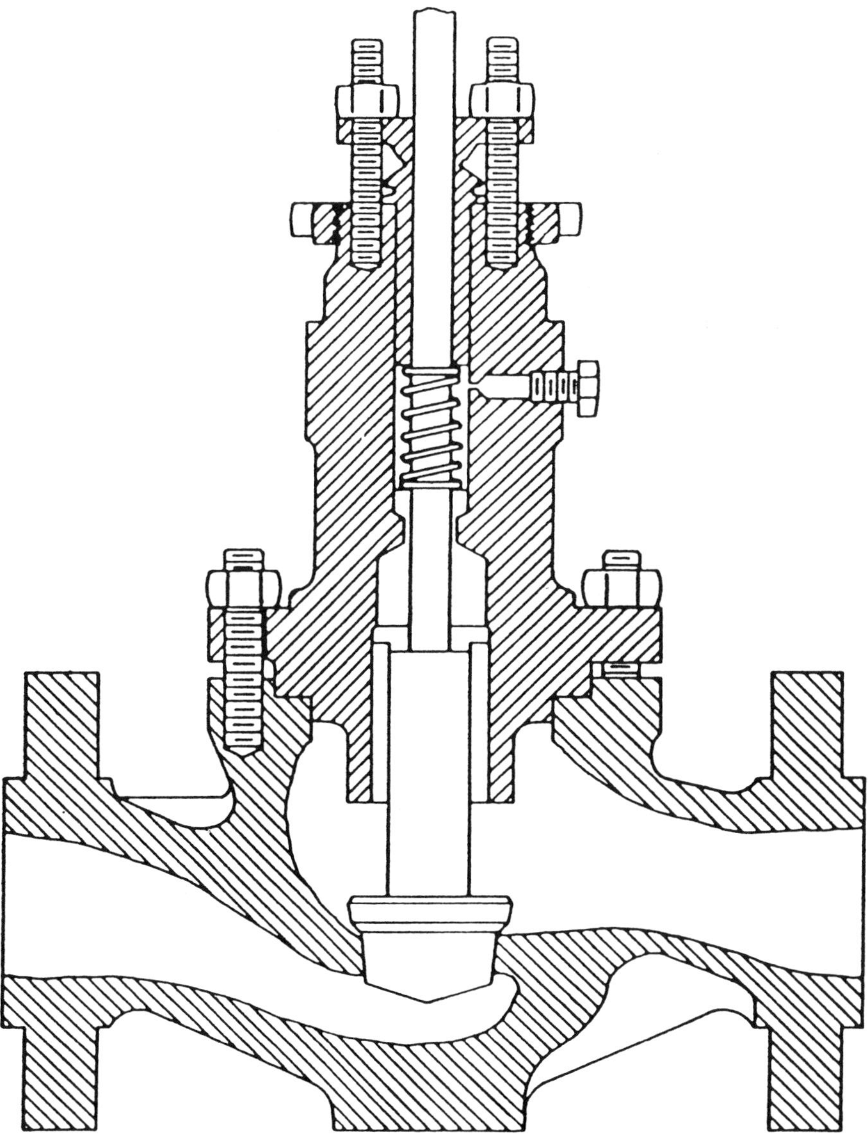

Figure 42 Globe valve.

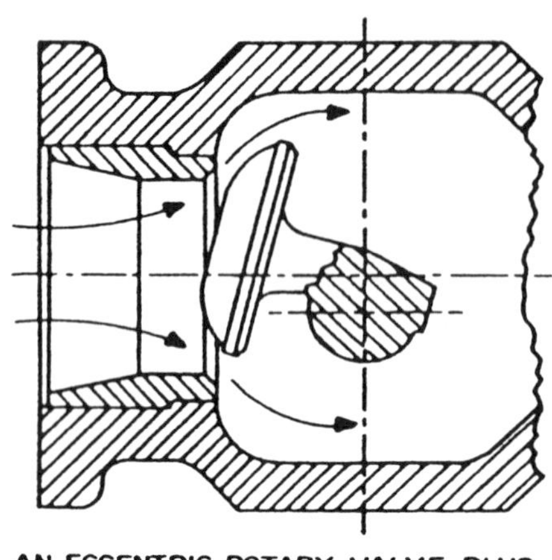

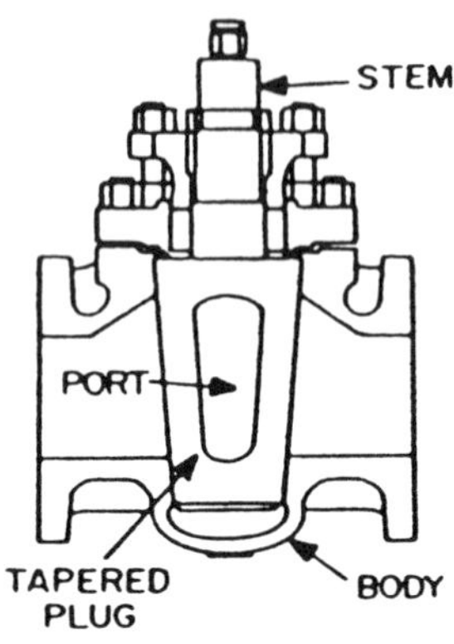

Figure 43 Plug valves.

plug is similar to the segmented ball valve. Both newer designs provide large flow for the valve size and high-pressure recovery, yet have reduced friction through most of the valve travel.

CONTROL VALVE ACTUATORS

Control valve actuators are available in a variety of types and sizes. In this section, some of the more common actuators are discussed. Included are pneumatic diaphragms and quarter-turn actuators.

Pneumatic Diaphragm

The energy of a compressible fluid, usually air, acts on a flexible member, the diaphragm, to provide linear motion of the actuator stem.

Pneumatic diaphragm actuators react to external pressure sources which may be the output pressure of a controller, an electric to pneumatic converter, a manual loader, or a positioner. Diaphragm actuators are available in two types, the spring and diaphragm and the springless diaphragm. A spring and diaphragm actuator is shown in Figure 44.

Characteristics of a pneumatic actuator are

- Pneumatic actuator offers high reliability at low cost and is easy to maintain.
- Fast acting.
- Available for either fail-open or fail-closed action.
- Normally uses air signal ranges of 3–15 psig.
- Has limited thrust capability. Normally limited to valves 8 in and smaller.

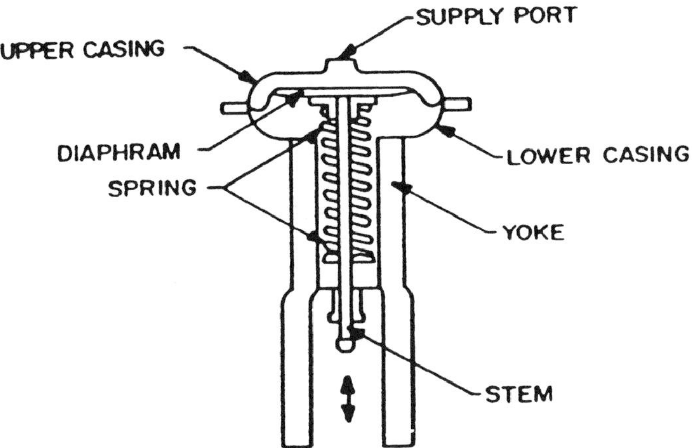

Figure 44 Linear pneumatic diaphragm actuator.

Control valves usually have relatively long stem travel, and stem position should be linearly proportional to the instrument output pressure (usually 3–15 psig) on the diaphragm. For a spring and diaphragm actuator, pressure on the diaphragm causes the stem to move. The force equals the pressure times the diaphragm area. This force is opposed by a spring. The valve and spring design permits a linear relationship between instrument air pressure and stem position. This assumes that the process flow force on the valve does not change over time.

The disadvantage of a spring is that its force is a constant at any one position. It cannot overcome varying friction and differential pressure forces. In a springless diaphragm actuator, the pressure on both sides of the diaphragm varies so that the sum of the two always equals the supply pressure. With a positioner, any reasonable supply pressure can be used. Pneumatic diaphragm-actuated modulating valves are typically butterfly, ball, and plug type valves or bladder type pinch valves. Remote position control from an external source is usually from pneumatic instruments. Current-to-pneumatic converters can be used for electrically based control systems.

Quarter-turn actuators are not a class of actuators. However, their use is common enough to include here. Motor gear train, piston, and pneumatic actuators can have a problem stopping a valve in the exact open and closed position. Limit switches can be used to operate the motor or pilot solenoids but it is difficult to set and keep limit switches in adjustment.

Characteristics of quarter-turn actuators are

- Either piston or diaphragm actuators can be used.
- Use where full closing of a valve to avoid leakage is important.
- Appropriate for butterfly, ball, and plug valves.

Two types of quarter-turn actuators are the rack and pinion type and the linkage type. For the rack and pinion type, a piston or diaphragm actuator moves a rack back and forth past a pinion. The linear motion of the piston is converted to circular motion. Precision adjusting screws are located at both ends of the cylinder to stop the piston at the desired position. There is no reliance on limit switches to stop the valve in the correct position.

A double-acting quarter-turn piston actuator is shown in Figure 45.

For the linkage-type actuator, the output force of the piston or diaphragm actuator is obtained by transferring the force through an arm to convert it to torque.

Figure 46 shows a double-acting piston linkage-type actuator and Figure 47 shows a diaphragm-actuated linkage-type actuator.

Both open/close and modulating operation is available for these actuators. Controls would be the same as that for the piston and diaphragm actuators discussed previously.

COMMON PUMP PROBLEMS

The following are common problems experienced in the field. These points should be checked to avoid difficulties and expensive service.

Failure to Deliver Liquid

- Pump is not primed or properly vented.
- High spot or air pocket in suction line which must be vented or preferably eliminated.
- Total head of system beyond pump's rating.

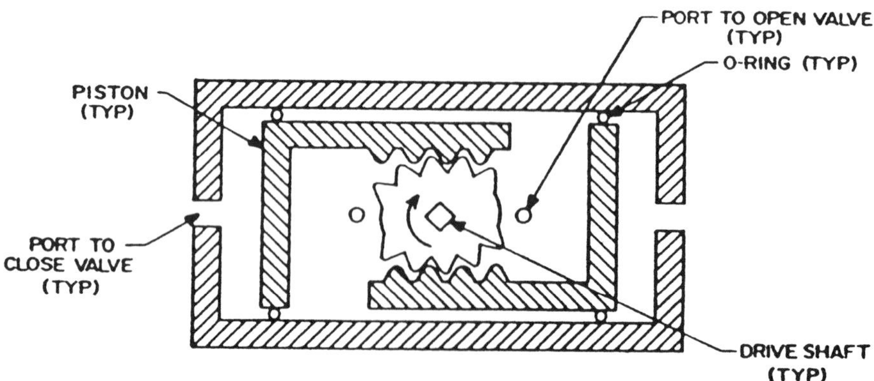

Figure 45 Double-acting piston, rack and pinion actuator.

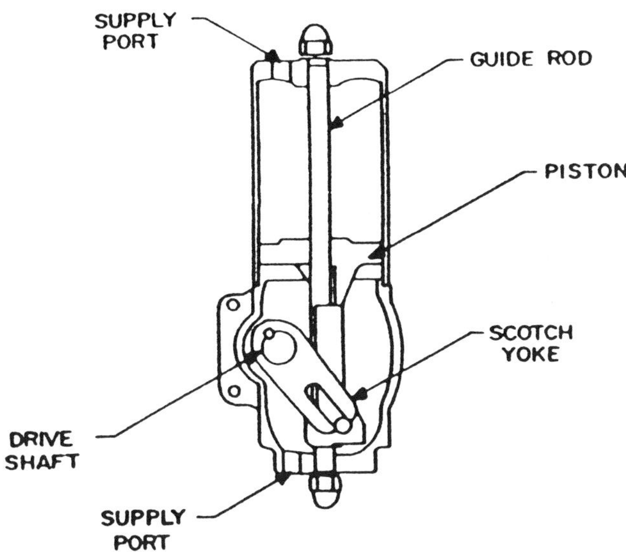

Figure 46 Double-acting piston, linkage actuator.

- Speed too low.
- Inadequate submergence (head on suction) where liquid handled is hot. (Consult manufacturer for the minimum submergence required.)
- Excessive suction life. (Consult manufacturer for maximum net suction lift which would include friction and other losses.)
- Impeller clogged.
- Incorrect direction of rotation Impeller mounted backwards on shaft.

Capacity Below Rating

Check previous points and also

- Air leaks in suction line or stuffing box
- Entrained air or gas in liquid being handled—crack pump casing vent
- Worn wearing rings
- Damaged impeller
- Defective packing
- Foot valve or suction line either too small or not sufficiently submerged
- Discharge valve not properly adjusted

Pressure Below Rating

Check previous points and also

- Pump delivering in excess of rated capacity—throttle, or if power is too high, consult manufacturer for reduced impeller diameter

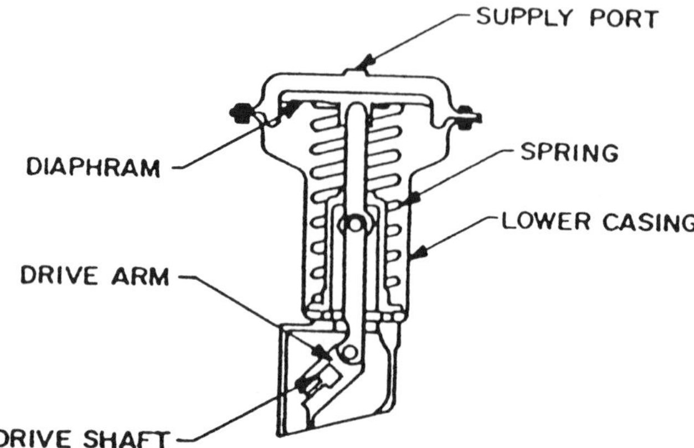

Figure 47 Diaphragm, rotary linkage actuator.

Loss of Prime

- Air leak in suction piping—add vent
- Air leak in stuffing box—add water seal
- Excessive suction lift—consult manufacturer
- Entrained gas in liquid—add vent line or install relief valve (at top of pump casing
- Water-sealing connection not functioning

Overheating of Pump at Shut-Off Condition

- Recirculating by-pass line plugged

Vibration and Noise

- Misalignment
- Unstable foundation
- Foreign material in pump causing unbalance
- Worn bearings
- Damaged impeller
- Defective packing
- Bent shaft
- Worn coupling
- Air bound—pump not fully primed
- Suction lift too high
- Pump delivering more than rated capacity
- Loose valve fittings in lines

Overheated Bearings

- Too much grease in ball bearing (or insufficient oil); poor alignment
- Damaged impeller or clearance rings
- Bent shaft
- Abnormal end thrust
- Dirt in bearings
- Excessive belt tension

Overload on Driver (Motor)

- Speed too high
- Pumping over capacity owing to existing head being lower than rated head
- Specific gravity or viscosity of liquid above design values
- Packing too tight or defective
- Wearing rings worn
- Shaft bent
- Poor alignment
- Rotor or wearing rings wedging or binding
- Consistency of paper stock being handled lower than design value
- Bearings improperly lubricated
- Motor improperly wired

Freezing of Liquid

- Pumps installed outdoors must be protected against freezing of the liquid pumped to prevent damage to pump parts.

APPENDIX

Useful Information and Data

A–1 Valences of Elements and Radicals

Hydrogen	H^+

Metal elements

Potassium	K^+
Sodium	Na^+
Silver	Ag^+
Mercurous	Hg^+
Mercuric	Hg^{2+}
Cuprous	Cu^+
Cupric	Cu^{2+}
Barium	Ba^{2+}
Calcium	Ca^{2+}
Lead	Pb^{2+}
Magnesium	Mg^{2+}
Nickel	Ni^{2+}
Zinc	Zn^{2+}
Ferrous	Fe^{2+}
Ferric	Fe^{3+}
Chromium	Cr^{3+}
Aluminum	Al^{3+}
Stannous	Sn^{2+}
Stannic	Sn^{4+}

Metal-acting radicals

Ammonium	NH_4^+
Methyl	CH_3^+
Ethyl	$C_2H_5^+$
Glyceryl	$C_3H_5^{3+}$

Oxygen	O^{2-}
Hydroxyl	OH^-

Acid radicals

Chloride	Cl^-
Bromide	Br^-
Iodide	I^-
Sulfide	S^{2-}
Chlorate	ClO_2^-
Chlorite	ClO_2^-
Hypochlorite	ClO^-
Perchlorate	ClO_4^-
Bromate	BrO_3^-
Bromite	BrO_2^-
Iodate	IO_3^-
Iodite	IO_2^-
Nitrate	NO_3^-
Nitrite	NO_2^-
Sulfate	SO_4^{2-}
Sulfite	SO_3^{2-}
Carbonate	CO_3^{2-}
Hydrogen carbonate	HCO_3^{2-}
Zincate	ZnO_2^{2-}
Stannite	SnO_2^{2-}
Phosphate	PO_4^{3-}
Monohydrogen phosphate	HPO_4^{2-}
Dihydrogen phosphate	$H_2PO_4^-$
Aluminate	AlO_3^{3-}

Organic acid radicals

Acetate	$C_2H_2O_2^-$
Palmitate	$C_{15}H_{31}CO_2^-$
Stearate	$C_{17}H_{35}CO_2^-$
Oleate	$C_{17}H_{33}CO_2^-$
Tartrate	$C_4H_4O_5^{2-}$
Oxalate	$C_2O_4^{2-}$

A–2 Metric Conversions

From/To	cm^3	liter	m^3	$in.^3$	ft^3	yd^3	fl oz	fl pt	fl qt	gal	bbl (oil)	bbl (liq)
cm^3	1	0.001	1×10^{-6}	0.06102	3.523×10^{-5}	1.31×10^{-6}	0.03381	0.00211	0.00106	2.64×10^{-4}	6.29×10^{-6}	8.39×10^{-6}
liter	1000	1	0.001	61.02	0.03532	0.00131	33.81	2.113	1.057	0.2642	0.00629	0.00839
m^3	1×10^{6}	1000	1	6.10×10^{4}	35.31	1.308	3.38×10^{4}	2113	1057	264.2	6.290	8.386
$in.^3$	16.39	0.01639	1.64×10^{-5}	1	5.79×10^{-4}	2.14×10^{-5}	0.5541	0.03463	0.01732	0.00433	1.03×10^{-4}	1.37×10^{-4}
ft^3	2.83×10^{4}	28.32	0.02832	1728	1	0.03704	957.5	59.84	29.92	7.481	0.1781	0.2375
yd^3	7.65×10^{5}	764.5	0.7646	4.67×10^{4}	27	1	2.59×10^{4}	1616	807.9	202.0	4.809	6.412
fl oz	29.57	0.02957	2.96×10^{-5}	1.805	0.00104	3.87×10^{-5}	1	0.06250	0.03125	0.00781	1.86×10^{-4}	2.48×10^{-4}
fl pt	473.2	0.4732	4.73×10^{-4}	28.88	0.01671	6.19×10^{-4}	16	1	0.5000	0.1250	0.00298	0.00397
fl qt	946.4	0.9463	9.46×10^{-4}	57.75	0.03342	0.00124	32	2	1	0.2500	0.00595	0.00794
gal	3785	3.785	0.00379	231.0	0.1337	0.00495	128	8	4	1	0.02381	0.03175
bbl (oil)	1.59×10^{5}	159.0	0.1590	9702	5.615	0.2079	5376	336	168	42	1	1.333
bbl (liq)	1.19×10^{5}	119.2	0.1192	7276	4.211	0.1560	4032	252	126	31.5	0.7500	1

A–3 Properties of Water

Temperature (°F)	Specific weight, γ (lb/ft^3)	Mass density, ρ (lb-s^2/ft^4)	Dynamic viscosity, $\mu \times 10^5$ (lb-s/ft^2)	Kinematic viscosity, $\nu \times 10^5$ (ft^2/s)	Surface energy, $\sigma \times 10^3$ (lb/ft)	Vapor pressure, p_v (lb/in.^{2})	Bulk modulus, $E \times 10^{-3}$ (lb/in.^{2})
32	62.42	1.940	3.746	1.931	5.18	0.09	290
40	62.43	1.938	3.229	1.664	5.14	0.12	295
50	62.41	1.936	2.735	1.410	5.09	0.18	300
60	62.37	1.934	2.359	1.217	5.04	0.26	312
70	62.30	1.931	2.050	1.059	5.00	0.36	320
80	62.22	1.927	1.799	0.930	4.92	0.51	323
90	62.11	1.923	1.595	0.826	4.86	0.70	326
100	62.00	1.918	1.424	0.739	4.80	0.95	329
110	61.86	1.913	1.284	0.667	4.73	1.24	331
120	61.71	1.908	1.168	0.609	4.65	1.69	333
130	61.55	1.902	1.069	0.558	4.60	2.22	332
140	61.38	1.896	0.981	0.514	4.54	2.89	330
150	61.20	1.890	0.905	0.476	4.47	3.72	328
160	61.00	1.896	0.838	0.442	4.41	4.74	326
170	60.80	1.890	0.780	0.413	4.33	5.99	322
180	60.58	1.883	0.726	0.385	4.26	7.51	318
190	60.36	1.876	0.678	0.362	4.19	9.34	313
200	60.12	1.868	0.637	0.341	4.12	11.52	308
212	59.83	1.860	0.593	0.319	4.04	14.7	300

A–4 U.S. National Drinking Water Regulations (May 1990 Safe Drinking Water Act)

Contaminant	SMCL[a]	Proposed SMCL[a]
Aluminum	—	0.05 mg/L
Chloride	250 mg/L	250 mg/L
Color	15 color units	15 color units
Copper	1 mg/L	1 mg/L
Corrositivity	Noncorrosive	Noncorrosive
Fluoride	2 mg/L	2 mg/L
Foaming agents	0.5 mg/L	0.5 mg/L
Iron	0.3 mg/L	0.3 mg/L
Manganese	0.05 mg/L	0.05 mg/L
Odor	3 (threshold odor number)	3 (threshold odor number)
pH	6.5–8.5	6.5–8.5
Silver	—	0.09 mg/L
Sulfate	250 mg/L	250 mg/L
Total dissolved solids (TDS)	500 mg/L	500 mg/L
Zinc	5 mg/L	5 mg/L

[a]Secondary maximum contaminant levels (SMCLs) are federally nonenforceable and establish limits for contaminants in drinking water which may affect the esthetic qualities and the public's acceptance of drinking water (e.g., taste and odor). These levels represent reasonable goals for drinking water quality. The states may establish higher or lower levels, which may be appropriate dependent on local conditions such as unavailability of alternate source waters or other compelling factors, and public health and welfare not being adversely affected.

A–5 Summary of National Primary Drinking Water Regulations (May 1990 Safe Drinking Water Act)

Contaminant	MCLG[a]	Proposed MCLG[a]	MCL[b]	Proposed MCL[b]
Microbiological contaminants				
Coliforms (total)	0		1/100 mL[c]	
Giardia Lamblia	0		TT[d]	
HPC	—		TT[d]	
Legionella	0		TT[d]	
Virus	0		TT[d]	
Turbidity: inorganic contaminants	—		1–5 NTU[e]	
Arsenic	—	—	0.05 mg/L	—
Asbestos	—	7 MF/L[f]	—	7 MFL
Barium	—	5 mg/L	1 mg/L	5 mg/L
Cadmium	—	0.005 mg/L	0.010 mg/L	0.005 mg/L
Chromium	—	0.1 mg/L	0.05 mg/L	0.1 mg/L
Cyanide	—	0.2 mg/L	—	0.2 mg/L

A–5 Continued

Contaminant	MCLG[a]	Proposed MCLG[a]	MCL[b]	Proposed MCL[b]
Fluoride	4.0 mg/L	—	4.0 mg/L	—
Lead	—	—	0.05 mg/L	—
Mercury	—	—	0.002 mg/L	—
Nickel	—	0.1 mg/L	—	0.1 mg/L
Nitrate[g]	—	10 mg/L	10 mg/L	10 mg/L
Nitrite[g]	—	1 mg/L	—	1 mg/L
Selenium	—	0.05 mg/L	0.01 mg/L	0.05 mg/L
Silver	—	—	0.05 mg/L	[h]
Turbidity: Organic contaminants				
2,4–D	—	0.07 mg/L	0.1 mg/L	0.07 mg/L
Endrin	—	0.002 mg/L	0.0002 mg/L	0.002 mg/L
Lindane	—	0.0002 mg/L	0.004 mg/L	0.0002 mg/L
Methoxychlor	—	0.4 mg/L	0.1 mg/L	0.4 mg/L
2,4,5-TP Silvex	—	0.05 mg/L	0.01 mg/L	0.05 mg/L
Benzene	0 mg/L	0 mg/L	0.005 mg/L	0.005 mg/L
Carbon tetrachloride	0 mg/L	0 mg/L	0.005 mg/L	0.005 mg/L
p-Dichlorobenzene	0.075 mg/L	0.075 mg/L	0.075 mg/L	0.075 mg/L
1,2-Dichloroethane	0 mg/L	0 mg/L	0.005 mg/L	0.005 mg/L
1,1-Dichlorethylene	0.007 mg/L	0.007 mg/L	0.007 mg/L	0.007 mg/L
1,1,1-Trichloroethane	0.20 mg/L	0.20 mg/L	0.20 mg/L	0.20 mg/L
Toluene	—	2 mg/L	—	2 mg/L
Trichlorethylene	0 mg/L	0 mg//L	0.005 mg/L	0.005 mg/L
Vinyl chloride	0 mg/L	0 mg/L	0.002 mg/L	0.002 mg/L
Xylenes	—	10 mg/L	—	10 mg/L
Total trihalomethanes (chloroform, bromoform, bromodichloromethane, dibromochloromethane)	—	—	0.10 mg/L	0.10 mg/L
Radionuclides				
Gross alpha particle activity	—	—	15 pCi/L	—
Gross beta partcile activity	—	—	4 mrem/yr	—
Radium 226 and 228 (total)	—	—	5 pCi/L	—

[a]Maximum contaminant level goal (MCLG) is a nonenforceable goal at which no known adverse health effects occur.

[b]Maximum contaminant level (MCL) is a federally enforceable standard.

[c]Revised regulations will be based on presence/absence concept rather than an estimate of coliform density, effective December, 1990.

[d]TT, Treatment technique requirements established in lieu of MCLs, effective December, 1990.

[e]Revised regulations will establish treatment technique requirements rather than an MCL for turbidity, effective December, 1990.

[f]MFL, Million fibers per liter longer than 10 μm.

[g]The MCLG and MCL for total nitrite and nitrate is 10 mg/L (as N).

[h]Deleted as primary regulation, proposed as secondary.

A–6 GLOSSARY OF WATER PURIFICATION TERMS

This glossary is not meant to be all-encompassing, but instead contains terms used in the different forms of purification and separation.

ASTM: American Society of Testing Materials—sets the standards for laboratory and electronics water.

absolute: Used in reference to the micrometer rating of cartridge or disc filters, indicating that all particles larger than a specified size will be trapped within or on the filter and will not pass through.

absorption: To take up a substance into the physical structure of a liquid or solid by physical or chemical action, but without chemical reaction.

acid rain: Rainfall above the natural pH range, caused by contact with atmospheric pollutants such as nitric and sulfuric oxides and carbon monoxide.

activated carbon: Granulated activated carbon used to remove tastes, odor, chlorine, chloramines, and some organics from water.

adsorption: The process by which molecules, colloids, or particles adhere to the surfaces by physical action but without chemical reaction.

aeration: The process of adding air to a water supply for the purpose of oxidation or mixing.

agglomerate: To bring together saller divisions to a larger mass.

alkalinity: Capacity of neutralizing acid, usually due to presence of bicarbonate or carbonate ions. Hydroxide, borate, silicate, or phosphate ions may contribute to alkalinity in treated waters.

angstrom: A unit of length equaling 10^{-10} meters, 10^{-4} micrometers, 10^{-8} centimeters and 4×10^{-9} inches. Symbol: Å.

anion: Negatively charged ion in an aqueous solution.

aquifer: Natural, undergound formation where mineral-bearing water is stored. Source of well water.

asbestos: A fibrous silicate material, chiefly calcium magnesium silicate; a noncombustible, non-conducting, and ehcmical-resistant material. A known lung carcinogen.

backwash: To reverse a solution's flow through a filtration system. Often used as a cleansing mechanism in sand and dual-media filters.

bacteria: Any of a class of microscopic single-celled organisms reproducing by fission or by spores. Characterized by round, rodlike spiral or filamentous bodies, often aggregated into colonies or mobile by means of flagella. Widely dispersed in soil, water, organic matter, and the bodies of plants and animals. Either autotrophic (self-sustaining, self-generative), saprophytic (derives nutrition from nonliving organic material already present in the environment), or parasitic (deriving nutrition from another living organism). Often symbiotic in humans, but sometimes pathogenic.

bactericide: Agent capable of destroying bacteria.
bacteriostat: Substance that inhibits bacterial growth and metabolism.
bar: Designation of pressure units. 1 bar = psi/14.5.
binders: In reference to cartridge filters, chemicals used to hold, or "bind," short fibers together in a filter.
blinding: In depth and surface filtration, a build-up of particulates on or within the filter, preventing fluid flow through the filter at normal pressures.
blowdown: In reference to boiler technology, the purge from the system of a small portion of concentrated boiler water in order to maintain a maximum level of dissolved and suspended solids in the system.
BOD (biochemical oxygen demand): A measure of the amount of oxygen required by bacteria for the biochemical degradation of organic material in a water sample.
bottled water: Commercial products sold in containers as pure water for drinking and domestic use.
candle turbidimeter: A device principally used to measure high-turbidity water with results expressed in Jackson turbidity units (JTU) or Formazine turbidity units (FTU). The JTU is measured with light scattering.
carbonate hardness: The hardness in a water caused by carbonates and bicarbonates of calcium and magnesium. The amount of hardness equivalent to the alkalinity formed and deposited when water is boiled. In boilers, carbonate hardness is readily removed by blowdown.
carcinogen: A substance or agent producing or inciting cancer.
cartridge: A filter device, usually disposable, filtering in the range of 0.1–100 μm, and usually 2–4 inches (51–102 mm) in diameter and 6–60 inches (152–1524 mm) in length.
caustic soda: Sodum hydroxide (NaOH), commonly known as lye. A commonly used chemical in water treatment.
cation: Positively charged ion in an aqueous solution.
chelating agents: A molecule, usually organic, which is soluble in water and undergoes reactions with metal ions to hold them in solution. A number of naturally occurring organic materials in water have chelating ability, such as humic acid and lignin. Owing to their chelating abilities, some organic materials interfere with water-softening processes.
chemical solution feeder: A pump used to meter chemicals, such as chlorine or polyphosphate, into a feed water supply.
chloramine: A combination of free chlorine and ammonia gas which retains its bactericidal qualities for a longer time than does free chlorine.
chlorination: The addition of small amounts of free chlorine, usually 0.2–2.0 ppm, to kill bacteria in a water supply.

chlorine: Chemical used for its qualities as a bleaching or oxidizing agent and disinfectant in water purification.

coagulant: Chemical added in water and wastewater applications to cause the formation of flocs that absorb, entrap, or otherwise bring together suspended matter so fine that it is defined as colloidal. Compounds of iron and aluminum are generally used to form flocs to allow removal of turbidity, bacteria, color, and other finely divided matter from water and wastewater.

coalescing: The separation of mixtures of immiscible fluids (such as oil and water) with different specific gravities. Can occur whenever two or more droplets collide and remain in contact and then become larger by passing through a coalescer. The enlarged drops then separate out of solution more rapidly.

COD (chemical oxygen demand): A measure of the organic matter content of a sample that can be oxidized by a strong chemical oxidant.

colloid: A substance of very fine particle size, typically between 0.1 and 0.001 μm in diameter suspended in liquid or dispersed in gas. Typically removable only by reverse osmosis, distillation, or ultrafiltration.

compaction: In crossflow filtration, the result of applied pressure compressing a reverse osmosis or ultrafiltration membrane which may result in a decline in flux.

compound: Composed of the union of separate elements, ingredients, or parts.

concentrate: In crossflow filtration, the portion of a feed stream which does not permeate the media but retains the ions, organics, and suspended particles which are rejected by the media.

concentration: The amount of material contained in a unit volume of fluid; the process of increasing the dissolved material per unit volume.

concentration polarization: In crossflow filtration, the formation of a more concentrated gradient of rejected material near the surface of the membrane causing either increased resistance to solvent transport, or an increase in local osmotic pressure, and possibly a change in rejection characteristics of the membrane.

condensate: Water obtained through evaporation and subsequent condensation. Normally the water resulting from condensing plant steam originally generated in a boiler. Water condensed in a water still operation is usually called distillate.

conductivity: The property of a substance's (in this case, water) ability to transmit electricity. The inverse of resistivity. Measured by a conductivity meter and described in microsiemen/cm or microhms/cm.

contaminant: A source of contamination, an impurity. Any substance in water which is not H_2O.

crossflow membrane filtration: A separation of the components of a fluid by semipermeable membranes through the application of pressure and tangential flow to the membrane surface. Includes the processes of reverse osmosis, ultrafiltration, nanofiltration, and microfiltration.

dalton: A unit of mass 1/16 the mass of oxygen, or 0.9997 mass unit. Named after John Dalton (1766–1844), the first theorist since Democritus (Greek, 5th century BC) to describe matter in terms of small particles—the founder of atomic theory. Nearly equivalent to 1 g of molecular weight of element or compound.

decarbonation: The process of removing CO_2 from water, typically using air contact towers.

degasification: The process of removing dissolved gasses in water, typically using a vacuum.

deionization (DI): Specially manufactured ion exchange resins which remove ionized salts from water. Can theoretically remove 100% of salts. Deionization typically does not remove organics, viruses, or bacteria, except through "accidental" trapping in the resin and specially made strong base anion resins which will remove gram-negative bacteria.

demineralization: The process of removing minerals from water, usually through deionization, reverse osmosis, or distillation.

detergent: A cleansing agent. Any of numerous synthetic water-soluble or liquid-organic preparations that are chemically different from soaps but resemble them in the ability to emulsify oils and hold dirt in suspension.

dialysis: A separation process dependent of different diffusion rates of solutes across a permeable membrane without pressure.

disinfection: To kill pathogenic organisms in a water supply or distribution system by means of heat, chemicals, or ultraviolet light.

dissolved solids: The residual material remaining after filtering the suspended material from water and evaporating the solution to a dry state.

distillate: The product water from distillation formed by condensing vapors.

distillation: The process of condensing steam from boiling water on a cool surface. Most contaminants do not vaporize and, therefore, do not pass to the distillate. Removes nearly 100% of all impurities.

effluent: The output stream exiting a treatment system.

electrodialysis: Dialysis that is conducted with the aid of an electromotive force applied to electrodes adjacent to both sides of the membrane.

endotoxin: A heat-resistant pyrogen, specifically a lipopolysaccharide found in the cell walls of viable and nonviable bacteria.

EPA: Environmental Protection Agency (United States): An organization that has set the potable water standards.

evaporation: Process where water passes from a liquid to a vapor state.

exhaustion: In water softening or ion exchange, the point where the resin can no longer exchange for additional ions.

FDA: Food and Drug Administration (United States).

feed: The input solution to a treatment/purification system, including the raw water supply prior to an treatment.

filter cake: The accumulated particles on a filter surface, usually from a slurry mixture.

filtrate: The portion of the feed stream which has passed through a filter.

floc: Coagulated groupings of formerly suspended particles which then settle by gravity.

flocculant: Chemical(s) which, when added to water, cause suspended particles to coagulate into larger groupings (flocs) which then settle by gravity.

flocculation: The process of agglomerating particles into larger groupings called flocs, which then settle by gravity.

flux: In crossflow filtration, the membrane throughput, usually expressed in volume per unit time per area, such as gallons per day per square foot or liters per hour per square meter.

fouling: In crossflow filtration, the reduction of flux that is attributed to a build-up of solids on the surface of the membrane.

FTU (formazine turbidity units): A measure of turbidity by a candle turbidimeter.

glassing: A form for silica scaling at high temperatures, usually in high-pressure boilers or stills.

gpd: Gallons per day.

grains per gallon (gpg): A unit of concentration equal to 17.1 milligrams per liter.

GRAS: Materials "generally regarded as safe," as listed by the FDA.

ground water: Water confined in permeable sand layers between rock or clay. All subsurface water.

hardness: The concentration of calcium and magnesium salts in water.

heavy metals: Metals having a high density or specific gravity of approximately 5 or higher. A generic term used to describe contaminants such as cadmium, lead, and mercury. Most are toxic to humans in low concentrations.

hemodialysis: The process of purifying a kidney patient's blood by means of dialysis membranes (see "dialysis").

hemolysis: Rupturing of red blood cells sometimes occurring during hemodialysis. May be caused by the presence of chloramines in the dialysis water supply.

high-purity water: Highly treated water with attention to microbiological reduction or elimination; the term commonly used in the pharmaceutical industry.

humic acid: A water-soluble organic compound composed of decayed vegetable matter which is leached into a water source by runoff. Present in most surface waters. Higher concentrations cause a brownish tint. Difficult to remove except by ultrafiltration or reverse osmosis.

hydrocarbon: An organic compound containing only carbon and hydrogen and often occurring in petroleum, natural gas, coal, and bitumens. Most successfully removed from water by coalescing (large volumes) or activated carbon (small volumes).

hydrogen sulfide (H_2S): A toxic gas that is detectable by a strong "rotten egg" odor. A common by-product of anaerobic bacteria.

hydrological cycle: The natural cycle of water as it passes through the environment by evaporation, condensation, precipitation, and retention in the oceans or on land.

injection: In water treatment, the introduction of a chemical or media into the process water for the purpose of altering its chemistry or filtering specific compounds.

ion: An atom or molecule which has lost or gained one or more electrons, thereby acquiring a net electric charge.

ion exchange: A process in which ions are preferentially adsorbed from a solution for equivalently charged ions attached to small solid structures (resin).

JTU (Jackson turbidity units): Turbidity test results from a candle turbidimeter.

LAL (Limulus amebocyte lysate): A reagent used in the detection of endotoxin, the pyrogen of greatest concern to the pharmaceutical industry. The LAL reagent is made from the blood of the horseshoe crab, *Limulus polyphemus*.

laminar flow: A flow in which rapid random fluctuations are absent.

LSI (Langelier saturation index): An expression of a calculation that allows the prediction of calcium carbonate precipitation at a specific condition, temperature, pH, TDS, hardness, and alkalinity.

medical device manufacturer: According to the FDA, a set of specific manufacturing and recordkeeping procedures which allows a manufacturer to be certified as a medical device manufacturer, the purpose of which is to assure physicians and patients that strict controls have been used and that component traceability is assured.

membrane: Highly engineered polymer film containing controlled distributions of pores. Membranes serve as a barrier permitting the passage of materials only up to a certain size, shape, or character. Membranes are used as the separation mechanism in reverse osmosis, electrodialysis, ultrafiltration, nanofiltration, and microfiltration, as disc filters in laboratories, and as pleated final filter cartridges, particularly in pharmaceutical and electronic applications.

mg/L: Milligrams of an element per liter of water.

microfiltration (MF): Filtration designed to remove particles and bacteria in the range of 0.1–2.0 or 3.0 μm in diameter.

micrometer: A metric unit of measurement equivalent to 10^{-6} meters or 10^{-4} centimeters. Symbol: μm.

mixed-bed: An ion exchange tank consisting of both cation and anion resin mixed together. Provides the most complete deionization of water, up to 18.3 MΩ/cm resistivity. Commonly used to polish water already treated by two-bed ion exchange tanks or reverse osmosis.

module: A membrane element combined with the membrane element housing.

molecule: The smallest physical unit of a substance, composed of one or more atoms, that retains the properties of that substance.

molecular weight (MW): The sum of the atomic weights of the constituents which make up a molecule. Often used to indicate size when referring to ultrafiltration of saccharide compounds (see dalton).

multiple-effect evaporation: Series-operation energy economizer system where heat from the steam generated (evaporated liquid) in the first stage is used to evaporate additional liquid in the second stage (by reducing system pressure), and so on, up to 10 or more effects.

nanofiltration (NF): A crossflow membrane separation process which removes particles in the 300–1000 molecular weight range, selected salts, and most organics.

nephelometer: A device used to measure mainly low-turbidity water with results expressed in nephelometric turbidity units (NTUs).

nominal: With regard to the micron rating of cartridge filters, refers to an approximate size particle, the vast majority of which will not pass through the filter. A small amount of particles this size or larger may pass through the filter.

noncarbonate hardness: Hardness caused by chlorides, sulfates, and nitrates of calcium and magnesium. Evaporation of waters containing these ions makes the water highly corrosive.

normal flow: The flow of the entire feedwater stream in a single direction directly through the filter media. The flow is generally "normal," or perpendicular, to the media.

NTU (nephelometric turbidity units): The result of passing a light beam through a water sample with a nephelometer to quantify low-turbidity water. The NTU is measured by light scattering.

osmosis: The spontaneous flow of water from a less concentrated solution to a more concentrated solution through a semipermeable membrane until energy equilibrium is achieved.

osmotic pressure: A measurement of the potential energy difference between solutions on either side of a semipermeable membrane. A factor in designing reverse osmosis equipment. The applied pressure must first overcome the osmotic pressure inherent in the chemical solution in order to get good purification and flux.

oxidize: To increase a molecule or ion in positive valence, losing electrons to an oxidizing agent.

oxidizing filters: Filters that use a catalytic media such as manganous oxides to oxidize iron and manganese and then filters the impurities from the water after they have been oxidized.

ozonator: A device which generates ozone by passing a high-voltage current through a chamber containing air or oxygen. Used as a disinfection system.

ozone (O_3): An unstable, highly reactive form of the oxygen formed by natural lightning or by passing a high-voltage electric charge through air. An excellent oxidizing agent and bactericide.

particle filtration (PF): Filtration rated in the range of 5–75 μm. Typically handled by cartridge filters.

particulate: Minute, separate pieces of matter.

permeable: Allowing some material to pass through.

permeate: That portion of the feed stream which passes through a membrane.

permeator: A hollow fine fiber membrane element consisting of thousands of hollow fibers.

pH: An expression of hydrogen ion concentration; specifically, the negative logarithm of the hydrogen ion concentration. The range 0–14, with 7 as neutral, 0–7 as acid, and 7–14 as alkaline (base).

phase: A state of matter, either solid, gaseous, or liquid.

polymer: A chemical compound with many repeating structural units produced by uniting many primary units, called monomers.

pore: An opening in a membrane or filter matrix.

porous: The ability of certain substances to pass fluids owing to an open physical structure.

ppb: Parts per billion, commonly considered equivalent to micrograms per liter (μg/L).

ppm: Parts per million, commonly considered equivalent to milligrams per liter (mg/L).

ppt: Parts per trillion, commonly considered equivalent to nanograms per liter (ng/L).

precipitate: An insoluble product of a chemical reaction of soluble compounds in water.

precipitation: The process of producing an insoluble reaction product from an aqueous chemical reaction, usually a crystalline compound that grows in size to be settleable.

precursors: Compounds such as humic acid which may lead to the creation of other compounds such as THMs.

psi: Pounds per square inch (pressure).

psig: Pounds per square inch gauge.

pyrogen: Any substance capable of producing a fever in mammals. Often an organic substance shed by bacteria during cell growth. Chemically and physically stable, they are not necessarily destroyed by conditions that kill bacteria.

reagent-grade water (ASTM): Water that meets the standards for reagent use promulgated by American Society of Testing and Materials. Four grades, RI through RIV, have been established depending upon intended use.

recirculation: (1) In crossflow membrane systems, the recycling of a portion of the concentrate stream to maintain a desirable flow across the membrane while the system is in operation. (2) In water system design, the continuous operation of the transfer pump to keep water flowing through the system above the use rate to reduce the hazard of bacterial growth.

regeneration: In ion exchange systems, the process of using either an acid, alkali, or salt solution to remove the accumulated cations or anions. The cation exchange resins take on hydrogen ions and the anion exchange resins take on hydroxide ions to restore themselves to the original hydrogen or hydroxide form when using strong acid and strong alkali solutions for the process.

rejection: In crossflow membrane systems, the process of retaining contaminants at the membrane that are larger than the membrane's pore sizes. In a membrane system, expressed as a percentage of the total presence of those contaminants.

resin: Specially manufactured polymer beads used in the ion exchange process to remove dissolved salts from water.

resistivity: The property of a substance (in this case, water) to resist the flow of electricity; the measurement of that resistance. The inverse of conductivity. Measured by a resistivity monitor, and described in Ω/cm.

reverse osmosis (RO): The separation of one component of a solution from another component by flowing the feed stream under pressure across a semipermeable membrane. RO removes ionized salts, colloids, and organics down to 150 molecular weight. Also called hyperfiltration.

saturation: The point at which a solution contains enough of a dissolved solid, liquid, or gas so that no more will dissolve into the solution at a given temperature and pressure.

scaling: The build-up of precipitated salts on a surface, such as pipes, tanks, or boiler condensate tubes.

SDI (silt density index): Test used to measure the level of suspended solids in feedwater for a reverse osmosis system.

semipermeable membrane: A membrane which allows a solvent such as water to pass through while rejecting certain dissolved or colloidal substances.

sepralator: A spiral-wound membrane cartridge or element in crossflow membrane systems. Modular and replaceable.

solutes: Matter dissolved in a solvent.

suspended solids (SS): Solid organic and inorganic particles in a solution that are held in suspension.

TDS (total dissolved solids): See dissolved solids.

THM (trihalogenated methane): Compounds formed by contact between free chlorine and certain organics to form materials similar to certain organic solvents. Considered a carcinogen.

TOC (total organic carbon): The amount of carbon bound in organic compounds in a water sample as determined by a standard laboratory test. The CO_2 is measured when a water sample is atomized in a combustion chamber.

traceability: In medical and pharmaceutical device manufacturing, the stringent recordkeeping on the use and origin of component materials.

transpires: The process of a plant giving off water directly to the air.

TS (total solids): The sum of total dissolved solids and total suspended solids.

TSS (total suspended solids): The residual matter which can be removed from a solution by filtration.

turbidity: A suspension of fine particles in water that causes cloudiness and will not readily settle because of small particle size.

turbidity units: Measurement of the relative ability of a solution to allow a light beam to pass through it.

two-bed: A pairing of cation and anion exchange tanks, typically operating in series. Best used for the deionization of relatively high volumes of water. Capable of product water resistivity of up to 1 MΩ/cm.

ultrafiltration (UF): Separation of one component of a solution from another component by means of pressure and flow exerted on a semipermeable membrane, with membrane pore sizes ranging from 10 Å to 0.2 μm. Typically rejects organics over 1000 MW while passing ions and small organics.

ultrapure water: Highly treated water of high resistivity and no organics; usually used in the semiconductor and pharmaceutical industries.

ultraviolet (UV): Radiation having a wave length shorter than visible light but no longer than x-rays. Ultraviolet light with a wavelength of 254 nm is used to kill bacteria.

USP: The United States Pharmacopeia, which publishes standards for the pharmaceutical industry, including those for water quality. Was established by the US Congress in 1884 to control make-up of drugs.

validation: In the pharmaceutical industry, the mandating of specific testing and recordkeeping procedures to ensure compliance not only with a specific quality but with a specific means to achieve that quality.

vaporize: To convert a liquid into a vapor.

virus: Any of a large group of submicroscopic infective agents capable of growth and multiplication only in living cells of a host.

VOC (volatile organic compound): Synthetic organic compounds which easily volatilize. Many are suspected carcinogens.

volatile: Readily passing off by evaporation.

WFI (water for injection): High-purity water intended for use as a solvent for the preparation of parenteral (injectable) solutions. Must meet specifications as listed in the USP.

WHO (World Health Organization, part of the United Nations): An international organization that has set the standards for potable water.

A–7 Wastewater-Treatment Processes

Pollutant removal	Treatment process and efficiency range (%)
Coarse solids	Screening (90–100)
Suspended solids	Coagulation (70–80)
	Flotation (40–60)
	Microstraining (50–60)
	Sedimentation (50–60)
Soluble organics	Biological (40–60)
	Activated carbon (65–75)
Oils	Activated carbon (90–95)
	Air flotation (80–90)
	Gravity separation (70–80)
Acids and bases	Neutralization (90–95)
Pathogens	Chlorination (98–99)
	Ozone (98–99)
Colloidal solids	Coagulation (70–80)
	Filtration (80–90)
	Microstraining (30–40)
	Ultrafiltration (95–98)
Ammonia	Air stripping (75–85)
	Ion exchange (90–95)
	Nitrification (75–85)
$NO–NO_2$	Denitrification (85–95)
Trace organics	Activated carbon (90–95)
	Biological systems (25–35)
	Air stripping (90–98)
	Ozone (70–80)
Heavy metals and inorganics	Activated carbon (to 95)
	(Chemical precipitation) (90–95)
	Ion exchange (90–95)
	Membrane processes (90–95)
	Ultrafiltration (0–75)

A–8 Removal Efficiencies of Selected Treatment Processes for Various Pollutants (%)

Processes	Pollutants											
	Coarse solids	Suspended solids	Soluble organics	Oils	Acids and bases	Pathogens	Colloidal solids	NH_3	NO_2–NO_3	P	Trace organics	Inorganics and heavy metals
Activated carbon			65–75	90–95							90–95	0–95
Aerobic biosystems			50–60							25–35	25–35	
Air stripping								75–85			90–98	
Anaerobic biosystems			40–50								25–35	
Chemical precipitation										90–95		90–95
Chlorination						98–99						
Coagulation and flocculation		70–80					70–80					
Denitrification									85–95			
Dissolved air flotation				80–90								
Electrodialysis											50–60	85–90
Filtration				80–90			80–90					
Flotation		40–60										
Gravity separation				70–80								
Ion exchange								90–95	90–95	90–95	70–80	90–95
Microstraining		50–60					30–40					
Neutralization					90–95							
Nitrification								75–85				
Ozonation						98–99						
Reverse osmosis											70–80	90–95
Screening	90–100											
Sedimentation		50–60										
Ultrafiltration							95–98				0–75	0–75

A–9 Typical Parameters that May Be Required in Wastewater Effluent Testing

pH	Magnesium	Zinc
Alkalinity or acidity	Hardness	Hexavalent chromium
Color	Aluminum	Total chromium
Conductivity	Iron	Total phosphate
Turbidity	Copper	Total nitrogen
Temperature	Fluoride	Ammonium nitrate
Suspended solids	Manganese	Oil and grease
Settleable solids	Cyanide	Phenol
Total organic carbon (TOC)	Nickel	Surfactants
Biochemical oxygen demand (BOD)	Lead	Chlorinated hydrocarbons
Chemical oxygen demand (COD)	Cadmium	Pesticides
Sulfate	Mercury	Fecal streptococci bacteria
Calcium	Arsenic	Coliform bacteria

A–10 SPECIFIC GRAVITY CONVERSION

Relationship Between Baumé Degrees and Specific Gravity for Liquids Lighter than Water

$$\text{Fomula—Specific Gravity} = \frac{140}{130 + {}^{\circ}\text{Baumé}}$$

Baumé degrees	Specific gravity (60°/60°F)	Baumé degrees	Specific gravity (60°/60°F)	Baumé degrees	Specific gravity (60°/60°F)	Baumé degrees	Specific gravity (60°/60°F)
10	1.00000	30	0.87500	50	0.77778	70	0.70000
11	0.99291	31	0.86957	51	0.77348	71	0.69652
12	0.98592	32	0.86420	52	0.76923	72	0.69307
13	0.97902	33	0.85890	53	0.76503	73	0.68966
14	0.97222	34	0.85366	54	0.76087	74	0.68627
15	0.96552	35	0.84848	55	0.75676	75	0.68293
16	0.95890	36	0.84337	56	0.75269	76	0.67961
17	0.95238	37	0.83832	57	0.74866	77	0.67633
18	0.94595	38	0.83333	58	0.74468	78	0.67308
19	0.93960	39	0.82840	59	0.74074	79	0.66986
20	0.93333	40	0.82353	60	0.73684	80	0.66667
21	0.92715	41	0.81871	61	0.73298	81	0.66351
22	0.92105	42	0.81395	62	0.72917	82	0.66038
23	0.91503	43	0.80925	63	0.72539	83	0.65728
24	0.90909	44	0.80460	64	0.72165	84	0.65421
25	0.90323	45	0.80000	65	0.71795	85	0.65117
26	0.89744	46	0.79545	66	0.71428	86	0.64815
27	0.89172	47	0.79096	67	0.71066	87	0.64516
28	0.88608	48	0.78652	68	0.70707	88	0.64220
29	0.88050	49	0.78212	69	0.70352	89	0.63927

Relationship Between Baumé Degrees and Specific Gravity for Liquids Heavier than Water

$$\text{Specific Gravity} = \frac{145}{145 - °\text{Baumé}}$$

Baumé degrees	Specific gravity (60°/60°F)	Baumé degrees	Specific gravity (60°/60°F)	Baumé degrees	Specific gravity (60°/60°F)	Baumé degrees	Specific gravity (60°/60°F)
0	1.00000	20	1.16000	40	1.38095	60	1.70588
1	1.00694	21	1.16935	41	1.39423	61	1.72619
2	1.01399	22	1.17886	42	1.40777	62	1.74699
3	1.02113	23	1.18852	43	1.42157	63	1.76829
4	1.02837	24	1.19835	44	1.43564	64	1.79012
5	1.03571	25	1.20833	45	1.45000	65	1.81250
6	1.04317	26	1.21849	46	1.46465	66	1.83544
7	1.05072	27	1.22881	47	1.47959	67	1.85897
8	1.05839	28	1.23932	48	1.49485	68	1.88312
9	1.06618	29	1.25000	49	1.51042	69	1.90789
10	1.07407	30	1.26087	50	1.52632	70	1.93333
11	1.08209	31	1.27193	51	1.54255	71	1.95946
12	1.09023	32	1.28319	52	1.55914	72	1.98630
13	1.09848	33	1.29464	53	1.57609	73	2.01389
14	1.10687	34	1.30631	54	1.59341	74	2.04225
15	1.11538	35	1.31818	55	1.61111	75	2.07143
16	1.12403	36	1.33028	56	1.62921	76	2.10145
17	1.13281	37	1.34259	57	1.64773	77	2.13235
18	1.14173	38	1.35514	58	1.66667	78	2.16418
19	1.15079	39	1.36792	59	1.68605	79	2.19697

Relation of API Hydrometer Scale to Specific Gravity

Degrees API	Specific gravity	Degrees API	Specific gravity	Degrees API	Specific gravity	Degrees API	Specific gravity
10	1.0000	30	0.8762	50	0.7796	70	0.7022
11	0.9930	31	0.8708	51	0.7753	71	0.6987
12	0.9861	32	0.8654	52	0.7711	72	0.6953
13	0.9792	33	0.8602	53	0.7669	73	0.6919
14	0.9725	34	0.8550	54	0.7628	74	0.6886
15	0.9659	35	0.8498	55	0.7587	75	0.6852
16	0.9593	36	0.8448	56	0.7547	76	0.6819
17	0.9529	37	0.8398	57	0.7507	77	0.6787
18	0.9465	38	0.8348	58	0.7467	78	0.6754
19	0.9402	39	0.8299	59	0.7428	79	0.6722
20	0.9340	40	0.8251	60	0.7389	80	0.6690
21	0.9279	41	0.8203	61	0.7351	81	0.6659
22	0.9218	42	0.8156	62	0.7313	82	0.6628
23	0.9159	43	0.8109	63	0.7275	83	0.6597
24	0.9100	44	0.8063	64	0.7238	84	0.6566
25	0.9042	45	0.8017	65	0.7201	85	0.6536
26	0.8984	46	0.7972	66	0.7165	86	0.6506
27	0.8927	47	0.7927	67	0.7128	87	0.6476
28	0.8871	48	0.7883	68	0.7093	88	0.6446
29	0.8816	49	0.7839	69	0.7057	89	0.6417

A–11 Decimal Equivalents

Fraction	Decimal	Fraction	Decimal
1/64	0.015625	33/64	0.515625
1/32	0.03125	17/32	0.53125
3/64	0.046875	35/64	0.546875
1/16	0.0625	9/16	0.5625
5/64	0.078125	37/64	0.578125
3/32	0.09375	19/32	0.59375
7/64	0.109375	39/64	0.609375
1/8	0.125	5/8	0.625
9/64	0.140625	41/64	0.640625
5/32	0.15625	21/32	0.65625
11/64	0.171875	43/64	0.671875
3/16	0.1875	11/16	0.6875
13/64	0.203125	45/64	0.703125
7/32	0.21875	23/32	0.71875
15/64	0.234375	47/64	0.734375
1/4	0.250	3/4	0.750
17/64	0.265625	49/64	0.765625
9/32	0.28125	25/32	0.78125
19/64	0.296875	51/64	0.796875
5/16	0.3125	13/16	0.8125
21/64	0.328125	53/64	0.828125
11/32	0.34375	27/32	0.84375
23/64	0.359375	55/64	0.859375
3/8	0.375	7/8	0.875
25/64	0.390625	57/64	0.890625
13/32	0.40625	29/32	0.90625
27/64	0.421875	59/64	0.921875
7/16	0.4375	15/16	0.9375
29/64	0.453125	61/64	0.953125
15/32	0.46875	31/32	0.96875
31/64	0.484375	63/64	0.984375
1/2	0.500	1	1.000

A–12 NOMENCLATURE AND SI CONVERSION TABLES

Metric units recommended for general use are given under the heading "API preferred metric unit." Other units that also may be needed are shown in the "other allowable" column. Preferred units do not preclude the use of other multiples or submultiples, as the choice of such unit-multiple is governed by the magnitude of the numerical value.

Notation used conforms to SI practice; that is, groups of three digits to the left or right of the decimal marker are separated by spaces. No commas or other triad spacers are used. E-notation is used for convenience because it is a standard method of display in many calculators and because of the inability of computers to print out or transmit superscripts. An asterisk (*) denotes that all the succeeding digits would be zeroes. If a conversion factor ends in zero but does not have an asterisk, then any subsequent digits would not necessarily be zeroes.

For example,

3.055 0 E+00 = $3.056\ 0 \times 10^{0}$ = 3.056 0
3.056* E–01 = $3.056\ 000 \times 10^{-1}$ = 0.305 600 0
9.290 304 E+02 = $9.290\ 304 \times 10^{2}$ = 929.030 4

Nomenclature for Conversion Tables

Symbol	Name	Quantity	Definition: In terms of other units	Definition: In terms of base units	Type of unit
A	Ampere	Electric current	—	—	Base
a	Annum (year)	Time	365 d	$3.153\ 600 \times 10^7$s	Allowable
bar	Bar	Pressure	10^5 Pa	10^5 kg/(m · s^2)	Allowable
Bq	Becquerel	Activity (of a radionuclide)	—	1 s^{-1}	Derived
C	Coulomb	Quantity of electricity, electric charge	—	1 A · s	Derived
°C	Degree celsius	Celsius temperature	—	—	Derived
cd	Candela	Luminous intensity	—	—	Base
cP	Centipoise	Dynamic viscosity	1 mPa · s	10^{-3} kg/(m · s)	Allowable
cSt	Centistokes	Kinematic viscosity	1 mm^2/s	10^{-6} m^2/s	Allowable
d	Day	Time	24 hr	8.640×10^4s	Allowable
F	Farad	Capacitance	1 C/V	1 s^4 A^2/(m^2 · kg)	Derived
g	Gram	Mass	—	10^{-3} kg	Allowable (submultiple of base unit)
Gy	Gray	Absorbed dose, specific energy imparted, kerma, absorbed dose index	1 J/kg	1 m^2/S^2	Derived
h	Hour	Time	60 min	3.6×10^3	Allowable
H	Henry	Inductance	1 Wb/A	1 m^2 · kg/(s^2 · A^2)	Derived
ha	Hectare	Area	—	10^4 m^2	Allowable
Hz	Hertz	Frequency (of periodic phenomenon)	—	1 s^{-1}	Derived
J	Joule	Work, energy, quantity of heat	1 N ·m	1 m^2 · kg/s^2	Derived
K	Kelvin	Thermodynamic temperature	—	—	Base
kg	Kilogram	Mass	—	—	Base
kn	Knot	Velocity	1.852 km/h	5.144 444 m^2/s	Allowable

Continued

Symbol	Name	Quantity	Definition		Type of unit
			In terms of other units	In terms of base units	
L	Liter	Volume	1 dm^3	10^{-3} m^3	Allowable
lm	Lumen	Luminous flux	—	1 cd · sr	Derived
lx	Lux	Illumination	1 lm/m^2	1 cd · sr/m^2	Derived
m	Meter	Length	—	—	Base
min	Minute	Time	—	60 s	Allowable
mol	Mole	Amount of substance	—	—	Base
N	Newton	Force	—	1 m · kg/s^2	Derived
naut. mi	Nautical mile	Distance	1.852 km	1.852 × 10^3 m	Allowable
Pa	Pascal	Pressure	1 N/m^2	1 kg/(m · s^2)	Derived
r	Revolution	Angular displacement	360°	2π rad	Allowable
rad	Radian	Plane angle	—	—	Supplementary
S	Siemens conductance	1 A/V	1 s^3 · $A^2/(m^2$ · kg)	Derived	
s	Second	Time	—	—	Base
sr	Steradian	Solid angle	—	—	Supplementary
Sv	Sievert	Dose equivalent	1 J/kg	1 m^2/s^2	Derived
t	Metric ton	Mass	1 Mg	10^3kg	Allowable
T	Tesla	Magnetic flux density	1 Wb/m^2	1 kg/(s^2 · A)	Derived
V	Volt	Electric potential, potential difference, electromotive force	1 W/A	1 m^2 · kg/(s^3 · A)	Derived
W	Watt	Power, radiant flux	1 J/s	1 m^2 · kg/s^3	Derived
Wb	Weber	Magnetic flux	1 V/s	1 m^2 ·kg/(s^2 · A)	Derived
Ω	Ohm	Electric resistance	1 V/A	1 m^2 · kg/(s^3 · A^2)	Derived
°	Degree	Plane angle	—	π/180 rad	Allowable

Recommended SI Units and Conversion Factors for Space and Time

Quantity	SI unit	Customary unit	Metric unit		Conversion factor (multiply quantity expressed in customary units by factor to get metric equivalent)	
			API preferred	Other allowable		
Length	m	naut mi	km		1.852*	E+00
				naut mi	1	
		mi	km		1.609 344*	E+00
		mi (US statute)	km		1.609 347	E+00
		Chain	m		2.011 684	E+01
		Rod	m		5.029 210	E+00
		Fathom	m		1.828 804	E+00
		m	m		1	
		yard	m		9.144*	E–01
		ft	m		3.048*	E–01
		ft (US survey)	m		3.048 006	E–01
		link	m		2.011 684	E–01
		in.	mm		2.54*	E+01
				cm	2.54*	E+00
		cm	mm		1.0*	E+01
				cm	1	
		mm	mm		1	
		mil	μm		2.54*	E+01
		micron (μ)	μm		1	
Surface texture	m	μin.	μm		2.54*	E–02
		nm	μm		1.0*	E–03

Continued

Quantity	SI unit	Customary unit	Metric unit: API preferred	Metric unit: Other allowable	Conversion factor (multiply quantity expressed in customary units by factor to get metric equivalent)	
Length/length	m/m	ft/mi	m/km		1.893 939	E–01
Length/volume	m/m^3	ft/US gal	m/m^3		8.051 964	E+01
		ft/ft^3	m/m^3		1.076 391	E+01
		ft/bbl	m/m^3		1.917 134	E+00
Length/temperature	m/K	see Temperature, Pressure, Vacuum				
Area	m^2	mi^2	km^2		2.589 988	E+00
		mi^2 (US statute)	km^2		2.589 998	E+00
		ha	m^2		1.0*	E+04
		acre	ha		4.046 873	E–01
				m^2	4.046 873	E+03
		sq chain	m^2		4.046 873	E+02
		sq rod	m^2		2.529 295	E+01
Area		yd^2	m^2		8.361 274	E–01
		ft^2	m^2		9.290 304*	E–02
		ft^2 (US survey)	m^2		9.290 341	E–02
		$in.^2$	mm^2		6.451 6*	E+02
				cm^2	6.451 6*	E+00
		cm^2	mm^2		1.0*	E+02
			cm^2	1		
		mm^2	mm^2		1	
Area/volume	m^2/m^3	$ft^2/in.^3$	m^2/cm^3		5.669 291	E–03

Area/mass	m^2/kg	cm^2/g	m^2/kg		1.0*	E–01
Volume, capacity	m^3	m^3	km^3		4.168 182	E+00
		acre · ft	m^3		1.233 489	E+03
				ha · m	1.233 489	E–01
		m^3	m^3		1	
		yd^3	m^3		7.645 549	E–01
		bbl (42 US gal)	m^3		1.589 873	E–01
		ft^3	m^3		2.831 685	E–02
			dm^3	L	2.831 685	E–01
		Can gal	m^3		4.546 09*	E–03
			dm^3	L	4.546 09*	E+00
		UK gal	m^3		4.546 092	E–03
			dm^3	L	4.546 092	E+00
		US gal	m^3		3.785 412	E–03
			dm^3	L	3.785 412	E+00
		L	dm^3	L	1	
		UK qt	dm^3	L	1.136 523	E+00
		US qt	dm^3	L	9.463 529	E–01
		UK pt	dm^3	L	5.682 615	E–01
		US pt	dm^3	L	4.731 765	E–001
Volume, capacity	m^3	UK fl oz	cm^3		2.841 308	E+01
		US fl oz	cm^3		2.957 353	E+01
		$in.^3$	cm^3		1.638 706	E+01
		mL	cm^3		1	
Volume/length	m^3/m	bbl/in.	m^3/m		6.259 342	E+00
(linear displacement)		bbl/ft	m^3/m		5.216 119	E–01
		ft^3/ft	m^3/m		9.290 304*	E–02
		U.S. gal/ft	dm^3/m	L/m	1.241 933	E+01
Volume/mass	m^3/kg	See Density, Specific Volume, Concentration, Dosage				

Continued

Quantity	SI unit	Customary unit	Metric unit: API preferred	Metric unit: Other allowable	Conversion factor (multiply quantity expressed in customary units by factor to get metric equivalent)	
Plane angle	rad	rad	rad		1	
		deg (°)	rad		1.745 329	E–02
				°	1	
		min (’)	rad		2.908 882	E–04
				’	1	
		sec (”)	rad		4.848 137	E–06
				”	1	
Solid angle	sr	sr	sr		1	
Time	s	million years (MY)	Ma		1	
		yr	a		1	
		wk	d		7.0*	E+00
		d	d		1	
		h	h		1	
				min	6.0*	E+01
		min	s		6.0*	E+01
				h	1.666 667	E–02
				min	1	
		s	s		1	
		millimicrosecond	ns		1	

Recommended SI Units and Conversion Factors for Mass and Amount of Substance

Quantity	SI unit	Customary unit	Metric unit: API preferred	Metric unit: Other allowable	Conversion factor (multiply quantity expressed in customary units by factor to get metric equivalent)	
Mass	kg	UK ton (long ton)	Mg	t	1.016 047	E+00
		US ton (short ton)	Mg	t	9.071 847	E–01
		UK cwt	kg		5.080 235	E+01
		US cwt	kg		4.535 924	E+01
		kg	kg		1	
		lb	kg		4.535 924	E–01
		oz (troy)	g		3.110 348	E+01
		oz (avdp)	g		2.834 952	E+01
		g	g		1	
		grain	mg		6.479 891	E+01
		mg	mg		1	
		μg	μg		1	
Mass/length	kg/m	see Mechanics				
Mass/area	kg/m^2	see Mechanics				
Mass/volume	kg/m^3	see Density, Specific Volum, Concentration, Dosage				
Mass/mass	kg/kg	see Density, Specific Volume, Concentration, Dosage				
Amount of substance	mol	ft^3 (60°F, 1 atm)	kmol		1.195 29	E–03
		ft^3 (60°F, 14.73 lbf/in.^{2})	kmol		1.198 06	E–03
		m^3 (0°C, 1 atm)	kmol		4.461 53	E–02
		m^3 (15°C, 1 atm)	kmol		4.229 28	E–02
		m^3 (20°C, 1 atm)	kmol		4.157 15	E–02
		m^3 (25°C, 1 atm)	kmol		4.087 43	E–02

Recommended SI Units and Conversion Factors for Heating Value, Entropy, and Heat Capacity

Quantity	SI unit	Customary unit	Metric unit		Conversion factor (multiply quantity expressed in customary units by factor to get metric equivalent)	
			API preferred	Other allowable		
Heating value (mass basis)	J/kg	Btu/lb	MJ/kg		2.326 000	E–03
			kJ/kg	J/g	2.326 000	E+00
				kW · h/kg	6.461 112	E–04
		cal/g	kJ/kg	J/g	4.184*	E+00
		cal/lb	J/kg		9.224 141	E+00
Heating value (mole basis)	J/mol	kcal/g mol	kJ/kmol		4.184*	E+03
		Btu/lb mol	MJ/kmol		2.326 000	E–03
			kJ/kmol		2.326 000	E+00
Heating value (volume basis-solids and liquids)	J/m^3	therm/US gal	MJ/m^3	kJ/dm^3	2.787 163	E+04
			kJ/m^3		2.787 163	E+07
				$kW \cdot h/dm^3$	7.742 119	E+00
		therm/UK gal	MJ/m^3	kJ/dm^3	2.320 798	E+04
			kJ/m^3		2.320 798	E+07
				$kW \cdot h/dm^3$	6.446 660	E+00
		therm/Can gal	MJ/m^3	kJ/dm^3	2.320 799	E+04
			kJ/m^3		2.320 799	E+07
				$kW \cdot h/dm^3$	6.446 663	E+00
		Btu/US gal	MJ/m^3	kJ/dm^3	2.787 163	E–01
			kJ/m^3		2.787 163	E+02
				$kW \cdot h/m^3$	7.742 119	E–02
		Btu/UK gal	MJ/m^3	kJ/dm^3	2.320 800	E–01
			kJ/m^3		2.320 800	E+02
				$kW \cdot h/m^3$	6.446 660	E–02

		Btu/Can gal	MJ/m^3	kJ/dm^3	2.320 799	E–01
			kJ/m^3		2.320 799	E+02
				$kW \cdot h/m^3$	6.446 663	E–02
		Btu/ft^3	MJ/m^3	kJ/dm^3	3.725 895	E–02
			kJ/m^3		3.725 895	E+01
				$kW \cdot h/m^3$	1.034 971	E–02
		$kcal/m^3$	MJ/m^3	kJ/dm^3	4.184*	E–03
			kJ/m^3		4.184*	E+00
		cal/mL	MJ/m^3		4.184*	E+00
		ft · lbf/US gal	kJ/m^3		3.581 692	E–01
Heating value	J/m^3	cal/mL	kJ/m^3	J/dm^3	4.184*	E+03
(volume basis—gases)		$kcal/m^3$	kJ/m^3	J/dm^3	4.184*	E+00
		Btu/ft^3	kJ/m^3	J/dm^3	3.725 895	E+01
Specific entropy	J/(kg · K)	Btu/(lb · °R)	kJ/(kg · K)	J/(g ·K)	4.186 8*	E+00
		cal/(g · K)	kJ/(kg · K)	J/(g ·K)	4.184*	E+00
		kcal/(kg · °C)	kJ/(kg · K)	J/(g · K)	4.184*	E+00
Specific heat	J/(kg · K)	kW · h/(kg · °C)	kJ/(kg · K)	J/(g · °C)	3.6*	E+03
Capacity		Btu/(lb · °F)	kJ/(kg · K)	J/(g · °C)	4.186 8 *	E+00
(mass basis)		kcal/(kg · °C)	kJ/(kg · K)	J/(g · °C)	4.184*	E+00
Molar heat	J/(mol · K)	Btu/(lb mol · °F)	kJ/(kmol · K)	J/(g · °C)	4.186 8*	E+00
Capacity		cal/(g mol · °C)	kJ/(kmol · K)	J/(g · °C)	4.184*	E+00

Recommended SI Units and Conversion Factors for Temperature, Pressure, and Vacuum

Quantity	SI unit	Customary unit	Metric unit: API preferred	Metric unit: Other allowable	Conversion factor (multiply quantity expressed in customary units by factor to get metric equivalent)	
Temperature (absolute)	K	°R	K		5/9	
		K	K		1	
Temperature (traditional)	K	°F	°C		(°F – 32)/1.8	
		°C	°C		1	
Temperature (difference)	K	°F	K	°C	5/9	
		°C	K	°C	1	
Temperature/length (geothermal gradient)	K/m	°F/100 ft	mK/m		1.822 689	E+01
Length/temperature (geothermal step)	m/K	ft/°F	m/K		5.486 4*	E–01
Pressure	Pa	atm (14.696 lbf/in.^{2}	MPa		1.013 250*	E–01
		or 760 mm Hg at 0°C)	kPa		1.013 250*	E+02
				bar	1.013 250*	E+00
		bar	MPa		1.0*	E–01
			kPa		1.0*	E+02
				bar	1	
		at (kgf/cm^2)	Mpa		9.806 650*	E–02
		(technical atmosphere)	kPa		9.806 650	E+01
				bar	9.806 650*	E–01
		lbf/in.^{2} (psi)	MPa		6.894 757	E–03
			kPa		6.894 757	E+00
				bar	6.894 757	E–02

		in. Hg at 60°F	kPa	3.376 85	E+00
		in. Hg at 32°F	kPa	3.386 38	E+00
		in. H_2O at 39.2°F	kPa	2.490 82	E–01
		in. H_2O at 60°F	kPa	2.488 4	E–01
		mm Hg at 0°C (torr)	kPa	1.333 22	E–01
		cm H_2O at 4°C	kPa	9.806 38	E–02
		lbf/ft^2(psf)	kPa	4.788 026	E–02
		μm Hg at 0°C	Pa	1.333 22	E–01
		μbar	Pa	1.0*	E–01
		dyn/cm^2	Pa	1.0*	E–01
Vacuum, draft	Pa	in. Hg at 60°F	kPa	3.376 85	E+00
		in. H_2O at 39.2°F	kPa	2.490 82	E–01
		in. H_2O at 60°F	kPa	2.488 4	E–01
		mm Hg at 0°C (torr)	kPa	1.333 22	E–01
		cm H_2O at 4°C	kPa	9.806 38	E–02
Liquid head	m	ft	m	3.048*	E–01
		in.	mm	2.54*	E+01
Pressure drop/length	Pa/m	psi/ft	kPa/m	2.262 059	E+01
		psi/100 ft	kPa/m	2.262 059	E–01
		psi/mi	kPa/km	4.284 203	E+00

Recommended SI Units and Conversion Factors for Density, Specific Volume, Concentration, and Dosage

Quantity	SI unit	Customary unit	Metric unit: API preferred	Metric unit: Other allowable	Conversion factor (multiply quantity expressed in customary units by factor to get metric equivalent)	
Density (gases)	kg/m^3	lb/ft^3	kg/m^3		1.601 846	E+01
Density (liquids)	kg/m^3	lb/US gal	kg/m^3		1.198 264	E+02
				kg/dm^3	1.198 264	E–01
		lb/UK gal	kg/m^3		9.977 633	E+01
				kg/dm^3	9.977 633	E–02
		lb/ft^3	kg/m^3		1.601 846	E+01
				kg/dm^3	1.601 846	E–02
		g/cm^3	kg/m^3		1.0*	E+03
				kg/dm^3	1	
		kg/L	kg/m^3		1.0*	E+03
		°API	kg/m^3		Use tables	
Density (solids)	kg/m^3	lb/ft^3	kg/m^3		1.601 846	E+01
				kg/dm^3	1.601 846	E–02
Specific volume (gases)	m^3/kg	ft^3/lb	m^3/kg		6.242 796	E–02
				dm^3/kg	6.242 796	E+01
Specific volume (liquids)	m^3/kg	ft^3/lb	m^3/kg		6.242 796	E–02
				dm^3/kg	6.242 796	E+01
		UK gal/lb	m^3/kg		1.022 242	E–02
				dm^3/kg	1.022 242	E+01
		US gal/lb	m^3/kg		8.345 404	E–03
				dm^3/kg	8.345 404	E+00
Molar volume	m^3/mol	L/g mol	$m^3/kmol$		1	
		ft^3/lb mol	$m^3/kmol$		6.242 796	E–02

Specific volume	m^3/kg	bbl/US ton	m^3/Mg	m^3/t	1.752 535	E–01
(clay yield)		bbl/UK ton	m^3/Mg	m^3/t	1.564 763	E–01
Yield (shale distillation)	m^3/kg	bbl/US ton	dm^3/Mg	dm^3/t	1.752 535	E+02
		bbl/UK ton	dm^3/Mg	dm^3/t	1.564 763	E+02
		U.S. gal/US ton	dm^3/Mg	dm^3/t	4.172 702	E+00
		U.S. gal/UK ton	dm^3/Mg	dm^3/t	3.725 627	E+00
Concentration	kg/kg	wt%	kg/kg		1.0*	E–02
(mass/mass)			g/kg		1.0*	E+01
		wt ppm	mg/kg		1	
Concentration	kg/m^3	lb/bbl	kg/m^3	g/dm^3	2.853 010	E+00
(mass/volume)		g/US gal	kg/m^3		2.641 720	E–01
		g/UK gal	kg/m^3		2.199 692	E–01
		lb/1000 US gal	g/m^3	mg/dm^3	1.198 264	E+02
		lb/1000 UK gal	g/m^3	mg/dm^3	9.977 633	E+01
		gr/US gal	g/m^3	mg/dm^3	1.711 806	E+01
		lb/1000 bbl	g/m^3	mg/dm^3	2.853 010	E+00
		mg/US gal	g/m^3	mg/dm^3	2.641 720	E–01
		gr/100 ft^3	mg/m^3		2.288 352	E+01
		gr/ft^3	mg/m^3		2.288 352	E+03
Concentration	m^3/m^3	bbl/bbl	m^3/m^3		1	
(volume/volume)		ft^3/ft^3	m^3/m^3		1	
		bbl/(acre · ft)	dm^3/m^3	L/m^3	1.288 923	E–01
		UK gal/ft^3	dm^3/m^3	L/m^3	1.605 437	E+02
		US gal/ft^3	dm^3/m^3	L/m^3	1.336 806	E+02
		mL/US gal	dm^3/m^3	L/m^3	2.641 720	E–01
		mL/UK gal	dm^3/m^3	L/m^3	2.199 692	E–01
		Vol %	m^3/m^3		1.0*	E–02
		Vol ppm	cm^3/m^3		1	
			dm^3/m^3	L/m^3	1.0*	E–03
		UK gal/1000 bbl	cm^3/m^3		2.859 406	E+01
		US gal/1000 bbl	cm^3/m^3		2.380 952	E+01

Continued

Quantity	SI unit	Customary unit	Metric unit: API preferred	Metric unit: Other allowable	Conversion factor (multiply quantity expressed in customary units by factor to get metric equivalent)	
Concentration (mole/volume)	mol/m^3	lb mol/US gal	kmol/m^3		1.198 264	E+02
		lb mol/UK gal	kmol/m^3		9.977 633	E+01
		lb mol/ft^3	kmol/m^3		1.601 846	E+01
		std ft^3(60°F, 1 atm)/bbl	kmol/m^3		7.518 18	E–03
Concentration (volume/mole)	m^3/mol	US gal/10000 std ft^3 (60°F/60°F)	dm^3/kmol	L/kmol	3.166 93	E+00
		bbl/million std ft^3 (60°F/60°F)	dm^3/kmol	L/kmol	1.330 11	E–01

Recommended SI Units and Conversion Factors for Facility Throughput and Capacity

Quantity	SI unit	Customary unit	Metric unit: API preferred	Metric unit: Other allowable	Conversion factor (multiply quantity expressed in customary units by factor to get metric equivalent)	
Throughput (mass basis)	kg/s	million lb/yr	Mg/a	t/a	4.535 924	E+02
		UK ton/yr	Mg/a	t/a	1.016 047	E+00
		US ton/yr	Mg/a	t/a	9.071 847	E–01
		UK ton/d	Mg/d	t/d	1.016 047	E+00
				t/h, Mg/h	4.233 529	E–02
		US ton/d	Mg/d	t/d	9.071 847	E–01
				t/h, Mg/h	3.779 936	E–02
		UK ton/h	Mg/h	t/h	1.016 047	E+00
		US ton/h	Mg/h	t/h	9.071 847	E–01
		lb/h	kg/h		4.535 924	E–01
Throughput (volume basis)	m^3/s	bbl/d	m^3/a		5.803 036	E+01
			m^3/d		1.589 873	E–01
			m^3/h		6.624 471	E–03
		ft^3/d	m^3/h		1.179 869	E–03
		bbl/h	m^3/h		1.589 873	E–01
		ft^3/h	m^3/h		2.831 685	E–02
		UK gal/h	m^3/h		4.546 092	E–03
				L/h	4.546 092	E+00
		US gal/h	m^3/h		3.785 412	E–03
				L/h	3.785 412	E+00
		UK gal/min	m^3/h		2.727 655	E–01
				L/min	4.546 092	E+00
		US gal/min	m^3/h		2.271 247	E–01
				L/min	3.785 412	E+00
Throughput (mole basis)	mol/s	lb mol/h	kmol/h		4.535 924	E–01
				kmol/s	1.259 979	E–04
Pipeline capacity	m^3/m	bbl/mi	m^3/km		9.879 013	E–02

Recommended SI Units and Conversion Factors for Flowrate

Quantity	SI unit	Customary unit	Metric unit		Conversion factor (multiply quantity expressed in customary units by factor to get metric equivalent)	
			API preferred	Other allowable		
Flowrate (mass basis)	kg/s	UK ton/min	kg/s		1.693 412	E+01
		US ton/min	kg/s		1.511 975	E+01
		UK ton/h	kg/s		2.822 353	E–01
		US ton/h	kg/s		2.519 958	E–01
		million lb/d	kg/s		5.249 912	E+00
		UK ton/d	kg/s		1.175 980	E–02
		US ton/d	kg/s		1.049 982	E–02
		million lb/yr	kg/s		1.438 332	E–02
		UK ton/yr	kg/s		3.221 864	E–05
		US ton/yr	kg/s		2.876 664	E–05
		lb/s	kg/s		4.535 924	E–01
		lb/min	kg/s		7.559 873	E–03
		lb/h	kg/s		1.259 979	E–04
Flowrate (volume basis)	m^3/s	bbl/d	dm^3/s		1.840 131	E–03
		ft^3/d	dm^3/s		3.277 413	E–04
		bbl/h	dm^3/s		4.416 314	E–02
		ft^3/h	dm^3/s		7.865 791	E–03
		UK gal/h	dm^3/s		1.262 803	E–03
		US gal/h	dm^3/s		1.051 503	E–03
		UK gal/min	dm^3/s		7.576 820	E–02
		US gal/min	dm^3/s		6.309 020	E–02
		ft^3/min	dm^3/s		4.719 474	E–01
		ft^3/s	dm^3/s		2.831 685	E+01

Flowrate (mole basis)	mol/s	lb mol/s	kmol/s		4.535 924	E–01
		lb mol/h	kmol/s		1.259 979	E–04
		million scf/sd	kmol/s		1.383 449	E–02
Flowrate/length (mass basis)	kg/(s · m)	lb/(s · ft)	kg/(s · m)		1.488 164	E+00
		lb/(h · ft)	kg/(s · m)		4.133 789	E–04
Flowrate/length (volume basis)	m^2/s	UK gal/(min · ft)	m^2/s	$m^3/(s \cdot m)$	2.485 833	E–04
		US gal/(min · ft)	m^2/s	$m^3/(s \cdot m)$	2.069 888	E–04
		UK gal/(h · in.)	m^2/s	$m^3/(s \cdot m)$	4.971 667	E–05
		US gal/(h · in.)	m^2/s	$m^3/(s \cdot m)$	4.139 776	E–05
		UK gal/(h · ft)	m^2/s	$m^3/(s \cdot m)$	4.143 055	E–06
		US gal/(h · ft)	m^2/s	$m^3/(s \cdot m)$	3.449 814	E–06
Flowrate/area (mass basis)	$kg/(s \cdot m^2)$	$lb/(s \cdot ft^2)$	$kg/(s \cdot m^2)$		4.882 428	E+00
		$lb/(h \cdot ft^2)$	$kg/(s \cdot m^2)$		1.356 230	E–03
Flowrate/area (volume basis)	m/s	$ft^3/(s \cdot ft^2)$	m/s	$m^3/(s \cdot m^2)$	3.048*	E–01
		$ft^3/(min \cdot ft^2)$	m/s	$m^3/(s \cdot m^2)$	5.08*	E–03
		UK gal/(h · in.^{2})	m/s	$m^3/(s \cdot m^2)$	1.957 349	E–03
		US gal/(h · in.^{2})	m/s	$m^3/(s \cdot m^2)$	1.629 833	E–03
		UK gal/(min · ft^2)	m/s	$m^3/(s \cdot m^2)$	8.155 621	E–04
		US gal/(min · ft^2)	m/s	$m^3/(s \cdot m^2)$	6.790 972	E–04
		UK gal/(h · ft^2)	m/s	$m^3/(s \cdot m^2)$	1.359 270	E–05
		US gal/(h · ft^2)	m/s	$m^3/(s \cdot m^2)$	1.131 829	E–05
Flowrate/pressure drop (productivity index)	$m^3/(s \cdot Pa)$	bbl/(d · psi)	$m^3/(d \cdot kPa)$		2.305 916	E–02

Recommended SI Units and Conversion Factors for Energy, Work, and Quantity of Heat, Power

Quantity	SI unit	Customary unit	Metric unit		Conversion factor (multiply quantity expressed in customarry units by factor to get metric equivalent)	
			API preferred	Other Allowable		
Energy, work, quantity of heat	J	quad	EJ		1.055 056	E+00
			TW · h		2.930 711	E+02
		therm	MJ		1.055 056	E–02
			kJ		1.055 056	E+05
				kW · h	2.930 711	E+01
		US tonf · mi	MJ		1.431 744	E+01
		hp · h	MJ		2.684 520	E+00
			kJ		2.684 520	E+03
				kW · h	7.456 999	E–01
		ch · h or CV · h	MJ		2.647 796	E+00
			kJ		2.647 796	E+03
			kW · h	7.354 99	E–01	
		kW · h	MJ		3.6*	E+00
			kJ		3.6*	E+03
		Chu	kJ		1.899 101	E+00
				kW · h	5.275 280	E–04
		Btu	kJ		1.055 056	E+00
				kW · h	2.930 711	E–04
		kcal	kJ		4.184*	E+00
		cal	kJ		4.184*	E–03
		ft · lbf	kJ		1.355 818	E–03
		J	kJ		1.0*	E–03

		lb · ft^2/s^2(ft · pdl)	kJ		4.214 011	E–05
		erg	J		1.0*	E–07
Impact energy	J	kgf · m	J		9.806 650*	E+00
		ft · lbf	J		1.355 818	E+00
Work/length	J/m	US tonf · mi/ft	MJ/m		4.697 322	E+01
Surface energy	J/m^2	erg/cm^2	mJ/m^2		1.0*	E+00
Power	W	quad/yr	EJ/a		1.055 056	E+00
				GW	3.345 561	E+01
		erg/a	TW		3.170 979	E–27
			GW		3.170 979	E–24
		million Btu/h	MW		2.930 711	E–01
		ton of refrigeration	kW		3.516 853	E+00
		Btu/s	kW		1.055 056	E+00
		kW	kW		1	
		hydraulic horse-power-hhp	kW		7.460 43	E–01
		hp (electric)	kW		7.46*	E–01
		hp (550 ft · lbf/s)	kW		7.456 999	E–01
		ch or CV	kW		7.354 99	E–01
		Btu/min	kW		1.758 427	E–02
		ft · lbf/s	kW		1.355 818	E–03
		kcal/h	W		1.162 222	E+00
		Btu/h	W		2.930 711	E–01
		ft · lbf/min	W		2.259 697	E–02
Power/area	W/m^3	Btu/(s · ft^2)	kW/m^2		1.135 653	E–01
		cal/(h · cm^2)	kW/m^2		1.162 222	E–02
		Btu/(h · ft^2)	kW/m^3		3.154 591	E–03
Heat flow unit (hfu; geothermics)		μcal/(s · cm^2)	mW/m^2		4.184*	E+01
Heat release rate, mixing power	W/m^3	hp/ft^3	kW/m^3		2.633 414	E+01
		cal/(h · cm^3)	kW/m^3		1.162 222	E+00
		Btu/(s · ft^3)	kW/m^3		3.725 895	E+01
		Btu/(h · ft^3)	kW/m^3		1.034 971	E–02

Continued

Quantity	SI unit	Customary unit	Metric unit		Conversion factor (multiply quantity expressed in customary units by factor to get metric equivalent)	
			API preferred	Other Allowable		
Heat generation unit-hgu (radioactive rocks)		cal/(s · cm^3)	$\mu W/m^3$		4.184*	E+12
Cooling duty (machinery)	W/W	Btu/(bhp · h	W/kW		3.930 148	E–01
Mass fuel consumption	kg/J	lb/(hp · h)	mg/J	kg/MJ	1.689 659	E–01
				kg/(kW · h)	6.082 774	E–01
Volume fuel consumption	m^3/J	m^3/(kW · h)	dm^3/MJ	mm^3/J	2.777 778	E+02
				dm^3/(kW · h)	1.0*	E+03
		US gal/(hp · h)	dm^3/MJ	mm^3/J	1.410 089	E+00
				dm^3/(kW · h)	5.076 321	E+00
		UK pt/(hp · h)	dm^3/MJ	mm^3/J	2.116 809	E–01
				dm^3/(kW · h)	7.620 512	E–01
Fuel consumption (automotive)	m^3/m	UK gal/mi	dm^3/100 km	L/100 km	2.824 811	E+02
		US gal/mi	dm^3/100 km	L/100 km	2.352 146	E+02
		mi/US gal	km/dm^3	km/L	4.251 437	E–01
		mi/UK gal	km/dm^3	km/L	3.540 060	E–01

Recommended SI Units and Conversion Factors for Mechanics

Quantity	SI unit	Customary unit	Metric unit		Conversion factor (multiply quantity expressed in customary units by factor to get metric equivalent)	
			API preferred	Other allowable		
Velocity (linear), speed	m/s	knot	km/h		1.852	E+00
				knot	1	
		mi/h	km/h		1.609 344*	E+00
		m/s	m/s		1	
		ft/s	m/s		3.048*	E–01
				cm/s	3.048*	E+01
				m/ms	3.048*	E–04
		ft/min	m/s		5.08*	E–03
				cm/s	5.08*	E–01
		ft/h	mm/s		8.466 667	E–02
				cm/s	8.466 667	E–03
		ft/d	mm/s		3.527 778	E–03
				m/d	3.048*	E–01
		in/s	mm/s		2.54*	E+01
				cm/s	2.54*	E+00
		in/min	mm/s		4.233 333	E–01
				cm/s	4.233 333	E–02
Velocity (angular)	rad/s	r/min	rad/s		1.047 198	E–01
		r/min		r/min	1	
		r/s	rad/s		6.283 185	E+00
		r/s		r/s	1	
		deg/min	rad/s		2.908 882	E–04
		deg/s	rad/s		1.745 329	E–02

Continued

Quantity	SI unit	Customary unit	Metric unit: API preferred	Metric unit: Other allowable	Conversion factor (multiply quantity expressed in customary units by factor to get metric equivalent)	
Reciprocal velocity	μs/m	μs/ft	μs/m		3.280 840	E+00
Acceleration (linear)	m/s^2	ft/s^2	m/s^2		3.048*	E–01
		gal (cm/s^2)	m/s^2		1.0*	E–02
Acceleration (angular)	rad/s^2	rad/s^2	rad/s^2		1	
		rpm/s	rad/s^2		1.047 198	E–01
Corrosion rate	mm/a	in./yr (ipy)	mm/a		2.54*	E+01
Momentum	kg · m/s	lb · ft/s	kg · m/s		1.382 550	E–01
Force	N	UK tonf	kN		9.964 016	E+00
		US tonf	kN		8.896 443	E+00
		kgf (kp)	N		9.806 650*	E+00
		lbf	N		4.448 222	E+00
		N	N		1	
		pdl	mN		1.382 550	E+02
		dyn	mN		1.0*	E–02
Bending moment, torque	N · m	US tonf · m	kN · m		2.711 636	E+00
		kgf · m	N · m		9.806 650*	E+00
		lbf · ft	N · m		1.355 818	E+00
		lbf · in.	N · m		1.129 848	E–01
		pdl · ft	N · m		4.214 011	E–02
Bending movement, length	N · m/m	lbf · ft/in.	N · m/m		5.337 866	E+01
		kgf · m/m	N · m/m		9.806 650*	E+00
	lbf · in./in.	N · m/m			4.448 222	E+00
Moment of inertia	$kg \cdot m^2$	$lb \cdot ft^2$	$kg \cdot m^2$		4.214 011	E–02
		$in.^4$	cm^4		4.162 314	E+01

Stress	Pa	US tonf/in.^{2}	MPa	N/mm^2	1.378 951	E+00
		kgf/mm^2	MPa	N/mm^2	9.806 650*	E+00
		US tonf/ft^2	MPa	N/mm^2	9.576 052	E–02
		lbf/in.^{2} (psi)	MPa	N/mm^2	6.894 757	E–03
		lbf/ft^2 (psf)	kPa		4.788 026	E–02
		dyn/cm^2	Pa		1.0*	E–01
Yield point, gel strength (drilling fluid)	lbf/100 ft^2	Pa			4.788 026	E+01
Mass/length	kg/m	lb/ft	kg/m		1.488 164	E+00
Mass/area structural loading,	kg/m^2	US ton/ft^2	Mg/m^2	t/m^2	9.764 855	E+00
bearing capacity (mass basis)		lb/ft^2	kg/m^2		4.882 428	E+00
Modulus of elasticity	Pa	lbf/in^2 (psi)	MPa	N/mm^2	6.894 757	E–03
Section modulus	m^3	in.^{3}	cm^3		1.638 706	E+01
Coefficient of thermal expansion	m/(m · K)	in./(in°F)	mm/(mm · K)	mm/(mm · °C)	5.555 556	E–01

Recommended SI Units and Conversion Factors for Transport Properties

Quantity	SI unit	Customary unit	Metric unit: API preferred	Metric unit: Other allowable	Conversion factor (multiply quantity expressed in customary units by factor to get metric equivalent)	
Diffusivity	m^2/s	ft^2/s	mm^2/s		9.290 304*	E+04
		cm^2/s	mm^2/s		1.0*	E+02
		ft^2/h	mm^2/s		2.580 64*	E+01
Thermal resistance	$K \cdot m^2/W$	$°C \cdot m^2 \cdot h/kcal$	$K \cdot m^2/kW$		8.604 208	E+02
		$°F \cdot ft^2 \cdot h/Btu$	$K \cdot m^2/kW$		1.761 102	E+02
Heat flux	W/m^2	$Btu/(h \cdot ft^2)$	kW/m^2		3.154 591	E–03
Thermal conductivity	$W/(m \cdot K)$	$cal/(s \cdot cm^2 \cdot °C/cm)$	$W/(m \cdot K)$	$W/(m^2 \cdot °C/m)$	4.184*	E+02
		$Btu/(h \cdot ft^2 \cdot °F/ft)$	$W/(m \cdot K)$	$W/(m^2 \cdot °C/m)$	1.730 735	E+00
		$kcal/(h \cdot m^2 \cdot °C/m)$	$W/(m \cdot K)$	$W/(m^2 \cdot °C/m)$	1.162 222	E+00
		$Btu/(h \cdot ft^2 \cdot °F/in.)$	$W/(m \cdot K)$	$W/(m^2 \cdot °C/m)$	1.422 279	E–01
		$cal/(h \cdot cm^2 \cdot °C/cm)$	$W/(m \cdot K)$	$W/(m^2 \cdot °C/m)$	1.162 222	E–01
Heat Transfer Coefficient	$W/(m^2 \cdot K)$	$cal/(s \cdot cm^2 \cdot °C)$	$kW/(m^2 \cdot K)$		4.184*	E+01
		$Btu/(s \cdot ft^2 \cdot °F)$	$kW/(m^2 \cdot K)$		2.044 175	E+01
		$cal/(h \cdot cm^2 \cdot °C)$	$kW/(m^2 \cdot K)$		1.162 222	E–02
		$Btu/(h \cdot ft^2 \cdot °F)$	$kW/(m^2 \cdot K)$		5.678 263	E–03
		$Btu/(h \cdot ft^2 \cdot °R)$	$kW/(m^2 \cdot K)$		5.678 263	E–03
		$kcal/(h \cdot m^2 \cdot °C)$	$kW/(m^2 \cdot K)$		1.162 222	E–03
Volumetric heat transfer coefficient	$W/(m^3 \cdot K)$	$Btu/(s \cdot ft^3 \cdot °F)$	$kW/(m^3 \cdot K)$		6.706 611	E+01
		$Btu/(h \cdot ft^3 \cdot °F)$	$kW/(m^3 \cdot K)$		1.862 947	E–02
Surface tension	N/m	dyn/cm	mN/m		1.0*	E+00

Viscosity (dynamic)	Pa · s	lbf · s/in.^{2}	mPa · s	cP	6.894 757	E+06
		lbf · s/ft^2	mPa · s	cP	4.788 026	E+04
		kgf · s/m^2	mPa · s	cP	9.806 650*	E+03
		dyn · s/cm^2	Pa · s		1.0*	E–01
			mPa · s	cP	1.0*	E+02
		P	Pa · s		1.0*	E–01
			mPa · s	cP	1.0*	E+02
		cP	mPa · s	cP	1.0*	E+00
Viscosity (kinematic)	m^2/s	ft^2/s	mm^2/s	cSt	9.290 304*	E+04
		in.^{2}/s	mm^2/s	cSt	6.451 6*	E+02
		ft^2/h	mm^2/s	cSt	2.580 64*	E+01
		m^2/h	mm^2/s	cSt	2.777 778	E+02
		cm^2/s	mm^2/s	cSt	1.0*	E+02
		St	mm^2/s	cSt	1.0*	E+02
		cSt	mm^2/s	cSt	1.0*	E+00
Permeability	m^2	D	μm^2	D	1.0*	E+00
		mD	μm^2	D	1.0*	E–03

Recommended SI Units and Conversion Factors for Electricity and Magnetism

Quantity	SI unit	Customary unit	Metric unit: API preferred	Metric unit: Other allowable	Conversion factor (multiply quantity expressed in customary units by factor to get metric equivalent)
Admittance	S	S	S		1
Capacitance	F	μF	μF		1
Capacity, storage battery	C	A · h	kC		3.6* E+00
Charge density	C/m³	C/mm³	C/mm³		1
Conductance	S	S	S		1
		Ω	S		1
Conductivity	S/m	S/m	S/m		
		mΩ/m	mS/m		1
Current density	A/m²	A/mm²	A/mm²		1
		A/in²	A/mm²		1.550 003 E–03
Displacement	C/m²	C/cm²	C/cm²		1
Electric charge	C	C	C		1
Electric current	A	A	A		1
electric dipole moment	C · m	C · m	C · m		1
Electric field strength	V/m	V/m	V/m		1
Electric flux	C	C	C		1
Electric polarization	C/m²	C/cm²	C/cm²		1
Electric potential	V	V	V		1
		mV	mV		1
Electromagnetic moment	A · m²	A · m²	A · m²		1
Electromotive force	V	V	V		1
Flux of displacement	C	C	C		1
Frequency	Hz	cycles/s	Hz		1
Impedance	Ω	Ω	Ω		1
Linear current density	A/m	A/mm	A/mm		1
Magnetic dipole moment	Wb · m	Wb · m	Wb · m		1

Magnetic field strength	A/m	A/mm	A/mm	1	
		oersted	A/m	7.957 747	E+01
Magnetic flux	Wb	mWb	mWb	1	
		maxwell	nWb	1.0*	E+01
Magnetic flux density	T	mT	mT	1	
		gauss	T	1.0*	E–04
		gamma	nT	1	
Magnetic induction	T	mT	mT	1	
Magnetic moment	$A \cdot m^2$	$A \cdot m^2$	$A \cdot m^2$	1	
Magnetic polarization	T	mT	mT	1	
Magnetic potential difference	A	A	A	1	
Magnetic vector potential	Wb/m	Wb/mm	Wb/mm	1	
Magnetization	A/m	A/mm	A/mm	1	
Modulus of admittance	S	S	S	1	
Modulus of impedance	Ω	Ω	Ω	1	
Mutual inductance	H	H	H	1	
Permeability	H/m	μH/m	μH/m	1	
Permeance	H	H	H	1	
Permittivity	F/m	μF/m	μF/m	1	
Potential difference	V	V	V	1	
Quantity of electricity	C	C	C	1	
Reactance	Ω	Ω	Ω	1	
Reluctance	H^{-1}	H^{-1}	H^{-1}	1	
Resistance	Ω	Ω	Ω	1	
Resistivity	$\Omega \cdot m$	$\Omega \cdot cm$	$\Omega \cdot cm$	1	
		$\Omega \cdot m$	$\Omega \cdot m$	1	
Self inductance	H	mH	mH	1	
Surface density of charge	C/m^2	mC/m^2	mC/m^2	1	
Susceptance	S	S	S	1	
Volume density of charge	C/m^3	C/mm^3	C/mm^3	1	

Recommended SI Units and Conversion Factors for Acoustics, Light, and Radiation

Quantity	SI unit	Customary unit	Metric unit: API preferred	Metric unit: Other allowable	Conversion factor (multiply quantity expressed in customary units by factor to get metric equivalent)	
Acoustical energy	J	J	J		1	
Acoustical intensity	W/m^2	W/cm^2	W/m^2		1.0*	E+04
Acoustical power	W	W	W		1	
Activity (of a radionuclide)	Bq	curie	Bq		3.7*	E+10
Illuminance, illumination	lx	footcandle (fc)	lx		1.076 391	E+01
		lm/ft^2	lx		1.076 391	E+01
Irradiance	W/m^2	W/m^2	W/m^2		1	
Light exposure	lx · s	fc · s	lx · s		1.076 391	E+01
Luminance	cd/m^2	cd/m^2	cd/m^2		1	
		footlambert (fL)	cd/m^2		3.426 259	E+00
Luminous efficacy	lm/W	lm/W	lm/W		1	
Luminous exitance	lm/m^2	lm/m^2	lm/m^2		1	
		lm/ft^2	lm/m^2		1.076 391	E+01
Luminous flux	lm	lm	lm		1	
Luminous intensity	cd	cd	cd		1	
Quantity of light	lm · s	talbot	lm · s		1	
Radiance	W/(m^2 · sr)	W/(m^2 · sr)	W/(m^2 · sr)		1	
Radiance energy	J	J	J		1	
Radiant flux	W	W	W		1	
Radiant intensity	W/sr	W/sr	W/sr		1	
Radiant power	W	W	W		1	
Sound pressure	Pa	Pa	Pa		1	
Wavelength	m	Å	nm		1.0*	E–01

Index